D0151108

CALCULUS

CALCULUS

Third Edition

Michael Spivak

CALCULUS
Third Edition

Library of Congress Catalog Card Number 80-82517

Publish or Perish, Inc.
PMB 377
1302 Waugh Drive
Houston, Texas 77019-4944

Manufactured in the United States of America
ISBN 0-914098-89-6

Dedicated to the Memory of Y. P.

PREFACE

I hold every man a debtor
to his profession,
from the which as men of course
doe seeke to receive countenance and profit,
so ought they of duty to endeavour
themselves by way of amends,
to be a help and
ornament thereunto.

FRANCIS BACON

PREFACE TO THE FIRST EDITION

Every aspect of this book was influenced by the desire to present calculus not merely as a prelude to but as the first real encounter with mathematics. Since the foundations of analysis provided the arena in which modern modes of mathematical thinking developed, calculus ought to be the place in which to expect, rather than avoid, the strengthening of insight with logic. In addition to developing the students' intuition about the beautiful concepts of analysis, it is surely equally important to persuade them that precision and rigor are neither deterrents to intuition, nor ends in themselves, but the natural medium in which to formulate and think about mathematical questions.

This goal implies a view of mathematics which, in a sense, the entire book attempts to defend. No matter how well particular topics may be developed, the goals of this book will be realized only if it succeeds as a whole. For this reason, it would be of little value merely to list the topics covered, or to mention pedagogical practices and other innovations. Even the cursory glance customarily bestowed on new calculus texts will probably tell more than any such extended advertisement, and teachers with strong feelings about particular aspects of calculus will know just where to look to see if this book fulfills their requirements.

A few features do require explicit comment, however. Of the twenty-nine chapters in the book, two (starred) chapters are optional, and the three chapters comprising Part V have been included only for the benefit of those students who might want to examine on their own a construction of the real numbers. Moreover, the appendices to Chapters 3 and 11 also contain optional material.

The order of the remaining chapters is intentionally quite inflexible, since the purpose of the book is to present calculus as the evolution of one idea, not as a collection of "topics." Since the most exciting concepts of calculus do not appear until Part III, it should be pointed out that Parts I and II will probably require less time than their length suggests—although the entire book covers a one-year course, the chapters are not meant to be covered at any uniform rate. A rather natural dividing point does occur between Parts II and III, so it is possible to reach differentiation and integration even more quickly by treating Part II very briefly, perhaps returning later for a more detailed treatment. This arrangement corresponds to the traditional organization of most calculus courses, but I feel that it will only diminish the value of the book for students who have seen a small amount of calculus previously, and for bright students with a reasonable background.

The problems have been designed with this particular audience in mind. They range from straightforward, but not overly simple, exercises which develop basic techniques and test understanding of concepts, to problems of considerable difficulty and, I hope, of comparable interest. There are about 625 problems in all. Those which emphasize manipulations usually contain many example, numbered with small Roman numerals, while small letters are used to label interrelated parts in other problems. Some indication of relative difficulty is provided by a system of starring and double starring, but there are so many criteria for judging difficulty, and so many hints have been provided, especially for harder problems, that this guide is not completely reliable. Many problems are so difficult, especially if the hints are not consulted, that the best of students will probably have to attempt only those which especially interest them; from the less difficult problems it should be easy to select a portion which will keep a good class busy, but not frustrated. The answer section contains solutions to about half the examples from an assortment of problems that should provided a good test of technical competence. A separate answer book contains the solutions of the other parts of these problems, and of all the other problems as well. Finally, there is a Suggested Reading list, to which the problems often refer, and a glossary of symbols.

I am grateful for the opportunity to mention the many people to whom I owe my thanks. Jane Bjorkgren performed prodigious feats of typing that compensated for my fitful production of the manuscript. Richard Serkey helped collect the material which provides historical sidelights in the problems, and Richard Weiss supplied the answers appearing in the back of the book. I am especially grateful to my friends Michael Freeman, Jay Goldman, Anthony Phillips, and Robert Wells for the care with which they read, and the relentlessness with which they criticized, a preliminary version of the book. Needles to say, they are not responsible for the deficiencies which remain, especially since I sometimes rejected suggestions which would have made the book appear suitable for a larger group of students. I must express my admiration for the editors and staff of W. A. Benjamin, Inc., who were always eager to increase the appeal of the book, while recognizing the audience for which it was intended.

The inadequacies which preliminary editions always involve were gallantly endured by a rugged group of freshmen in the honors mathematics course at Brandeis University during the academic year 1965–1966. About half of this course was devoted to algebra and topology, while the other half covered calculus, with the preliminary edition as the text. It is almost obligatory in such circumstances to report that the preliminary version was a gratifying success. This is always safe—after all, the class is unlikely to rise up in a body and protest publicly—but the students themselves, it seems to me, deserve the right to assign credit for the thoroughness with which they absorbed an impressive amount of mathematics. I am content to hope that some other students will be able to use the book to such good purpose, and with such enthusiasm.

Waltham, Massachusetts
February 1967

MICHAEL SPIVAK

PREFACE TO THE SECOND EDITION

I have often been told that the title of this book should really be something like "An Introduction to Analysis," because the book is usually used in courses where the students have already learned the mechanical aspects of calculus—such courses are standard in Europe, and they are becoming more common in the United States. After thirteen years it seems too late to change the title, but other changes, in addition to the correction of numerous misprints and mistakes, seemed called for. There are now separate Appendices for many topics that were previously slighted: polar coordinates, uniform continuity, parameterized curves, Riemann sums, and the use of integrals for evaluating lengths, volumes and surface areas. A few topics, like manipulations with power series, have been discussed more thoroughly in the text, and there are also more problems on these topics, while other topics, like Newton's method and the trapezoid rule and Simpson's rule, have been developed in the problems. There are in all about 160 new problems, many of which are intermediate in difficulty between the few routine problems at the beginning of each chapter and the more difficult ones that occur later.

Most of the new problems are the work of Ted Shifrin. Frederick Gordon pointed out several serious mistakes in the original problems, and supplied some non-trivial corrections, as well as the neat proof of Theorem 12-2, which took two Lemmas and two pages in the first edition. Joseph Lipman also told me of this proof, together with the similar trick for the proof of the last theorem in the Appendix to Chapter 11, which went unproved in the first edition. Roy O. Davies told me the trick for Problem 11-66, which previously was proved only in Problem 20-8 [21-8 in the third edition], and Marina Ratner suggested several interesting problems, especially ones on uniform continuity and infinite series. To all these people go my thanks, and the hope that in the process of fashioning the new edition their contributions weren't too badly botched.

MICHAEL SPIVAK

PREFACE TO THE THIRD EDITION

The most significant change in this third edition is the inclusion of a new (starred) Chapter 17 on planetary motion, in which calculus is employed for a substantial physics problem.

In preparation for this, the old Appendix to Chapter 4 has been replaced by three Appendices: the first two cover vectors and conic sections, while polar coordinates are now deferred until the third Appendix, which also discusses the polar coordinate equations of the conic sections. Moreover, the Appendix to Chapter 12 has been extended to treat vector operations on vector-valued curves.

Another large change is merely a rearrangement of old material: "The Cosmopolitan Integral," previously a second Appendix to Chapter 13, is now an Appendix to the chapter on "Integration in Elementary Terms" (previously Chapter 18, now Chapter 19); moreover, those problems from that chapter which used the material from that Appendix now appear as problems in the newly placed Appendix.

A few other changes and renumbering of Problems result from corrections, and elimination of incorrect problems.

I was both startled and somewhat dismayed when I realized that after allowing 13 years to elapse between the first and second editions of the book, I have allowed another 14 years to elapse before this third edition. During this time I seem to have accumulated a not-so-short list of corrections, but no longer have the original communications, and therefore cannot properly thank the various individuals involved (who by now have probably lost interest anyway). I have had time to make only a few changes to the Suggested Reading, which after all these years probably requires a complete revision; this will have to wait until the next edition, which I hope to make in a more timely fashion.

MICHAEL SPIVAK

CONTENTS

CALCULUS

PART 1
PROLOGUE

To be conscious that
you are ignorant is a great step
to knowledge.

BENJAMIN DISRAELI

CHAPTER 1 BASIC PROPERTIES OF NUMBERS

The title of this chapter expresses in a few words the mathematical knowledge required to read this book. In fact, this short chapter is simply an explanation of what is meant by the "basic properties of numbers," all of which—addition and multiplication, subtraction and division, solutions of equations and inequalities, factoring and other algebraic manipulations—are already familiar to us. Nevertheless, this chapter is not a review. Despite the familiarity of the subject, the survey we are about to undertake will probably seem quite novel; it does not aim to present an extended review of old material, but to condense this knowledge into a few simple and obvious properties of numbers. Some may even seem too obvious to mention, but a surprising number of diverse and important facts turn out to be consequences of the ones we shall emphasize.

Of the twelve properties which we shall study in this chapter, the first nine are concerned with the fundamental operations of addition and multiplication. For the moment we consider only addition: this operation is performed on a pair of numbers—the sum $a + b$ exists for any two given numbers a and b (which may possibly be the same number, of course). It might seem reasonable to regard addition as an operation which can be performed on several numbers at once, and consider the sum $a_1 + \cdots + a_n$ of n numbers $a_1, \ldots, a_n$ as a basic concept. It is more convenient, however, to consider addition of pairs of numbers only, and to define other sums in terms of sums of this type. For the sum of three numbers a, b, and c, this may be done in two different ways. One can first add b and c, obtaining $b + c$, and then add a to this number, obtaining $a + (b + c)$; or one can first add a and b, and then add the sum $a + b$ to c, obtaining $(a + b) + c$. Of course, the two compound sums obtained are equal, and this fact is the very first property we shall list:

(P1) If a, b, and c are any numbers, then

$$a + (b + c) = (a + b) + c.$$

The statement of this property clearly renders a separate concept of the sum of three numbers superfluous; we simply agree that $a + b + c$ denotes the number $a + (b + c) = (a + b) + c$. Addition of four numbers requires similar, though slightly more involved, considerations. The symbol $a + b + c + d$ is defined to mean

$$\begin{array}{lll} & (1) & ((a + b) + c) + d, \\ \text{or} & (2) & (a + (b + c)) + d, \\ \text{or} & (3) & a + ((b + c) + d), \\ \text{or} & (4) & a + (b + (c + d)), \\ \text{or} & (5) & (a + b) + (c + d). \end{array}$$

This definition is unambiguous since these numbers are all equal. Fortunately, *this* fact need not be listed separately, since it follows from the property P1 already listed. For example, we know from P1 that

$$(a+b)+c = a+(b+c),$$

and it follows immediately that (1) and (2) are equal. The equality of (2) and (3) is a direct consequence of P1, although this may not be apparent at first sight (one must let $b+c$ play the role of b in P1, and d the role of c). The equalities (3) = (4) = (5) are also simple to prove.

It is probably obvious that an appeal to P1 will also suffice to prove the equality of the 14 possible ways of summing five numbers, but it may not be so clear how we can reasonably arrange a proof that this is so without actually listing these 14 sums. Such a procedure is feasible, but would soon cease to be if we considered collections of six, seven, or more numbers; it would be totally inadequate to prove the equality of all possible sums of an arbitrary finite collection of numbers $a_1, \dots, a_n$. This fact may be taken for granted, but for those who would like to worry about the proof (and it is worth worrying about once) a reasonable approach is outlined in Problem 24. Henceforth, we shall usually make a tacit appeal to the results of this problem and write sums $a_1 + \cdots + a_n$ with a blithe disregard for the arrangement of parentheses.

The number 0 has one property so important that we list it next:

(P2) If a is any number, then

$$a+0 = 0+a = a.$$

An important role is also played by 0 in the third property of our list:

(P3) For every number a, there is a number $-a$ such that

$$a+(-a) = (-a)+a = 0.$$

Property P2 ought to represent a distinguishing characteristic of the number 0, and it is comforting to note that we are already in a position to prove this. Indeed, if a number x satisfies

$$a+x = a$$

for any one number a, then $x = 0$ (and consequently this equation also holds for all numbers a). The proof of this assertion involves nothing more than subtracting a from both sides of the equation, in other words, adding $-a$ to both sides; as the following detailed proof shows, all three properties P1–P3 must be used to justify this operation.

$$\begin{array}{ll} \text{If} & a+x = a, \\ \text{then} & (-a)+(a+x) = (-a)+a = 0; \\ \text{hence} & ((-a)+a)+x = 0; \\ \text{hence} & 0+x = 0; \\ \text{hence} & x = 0. \end{array}$$

As we have just hinted, it is convenient to regard subtraction as an operation derived from addition: we consider $a - b$ to be an abbreviation for $a + (-b)$. It is then possible to find the solution of certain simple equations by a series of steps (each justified by P1, P2, or P3) similar to the ones just presented for the equation $a + x = a$. For example:

$$\begin{array}{ll} \text{If} & x+3=5, \\ \text{then} & (x+3)+(-3)=5+(-3); \\ \text{hence} & x+(3+(-3))=5-3=2; \\ \text{hence} & x+0=2; \\ \text{hence} & x=2. \end{array}$$

Naturally, such elaborate solutions are of interest only until you become convinced that they can always be supplied. In practice, it is usually just a waste of time to solve an equation by indicating so explicitly the reliance on properties P1, P2, and P3 (or any of the further properties we shall list).

Only one other property of addition remains to be listed. When considering the sums of three numbers a, b, and c, only two sums were mentioned: $(a + b) + c$ and $a + (b + c)$. Actually, several other arrangements are obtained if the order of a, b, and c is changed. That these sums are all equal depends on

(P4) If a and b are any numbers, then

$$a + b = b + a.$$

The statement of P4 is meant to emphasize that although the operation of addition of pairs of numbers might conceivably depend on the order of the two numbers, in fact it does not. It is helpful to remember that not all operations are so well behaved. For example, subtraction does not have this property: usually $a - b \neq b - a$. In passing we might ask just when $a - b$ does equal $b - a$, and it is amusing to discover how powerless we are if we rely only on properties P1–P4 to justify our manipulations. Algebra of the most elementary variety shows that $a - b = b - a$ only when $a = b$. Nevertheless, it is impossible to derive this fact from properties P1–P4; it is instructive to examine the elementary algebra carefully and determine which step(s) cannot be justified by P1–P4. We will indeed be able to justify all steps in detail when a few more properties are listed. Oddly enough, however, the crucial property involves multiplication.

The basic properties of multiplication are fortunately so similar to those for addition that little comment will be needed; both the meaning and the consequences should be clear. (As in elementary algebra, the product of a and b will be denoted by $a \cdot b$, or simply ab.)

(P5) If a, b, and c are any numbers, then

$$a \cdot (b \cdot c) = (a \cdot b) \cdot c.$$

(P6) If a is any number, then

$$a \cdot 1 = 1 \cdot a = a.$$

Moreover, $1 \neq 0$.

(The assertion that $1 \neq 0$ may seem a strange fact to list, but we have to list it, because there is no way it could possibly be proved on the basis of the other properties listed—these properties would all hold if there were only one number, namely, 0.)

(P7) For every number $a \neq 0$, there is a number a^{-1} such that

$$a \cdot a^{-1} = a^{-1} \cdot a = 1.$$

(P8) If a and b are any numbers, then

$$a \cdot b = b \cdot a.$$

One detail which deserves emphasis is the appearance of the condition $a \neq 0$ in P7. This condition is quite necessary; since $0 \cdot b = 0$ for all numbers b, there is *no* number 0^{-1} satisfying $0 \cdot 0^{-1} = 1$. This restriction has an important consequence for division. Just as subtraction was defined in terms of addition, so division is defined in terms of multiplication: The symbol a/b means $a \cdot b^{-1}$. Since 0^{-1} is meaningless, $a/0$ is also meaningless—division by 0 is *always* undefined.

Property P7 has two important consequences. If $a \cdot b = a \cdot c$, it does not necessarily follow that $b = c$; for if $a = 0$, then both $a \cdot b$ and $a \cdot c$ are 0, no matter what b and c are. However, if $a \neq 0$, then $b = c$; this can be deduced from P7 as follows:

$$\begin{array}{ll} \text{If} & a \cdot b = a \cdot c \text{ and } a \neq 0, \\ \text{then} & a^{-1} \cdot (a \cdot b) = a^{-1} \cdot (a \cdot c); \\ \text{hence} & (a^{-1} \cdot a) \cdot b = (a^{-1} \cdot a) \cdot c; \\ \text{hence} & 1 \cdot b = 1 \cdot c; \\ \text{hence} & b = c. \end{array}$$

It is also a consequence of P7 that if $a \cdot b = 0$, then either $a = 0$ or $b = 0$. In fact,

$$\begin{array}{ll} \text{if} & a \cdot b = 0 \text{ and } a \neq 0, \\ \text{then} & a^{-1} \cdot (a \cdot b) = 0; \\ \text{hence} & (a^{-1} \cdot a) \cdot b = 0; \\ \text{hence} & 1 \cdot b = 0; \\ \text{hence} & b = 0. \end{array}$$

(It may happen that $a = 0$ and $b = 0$ are both true; this possibility is not excluded when we say "either $a = 0$ or $b = 0$"; in mathematics "or" is always used in the sense of "one or the other, or both.")

This latter consequence of P7 is constantly used in the solution of equations. Suppose, for example, that a number x is known to satisfy

$$(x - 1)(x - 2) = 0.$$

Then it follows that either $x - 1 = 0$ or $x - 2 = 0$; hence $x = 1$ or $x = 2$.

On the basis of the eight properties listed so far it is still possible to prove very little. Listing the next property, which combines the operations of addition and multiplication, will alter this situation drastically.

(P9) If a, b, and c are any numbers, then

$$a \cdot (b + c) = a \cdot b + a \cdot c.$$

(Notice that the equation $(b + c) \cdot a = b \cdot a + c \cdot a$ is also true, by P8.)

As an example of the usefulness of P9 we will now determine just when $a - b = b - a$:

$$\begin{array}{ll} \text{If} & a - b = b - a, \\ \text{then} & (a - b) + b = (b - a) + b = b + (b - a); \\ \text{hence} & a = b + b - a; \\ \text{hence} & a + a = (b + b - a) + a = b + b. \\ \text{Consequently} & a \cdot (1 + 1) = b \cdot (1 + 1), \\ \text{and therefore} & a = b. \end{array}$$

A second use of P9 is the justification of the assertion $a \cdot 0 = 0$ which we have already made, and even used in a proof on page 6 (can you find where?). This fact was not listed as one of the basic properties, even though no proof was offered when it was first mentioned. With P1–P8 alone a proof was not possible, since the number 0 appears only in P2 and P3, which concern addition, while the assertion in question involves multiplication. With P9 the proof is simple, though perhaps not obvious: We have

$$\begin{aligned} a \cdot 0 + a \cdot 0 &= a \cdot (0 + 0) \\ &= a \cdot 0; \end{aligned}$$

as we have already noted, this immediately implies (by adding $-(a \cdot 0)$ to both sides) that $a \cdot 0 = 0$.

A series of further consequences of P9 may help explain the somewhat mysterious rule that the product of two negative numbers is positive. To begin with, we will establish the more easily acceptable assertion that $(-a) \cdot b = -(a \cdot b)$. To prove this, note that

$$\begin{aligned} (-a) \cdot b + a \cdot b &= [(-a) + a] \cdot b \\ &= 0 \cdot b \\ &= 0. \end{aligned}$$

It follows immediately (by adding $-(a \cdot b)$ to both sides) that $(-a) \cdot b = -(a \cdot b)$. Now note that

$$\begin{aligned} (-a) \cdot (-b) + [-(a \cdot b)] &= (-a) \cdot (-b) + (-a) \cdot b \\ &= (-a) \cdot [(-b) + b] \\ &= (-a) \cdot 0 \\ &= 0. \end{aligned}$$

Consequently, adding $(a \cdot b)$ to both sides, we obtain

$$(-a) \cdot (-b) = a \cdot b.$$

The fact that the product of two negative numbers is positive is thus a consequence of P1–P9. In other words, *if we want P1 to P9 to be true, the rule for the product of two negative numbers is forced upon us.*

The various consequences of P9 examined so far, although interesting and important, do not really indicate the significance of P9; after all, we could have listed each of these properties separately. Actually, P9 is the justification for almost all algebraic manipulations. For example, although we have shown how to solve the equation

$$(x - 1)(x - 2) = 0,$$

we can hardly expect to be presented with an equation in this form. We are more likely to be confronted with the equation

$$x^2 - 3x + 2 = 0.$$

The "factorization" $x^2 - 3x + 2 = (x - 1)(x - 2)$ is really a triple use of P9:

$$\begin{aligned}(x - 1) \cdot (x - 2) &= x \cdot (x - 2) + (-1) \cdot (x - 2)\\ &= x \cdot x + x \cdot (-2) + (-1) \cdot x + (-1) \cdot (-2)\\ &= x^2 + x[(-2) + (-1)] + 2\\ &= x^2 - 3x + 2.\end{aligned}$$

A final illustration of the importance of P9 is the fact that this property is actually used every time one multiplies arabic numerals. For example, the calculation

$$\begin{array}{r} 13\\ \times 24\\ \hline 52\\ 26\ \\ \hline 312 \end{array}$$

is a concise arrangement for the following equations:

$$\begin{aligned}13 \cdot 24 &= 13 \cdot (2 \cdot 10 + 4)\\ &= 13 \cdot 2 \cdot 10 + 13 \cdot 4\\ &= 26 \cdot 10 + 52.\end{aligned}$$

(Note that moving 26 to the left in the above calculation is the same as writing $26 \cdot 10$.) The multiplication $13 \cdot 4 = 52$ uses P9 also:

$$\begin{aligned}13 \cdot 4 &= (1 \cdot 10 + 3) \cdot 4\\ &= 1 \cdot 10 \cdot 4 + 3 \cdot 4\\ &= 4 \cdot 10 + 12\\ &= 4 \cdot 10 + 1 \cdot 10 + 2\\ &= (4 + 1) \cdot 10 + 2\\ &= 5 \cdot 10 + 2\\ &= 52.\end{aligned}$$

The properties P1–P9 have descriptive names which are not essential to remember, but which are often convenient for reference. We will take this opportunity to list properties P1–P9 together and indicate the names by which they are commonly designated.

(P1)	(Associative law for addition)	$a + (b + c) = (a + b) + c$.
(P2)	(Existence of an additive identity)	$a + 0 = 0 + a = a$.
(P3)	(Existence of additive inverses)	$a + (-a) = (-a) + a = 0$.
(P4)	(Commutative law for addition)	$a + b = b + a$.
(P5)	(Associative law for multiplication)	$a \cdot (b \cdot c) = (a \cdot b) \cdot c$.
(P6)	(Existence of a multiplicative identity)	$a \cdot 1 = 1 \cdot a = a; \quad 1 \neq 0$.
(P7)	(Existence of multiplicative inverses)	$a \cdot a^{-1} = a^{-1} \cdot a = 1$, for $a \neq 0$.
(P8)	(Commutative law for multiplication)	$a \cdot b = b \cdot a$.
(P9)	(Distributive law)	$a \cdot (b + c) = a \cdot b + a \cdot c$.

The three basic properties of numbers which remain to be listed are concerned with inequalities. Although inequalities occur rarely in elementary mathematics, they play a prominent role in calculus. The two notions of inequality, $a < b$ (a is less than b) and $a > b$ (a is greater than b), are intimately related: $a < b$ means the same as $b > a$ (thus $1 < 3$ and $3 > 1$ are merely two ways of writing the same assertion). The numbers a satisfying $a > 0$ are called **positive**, while those numbers a satisfying $a < 0$ are called **negative**. While positivity can thus be defined in terms of $<$, it is possible to reverse the procedure: $a < b$ can be defined to mean that $b - a$ is positive. In fact, it is convenient to consider the collection of all positive numbers, denoted by P, as the basic concept, and state all properties in terms of P:

(P10) (Trichotomy law) For every number a, one and only one of the following holds:

(i) $a = 0$,
(ii) a is in the collection P,
(iii) $-a$ is in the collection P.

(P11) (Closure under addition) If a and b are in P, then $a + b$ is in P.

(P12) (Closure under multiplication) If a and b are in P, then $a \cdot b$ is in P.

These three properties should be complemented with the following definitions:

$$\begin{aligned} a > b &\quad \text{if} \quad a - b \text{ is in } P; \\ a < b &\quad \text{if} \quad b > a; \\ a \geq b &\quad \text{if} \quad a > b \text{ or } a = b; \\ a \leq b &\quad \text{if} \quad a < b \text{ or } a = b.^* \end{aligned}$$

Note, in particular, that $a > 0$ if and only if a is in P.

All the familiar facts about inequalities, however elementary they may seem, are consequences of P10–P12. For example, if a and b are any two numbers, then precisely one of the following holds:

(i) $a - b = 0$,
(ii) $a - b$ is in the collection P,
(iii) $-(a - b) = b - a$ is in the collection P.

Using the definitions just made, it follows that precisely one of the following holds:

(i) $a = b$,
(ii) $a > b$,
(iii) $b > a$.

A slightly more interesting fact results from the following manipulations. If $a < b$, so that $b - a$ is in P, then surely $(b + c) - (a + c)$ is in P; thus, if $a < b$, then $a + c < b + c$. Similarly, suppose $a < b$ and $b < c$. Then

$$\begin{aligned} & b - a \text{ is in } P, \\ \text{and} \quad & c - b \text{ is in } P, \\ \text{so} \quad & c - a = (c - b) + (b - a) \text{ is in } P. \end{aligned}$$

This shows that if $a < b$ and $b < c$, then $a < c$. (The two inequalities $a < b$ and $b < c$ are usually written in the abbreviated form $a < b < c$, which has the third inequality $a < c$ almost built in.)

The following assertion is somewhat less obvious: If $a < 0$ and $b < 0$, then $ab > 0$. The only difficulty presented by the proof is the unraveling of definitions. The symbol $a < 0$ means, by definition, $0 > a$, which means $0 - a = -a$ is in P. Similarly $-b$ is in P, and consequently, by P12, $(-a)(-b) = ab$ is in P. Thus $ab > 0$.

The fact that $ab > 0$ if $a > 0$, $b > 0$ and also if $a < 0$, $b < 0$ has one special consequence: $a^2 > 0$ if $a \neq 0$. Thus squares of nonzero numbers are always positive, and in particular we have proved a result which might have seemed sufficiently elementary to be included in our list of properties: $1 > 0$ (since $1 = 1^2$).

*There is one slightly perplexing feature of the symbols $\geq$ and $\leq$. The statements

$$\begin{aligned} 1 + 1 &\leq 3 \\ 1 + 1 &\leq 2 \end{aligned}$$

are both true, even though we know that $\leq$ could be replaced by $<$ in the first, and by $=$ in the second. This sort of thing is bound to occur when $\leq$ is used with specific numbers; the usefulness of the symbol is revealed by a statement like Theorem 1—here equality holds for some values of a and b, while inequality holds for other values.

The fact that $-a > 0$ if $a < 0$ is the basis of a concept which will play an extremely important role in this book. For any number a, we define the **absolute value** $|a|$ of a as follows:

$$|a| = \begin{cases} a, & a \geq 0 \\ -a, & a \leq 0. \end{cases}$$

Note that $|a|$ is always positive, except when $a = 0$. For example, we have $|-3| = 3$, $|7| = 7$, $|1 + \sqrt{2} - \sqrt{3}| = 1 + \sqrt{2} - \sqrt{3}$, and $|1 + \sqrt{2} - \sqrt{10}| = \sqrt{10} - \sqrt{2} - 1$. In general, the most straightforward approach to any problem involving absolute values requires treating several cases separately, since absolute values are defined by cases to begin with. This approach may be used to prove the following very important fact about absolute values.

THEOREM 1 For all numbers a and b, we have

$$|a + b| \leq |a| + |b|.$$

PROOF We will consider 4 cases:

$$\begin{array}{lll} (1) & a \geq 0, & b \geq 0; \\ (2) & a \geq 0, & b \leq 0; \\ (3) & a \leq 0, & b \geq 0; \\ (4) & a \leq 0, & b \leq 0. \end{array}$$

In case (1) we also have $a + b \geq 0$, and the theorem is obvious; in fact,

$$|a + b| = a + b = |a| + |b|,$$

so that in this case equality holds.

In case (4) we have $a + b \leq 0$, and again equality holds:

$$|a + b| = -(a + b) = -a + (-b) = |a| + |b|.$$

In case (2), when $a \geq 0$ and $b \leq 0$, we must prove that

$$|a + b| \leq a - b.$$

This case may therefore be divided into two subcases. If $a + b \geq 0$, then we must prove that

$$\begin{aligned} & a + b \leq a - b, \\ \text{i.e.,} \quad & \quad\; b \leq -b, \end{aligned}$$

which is certainly true since $b \leq 0$ and hence $-b \geq 0$. On the other hand, if $a + b \leq 0$, we must prove that

$$\begin{aligned} & -a - b \leq a - b, \\ \text{i.e.,} \quad & \quad\; -a \leq a, \end{aligned}$$

which is certainly true since $a \geq 0$ and hence $-a \leq 0$.

Finally, note that case (3) may be disposed of with no additional work, by applying case (2) with a and b interchanged. ▌

Although this method of treating absolute values (separate consideration of various cases) is sometimes the only approach available, there are often simpler methods which may be used. In fact, it is possible to give a much shorter proof of Theorem 1; this proof is motivated by the observation that

$$|a| = \sqrt{a^2}.$$

(Here, and throughout the book, $\sqrt{x}$ denotes the *positive* square root of x; this symbol is defined only when $x \geq 0$.) We may now observe that

$$\begin{aligned}(|a+b|)^2 = (a+b)^2 &= a^2 + 2ab + b^2 \\ &\leq a^2 + 2|a| \cdot |b| + b^2 \\ &= |a|^2 + 2|a| \cdot |b| + |b|^2 \\ &= (|a| + |b|)^2.\end{aligned}$$

From this we can conclude that $|a+b| \leq |a| + |b|$ because $x^2 < y^2$ implies $x < y$, provided that x and y are both nonnegative; a proof of *this* fact is left to the reader (Problem 5).

One final observation may be made about the theorem we have just proved: a close examination of either proof offered shows that

$$|a+b| = |a| + |b|$$

if a and b have the same sign (i.e., are both positive or both negative), or if one of the two is 0, while

$$|a+b| < |a| + |b|$$

if a and b are of opposite signs.

We will conclude this chapter with a subtle point, neglected until now, whose inclusion is required in a conscientious survey of the properties of numbers. After stating property P9, we proved that $a - b = b - a$ implies $a = b$. The proof began by establishing that

$$a \cdot (1+1) = b \cdot (1+1),$$

from which we concluded that $a = b$. This result is obtained from the equation $a \cdot (1+1) = b \cdot (1+1)$ by dividing both sides by $1+1$. Division by 0 should be avoided scrupulously, and it must therefore be admitted that the validity of the argument depends on knowing that $1+1 \neq 0$. Problem 25 is designed to convince you that this fact cannot possibly be proved from properties P1–P9 alone! Once P10, P11, and P12 are available, however, the proof is very simple: We have already seen that $1 > 0$; it follows that $1+1 > 0$, and in particular $1+1 \neq 0$.

This last demonstration has perhaps only strengthened your feeling that it is absurd to bother proving such obvious facts, but an honest assessment of our present situation will help justify serious consideration of such details. In this chapter we have assumed that numbers are familiar objects, and that P1–P12 are merely explicit statements of obvious, well-known properties of numbers. It would be difficult, however, to justify this assumption. Although one learns how to "work with" numbers in school, just what numbers *are*, remains rather vague. A great deal of this book is devoted to elucidating the concept of numbers, and by the end of the book we will have become quite well acquainted with them. But it will be necessary to work with numbers throughout the book. It is therefore reasonable to admit frankly that we do not yet thoroughly understand numbers; we may still say that, in whatever way numbers are finally defined, they should certainly have properties P1–P12.

Most of this chapter has been an attempt to present convincing evidence that P1–P12 are indeed basic properties which we should assume in order to deduce other familiar properties of numbers. Some of the problems (which indicate the derivation of other facts about numbers from P1–P12) are offered as further evidence. It is still a crucial question whether P1–P12 actually account for *all* properties of numbers. As a matter of fact, we shall soon see that they do *not*. In the next chapter the deficiencies of properties P1–P12 will become quite clear, but the proper means for correcting these deficiencies is not so easily discovered. The crucial additional basic property of numbers which we are seeking is profound and subtle, quite unlike P1–P12. The discovery of this crucial property will require all the work of Part II of this book. In the remainder of Part I we will begin to see why some additional property is required; in order to investigate this we will have to consider a little more carefully what we mean by "numbers."

PROBLEMS

1. Prove the following:

(i) If $ax = a$ for some number $a \neq 0$, then $x = 1$.
(ii) $x^2 - y^2 = (x - y)(x + y)$.
(iii) If $x^2 = y^2$, then $x = y$ or $x = -y$.
(iv) $x^3 - y^3 = (x - y)(x^2 + xy + y^2)$.
(v) $x^n - y^n = (x - y)(x^{n-1} + x^{n-2}y + \cdots + xy^{n-2} + y^{n-1})$.
(vi) $x^3 + y^3 = (x + y)(x^2 - xy + y^2)$. (There is a particularly easy way to do this, using (iv), and it will show you how to find a factorization for $x^n + y^n$ whenever n is odd.)

2. What is wrong with the following "proof"? Let $x = y$. Then

$$\begin{aligned} x^2 &= xy, \\ x^2 - y^2 &= xy - y^2, \\ (x + y)(x - y) &= y(x - y), \\ x + y &= y, \\ 2y &= y, \\ 2 &= 1. \end{aligned}$$

3. Prove the following:

(i) $\dfrac{a}{b} = \dfrac{ac}{bc}$, if $b, c \neq 0$.

(ii) $\dfrac{a}{b} + \dfrac{c}{d} = \dfrac{ad + bc}{bd}$, if $b, d \neq 0$.

(iii) $(ab)^{-1} = a^{-1}b^{-1}$, if $a, b \neq 0$. (To do this you must remember the defining property of $(ab)^{-1}$.)

(iv) $\dfrac{a}{b} \cdot \dfrac{c}{d} = \dfrac{ac}{db}$, if $b, d \neq 0$.

(v) $\dfrac{a}{b} \Big/ \dfrac{c}{d} = \dfrac{ad}{bc}$, if $b, c, d \neq 0$.

(vi) If $b, d \neq 0$, then $\dfrac{a}{b} = \dfrac{c}{d}$ if and only if $ad = bc$. Also determine when $\dfrac{a}{b} = \dfrac{b}{a}$.

4. Find all numbers x for which

(i) $4 - x < 3 - 2x$.
(ii) $5 - x^2 < 8$.
(iii) $5 - x^2 < -2$.
(iv) $(x - 1)(x - 3) > 0$. (When is a product of two numbers positive?)
(v) $x^2 - 2x + 2 > 0$.
(vi) $x^2 + x + 1 > 2$.
(vii) $x^2 - x + 10 > 16$.
(viii) $x^2 + x + 1 > 0$.
(ix) $(x - \pi)(x + 5)(x - 3) > 0$.
(x) $(x - \sqrt[3]{2})(x - \sqrt{2}) > 0$.
(xi) $2^x < 8$.
(xii) $x + 3^x < 4$.
(xiii) $\dfrac{1}{x} + \dfrac{1}{1 - x} > 0$.
(xiv) $\dfrac{x - 1}{x + 1} > 0$.

5. Prove the following:

(i) If $a < b$ and $c < d$, then $a + c < b + d$.
(ii) If $a < b$, then $-b < -a$.
(iii) If $a < b$ and $c > d$, then $a - c < b - d$.
(iv) If $a < b$ and $c > 0$, then $ac < bc$.
(v) If $a < b$ and $c < 0$, then $ac > bc$.
(vi) If $a > 1$, then $a^2 > a$.
(vii) If $0 < a < 1$, then $a^2 < a$.
(viii) If $0 \leq a < b$ and $0 \leq c < d$, then $ac < bd$.
(ix) If $0 \leq a < b$, then $a^2 < b^2$. (Use (viii).)
(x) If $a, b \geq 0$ and $a^2 < b^2$, then $a < b$. (Use (ix), backwards.)

6. (a) Prove that if $0 \leq x < y$, then $x^n < y^n$, $n = 1, 2, 3, \ldots$.
(b) Prove that if $x < y$ and n is odd, then $x^n < y^n$.
(c) Prove that if $x^n = y^n$ and n is odd, then $x = y$.
(d) Prove that if $x^n = y^n$ and n is even, then $x = y$ or $x = -y$.

7. Prove that if $0 < a < b$, then

$$a < \sqrt{ab} < \frac{a+b}{2} < b.$$

Notice that the inequality $\sqrt{ab} \le (a+b)/2$ holds for all $a, b \ge 0$. A generalization of this fact occurs in Problem 2-22.

***8.** Although the basic properties of inequalities were stated in terms of the collection P of all positive numbers, and $<$ was defined in terms of P, this procedure can be reversed. Suppose that P10–P12 are replaced by

(P′10) For any numbers a and b one, and only one, of the following holds:

(i) $a = b$,
(ii) $a < b$,
(iii) $b < a$.

(P′11) For any numbers a, b, and c, if $a < b$ and $b < c$, then $a < c$.

(P′12) For any numbers a, b, and c, if $a < b$, then $a + c < b + c$.

(P′13) For any numbers a, b, and c, if $a < b$ and $0 < c$, then $ac < bc$.

Show that P10–P12 can then be deduced as theorems.

9. Express each of the following with at least one less pair of absolute value signs.

(i) $|\sqrt{2} + \sqrt{3} - \sqrt{5} + \sqrt{7}|$.
(ii) $|(|a+b| - |a| - |b|)|$.
(iii) $|(|a+b| + |c| - |a+b+c|)|$.
(iv) $|x^2 - 2xy + y^2|$.
(v) $|(|\sqrt{2} + \sqrt{3}| - |\sqrt{5} - \sqrt{7}|)|$.

10. Express each of the following without absolute value signs, treating various cases separately when necessary.

(i) $|a + b| - |b|$.
(ii) $|(|x| - 1)|$.
(iii) $|x| - |x^2|$.
(iv) $a - |(a - |a|)|$.

11. Find all numbers x for which

(i) $|x - 3| = 8$.
(ii) $|x - 3| < 8$.
(iii) $|x + 4| < 2$.
(iv) $|x - 1| + |x - 2| > 1$.
(v) $|x - 1| + |x + 1| < 2$.

(vi) $|x-1|+|x+1|<1$.
(vii) $|x-1|\cdot|x+1|=0$.
(viii) $|x-1|\cdot|x+2|=3$.

12. Prove the following:

(i) $|xy|=|x|\cdot|y|$.
(ii) $\left|\frac{1}{x}\right|=\frac{1}{|x|}$, if $x\neq 0$. (The best way to do this is to remember what $|x|^{-1}$ is.)
(iii) $\frac{|x|}{|y|}=\left|\frac{x}{y}\right|$, if $y\neq 0$.
(iv) $|x-y|\leq|x|+|y|$. (Give a very short proof.)
(v) $|x|-|y|\leq|x-y|$. (A very short proof is possible, if you write things in the right way.)
(vi) $|(|x|-|y|)|\leq|x-y|$. (Why does this follow immediately from (v)?)
(vii) $|x+y+z|\leq|x|+|y|+|z|$. Indicate when equality holds, and prove your statement.

13. The maximum of two numbers x and y is denoted by $\max(x,y)$. Thus $\max(-1,3)=\max(3,3)=3$ and $\max(-1,-4)=\max(-4,-1)=-1$. The minimum of x and y is denoted by $\min(x,y)$. Prove that

$$\max(x,y)=\frac{x+y+|y-x|}{2},$$
$$\min(x,y)=\frac{x+y-|y-x|}{2}.$$

Derive a formula for $\max(x,y,z)$ and $\min(x,y,z)$, using, for example

$$\max(x,y,z)=\max(x,\max(y,z)).$$

14. (a) Prove that $|a|=|-a|$. (The trick is not to become confused by too many cases. First prove the statement for $a\geq 0$. Why is it then obvious for $a\leq 0$?)
(b) Prove that $-b\leq a\leq b$ if and only if $|a|\leq b$. In particular, it follows that $-|a|\leq a\leq|a|$.
(c) Use this fact to give a new proof that $|a+b|\leq|a|+|b|$.

***15.** Prove that if x and y are not both 0, then

$$x^2+xy+y^2>0,$$
$$x^4+x^3y+x^2y^2+xy^3+y^4>0.$$

Hint: Use Problem 1.

***16.** (a) Show that

$$(x+y)^2=x^2+y^2 \quad \text{only when } x=0 \text{ or } y=0,$$
$$(x+y)^3=x^3+y^3 \quad \text{only when } x=0 \text{ or } y=0 \text{ or } x=-y.$$

(b) Using the fact that

$$x^2 + 2xy + y^2 = (x+y)^2 \geq 0,$$

show that $4x^2 + 6xy + 4y^2 > 0$ unless x and y are both 0.

(c) Use part (b) to find out when $(x+y)^4 = x^4 + y^4$.

(d) Find out when $(x+y)^5 = x^5+y^5$. Hint: From the assumption $(x+y)^5 = x^5+y^5$ you should be able to derive the equation $x^3+2x^2y+2xy^2+y^3 = 0$, if $xy \neq 0$. This implies that $(x+y)^3 = x^2y + xy^2 = xy(x+y)$.

You should now be able to make a good guess as to when $(x+y)^n = x^n + y^n$; the proof is contained in Problem 11-57.

17. (a) Find the smallest possible value of $2x^2 - 3x + 4$. Hint: "Complete the square," i.e., write $2x^2 - 3x + 4 = 2(x - 3/4)^2 + ?$

(b) Find the smallest possible value of $x^2 - 3x + 2y^2 + 4y + 2$.

(c) Find the smallest possible value of $x^2 + 4xy + 5y^2 - 4x - 6y + 7$.

18. (a) Suppose that $b^2 - 4c \geq 0$. Show that the numbers

$$\frac{-b + \sqrt{b^2 - 4c}}{2}, \qquad \frac{-b - \sqrt{b^2 - 4c}}{2}$$

both satisfy the equation $x^2 + bx + c = 0$.

(b) Suppose that $b^2 - 4c < 0$. Show that there are no numbers x satisfying $x^2 + bx + c = 0$; in fact, $x^2 + bx + c > 0$ for all x. Hint: Complete the square.

(c) Use this fact to give another proof that if x and y are not both 0, then $x^2 + xy + y^2 > 0$.

(d) For which numbers α is it true that $x^2 + \alpha xy + y^2 > 0$ whenever x and y are not both 0?

(e) Find the smallest possible value of $x^2 + bx + c$ and of $ax^2 + bx + c$, for $a > 0$.

19. The fact that $a^2 \geq 0$ for all numbers a, elementary as it may seem, is nevertheless the fundamental idea upon which most important inequalities are ultimately based. The great-granddaddy of all inequalities is the *Schwarz inequality*:

$$x_1y_1 + x_2y_2 \leq \sqrt{x_1{}^2 + x_2{}^2}\sqrt{y_1{}^2 + y_2{}^2}.$$

(A more general form occurs in Problem 2-21.) The three proofs of the Schwarz inequality outlined below have only one thing in common—their reliance on the fact that $a^2 \geq 0$ for all a.

(a) Prove that if $x_1 = \lambda y_1$ and $x_2 = \lambda y_2$ for some number λ, then equality holds in the Schwarz inequality. Prove the same thing if $y_1 = y_2 = 0$. Now suppose that y_1 and y_2 are not both 0, and that there is no number

λ such that $x_1 = \lambda y_1$ and $x_2 = \lambda y_2$. Then

$$\begin{aligned} 0 &< (\lambda y_1 - x_1)^2 + (\lambda y_2 - x_2)^2 \\ &= \lambda^2({y_1}^2 + {y_2}^2) - 2\lambda(x_1 y_1 + x_2 y_2) + ({x_1}^2 + {x_2}^2). \end{aligned}$$

Using Problem 18, complete the proof of the Schwarz inequality.

(b) Prove the Schwarz inequality by using $2xy \le x^2 + y^2$ (how is this derived?) with

$$x = \frac{x_i}{\sqrt{{x_1}^2 + {x_2}^2}}, \qquad y = \frac{y_i}{\sqrt{{y_1}^2 + {y_2}^2}},$$

first for $i = 1$ and then for $i = 2$.

(c) Prove the Schwarz inequality by first proving that

$$({x_1}^2 + {x_2}^2)({y_1}^2 + {y_2}^2) = (x_1 y_1 + x_2 y_2)^2 + (x_1 y_2 - x_2 y_1)^2.$$

(d) Deduce, from each of these three proofs, that equality holds only when $y_1 = y_2 = 0$ or when there is a number λ such that $x_1 = \lambda y_1$ and $x_2 = \lambda y_2$.

In our later work, three facts about inequalities will be crucial. Although proofs will be supplied at the appropriate point in the text, a personal assault on these problems is infinitely more enlightening than a perusal of a completely worked-out proof. The statements of these propositions involve some weird numbers, but their basic message is very simple: if x is close enough to x_0, and y is close enough to y_0, then $x+y$ will be close to $x_0 + y_0$, and xy will be close to $x_0 y_0$, and $1/y$ will be close to $1/y_0$. The symbol "ε" which appears in these propositions is the fifth letter of the Greek alphabet ("epsilon"), and could just as well be replaced by a less intimidating Roman letter; however, tradition has made the use of ε almost sacrosanct in the contexts to which these theorems apply.

20. Prove that if

$$|x - x_0| < \frac{\varepsilon}{2} \quad \text{and} \quad |y - y_0| < \frac{\varepsilon}{2},$$

then

$$\begin{aligned} |(x + y) - (x_0 + y_0)| &< \varepsilon, \\ |(x - y) - (x_0 - y_0)| &< \varepsilon. \end{aligned}$$

***21.** Prove that if

$$|x - x_0| < \min\left(\frac{\varepsilon}{2(|y_0| + 1)}, 1\right) \quad \text{and} \quad |y - y_0| < \frac{\varepsilon}{2(|x_0| + 1)},$$

then $|xy - x_0 y_0| < \varepsilon$.

(The notation "min" was defined in Problem 13, but the formula provided by that problem is irrelevant at the moment; the first inequality in the hypothesis just means that

$$|x - x_0| < \frac{\varepsilon}{2(|y_0| + 1)} \quad \text{and} \quad |x - x_0| < 1;$$

at one point in the argument you will need the first inequality, and at another point you will need the second. One more word of advice: since the hypotheses only provide information about $x - x_0$ and $y - y_0$, it is almost a foregone conclusion that the proof will depend upon writing $xy - x_0y_0$ in a way that involves $x - x_0$ and $y - y_0$.)

***22.** Prove that if $y_0 \neq 0$ and

$$|y - y_0| < \min\left(\frac{|y_0|}{2}, \frac{\varepsilon |y_0|^2}{2}\right),$$

then $y \neq 0$ and

$$\left|\frac{1}{y} - \frac{1}{y_0}\right| < \varepsilon.$$

***23.** Replace the question marks in the following statement by expressions involving ε, x_0, and y_0 so that the conclusion will be true:

If $y_0 \neq 0$ and

$$|y - y_0| < ? \quad \text{and} \quad |x - x_0| < ?$$

then $y \neq 0$ and

$$\left|\frac{x}{y} - \frac{x_0}{y_0}\right| < \varepsilon.$$

This problem is trivial in the sense that its solution follows from Problems 21 and 22 with almost no work at all (notice that $x/y = x \cdot 1/y$). The crucial point is not to become confused; decide which of the two problems should be used first, and don't panic if your answer looks unlikely.

***24.** This problem shows that the actual placement of parentheses in a sum is irrelevant. The proofs involve "mathematical induction"; if you are not familiar with such proofs, but still want to tackle this problem, it can be saved until after Chapter 2, where proofs by induction are explained.

Let us agree, for definiteness, that $a_1 + \cdots + a_n$ will denote

$$a_1 + (a_2 + (a_3 + \cdots + (a_{n-2} + (a_{n-1} + a_n))) \cdots).$$

Thus $a_1 + a_2 + a_3$ denotes $a_1 + (a_2 + a_3)$, and $a_1 + a_2 + a_3 + a_4$ denotes $a_1 + (a_2 + (a_3 + a_4))$, etc.

(a) Prove that

$$(a_1 + \cdots + a_k) + a_{k+1} = a_1 + \cdots + a_{k+1}.$$

Hint: Use induction on k.

(b) Prove that if $n \geq k$, then

$$(a_1 + \cdots + a_k) + (a_{k+1} + \cdots + a_n) = a_1 + \cdots + a_n.$$

Hint: Use part (a) to give a proof by induction on k.

(c) Let $s(a_1, \ldots, a_k)$ be some sum formed from $a_1, \ldots, a_k$. Show that

$$s(a_1, \ldots, a_k) = a_1 + \cdots + a_k.$$

Hint: There must be two sums $s'(a_1, \ldots, a_l)$ and $s''(a_{l+1}, \ldots, a_k)$ such that

$$s(a_1, \ldots, a_k) = s'(a_1, \ldots, a_l) + s''(a_{l+1}, \ldots, a_k).$$

25. Suppose that we interpret "number" to mean either 0 or 1, and $+$ and $\cdot$ to be the operations defined by the following two tables.

$+$	0	1
0	0	1
1	1	0

$\cdot$	0	1
0	0	0
1	0	1

Check that properties P1–P9 all hold, even though $1 + 1 = 0$.

CHAPTER 2 NUMBERS OF VARIOUS SORTS

In Chapter 1 we used the word "number" very loosely, despite our concern with the basic properties of numbers. It will now be necessary to distinguish carefully various kinds of numbers.

The simplest numbers are the "counting numbers"

$$1, 2, 3, \ldots.$$

The fundamental significance of this collection of numbers is emphasized by its symbol $\mathbf{N}$ (for **natural numbers**). A brief glance at P1–P12 will show that our basic properties of "numbers" do not apply to $\mathbf{N}$—for example, P2 and P3 do not make sense for $\mathbf{N}$. From this point of view the system $\mathbf{N}$ has many deficiencies. Nevertheless, $\mathbf{N}$ is sufficiently important to deserve several comments before we consider larger collections of numbers.

The most basic property of $\mathbf{N}$ is the principle of "mathematical induction." Suppose $P(x)$ means that the property P holds for the number x. Then the principle of mathematical induction states that $P(x)$ is true for all natural numbers x provided that

(1) $P(1)$ is true.

(2) Whenever $P(k)$ is true, $P(k+1)$ is true.

Note that condition (2) merely asserts the truth of $P(k+1)$ under the assumption that $P(k)$ is true; this suffices to ensure the truth of $P(x)$ for all x, if condition (1) also holds. In fact, if $P(1)$ is true, then it follows that $P(2)$ is true (by using (2) in the special case $k = 1$). Now, since $P(2)$ is true it follows that $P(3)$ is true (using (2) in the special case $k = 2$). It is clear that each number will eventually be reached by a series of steps of this sort, so that $P(k)$ is true for all numbers k.

A favorite illustration of the reasoning behind mathematical induction envisions an infinite line of people,

$$\text{person number 1, person number 2, person number 3, } \ldots.$$

If each person has been instructed to tell any secret he hears to the person behind him (the one with the next largest number) and a secret is told to person number 1, then clearly every person will eventually learn the secret. If $P(x)$ is the assertion that person number x will learn the secret, then the instructions given (to tell all secrets learned to the next person) assures that condition (2) is true, and telling the secret to person number 1 makes (1) true. The following example is a less facetious use of mathematical induction. There is a useful and striking formula which expresses the sum of the first n numbers in a simple way:

$$1 + \cdots + n = \frac{n(n+1)}{2}.$$

To prove this formula, note first that it is clearly true for $n = 1$. Now *assume* that for some natural number k we have

$$1 + \cdots + k = \frac{k(k+1)}{2}.$$

Then

$$\begin{aligned} 1 + \cdots + k + (k+1) &= \frac{k(k+1)}{2} + k + 1 \\ &= \frac{k(k+1) + 2k + 2}{2} \\ &= \frac{k^2 + 3k + 2}{2} \\ &= \frac{(k+1)(k+2)}{2}, \end{aligned}$$

so the formula is also true for $k + 1$. By the principle of induction this proves the formula for all natural numbers n. This particular example illustrates a phenomenon that frequently occurs, especially in connection with formulas like the one just proved. Although the proof by induction is often quite straightforward, the method by which the formula was discovered remains a mystery. Problems 5 and 6 indicate how some formulas of this type may be derived.

The principle of mathematical induction may be formulated in an equivalent way without speaking of "properties" of a number, a term which is sufficiently vague to be eschewed in a mathematical discussion. A more precise formulation states that if A is any collection (or "set"—a synonymous mathematical term) of natural numbers and

(1) 1 is in A,
(2) $k + 1$ is in A whenever k is in A,

then A is the set of all natural numbers. It should be clear that this formulation adequately replaces the less formal one given previously—we just consider the set A of natural numbers x which satisfy $P(x)$. For example, suppose A is the set of natural numbers n for which it is true that

$$1 + \cdots + n = \frac{n(n+1)}{2}.$$

Our previous proof of this formula showed that A contains 1, and that $k + 1$ is in A, if k is. It follows that A is the set of all natural numbers, i.e., that the formula holds for all natural numbers n.

There is yet another rigorous formulation of the principle of mathematical induction, which looks quite different. If A is any collection of natural numbers, it is tempting to say that A must have a smallest member. Actually, this statement can fail to be true in a rather subtle way. A particularly important set of natural numbers is the collection A that contains no natural numbers at all, the "empty collection" or "null set,"* denoted by $\emptyset$. The null set $\emptyset$ is a collection of natural numbers that has no smallest member—in fact, it has no members at all. This is the only possible exception, however; if A is a nonnull set of natural numbers, then A has a least member. This "intuitively obvious" statement, known as the "well-ordering principle," can be proved from the principle of induction as follows. Suppose that the set A has no least member. Let B be the set of natural numbers n such that $1, \dots, n$ are all *not* in A. Clearly 1 is in B (because if 1 were in A, then A would have 1 as smallest member). Moreover, if $1, \dots, k$ are not in A, surely $k+1$ is not in A (otherwise $k+1$ would be the smallest member of A), so $1, \dots, k+1$ are all not in A. This shows that if k is in B, then $k+1$ is in B. It follows that every number n is in B, i.e., the numbers $1, \dots, n$ are *not* in A for any natural number n. Thus $A = \emptyset$, which completes the proof.

It is also possible to prove the principle of induction from the well-ordering principle (Problem 10). Either principle may be considered as a basic assumption about the natural numbers.

There is still another form of induction which should be mentioned. It sometimes happens that in order to prove $P(k+1)$ we must assume not only $P(k)$, but also $P(l)$ for all natural numbers $l \le k$. In this case we rely on the "principle of complete induction": If A is a set of natural numbers and

(1) 1 is in A,

(2) $k+1$ is in A if $1, \dots, k$ are in A,

then A is the set of all natural numbers.

Although the principle of complete induction may appear much stronger than the ordinary principle of induction, it is actually a consequence of that principle. The proof of this fact is left to the reader, with a hint (Problem 11). Applications will be found in Problems 7, 17, 20 and 22.

Closely related to proofs by induction are "recursive definitions." For example, the number $n!$ (read "n factorial") is defined as the product of all the natural numbers less than or equal to n:

$$n! = 1 \cdot 2 \cdot \ldots \cdot (n-1) \cdot n.$$

This can be expressed more precisely as follows:

$$\begin{aligned} &(1) \quad 1! = 1 \\ &(2) \quad n! = n \cdot (n-1)!. \end{aligned}$$

This form of the definition exhibits the relationship between $n!$ and $(n-1)!$ in an

*Although it may not strike you as a collection, in the ordinary sense of the word, the null set arises quite naturally in many contexts. We frequently consider the set A, consisting of all x satisfying some property P; often we have no guarantee that P is satisfied by *any* number, so that A might be $\emptyset$—in fact often one proves that P is always false by showing that $A = \emptyset$.

explicit way that is ideally suited for proofs by induction. Problem 23 reviews a definition already familiar to you, which may be expressed more succinctly as a recursive definition; as this problem shows, the recursive definition is really necessary for a rigorous proof of some of the basic properties of the definition.

One definition which may not be familiar involves some convenient notation which we will constantly be using. Instead of writing

$$a_1 + \cdots + a_n,$$

we will usually employ the Greek letter Σ (capital sigma, for "sum") and write

$$\sum_{i=1}^{n} a_i.$$

In other words, $\sum_{i=1}^{n} a_i$ denotes the sum of the numbers obtained by letting $i = 1, 2, \ldots, n$. Thus

$$\sum_{i=1}^{n} i = 1 + 2 + \cdots + n = \frac{n(n+1)}{2}.$$

Notice that the letter i really has nothing to do with the number denoted by $\sum_{i=1}^{n} i$, and can be replaced by any convenient symbol (except n, of course!):

$$\sum_{j=1}^{n} j = \frac{n(n+1)}{2},$$

$$\sum_{j=1}^{i} j = \frac{i(i+1)}{2},$$

$$\sum_{n=1}^{j} n = \frac{j(j+1)}{2}.$$

To define $\sum_{i=1}^{n} a_i$ precisely really requires a recursive definition:

$$\text{(1)} \quad \sum_{i=1}^{1} a_i = a_1,$$

$$\text{(2)} \quad \sum_{i=1}^{n} a_i = \sum_{i=1}^{n-1} a_i + a_n.$$

But only purveyors of mathematical austerity would insist too strongly on such precision. In practice, all sorts of modifications of this symbolism are used, and no one ever considers it necessary to add any words of explanation. The symbol

$$\sum_{\substack{i=1\\ i\neq 4}}^{n} a_i,$$

for example, is an obvious way of writing

$$a_1 + a_2 + a_3 + a_5 + a_6 + \cdots + a_n,$$

or more precisely,

$$\sum_{i=1}^{3} a_i + \sum_{i=5}^{n} a_i.$$

The deficiencies of the natural numbers which we discovered at the beginning of this chapter may be partially remedied by extending this system to the set of **integers**

$$\dots, -2, -1, 0, 1, 2, \dots.$$

This set is denoted by **Z** (from German "Zahl," number). Of properties P1–P12, only P7 fails for **Z**.

A still larger system of numbers is obtained by taking quotients m/n of integers (with $n \neq 0$). These numbers are called **rational numbers**, and the set of all rational numbers is denoted by **Q** (for "quotients"). In this system of numbers all of P1–P12 are true. It is tempting to conclude that the "properties of numbers," which we studied in some detail in Chapter 1, refer to just one set of numbers, namely, **Q**. There is, however, a still larger collection of numbers to which properties P1–P12 apply—the set of all **real numbers**, denoted by **R**. The real numbers include not only the rational numbers, but other numbers as well (the **irrational numbers**) which can be represented by infinite decimals; π and $\sqrt{2}$ are both examples of irrational numbers. The proof that π is irrational is not easy—we shall devote all of Chapter 16 of Part III to a proof of this fact. The irrationality of $\sqrt{2}$, on the other hand, is quite simple, and was known to the Greeks. (Since the Pythagorean theorem shows that an isosceles right triangle, with sides of length 1, has a hypotenuse of length $\sqrt{2}$, it is not surprising that the Greeks should have investigated this question.) The proof depends on a few observations about the natural numbers. Every natural number n can be written either in the form $2k$ for some integer k, or else in the form $2k + 1$ for some integer k (this "obvious" fact has a simple proof by induction (Problem 8)). Those natural numbers of the form $2k$ are called **even**; those of the form $2k + 1$ are called **odd**. Note that even numbers have even squares, and odd numbers have odd squares:

$$(2k)^2 = 4k^2 = 2 \cdot (2k^2),$$
$$(2k+1)^2 = 4k^2 + 4k + 1 = 2 \cdot (2k^2 + 2k) + 1.$$

In particular it follows that the converse must also hold: if n^2 is even, then n is even; if n^2 is odd, then n is odd. The proof that $\sqrt{2}$ is irrational is now quite simple. Suppose that $\sqrt{2}$ were rational; that is, suppose there were natural numbers p

and q such that

$$\left(\frac{p}{q}\right)^2 = 2.$$

We can assume that p and q have no common divisor (since all common divisors could be divided out to begin with). Now we have

$$p^2 = 2q^2.$$

This shows that p^2 is even, and consequently p must be even; that is, $p = 2k$ for some natural number k. Then

$$p^2 = 4k^2 = 2q^2,$$

so

$$2k^2 = q^2.$$

This shows that q^2 is even, and consequently that q is even. Thus both p and q are even, contradicting the fact that p and q have no common divisor. This contradiction completes the proof.

It is important to understand precisely what this proof shows. We have demonstrated that there is no rational number x such that $x^2 = 2$. This assertion is often expressed more briefly by saying that $\sqrt{2}$ is irrational. Note, however, that the use of the symbol $\sqrt{2}$ implies the existence of *some* number (necessarily irrational) whose square is 2. We have not proved that such a number exists and we can assert confidently that, at present, a proof is *impossible* for us. Any proof at this stage would have to be based on P1–P12 (the only properties of **R** we have mentioned); since P1–P12 are also true for **Q** the exact same argument would show that there is a rational number whose square is 2, and this we know is false. (Note that the reverse argument will not work—our proof that there is no rational number whose square is 2 cannot be used to show that there is no real number whose square is 2, because our proof used not only P1–P12 but also a special property of **Q**, the fact that every number in **Q** can be written p/q for integers p and q.)

This particular deficiency in our list of properties of the real numbers could, of course, be corrected by adding a new property which asserts the existence of square roots of positive numbers. Resorting to such a measure is, however, neither aesthetically pleasing nor mathematically satisfactory; we would still not know that every number has an nth root if n is odd, and that every positive number has an nth root if n is even. Even if we assumed this, we could not prove the existence of a number x satisfying $x^5 + x + 1 = 0$ (even though there does happen to be one), since we do not know how to write the solution of the equation in terms of nth roots (in fact, it is known that the solution cannot be written in this form). And, of course, we certainly do not wish to assume that all equations have solutions, since this is false (no real number x satisfies $x^2 + 1 = 0$, for example). In fact, this direction of investigation is not a fruitful one. The most useful hints about the property distinguishing **R** from **Q**, the most compelling evidence for the necessity of elucidating this property, do not come from the study of numbers alone. In order to study the properties of the real numbers in a more profound way, we must study more than the real numbers. At this point we must begin with the foundations of calculus, in particular the fundamental concept on which calculus is based—functions.

PROBLEMS

1. Prove the following formulas by induction.

(i) $1^2 + \cdots + n^2 = \dfrac{n(n+1)(2n+1)}{6}$.

(ii) $1^3 + \cdots + n^3 = (1 + \cdots + n)^2$.

2. Find a formula for

(i) $\displaystyle\sum_{i=1}^{n}(2i-1) = 1 + 3 + 5 + \cdots + (2n-1)$.

(ii) $\displaystyle\sum_{i=1}^{n}(2i-1)^2 = 1^2 + 3^2 + 5^2 + \cdots + (2n-1)^2$.

Hint: What do these expressions have to do with $1 + 2 + 3 + \cdots + 2n$ and $1^2 + 2^2 + 3^2 + \cdots + (2n)^2$?

3. If $0 \le k \le n$, the "binomial coefficient" $\binom{n}{k}$ is defined by

$$\binom{n}{k} = \frac{n!}{k!(n-k)!} = \frac{n(n-1)\cdots(n-k+1)}{k!}, \quad \text{if } k \neq 0, n$$

$$\binom{n}{0} = \binom{n}{n} = 1.$$ (This becomes a special case of the first formula if we define $0! = 1$.)

(a) Prove that

$$\binom{n+1}{k} = \binom{n}{k-1} + \binom{n}{k}.$$

(The proof does not require an induction argument.)

This relation gives rise to the following configuration, known as "Pascal's triangle"—a number not on one of the sides is the sum of the two numbers above it; the binomial coefficient $\binom{n}{k}$ is the $(k+1)$st number in the $(n+1)$st row.

$$\begin{array}{ccccccccccc} &&&&& 1 &&&&& \\ &&&& 1 && 1 &&&& \\ &&& 1 && 2 && 1 &&& \\ && 1 && 3 && 3 && 1 && \\ & 1 && 4 && 6 && 4 && 1 & \\ 1 && 5 && 10 && 10 && 5 && 1 \\ &&&&& \cdots &&&&& \end{array}$$

(b) Notice that all the numbers in Pascal's triangle are natural numbers. Use part (a) to prove by induction that $\binom{n}{k}$ is always a natural number. (Your entire proof by induction will, in a sense, be summed up in a glance by Pascal's triangle.)

(c) Give another proof that $\binom{n}{k}$ is a natural number by showing that $\binom{n}{k}$ is the number of sets of exactly k integers each chosen from 1, $\dots, n$.

(d) Prove the "binomial theorem": If a and b are any numbers and n is a natural number, then

$$(a+b)^n = a^n + \binom{n}{1}a^{n-1}b + \binom{n}{2}a^{n-2}b^2 + \cdots + \binom{n}{n-1}ab^{n-1} + b^n$$
$$= \sum_{j=0}^{n} \binom{n}{j} a^{n-j}b^j.$$

(e) Prove that

(i) $\displaystyle\sum_{j=0}^{n} \binom{n}{j} = \binom{n}{0} + \cdots + \binom{n}{n} = 2^n.$

(ii) $\displaystyle\sum_{j=0}^{n} (-1)^j \binom{n}{j} = \binom{n}{0} - \binom{n}{1} + \cdots \pm \binom{n}{n} = 0.$

(iii) $\displaystyle\sum_{l \text{ odd}} \binom{n}{l} = \binom{n}{1} + \binom{n}{3} + \cdots = 2^{n-1}.$

(iv) $\displaystyle\sum_{l \text{ even}} \binom{n}{l} = \binom{n}{0} + \binom{n}{2} + \cdots = 2^{n-1}.$

4. (a) Prove that

$$\sum_{k=0}^{l} \binom{n}{k}\binom{m}{l-k} = \binom{n+m}{l}.$$

Hint: Apply the binomial theorem to $(1+x)^n(1+x)^m$.

(b) Prove that

$$\sum_{k=0}^{n} \binom{n}{k}^2 = \binom{2n}{n}.$$

5. (a) Prove by induction on n that

$$1 + r + r^2 + \cdots + r^n = \frac{1 - r^{n+1}}{1 - r}$$

if $r \neq 1$ (if $r = 1$, evaluating the sum certainly presents no problem).

(b) Derive this result by setting $S = 1+r+\cdots+r^n$, multiplying this equation by r, and solving the two equations for S.

6. The formula for $1^2 + \cdots + n^2$ may be derived as follows. We begin with the formula

$$(k+1)^3 - k^3 = 3k^2 + 3k + 1.$$

Writing this formula for $k = 1, \ldots, n$ and adding, we obtain

$$\begin{aligned}
2^3 - 1^3 &= 3 \cdot 1^2 + 3 \cdot 1 + 1 \\
3^3 - 2^3 &= 3 \cdot 2^2 + 3 \cdot 2 + 1 \\
&\ \ \vdots \\
(n+1)^3 - n^3 &= 3 \cdot n^2 + 3 \cdot n + 1 \\
\hline
(n+1)^3 - 1 &= 3[1^2 + \cdots + n^2] + 3[1 + \cdots + n] + n.
\end{aligned}$$

Thus we can find $\sum_{k=1}^{n} k^2$ if we already know $\sum_{k=1}^{n} k$ (which could have been found in a similar way). Use this method to find

(i) $1^3 + \cdots + n^3$.

(ii) $1^4 + \cdots + n^4$.

(iii) $\dfrac{1}{1 \cdot 2} + \dfrac{1}{2 \cdot 3} + \cdots + \dfrac{1}{n(n+1)}$.

(iv) $\dfrac{3}{1^2 \cdot 2^2} + \dfrac{5}{2^2 \cdot 3^2} + \cdots + \dfrac{2n+1}{n^2(n+1)^2}$.

***7.** Use the method of Problem 6 to show that $\sum_{i=1}^{n} k^p$ can always be written in the form

$$\frac{n^{p+1}}{p+1} + An^p + Bn^{p-1} + Cn^{p-2} + \cdots .$$

(The first 10 such expressions are

$$\sum_{k=1}^{n} k = \tfrac{1}{2}n^2 + \tfrac{1}{2}n$$

$$\sum_{k=1}^{n} k^2 = \tfrac{1}{3}n^3 + \tfrac{1}{2}n^2 + \tfrac{1}{6}n$$

$$\sum_{k=1}^{n} k^3 = \tfrac{1}{4}n^4 + \tfrac{1}{2}n^3 + \tfrac{1}{4}n^2$$

$$\sum_{k=1}^{n} k^4 = \tfrac{1}{5}n^5 + \tfrac{1}{2}n^4 + \tfrac{1}{3}n^3 - \tfrac{1}{30}n$$

$$\sum_{k=1}^{n} k^5 = \tfrac{1}{6}n^6 + \tfrac{1}{2}n^5 + \tfrac{5}{12}n^4 - \tfrac{1}{12}n^2$$

$$\sum_{k=1}^{n} k^6 = \tfrac{1}{7}n^7 + \tfrac{1}{2}n^6 + \tfrac{1}{2}n^5 - \tfrac{1}{6}n^3 + \tfrac{1}{42}n$$

$$\sum_{k=1}^{n} k^7 = \tfrac{1}{8}n^8 + \tfrac{1}{2}n^7 + \tfrac{7}{12}n^6 - \tfrac{7}{24}n^4 + \tfrac{1}{12}n^2$$

$$\sum_{k=1}^{n} k^8 = \tfrac{1}{9}n^9 + \tfrac{1}{2}n^8 + \tfrac{2}{3}n^7 - \tfrac{7}{15}n^5 + \tfrac{2}{9}n^3 - \tfrac{1}{30}n$$

$$\sum_{k=1}^{n} k^9 = \tfrac{1}{10}n^{10} + \tfrac{1}{2}n^9 + \tfrac{3}{4}n^8 - \tfrac{7}{10}n^6 + \tfrac{1}{2}n^4 - \tfrac{3}{20}n^2$$

$$\sum_{k=1}^{n} k^{10} = \tfrac{1}{11}n^{11} + \tfrac{1}{2}n^{10} + \tfrac{5}{6}n^9 - 1n^7 + 1n^5 - \tfrac{1}{2}n^3 + \tfrac{5}{66}n.$$

Notice that the coefficients in the second column are always $\frac{1}{2}$, and that after the third column the powers of n with nonzero coefficients decrease by 2 until n^2 or n is reached. The coefficients in all but the first two columns seem to be rather haphazard, but there actually is some sort of pattern; finding it may be regarded as a super-perspicacity test. See Problem 27-17 for the complete story.)

8. Prove that every natural number is either even or odd.

9. Prove that if a set A of natural numbers contains n_0 and contains $k+1$ whenever it contains k, then A contains all natural numbers $\geq n_0$.

10. Prove the principle of mathematical induction from the well-ordering principle.

11. Prove the principle of complete induction from the ordinary principle of induction. Hint: If A contains 1 and A contains $n+1$ whenever it contains $1, \ldots, n$, consider the set B of all k such that $1, \ldots, k$ are all in A.

12. (a) If a is rational and b is irrational, is $a+b$ necessarily irrational? What if a and b are both irrational?
(b) If a is rational and b is irrational, is ab necessarily irrational? (Careful!)
(c) Is there a number a such that a^2 is irrational, but a^4 is rational?

(d) Are there two irrational numbers whose sum and product are both rational?

13. (a) Prove that $\sqrt{3}$, $\sqrt{5}$, and $\sqrt{6}$ are irrational. Hint: To treat $\sqrt{3}$, for example, use the fact that every integer is of the form $3n$ or $3n+1$ or $3n+2$. Why doesn't this proof work for $\sqrt{4}$?
(b) Prove that $\sqrt[3]{2}$ and $\sqrt[3]{3}$ are irrational.

14. Prove that

(a) $\sqrt{2}+\sqrt{6}$ is irrational.
(b) $\sqrt{2}+\sqrt{3}$ is irrational.

15. (a) Prove that if $x = p+\sqrt{q}$ where p and q are rational, and m is a natural number, then $x^m = a+b\sqrt{q}$ for some rational a and b.
(b) Prove also that $(p-\sqrt{q})^m = a-b\sqrt{q}$.

16. (a) Prove that if m and n are natural numbers and $m^2/n^2 < 2$, then $(m+2n)^2/(m+n)^2 > 2$; show, moreover, that

$$\frac{(m+2n)^2}{(m+n)^2} - 2 < 2 - \frac{m^2}{n^2}.$$

(b) Prove the same results with all inequality signs reversed.
(c) Prove that if $m/n < \sqrt{2}$, then there is another rational number m'/n' with $m/n < m'/n' < \sqrt{2}$.

***17.** It seems likely that $\sqrt{n}$ is irrational whenever the natural number n is not the square of another natural number. Although the method of Problem 13 may actually be used to treat any particular case, it is not clear in advance that it will always work, and a proof for the general case requires some extra information. A natural number p is called a **prime number** if it is impossible to write $p = ab$ for natural numbers a and b unless one of these is p, and the other 1; for convenience we also agree that 1 is *not* a prime number. The first few prime numbers are 2, 3, 5, 7, 11, 13, 17, 19. If $n > 1$ is not a prime, then $n = ab$, with a and b both $< n$; if either a or b is not a prime it can be factored similarly; continuing in this way proves that we can write n as a product of primes. For example, $28 = 4 \cdot 7 = 2 \cdot 27$.

(a) Turn this argument into a rigorous proof by complete induction. (To be sure, any reasonable mathematician would accept the informal argument, but this is partly because it would be obvious to her how to state it rigorously.)

A fundamental theorem about integers, which we will not prove here, states that this factorization is unique, except for the order of the factors. Thus, for example, 28 can never be written as a product of primes one of which is 3, nor can it be written in a way that involves 2 only once (now you should appreciate why 1 is not allowed as a prime).

(b) Using this fact, prove that $\sqrt{n}$ is irrational unless $n = m^2$ for some natural number m.

(c) Prove more generally that $\sqrt[k]{n}$ is irrational unless $n = m^k$.

(d) No discussion of prime numbers should fail to allude to Euclid's beautiful proof that there are infinitely many of them. Prove that there cannot be only finitely many prime numbers $p_1, p_2, p_3, \dots, p_n$ by considering $p_1 \cdot p_2 \cdot \dots \cdot p_n + 1$.

***18.** (a) Prove that if x satisfies

$$x^n + a_{n-1}x^{n-1} + \cdots + a_0 = 0,$$

for some integers $a_{n-1}, \dots, a_0$, then x is irrational unless x is an integer. (Why is this a generalization of Problem 17?)

(b) Prove that $\sqrt{6} - \sqrt{2} - \sqrt{3}$ is irrational.

(c) Prove that $\sqrt{2} + \sqrt[3]{2}$ is irrational. Hint: Start by working out the first 6 powers of this number.

19. Prove Bernoulli's inequality: If $h > -1$, then

$$(1+h)^n \geq 1 + nh.$$

Why is this trivial if $h > 0$?

20. The Fibonacci sequence $a_1, a_2, a_3, \dots$ is defined as follows:

$$\begin{aligned} a_1 &= 1, \\ a_2 &= 1, \\ a_n &= a_{n-1} + a_{n-2} \qquad \text{for } n \geq 3. \end{aligned}$$

This sequence, which begins $1, 1, 2, 3, 5, 8, \dots$, was discovered by Fibonacci (circa 1175–1250), in connection with a problem about rabbits. Fibonacci assumed that an initial pair of rabbits gave birth to one new pair of rabbits per month, and that after two months each new pair behaved similarly. The number a_n of pairs born in the nth month is $a_{n-1} + a_{n-2}$, because a pair of rabbits is born for each pair born the previous month, and moreover each pair born two months ago now gives birth to another pair. The number of interesting results about this sequence is truly amazing—there is even a Fibonacci Association which publishes a journal, *The Fibonacci Quarterly*. Prove that

$$a_n = \frac{\left(\dfrac{1+\sqrt{5}}{2}\right)^n - \left(\dfrac{1-\sqrt{5}}{2}\right)^n}{\sqrt{5}}.$$

One way of deriving this astonishing formula is presented in Problem 24-15.

21. The Schwarz inequality (Problem 1-19) actually has a more general form:

$$\sum_{i=1}^n x_i y_i \leq \sqrt{\sum_{i=1}^n x_i^2}\sqrt{\sum_{i=1}^n y_i^2}.$$

Give three proofs of this, analogous to the three proofs in Problem 1-19.

22. The result in Problem 1-7 has an important generalization: If $a_1, \ldots, a_n \geq 0$, then the "arithmetic mean"

$$A_n = \frac{a_1 + \cdots + a_n}{n}$$

and "geometric mean"

$$G_n = \sqrt[n]{a_1 \ldots a_n}$$

satisfy

$$G_n \leq A_n.$$

(a) Suppose that $a_1 < A_n$. Then some a_i satisfies $a_i > A_n$; for convenience, say $a_2 > A_n$. Let $\bar{a}_1 = A_n$ and let $\bar{a}_2 = a_1 + a_2 - \bar{a}_1$. Show that

$$\bar{a}_1\bar{a}_2 \geq a_1a_2.$$

Why does repeating this process enough times eventually prove that $G_n \leq A_n$? (This is another place where it is a good exercise to provide a formal proof by induction, as well as an informal reason.) When does equality hold in the formula $G_n \leq A_n$?

The reasoning in this proof is related to another interesting proof.

(b) Using the fact that $G_n \leq A_n$ when $n = 2$, prove, by induction on k, that $G_n \leq A_n$ for $n = 2^k$.

(c) For a general n, let $2^m > n$. Apply part (b) to the 2^m numbers

$$a_1, \ldots, a_n, \underbrace{A_n, \ldots, A_n}_{2^m - n \text{ times}}$$

to prove that $G_n \leq a_n$.

23. The following is a recursive definition of a^n:

$$\begin{aligned} a^1 &= a, \\ a^{n+1} &= a^n \cdot a. \end{aligned}$$

Prove, by induction, that

$$\begin{aligned} a^{n+m} &= a^n \cdot a^m, \\ (a^n)^m &= a^{nm}. \end{aligned}$$

(Don't try to be fancy: use either induction on n or induction on m, not both at once.)

24. Suppose we know properties P1 and P4 for the natural numbers, but that multiplication has never been mentioned. Then the following can be used as a recursive definition of multiplication:

$$\begin{aligned} 1 \cdot b &= b, \\ (a+1) \cdot b &= a \cdot b + b. \end{aligned}$$

Prove the following (in the order suggested!):

$$a \cdot (b + c) = a \cdot b + a \cdot c \text{ (use induction on } a\text{)},$$
$$a \cdot 1 = a,$$
$$a \cdot b = b \cdot a \text{ (you just finished proving the case } b = 1\text{)}.$$

25. In this chapter we began with the natural numbers and gradually built up to the real numbers. A completely rigorous discussion of this process requires a little book in itself (see Part V). No one has ever figured out how to get to the real numbers without going through this process, but if we do accept the real numbers as given, then the natural numbers can be *defined* as the real numbers of the form 1, $1+1$, $1+1+1$, etc. The whole point of this problem is to show that there is a rigorous mathematical way of saying "etc."

(a) A set A of real numbers is called **inductive** if

(1) 1 is in A,
(2) $k + 1$ is in A whenever k is in A.

Prove that

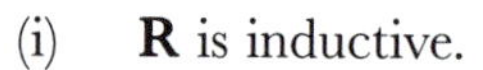

(i) **R** is inductive.
(ii) The set of positive real numbers is inductive.
(iii) The set of positive real numbers unequal to $\frac{1}{2}$ is inductive.
(iv) The set of positive real numbers unequal to 5 is not inductive.
(v) If A and B are inductive, then the set C of real numbers which are in both A and B is also inductive.

(b) A real number n will be called a **natural number** if n is in *every* inductive set.

(i) Prove that 1 is a natural number.
(ii) Prove that $k + 1$ is a natural number if k is a natural number.

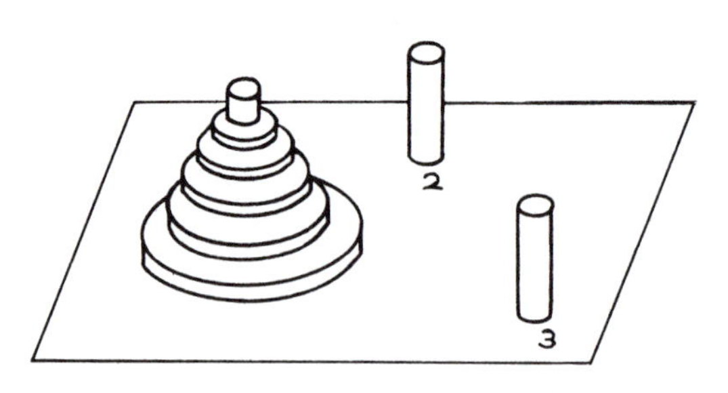

FIGURE 1

26. There is a puzzle consisting of three spindles, with n concentric rings of decreasing diameter stacked on the first (Figure 1). A ring at the top of a stack may be moved from one spindle to another spindle, provided that it is not placed on top of a smaller ring. For example, if the smallest ring is moved to spindle 2 and the next-smallest ring is moved to spindle 3, then the smallest ring may be moved to spindle 3 also, on top of the next-smallest. Prove that the entire stack of n rings can be moved onto spindle 3 in $2^n - 1$ moves, and that this cannot be done in fewer than $2^n - 1$ moves.

***27.** University B. once boasted 17 tenured professors of mathematics. Tradition prescribed that at their weekly luncheon meeting, faithfully attended by all 17, any members who had discovered an error in their published work should make an announcement of this fact, and promptly resign. Such an announcement had never actually been made, because no professor was aware of any errors in her or his work. This is not to say that no errors existed, however. In fact, over the years, in the work of every member of the department at least one error had been found, by some other member of the department. This error had been mentioned to all other members of the department, but the actual author of the error had been kept ignorant of the fact, to forestall any resignations.

One fateful year, the department was augmented by a visitor from another university, one Prof. X, who had come with hopes of being offered a permanent position at the end of the academic year. Naturally, he was apprised, by various members of the department, of the published errors which had been discovered. When the hoped-for appointment failed to materialize, Prof. X obtained his revenge at the last luncheon of the year. "I have enjoyed my visit here very much," he said, "but I feel that there is one thing that I have to tell you. At least one of you has published an incorrect result, which has been discovered by others in the department." What happened the next year?

****28.** After figuring out, or looking up, the answer to Problem 27, consider the following: Each member of the department already knew what Prof. X asserted, so how could his saying it change anything?

PART 2
FOUNDATIONS

The statement is so frequently made
that the differential calculus deals with
continuous magnitude, and yet
an explanation of this continuity is
nowhere given;
even the most rigorous expositions
of the differential calculus do not base
their proofs upon continuity but,
with more or less consciousness of the fact,
they either appeal to geometric notions
or those suggested by geometry,
or depend upon theorems which are never
established in a purely arithmetic manner.
Among these, for example,
belongs the above-mentioned theorem,
and a more careful investigation
convinced me that this theorem, or
any one equivalent to it, can be regarded
in some way as a sufficient basis
for infinitesimal analysis.
It then only remained to discover its true
origin in the elements of arithmetic
and thus at the same time
to secure a real definition of
the essence of continuity.
I succeeded Nov. 24, 1858, and
a few days afterward I communicated
the results
of my meditations to my dear friend
Durège with whom I had a long
and lively discussion.

RICHARD DEDEKIND

CHAPTER 3 FUNCTIONS

Undoubtedly the most important concept in all of mathematics is that of a function—in almost every branch of modern mathematics functions turn out to be the central objects of investigation. It will therefore probably not surprise you to learn that the concept of a function is one of great generality. Perhaps it will be a relief to learn that, for the present, we will be able to restrict our attention to functions of a very special kind; even this small class of functions will exhibit sufficient variety to engage our attention for quite some time. We will not even begin with a proper definition. For the moment a provisional definition will enable us to discuss functions at length, and will illustrate the intuitive notion of functions, as understood by mathematicians. Later, we will consider and discuss the advantages of the modern mathematical definition. Let us therefore begin with the following:

PROVISIONAL DEFINITION A function is a rule which assigns, to each of certain real numbers, some other real number.

The following examples of functions are meant to illustrate and amplify this definition, which, admittedly, requires some such clarification.

Example 1 The rule which assigns to each number the square of that number.

Example 2 The rule which assigns to each number y the number

$$\frac{y^3 + 3y + 5}{y^2 + 1}.$$

Example 3 The rule which assigns to each number $c \neq 1, -1$ the number

$$\frac{c^3 + 3c + 5}{c^2 - 1}.$$

Example 4 The rule which assigns to each number x satisfying $-17 \le x \le \pi/3$ the number x^2.

Example 5 The rule which assigns to each number a the number 0 if a is irrational, and the number 1 if a is rational.

Example 6 The rule which assigns

to 2 the number 5,

to 17 the number $\dfrac{36}{\pi}$,

$$\text{to } \frac{\pi^2}{17} \text{ the number } 28,$$

$$\text{to } \frac{36}{\pi} \text{ the number } 28,$$

and to any $y \neq 2, 17, \pi^2/17$, or $36/\pi$, the number 16 if y is of the form $a + b\sqrt{2}$ for a, b in $\mathbf{Q}$.

Example 7 The rule which assigns to each number t the number $t^3 + x$. (This rule depends, of course, on what the number x is, so we are really describing infinitely many different functions, one for each number x.)

Example 8 The rule which assigns to each number z the number of 7's in the decimal expansion of z, if this number is finite, and $-\pi$ if there are infinitely many 7's in the decimal expansion of z.

One thing should be abundantly clear from these examples—a function is *any* rule that assigns numbers to certain other numbers, not just a rule which can be expressed by an algebraic formula, or even by one uniform condition which applies to every number; nor is it necessarily a rule which you, or anybody else, can actually apply in practice (no one knows, for example, what rule 8 associates to π). Moreover, the rule may neglect some numbers and it may not even be clear to which numbers the function applies (try to determine, for example, whether the function in Example 6 applies to π). The set of numbers to which a function *does* apply is called the *domain* of the function.

Before saying anything else about functions we badly need some notation. Since throughout this book we shall frequently be talking about functions (indeed we shall hardly ever talk about anything else) we need a convenient way of naming functions, and of referring to functions in general. The standard practice is to denote a function by a letter. For obvious reasons the letter "f" is a favorite, thereby making "g" and "h" other obvious candidates, but any letter (or any reasonable symbol, for that matter) will do, not excluding "x" and "y", although these letters are usually reserved for indicating numbers. If f is a function, then the number which f associates to a number x is denoted by $f(x)$—this symbol is read "f of x" and is often called the **value of f at x**. Naturally, if we denote a function by x, some other letter must be chosen to denote the number (a perfectly legitimate, though perverse, choice would be "f," leading to the symbol $x(f)$). Note that the symbol $f(x)$ makes sense only for x in the domain of f; for other x the symbol $f(x)$ is not defined.

If the functions defined in Examples 1–8 are denoted by $f, g, h, r, s, \theta, \alpha_x$, and y, then we can rewrite their definitions as follows:

(1) $f(x) = x^2$ for all x.

(2) $g(y) = \dfrac{y^3 + 3y + 5}{y^2 + 1}$ for all y.

(3) $h(c) = \dfrac{c^3 + 3c + 5}{c^2 - 1}$ for all $c \neq 1, -1$.

(4) $r(x) = x^2$ for all x such that $-17 \le x \le \pi/3$.

(5) $s(x) = \begin{cases} 0, & x \text{ irrational} \\ 1, & x \text{ rational.} \end{cases}$

(6) $\theta(x) = \begin{cases} 5, & x = 2 \\ \dfrac{36}{\pi}, & x = 17 \\ 28, & x = \dfrac{\pi^2}{17} \\ 28, & x = \dfrac{36}{\pi} \\ 16, & x \ne 2,\ 17,\ \dfrac{\pi^2}{17}, \text{ or } \dfrac{36}{\pi}, \text{ and } x = a + b\sqrt{2} \text{ for } a, b \text{ in } \mathbf{Q}. \end{cases}$

(7) $\alpha_x(t) = t^3 + x$ for all numbers t.

(8) $y(x) = \begin{cases} n, & \text{exactly } n \text{ 7's appear in the decimal expansion of } x \\ -\pi, & \text{infinitely many 7's appear in the decimal expansion of } x. \end{cases}$

These definitions illustrate the common procedure adopted for defining a function f—indicating what $f(x)$ is for every number x in the domain of f. (Notice that this is exactly the same as indicating $f(a)$ for every number a, or $f(b)$ for every number b, etc.) In practice, certain abbreviations are tolerated. Definition (1) could be written simple

$$(1) \quad f(x) = x^2$$

the qualifying phrase "for all x" being understood. Of course, for definition (4) the only possible abbreviation is

$$(4) \quad r(x) = x^2, \qquad -17 \le x \le \pi/3.$$

It is usually understood that a definition such as

$$k(x) = \frac{1}{x} + \frac{1}{x-1}, \qquad x \ne 0, 1$$

can be shortened to

$$k(x) = \frac{1}{x} + \frac{1}{x-1};$$

in other words, *unless the domain is explicitly restricted further, it is understood to consist of all numbers for which the definition makes any sense at all.*

You should have little difficulty checking the following assertions about the functions defined above:

$$f(x+1) = f(x) + 2x + 1;$$
$$g(x) = h(x) \text{ if } x^3 + 3x + 5 = 0;$$
$$r(x+1) = r(x) + 2x + 1 \text{ if } -17 \le x \le \frac{\pi}{3} - 1;$$

$$s(x+y) = s(x) \text{ if } y \text{ is rational;}$$

$$\theta\left(\frac{\pi^2}{17}\right) = \theta\left(\frac{36}{\pi}\right);$$

$$\alpha_x(x) = x \cdot [f(x)+1];$$

$$y\left(\frac{1}{3}\right) = 0, \quad y\left(\frac{7}{9}\right) = -\pi.$$

If the expression $f(s(a))$ looks unreasonable to you, then you are forgetting that $s(a)$ is a number like any other number, so that $f(s(a))$ makes sense. As a matter of fact, $f(s(a)) = s(a)$ for all a. Why? Even more complicated expressions than $f(s(a))$ are, after a first exposure, no more difficult to unravel. The expression

$$f(r(s(\theta(\alpha_3(y(\tfrac{1}{3})))))),$$

formidable as it appears, may be evaluated quite easily with a little patience:

$$\begin{aligned} &f(r(s(\theta(\alpha_3(y(\tfrac{1}{3})))))) \\ &= f(r(s(\theta(\alpha_3(0))))) \\ &= f(r(s(\theta(3)))) \\ &= f(r(s(16))) \\ &= f(r(1)) \\ &= f(1) \\ &= 1. \end{aligned}$$

The first few problems at the end of this chapter give further practice manipulating this symbolism.

The function defined in (1) is a rather special example of an extremely important class of functions, the polynomial functions. A function f is a **polynomial function** if there are real numbers $a_0, \ldots, a_n$ such that

$$f(x) = a_n x^n + a_{n-1}x^{n-1} + \cdots + a_2x^2 + a_1x + a_0, \qquad \text{for all } x$$

(when $f(x)$ is written in this form it is usually tacitly assumed that $a_n \neq 0$). The highest power of x with a nonzero coefficient is called the **degree** of f; for example, the polynomial function f defined by $f(x) = 5x^6 + 137x^4 - \pi$ has degree 6.

The functions defined in (2) and (3) belong to a somewhat larger class of functions, the **rational functions**; these are the functions of the form p/q where p and q are polynomial functions (and q is not the function which is always 0). The rational functions are themselves quite special examples of an even larger class of functions, very thoroughly studied in calculus, which are simpler than many of the functions first mentioned in this chapter. The following are examples of this kind of function:

(9) $f(x) = \dfrac{x + x^2 + x\sin^2 x}{x\sin x + x\sin^2 x}$

(10) $f(x) = \sin(x^2)$.

(11) $f(x) = \sin(\sin(x^2))$.

$$(12)\quad f(x) = \sin^2(\sin(\sin^2(x \sin^2 x^2))) \cdot \sin\left(\frac{x + \sin(x \sin x)}{x + \sin x}\right).$$

By what criterion, you may feel impelled to ask, can such functions, especially a monstrosity like (12), be considered simple? The answer is that they can be built up from a few simple functions using a few simple means of combining functions. In order to construct the functions (9)–(12) we need to start with the "identity function" I, for which $I(x) = x$, and the "sine function" sin, whose value $\sin(x)$ at x is often written simple $\sin x$. The following are some of the important ways in which functions may be combined to produce new functions.

If f and g are any two functions, we can define a new function $f + g$, called the **sum** of f and g, by the equation

$$(f + g)(x) = f(x) + g(x).$$

Note that according to the conventions we have adopted, the domain of $f + g$ consists of all x for which "$f(x) + g(x)$" makes sense, i.e., the set of all x in both domain f and domain g. If A and B are any two sets, then $A \cap B$ (read "A intersect B" or "the intersection of A and B") denotes the set of x in both A and B; this notation allows us to write $\text{domain}(f + g) = \text{domain } f \cap \text{domain } g$.

In a similar vein, we define the **product** $f \cdot g$ and the **quotient** $\dfrac{f}{g}$ (or f/g) of f and g by

$$(f \cdot g)(x) = f(x) \cdot g(x)$$

and

$$\left(\frac{f}{g}\right)(x) = \frac{f(x)}{g(x)}.$$

Moreover, if g is a function and c is a number, we define a new function $c \cdot g$ by

$$(c \cdot g)(x) = c \cdot g(x).$$

This becomes a special case of the notation $f \cdot g$ if we agree that the symbol c should also represent the function f defined by $f(x) = c$; such a function, which has the same value for all numbers x, is called a **constant function**.

The domain of $f \cdot g$ is $\text{domain } f \cap \text{domain } g$, and the domain of $c \cdot g$ is simply the domain of g. On the other hand, the domain of f/g is rather complicated—it may be written $\text{domain } f \cap \text{domain } g \cap \{x : g(x) \neq 0\}$, the symbol $\{x : g(x) \neq 0\}$ denoting the set of numbers x such that $g(x) \neq 0$. In general, $\{x : \ldots\}$ denotes the set of all x such that "$\ldots$" is true. Thus $\{x : x^3 + 3 < 11\}$ denotes the set of all numbers x such that $x^3 < 8$, and consequently $\{x : x^3 + 3 < 11\} = \{x : x < 2\}$. Either of these symbols could just as well have been written using y everywhere instead of x. Variations of this notation are common, but hardly require any discussion. Any one can guess that $\{x > 0 : x^3 < 8\}$ denotes the set of positive numbers whose cube is less than 8; it could be expressed more formally as $\{x : x > 0 \text{ and } x^3 < 8\}$. Incidentally, this set is equal to the set $\{x : 0 < x < 2\}$. One variation is slightly less transparent, but very standard. The set $\{1, 3, 2, 4\}$, for example, contains just the four numbers 1, 2, 3, and 4; it can also be denoted by $\{x : x = 1 \text{ or } x = 3 \text{ or } x = 2 \text{ or } x = 4\}$.

Certain facts about the sum, product, and quotient of functions are obvious consequences of facts about sums, products, and quotients of numbers. For example, it is very easy to prove that

$$(f + g) + h = f + (g + h).$$

The proof is characteristic of almost every proof which demonstrates that two functions are equal—the two functions must be shown to have the same domain, and the same value at any number in the domain. For example, to prove that $(f+g)+h = f+(g+h)$, note that unraveling the definition of the two sides gives

$$\begin{aligned} [(f + g) + h](x) &= (f + g)(x) + h(x) \\ &= [f(x) + g(x)] + h(x) \end{aligned}$$

and

$$\begin{aligned} [f + (g + h)](x) &= f(x) + (g + h)(x) \\ &= f(x) + [g(x) + h(x)], \end{aligned}$$

and the equality of $[f(x)+g(x)]+h(x)$ and $f(x)+[g(x)+h(x)]$ is a fact about numbers. In this proof the equality of the two domains was not explicitly mentioned because this is obvious, as soon as we begin to write down these equations; the domain of $(f + g) + h$ and of $f + (g + h)$ is clearly domain $f \cap$ domain $g \cap$ domain h. We naturally write $f + g + h$ for $(f + g) + h = f + (g + h)$, precisely as we did for numbers.

It is just as easy to prove that $(f \cdot g) \cdot h = f \cdot (g \cdot h)$, and this function is denoted by $f \cdot g \cdot h$. The equations $f + g = g + f$ and $f \cdot g = g \cdot f$ should also present no difficulty.

Using the operations $+$, $\cdot$, $/$ we can now express the function f defined in (9) by

$$f = \frac{I + I \cdot I + I \cdot \sin \cdot \sin}{I \cdot \sin + I \cdot \sin \cdot \sin}.$$

It should be clear, however, that we cannot express function (10) this way. We require yet another way of combining functions. This combination, the composition of two functions, is by far the most important.

If f and g are any two functions, we define a new function $f \circ g$, the **composition** of f and g, by

$$(f \circ g)(x) = f(g(x));$$

the domain of $f \circ g$ is $\{x : x \text{ is in domain } g \text{ and } g(x) \text{ is in domain } f\}$. The symbol "$f \circ g$" is often read "$f$ circle g." Compared to the phrase "the composition of f and g" this has the advantage of brevity, of course, but there is another advantage of far greater import: there is much less chance of confusing $f \circ g$ with $g \circ f$, and these must *not* be confused, since they are not usually equal; in fact, almost any f and g chosen at random will illustrate this point (try $f = I \cdot I$ and $g = \sin$, for example). Lest you become too apprehensive about the operation of composition, let us hasten to point out that composition *is* associative:

$$(f \circ g) \circ h = f \circ (g \circ h)$$

(and the proof is a triviality); this function is denoted by $f \circ g \circ h$. We can now write the functions (10), (11), (12) as

(10) $f = \sin \circ (I \cdot I)$,

(11) $f = \sin \circ \sin \circ (I \cdot I)$,

(12) $f = (\sin \cdot \sin) \circ \sin \circ (\sin \cdot \sin) \circ (I \cdot [(\sin \cdot \sin) \circ (I \cdot I)]) \cdot \sin \circ \left(\dfrac{I + \sin \circ (I \cdot \sin)}{I + \sin} \right).$

One fact has probably already become clear. Although this method of writing functions reveals their "structure" very clearly, it is hardly short or convenient. The shortest name for the function f such that $f(x) = \sin(x^2)$ for all x unfortunately seems to be "the function f such that $f(x) = \sin(x^2)$ for all x." The need for abbreviating this clumsy description has been clear for two hundred years, but no reasonable abbreviation has received universal acclaim. At present the strongest contender for this honor is something like

$$x \to \sin(x^2)$$

(read "x goes to $\sin(x^2)$" or just "x arrow $\sin(x^2)$"), but it is hardly popular among writers of calculus textbooks. In this book we will tolerate a certain amount of ellipsis, and speak of "the function $f(x) = \sin(x^2)$." Even more popular is the quite drastic abbreviation: "the function $\sin(x^2)$." For the sake of precision we will never use this description, which, strictly speaking, confuses a number and a function, but it is so convenient that you will probably end up adopting it for personal use. As with any convention, utility is the motivating factor, and this criterion is reasonable so long as the slight logical deficiencies cause no confusion. On occasion, confusion *will* arise unless a more precise description is used. For example, "the function $x + t^3$" is an ambiguous phrase; it could mean either

$$x \to x + t^3, \text{ i.e., the function } f \text{ such that } f(x) = x + t^3 \text{ for all } x$$

or

$$t \to x + t^3, \text{ i.e., the function } f \text{ such that } f(t) = x + t^3 \text{ for all } t.$$

As we shall see, however, for many important concepts associated with functions, calculus has a notation which contains the "$x \to$" built in.

By now we have made a sufficiently extensive investigation of functions to warrant reconsidering our definition. We have defined a function as a "rule," but it is hardly clear what this means. If we ask "What happens if you break this rule?" it is not easy to say whether this question is merely facetious or actually profound.

A more substantial objection to the use of the word "rule" is that

$$f(x) = x^2$$

and

$$f(x) = x^2 + 3x + 3 - 3(x + 1)$$

are certainly *different* rules, if by a rule we mean the actual instructions given for determining $f(x)$; nevertheless, we want

$$f(x) = x^2$$

and

$$f(x) = x^2 + 3x + 3 - c(x + 1)$$

to define the same function. For this reason, a function is sometimes defined as an "association" between numbers; unfortunately the word "association" escapes the objections raised against "rule" only because it is even more vague.

There is, of course, a satisfactory way of defining functions, or we should never have gone to the trouble of criticizing our original definition. But a satisfactory definition can never be constructed by finding synonyms for English words which are troublesome. The definition which mathematicians have finally accepted for "function" is a beautiful example of the means by which intuitive ideas have been incorporated into rigorous mathematics. The correct question to ask about a function is not "What is a rule?" or "What is an association?" but "What does one have to know about a function in order to know all about it?" The answer to the last question is easy—for each number x one needs to know the number $f(x)$; we can imagine a table which would display all the information one could desire about the function $f(x) = x^2$:

x	$f(x)$
1	1
-1	1
2	4
-2	4
$\sqrt{2}$	2
$-\sqrt{2}$	2
π	π^2
$-\pi$	π^2

It is not even necessary to arrange the numbers in a table (which would actually be impossible if we wanted to list all of them). Instead of a two column array we can consider various pairs of numbers

$$(1, 1),\ (-1, 1),\ (2, 4),\ (-2, 4),\ (\pi, \pi^2),\ (\sqrt{2}, 2), \ldots$$

simply collected together into a set.* To find $f(1)$ we simply take the second number of the pair whose first member is 1; to find $f(\pi)$ we take the second number of the pair whose first member is π. We seem to be saying that a function might as well be defined as a collection of pairs of numbers. For example, if we were given the following collection (which contains just 5 pairs):

$$f = \{(1,7),\ (3,7),\ (5,3),\ (4,8),\ (8,4)\},$$

then $f(1) = 7$, $f(3) = 7$, $f(5) = 3$, $f(4) = 8$, $f(8) = 4$ and 1, 3, 4, 5, 8 are the only numbers in the domain of f. If we consider the collection

$$f = \{(1,7),\ (3,7),\ (2,5),\ (1,8),\ (8,4)\},$$

then $f(3) = 7$, $f(2) = 5$, $f(8) = 4$; but it is impossible to decide whether $f(1) = 7$ or $f(1) = 8$. In other words, a function cannot be defined to be any old collection of pairs of numbers; we must rule out the possibility which arose in this case. We are therefore led to the following definition.

DEFINITION

A **function** is a collection of pairs of numbers with the following property: if (a, b) and (a, c) are both in the collection, then $b = c$; in other words, the collection must not contain two different pairs with the same first element.

This is our first full-fledged definition, and illustrates the format we shall always use to define significant new concepts. These definitions are so important (at least as important as theorems) that it is essential to know when one is actually at hand, and to distinguish them from comments, motivating remarks, and casual explanations. They will be preceded by the word **DEFINITION**, contain the term being defined in boldface letters, and constitute a paragraph unto themselves.

There is one more definition (actually defining two things at once) which can now be made rigorously:

DEFINITION

If f is a function, the **domain** of f is the set of all a for which there is some b such that (a, b) is in f. If a is in the domain of f, it follows from the definition of a function that there is, in fact, a *unique* number b such that (a, b) is in f. This unique b is denoted by $\boldsymbol{f(a)}$.

With this definition we have reached our goal: the important thing about a function f is that a number $f(x)$ is determined for each number x in its domain. You may feel that we have also reached the point where an intuitive definition has been replaced by an abstraction with which the mind can hardly grapple. Two consolations may be offered. First, although a function has been defined as a

*The pairs occurring here are often called "ordered pairs," to emphasize that, for example, $(2,4)$ is not the same pair as $(4,2)$. It is only fair to warn that we are going to define functions in terms of ordered pairs, another undefined term. Ordered pairs *can* be defined, however, and an appendix to this chapter has been provided for skeptics.

collection of pairs, there is nothing to stop you from *thinking* of a function as a rule. Second, neither the intuitive nor the formal definition indicates the best way of thinking about functions. The best way is to draw pictures; but this requires a chapter all by itself.

PROBLEMS

1. Let $f(x) = 1/(1+x)$. What is

(i) $f(f(x))$ (for which x does this make sense?).
(ii) $f\left(\dfrac{1}{x}\right)$.
(iii) $f(cx)$.
(iv) $f(x+y)$.
(v) $f(x)+f(y)$.
(vi) For which numbers c is there a number x such that $f(cx) = f(x)$. Hint: There are a lot more than you might think at first glance.
(vii) For which numbers c is it true that $f(cx) = f(x)$ for two different numbers x?

2. Let $g(x) = x^2$, and let

$$h(x) = \begin{cases} 0, & x \text{ rational} \\ 1, & x \text{ irrational.} \end{cases}$$

(i) For which y is $h(y) \leq y$?
(ii) For which y is $h(y) \leq g(y)$?
(iii) What is $g(h(z)) - h(z)$?
(iv) For which w is $g(w) \leq w$?
(v) For which ε is $g(g(\varepsilon)) = g(\varepsilon)$?

3. Find the domain of the functions defined by the following formulas.

(i) $f(x) = \sqrt{1-x^2}$.
(ii) $f(x) = \sqrt{1-\sqrt{1-x^2}}$.
(iii) $f(x) = \dfrac{1}{x-1} + \dfrac{1}{x-2}$.
(iv) $f(x) = \sqrt{1-x^2} + \sqrt{x^2-1}$.
(v) $f(x) = \sqrt{1-x} + \sqrt{x-2}$.

4. Let $S(x) = x^2$, let $P(x) = 2^x$, and let $s(x) = \sin x$. Find each of the following. In each case you answer should be a *number*.

(i) $(S \circ P)(y)$.
(ii) $(S \circ s)(y)$.
(iii) $(S \circ P \circ s)(t) + (s \circ P)(t)$.
(iv) $s(t^3)$.

5. Express each of the following functions in terms of S, P, s, using only $+$, $\cdot$, and $\circ$ (for example, the answer to (i) is $P \circ s$). In each case your answer should be a *function.*

(i) $f(x) = 2^{\sin x}$.
(ii) $f(x) = \sin 2^x$.
(iii) $f(x) = \sin x^2$.
(iv) $f(x) = \sin^2 x$ (remember that $\sin^2 x$ is an abbreviation for $(\sin x)^2$).
(v) $f(t) = 2^{2^t}$. (Note: a^{b^c} *always* means $a^{(b^c)}$; this convention is adopted because $(a^b)^c$ can be written more simply as a^{bc}.)
(vi) $f(u) = \sin(2^u + 2^{u^2})$.
(vii) $f(y) = \sin\left(\sin\left(\sin(2^{2^{2^{\sin y}}})\right)\right)$.
(viii) $f(a) = 2^{\sin^2 a} + \sin(a^2) + 2^{\sin(a^2 + \sin a)}$.

Polynomial functions, because they are simple, yet flexible, occupy a favored role in most investigations of functions. The following two problems illustrate their flexibility, and guide you through a derivation of their most important elementary properties.

6. (a) If $x_1, \ldots, x_n$ are distinct numbers, find a polynomial function f_i of degree $n-1$ which is 1 at x_i and 0 at x_j for $j \neq i$. Hint: the product of all $(x - x_j)$ for $j \neq i$, is 0 at x_j if $j \neq i$. (This product is usually denoted by

$$\prod_{\substack{j=1 \\ j \neq i}}^{n} (x - x_j),$$

the symbol Π (capital pi) playing the same role for products that Σ plays for sums.)

(b) Now find a polynomial function f of degree $n-1$ such that $f(x_i) = a_i$, where $a_1, \ldots, a_n$ are given numbers. (You should use the functions f_i from part (a). The formula you will obtain is called the "Lagrange interpolation formula.")

7. (a) Prove that for any polynomial function f, and any number a, there is a polynomial function g, and a number b, such that $f(x) = (x-a)g(x)+b$ for all x. (The idea is simply to divide $(x - a)$ into $f(x)$ by long division, until a constant remainder is left. For example, the calculation

$$\begin{array}{r|llll}
 & x^2 & +x & & -2 \\ \hline
x-1 & x^3 & & & -3x+1 \\
 & x^3 & -x^2 & & \\ \cline{2-4}
 & & x^2 & -3x & \\
 & & x^2 & -x & \\ \cline{3-4}
 & & & & -2x+1 \\
 & & & & -2x+2 \\ \cline{5-5}
 & & & & -1
\end{array}$$

shows that $x^3 - 3x + 1 = (x - 1)(x^2 + x - 2) - 1$. A formal proof is possible by induction on the degree of f.)

(b) Prove that if $f(a) = 0$, then $f(x) = (x - a)g(x)$ for some polynomial function g. (The converse is obvious.)

(c) Prove that if f is a polynomial function of degree n, then f has at most n roots, i.e., there are at most n numbers a with $f(a) = 0$.

(d) Show that for each n there is a polynomial function of degree n with n roots. If n is even find a polynomial function of degree n with no roots, and if n is odd find one with only one root.

8. For which numbers a, b, c, and d will the function

$$f(x) = \frac{ax + b}{cx + d}$$

satisfy $f(f(x)) = x$ for all x?

9. (a) If A is any set of real numbers, define a function C_A as follows:

$$C_A(x) = \begin{cases} 1, & x \text{ in } A \\ 0, & x \text{ not in } A. \end{cases}$$

Find expressions for $C_{A\cap B}$ and $C_{A\cup B}$ and $C_{\mathbf{R}-A}$, in terms of C_A and C_B. (The symbol $A \cap B$ was defined in this chapter, but the other two may be new to you. They can be defined as follows:

$$A \cup B = \{x : x \text{ is in } A \text{ or } x \text{ is in } B\},$$
$$\mathbf{R} - A = \{x : x \text{ is in } \mathbf{R} \text{ but } x \text{ is not in } A\}.)$$

(b) Suppose f is a function such that $f(x) = 0$ or 1 for each x. Prove that there is a set A such that $f = C_A$.

(c) Show that $f = f^2$ if and only if $f = C_A$ for some set A.

10. (a) For which functions f is there a function g such that $f = g^2$? Hint: You can certainly answer this question if "function" is replaced by "number."

(b) For which functions f is there a function g such that $f = 1/g$?

*(c) For which functions b and c can we find a function x such that

$$(x(t))^2 + b(t)x(t) + c(t) = 0$$

for all numbers t?

*(d) What conditions must the functions a and b satisfy if there is to be a function x such that

$$a(t)x(t) + b(t) = 0$$

for all numbers t? How many such functions x will there be?

11. (a) Suppose that H is a function and y is a number such that $H(H(y)) = y$. What is

$$\underbrace{H(H(H(\cdots(H(y)\cdots)}_{80 \text{ times}}?$$

(b) Same question if 80 is replaced by 81.
(c) Same question if $H(H(y)) = H(y)$.
*(d) Find a function H such that $H(H(x)) = H(x)$ for all numbers x, and such that $H(1) = 36$, $H(2) = \pi/3$, $H(13) = 47$, $H(36) = 36$, $H(\pi/3) = \pi/3$, $H(47) = 47$. (Don't try to "solve" for $H(x)$; there are many functions H with $H(H(x)) = H(x)$. The extra conditions on H are supposed to suggest a way of finding a suitable H.)
*(e) Find a function H such that $H(H(x)) = H(x)$ for all x, and such that $H(1) = 7$, $H(17) = 18$.

12. A function f is **even** if $f(x) = f(-x)$ and **odd** if $f(x) = -f(-x)$. For example, f is even if $f(x) = x^2$ or $f(x) = |x|$ or $f(x) = \cos x$, while f is odd if $f(x) = x$ or $f(x) = \sin x$.

(a) Determine whether $f + g$ is even, odd, or not necessarily either, in the four cases obtained by choosing f even or odd, and g even or odd. (Your answers can most conveniently be displayed in a 2×2 table.)
(b) Do the same for $f \cdot g$.
(c) Do the same for $f \circ g$.
(d) Prove that every even function f can be written $f(x) = g(|x|)$, for infinitely many functions g.

***13.** (a) Prove that any function f with domain $\mathbf{R}$ can be written $f = E + O$, where E is even and O is odd.
(b) Prove that this way of writing f is unique. (If you try to do part (b) first, by "solving" for E and O you will probably find the solution to part (a).)

14. If f is any function, define a new function $|f|$ by $|f|(x) = |f(x)|$. If f and g are functions, define two new functions, $\max(f, g)$ and $\min(f, g)$, by

$$\max(f, g)(x) = \max(f(x), g(x)),$$
$$\min(f, g)(x) = \min(f(x), g(x)).$$

Find an expression for $\max(f, g)$ and $\min(f, g)$ in terms of $|\ \ |$.

15. (a) Show that $f = \max(f, 0) + \min(f, 0)$. This particular way of writing f is fairly useful; the functions $\max(f, 0)$ and $\min(f, 0)$ are called the **positive** and **negative parts** of f.
(b) A function f is called **nonnegative** if $f(x) \geq 0$ for all x. Prove that any function f can be written $f = g - h$, where g and h are nonnegative, in infinitely many ways. (The "standard way" is $g = \max(f, 0)$ and $h = -\min(f, 0)$.) Hint: Any *number* can certainly be written as the difference of two nonnegative *numbers* in infinitely many ways.

***16.** Suppose f satisfies $f(x + y) = f(x) + f(y)$ for all x and y.

(a) Prove that $f(x_1 + \cdots + x_n) = f(x_1) + \cdots + f(x_n)$.
(b) Prove that there is some number c such that $f(x) = cx$ for all *rational* numbers x (at this point we're not trying to say anything about $f(x)$ for irrational x). Hint: First figure out what c must be. Now prove that

$f(x) = cx$, first when x is a natural number, then when x is an integer, then when x is the reciprocal of an integer and, finally, for all rational x.

***17.** If $f(x) = 0$ for all x, then f satisfies $f(x+y) = f(x) + f(y)$ for all x and y, and also $f(x \cdot y) = f(x) \cdot f(y)$ for all x and y. Now suppose that f satisfies these two properties, but that $f(x)$ is not always 0. Prove that $f(x) = x$ for all x, as follows:

(a) Prove that $f(1) = 1$.
(b) Prove that $f(x) = x$ if x is rational.
(c) Prove that $f(x) > 0$ if $x > 0$. (This part is tricky, but if you have been paying attention to the philosophical remarks accompanying the problems in the last two chapters, you will know what to do.)
(d) Prove that $f(x) > f(y)$ if $x > y$.
(e) Prove that $f(x) = x$ for all x. Hint: Use the fact that between any two numbers there is a rational number.

***18.** Precisely what conditions must f, g, h, and k satisfy in order that $f(x)g(y) = h(x)k(y)$ for all x and y?

***19.** (a) Prove that there do *not* exist functions f and g with either of the following properties:

(i) $f(x) + g(y) = xy$ for all x and y.
(ii) $f(x) \cdot g(y) = x + y$ for all x and y.

Hint: Try to get some information about f or g by choosing particular values of x and y.

(b) Find functions f and g such that $f(x+y) = g(xy)$ for all x and y.

***20.** (a) Find a function f, other than a constant function, such that $|f(y) - f(x)| \leq |y - x|$.
(b) Suppose that $f(y) - f(x) \leq (y-x)^2$ for all x and y. (Why does this imply that $|f(y) - f(x)| \leq (y-x)^2$?) Prove that f is a constant function. Hint: Divide the interval from x to y into n equal pieces.

21. Prove or give a counterexample for each of the following assertions:

(a) $f \circ (g + h) = f \circ g + f \circ h$.
(b) $(g + h) \circ f = g \circ f + h \circ f$.
(c) $\dfrac{1}{f \circ g} = \dfrac{1}{f} \circ g$.
(d) $\dfrac{1}{f \circ g} = f \circ \left(\dfrac{1}{g}\right)$.

22. (a) Suppose $g = h \circ f$. Prove that if $f(x) = f(y)$, then $g(x) = g(y)$.
(b) Conversely, suppose that f and g are two functions such that $g(x) = g(y)$ whenever $f(x) = f(y)$. Prove that $g = h \circ f$ for some function h. Hint: Just try to define $h(z)$ when z is of the form $z = f(x)$ (these are the only z that matter) and use the hypotheses to show that your definition will not run into trouble.

23. Suppose that $f \circ g = I$, where $I(x) = x$. Prove that

(a) if $x \neq y$, then $g(x) \neq g(y)$;
(b) every number b can be written $b = f(a)$ for some number a.

***24.** (a) Suppose g is a function with the property that $g(x) \neq g(y)$ if $x \neq y$. Prove that there is a function f such that $f \circ g = I$.
(b) Suppose that f is a function such that every number b can be written $b = f(a)$ for some number a. Prove that there is a function g such that $f \circ g = I$.

***25.** Find a function f such that $g \circ f = I$ for some g, but such that there is no function h with $f \circ h = I$.

***26.** Suppose $f \circ g = I$ and $h \circ f = I$. Prove that $g = h$. Hint: Use the fact that composition is associative.

27. (a) Suppose $f(x) = x+1$. Are there any functions g such that $f \circ g = g \circ f$?
(b) Suppose f is a constant function. For which functions g does $f \circ g = g \circ f$?
(c) Suppose that $f \circ g = g \circ f$ for *all* functions g. Show that f is the identity function, $f(x) = x$.

28. (a) Let F be the set of all functions whose domain is **R**. Prove that, using $+$ and $\cdot$ as defined in this chapter, all of properties P1–P9 except P7 hold for F, provided 0 and 1 are interpreted as constant functions.
(b) Show that P7 does not hold.
*(c) Show that P10–P12 cannot hold. In other words, show that there is no collection P of functions in F, such that P10–P12 hold for P. (It is sufficient, and will simplify things, to consider only functions which are 0 except at two points x_0 and x_1.)
(d) Suppose we define $f < g$ to mean that $f(x) < g(x)$ for all x. Which of P′10–P′13 (in Problem 1-8) now hold?
(e) If $f < g$, is $h \circ f < h \circ g$? Is $f \circ h < g \circ h$?

APPENDIX. ORDERED PAIRS

Not only in the definition of a function, but in other parts of the book as well, it is necessary to use the notion of an ordered pair of objects. A definition has not yet been given, and we have never even stated explicitly what properties an ordered pair is supposed to have. The one property which we will require states formally that the ordered pair (a, b) should be determined by a and b, and the order in which they are given:

$$\text{if } (a, b) = (c, d), \text{ then } a = c \text{ and } b = d.$$

Ordered pairs may be treated most conveniently by simply introducing (a, b) as an undefined term and adopting the basic property as an axiom—since this property is the only significant fact about ordered pairs, there is not much point worrying about what an ordered pair "really" is. Those who find this treatment satisfactory need read no further.

The rest of this short appendix is for the benefit of those readers who will feel uncomfortable unless ordered pairs are somehow defined so that this basic property becomes a theorem. There is no point in restricting our attention to ordered pairs of numbers; it is just as reasonable, and just as important, to have available the notion of an ordered pair of any two mathematical objects. This means that our definition ought to involve only concepts common to all branches of mathematics. The one common concept which pervades all areas of mathematics is that of a set, and ordered pairs (like everything else in mathematics) can be defined in this context; an ordered pair will turn out to be a set of a rather special sort.

The set $\{a, b\}$, containing the two elements a and b, is an obvious first choice, but will not do as a definition for (a, b), because there is no way of determining from $\{a, b\}$ which of a or b is meant to be the first element. A more promising candidate is the rather startling set:

$$\{\{a\}, \{a, b\}\}.$$

This set has two members, both of which are *themselves* sets; one member is the set $\{a\}$, containing the single member a, the other is the set $\{a, b\}$. Shocking as it may seem, we are going to define (a, b) to be this set. The justification for this choice is given by the theorem immediately following the definition—the definition works, and there really isn't anything else worth saying.

DEFINITION $\quad (\mathbf{a}, \mathbf{b}) = \{\{a\}, \{a, b\}\}.$

THEOREM 1 If $(a, b) = (c, d)$, then $a = c$ and $b = d$.

PROOF The hypothesis means that

$$\{\{a\}, \ \{a, b\}\} = \{\{c\}, \ \{c, d\}\}.$$

Now $\{\{a\}, \ \{a, b\}\}$ contains just two members, $\{a\}$ and $\{a, b\}$; and a is the only common element of these two members of $\{\{a\}, \ \{a, b\}\}$. Similarly, c is the unique common member of both members of $\{\{c\}, \{c,d\}\}$. Therefore $a = c$. We therefore have

$$\{\{a\}, \{a,b\}\} = \{\{a\}, \{a,d\}\},$$

and only the proof that $b = d$ remains. It is convenient to distinguish 2 cases.

Case 1. $b = a$. In this case, $\{a,b\} = \{a\}$, so the set $\{\{a\}, \{a,b\}\}$ really has only one member, namely, $\{a\}$. The same must be true of $\{\{a\}, \{a,d\}\}$, so $\{a,d\} = \{a\}$, which implies that $d = a = b$.

Case 2. $b \neq a$. In this case, b is in one member of $\{\{a\}, \{a,b\}\}$ but not in the other. It must therefore be true that b is in one member of $\{\{a\}, \{a,d\}\}$ but not in the other. This can happen only if b is in $\{a,d\}$, but b is not in $\{a\}$; thus $b = a$ or $b = d$, but $b \neq a$; so $b = d$. ▮

CHAPTER 4 GRAPHS

Mention the real numbers to a mathematician and the image of a straight line will probably form in her mind, quite involuntarily. And most likely she will neither banish nor too eagerly embrace this mental picture of the real numbers. "Geometric intuition" will allow her to interpret statements about numbers in terms of this picture, and may even suggest methods of proving them. Although the properties of the real numbers which were studied in Part I are not greatly illuminated by a geometric picture, such an interpretation will be a great aid in Part II.

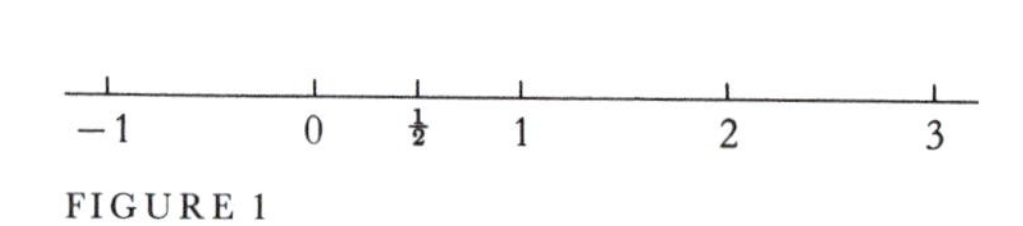

FIGURE 1

You are probably already familiar with the conventional method of considering the straight line as a picture of the real numbers, i.e., of associating to each real number a point on a line. To do this (Figure 1) we pick, arbitrarily, a point which we label 0, and a point to the right, which we label 1. The point twice as far to the right is labeled 2, the point the same distance from 0 to 1, but to the left of 0, is labeled -1, etc. With this arrangement, if $a < b$, then the point corresponding to a lies to the left of the point corresponding to b. We can also draw rational numbers, such as $\frac{1}{2}$, in the obvious way. It is usually taken for granted that the irrational numbers also somehow fit into this scheme, so that every real number can be drawn as a point on the line. We will not make too much fuss about justifying this assumption, since this method of "drawing" numbers is intended solely as a method of picturing certain abstract ideas, and our proofs will never rely on these pictures (although we will frequently use a picture to suggest or help explain a proof). Because this geometric picture plays such a prominent, albeit inessential role, geometric terminology is frequently employed when speaking of numbers—thus a number is sometimes called a *point*, and $\mathbf{R}$ is often called the *real line*.

The number $|a-b|$ has a simple interpretation in terms of this geometric picture: it is the distance between a and b, the length of the line segment which has a as one end point and b as the other. This means, to choose an example whose frequent occurrence justifies special consideration, that the set of numbers x which satisfy $|x-a| < \varepsilon$ may be pictured as the collection of points whose distance from a is less than ε. This set of points is the "interval" from $a-\varepsilon$ to $a+\varepsilon$, which may also be described as the points corresponding to numbers x with $a-\varepsilon < x < a+\varepsilon$ (Figure 2).

FIGURE 2

Sets of numbers which correspond to intervals arise so frequently that it is desirable to have special names for them. The set $\{x : a < x < b\}$ is denoted by (a, b) and called the **open interval** from a to b. This notation naturally creates some ambiguity, since (a, b) is also used to denote a pair of numbers, but in context it is always clear (or can easily be made clear) whether one is talking about a pair or an interval. Note that if $a \geq b$, then $(a, b) = \emptyset$, the set with no elements; in prac-

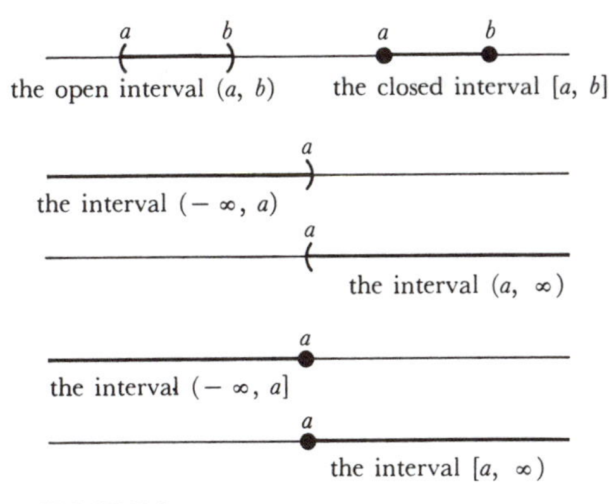

FIGURE 3

tice, however, it is almost always assumed (explicitly if one has been careful, and implicitly otherwise), that whenever an interval (a, b) is mentioned, the number a is less than b.

The set $\{x : a \leq x \leq b\}$ is denoted by $[a, b]$ and is called the **closed interval** from a to b. This symbol is usually reserved for the case $a < b$, but it is sometimes used for $a = b$, also. The usual pictures for the intervals (a, b) and $[a, b]$ are shown in Figure 3; since no reasonably accurate picture could ever indicate the difference between the two intervals, various conventions have been adopted. Figure 3 also shows certain "infinite" intervals. The set $\{x : x > a\}$ is denoted by (a, ∞), while the set $\{x : x \geq a\}$ is denoted by $[a, \infty)$; the sets $(-\infty, a)$ and $(-\infty, a]$ are defined similarly. At this point a standard warning must be issued: the symbols ∞ and $-\infty$, though usually read "infinity" and "minus infinity," are *purely* suggestive; there is no number "∞" which satisfies $\infty \geq a$ for all numbers a. While the symbols ∞ and $-\infty$ will appear in many contexts, it is always necessary to define these uses in ways that refer only to numbers. The set $\mathbf{R}$ of all real numbers is also considered to be an "interval," and is sometimes denoted by $(-\infty, \infty)$.

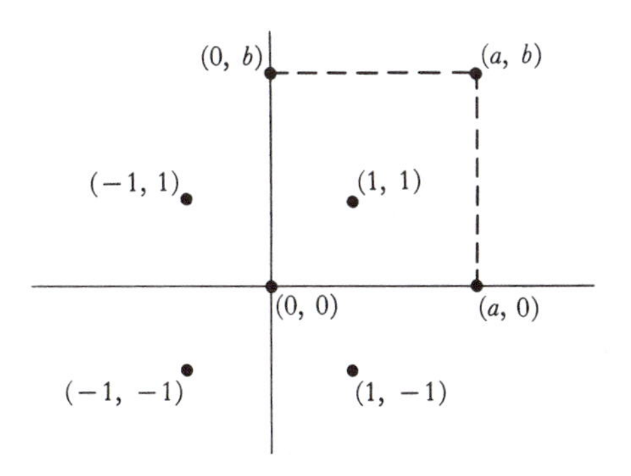

FIGURE 4

Of even greater interest to us than the method of drawing numbers is a method of drawing pairs of numbers. This procedure, probably also familiar to you, requires a "coordinate system," two straight lines intersecting at right angles. To distinguish these straight lines, we call one the *horizontal axis*, and one the *vertical axis*. (More prosaic terminology, such as the "first" and "second" axes, is probably preferable from a logical point of view, but most people hold their books, or at least their blackboards, in the same way, so that "horizontal" and "vertical" are more descriptive.) Each of the two axes could be labeled with real numbers, but we can also label points on the horizontal axis with pairs $(a, 0)$ and points on the vertical axis with pairs $(0, b)$, so that the intersection of the two axes, the "origin" of the coordinate system, is labeled $(0, 0)$. Any pair (a, b) can now be drawn as in Figure 4, lying at the vertex of the rectangle whose other three vertices are labeled $(0, 0)$, $(a, 0)$, and $(0, b)$. The numbers a and b are called the *first* and *second coordinates*, respectively, of the point determined in this way.

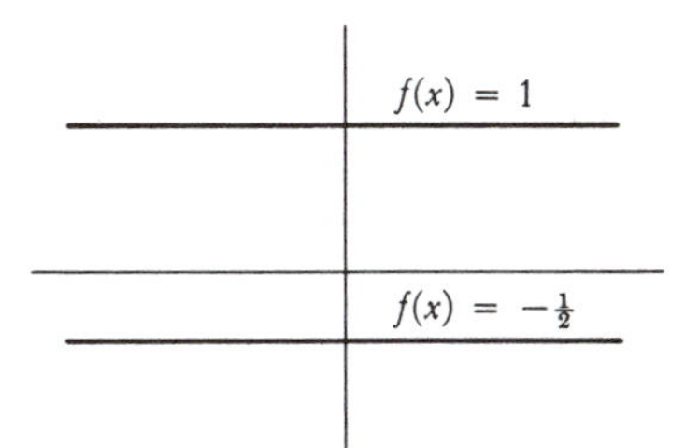

FIGURE 5

Our real concern, let us recall, is a method of drawing functions. Since a function is just a collection of pairs of numbers, we can draw a function by drawing each of the pairs in the function. The drawing obtained in this way is called the **graph** of the function. In other words, the graph of f contains all the points corresponding to pairs $(x, f(x))$. Since most functions contain infinitely many pairs, drawing the graph promises to be a laborious undertaking, but, in fact, many functions have graphs which are quite easy to draw.

Not surprisingly, the simplest functions of all, the constant functions $f(x) = c$, have the simplest graphs. It is easy to see that the graph of the function $f(x) = c$ is a straight line parallel to the horizontal axis, at distance c from it (Figure 5).

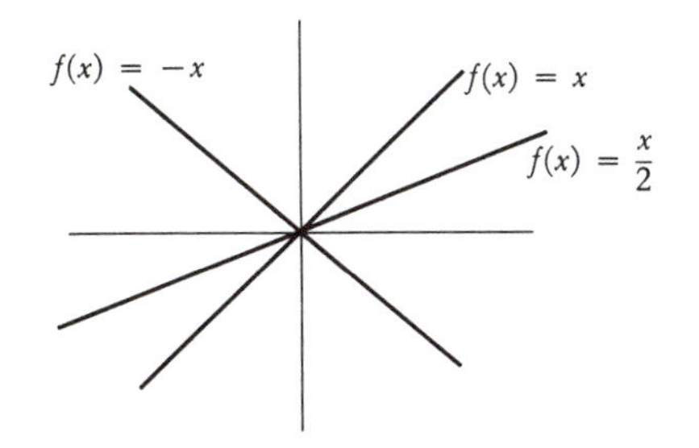

FIGURE 6

The functions $f(x) = cx$ also have particularly simple graphs—straight lines through $(0, 0)$, as in Figure 6. A proof of this fact is indicated in Figure 7:

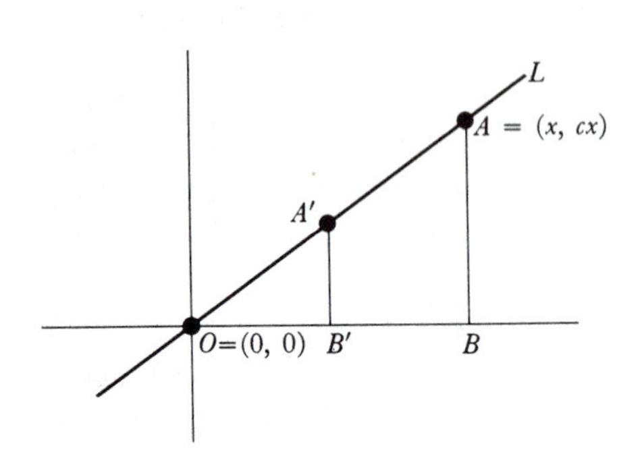

FIGURE 7

Let x be some number not equal to 0, and let L be the straight line which passes through the origin O, corresponding to $(0, 0)$, and through the point A, corresponding to (x, cx). A point A', with first coordinate y, will lie on L when the triangle $A'B'O$ is similar to the triangle ABO, thus when

$$\frac{A'B'}{OB'} = \frac{AB}{OB} = c;$$

this is precisely the condition that A' corresponds to the pair (y, cy), i.e., that A' lies on the graph of f. The argument has implicitly assumed that $c > 0$, but the other cases are treated easily enough. The number c, which measures the ratio of the sides of the triangles appearing in the proof, is called the *slope* of the straight line, and a line parallel to this line is also said to have slope c.

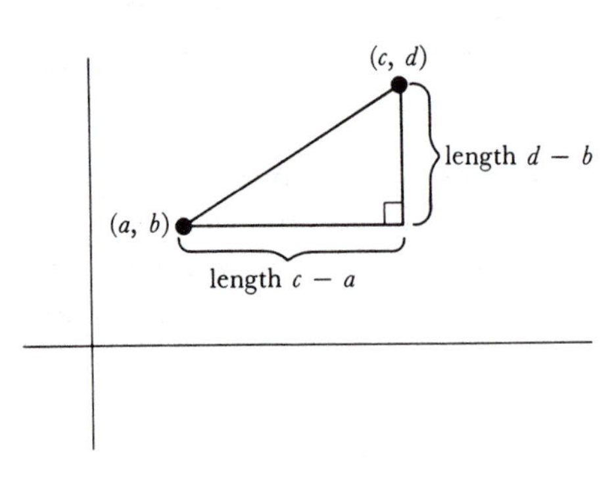

FIGURE 8

This demonstration has neither been labeled nor treated as a formal proof. Indeed, a rigorous demonstration would necessitate a digression which we are not at all prepared to follow. The rigorous proof of *any* statement connecting geometric and algebraic concepts would first require a real proof (or a precisely stated assumption) that the points on a straight line correspond in an exact way to the real numbers. Aside from this, it would be necessary to develop plane geometry as precisely as we intend to develop the properties of real numbers. Now the detailed development of plane geometry is a beautiful subject, but it is by no means a prerequisite for the study of calculus. We shall use geometric pictures only as an aid to intuition; for our purposes (and for most of mathematics) it is perfectly satisfactory to *define* the plane to be the set of all pairs of real numbers, and to *define* straight lines as certain collections of pairs, including, among others, the collections $\{(x, cx) : x \text{ a real number}\}$. To provide this artificially constructed geometry with all the structure of geometry studied in high school, one more definition is required. If (a, b) and (c, d) are two points in the plane, i.e., pairs of real numbers, we *define* the **distance** between (a, b) and (c, d) to be

$$\sqrt{(a-c)^2 + (b-d)^2}.$$

If the motivation for this definition is not clear, Figure 8 should serve as adequate explanation—with this definition the Pythagorean theorem has been built into our geometry.*

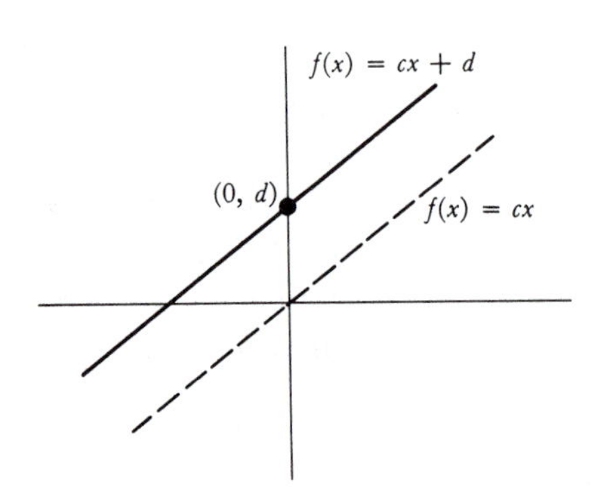

FIGURE 9

Reverting once more to our informal geometric picture, it is not hard to see (Figure 9) that the graph of the function $f(x) = cx + d$ is a straight line with slope c, passing through the point $(0, d)$. For this reason, the functions $f(x) = cx + d$ are called **linear functions**. Simple as they are, linear functions occur frequently, and you should feel comfortable working with them. The following is a typical problem whose solution should not cause any trouble. Given two distinct points (a, b) and (c, d), find the linear function f whose graph goes through (a, b) and (c, d). This amounts to saying that $f(a) = b$ and $f(c) = d$. If

*The fastidious reader might object to this definition on the grounds that nonnegative numbers are not yet known to have square roots. This objection is really unanswerable at the moment—the definition will just have to be accepted with reservations, until this little point is settled.

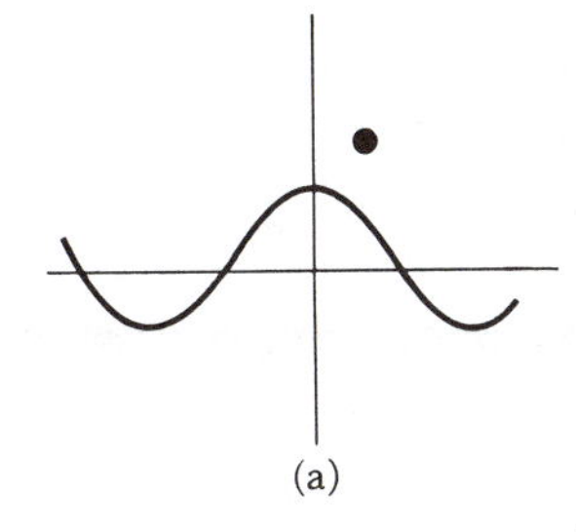

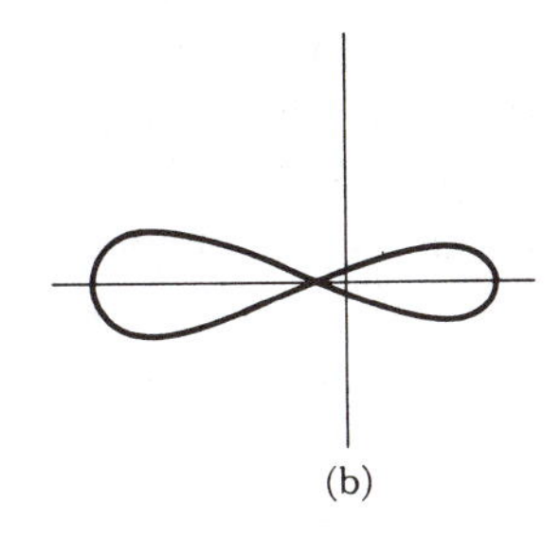

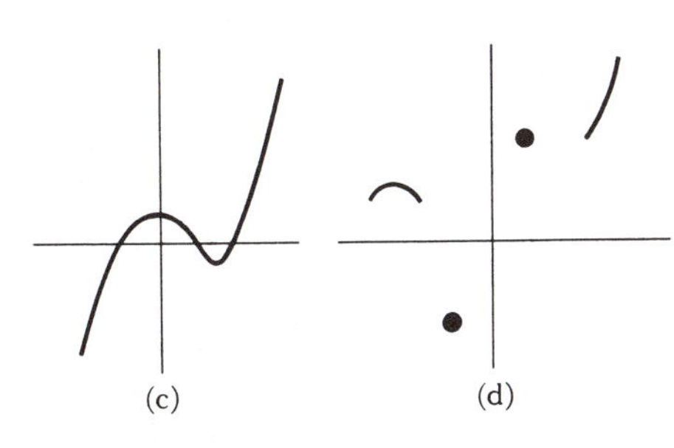

FIGURE 10

f is to be of the form $f(x) = \alpha x + \beta$, then we must have

$$\begin{aligned} \alpha a + \beta &= b, \\ \alpha c + \beta &= d; \end{aligned}$$

therefore $\alpha = (d - b)/(c - a)$ and $\beta = b - [(d - b)/(c - a)]a$, so

$$f(x) = \frac{d-b}{c-a}x + b - \frac{d-b}{c-a}a = \frac{d-b}{c-a}(x - a) + b,$$

a formula most easily remembered by using the "point-slope form" (see Problem 6).

Of course, this solution is possible only if $a \neq c$; the graphs of linear functions account only for the straight lines which are not parallel to the vertical axis. The vertical straight lines are not the graph of *any* function at all; in fact, the graph of a function can never contain even two distinct points on the same vertical line. This conclusion is immediate from the definition of a function—two points on the same vertical line correspond to pairs of the form (a, b) and (a, c) and, by definition, a function cannot contain (a, b) and (a, c) if $b \neq c$. Conversely, if a set of points in the plane has the property that no two points lie on the same vertical line, then it is surely the graph of a function. Thus, the first two sets in Figure 10 are not graphs of functions and the last two are; notice that the fourth is the graph of a function whose domain is not all of $\mathbf{R}$, since some vertical lines have no points on them at all.

After the linear functions the simplest is perhaps the function $f(x) = x^2$. If we draw some of the pairs in f, i.e., some of the pairs of the form (x, x^2), we obtain a picture like Figure 11.

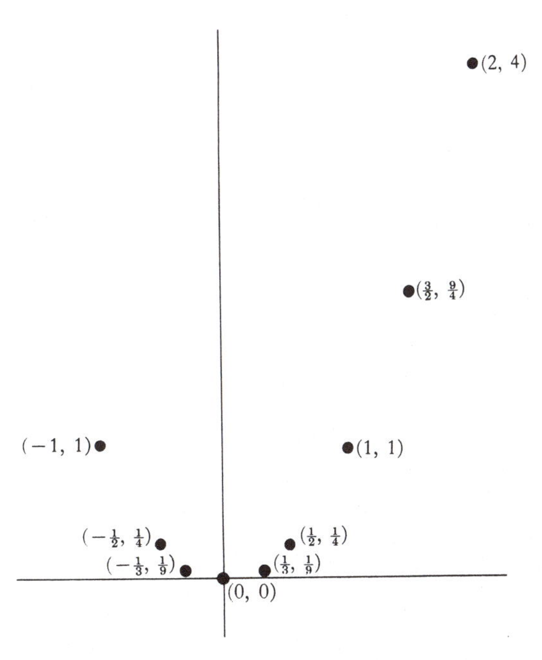

FIGURE 11

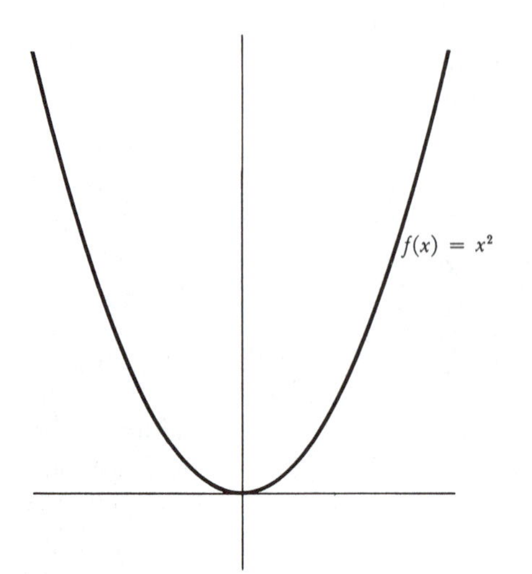

FIGURE 12

It is not hard to convince yourself that all the pairs (x, x^2) lie along a curve like the one shown in Figure 12; this curve is known as a **parabola**.

Since a graph is just a drawing on paper, made (in this case) with printer's ink, the question "Is this what the graph really looks like?" is hard to phrase in any sensible manner. No drawing is ever *really* correct since the line has thickness. Nevertheless, there are some questions which one *can* ask: for example, how can you be sure that the graph does not look like one of the drawings in Figure 13? It is easy to see, and even to prove, that the graph cannot look like (a); for if $0 < x < y$, then $x^2 < y^2$, so the graph should be higher at y than at x, which is not the case in (a) . It is also easy to see, simply by drawing a very accurate graph, first plotting many pairs (x, x^2), that the graph cannot have a large "jump" as in (b) or a "corner" as in (c). In order to prove these assertions, however, we first need to say, in a mathematical way, what it means for a function not to have a "jump" or "corner"; these ideas already involve some of the fundamental concepts of calculus. Eventually we will be able to define them rigorously, but meanwhile you may amuse yourself by attempting to define these concepts, and then examining your definitions critically. Later these definitions may be compared with the ones mathematicians have agreed upon. If they compare favorably, you are certainly to be congratulated!

The functions $f(x) = x^n$, for various natural numbers n, are sometimes called **power functions**. Their graphs are most easily compared as in Figure 14, by drawing several at once.

The power functions are only special cases of polynomial functions, introduced in the previous chapter. Two particular polynomial functions are graphed in

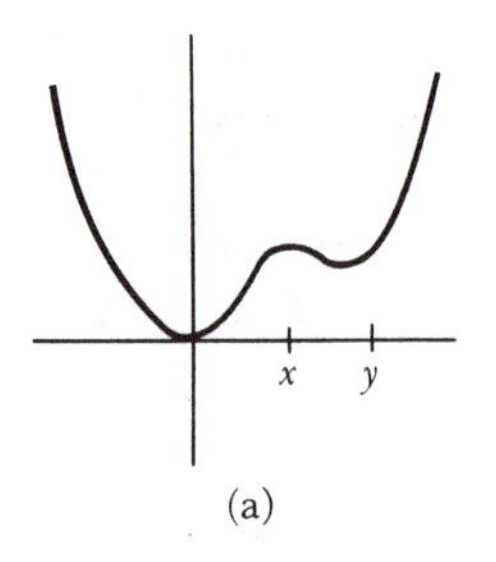

(a)

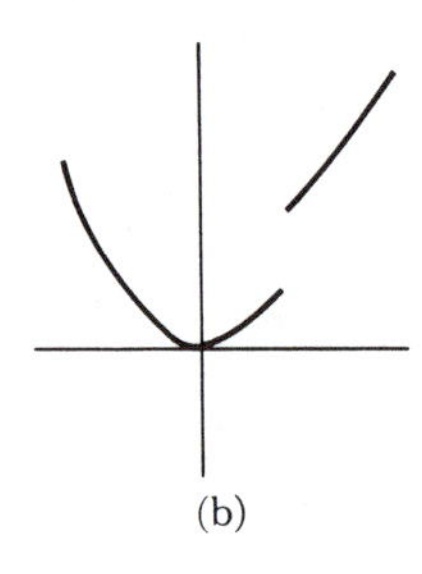

(b)

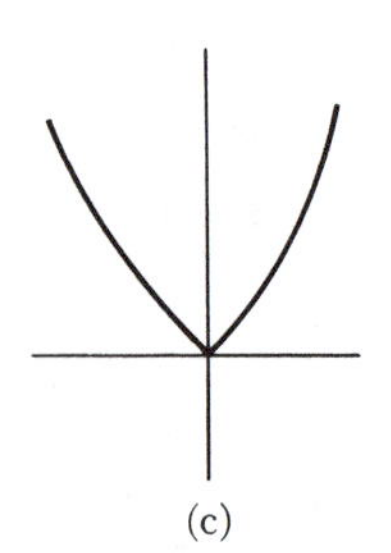

(c)

FIGURE 13

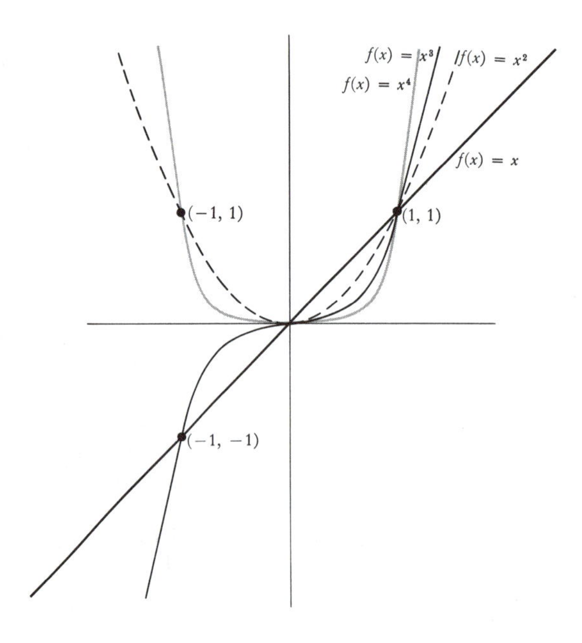

FIGURE 14

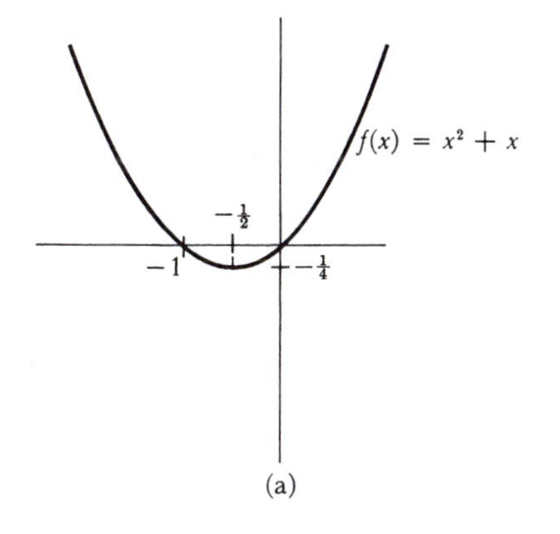

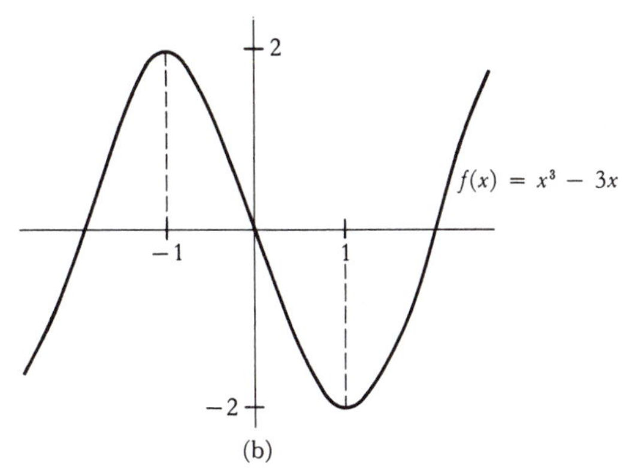

FIGURE 15

Figure 15, while Figure 16 is meant to give a general idea of the graph of the polynomial function

$$f(x) = a_n x^n + a_{n-1} x^{n-1} + \cdots + a_0,$$

in the case $a_n > 0$.

In general, the graph of f will have at most $n-1$ "peaks" or "valleys" (a "peak" is a point like $(x, f(x))$ in Figure 16, while a "valley" is a point like $(y, f(y))$. The number of peaks and valleys may actually be much smaller (the power functions, for example, have at most one valley). Although these assertions are easy to make, we will not even contemplate giving proofs until Part III (once the powerful methods of Part III are available, the proofs will be very easy).

Figure 17 illustrates the graphs of several rational functions. The rational functions exhibit even greater variety than the polynomial functions, but their behavior will also be easy to analyze once we can use the derivative, the basic tool of Part III.

Many interesting graphs can be constructed by "piecing together" the graphs of functions already studied. The graph in Figure 18 is made up entirely of straight lines. The function f with this graph satisfies

$$f\left(\frac{1}{n}\right) = (-1)^{n+1},$$

$$f\left(\frac{-1}{n}\right) = (-1)^{n+1},$$

$$f(x) = 1, \qquad |x| \geq 1,$$

and is a linear function on each interval $[1/(n+1), 1/n]$ and $[-1/n, -1/(n+1)]$. (The number 0 is not in the domain of f.) Of course, one can write out an explicit formula for $f(x)$, when x is in $[1/(n+1), 1/n]$; this is a good exercise in the use of linear functions, and will also convince you that a picture is worth a thousand words.

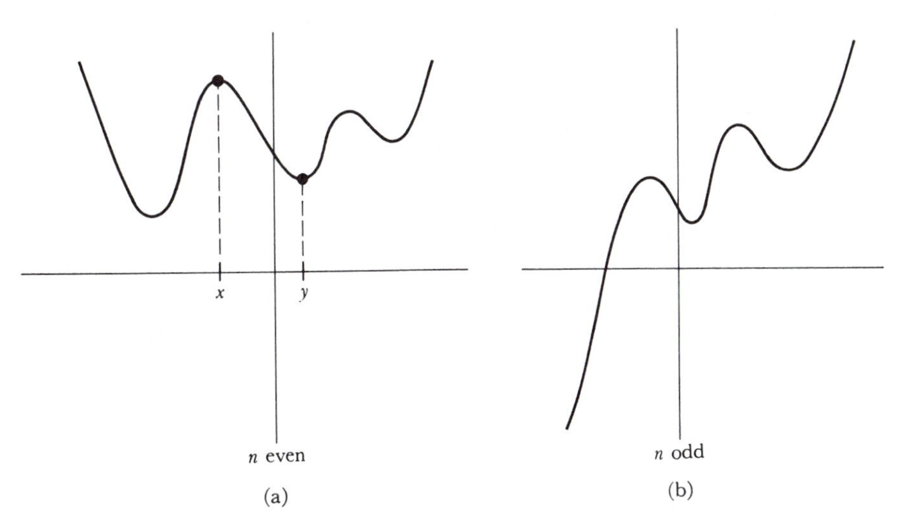

FIGURE 16

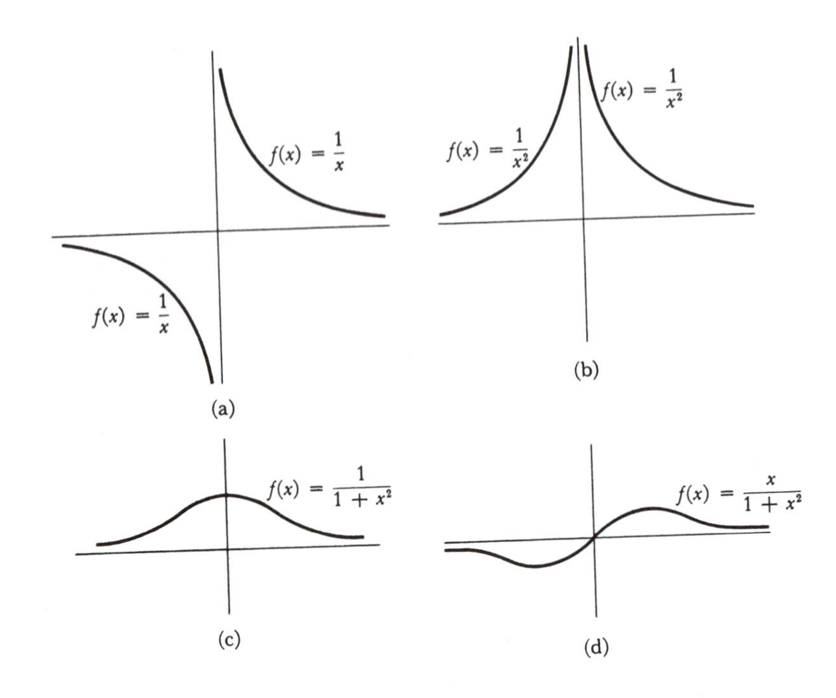

FIGURE 17

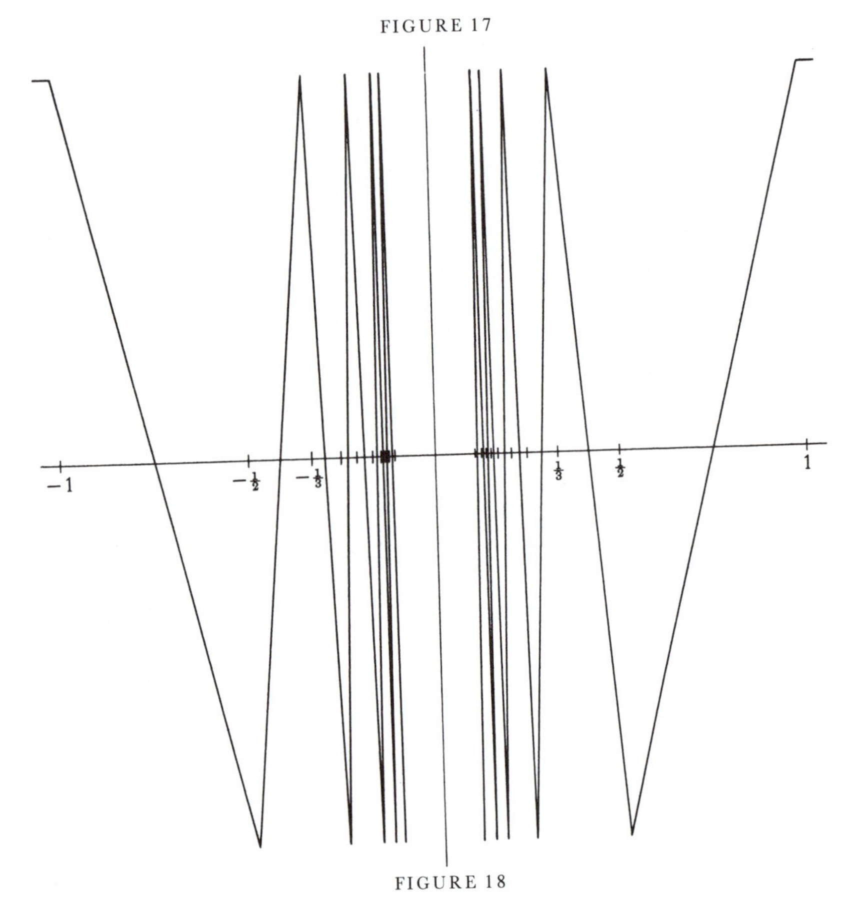

FIGURE 18

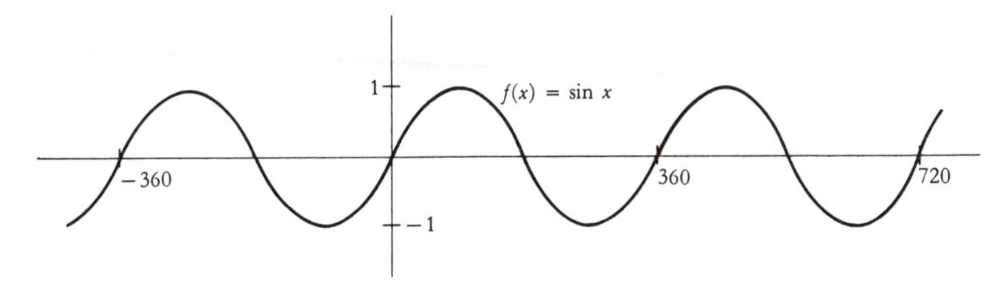

FIGURE 19

It is actually possible to define, in a much simpler way, a function which exhibits this same property of oscillating infinitely often near 0, by using the sine function. In Chapter 15 we will discuss this function in detail, and radian measure in particular; for the time being it will be easiest to use degree measurements for angles. The graph of the sine function is shown in Figure 19 (the scale on the horizontal axis has been altered so that the graph will be clearer; radian measure has, besides important mathematical properties, the additional advantage that such changes are unnecessary).

Now consider the function $f(x) = \sin 1/x$. The graph of f is shown in Figure 20. To draw this graph it helps to first observe that

$$\begin{aligned} f(x) &= 0 \qquad \text{for } x = \frac{1}{180}, \frac{1}{360}, \frac{1}{540}, \dots, \\ f(x) &= 1 \qquad \text{for } x = \frac{1}{90}, \frac{1}{90+360}, \frac{1}{90+720}, \dots, \\ f(x) &= -1 \qquad \text{for } x = \frac{1}{270}, \frac{1}{270+360}, \frac{1}{270+720}, \dots. \end{aligned}$$

Notice that when x is large, so that $1/x$ is small, $f(x)$ is also small; when x is

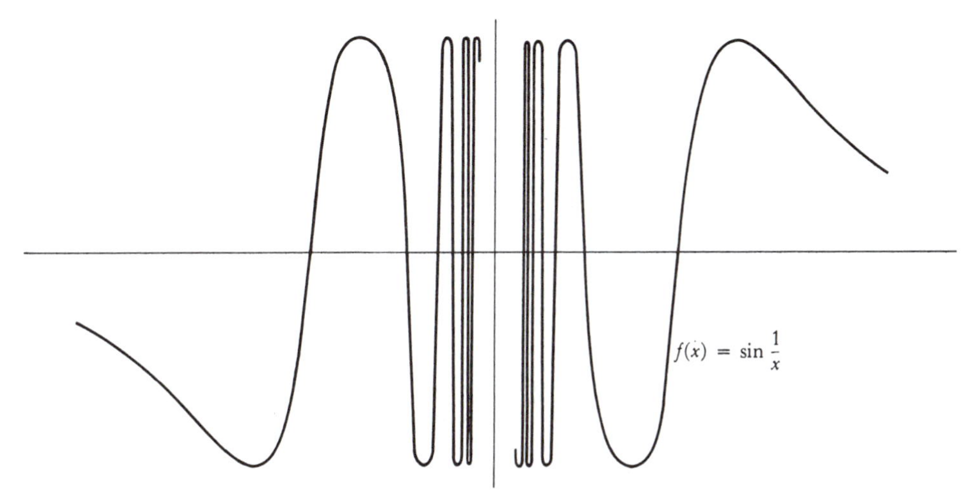

FIGURE 20

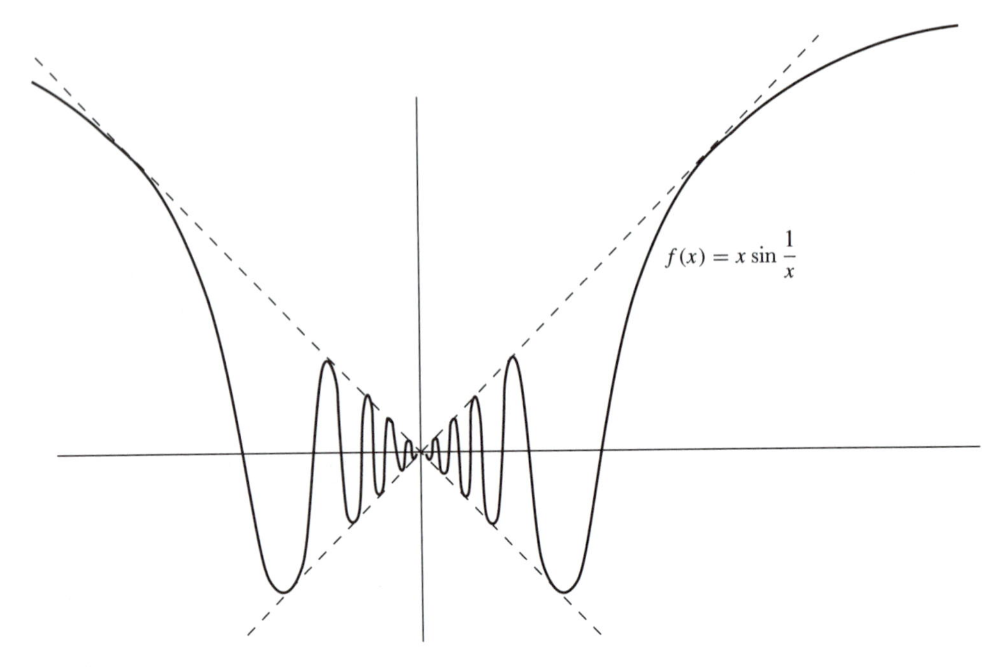

FIGURE 21

"large negative," that is, when $|x|$ is large for negative x, again $f(x)$ is close to 0, although $f(x) < 0$.

An interesting modification of this function is $f(x) = x \sin 1/x$. The graph of this function is sketched in Figure 21. Since $\sin 1/x$ oscillates infinitely often near 0 between 1 and -1, the function $f(x) = x \sin 1/x$ oscillates infinitely often between x and $-x$. The behavior of the graph for x large or large negative is harder to

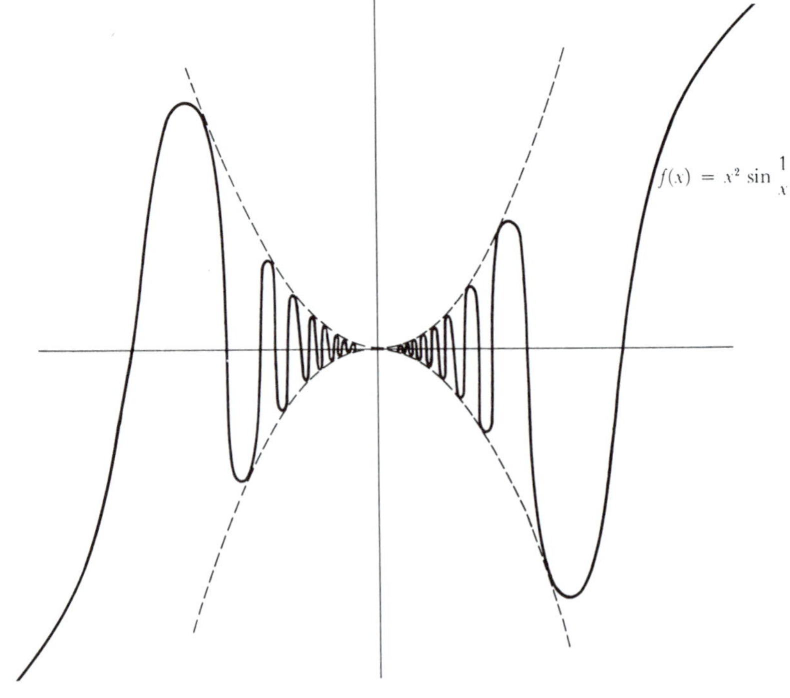

FIGURE 22

analyze. Since $\sin 1/x$ is getting close to 0, while x is getting larger and larger, there seems to be no telling what the product will do. It *is* possible to decide, but this is another question that is best deferred to Part III. The graph of $f(x) = x^2 \sin 1/x$ has also been illustrated (Figure 22).

For these infinitely oscillating functions, it is clear that the graph cannot hope to be really "accurate." The best we can do is to show part of it, and leave out the part near 0 (which is the interesting part). Actually, it is easy to find much simpler functions whose graphs cannot be "accurately" drawn. The graphs of

$$f(x) = \begin{cases} x^2, & x < 1 \\ 2, & x \geq 1 \end{cases} \qquad \text{and} \qquad g(x) = \begin{cases} x^2, & x \leq 1 \\ 2, & x > 1 \end{cases}$$

can only be distinguished by some convention similar to that used for open and closed intervals (Figure 23).

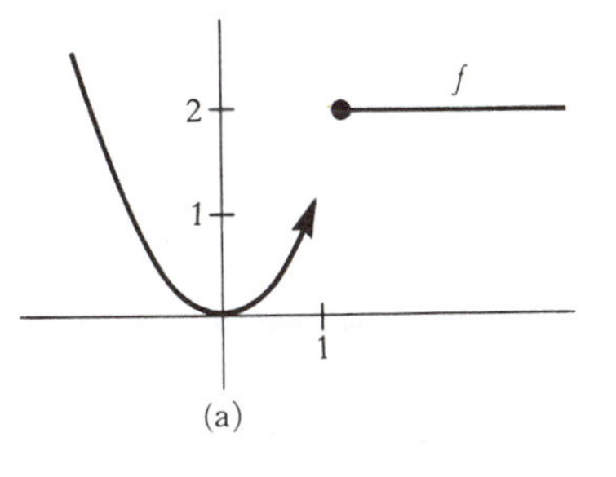

(a)

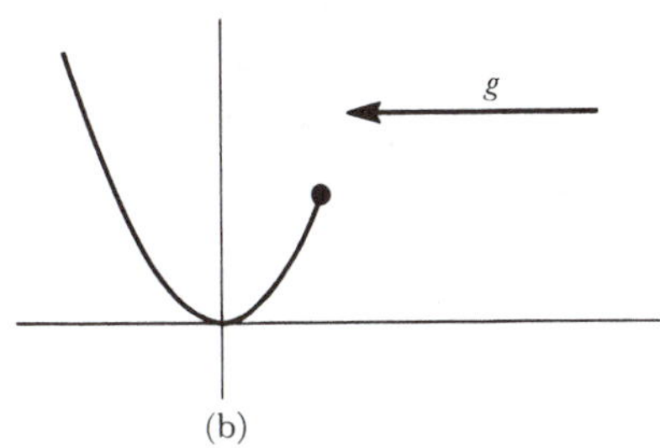

(b)

FIGURE 23

Out last example is a function whose graph is spectacularly nondrawable:

$$f(x) = \begin{cases} 0, & x \text{ irrational} \\ 1, & x \text{ rational.} \end{cases}$$

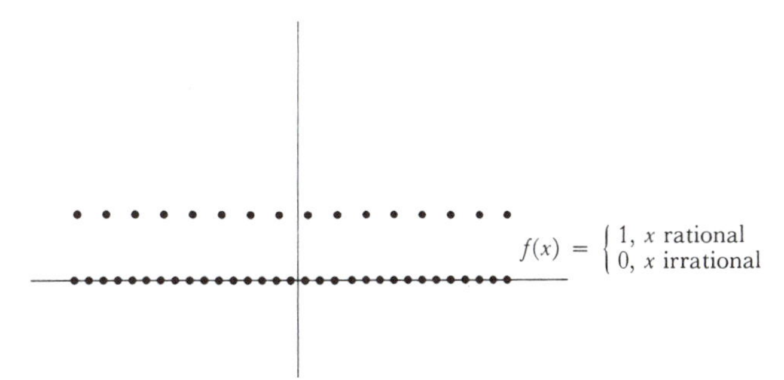

FIGURE 24

The graph of f must contain infinitely many points on the horizontal axis and also infinitely many points on a line parallel to the horizontal axis, but it must not contain either of these lines entirely. Figure 24 shows the usual textbook picture of the graph. To distinguish the two parts of the graph, the dots are placed closer together on the line corresponding to irrational x. (There is actually a mathematical reason behind this convention, but it depends on some sophisticated ideas, introduced in Problems 21-5 and 21-6.)

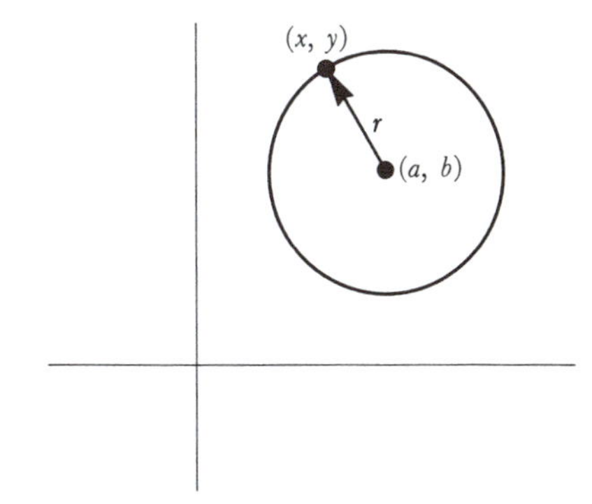

FIGURE 25

The peculiarities exhibited by some functions are so engrossing that it is easy to forget some of the simplest, and most important, subsets of the plane, which are not the graphs of functions. The most important example of all is the **circle**. A circle with center (a, b) and radius $r > 0$ contains, by definition, all the points (x, y) whose distance from (a, b) is equal to r. The circle thus consists (Figure 25) of all points (x, y) with

$$\sqrt{(x-a)^2 + (y-b)^2} = r$$

or

$$(x-a)^2 + (y-b)^2 = r^2.$$

The circle with center (0,0) and radius 1, often regarded as a sort of standard copy, is called the *unit circle*.

A close relative of the circle is the **ellipse**. This is defined as the set of points, the *sum* of whose distances from two "focus" points is a constant. (When the two foci are the same, we obtain a circle.) If, for convenience, the focus points are taken to be $(-c, 0)$ and $(c, 0)$, and the sum of the distances is taken to be $2a$ (the factor 2 simplifies some algebra), then (x, y) is on the ellipse if and only if

$$\sqrt{(x-(-c))^2+y^2}+\sqrt{(x-c)^2+y^2}=2a$$

or

$$\sqrt{(x+c)^2+y^2}=2a-\sqrt{(x-c)^2+y^2}$$

or

$$x^2+2cx+c^2+y^2=4a^2-4a\sqrt{(x-c)^2+y^2}+x^2-2cx+c^2+y^2$$

or

$$4(cx-a^2)=-4a\sqrt{(x-c)^2+y^2}$$

or

$$c^2x^2-2cxa^2+a^4=a^2(x^2-2cx+c^2+y^2)$$

or

$$(c^2-a^2)x^2-a^2y^2=a^2(c^2-a^2)$$

or

$$\frac{x^2}{a^2}+\frac{y^2}{a^2-c^2}=1.$$

This is usually written simply

$$\frac{x^2}{a^2}+\frac{y^2}{b^2}=1,$$

where $b = \sqrt{a^2-c^2}$ (since we must clearly choose $a > c$, it follows that $a^2 - c^2 > 0$). A picture of an ellipse is shown in Figure 26. The ellipse intersects the horizontal axis when $y = 0$, so that

$$\frac{x^2}{a^2}=1, \qquad x=\pm a,$$

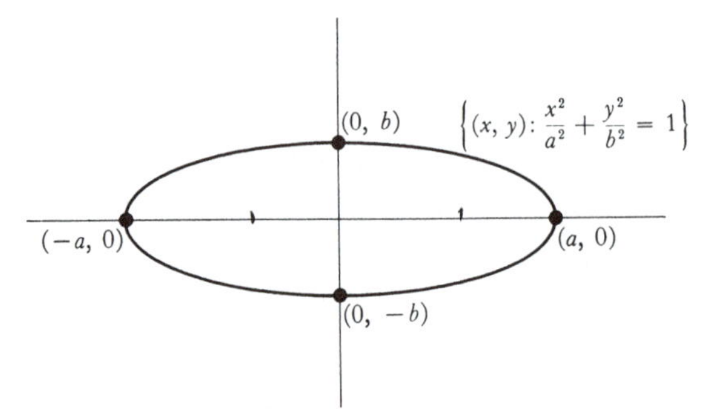

FIGURE 26

and it intersects the vertical axis when $x = 0$, so that

$$\frac{y^2}{b^2} = 1, \qquad y = \pm b.$$

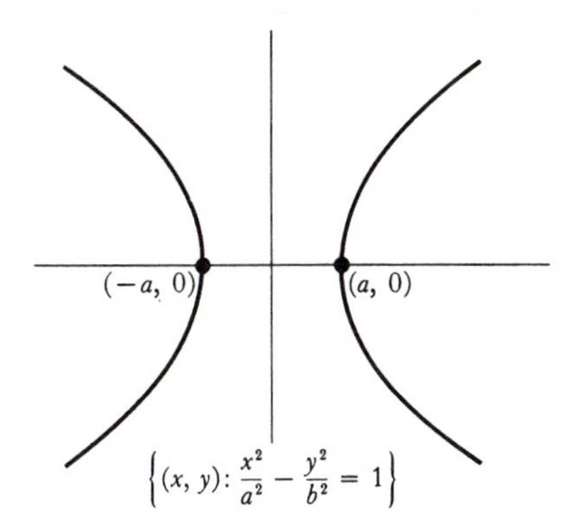

FIGURE 27

The **hyperbola** is defined analogously, except that we require the *difference* of the two distances to be constant. Choosing the points $(-c, 0)$ and $(c, 0)$ once again, and the constant difference as $2a$, we obtain, as the condition that (x, y) be on the hyperbola,

$$\sqrt{(x+c)^2 + y^2} - \sqrt{(x-c)^2 + y^2} = \pm 2a,$$

which may be simplified to

$$\frac{x^2}{a^2} + \frac{y^2}{a^2 - c^2} = 1.$$

In this case, however, we must clearly choose $c > a$, so that $a^2 - c^2 < 0$. If $b = \sqrt{c^2 - a^2}$, then (x, y) is on the hyperbola if and only if

$$\frac{x^2}{a^2} - \frac{y^2}{b^2} = 1.$$

The picture is shown in Figure 27. It contains two pieces, because the difference between the distances of (x, y) from $(-c, 0)$ and $(c, 0)$ may be taken in two different orders. The hyperbola intersects the horizontal axis when $y = 0$, so that $x = \pm a$, but it never intersects the vertical axis.

It is interesting to compare (Figure 28) the hyperbola with $a = b = \sqrt{2}$ and the graph of the function $f(x) = 1/x$. The drawings look quite similar, and the two sets are actually identical, except for a rotation through an angle of 45° (Problem 23).

Clearly no rotation of the plane will change circles or ellipses into the graphs of functions. Nevertheless, the study of these important geometric figures can often be reduced to the study of functions. Ellipses, for example, are made up of the

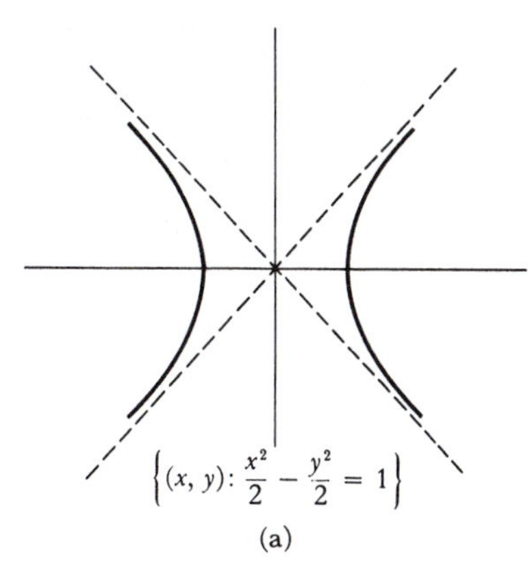

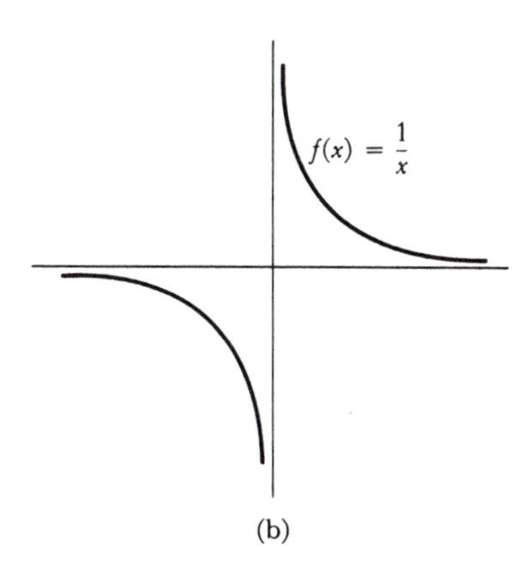

FIGURE 28

graphs of two functions,

$$f(x) = b\sqrt{1-(x^2/a^2)}, \qquad -a \le x \le a$$

and

$$g(x) = -b\sqrt{1-(x^2/a^2)}, \qquad -a \le x \le a.$$

Of course, there are many other pairs of functions with this same property. For example, we can take

$$f(x) = \begin{cases} b\sqrt{1-(x^2/a^2)}, & 0 < x \le a \\ -b\sqrt{1-(x^2/a^2)}, & -a \le x \le 0 \end{cases}$$

and

$$g(x) = \begin{cases} -b\sqrt{1-(x^2/a^2)}, & 0 < x \le a \\ b\sqrt{1-(x^2/a^2}, & -a \le x \le 0. \end{cases}$$

We could also choose

$$f(x) = \begin{cases} b\sqrt{1-(x^2/a^2)}, & x \text{ rational}, \ -a \le x \le a \\ -b\sqrt{1-(x^2/a^2)}, & x \text{ irrational}, -a \le x \le a \end{cases}$$

and

$$g(x) = \begin{cases} -b\sqrt{1-(x^2/a^2)}, & x \text{ rational}, \ -a \le x \le a \\ b\sqrt{1-(x^2/a^2)}, & x \text{ irrational}, -a \le x \le a. \end{cases}$$

But all these other pairs necessarily involve unreasonable functions which jump around. A proof, or even a precise statement of this fact, is too difficult at present. Although you have probably already begun to make a distinction between those functions with reasonable graphs, and those with unreasonable graphs, you may find it very difficult to state a reasonable definition of reasonable functions. A mathematical definition of this concept is by no means easy, and a great deal of this book may be viewed as successive attempts to impose more and more conditions that a "reasonable" function must satisfy. As we define some of these conditions, we will take time out to ask if we have really succeeded in isolating the functions which deserve to be called reasonable. The answer, unfortunately, will always be "no," or at best, a qualified "yes."

PROBLEMS

1. Indicate on a straight line the set of all x satisfying the following conditions. Also name each set, using the notation for intervals (in some cases you will also need the $\cup$ sign).

(i) $|x-3| < 1$.
(ii) $|x-3| \le 1$.
(iii) $|x-a| < \varepsilon$.
(iv) $|x^2-1| < \frac{1}{2}$.

(v) $\dfrac{1}{1+x^2} \geq \frac{1}{5}$.

(vi) $\dfrac{1}{1+x^2} \leq a$ (give an answer in terms of a, distinguishing various cases).

(vii) $x^2 + 1 \geq 2$.

(viii) $(x+1)(x-1)(x-2) > 0$.

2. There is a very useful way of describing the points of the closed interval $[a, b]$ (where we assume, as usual, that $a < b$).

(a) First consider the interval $[0, b]$, for $b > 0$. Prove that if x is in $[0, b]$, then $x = tb$ for some t with $0 \leq t \leq 1$. What is the significance of the number t? What is the mid-point of the interval $[0, b]$?

(b) Now prove that if x is in $[a, b]$, then $x = (1-t)a + tb$ for some t with $0 \leq t \leq 1$. Hint: This expression can also be written as $a + t(b - a)$. What is the midpoint of the interval $[a, b]$? What is the point $1/3$ of the way from a to b?

(c) Prove, conversely, that if $0 \leq t \leq 1$, then $(1-t)a + tb$ is in $[a, b]$.

(d) The points of the *open* interval (a, b) are those of the form $(1-t)a + tb$ for $0 < t < 1$.

3. Draw the set of all points (x, y) satisfying the following conditions. (In most cases your picture will be a sizable portion of a plane, not just a line or curve.)

(i) $x > y$.

(ii) $x + a > y + b$.

(iii) $y < x^2$.

(iv) $y \leq x^2$.

(v) $|x - y| < 1$.

(vi) $|x + y| < 1$.

(vii) $x + y$ is an integer.

(viii) $\dfrac{1}{x+y}$ is an integer.

(ix) $(x-1)^2 + (y-2)^2 < 1$.

(x) $x^2 < y < x^4$.

4. Draw the set of all points (x, y) satisfying the following conditions:

(i) $|x| + |y| = 1$.

(ii) $|x| - |y| = 1$.

(iii) $|x - 1| = |y - 1|$.

(iv) $|1 - x| = |y - 1|$.

(v) $x^2 + y^2 = 0$.

(vi) $xy = 0$.

(vii) $x^2 - 2x + y^2 = 4$.

(viii) $x^2 = y^2$.

5. Draw the set of all points (x, y) satisfying the following conditions:

(i) $x = y^2$.
(ii) $\frac{y^2}{a^2} - \frac{x^2}{b^2} = 1$.
(iii) $x = |y|$.
(iv) $x = \sin y$.

Hint: You already know the answers when x and y are interchanged.

6. (a) Show that the straight line through (a, b) with slope m is the graph of the function $f(x) = m(x - a) + b$. This formula, known as the "point-slope form" is far more convenient than the equivalent expression $f(x) = mx + (b - ma)$; it is immediately clear from the point-slope form that the slope is m, and that the value of f at a is b.

(b) For $a \neq c$, show that the straight line through (a, b) and (c, d) is the graph of the function

$$f(x) = \frac{d - b}{c - a}(x - a) + b.$$

(c) When are the graphs of $f(x) = mx + b$ and $g(x) = m'x + b'$ parallel straight lines?

7. (a) For any numbers A, B, and C, with A and B not both 0, show that the set of all (x, y) satisfying $Ax + By + C = 0$ is a straight line (possibly a vertical one). Hint: First decide when a vertical straight line is described.

(b) Show conversely that every straight line, including vertical ones, can be described as the set of all (x, y) satisfying $Ax + By + C = 0$.

(1, m)
(1, n)

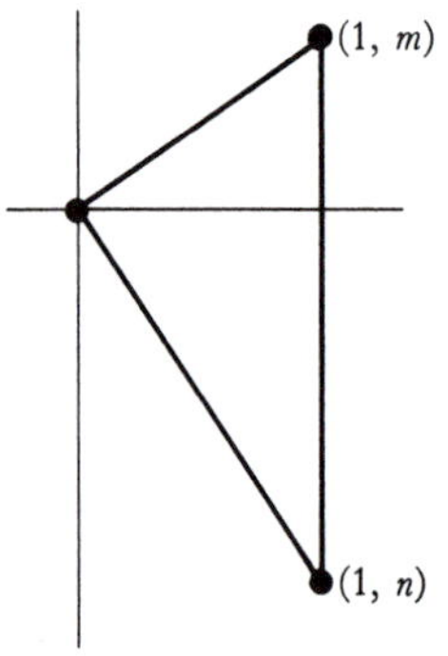

FIGURE 29

8. (a) Prove that the graphs of the functions

$$f(x) = mx + b,$$
$$g(x) = nx + c,$$

are perpendicular if $mn = -1$, by computing the squares of the lengths of the sides of the triangle in Figure 29. (Why is this special case, where the lines intersect at the origin, as good as the general case?)

(b) Prove that the two straight lines consisting of all (x, y) satisfying the conditions

$$Ax + By + C = 0,$$
$$A'x + B'y + C' = 0,$$

are perpendicular if and only if $AA' + BB' = 0$.

9. (a) Prove, using Problem 1-19, that

$$\sqrt{(x_1 + y_1)^2 + (x_2 + y_2)^2} \leq \sqrt{x_1{}^2 + x_2{}^2} + \sqrt{y_1{}^2 + y_2{}^2}.$$

(b) Prove that

$$\sqrt{(x_3 - x_1)^2 + (y_3 - y_1)^2} \le \sqrt{(x_2 - x_1)^2 + (y_2 - y_1)^2} + \sqrt{(x_3 - x_2)^2 + (y_3 - y_2)^2}.$$

Interpret this inequality geometrically (it is called the "triangle inequality"). When does strict inequality hold?

10. Sketch the graphs of the following functions, plotting enough points to get a good idea of the general appearance. (Part of the problem is to make a reasonable decision how many is "enough"; the queries posed below are meant to show that a little thought will often be more valuable than hundreds of individual points.)

(i) $f(x) = x + \dfrac{1}{x}$. (What happens for x near 0, and for large x? Where does the graph lie in relation to the graph of the identify function? Why does it suffice to consider only positive x at first?)

(ii) $f(x) = x - \dfrac{1}{x}$.

(iii) $f(x) = x^2 + \dfrac{1}{x^2}$.

(iv) $f(x) = x^2 - \dfrac{1}{x^2}$.

11. Describe the general features of the graph of f if

(i) f is even.
(ii) f is odd.
(iii) f is nonnegative.
(iv) $f(x) = f(x + a)$ for all x (a function with this property is called **periodic**, with **period** a.

12. Graph the functions $f(x) = \sqrt[m]{x}$ for $m = 1, 2, 3, 4$. (There is an easy way to do this, using Figure 14. Be sure to remember, however, that $\sqrt[m]{x}$ means the *positive* mth root of x when m is even; you should also note that there will be an important difference between the graphs when m is even and when m is odd.)

13. (a) Graph $f(x) = |x|$ and $f(x) = x^2$.
(b) Graph $f(x) = |\sin x|$ and $f(x) = \sin^2 x$. (There is an important difference between the graphs, which we cannot yet even describe rigorously. See if you can discover what it is; part (a) is meant to be a clue.)

14. Describe the graph of g in terms of the graph of f if

(i) $g(x) = f(x) + c$.
(ii) $g(x) = f(x + c)$. (It is easy to make a mistake here.)
(iii) $g(x) = cf(x)$.
(iv) $g(x) = f(cx)$. (Distinguish the cases $c = 0$, $c > 0$, $c < 0$.)
(v) $g(x) = f(1/x)$.
(vi) $g(x) = f(|x|)$.
(vii) $g(x) = |f(x)|$.
(viii) $g(x) = \max(f, 0)$.
(ix) $g(x) = \min(f, 0)$.
(x) $g(x) = \max(f, 1)$.

15. Draw the graph of $f(x) = ax^2 + bx + c$. Hint: Use the methods of Problem 1-18.

16. Suppose that A and C are not both 0. Show that the set of all (x, y) satisfying

$$Ax^2 + Bx + Cy^2 + Dy + E = 0$$

is either a parabola, an ellipse, or an hyperbola (or possibly $\emptyset$). Hint: The case $C = 0$ is essentially Problem 15, and the case $A = 0$ is just a minor variant. Now consider separately the cases where A and B are both positive or negative, and where one is positive while the other is negative.

17. The symbol $[x]$ denotes the largest integer which is $\leq x$. Thus, $[2.1] = [2] = 2$ and $[-0.9] = [-1] = -1$. Draw the graph of the following functions (they are all quite interesting, and several will reappear frequently in other problems).

(i) $f(x) = [x]$.
(ii) $f(x) = x - [x]$.
(iii) $f(x) = \sqrt{x - [x]}$.
(iv) $f(x) = [x] + \sqrt{x - [x]}$.
(v) $f(x) = \left[\dfrac{1}{x}\right]$.
(vi) $f(x) = \dfrac{1}{\left[\dfrac{1}{x}\right]}$.

18. Graph the following functions.

(i) $f(x) = \{x\}$, where $\{x\}$ is defined to be the distance from x to the nearest integer.
(ii) $f(x) = \{2x\}$.
(iii) $f(x) = \{x\} + \frac{1}{2}\{2x\}$.
(iv) $f(x) = \{4x\}$.
(v) $f(x) = \{x\} + \frac{1}{2}\{2x\} + \frac{1}{4}\{4x\}$.

Many functions may be described in terms of the decimal expansion of a number. Although we will not be in a position to describe infinite decimals rigorously until Chapter 23, your intuitive notion of infinite decimals should suffice to carry you through the following problem, and others which occur before Chapter 23. There is one ambiguity about infinite decimals which must be eliminated: Every decimal ending in a string of 9's is equal to another ending in a string of 0's (e.g., $1.23999\ldots = 1.24000\ldots$). We will always use the one ending in 9's.

***19.** Describe as best you can the graphs of the following functions (a complete picture is usually out of the question).

(i) $f(x) =$ the 1st number in the decimal expansion of x.
(ii) $f(x) =$ the 2nd number in the decimal expansion of x.
(iii) $f(x) =$ the number of 7's in the decimal expansion of x if this number is finite, and 0 otherwise.
(iv) $f(x) = 0$ if the number of 7's in the decimal expansion of x is finite, and 1 otherwise.
(v) $f(x) =$ the number obtained by replacing all digits in the decimal expansion of x which come after the first 7 (if any) by 0.
(vi) $f(x) = 0$ if 1 never appears in the decimal expansion of x, and n if 1 first appears in the nth place.

***20.** Let

$$f(x) = \begin{cases} 0, & x \text{ irrational} \\ \dfrac{1}{q}, & x = \dfrac{p}{q} \text{ rational in lowest terms.} \end{cases}$$

(A number p/q is in **lowest terms** if p and q are integers with no common factor, and $q > 0$). Draw the graph of f as well as you can (don't sprinkle points randomly on the paper; consider first the rational numbers with $q = 2$, then those with $q = 3$, etc.).

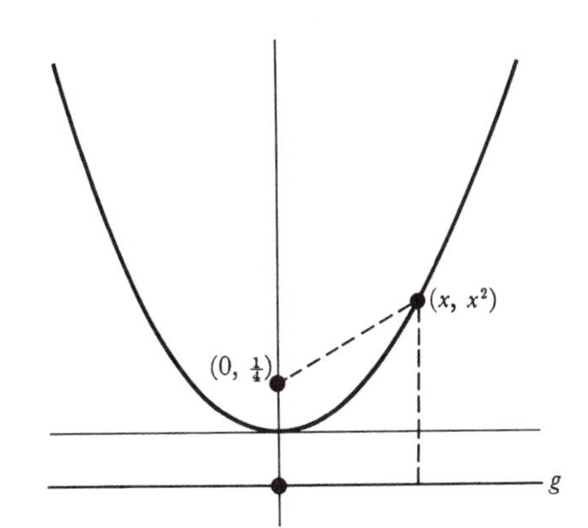

FIGURE 30

21. (a) The points on the graph of $f(x) = x^2$ are the ones of the form (x, x^2). Prove that each such point is equidistant from the point $(0, \frac{1}{4})$ and the graph of $g(x) = -\frac{1}{4}$. (See Figure 30.)

(b) Given a point $P = (\alpha, \beta)$ and a horizontal line L, the graph of $g(x) = \gamma$, show that the set of all points (x, y) equidistant from P and L is the graph of a function of the form $f(x) = ax^2 + bx + c$.

***22.** (a) Show that the square of the distance from (c, d) to (x, mx) is

$$x^2(m^2 + 1) + x(-2md - 2c) + d^2 + c^2.$$

Using Problem 1-18 to find the minimum of these numbers, show that the distance from (c, d) to the graph of $f(x) = mx$ is

$$|cm - d|/\sqrt{m^2 + 1}.$$

(b) Find the distance from (c, d) to the graph of $f(x) = mx + b$. (Reduce this case to part (a).)

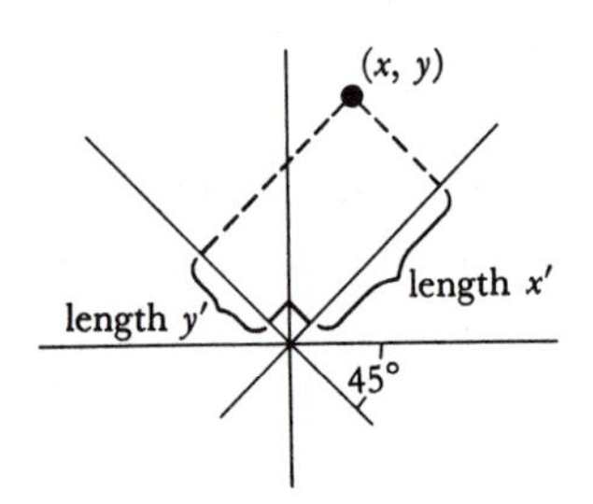

FIGURE 31

***23.** (a) Using Problem 22, show that the numbers x' and y' indicated in Figure 31 are given by

$$x' = \frac{1}{\sqrt{2}}x + \frac{1}{\sqrt{2}}y,$$

$$y' = -\frac{1}{\sqrt{2}}x + \frac{1}{\sqrt{2}}y.$$

(b) Show that the set of all (x, y) with $(x'/\sqrt{2})^2 - (y'/\sqrt{2})^2 = 1$ is the same as the set of all (x, y) with $xy = 1$.

APPENDIX 1. VECTORS

Suppose that v is a point in the plane; in other words, v is a pair of numbers

$$v = (v_1, v_2).$$

For convenience, we will use this convention that subscripts indicate the first and second pairs of a point that has been described by a single letter. Thus, if we mention the points w and z, it will be understood that w is the pair (w_1, w_2), while z is the pair (z_1, z_2).

Instead of the actual pair of numbers (v_1, v_2), we often picture v as an arrow from the origin O to this point (Figure 1), and we refer to these arrows as *vectors* in the plane. Of course, we've haven't really said anything new yet, we've simply introduced an alternate term for a point of the plane, and another mental picture. The real point of the new terminology is to emphasize that we are going to do some new things with points in the plane.

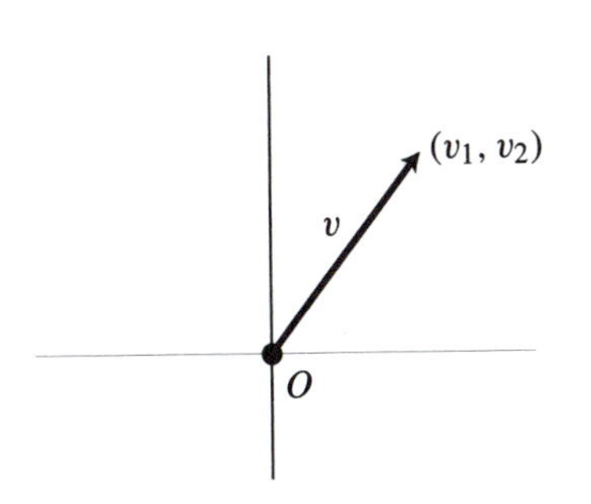

FIGURE 1

For example, suppose that we have two vectors (i.e., points) in the plane,

$$v = (v_1, v_2), \qquad w = (w_1, w_2).$$

Then we can define a new vector (a new point of the plane) $v + w$ by the equation

$$(1) \qquad v + w = (v_1 + w_1, v_2 + w_2).$$

Notice that all the letters on the right side of this equation are numbers, and the + sign is just our usual addition of numbers. On the other hand, the + sign on the left side is new: previously, the sum of two points in the plane wasn't defined, and we've simply used equation (1) as a *definition*.

A very fussy mathematician might want to use some new symbol for this newly defined operation, like

$$v \mathbf{+} w, \qquad \text{or perhaps} \qquad v \oplus w,$$

but there's really no need to insist on this; since $v + w$ hasn't been defined before, there's no possibility of confusion, so we might as well keep the notation simple.

Of course, any one can make new notation; for example, since it's our definition, we could just as well have defined $v + w$ as $(v_1 + w_1 \cdot w_2,\ v_2 + {w_1}^2)$, or by some other equally weird formula. The real question is, does our new construction have any particular significance?

Figure 2 shows two vectors v and w, as well as the point

$$(v_1 + w_1, v_2 + w_2),$$

which, for the moment, we have simply indicated in the usual way, without drawing an arrow. Note that it is easy to compute the slope of the line L between v and our new point: as indicated in Figure 2, this slope is just

$$\frac{(v_2 + w_2) - v_2}{(v_1 + w_1) - v_1} = \frac{w_2}{w_1},$$

and this, of course, is the slope of our vector w, from the origin O to (w_1, w_2). In other words, the line L is parallel to w.

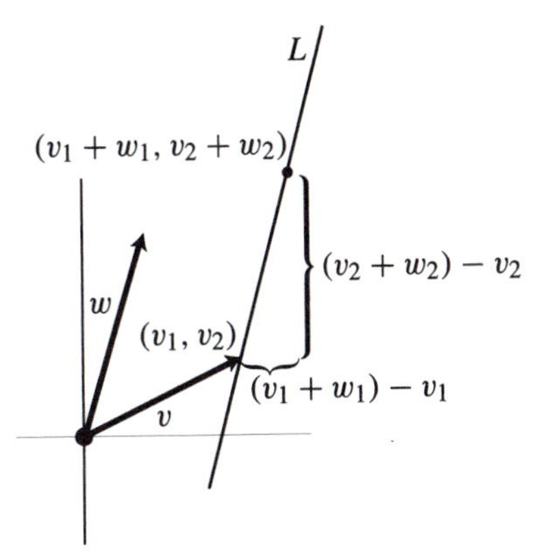

FIGURE 2

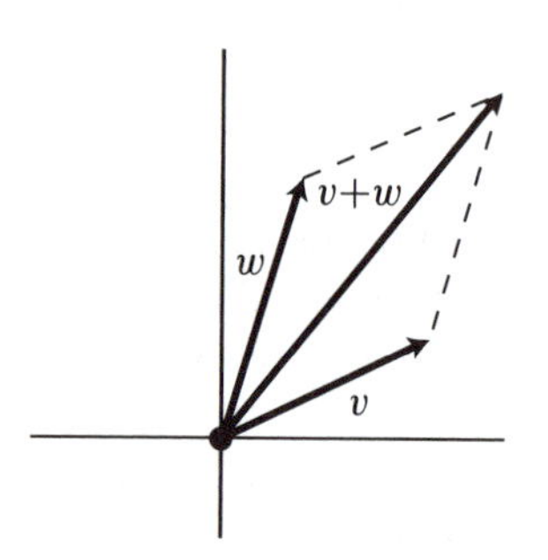

FIGURE 3

Similarly, the slope of the line M between (w_1, w_2) and our new point is

$$\frac{(v_2 + w_2) - w_2}{(v_1 + w_1) - v_2} = \frac{v_2}{v_1},$$

which is the slope of the vector v; so M is parallel to v. In short, the new point $v + w$ lies on the parallelogram having v and w as sides. When we draw $v + w$ as an arrow (Figure 3), it points along the diagonal of this parallelogram. In physics, vectors are used to symbolize forces, and the sum of two vectors represents the resultant force when two different forces are applied simultaneously to the same object.

Figure 4 shows another way of visualizing the sum $v+w$. If we use "w" to denote an arrow parallel to w, and having the same length, but starting at v instead of at the origin, then $v + w$ is the vector from O to the final endpoint; thus we get to $v + w$ by first following v, and then following w.

Many of the properties of $+$ for ordinary numbers also hold for this new $+$ for vectors. For example, the "commutative law"

$$v + w = w + v,$$

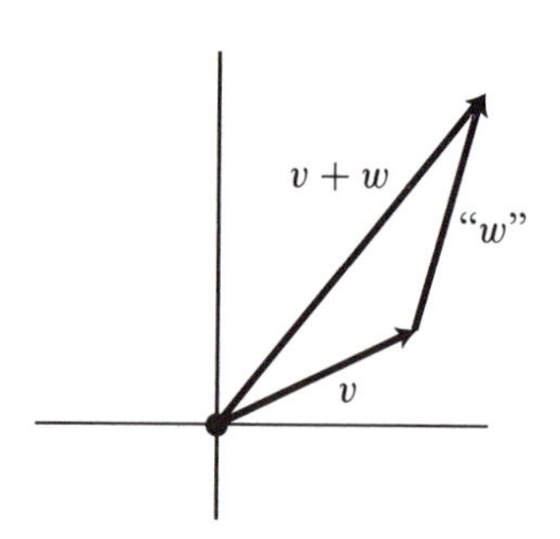

FIGURE 4

is obvious from the geometric picture, since the parallelogram spanned by v and w is the same as the parallelogram spanned by w and v. It is also easily checked analytically, since it states that

$$(v_1 + w_1, v_2 + w_2) = (w_1 + v_1, w_2 + v_2),$$

and thus simply depends on the commutative law for numbers:

$$v_1 + w_1 = w_1 + v_1,$$
$$v_2 + w_2 = w_2 + v_2.$$

Similarly, unraveling definitions, we find the "associative law"

$$[v + w] + z = v + [w + z].$$

Figure 5 indicates a method of finding $v + w + z$.

The origin $O = (0, 0)$ is an "additive identity,"

$$O + v = v + O = v,$$

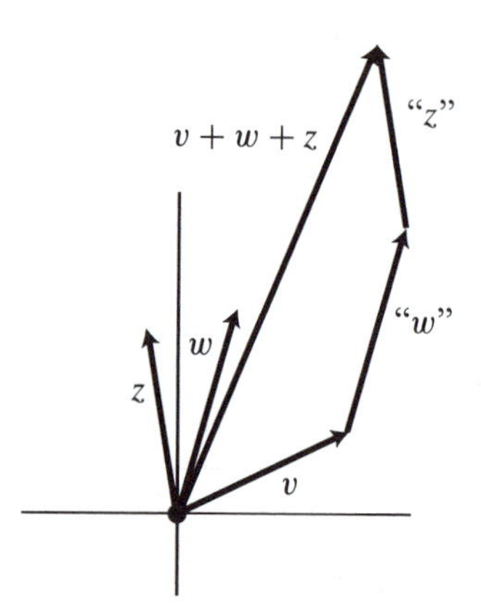

FIGURE 5

and if we define

$$-v = (-v_1, -v_2),$$

then we also have

$$v + (-v) = -v + v = O.$$

Naturally we can also define

$$w - v = w + (-v),$$

exactly as with numbers; equivalently,

$$w - v = (w_1 - v_1, w_2 - v_2).$$

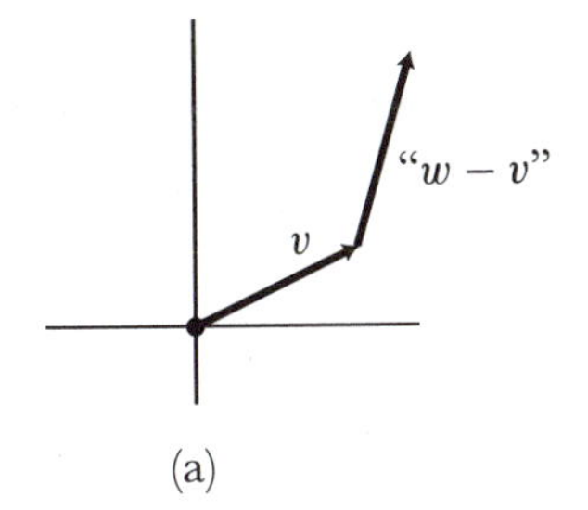

(a)

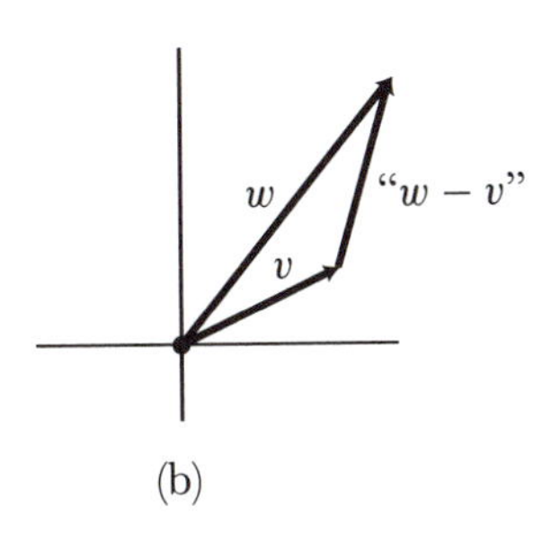

(b)

FIGURE 6

Just as with numbers, our definition of $w - v$ simply means that it satisfies

$$v + (w - v) = w.$$

Figure 6(a) shows v and an arrow "$w - v$" that is parallel to $w - v$ but that starts at the endpoint of v. As we established with Figure 4, the vector from the origin to the endpoint of this arrow is just $v + (w - v) = w$ (Figure 6(b)). In other words, we can picture $w - v$ geometrically as the arrow that goes from v to w (except that it must then be moved back to the origin).

There is also a way of multiplying a number by a vector: For a number a and a vector $v = (v_1, v_2)$, we define

$$a \cdot v = (av_1, av_2)$$

(We sometimes simply write av instead of $a \cdot v$; of course, it is then especially important to remember that v denotes a vector, rather than a number.) The vector $a \cdot v$ points in the same direction as v when $a > 0$ and in the opposite direction when $a < 0$ (Figure 7).

You can easily check the following formulas:

$$\begin{aligned} a \cdot (b \cdot v) &= (ab) \cdot v, \\ 1 \cdot v &= v, \\ 0 \cdot v &= O, \\ -1 \cdot v &= -v. \end{aligned}$$

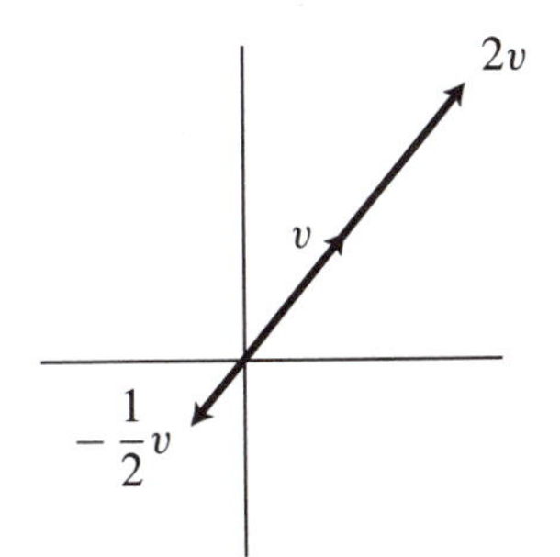

FIGURE 7

Notice that we have only defined a product of a number and a vector, we have not defined a way of 'multiplying' two vectors to get another vector.* However, there are various ways of 'multiplying' vectors to get numbers, which are explored in the following problems.

PROBLEMS

1. Given a point v of the plane, let $R_\theta(v)$ be the result of rotating v around the origin through an angle of θ (Figure 8). The aim of this problem is to obtain a formula for R_θ, with minimal calculation.

(a) Show that

$$\begin{aligned} R_\theta(1, 0) &= (\cos\theta, \sin\theta), \\ R_\theta(0, 1) &= (-\sin\theta, \cos\theta). \end{aligned}$$

(b) Explain why we have

$$\begin{aligned} R_\theta(v + w) &= R_\theta(v) + R_\theta(w), \\ R_\theta(a \cdot w) &= a \cdot R_\theta(w). \end{aligned}$$

(c) Now show that for any point (x, y) we have

$$R_\theta(x, y) = (x\cos\theta - y\sin\theta, x\sin\theta + y\cos\theta).$$

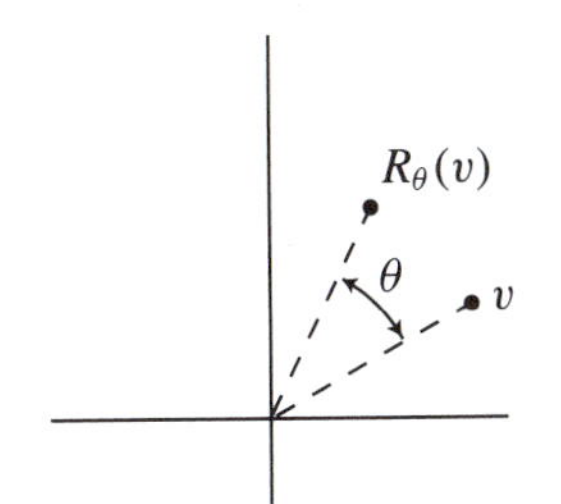

FIGURE 8

*If you jump to Chapter 25, you'll find that there is an important way of defining a product, but this is something very special for the plane—it doesn't work for vectors in 3-space, for example, even though the other constructions do.

(d) Use this result to give another solution to Problem 4-23.

2. Given v and w, we define the *number*

$$v \bullet w = v_1 w_1 + v_2 w_2;$$

this is often called the 'dot product' or 'scalar product' of v and w ('scalar' being a rather old-fashioned word for a number, as opposed to a vector).

(a) Given v, find a vector w such that $v \bullet w = 0$. Now describe the set of all such vectors w.

(b) Show that

$$v \bullet w = w \bullet v$$
$$v \bullet (w + z) = v \bullet w + v \bullet z$$

and that

$$a \cdot (v \bullet w) = (a \cdot v) \bullet w = v \bullet (a \cdot w).$$

Notice that the last of these equations involves *three* products: the dot product $\bullet$ of two vectors; the product $\cdot$ of a number and a vector; and the ordinary product $\cdot$ of two numbers.

(c) Show that $v \bullet v \geq 0$, and that $v \bullet v = 0$ only when $v = O$. Hence we can define the *norm* $\|v\|$ as

$$\|v\| = \sqrt{v \bullet v},$$

which will be 0 only for $v = O$. What is the geometric interpretation of the norm?

(d) Prove that

$$\|v + w\| \leq \|v\| + \|w\|,$$

and that equality holds if and only if $v = 0$ or $w = 0$ or $w = a \cdot v$ for some number $a > 0$.

(e) Show that

$$v \bullet w = \frac{\|v + w\|^2 - \|v - w\|^2}{4}.$$

3. (a) Let R_θ be rotation by an angle of θ (Problem 1). Show that

$$R_\theta(v) \bullet R_\theta(w) = v \bullet w.$$

(b) Let $e = (1, 0)$ be the vector of length 1 pointing along the first axis, and let $w = (\cos\theta, \sin\theta)$; this is a vector of length 1 that makes an angle of θ with the first axis (compare Problem 1). Calculate that

$$e \bullet w = \cos\theta.$$

Conclude that in general

$$v \bullet w = \|v\| \cdot \|w\| \cdot \cos\theta,$$

where θ is the angle between v and w.

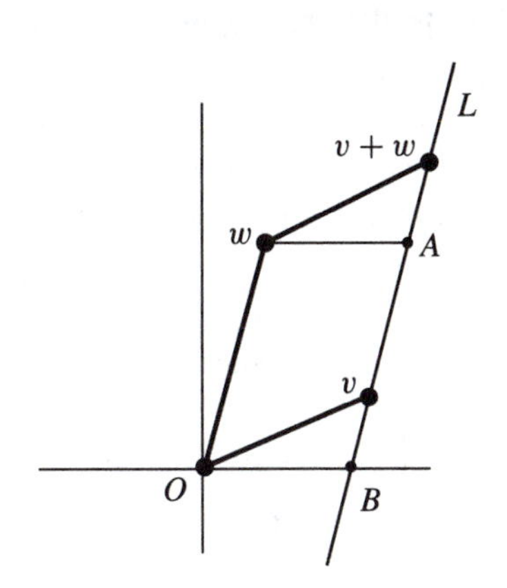

FIGURE 9

4. Given two vectors v and w, we'd expect to have a simple formula, involving the coordinates v_1, v_2, w_1, w_2, for the area of the parallelogram they span. Figure 9 indicates a strategy for finding such a formula: since the triangle with vertices $w, A, v + w$ is congruent to the triangle OBv, we can reduce the problem to an easier one where one side of the parallelogram lies along the horizontal axis:

(a) The line L passes through v and is parallel to w, so has slope w_2/w_1. Conclude that the point B has coordinate

$$\frac{v_1 w_2 - w_1 v_2}{w_2},$$

and that the parallelogram therefore has area

$$\det(v, w) = v_1 w_2 - w_1 v_2.$$

This formula, which defines the *determinant* det, certainly seems to be simple enough, but it can't really be true that $\det(v, w)$ always gives the area. After all, we clearly have

$$\det(w, v) = -\det(v, w),$$

so sometimes det will be negative! Indeed, it is easy to see that our "derivation" made all sorts of assumptions (that w_2 was positive, that B had a positive coordinate, etc.) Nevertheless, it seems likely that $\det(v, w)$ is $\pm$ the area; the next problem gives an independent proof.

5. (a) If v points along the positive horizontal axis, show that $\det(v, w)$ is the area of the parallelogram spanned by v and w for w above the horizontal axis ($w_2 > 0$), and the negative of the area for w below this axis.

(b) If R_θ is rotation by an angle of θ (Problem 1), show that

$$\det(R_\theta v, R_\theta w) = \det(v, w).$$

Conclude that $\det(v, w)$ is the area of the parallelogram spanned by v and w when the rotation from v to w is counterclockwise, and the negative of the area when it is clockwise.

6. Show that

$$\det(v, w + z) = \det(v, w) + \det(v, z)$$
$$\det(v + w, z) = \det(v, z) + \det(w, z)$$

and that

$$a \det(v, w) = \det(a \cdot v, w) = \det(v, a \cdot w).$$

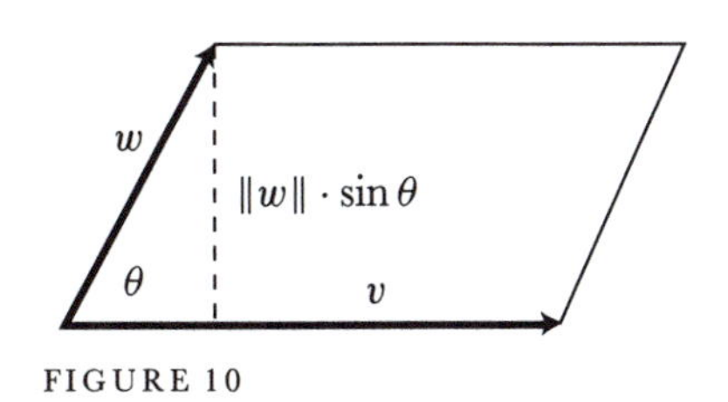

FIGURE 10

7. Using the method of Problem 3, show that

$$\det(v, w) = \|v\| \cdot \|w\| \cdot \sin\theta,$$

which is also obvious from the geometric interpretation (Figure 10).

APPENDIX 2. THE CONIC SECTIONS

Although we will be concerned almost exclusively with figures in the plane, defined formally as the set of all pairs of real numbers, in this Appendix we want to consider three-dimensional space, which we can describe in terms of triples of real numbers, using a "three-dimensional coordinate system," consisting of three straight lines intersecting at right angles (Figure 1). Our *horizontal* and *vertical* axes now mutate to two axes in a horizontal plane, with the third axis perpendicular to both.

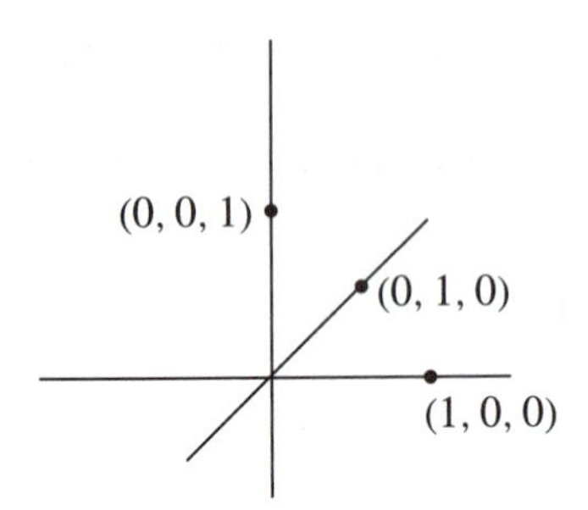

FIGURE 1

One of the simplest subsets of this three-dimensional space is the (infinite) *cone* illustrated in Figure 2; this cone may be produced by rotating a "generating line," of slope C say, around the third axis.

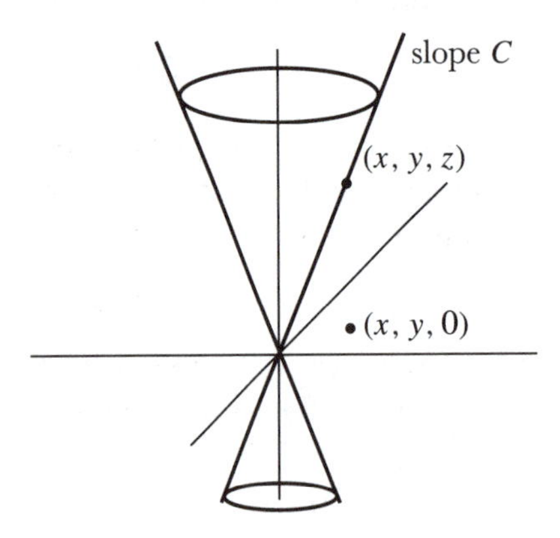

FIGURE 2

For any given first two coordinates x and y, the point $(x, y, 0)$ in the horizontal plane has distance $\sqrt{x^2 + y^2}$ from the origin, and thus

(1) (x, y, z) is in the cone if and only if $z = \pm C\sqrt{x^2 + y^2}$.

We can descend from these three-dimensional vistas to the more familiar two-dimensional one by asking what happens when we intersect this cone with some plane P (Figure 3).

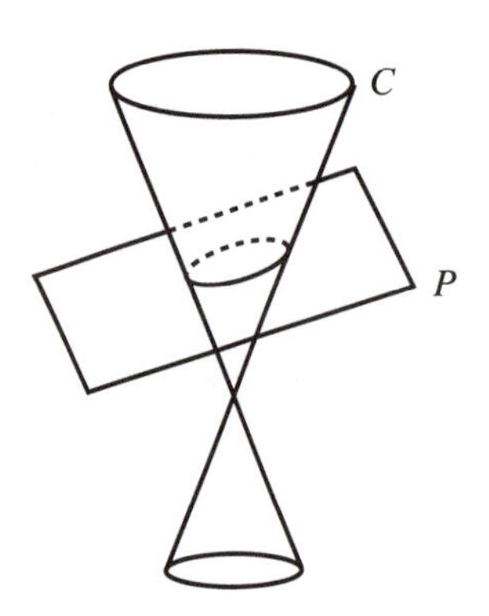

FIGURE 3

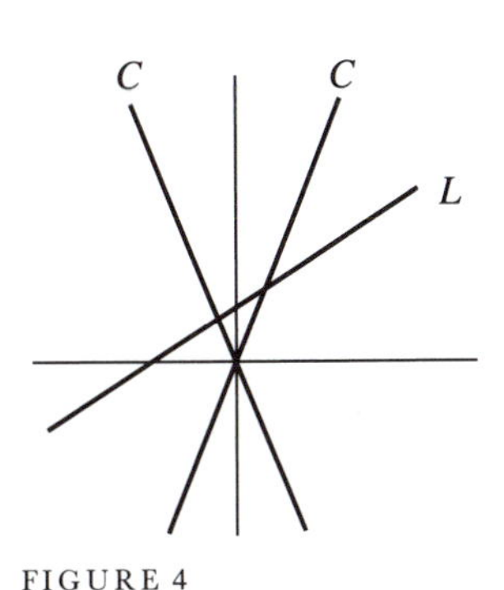

FIGURE 4

If the plane is parallel to the horizontal plane, there's certainly no mystery—the intersection is just a circle. Otherwise, the plane P intersects the horizontal plane in a straight line. We can make things a lot simpler for ourselves if we rotate everything so that this intersection line points straight out from the plane of the paper, while the first axis is in the usual position that we are familiar with. The plane P is thus viewed "straight on," so that all we see (Figure 4) is its intersection L with the plane of the first and third axes; from this view-point the cone itself simply appears as two straight lines.

In the plane of the first and third axes, the line L can be described as the collection of all points of the form

$$(x, Mx + B),$$

where M is the slope of L. For an arbitrary point (x, y, z) it follows that

$$(2) \qquad (x, y, z) \text{ is in the plane } P \text{ if and only if } z = Mx + B.$$

Combining (1) and (2), we see that (x, y, z) is in the intersection of the cone and the plane if and only if

$$(*) \qquad Mx + B = \pm C\sqrt{x^2 + y^2}.$$

FIGURE 5

Now we have to choose coordinate axes in the plane P. We can choose L as the first axis, measuring distances from the intersection Q with the horizontal plane (Figure 5); for the second axis we just choose the line through Q parallel to our original second axis. If the first coordinate of a point in P with respect to these axes is x, then the first coordinate of this point with respect to the original axes can be written in the form

$$\alpha x + \beta$$

for some α and β. On the other hand, if the second coordinate of the point with respect to these axes is y, then y is also the second coordinate with respect to the original axes.

Consequently, $(*)$ says that the point lies on the intersection of the plane and the cone if and only if

$$M(\alpha x + \beta) + B = \pm C\sqrt{(\alpha x + \beta)^2 + y^2}.$$

Although this looks fairly complicated, after squaring we can write this as

$$\alpha^2 C^2 y^2 + \alpha^2(M^2 - A^2)x^2 + Ex + F = 0$$

for some E and F that we won't bother writing out. Dividing by α^2 simplifies this to

$$C^2 y^2 + (C^2 - M^2)x^2 + Gx + H = 0.$$

Now Problem 4-16 indicates that this is either a parabola, an ellipse, or an hyperbola. In fact, looking a little more closely at the solution (and interchanging the roles of x and y), we see that the values of G and H are irrelevant:

(1) If $M = \pm C$ we obtain a parabola;
(2) If $C^2 > M^2$ we obtain an ellipse;
(3) If $C^2 < M^2$ we obtain an hyperbola.

These analytic conditions are easy to interpret geometrically (Figure 6):

(1) If our plane is parallel to one of the generating lines of the cone we obtain a parabola;
(2) If our plane slopes less than the generating line of the cone (so that our intersection omits one half of the cone) we obtain an ellipse;
(3) If our plane slopes more than the generating line of the cone we obtain an hyperbola.

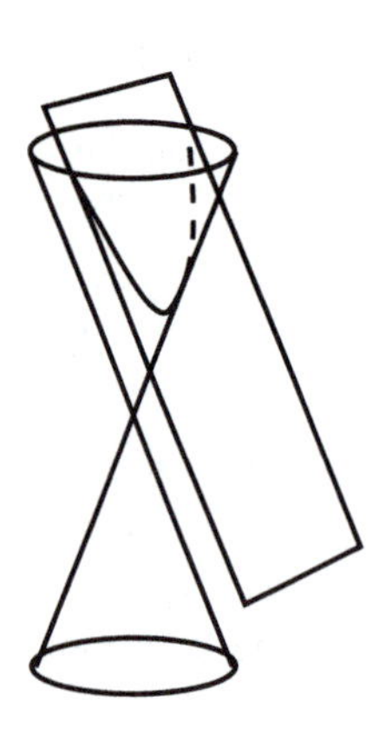

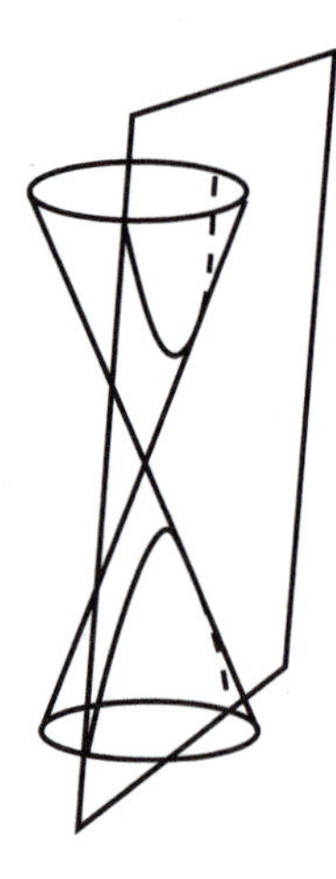

FIGURE 6

In fact, the very names of these "conic sections" are related to this description. The word *parabola* comes from a Greek root meaning 'alongside,' the same root that appears in parable, not to mention paradigm, paradox, paragon, paragraph, paralegal, parallax, parallel, even parachute. *Ellipse* comes from a Greek root meaning 'defect,' or omission, as in ellipsis (an omission, ... or the dots that indicate it). And *hyperbola* comes from a Greek root meaning 'throwing beyond,' or excess. With the currency of words like hyperactive, hypersensitive, and hyperventilate, not to mention hype, one can probably say, without risk of hyperbole, that this root is familiar to almost everyone.*

PROBLEMS

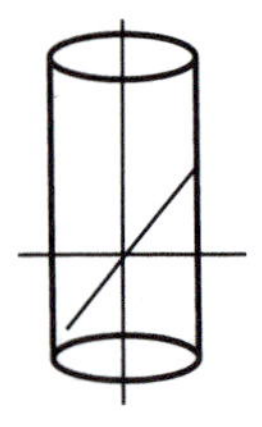

FIGURE 7

1. Consider a cylinder with a generator perpendicular to the horizontal plane (Figure 7); the only requirement for a point (x, y, z) to lie on this cylinder is

*Although the correspondence between these roots and the geometric picture correspond so beautifully, for the sake of dull accuracy it has to be reported that the Greeks originally applied the words to describe features of certain equations involving the conic sections.

that (x, y) lies on a circle:

$$x^2 + y^2 = C^2.$$

Show that the intersection of a plane with this cylinder can be described by an equation of the form

$$(\alpha x + \beta)^2 + y^2 = C^2.$$

What possibilities are there?

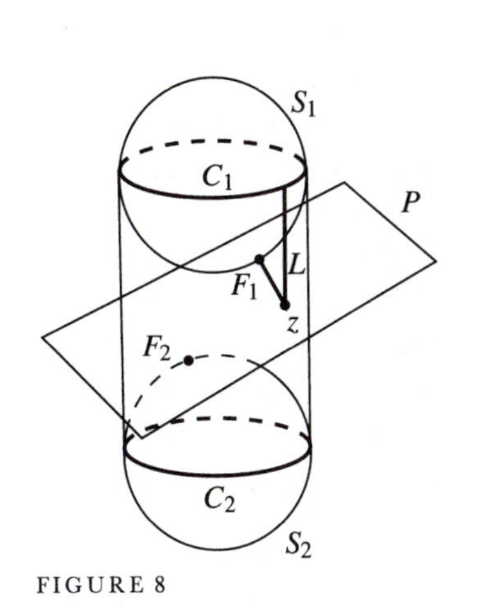

FIGURE 8

2. In Figure 8, the sphere S_1 has the same diameter as the cylinder, so that its equator C_1 lies along the cylinder; it is also tangent to the plane P at F_1. Similarly, the equator C_2 of S_2 lies along the cylinder, and S_2 is tangent to P at F_2.

(a) Let z be any point on the intersection of P and the cylinder. Explain why the length of the line from z to F_1 is equal to the length of the vertical line L from z to C_1.

(b) By proving a similar fact for the length of the line from z to F_2, show that the distance from z to F_1 plus the distance from z to F_2 is a constant, so that the intersection is an ellipse, with foci F_1 and F_2.

3. Similarly, use Figure 9 to prove geometrically that the intersection of a plane and a cone is an ellipse.

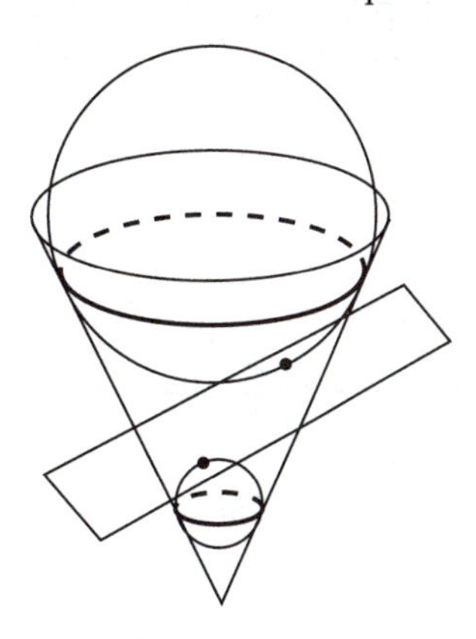

FIGURE 9

APPENDIX 3. POLAR COORDINATES

FIGURE 1

In this chapter we've been acting all along as if there's only one way to label points in the plane with pairs of numbers. Actually, there are many different ways, each giving rise to a different "coordinate system." The usual coordinates of a point are called its cartesian coordinates, after the French mathematician and philosopher René Descartes (1596–1650), who first introduced the idea of coordinate systems. In many situations it is more convenient to introduce polar coordinates, which are illustrated in Figure 1. To the point P we assign the polar coordinates (r, θ), where r is the distance from the origin O to P, and θ is the angle between the horizontal axis and the line from O to P. This angle can be measured either in degrees or in radians (Chapter 15), but in either case θ is not determined unambiguously. For example, with degree measurement, points on the right side of the horizontal axis could have either $\theta = 0$ or $\theta = 360$; moreover, θ is completely ambiguous at the origin O. So it is necessary to exclude some ray through the origin if we want to assign a unique pair (r, θ) to each point under consideration.

On the other hand, there is no problem associating a unique point to any pair (r, θ). In fact, we can even associate a point to (r, θ) when $r < 0$, according to the scheme indicated in Figure 2. Thus, it always makes sense to talk about "the point with polar coordinates (r, θ)," even though there is some ambiguity when we talk about "the polar coordinates" of a given point.

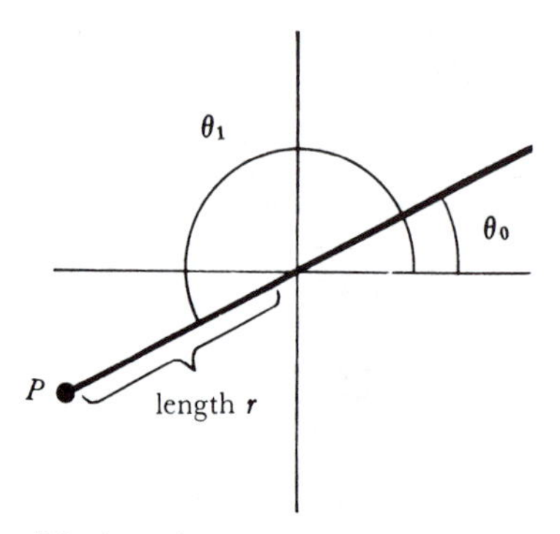

P is the point with polar coordinates (r, θ_1) and also the point with polar coordinates $(-r, \theta_0)$.

FIGURE 2

It is clear from Figure 1 (and Figure 2) that the point with polar coordinates (r, θ) has cartesian coordinates (x, y) given by

$$x = r\cos\theta, \qquad y = r\sin\theta.$$

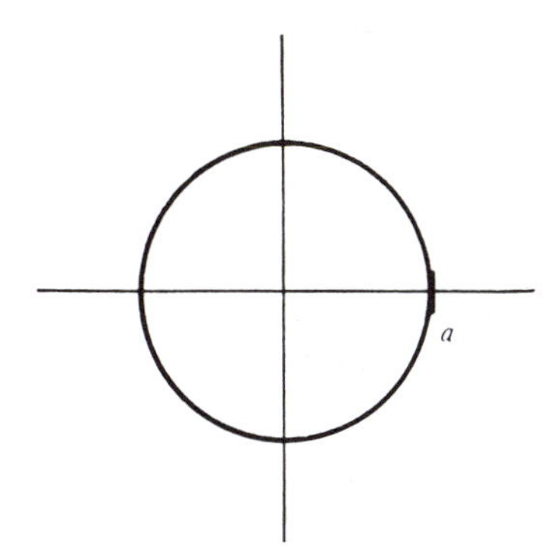

FIGURE 3

Conversely, if a point has cartesian coordinates (x, y), then (any of) its polar coordinates (r, θ) satisfy

$$r = \pm\sqrt{x^2 + y^2}$$
$$\tan\theta = \frac{y}{x} \qquad \text{if } x \neq 0.$$

Now suppose that f is a function. Then by the **graph of f in polar coordinates** we mean the collection of all points P with polar coordinates (r, θ) satisfying $r = f(\theta)$. In other words, the graph of f in polar coordinates is the collection of all points with polar coordinates $(f(\theta), \theta)$. No special significance should be attached to the fact that we are considering pairs $(f(\theta), \theta)$, with $f(\theta)$ first, as opposed to pairs $(x, f(x))$ in the usual graph of f; it is purely a matter of convention that r is considered the first polar coordinate and θ is considered the second.

The graph of f in polar coordinates is often described as "the graph of the equation $r = f(\theta)$." For example, suppose that f is a constant function, $f(\theta) = a$ for all θ. The graph of the equation $r = a$ is simply a circle with center O and radius a (Figure 3). This example illustrates, in a rather blatant way, that polar coordinates are likely to make things simpler in situations that involve symmetry with respect to the origin O.

The graph of the equation $r = \theta$ is shown in Figure 4. The solid line corresponds to all values of $\theta \geq 0$, while the dashed line corresponds to values of $\theta \leq 0$.

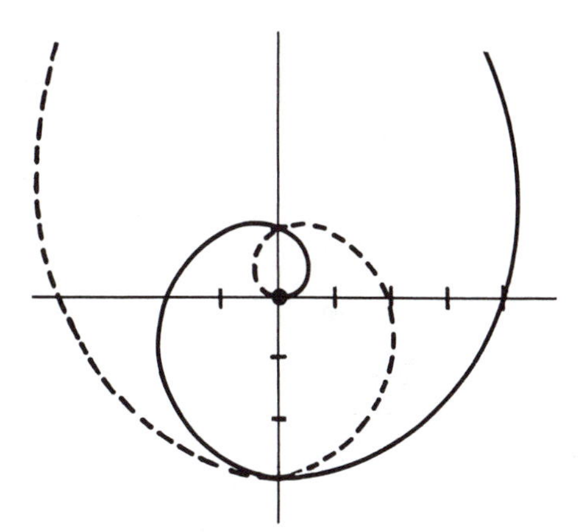

FIGURE 4 **Spiral of Archimedes**

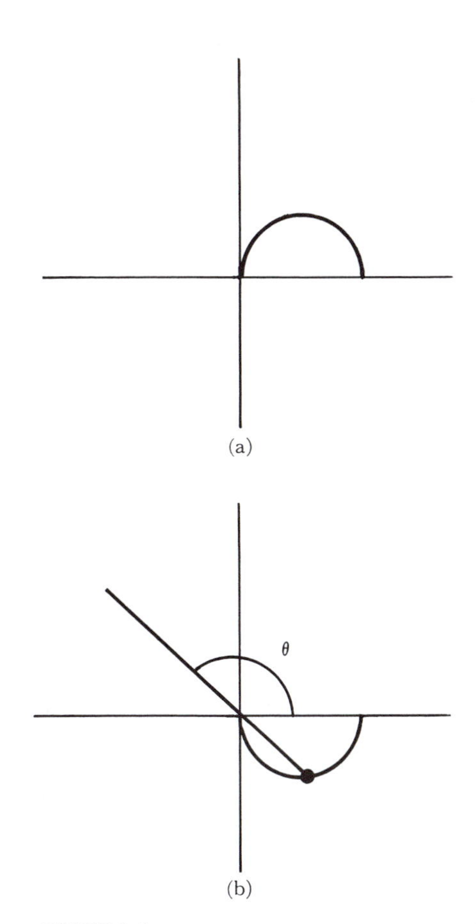

FIGURE 5

As another example involving both positive and negative r, consider the graph of the equation $r = \cos\theta$. Figure 5(a) shows the part that corresponds to $0 \leq \theta \leq 90$ [with θ in degrees]. Figure 5(b) shows the part corresponding to $90 \leq \theta \leq 180$; here $r < 0$. You can check that no new points are added for $\theta > 180$ or $\theta < 0$. It is easy to describe this same graph in terms of the cartesian coordinates of its points. Since the polar coordinates of any point on the graph satisfy

$$r = \cos\theta,$$

and hence

$$r^2 = r\cos\theta,$$

its cartesian coordinates satisfy the equation

$$x^2 + y^2 = x$$

which describes a circle (Problem 3-16). [Conversely, it is clear that if the cartesian coordinates of a point satisfy $x^2 + y^2 = x$, then it lies on the graph of the equation $r = \cos\theta$.]

Although we've now gotten a circle in two different ways, we might well be hesitant about trying to find the equation of an ellipse in polar coordinates. But it turns out that we can get a very nice equation if we choose one of the *foci* as the origin. Figure 6 shows an ellipse with one focus at O, with the sum of the distances of all points from O and the other focus $\mathbf{f}$ being $2a$. We've chosen $\mathbf{f}$ to the left of O, with coordinates written as

$$(-2\varepsilon a, 0).$$

(We have $0 \le \varepsilon < 1$, since we must have $2a >$ distance from $\mathbf{f}$ to O).

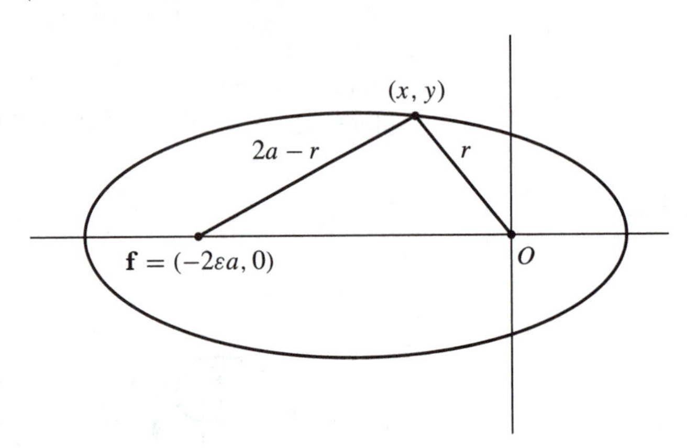

FIGURE 6

The distance r from (x, y) to O is given by

$$(1) \qquad r^2 = x^2 + y^2.$$

By assumption, the distance from (x, y) to $\mathbf{f}$ is $2a - r$, hence

$$(2a - r)^2 = (x - [-2\varepsilon a])^2 + y^2,$$

or

$$(2) \qquad 4a^2 - 4ar + r^2 = x^2 + 4\varepsilon ax + 4\varepsilon^2 a^2 + y^2.$$

Subtracting (1) from (2), and dividing by $4a$, we get

$$a - r = \varepsilon x + \varepsilon^2 a,$$

or

$$\begin{aligned} r &= a - \varepsilon x - \varepsilon^2 a \\ &= (1-\varepsilon^2)a - \varepsilon x, \end{aligned}$$

which we can write as

$$r = \Lambda - \varepsilon x, \qquad \text{for } \Lambda = (1-\varepsilon^2)a. \tag{3}$$

Substituting $r\cos\theta$ for x, we have

$$\begin{aligned} r &= \Lambda - \varepsilon r \cos\theta, \\ r(1+\varepsilon\cos\theta) &= \Lambda, \end{aligned}$$

and thus

$$r = \frac{\Lambda}{1+\varepsilon\cos\theta}. \tag{4}$$

In Chapter 4 we found that

$$\frac{x^2}{a^2} + \frac{y^2}{b^2} = 1 \tag{5}$$

is the equation in cartesian coordinates for an ellipse with $2a$ as the sum of the distances to the foci, but with the foci at $(-c, 0)$ and $(c, 0)$, where

$$b = \sqrt{a^2 - c^2}.$$

Since the distance between the foci is $2c$, this corresponds to the ellipse (4) when we take $c = \varepsilon a$ or $\varepsilon = c/a$ (with equation (3) determining Λ). Conversely, given the ellipse described by (4), for the corresponding equation (5) the value of a is determined by (3),

$$a = \frac{\Lambda}{1-\varepsilon^2},$$

and again using $c = \varepsilon a$, we get

$$b = \sqrt{a^2 - c^2} = \sqrt{a^2 - \varepsilon^2 a^2} = a\sqrt{1-\varepsilon^2} = \frac{\Lambda}{\sqrt{1-\varepsilon^2}}.$$

Thus, we can obtain a and b, the lengths of the major and minor axes, immediately from ε and Λ.

The number

$$\varepsilon = \frac{c}{a} = \frac{\sqrt{a^2 - b^2}}{a} = \sqrt{1 - \left(\frac{b}{a}\right)^2},$$

the *eccentricity* of the ellipse, determines the "shape" of the ellipse (the ratio of the major and minor axes), while the number Λ determines its "size," as shown by (4).

PROBLEMS

1. If two points have polar coordinates (r_1, θ_1) and (r_2, θ_2), show that the distance d between them is given by

$$d^2 = {r_1}^2 + {r_2}^2 - 2r_1r_2\cos(\theta_1 - \theta_2).$$

What does this say geometrically?

2. Describe the general features of the graph of f in polar coordinates if

(i) f is even.
(ii) f is odd.
(iii) $f(\theta) = f(\theta + 180)$ [when θ is measured in degrees].

3. Sketch the graphs of the following equations.

(i) $r = a\sin\theta$.
(ii) $r = a\sec\theta$. Hint: It is a very simple graph!
(iii) $r = \cos 2\theta$. Good luck on this one!
(iv) $r = \cos 3\theta$.
(v) $r = |\cos 2\theta|$.
(vi) $r = |\cos 3\theta|$.

4. Find equations for the cartesian coordinates of points on the graphs (i), (ii) and (iii) in Problem 3.

5. Consider a hyperbola, where the difference of the distance between the two foci is the constant $2a$, and choose one focus at O and the other at $(-2\varepsilon a, 0)$. (In this case, we must have $\varepsilon > 1$). Show that we obtain the exact same equation in polar coordinates

$$r = \frac{\Lambda}{1 + \varepsilon\cos\theta}$$

as we obtained for an ellipse.

6. Consider the set of points (x, y) such that the distance (x, y) to O is equal to the distance from (x, y) to the line $y = a$ (Figure 7). Show that the distance to the line is $a - r\cos\theta$, and conclude that the equation can be written

$$a = r(1 + \cos\theta).$$

Notice that this equation for a parabola is again of the same from as (4).

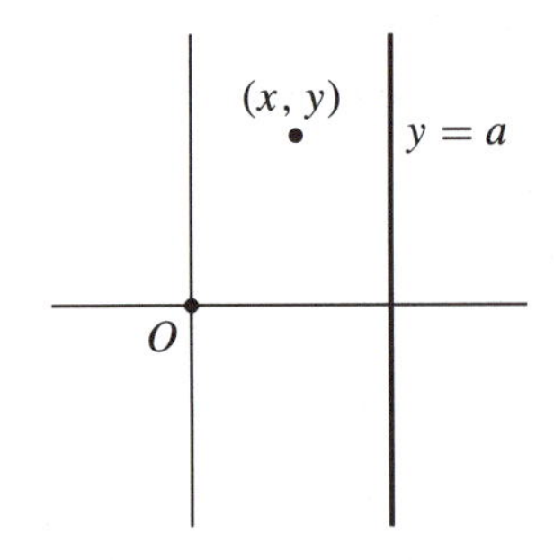

FIGURE 7

7. Now, for any Λ and ε, consider the graph in polar coordinates of the equation (4), which implies (3). Show that the points satisfying this equation satisfy

$$(1 - \varepsilon^2)x^2 + y^2 = \Lambda^2 - 2\Lambda\varepsilon x.$$

Using Problem 4-16, show that this is an ellipse for $\varepsilon < 1$, a parabola for $\varepsilon = 1$, and a hyperbola for $\varepsilon > 1$.

8. (a) Sketch the graph of the *cardioid* $r = 1 - \sin\theta$.
(b) Show that it is also the graph of $r = -1 - \sin\theta$.
(c) Show that it can be described by the equation

$$x^2 + y^2 = \sqrt{x^2 + y^2} - y,$$

and conclude that it can be described by the equation

$$(x^2 + y^2 + y)^2 = x^2 + y^2$$

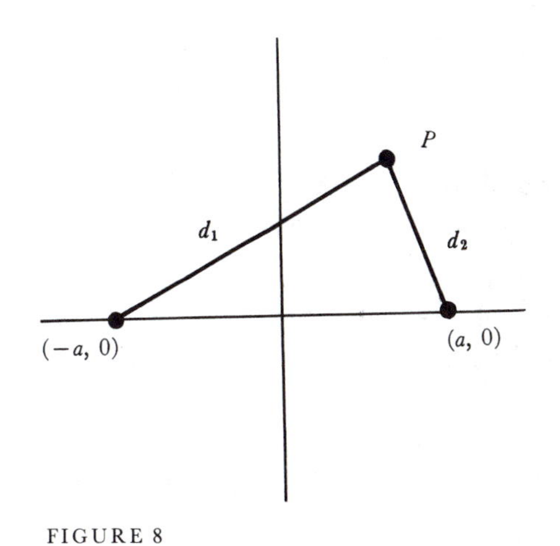

FIGURE 8

9. Sketch the graphs of the following equations.

(i) $r = 1 - \frac{1}{2}\sin\theta$.
(ii) $r = 1 - 2\sin\theta$.
(iii) $r = 2 + \cos\theta$.

10. (a) Sketch the graph of the *lemniscate*

$$r^2 = 2a^2 \cos 2\theta.$$

(b) Find an equation for its cartesian coordinates.
(c) Show that it is the collection of all points P in Figure 8 satisfying $d_1 d_2 = a^2$.
(d) Make a guess about the shape of the curves formed by the set of all P satisfying $d_1 d_2 = b$, when $b > a^2$ and when $b < a^2$.

CHAPTER 5 LIMITS

The concept of a limit is surely the most important, and probably the most difficult one in all of calculus. The goal of this chapter is the definition of limits, but we are, once more, going to begin with a provisional definition; what we shall define is not the word "limit" but the notion of a function approaching a limit.

PROVISIONAL DEFINITION

The function f approaches the limit l near a, if we can make $f(x)$ as close as we like to l by requiring that x be sufficiently close to, but unequal to, a.

Of the six functions graphed in Figure 1, only the first three approach l at a. Notice that although $g(a)$ is not defined, and $h(a)$ is defined "the wrong way," it is still true that g and h approach l near a. This is because we explicitly ruled out, in our definition, the necessity of ever considering the value of the function at a—it is only necessary that $f(x)$ should be close to l for x close to a, but *unequal to* a. We are simply not interested in the value of $f(a)$, or even in the question of whether $f(a)$ is defined.

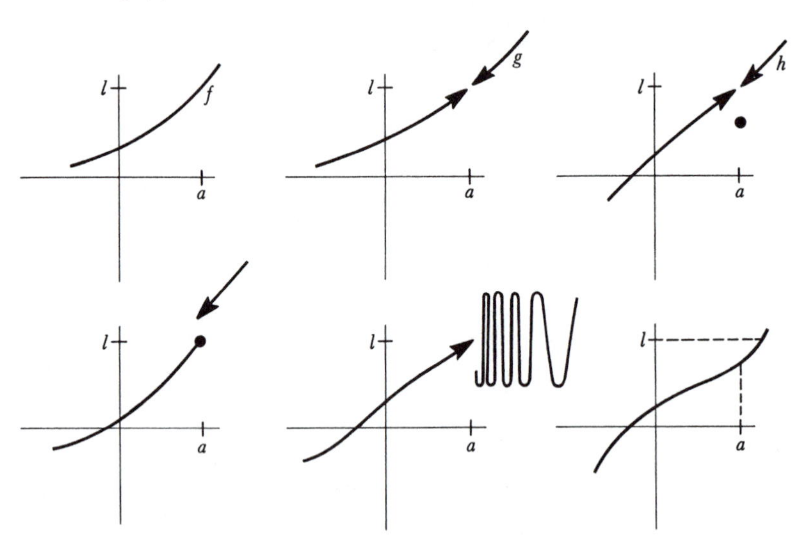

FIGURE 1

One convenient way of picturing the assertion that f approaches l near a is provided by a method of drawing functions that was not mentioned in Chapter 4. In this method, we draw two straight lines, each representing $\mathbf{R}$, and arrows from a point x in one, to $f(x)$ in the other. Figure 2 illustrates such a picture for two different functions.

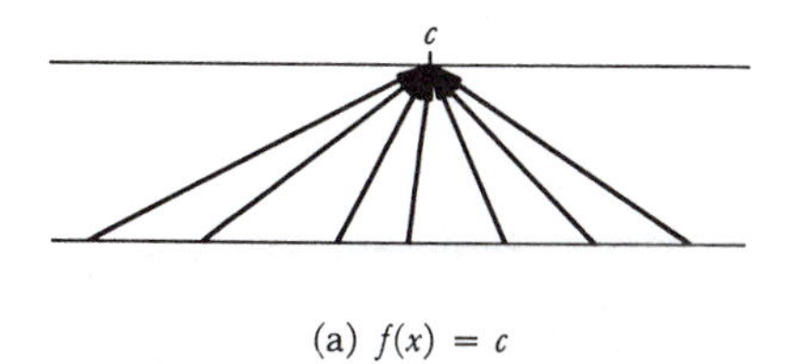

(a) $f(x) = c$

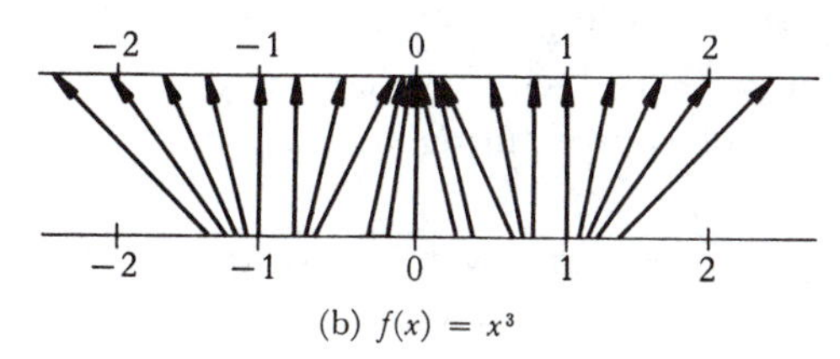

(b) $f(x) = x^3$

FIGURE 2

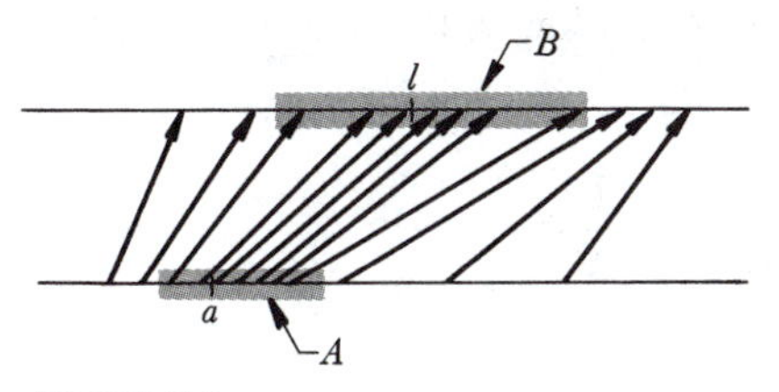

FIGURE 3

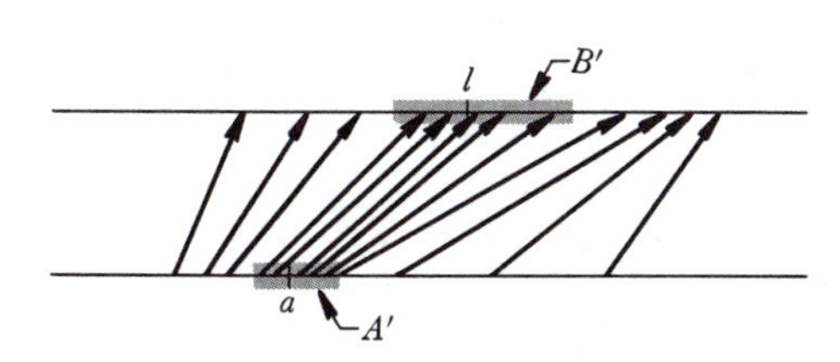

FIGURE 4

Now consider a function f whose drawing looks like Figure 3. Suppose we ask that $f(x)$ be close to l, say within the open interval B which has been drawn in Figure 3. This can be guaranteed if we consider only the numbers x in the interval A of Figure 3. (In this diagram we have chosen the largest interval which will work; any smaller interval containing a could have been chosen instead.) If we choose a smaller interval B' (Figure 4) we will, usually, have to choose a smaller A', but no matter how small we choose the open interval B, there is always supposed to be some open interval A which works.

A similar pictorial interpretation is possible in terms of the graph of f, but in this case the interval B must be drawn on the vertical axis, and the set A on the horizontal axis. The fact that $f(x)$ is in B when x is in A means that the part of the graph lying over A is contained in the region which is bounded by the horizontal lines through the end points of B; compare Figure 5(a), where a valid interval A has been chosen, with Figure 5(b), where A is too large.

In order to apply our definition to a particular function, let us consider $f(x) = x \sin 1/x$ (Figure 6). Despite the erratic behavior of this function near 0 it is clear, at least intuitively, that f approaches 0 near 0, and it is certainly to be hoped that our definition will allow us to reach the same conclusion. In the case we are considering, both a and l of the definition are 0, so we must ask if we can get $f(x) = x \sin 1/x$ as close to 0 as desired if we require that x be sufficiently close to 0, but $\neq 0$. To be specific, suppose we wish to get $x \sin 1/x$ within $\frac{1}{10}$ of 0. This means we want

$$-\frac{1}{10} < x \sin \frac{1}{x} < \frac{1}{10},$$

or, more succinctly, $|x \sin 1/x| < \frac{1}{10}$. Now this is easy. Since

$$\left|\sin \frac{1}{x}\right| \leq 1, \qquad \text{for all } x \neq 0,$$

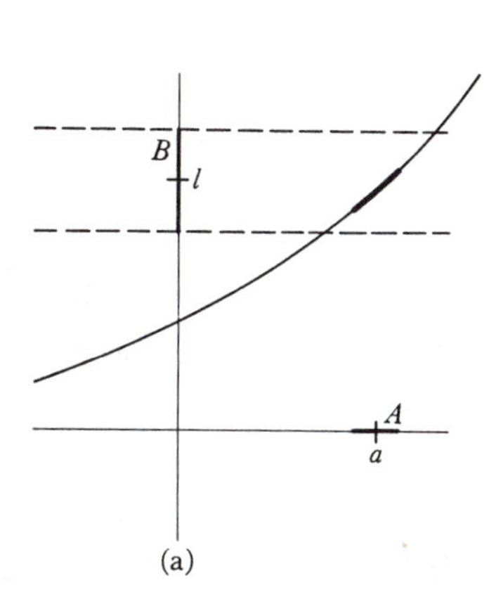

(a)

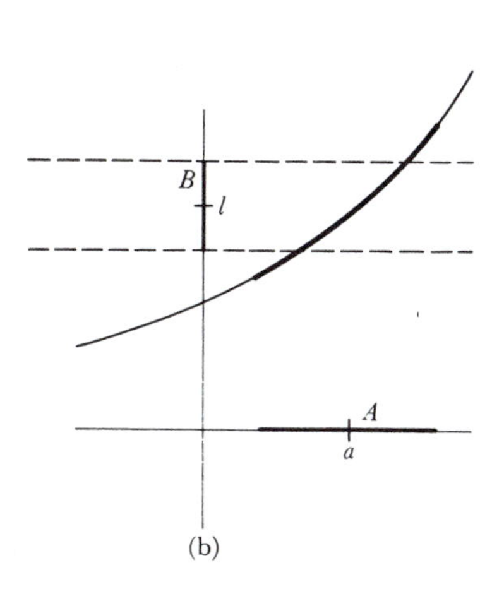

(b)

FIGURE 5

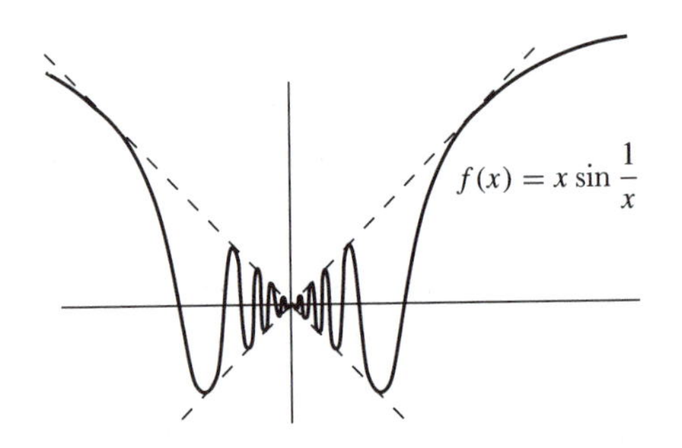

FIGURE 6

we have

$$\left| x \sin \frac{1}{x} \right| \le |x|, \qquad \text{for all } x \ne 0.$$

This means that if $|x| < \frac{1}{10}$ and $x \ne 0$, then $|x \sin 1/x| < \frac{1}{10}$; in other words, $x \sin 1/x$ is within $\frac{1}{10}$ of 0 provided that x is within $\frac{1}{10}$ of 0, but $\ne 0$. There is nothing special about the number $\frac{1}{10}$; it is just as easy to guarantee that $|f(x)-0| < \frac{1}{100}$—simply require that $|x| < \frac{1}{100}$, but $x \ne 0$. In fact, if we take any positive number ε we can make $|f(x) - 0| < \varepsilon$ simply by requiring that $|x| < \varepsilon$, and $x \ne 0$.

For the function $f(x) = x^2 \sin 1/x$ (Figure 7) it seems even clearer that f approaches 0 near 0. If, for example, we want

$$\left| x^2 \sin \frac{1}{x} \right| < \frac{1}{10},$$

then we certainly need only require that $|x| < \frac{1}{10}$ and $x \ne 0$, since this implies that $|x^2| < \frac{1}{100}$ and consequently

$$\left| x^2 \sin \frac{1}{x} \right| \le |x^2| < \frac{1}{100} < \frac{1}{10}.$$

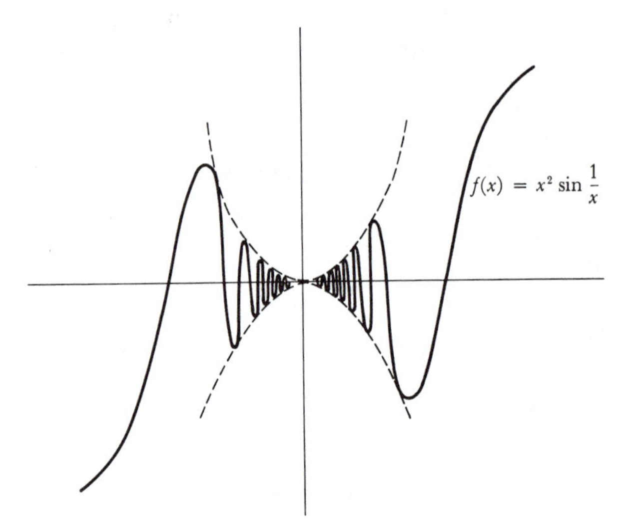

FIGURE 7

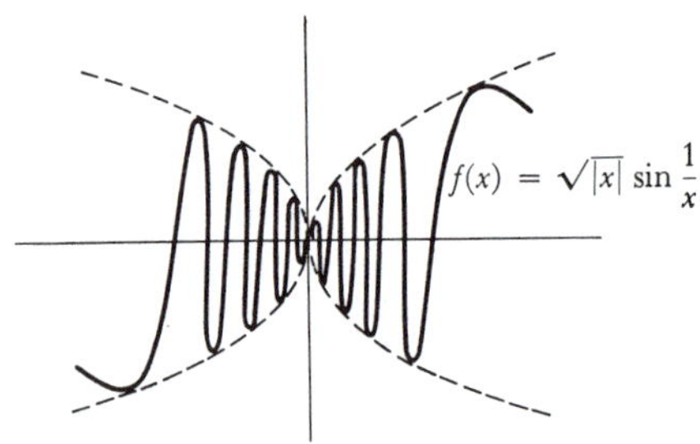

FIGURE 8

(We could do even better, and allow $|x| < 1/\sqrt{10}$ and $x \ne 0$, but there is no particular virtue in being as economical as possible.) In general, if $\varepsilon > 0$, to ensure that

$$\left| x^2 \sin \frac{1}{x} \right| < \varepsilon,$$

we need only require that

$$|x| < \varepsilon \quad \text{and} \quad x \ne 0,$$

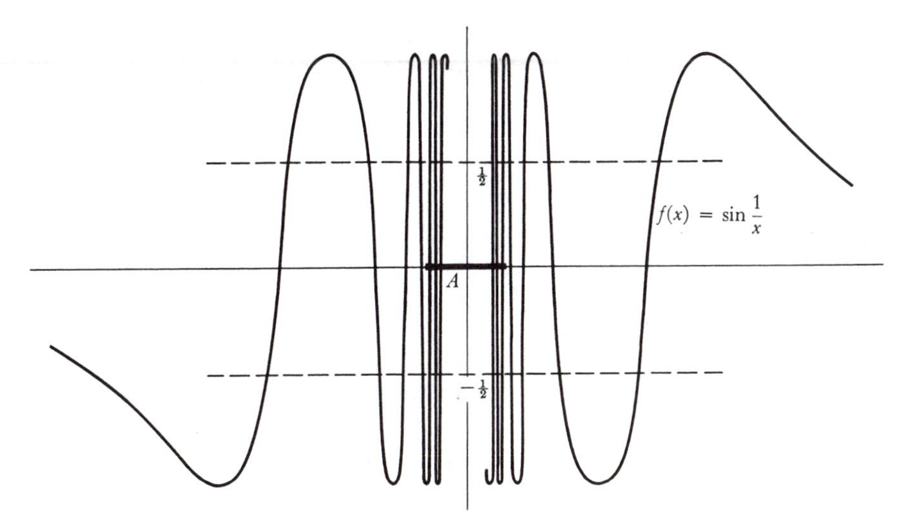

FIGURE 9

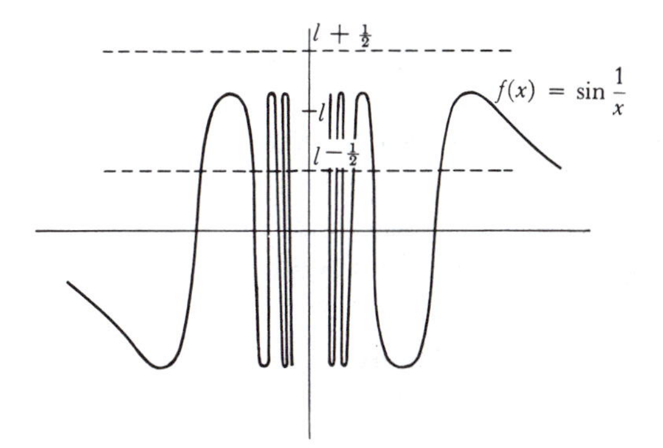

FIGURE 10

provided that $\varepsilon \leq 1$. If we are given an ε which is greater than 1 (it might be, even thought it is "small" ε's which are of interest), then it does not suffice to require that $|x| < \varepsilon$, but it certainly suffices to require that $|x| < 1$ and $x \neq 0$.

As a third example, consider the function $f(x) = \sqrt{|x|} \sin 1/x$ (Figure 8). In order to make $|\sqrt{|x|} \sin 1/x| < \varepsilon$ we can require that

$$|x| < \varepsilon^2 \quad \text{and} \quad x \neq 0$$

(the algebra is left to you).

Finally, let us consider the function $f(x) = \sin 1/x$ (Figure 9). For this function it is *false* that f approaches 0 near 0. This amounts to saying that it is not true for every number $\varepsilon > 0$ that we can get $|f(x) - 0| < \varepsilon$ by choosing x sufficiently small, and $\neq 0$. To show this we simply have to find *one* $\varepsilon > 0$ for which the condition $|f(x) - 0| < \varepsilon$ cannot be guaranteed, no matter how small we require $|x|$ to be. In fact, $\varepsilon = \frac{1}{2}$ will do: it is impossible to ensure that $|f(x)| < \frac{1}{2}$ no matter how small we require $|x|$ to be; for if A is any interval containing 0, there is some number $x = 1/(90 + 360n)$ which is in this interval, and for this x we have $f(x) = 1$.

This same argument can be used (Figure 10) to show that f does not approach *any* number near 0. To show this we must again find, for any particular number l, some number $\varepsilon > 0$ so that $|f(x) - l| < \varepsilon$ is *not* true, no matter how small x is required to be. The choice $\varepsilon = \frac{1}{2}$ works for any number l; that is, no matter how small we require $|x|$ to be, we cannot ensure that $|f(x) - l| < \frac{1}{2}$. The reason is, that for any interval A containing 0 there is some $x_1 = 1/(90 + 360n)$ in this interval, so that

$$f(x_1) = 1,$$

and also some $x_2 = 1/(270 + 360m)$ in this interval, so that

$$f(x_2) = -1.$$

But the interval from $l - \frac{1}{2}$ to $l + \frac{1}{2}$ cannot contain both -1 and 1, since its total length is only 1; so we cannot have

$$|1 - l| < \tfrac{1}{2} \text{ and also } |-1 - l| < \tfrac{1}{2},$$

no matter what l is.

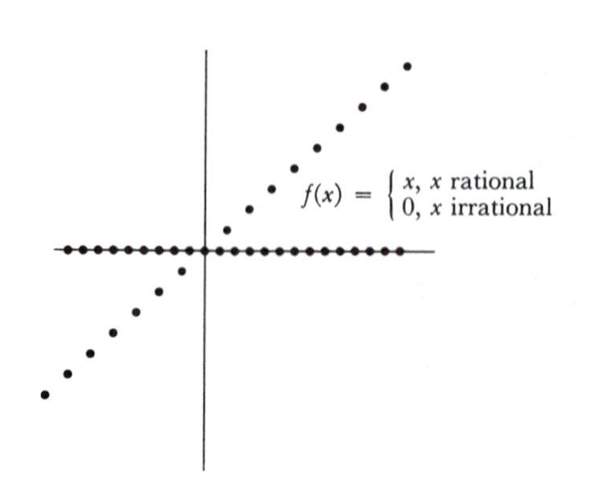

FIGURE 11

The phenomenon exhibited by $f(x) = \sin 1/x$ near 0 can occur in many ways. If we consider the function

$$f(x) = \begin{cases} 0, & x \text{ irrational} \\ 1, & x \text{ rational}, \end{cases}$$

then, no matter what a is, f does not approach any number l near a. In fact, we cannot make $|f(x) - l| < \frac{1}{4}$ no matter how close we bring x to a, because in any interval around a there are numbers x with $f(x) = 0$, and also numbers x with $f(x) = 1$, so that we would need $|0 - l| < \frac{1}{4}$ and also $|1 - l| < \frac{1}{4}$.

An amusing variation on this behavior is presented by the function shown in Figure 11:

$$f(x) = \begin{cases} x, & x \text{ rational} \\ 0, & x \text{ irrational}. \end{cases}$$

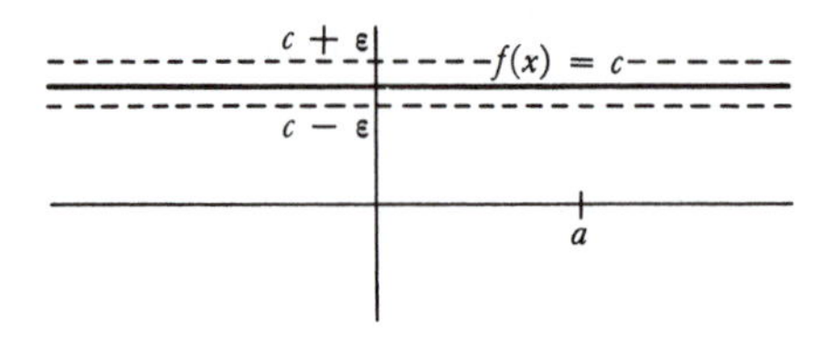

FIGURE 12

The behavior of this function is "opposite" to that of $g(x) = \sin 1/x$; it approaches 0 at 0, but does not approach any number at a, if $a \neq 0$. By now you should have no difficulty convincing yourself that this is true.

As a contrast to the functions considered so far, which have been quite pathological, we will now examine some of the simplest functions.

If $f(x) = c$, then f approaches c near a, for every number a. In fact, to ensure that $|f(x) - c| < \varepsilon$ one does not need to restrict x to be near a at all; the condition is automatically satisfied (Figure 12).

As a slight variation, let f be the function shown in Figure 13:

$$f(x) = \begin{cases} -1, & x < 0 \\ 1, & x > 0. \end{cases}$$

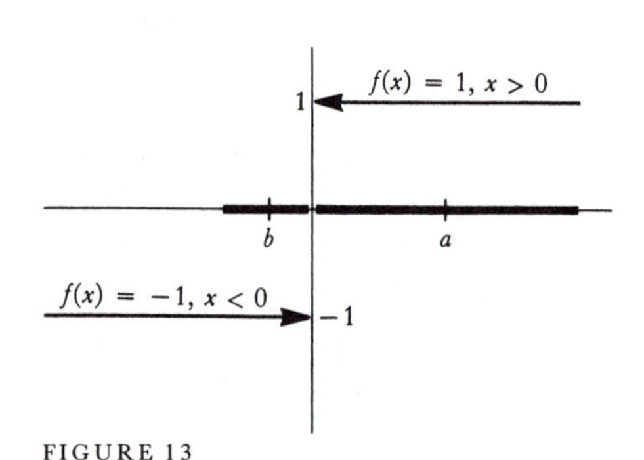

FIGURE 13

If $a > 0$, then f approaches 1 near a: indeed, to ensure that $|f(x) - 1| < \varepsilon$ it certainly suffices to require that $|x - a| < a$, since this implies

$$\begin{aligned} -a &< x - a \\ \text{or} \quad 0 &< x \end{aligned}$$

so that $f(x) = 1$. Similarly, if $b < 0$, then f approaches -1 near b: to ensure that $|f(x) - (-1)| < \varepsilon$ it suffices to require that $|x - b| < -b$. Finally, as you may easily check, f does not approach any number near 0.

The function $f(x) = x$ is easily dealt with. Clearly f approaches a near a: to ensure that $|f(x) - a| < \varepsilon$ we just have to require that $|x - a| < \varepsilon$.

The function $f(x) = x^2$ requires a little more work. To show that f approaches a^2 near a, we must decide how to ensure that

$$|x^2 - a^2| < \varepsilon.$$

Factoring looks like the most promising procedure: we want

$$|x - a| \cdot |x + a| < \varepsilon.$$

Obviously the factor $|x + a|$ is the one that will cause trouble. On the other hand, there is no need to make $|x + a|$ particularly small; as long as we know *some* bound on the values of $|x + a|$ we will be in good shape. For example, if $|x + a| < 1{,}000{,}000$, then we will just need to require that $|x - a| < \varepsilon/1{,}000{,}000$. Therefore, to begin with, let us require that $|x - a| < 1$ (any positive number other than 1 would do just as well); presumably this will ensure that x is not too large, and consequently that $|x + a|$ is not too large. As a matter of fact, Problem 1-12 shows that

$$|x| - |a| \le |x - a| < 1,$$

so

$$|x| < 1 + |a|,$$

and consequently

$$|x + a| \le |x| + |a| < 2|a| + 1.$$

Now we need only the additional requirement that $|x - a| < \varepsilon/(2|a| + 1)$. In other words,

$$\text{if } |x - a| < \min\left(1, \frac{\varepsilon}{2|a| + 1}\right), \quad \text{then } |x^2 - a^2| < \varepsilon.$$

Naturally, $\min(1, \varepsilon/(2|a| + 1))$ will just be $\varepsilon/(2|a| + 1)$ for small ε.

Precisely the same sort of trick will show that if $f(x) = x^3$, then f approaches a^3 near a. In fact,

$$\text{if } |x - a| < \min\left(1, \frac{\varepsilon}{(1 + |a|)^2 + |a|(1 + |a|) + |a|^2}\right), \quad \text{then } |x^3 - a^3| < \varepsilon.$$

The proof of this assertion will show where the weird denominator comes from: If $|x - a| < 1$, then $|x| < |a| + 1$, and consequently

$$\begin{aligned} |x^2 + ax + a^2| &\le |x|^2 + |a| \cdot |x| + |a|^2 \\ &< (1 + |a|)^2 + |a|(1 + |a|) + |a|^2. \end{aligned}$$

Therefore

$$\begin{aligned} |x^3 - a^3| &= |x - a| \cdot |x^2 + ax + a^2| \\ &< \frac{\varepsilon}{(1 + |a|)^2 + |a|(1 + |a|) + |a|^2} \cdot [(1 + |a|)^2 + |a|(1 + |a|) + |a|^2] \\ &= \varepsilon. \end{aligned}$$

The time has now come to point out that of the many demonstrations about limits which we have given, not one has been a real proof. The fault lies not with our reasoning, but with our definition. If our provisional definition of a function was open to criticism, our provisional definition of approaching a limit is even more vulnerable. This definition is simply not sufficiently precise to be used in proofs. It is hardly clear how one "makes" $f(x)$ close to l (whatever "close" means) by "requiring" x to be sufficiently close to a (however close "sufficiently" close is supposed to be). Despite the criticisms of our definition you may feel (I certainly hope you do) that our arguments were nevertheless quite convincing. In order to present any sort of argument at all, we have been practically forced to invent the real definition. It is possible to arrive at this definition in several steps, each one clarifying some obscure phrase which still remains. Let us begin, once again, with the provisional definition:

> The function f approaches the limit l near a, if we can make $f(x)$ as close as we like to l by requiring that x be sufficiently close to, but unequal to, a.

The very first change which we made in this definition was to note that making $f(x)$ close to l meant making $|f(x) - l|$ small, and similarly for x and a:

> The function f approaches the limit l near a, if we can make $|f(x) - l|$ as small as we like by requiring that $|x - a|$ be sufficiently small, and $x \neq a$.

The second, more crucial, change was to note that making $|f(x) - l|$ "as small as we like" means making $|f(x) - l| < \varepsilon$ for any $\varepsilon > 0$ that happens to be given us:

> The function f approaches the limit l near a, if for every number $\varepsilon > 0$ we can make $|f(x) - l| < \varepsilon$ by requiring that $|x - a|$ be sufficiently small, and $x \neq a$.

There is a common pattern to all the demonstrations about limits which we have given. For each number $\varepsilon > 0$ we found some other positive number, δ say, with the property that if $x \neq a$ and $|x - a| < \delta$, then $|f(x) - l| < \varepsilon$. For the function $f(x) = x \sin 1/x$ (with $a = 0$, $l = 0$), the number δ was just the number ε; for $f(x) = \sqrt{|x|} \sin 1/x$, it was ε^2; for $f(x) = x^2$ it was the minimum of 1 and $\varepsilon/(2|a| + 1)$. In general, it may not be at all clear how to find the number δ, given ε, but it is the condition $|x - a| < \delta$ which expresses how small "sufficiently" small must be:

> The function f approaches the limit l near a, if for every $\varepsilon > 0$ there is some $\delta > 0$ such that, for all x, if $|x - a| < \delta$ and $x \neq a$, then $|f(x) - l| < \varepsilon$.

This is practically the definition we will adopt. We will make only one trivial change, noting that "$|x - a| < \delta$ and $x \neq a$" can just as well be expressed "$0 < |x - a| < \delta$."

DEFINITION

> The function $\boldsymbol{f}$ **approaches the limit** $\boldsymbol{l}$ **near** $\boldsymbol{a}$ means: for every $\varepsilon > 0$ there is some $\delta > 0$ such that, for all x, if $0 < |x - a| < \delta$, then $|f(x) - l| < \varepsilon$.

This definition is so important (*everything* we do from now on depends on it) that proceeding any further without knowing it is hopeless. If necessary memorize it, like a poem! That, at least, is better than stating it incorrectly; if you do this you are doomed to give incorrect proofs. A good exercise in giving correct proofs is to review every fact already demonstrated about functions approaching limits, giving real proofs of each. This requires writing down the correct definition of what you are proving, but not much more—all the algebraic work has been done already. When proving that f does *not* approach l at a, be sure to negate the definition correctly:

If it is *not* true that

> for every $\varepsilon > 0$ there is some $\delta > 0$ such that, for all x, if $0 < |x - a| < \delta$, then $|f(x) - l| < \varepsilon$,

then

> there is *some* $\varepsilon > 0$ such that for *every* $\delta > 0$ there is *some* x which satisfies $0 < |x - a| < \delta$ but not $|f(x) - l| < \varepsilon$.

Thus, to show that the function $f(x) = \sin 1/x$ does not approach 0 near 0, we consider $\varepsilon = \frac{1}{2}$ and note that for every $\delta > 0$ there is some x with $0 < |x - 0| < \delta$ but not $|\sin 1/x - 0| < \frac{1}{2}$—namely, an x of the form $1/(90 + 360n)$, where n is so large that $1/(90 + 360n) < \delta$.

As an illustration of the use of the definition of a function approaching a limit, we have reserved the function shown in Figure 14, a standard example, but one of the most complicated:

$$f(x) = \begin{cases} 0, & x \text{ irrational, } 0 < x < 1 \\ 1/q, & x = p/q \text{ in lowest terms, } 0 < x < 1. \end{cases}$$

(Recall that p/q is in lowest terms if p and q are integers with no common factor and $q > 0$.)

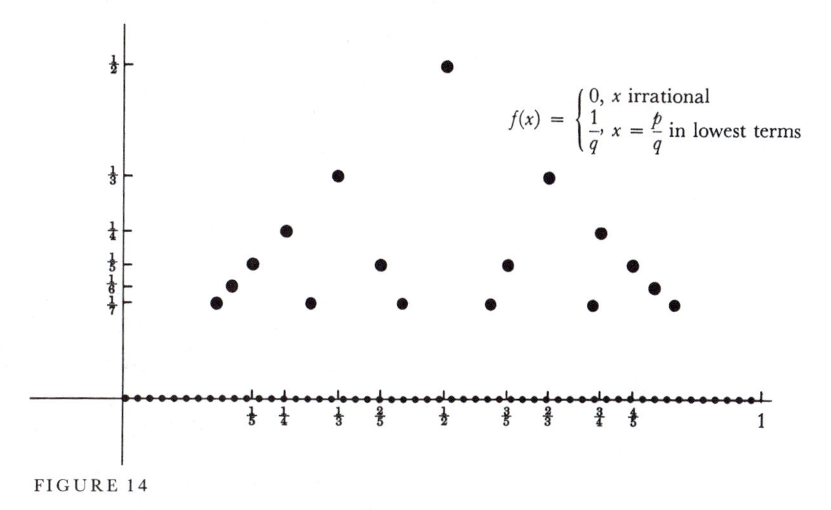

FIGURE 14

For any number a, with $0 < a < 1$, the function f approaches 0 at a. To prove this, consider any number $\varepsilon > 0$. Let n be a natural number so large that $1/n \leq \varepsilon$. Notice that the only numbers x for which $|f(x) - 0| < \varepsilon$ could be false are:

$$\frac{1}{2}; \frac{1}{3}, \frac{2}{3}; \frac{1}{4}, \frac{3}{4}; \frac{1}{5}, \frac{2}{5}, \frac{3}{5}, \frac{4}{5}; \ldots; \frac{1}{n}, \ldots, \frac{n-1}{n}.$$

(If a is rational, then a might be one of these numbers.) However many of these numbers there may be, there are, at any rate, only finitely many. Therefore, of all these numbers, one is closest to a; that is, $|p/q - a|$ is smallest for one p/q among these numbers. (If a happens to be one of these numbers, then consider only the values $|p/q - a|$ for $p/q \neq a$.) This closest distance may be chosen as the δ. For if $0 < |x - a| < \delta$, then x is *not* one of

$$\frac{1}{2}, \ldots, \frac{n-1}{n}$$

and therefore $|f(x) - 0| < \varepsilon$ *is* true. This completes the proof. Note that our description of the δ which works for a given ε is completely adequate—there is no reason why we must give a formula for δ in terms of ε.

Armed with our definition, we are now prepared to prove our first theorem; you have probably assumed the result all along, which is a very reasonable thing to do. This theorem is really a test case for our definition: if the theorem could not be proved, our definition would be useless.

THEOREM 1 A function cannot approach two different limits near a. In other words, if f approaches l near a, and f approaches m near a, then $l = m$.

PROOF Since this is our first theorem about limits it will certainly be necessary to translate the hypotheses according to the definition.

Since f approaches l near a, we know that for any $\varepsilon > 0$ there is some number $\delta_1 > 0$ such that, for all x,

$$\text{if } 0 < |x - a| < \delta_1, \text{ then } |f(x) - l| < \varepsilon.$$

We also know, since f approaches m near a, that there is some $\delta_2 > 0$ such that, for all x,

$$\text{if } 0 < |x - a| < \delta_2, \text{ then } |f(x) - m| < \varepsilon.$$

We have had to use two numbers, δ_1 and δ_2, since there is no guarantee that the δ which works in one definition will work in the other. But, in fact, it is now easy to conclude that for any $\varepsilon > 0$ there is some $\delta > 0$ such that, for all x,

$$\text{if } 0 < |x - a| < \delta, \text{ then } |f(x) - l| < \varepsilon \text{ and } |f(x) - m| < \varepsilon;$$

we simply choose $\delta = \min(\delta_1, \delta_2)$.

To complete the proof we just have to pick a particular $\varepsilon > 0$ for which the two conditions

$$|f(x) - l| < \varepsilon \quad \text{and} \quad |f(x) - m| < \varepsilon$$

cannot both hold, if $l \neq m$. The proper choice is suggested by Figure 15. If $l \neq m$, so that $|l - m| > 0$, we can choose $|l - m|/2$ as our ε. It follows that there is a $\delta > 0$ such that, for all x,

$$\text{if } 0 < |x - a| < \delta, \text{ then } |f(x) - l| < \frac{|l - m|}{2}$$
$$\text{and } |f(x) - m| < \frac{|l - m|}{2}.$$

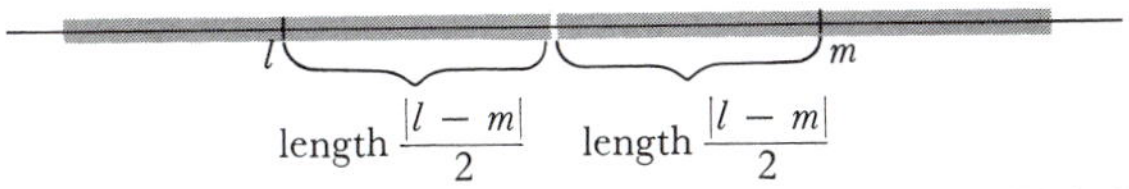

FIGURE 15

This implies that for $0 < |x - a| < \delta$ we have

$$\begin{aligned} |l - m| = |l - f(x) + f(x) - m| &\leq |l - f(x)| + |f(x) - m| \\ &< \frac{|l - m|}{2} + \frac{|l - m|}{2} \\ &= |l - m|, \end{aligned}$$

a contradiction. ▮

The number l which f approaches near a is denoted by $\lim_{x \to a} f(x)$ (read: the limit of $f(x)$ as x approaches a). This definition is possible only because of Theorem 1, which ensures that $\lim_{x \to a} f(x)$ never has to stand for two different numbers. The equation

$$\lim_{x \to a} f(x) = l$$

has exactly the same meaning as the phrase

$$f \text{ approaches } l \text{ near } a.$$

The possibility still remains that f does not approach l near a, for any l, so that $\lim_{x \to a} f(x) = l$ is false for every number l. This is usually expressed by saying that "$\lim_{x \to a} f(x)$ does not exist."

Notice that our new notation introduces an extra, utterly irrelevant letter x, which could be replaced by t, y, or any other letter which does not already appear—the symbols

$$\lim_{x \to a} f(x), \qquad \lim_{t \to a} f(t), \qquad \lim_{y \to a} f(y),$$

all denote precisely the same number, which depends on f and a, and has nothing to do with x, t, or y (these letters, in fact, do not denote anything at all). A more logical symbol would be something like $\lim_a f$, but this notation, despite its brevity, is so infuriatingly rigid that almost no one has seriously tried to use it. The notation $\lim_{x \to a} f(x)$ is much more useful because a function f often has no simple name, even though it might be possible to express $f(x)$ by a simple formula involving x. Thus, the short symbol

$$\lim_{x \to a} (x^2 + \sin x)$$

could be paraphrased only by the awkward expression

$$\lim_a f, \text{ where } f(x) = x^2 + \sin x.$$

Another advantage of the standard symbolism is illustrated by the expressions

$$\lim_{x \to a} x + t^3,$$
$$\lim_{t \to a} x + t^3.$$

The first means the number which f approaches near a when

$$f(x) = x + t^3, \quad \text{for all } x;$$

the second means the number which f approaches near a when

$$f(t) = x + t^3, \quad \text{for all } t.$$

You should have little difficulty (especially if you consult Theorem 2) proving that

$$\lim_{x \to a} x + t^3 = a + t^3,$$
$$\lim_{t \to a} x + t^3 = x + a^3.$$

These examples illustrate the main advantage of our notation, which is its flexibility. In fact, the notation $\lim_{x \to a} f(x)$ is so flexible that there is some danger of forgetting what it really means. Here is a simple exercise in the use of this notation, which will be important later: first interpret precisely, and then prove the equality of the expressions

$$\lim_{x \to a} f(x) \quad \text{and} \quad \lim_{h \to 0} f(a + h).$$

An important part of this chapter is the proof of a theorem which will make it easy to find many limits. The proof depends upon certain properties of inequalities and absolute values, hardly surprising when one considers the definition of limit. Although these facts have already been stated in Problems 1-20, 1-21, and 1-22, because of their importance they will be presented once again, in the form of a lemma (a lemma is an auxiliary theorem, a result that justifies its existence only by virtue of its prominent role in the proof of another theorem). The lemma says, roughly, that if x is close to x_0, and y is close to y_0, then $x + y$ will be close to $x_0 + y_0$, and xy will be close to $x_0 y_0$, and $1/y$ will be close to $1/y_0$. This intuitive statement is much easier to remember than the precise estimates of the lemma, and it is not unreasonable to read the proof of Theorem 2 first, in order to see just how these estimates are used.

LEMMA (1) If

$$|x - x_0| < \frac{\varepsilon}{2} \text{ and } |y - y_0| < \frac{\varepsilon}{2},$$

then

$$|(x + y) - (x_0 + y_0)| < \varepsilon.$$

(2) If

$$|x - x_0| < \min\left(1, \frac{\varepsilon}{2(|y_0| + 1)}\right) \quad \text{and} \quad |y - y_0| < \frac{\varepsilon}{2(|x_0| + 1)},$$

then

$$|xy - x_0 y_0| < \varepsilon.$$

(3) If $y_0 \neq 0$ and

$$|y - y_0| < \min\left(\frac{|y_0|}{2}, \frac{\varepsilon |y_0|^2}{2}\right),$$

then $y \neq 0$ and

$$\left|\frac{1}{y} - \frac{1}{y_0}\right| < \varepsilon.$$

PROOF (1)

$$\begin{aligned} |(x + y) - (x_0 + y_0)| &= |(x - x_0) + (y - y_0)| \\ &\leq |x - x_0| + |y - y_0| < \frac{\varepsilon}{2} + \frac{\varepsilon}{2} = \varepsilon. \end{aligned}$$

(2) Since $|x - x_0| < 1$ we have

$$|x| - |x_0| \leq |x - x_0| < 1,$$

so that

$$|x| < 1 + |x_0|.$$

Thus

$$\begin{aligned} |xy - x_0 y_0| &= |x(y - y_0) + y_0(x - x_0)| \\ &\leq |x| \cdot |y - y_0| + |y_0| \cdot |x - x_0| \\ &< (1 + |x_0|) \cdot \frac{\varepsilon}{2(|x_0| + 1)} + |y_0| \cdot \frac{\varepsilon}{2(|y_0| + 1)} \\ &< \frac{\varepsilon}{2} + \frac{\varepsilon}{2} = \varepsilon. \end{aligned}$$

(3) We have

$$|y_0| - |y| \leq |y - y_0| < \frac{y_0}{2},$$

so $|y| > |y_0|/2$. In particular, $y \neq 0$, and

$$\frac{1}{|y|} < \frac{2}{|y_0|}.$$

Thus

$$\left|\frac{1}{y} - \frac{1}{y_0}\right| = \frac{|y_0 - y|}{|y| \cdot |y_0|} < \frac{2}{|y_0|} \cdot \frac{1}{|y_0|} \cdot \frac{\varepsilon |y_0|^2}{2} = \varepsilon. \blacksquare$$

THEOREM 2 If $\lim_{x\to a} f(x) = l$ and $\lim_{x\to a} g(x) = m$, then

$$(1) \quad \lim_{x\to a}(f+g)(x) = l + m;$$

$$(2) \quad \lim_{x\to a}(f \cdot g)(x) = l \cdot m.$$

Moreover, if $m \neq 0$, then

$$(3) \quad \lim_{x\to a}\left(\frac{1}{g}\right)(x) = \frac{1}{m}.$$

PROOF The hypothesis means that for every $\varepsilon > 0$ there are $\delta_1, \delta_2 > 0$ such that, for all x,

$$\begin{aligned} &\text{if } 0 < |x-a| < \delta_1, \text{ then } |f(x) - l| < \varepsilon, \\ \text{and } &\text{if } 0 < |x-a| < \delta_2, \text{ then } |g(x) - m| < \varepsilon. \end{aligned}$$

This means (since, after all, $\varepsilon/2$ is also a positive number) that there are $\delta_1, \delta_2 > 0$ such that, for all x,

$$\begin{aligned} &\text{if } 0 < |x-a| < \delta_1, \text{ then } |f(x) - l| < \frac{\varepsilon}{2}, \\ \text{and } &\text{if } 0 < |x-a| < \delta_2, \text{ then } |g(x) - m| < \frac{\varepsilon}{2}. \end{aligned}$$

Now let $\delta = \min(\delta_1, \delta_2)$. If $0 < |x-a| < \delta$, then $0 < |x-a| < \delta_1$ and $0 < |x-a| < \delta_2$ are both true, so both

$$|f(x) - l| < \frac{\varepsilon}{2} \quad \text{and} \quad |g(x) - m| < \frac{\varepsilon}{2}$$

are true. But by part (1) of the lemma this implies that $|(f+g)(x) - (l+m)| < \varepsilon$. This proves (1).

To prove (2) we proceed similarly, after consulting part (2) of the lemma. If $\varepsilon > 0$ there are $\delta_1, \delta_2 > 0$ such that, for all x,

$$\begin{aligned} &\text{if } 0 < |x-a| < \delta_1, \text{ then } |f(x) - l| < \min\left(1, \frac{\varepsilon}{2(|m|+1)}\right), \\ \text{and } &\text{if } 0 < |x-a| < \delta_2, \text{ then } |g(x) - m| < \frac{\varepsilon}{2(|l|+1)}. \end{aligned}$$

Again let $\delta = \min(\delta_1, \delta_2)$. If $0 < |x-a| < \delta$, then

$$|f(x) - l| < \min\left(1, \frac{\varepsilon}{2(|m|+1)}\right) \quad \text{and} \quad |g(x) - m| < \frac{\varepsilon}{2(|l|+1)}.$$

So, by the lemma, $|(f \cdot g)(x) - l \cdot m| < \varepsilon$, and this proves (2).

Finally, if $\varepsilon > 0$ there is a $\delta > 0$ such that, for all x,

$$\text{if } 0 < |x-a| < \delta, \text{ then } |g(x) - m| < \min\left(\frac{|m|}{2}, \frac{\varepsilon|m|^2}{2}\right).$$

But according to part (3) of the lemma this means, first, that $g(x) \neq 0$, so $(1/g)(x)$ makes sense, and second that

$$\left|\left(\frac{1}{g}\right)(x) - \frac{1}{m}\right| < \varepsilon.$$

This proves (3). ▮

Using Theorem 2 we can prove, trivially, such facts as

$$\lim_{x \to a} \frac{x^3 + 7x^5}{x^2 + 1} = \frac{a^3 + 7a^5}{a^2 + 1},$$

without going through the laborious process of finding a δ, given an ε. We must begin with

$$\lim_{x \to a} 7 = 7,$$
$$\lim_{x \to a} 1 = 1,$$
$$\lim_{x \to a} x = a,$$

but these are easy to prove directly. If we *want* to find the δ, however, the proof of Theorem 2 amounts to a prescription for doing this. Suppose, to take a simpler example, that we want to find a δ such that, for all x,

$$\text{if } 0 < |x - a| < \delta, \text{ then } |x^2 + x - (a^2 + a)| < \varepsilon.$$

Consulting the proof of Theorem 2(1), we see that we must first find δ_1 and $\delta_2 > 0$ such that, for all x,

$$\text{if } 0 < |x - a| < \delta_1, \text{ then } |x^2 - a^2| < \frac{\varepsilon}{2}$$
$$\text{and} \quad \text{if } 0 < |x - a| < \delta_2, \text{ then } |x - a| < \frac{\varepsilon}{2}.$$

Since we have already given proofs that $\lim_{x \to a} x^2 = a^2$ and $\lim_{x \to a} x = a$, we know how to do this:

$$\delta_1 = \min\left(1, \frac{\frac{\varepsilon}{2}}{2|a| + 1}\right),$$
$$\delta_2 = \frac{\varepsilon}{2}.$$

Thus we can take

$$\delta = \min(\delta_1, \delta_2) = \min\left(\min\left(1, \frac{\frac{\varepsilon}{2}}{2|a| + 1}\right), \frac{\varepsilon}{2}\right).$$

If $a \neq 0$, the same method can be used to find a $\delta > 0$ such that, for all x,

$$\text{if } 0 < |x - a| < \delta, \text{ then } \left| \frac{1}{x^2} - \frac{1}{a^2} \right| < \varepsilon.$$

The proof of Theorem 2(3) shows that the second condition will follow if we find a $\delta > 0$ such that, for all x,

$$\text{if } 0 < |x - a| < \delta, \text{ then } |x^2 - a^2| < \min\left(\frac{|a|^2}{2}, \frac{\varepsilon|a|^4}{2}\right).$$

Thus we can take

$$\delta = \min\left(1, \frac{\min\left(\dfrac{|a|^2}{2}, \dfrac{\varepsilon|a|^4}{2}\right)}{2|a| + 1}\right).$$

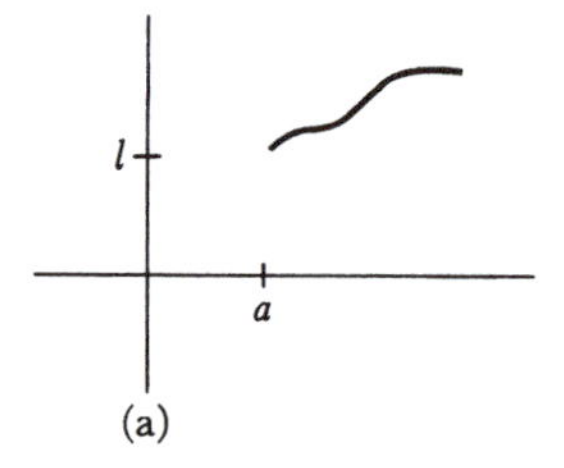

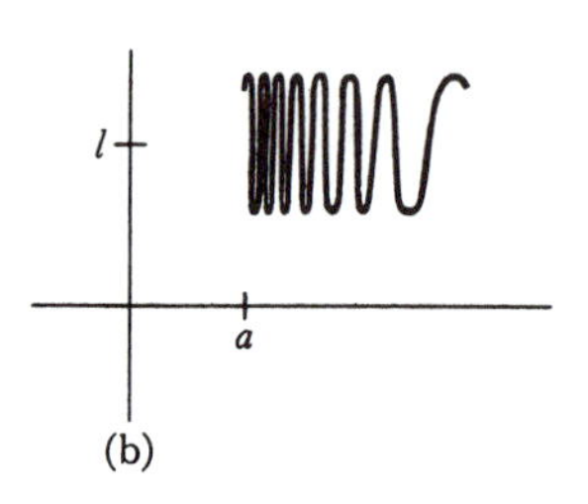

FIGURE 16

Naturally, these complicated expressions for δ can be simplified considerably, after they have been derived.

One technical detail in the proof of Theorem 2 deserves some discussion. In order for $\lim_{x \to a} f(x)$ to be defined it is, as we know, not necessary for f to be defined at a, nor is it necessary for f to be defined at all points $x \neq a$. However, there must be some $\delta > 0$ such that $f(x)$ is defined for x satisfying $0 < |x - a| < \delta$; otherwise the clause

$$\text{“if } 0 < |x - a| < \delta, \text{ then } |f(x) - l| < \varepsilon\text{”}$$

would make no sense at all, since the symbol $f(x)$ would make no sense for some x's. If f and g are two functions for which the definition makes sense, it is easy to see that the same is true for $f + g$ and $f \cdot g$. But this is not so clear for $1/g$, since $1/g$ is undefined for x with $g(x) = 0$. However, this fact gets established in the proof of Theorem 2(3).

There are times when we would like to speak of the limit which f approaches at a, even though there is no $\delta > 0$ such that $f(x)$ is defined for x satisfying $0 < |x - a| < \delta$. For example, we want to distinguish the behavior of the two functions shown in Figure 16, even though they are not defined for numbers less than a. For the function of Figure 16(a) we write

$$\lim_{x \to a^+} f(x) = l \quad \text{or} \quad \lim_{x \downarrow a} f(x) = l.$$

(The symbols on the left are read: the limit of $f(x)$ as x approaches a from above.) These “limits from above” are obviously closely related to ordinary limits, and the definition is very similar: $\lim_{x \to a^+} f(x) = l$ means that for every $\varepsilon > 0$ there is a $\delta > 0$ such that, for all x,

$$\text{if } 0 < x - a < \delta, \text{ then } |f(x) - l| < \varepsilon.$$

(The condition “$0 < x - a < \delta$” is equivalent to “$0 < |x - a| < \delta$ and $x > a$.”)

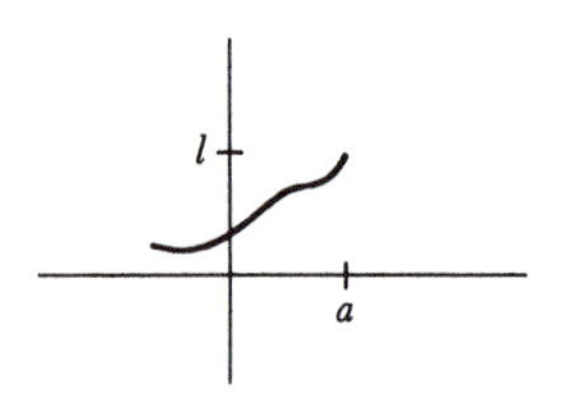

FIGURE 17

“Limits from below” (Figure 17) are defined similarly: $\lim_{x \to a^-} f(x) = l$ (or

$\lim_{x \uparrow a} f(x) = l$) means that for every $\varepsilon > 0$ there is a $\delta > 0$ such that, for all x,

$$\text{if } 0 < a - x < \delta, \text{ then } |f(x) - l| < \varepsilon.$$

It is quite possible to consider limits from above and below even if f is defined for numbers both greater and less than a. Thus, for the function f of Figure 13, we have

$$\lim_{x \to 0^+} f(x) = 1 \quad \text{and} \quad \lim_{x \to 0^-} f(x) = -1.$$

It is an easy exercise (Problem 29) to show that $\lim_{x \to a} f(x)$ exists if and only if $\lim_{x \to a^+} f(x)$ and $\lim_{x \to a^-} f(x)$ both exist and are equal.

Like the definitions of limits from above and below, which have been smuggled into the text informally, there are other modifications of the limit concept which will be found useful. In Chapter 4 it was claimed that if x is large, then $\sin 1/x$ is close to 0. This assertion is usually written

$$\lim_{x \to \infty} \sin 1/x = 0.$$

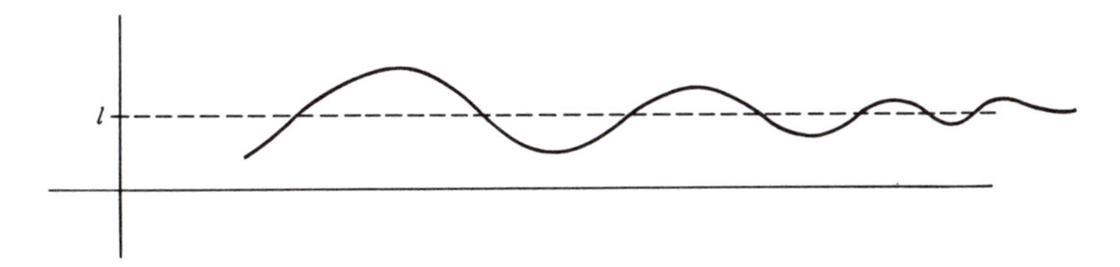

FIGURE 18

The symbol $\lim_{x \to \infty} f(x)$ is read "the limit of $f(x)$ as x approaches ∞," or "as x becomes infinite," and a limit of the form $\lim_{x \to \infty} f(x)$ is often called a limit at infinity. Figure 18 illustrates a general situation where $\lim_{x \to \infty} f(x) = l$. Formally, $\lim_{x \to \infty} f(x) = l$ means that for every $\varepsilon > 0$ there is a number N such that, for all x,

$$\text{if } x > N, \text{ then } |f(x) - l| < \varepsilon.$$

The analogy with the definition of ordinary limits should be clear: whereas the condition "$0 < |x - a| < \delta$" expresses the fact that x is close to a, the condition "$x > N$" expresses the fact that x is large.

We have spent so little time on limits from above and below, and at infinity, because the general philosophy behind the definitions should be clear if you understand the definition of ordinary limits (which are by far the most important). Many exercises on these definitions are provided in the Problems, which also contain several other types of limits which are occasionally useful.

PROBLEMS

1. Find the following limits. (These limits all follow, after some algebraic manipulations, from the various parts of Theorem 2; be sure you know which ones are used in each case, but don't bother listing them.)

(i) $\lim\limits_{x \to 1} \dfrac{x^2 - 1}{x + 1}$.

(ii) $\lim\limits_{x \to 2} \dfrac{x^3 - 8}{x - 2}$.

(iii) $\lim\limits_{x \to 3} \dfrac{x^3 - 8}{x - 2}$.

(iv) $\lim\limits_{x \to y} \dfrac{x^n - y^n}{x - y}$.

(v) $\lim\limits_{y \to x} \dfrac{x^n - y^n}{x - y}$.

(vi) $\lim\limits_{h \to 0} \dfrac{\sqrt{a + h} - \sqrt{a}}{h}$.

2. Find the following limits.

(i) $\lim\limits_{x \to 1} \dfrac{1 - \sqrt{x}}{1 - x}$.

(ii) $\lim\limits_{x \to 0} \dfrac{1 - \sqrt{1 - x^2}}{x}$.

(iii) $\lim\limits_{x \to 0} \dfrac{1 - \sqrt{1 - x^2}}{x^2}$.

3. In each of the following cases, find a δ such that $|f(x) - l| < \varepsilon$ for all x satisfying $0 < |x - a| < \delta$.

(i) $f(x) = x^4$; $l = a^4$.

(ii) $f(x) = \dfrac{1}{x}$; $a = 1$, $l = 1$.

(iii) $f(x) = x^4 + \dfrac{1}{x}$; $a = 1$, $l = 2$.

(iv) $f(x) = \dfrac{x}{1 + \sin^2 x}$; $a = 0$, $l = 0$.

(v) $f(x) = \sqrt{|x|}$; $a = 0$, $l = 0$.

(vi) $f(x) = \sqrt{x}$; $a = 1$, $l = 1$.

4. For each of the functions in Problem 4-17, decide for which numbers a the limit $\lim\limits_{x \to a} f(x)$ exists.

***5.** (a) Do the same for each of the functions in Problem 4-19.
(b) Same problem, if we use infinite decimals ending in a string of 0's instead of those ending in a string of 9's.

6. Suppose the functions f and g have the following property: for all $\varepsilon > 0$ and all x,

$$\text{if } 0 < |x-2| < \sin^2\left(\frac{\varepsilon^2}{9}\right) + \varepsilon, \text{ then } |f(x) - 2| < \varepsilon,$$

$$\text{if } 0 < |x-2| < \varepsilon^2, \text{ then } |g(x) - 4| < \varepsilon.$$

For each $\varepsilon > 0$ find a $\delta > 0$ such that, for all x,

(i) if $0 < |x-2| < \delta$, then $|f(x) + g(x) - 6| < \varepsilon$.

(ii) if $0 < |x-2| < \delta$, then $|f(x)g(x) - 8| < \varepsilon$.

(iii) if $0 < |x-2| < \delta$, then $\left|\dfrac{1}{g(x)} - \dfrac{1}{4}\right| < \varepsilon$.

(iv) if $0 < |x-2| < \delta$, then $\left|\dfrac{f(x)}{g(x)} - \dfrac{1}{2}\right| < \varepsilon$.

7. Give an example of a function f for which the following assertion is *false*: If $|f(x) - l| < \varepsilon$ when $0 < |x-a| < \delta$, then $|f(x) - l| < \varepsilon/2$ when $0 < |x-a| < \delta/2$.

8. (a) If $\lim_{x\to a} f(x)$ and $\lim_{x\to a} g(x)$ do not exist, can $\lim_{x\to a} [f(x) + g(x)]$ or $\lim_{x\to a} f(x)g(x)$ exist?

(b) If $\lim_{x\to a} f(x)$ exists and $\lim_{x\to a} [f(x) + g(x)]$ exists, must $\lim_{x\to a} g(x)$ exist?

(c) If $\lim_{x\to a} f(x)$ exists and $\lim_{x\to a} g(x)$ does not exist, can $\lim_{x\to a} [f(x) + g(x)]$ exist?

(d) If $\lim_{x\to a} f(x)$ exists and $\lim_{x\to a} f(x)g(x)$ exists, does it follow that $\lim_{x\to a} g(x)$ exists?

9. Prove that $\lim_{x\to a} f(x) = \lim_{h\to 0} f(a+h)$. (This is mainly an exercise in understanding what the terms mean.)

10. (a) Prove that $\lim_{x\to a} f(x) = l$ if and only if $\lim_{x\to a} [f(x) - l] = 0$. (First see why the assertion is obvious; then provide a rigorous proof. In this chapter most problems which ask for proofs should be treated in the same way.)

(b) Prove that $\lim_{x\to 0} f(x) = \lim_{x\to a} f(x-a)$.

(c) Prove that $\lim_{x\to 0} f(x) = \lim_{x\to 0} f(x^3)$.

(d) Give an example where $\lim_{x\to 0} f(x^2)$ exists, but $\lim_{x\to 0} f(x)$ does not.

11. Suppose there is a $\delta > 0$ such that $f(x) = g(x)$ when $0 < |x-a| < \delta$. Prove that $\lim_{x\to a} f(x) = \lim_{x\to a} g(x)$. In other words, $\lim_{x\to a} f(x)$ depends only on the values of $f(x)$ for x near a—this fact is often expressed by saying that limits are a "local property." (It will clearly help to use δ', or some other letter, instead of δ, in the definition of limits.)

12. (a) Suppose that $f(x) \le g(x)$ for all x. Prove that $\lim_{x\to a} f(x) \le \lim_{x\to a} g(x)$, provided that these limits exist.

(b) How can the hypotheses by weakened?

(c) If $f(x) < g(x)$ for all x, does it necessarily follow that $\lim_{x\to a} f(x) < \lim_{x\to a} g(x)$?

13. Suppose that $f(x) \le g(x) \le h(x)$ and that $\lim_{x\to a} f(x) = \lim_{x\to a} h(x)$. Prove that $\lim_{x\to a} g(x)$ exists, and that $\lim_{x\to a} g(x) = \lim_{x\to a} f(x) = \lim_{x\to a} h(x)$. (Draw a picture!)

***14.** (a) Prove that if $\lim_{x\to 0} f(x)/x = l$ and $b \ne 0$, then $\lim_{x\to 0} f(bx)/x = bl$. Hint: Write $f(bx)/x = b[f(bx)/bx]$.

(b) What happens if $b = 0$?

(c) Part (a) enables us to find $\lim_{x\to 0} (\sin 2x)/x$ in terms of $\lim_{x\to 0} (\sin x)/x$. Find this limit in another way.

15. Evaluate the following limits in terms of the number $\alpha = \lim_{x\to 0} (\sin x)/x$.

(i) $\displaystyle \lim_{x\to 0} \frac{\sin 2x}{x}$.

(ii) $\displaystyle \lim_{x\to 0} \frac{\sin ax}{\sin bx}$.

(iii) $\displaystyle \lim_{x\to 0} \frac{\sin^2 2x}{x}$.

(iv) $\displaystyle \lim_{x\to 0} \frac{\sin^2 2x}{x^2}$.

(v) $\displaystyle \lim_{x\to 0} \frac{1 - \cos x}{x^2}$.

(vi) $\displaystyle \lim_{x\to 0} \frac{\tan^2 x + 2x}{x + x^2}$.

(vii) $\displaystyle \lim_{x\to 0} \frac{x \sin x}{1 - \cos x}$.

(viii) $\displaystyle \lim_{h\to 0} \frac{\sin(x+h) - \sin x}{h}$.

(ix) $\displaystyle \lim_{x\to 1} \frac{\sin(x^2 - 1)}{x - 1}$.

(x) $\displaystyle \lim_{x\to 0} \frac{x^2(3 + \sin x)}{(x + \sin x)^2}$.

(xi) $\displaystyle \lim_{x\to 1} (x^2 - 1)^3 \sin\left(\frac{1}{x-1}\right)^3$.

16. (a) Prove that if $\lim_{x\to a} f(x) = l$, then $\lim_{x\to a} |f|(x) = |l|$.

(b) Prove that if $\lim_{x\to a} f(x) = l$ and $\lim_{x\to a} g(x) = m$, then $\lim_{x\to a} \max(f, g)(x) = \max(l, m)$ and similarly for min.

17. (a) Prove that $\lim_{x \to 0} 1/x$ does not exist, i.e., show that $\lim_{x \to 0} 1/x = l$ is false for every number l.
(b) Prove that $\lim_{x \to 1} 1/(x-1)$ does not exist.

18. Prove that if $\lim_{x \to a} f(x) = l$, then there is a number $\delta > 0$ and a number M such that $|f(x)| < M$ if $0 < |x - a| < \delta$. (What does this mean pictorially?) Hint: Why does it suffice to prove that $l-1 < f(x) < l+1$ for $0 < |x-a| < \delta$?

19. Prove that if $f(x) = 0$ for irrational x and $f(x) = 1$ for rational x, then $\lim_{x \to a} f(x)$ does not exist for any a.

***20.** Prove that if $f(x) = x$ for rational x, and $f(x) = -x$ for irrational x, then $\lim_{x \to a} f(x)$ does not exist if $a \neq 0$.

21. (a) Prove that if $\lim_{x \to 0} g(x) = 0$, then $\lim_{x \to 0} g(x) \sin 1/x = 0$.
(b) Generalize this fact as follows: If $\lim_{x \to 0} g(x) = 0$ and $|h(x)| \le M$ for all x, then $\lim_{x \to 0} g(x)h(x) = 0$. (Naturally it is unnecessary to do part (a) if you succeed in doing part (b); actually the statement of part (b) may make it easier than (a)—that's one of the values of generalization.)

22. Consider a function f with the following property: if g is any function for which $\lim_{x \to 0} g(x)$ does not exist, then $\lim_{x \to 0} [f(x) + g(x)]$ also does not exist. Prove that this happens if and only if $\lim_{x \to 0} f(x)$ *does* exist. Hint: This is actually very easy: the assumption that $\lim_{x \to 0} f(x)$ does not exist leads to an immediate contradiction if you consider the right g.

****23.** This problem is the analogue of Problem 22 when $f + g$ is replaced by $f \cdot g$. In this case the situation is considerably more complex, and the analysis requires several steps (those in search of an especially challenging problem can attempt an independent solution).

(a) Suppose that $\lim_{x \to 0} f(x)$ exists and is $\neq 0$. Prove that if $\lim_{x \to 0} g(x)$ does not exist, then $\lim_{x \to 0} f(x)g(x)$ also does not exist.
(b) Prove the same result if $\lim_{x \to 0} |f(x)| = \infty$. (The precise definition of this sort of limit is given in Problem 37.)
(c) Prove that if neither of these two conditions holds, then there is a function g such that $\lim_{x \to 0} g(x)$ does not exist, but $\lim_{x \to 0} f(x)g(x)$ does exist.
Hint: Consider separately the following two cases: (1) for some $\varepsilon > 0$ we have $|f(x)| > \varepsilon$ for all sufficiently small x. (2) For every $\varepsilon > 0$, there are arbitrarily small x with $|f(x)| < \varepsilon$. In the second case, begin by choosing points x_n with $|x_n| < 1/n$ and $|f(x_n)| < 1/n$.

***24.** Suppose that A_n is, for each natural number n, some *finite* set of numbers in $[0, 1]$, and that A_n and A_m have no members in common if $m \neq n$. Define

f as follows:

$$f(x) = \begin{cases} 1/n, & x \text{ in } A_n \\ 0, & x \text{ not in } A_n \text{ for any } n. \end{cases}$$

Prove that $\lim_{x \to a} f(x) = 0$ for all a in $[0, 1]$.

25. Explain why the following definitions of $\lim_{x \to a} f(x) = l$ are all correct: For every $\delta > 0$ there is an $\varepsilon > 0$ such that, for all x,

(i) if $0 < |x - a| < \varepsilon$, then $|f(x) - l| < \delta$.
(ii) if $0 < |x - a| < \varepsilon$, then $|f(x) - l| \le \delta$.
(iii) if $0 < |x - a| < \varepsilon$, then $|f(x) - l| < 5\delta$.
(iv) if $0 < |x - a| < \varepsilon/10$, then $|f(x) - l| < \delta$.

***26.** Give examples to show that the following definitions of $\lim_{x \to a} f(x) = l$ are *not* correct.

(a) For all $\delta > 0$ there is an $\varepsilon > 0$ such that if $0 < |x - a| < \delta$, then $|f(x) - l| < \varepsilon$.
(b) For all $\varepsilon > 0$ there is a $\delta > 0$ such that if $|f(x) - l| < \varepsilon$, then $0 < |x - a| < \delta$.

27. For each of the functions in Problem 4-17 indicate for which numbers a the one-sided limits $\lim_{x \to a^+} f(x)$ and $\lim_{x \to a^-} f(x)$ exist.

***28.** (a) Do the same for each of the functions in Problem 4-19.
(b) Also consider what happens if decimals ending in 0's are used instead of decimals ending in 9's.

29. Prove that $\lim_{x \to a} f(x)$ exists if $\lim_{x \to a^+} f(x) = \lim_{x \to a^-} f(x)$.

30. Prove that

(i) $\lim_{x \to 0^+} f(x) = \lim_{x \to 0^-} f(-x)$.
(ii) $\lim_{x \to 0} f(|x|) = \lim_{x \to 0^+} f(x)$.
(iii) $\lim_{x \to 0} f(x^2) = \lim_{x \to 0^+} f(x)$.

(These equations, and others like them, are open to several interpretations. They might mean only that the two limits are equal if they both exist; or that if a certain one of the limits exists, the other also exists and is equal to it; or that if either limit exists, then the other exists and is equal to it. Decide for yourself which interpretations are suitable.)

***31.** Suppose that $\lim_{x \to a^-} f(x) < \lim_{x \to a^+} f(x)$. (Draw a picture to illustrate this assertion.) Prove that there is some $\delta > 0$ such that $f(x) < f(y)$ whenever $x < a < y$ and $|x - a| < \delta$ and $|y - a| < \delta$. Is the converse true?

***32.** Prove that $\lim_{x\to\infty}(a_nx^n+\cdots+a_0)/(b_mx^m+\cdots+b_0)$ exists if and only if $m \geq n$. What is the limit when $m = n$? When $m > n$? Hint: the one easy limit is $\lim_{x\to\infty} 1/x^k = 0$; do some algebra so that this is the only information you need.

33. Find the following limits.

(i) $\lim_{x\to\infty} \dfrac{x+\sin^3 x}{5x+6}$.

(ii) $\lim_{x\to\infty} \dfrac{x\sin x}{x^2+5}$.

(iii) $\lim_{x\to\infty} \sqrt{x^2+x}-x$.

(iv) $\lim_{x\to\infty} \dfrac{x^2(1+\sin^2 x)}{(x+\sin x)^2}$.

34. Prove that $\lim_{x\to 0^+} f(1/x) = \lim_{x\to\infty} f(x)$.

35. Find the following limits in terms of the number $\alpha = \lim_{x\to 0}(\sin x)/x$.

(i) $\lim_{x\to\infty} \dfrac{\sin x}{x}$.

(ii) $\lim_{x\to\infty} x\sin\dfrac{1}{x}$.

36. Define "$\lim_{x\to-\infty} f(x) = l$."

(a) Find $\lim_{x\to-\infty}(a_nx^n+\cdots+a^0)/(b_mx^m+\cdots+b_0)$.

(b) Prove that $\lim_{x\to\infty} f(x) = \lim_{x\to-\infty} f(-x)$.

(c) Prove that $\lim_{x\to 0^-} f(1/x) = \lim_{x\to-\infty} f(x)$.

37. We define $\lim_{x\to a} f(x) = \infty$ to mean that for all N there is a $\delta > 0$ such that, for all x, if $0 < |x-a| < \delta$, then $f(x) > N$. (Draw an appropriate picture!)

(a) Show that $\lim_{x\to 3} 1/(x-3)^2 = \infty$.

(b) Prove that if $f(x) > \varepsilon > 0$ for all x, and $\lim_{x\to a} g(x) = 0$, then

$$\lim_{x\to a} f(x)/|g(x)| = \infty.$$

38. (a) Define $\lim_{x\to a^+} f(x) = \infty$, $\lim_{x\to a^-} f(x) = \infty$, and $\lim_{x\to a} f(x) = \infty$. (Or at least convince yourself that you could write down the definitions if you had the energy. How many other such symbols can you define?)

(b) Prove that $\lim_{x\to 0^+} 1/x = \infty$.

(c) Prove that $\lim_{x\to 0^+} f(x) = \infty$ if and only if $\lim_{x\to\infty} f(1/x) = \infty$.

39. Find the following limits, when they exist.

(i) $\lim_{x\to\infty} \dfrac{x^3+4x-7}{7x^2-x+1}$

(ii) $\lim_{x \to \infty} x(1 + \sin^2 x)$.

(iii) $\lim_{x \to \infty} x \sin^2 x$.

(iv) $\lim_{x \to \infty} x^2 \sin \frac{1}{x}$.

(v) $\lim_{x \to \infty} \sqrt{x^2 + 2x} - x$.

(vi) $\lim_{x \to \infty} x(\sqrt{x + 2} - \sqrt{x})$.

(vii) $\lim_{x \to \infty} \frac{\sqrt{|x|}}{x}$.

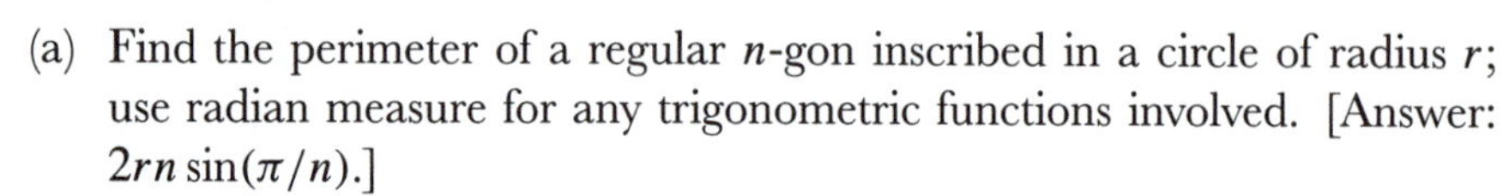

40. (a) Find the perimeter of a regular n-gon inscribed in a circle of radius r; use radian measure for any trigonometric functions involved. [Answer: $2rn \sin(\pi/n)$.]

(b) What value does this perimeter approach as n becomes very large?

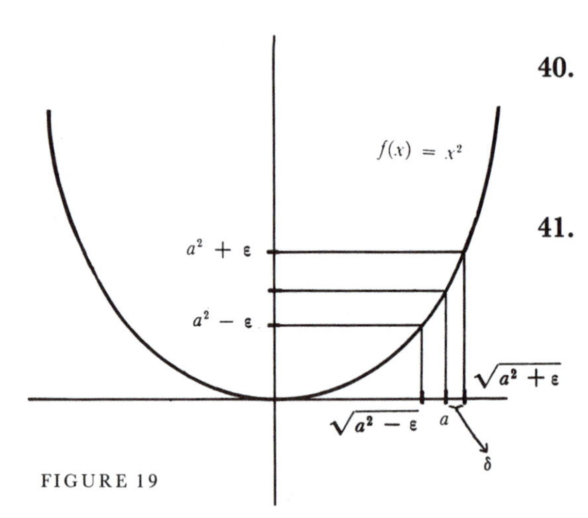

FIGURE 19

41. After sending the manuscript for the first edition of this book off to the printer, I thought of a much simpler way to prove that $\lim_{x \to a} x^2 = a^2$ and $\lim_{x \to a} x^3 = a^3$, without going through all the factoring tricks on page 95. Suppose, for example, that we want to prove that $\lim_{x \to a} x^2 = a^2$, where $a > 0$. Given $\varepsilon > 0$, we simply let δ be the minimum of $\sqrt{a^2 + \varepsilon} - a$ and $a - \sqrt{a^2 - \varepsilon}$ (see Figure 19); then $|x - a| < \delta$ implies that $\sqrt{a^2 - \varepsilon} < x < \sqrt{a^2 + \varepsilon}$, so $a^2 - \varepsilon < x^2 < a^2 + \varepsilon$, or $|x^2 - a^2| < \varepsilon$. It is fortunate that these pages had already been set, so that I couldn't make these changes, because this "proof" is completely fallacious. Wherein lies the fallacy?

CHAPTER 6 CONTINUOUS FUNCTIONS

If f is an arbitrary function, it is not necessarily true that

$$\lim_{x \to a} f(x) = f(a).$$

In fact, there are many ways this can fail to be true. For example, f might not even be defined at a, in which case the equation makes no sense (Figure 1).

FIGURE 1

Again, $\lim_{x \to a} f(x)$ might not exist (Figure 2). Finally, as illustrated in Figure 3, even if f is defined at a and $\lim_{x \to a} f(x)$ exists, the limit might not equal $f(a)$.

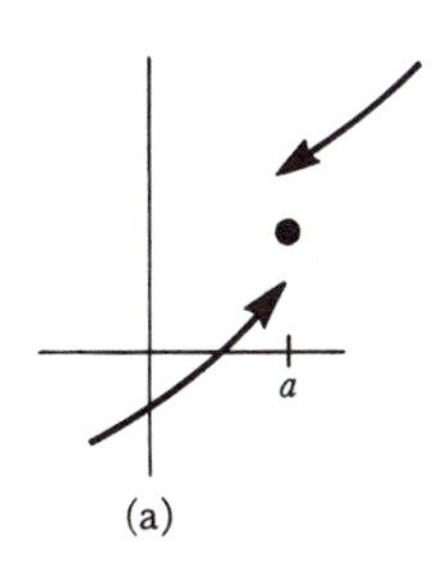

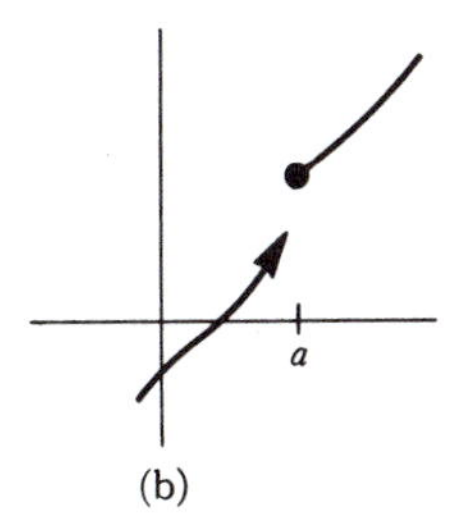

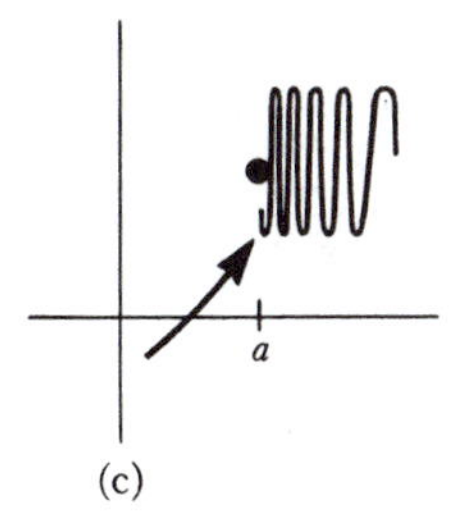

FIGURE 2

We would like to regard all behavior of this type as abnormal and honor, with some complimentary designation, functions which do not exhibit such peculiarities. The term which has been adopted is "continuous." Intuitively, a function f is continuous if the graph contains no breaks, jumps, or wild oscillations. Although this description will usually enable you to decide whether a function is continuous simply by looking at its graph (a skill well worth cultivating) it is easy to be fooled, and the precise definition is *very* important.

DEFINITION

The function f is **continuous at *a*** if

$$\lim_{x \to a} f(x) = f(a).$$

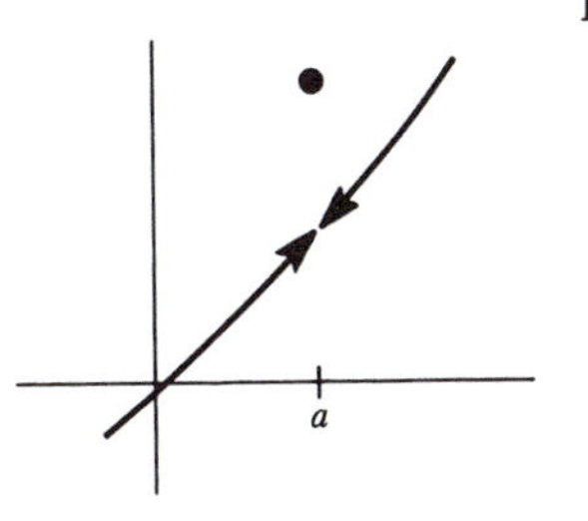

FIGURE 3

We will have no difficulty finding many examples of functions which are, or are not, continuous at some number a—every example involving limits provides an example about continuity, and Chapter 5 certainly provides enough of these.

The function $f(x) = \sin 1/x$ is not continuous at 0, because it is not even defined at 0, and the same is true of the function $g(x) = x \sin 1/x$. On the other hand, if we are willing to extend the second of these functions, that is, if we wish to define a new function G by

$$G(x) = \begin{cases} x \sin 1/x, & x \neq 0 \\ a, & x = 0, \end{cases}$$

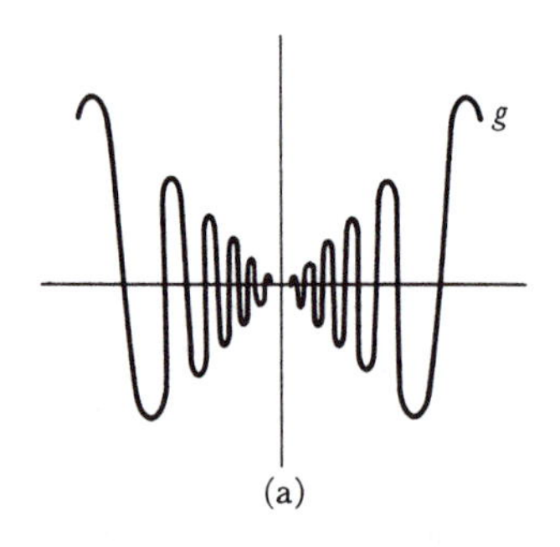

then the choice of $a = G(0)$ can be made in such a way that G will be continuous at 0—to do this we can (if fact, we must) define $G(0) = 0$ (Figure 4). This sort of extension is not possible for f; if we define

$$F(x) = \begin{cases} \sin 1/x, & x \neq 0 \\ a, & x = 0, \end{cases}$$

then F will not be continuous at 0, no matter what a is, because $\lim_{x \to 0} f(x)$ does not exist.

The function

$$f(x) = \begin{cases} x, & x \text{ rational} \\ 0, & x \text{ irrational} \end{cases}$$

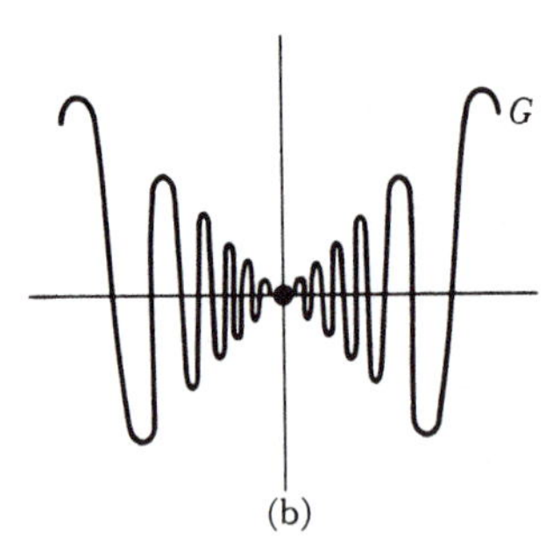

FIGURE 4

is not continuous at a, if $a \neq 0$, since $\lim_{x \to a} f(x)$ does not exist. However, $\lim_{x \to 0} f(x) = 0 = f(0)$, so f is continuous at precisely one point, 0.

The functions $f(x) = c$, $g(x) = x$, and $h(x) = x^2$ are continuous at all numbers a, since

$$\lim_{x \to a} f(x) = \lim_{x \to a} c = c = f(a),$$
$$\lim_{x \to a} g(x) = \lim_{x \to a} x = a = g(a),$$
$$\lim_{x \to a} h(x) = \lim_{x \to a} x^2 = a^2 = h(a).$$

Finally, consider the function

$$f(x) = \begin{cases} 0, & x \text{ irrational} \\ 1/q, & x = p/q \text{ in lowest terms.} \end{cases}$$

In Chapter 5 we showed that $\lim_{x \to a} f(x) = 0$ for all a. Since $0 = f(a)$ only when a is irrational, this function is continuous at a if a is irrational, but not if a is rational.

It is even easier to give examples of continuity if we prove two simple theorems.

THEOREM 1 If f and g are continuous at a, then

(1) $f + g$ is continuous at a,
(2) $f \cdot g$ is continuous at a.

Moreover, if $g(a) \neq 0$, then

(3) $1/g$ is continuous at a.

PROOF Since f and g are continuous at a,

$$\lim_{x \to a} f(x) = f(a) \quad \text{and} \quad \lim_{x \to a} g(x) = g(a).$$

By Theorem 2(1) of Chapter 5 this implies that

$$\lim_{x \to a} (f + g)(x) = f(a) + g(a) = (f + g)(a),$$

which is just the assertion that $f + g$ is continuous at a. The proofs of parts (2) and (3) are left to you. ▮

Starting with the functions $f(x) = c$ and $f(x) = x$, which are continuous at a, for every a, we can use Theorem 1 to conclude that a function

$$f(x) = \frac{b_n x^n + b_{n-1}x^{n-1} + \cdots + b_0}{c_m x^m + c_{m-1}x^{m-1} + \cdots + c_0}$$

is continuous at every point in its domain. But it is harder to get much further than that. When we discuss the sine function in detail it will be easy to prove that sin is continuous at a for all a; let us assume this fact meanwhile. A function like

$$f(x) = \frac{\sin^2 x + x^2 + x^4 \sin x}{\sin^{27} x + 4x^2 \sin^2 x}$$

can now be proved continuous at every point in its domain. But we are still unable to prove the continuity of a function like $f(x) = \sin(x^2)$; we obviously need a theorem about the composition of continuous functions. Before stating this theorem, the following point about the definition of continuity is worth noting. If we translate the equation $\lim_{x \to a} f(x) = f(a)$ according to the definition of limits, we obtain

for every $\varepsilon > 0$ there is $\delta > 0$ such that, for all x,
if $0 < |x - a| < \delta$, then $|f(x) - f(a)| < \varepsilon$.

But in this case, where the limit is $f(a)$, the phrase

$$0 < |x - a| < \delta$$

may be changed to the simpler condition

$$|x - a| < \delta,$$

since if $x = a$ it is certainly true that $|f(x) - f(a)| < \varepsilon$.

THEOREM 2 If g is continuous at a, and f is continuous at $g(a)$, then $f \circ g$ is continuous at a. (Notice that f is required to be continuous at $g(a)$, not at a.)

PROOF Let $\varepsilon > 0$. We wish to find a $\delta > 0$ such that for all x,

if $|x - a| < \delta$, then $|(f \circ g)(x) - (f \circ g)(a)| < \varepsilon$,
i.e., $|f(g(x)) - f(g(a))| < \varepsilon$.

We first use continuity of f to estimate how close $g(x)$ must be to $g(a)$ in order for this inequality to hold. Since f is continuous at $g(a)$, there is a $\delta' > 0$ such that for all y,

$$(1) \qquad \text{if } |y - g(a)| < \delta', \text{ then } |f(y) - f(g(a))| < \varepsilon.$$

In particular, this means that

$$(2) \qquad \text{if } |g(x) - g(a)| < \delta', \text{ then } |f(g(x)) - f(g(a))| < \varepsilon.$$

We now use continuity of g to estimate how close x must be to a in order for the inequality $|g(x) - g(a)| < \delta'$ to hold. The number δ' is a positive number just like any other positive number; we can therefore take δ' as the ε (!) in the definition of continuity of g at a. We conclude that there is a $\delta > 0$ such that, for all x,

$$(3) \qquad \text{if } |x - a| < \delta, \text{ then } |g(x) - g(a)| < \delta'.$$

Combining (2) and (3) we see that for all x,

$$\text{if } |x - a| < \delta, \text{ then } |f(g(x)) - f(g(a))| < \varepsilon. \blacksquare$$

We can now reconsider the function

$$f(x) = \begin{cases} x \sin 1/x, & x \neq 0 \\ 0, & x = 0. \end{cases}$$

We have already noted that f is continuous at 0. A few applications of Theorems 1 and 2, together with the continuity of sin, show that f is also continuous at a, for $a \neq 0$. Functions like $f(x) = \sin(x^2 + \sin(x + \sin^2(x^3)))$ should be equally easy for you to analyze.

The few theorems of this chapter have all been related to continuity of functions at a single point, but the concept of continuity doesn't begin to be really interesting until we focus our attention on functions which are continuous at all points of some interval. If f is continuous at x for all x in (a, b), then f is called **continuous on** (a, b). Continuity on a closed interval must be defined a little differently; a function f is called **continuous on** $[a, b]$ if

(1) f is continuous at x for all x in (a, b),
(2) $\lim_{x \to a^+} f(x) = f(a)$ and $\lim_{x \to b^-} f(x) = f(b)$.

Functions which are continuous on an interval are usually regarded as especially well behaved; indeed continuity might be specified as the first condition which a "reasonable" function ought to satisfy. A continuous function is sometimes described, intuitively, as one whose graph can be drawn without lifting your pencil from the paper. Consideration of the function

$$f(x) = \begin{cases} x \sin 1/x, & x \neq 0 \\ 0, & x = 0 \end{cases}$$

shows that this description is a little too optimistic, but it is nevertheless true that there are many important results involving functions which are continuous on an interval. There theorems are generally much harder than the ones in this chapter, but there is a simple theorem which forms a bridge between the two kinds of results. The hypothesis of this theorem requires continuity at only a single point, but the conclusion describes the behavior of the function on some interval containing the point. Although this theorem is really a lemma for later arguments, it is included here as a preview of things to come.

THEOREM 3 Suppose f is continuous at a, and $f(a) > 0$. Then there is a number $\delta > 0$ such that $f(x) > 0$ for all x satisfying $|x - a| < \delta$. Similarly, if $f(a) < 0$, then there is a number $\delta > 0$ such that $f(x) < 0$ for all x satisfying $|x - a| < \delta$.

PROOF Consider the case $f(a) > 0$. Since f is continuous at a, if $\varepsilon > 0$ there is a $\delta > 0$ such that, for all x,

$$\text{if } |x - a| < \delta, \text{ then } |f(x) - f(a)| < \varepsilon.$$

Since $f(a) > 0$ we can take $f(a)$ as the ε. Thus there is $\delta > 0$ so that for all x,

$$\text{if } |x - a| < \delta, \text{ then } |f(x) - f(a)| < f(a),$$

and this last inequality implies $f(x) > 0$.

A similar proof can be given in the case $f(a) < 0$; take $\varepsilon = -f(a)$. Or one can apply the first case to the function $-f$. ▌

PROBLEMS

1. For which of the following functions f is there a continuous function F with domain $\mathbf{R}$ such that $F(x) = f(x)$ for all x in the domain of f?

(i) $f(x) = \dfrac{x^2 - 4}{x - 2}$.

(ii) $f(x) = \dfrac{|x|}{x}$.

(iii) $f(x) = 0$, x irrational.

(iv) $f(x) = 1/q$, $x = p/q$ rational in lowest terms.

2. At which points are the functions of Problems 4-17 and 4-19 continuous?

3. (a) Suppose that f is a function satisfying $|f(x)| \le |x|$ for all x. Show that f is continuous at 0. (Notice that $f(0)$ must equal 0.)

(b) Give an example of such a function f which is not continuous at any $a \ne 0$.

(c) Suppose that g is continuous at 0 and $g(0) = 0$, and $|f(x)| \le |g(x)|$. Prove that f is continuous at 0.

4. Give an example of a function f such that f is continuous nowhere, but $|f|$ is continuous everywhere.

5. For each number a, find a function which is continuous at a, but not at any other points.

6. (a) Find a function f which is discontinuous at $1, \frac{1}{2}, \frac{1}{3}, \frac{1}{4}, \dots$ but continuous at all other points.
(b) Find a function f which is discontinuous at $1, \frac{1}{2}, \frac{1}{3}, \frac{1}{4}, \dots$, and at 0, but continuous at all other points.

7. Suppose that f satisfies $f(x+y) = f(x) + f(y)$, and that f is continuous at 0. Prove that f is continuous at a for all a.

8. Suppose that f is continuous at a and $f(a) = 0$. Prove that if $\alpha \neq 0$, then $f + \alpha$ is nonzero in some open interval containing a.

9. (a) Suppose f is *not* continuous at a. Prove that for some number $\varepsilon > 0$ there are numbers x arbitrarily close to a with $|f(x) - f(a)| > \varepsilon$. Illustrate graphically.
(b) Conclude that for some number $\varepsilon > 0$ *either* there are numbers x arbitrarily close to a with $f(x) < f(a) - \varepsilon$ *or* there are numbers x arbitrarily close to a with $f(x) > f(a) + \varepsilon$.

10. (a) Prove that if f is continuous at a, then so is $|f|$.
(b) Prove that every continuous f can be written $f = E + O$, where E is even and continuous and O is odd and continuous.
(c) Prove that if f and g are continuous, then so are $\max(f, g)$ and $\min(f, g)$.
(d) Prove that every continuous f can be written $f = g - h$, where g and h are nonnegative and continuous.

11. Prove Theorem 1(3) by using Theorem 2 and continuity of the function $f(x) = 1/x$.

***12.** (a) Prove that if f is continuous at l and $\lim_{x \to a} g(x) = l$, then $\lim_{x \to a} f(g(x)) = f(l)$. (You can go right back to the definitions, but it is easier to consider the function G with $G(x) = g(x)$ for $x \neq a$, and $G(a) = l$.)
(b) Show that if continuity of f at l is not assumed, then it is not generally true that $\lim_{x \to a} f(g(x)) = f(\lim_{x \to a} g(x))$. Hint: Try $f(x) = 0$ for $x \neq l$, and $f(l) = 1$.

13. (a) Prove that if f is continuous on $[a, b]$, then there is a function g which is continuous on $\mathbf{R}$, and which satisfies $g(x) = f(x)$ for all x in $[a, b]$. Hint: Since you obviously have a great deal of choice, try making g constant on $(-\infty, a]$ and $[b, \infty)$.
(b) Give an example to show that this assertion is false if $[a, b]$ is replaced by (a, b).

14. (a) Suppose that g and h are continuous at a, and that $g(a) = h(a)$. Define $f(x)$ to be $g(x)$ if $x \geq a$ and $h(x)$ if $x \leq a$. Prove that f is continuous at a.
(b) Suppose g is continuous on $[a, b]$ and h is continuous on $[b, c]$ and $g(b) = h(b)$. Let $f(x)$ be $g(x)$ for x in $[a, b]$ and $h(x)$ for x in $[b, c]$.

Show that f is continuous on $[a, c]$. (Thus, continuous functions can be "pasted together".)

15. (a) Prove the following version of Theorem 3 for "right-hand continuity": Suppose that $\lim_{x \to a^+} f(x) = f(a)$, and $f(a) > 0$. Then there is a number $\delta > 0$ such that $f(x) > 0$ for all x satisfying $0 \le x - a < \delta$. Similarly, if $f(a) < 0$, then there is a number $\delta > 0$ such that $f(x) < 0$ for all x satisfying $0 \le x - a < \delta$.

(b) Prove a version of Theorem 3 when $\lim_{x \to b^-} f(x) = f(b)$.

16. If $\lim_{x \to a} f(x)$ exists, but is $\neq f(a)$, then f is said to have a **removable discontinuity** at a.

(a) If $f(x) = \sin 1/x$ for $x \neq 0$ and $f(0) = 1$, does f have a removable discontinuity at 0? What if $f(x) = x \sin 1/x$ for $x \neq 0$, and $f(0) = 1$?

(b) Suppose f has a removable discontinuity at a. Let $g(x) = f(x)$ for $x \neq a$, and let $g(a) = \lim_{x \to a} f(x)$. Prove that g is continuous at a. (Don't work very hard; this is quite easy.)

(c) Let $f(x) = 0$ if x is irrational, and let $f(p/q) = 1/q$ if p/q is in lowest terms. What is the function g defined by $g(x) = \lim_{y \to x} f(y)$?

*(d) Let f be a function with the property that every point of discontinuity is a removable discontinuity. This means that $\lim_{y \to x} f(y)$ exists for all x, but f may be discontinuous at some (even infinitely many) numbers x. Define $g(x) = \lim_{y \to x} f(y)$. Prove that g is continuous. (This is not quite so easy as part (b).)

**(e) Is there a function f which is discontinuous at every point, and which has only removable discontinuities? (It is worth thinking about this problem now, but mainly as a test of intuition; even if you suspect the correct answer, you will almost certainly be unable to prove it at the present time. See Problem 22-33.)

CHAPTER 7 THREE HARD THEOREMS

This chapter is devoted to three theorems about continuous functions, and some of their consequences. The proofs of the three theorems themselves will not be given until the next chapter, for reasons which are explained at the end of this chapter.

THEOREM 1 If f is continuous on $[a, b]$ and $f(a) < 0 < f(b)$, then there is some x in $[a, b]$ such that $f(x) = 0$.

(Geometrically, this means that the graph of a continuous function which starts below the horizontal axis and ends above it must cross this axis at some point, as in Figure 1.)

THEOREM 2 If f is continuous on $[a, b]$, then f is bounded above on $[a, b]$, that is, there is some number N such that $f(x) \leq N$ for all x in $[a, b]$.

(Geometrically, this theorem means that the graph of f lies below some line parallel to the horizontal axis, as in Figure 2.)

THEOREM 3 If f is continuous on $[a, b]$, then there is some number y in $[a, b]$ such that $f(y) \geq f(x)$ for all x in $[a, b]$ (Figure 3).

These three theorems differ markedly from the theorems of Chapter 6. The hypotheses of those theorems always involved continuity at a single point, while

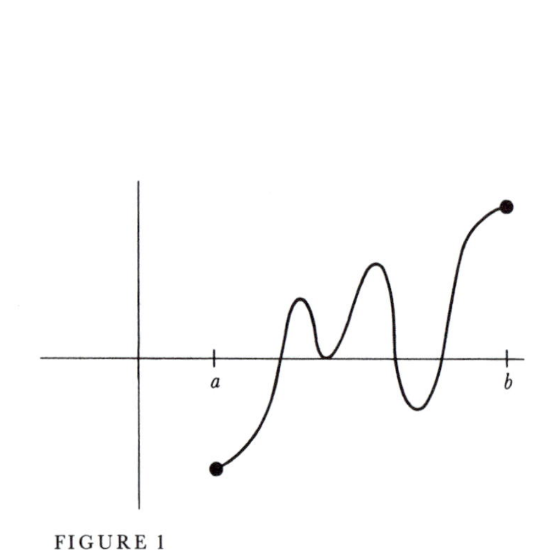

FIGURE 1

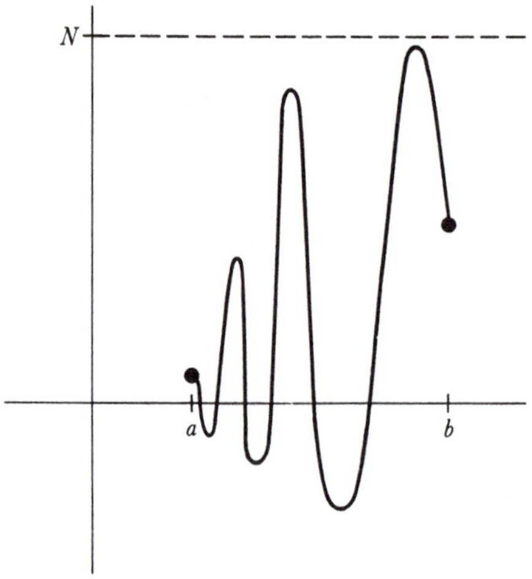

FIGURE 2

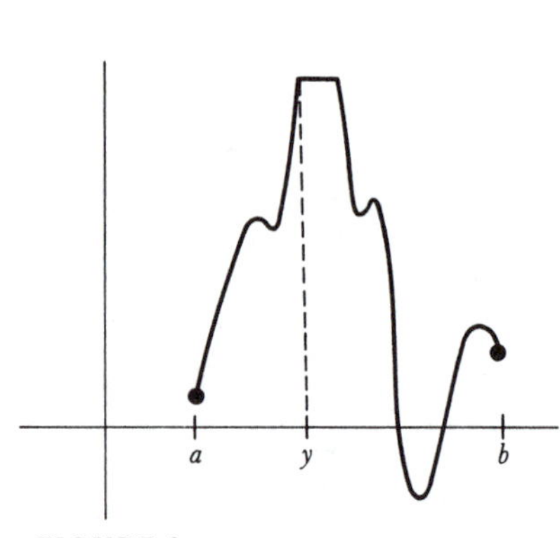

FIGURE 3

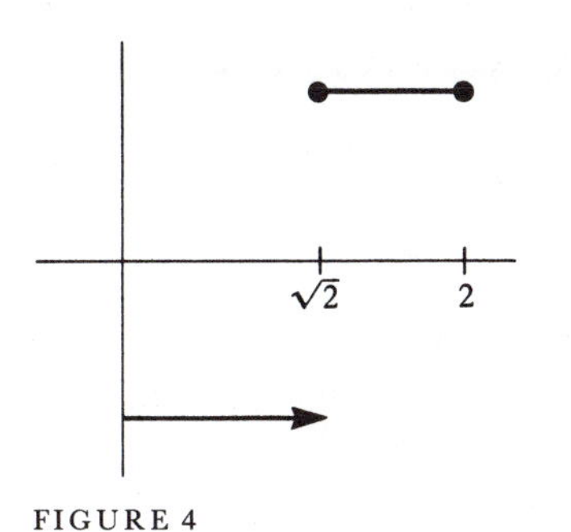

FIGURE 4

the hypotheses of the present theorems require continuity on a whole interval $[a, b]$—if continuity fails to hold at a single point, the conclusions may fail. For example, let f be the function shown in Figure 4,

$$f(x) = \begin{cases} -1, & 0 \le x < \sqrt{2} \\ 1, & \sqrt{2} \le x \le 2. \end{cases}$$

Then f is continuous at every point of $[0, 2]$ except $\sqrt{2}$, and $f(0) < 0 < f(2)$, but there is no point x in $[0, 2]$ such that $f(x) = 0$; the discontinuity at the single point $\sqrt{2}$ is sufficient to destroy the conclusion of Theorem 1.

Similarly, suppose that f is the function shown in Figure 5,

$$f(x) = \begin{cases} 1/x, & x \ne 0 \\ 0, & x = 0. \end{cases}$$

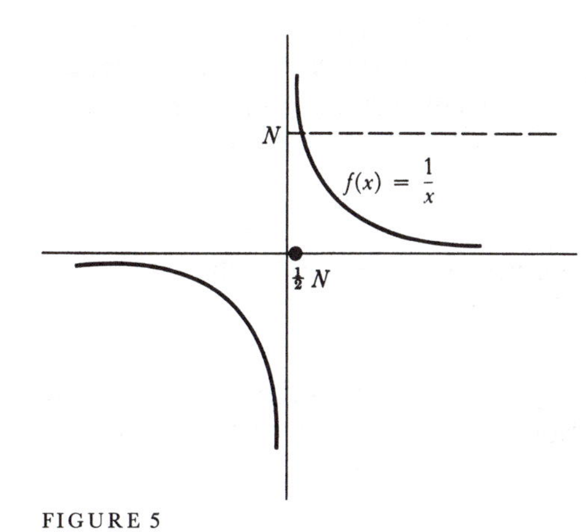

FIGURE 5

Then f is continuous at every point of $[0, 1]$ except 0, but f is not bounded above on $[0, 1]$. In fact, for any number $N > 0$ we have $f(1/2N) = 2N > N$.

This example also shows that the closed interval $[a, b]$ in Theorem 2 cannot be replaced by the open interval (a, b), for the function f is continuous on $(0, 1)$, but is not bounded there.

Finally, consider the function shown in Figure 6,

$$f(x) = \begin{cases} x^2, & x < 1 \\ 0, & x \ge 1. \end{cases}$$

On the interval $[0, 1]$ the function f is bounded above, so f does satisfy the conclusion of Theorem 2, even though f is not continuous on $[0, 1]$. But f does not satisfy the conclusion of Theorem 3—there is no y in $[0, 1]$ such that $f(y) \ge f(x)$ for all x in $[0, 1]$; in fact, it is certainly not true that $f(1) \ge f(x)$ for all x in $[0, 1]$ so we cannot choose $y = 1$, nor can we choose $0 \le y < 1$ because $f(y) < f(x)$ if x is any number with $y < x < 1$.

This example shows that Theorem 3 is considerably stronger than Theorem 2. Theorem 3 is often paraphrased by saying that a continuous function on a closed interval "takes on its maximum value" on that interval.

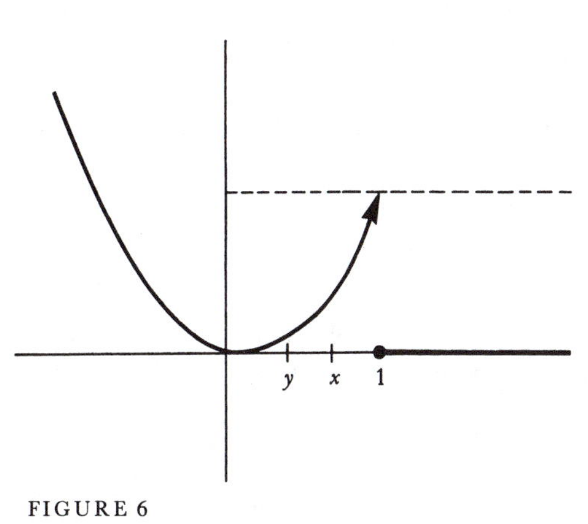

FIGURE 6

As a compensation for the stringency of the hypotheses of our three theorems, the conclusions are of a totally different order than those of previous theorems. They describe the behavior of a function, not just near a point, but on a whole interval; such "global" properties of a function are always significantly more difficult to prove than "local" properties, and are correspondingly of much greater power. To illustrate the usefulness of Theorems 1, 2, and 3, we will soon deduce some important consequences, but it will help to first mention some simple generalizations of these theorems.

THEOREM 4 If f is continuous on $[a, b]$ and $f(a) < c < f(b)$, then there is some x in $[a, b]$ such that $f(x) = c$.

PROOF Let $g = f - c$. Then g is continuous, and $g(a) < 0 < g(b)$. By Theorem 1, there is some x in $[a, b]$ such that $g(x) = 0$. But this means that $f(x) = c$. ▮

THEOREM 5 If f is continuous on $[a, b]$ and $f(a) > c > f(b)$, then there is some x in $[a, b]$ such that $f(x) = c$.

PROOF The function $-f$ is continuous on $[a, b]$ and $-f(a) < -c < -f(b)$. By Theorem 4 there is some x in $[a, b]$ such that $-f(x) = -c$, which means that $f(x) = c$. ▮

Theorems 4 and 5 together show that f takes on any value between $f(a)$ and $f(b)$. We can do even better than this: if c and d are in $[a, b]$, then f takes on any value between $f(c)$ and $f(d)$. The proof is simple: if, for example, $c < d$, then just apply Theorems 4 and 5 to the interval $[c, d]$. Summarizing, if a continuous function on an interval takes on two values, it takes on every value in between; this slight generalization of Theorem 1 is often called the Intermediate Value Theorem.

THEOREM 6 If f is continuous on $[a, b]$, then f is bounded below on $[a, b]$, that is, there is some number N such that $f(x) \geq N$ for all x in $[a, b]$.

PROOF The function $-f$ is continuous on $[a, b]$, so by Theorem 2 there is a number M such that $-f(x) \leq M$ for all x in $[a, b]$. But this means that $f(x) \geq -M$ for all x in $[a, b]$, so we can let $N = -M$. ▮

Theorems 2 and 6 together show that a continuous function f on $[a, b]$ is bounded on $[a, b]$, that is, there is a number N such that $|f(x)| \leq N$ for all x in $[a, b]$. In fact, since Theorem 2 ensures the existence of a number N_1 such that $f(x) \leq N_1$ for all x in $[a, b]$, and Theorem 6 ensures the existence of a number N_2 such that $f(x) \geq N_2$ for all x in $[a, b]$, we can take $N = \max(|N_1|, |N_2|)$.

THEOREM 7 If f is continuous on $[a, b]$, then there is some y in $[a, b]$ such that $f(y) \leq f(x)$ for all x in $[a, b]$.
(A continuous function on a closed interval takes on its minimum value on that interval.)

PROOF The function $-f$ is continuous on $[a, b]$; by Theorem 3 there is some y in $[a, b]$ such that $-f(y) \geq -f(x)$ for all x in $[a, b]$, which means that $f(y) \leq f(x)$ for all x in $[a, b]$. ▮

Now that we have derived the trivial consequences of Theorems 1, 2, and 3, we can begin proving a few interesting things.

THEOREM 8 Every positive number has a square root. In other words, if $\alpha > 0$, then there is some number x such that $x^2 = \alpha$.

PROOF Consider the function $f(x) = x^2$, which is certainly continuous. Notice that the statement of the theorem can be expressed in terms of f: "the number α has a square root" means that f takes on the value α. The proof of this fact about f will be an easy consequence of Theorem 4.

There is obviously a number $b > 0$ such that $f(b) > \alpha$ (as illustrated in Figure 7); in fact, if $\alpha > 1$ we can take $b = \alpha$, while if $\alpha < 1$ we can take $b = 1$. Since $f(0) < \alpha < f(b)$, Theorem 4 applied to $[0, b]$ implies that for some x (in $[0, b]$), we have $f(x) = \alpha$, i.e., $x^2 = \alpha$. ▮

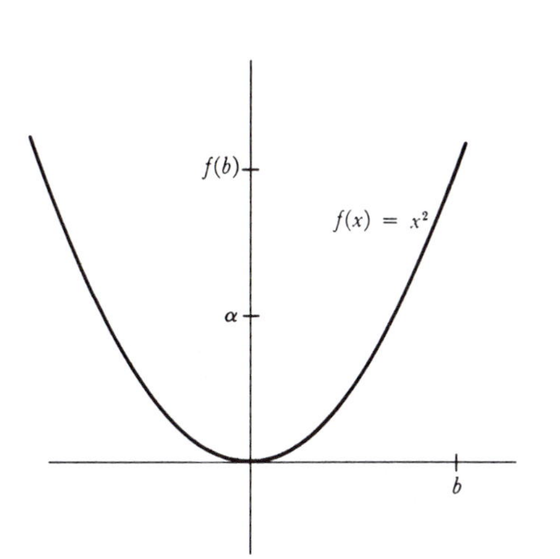

FIGURE 7

Precisely the same argument can be used to prove that a positive number has an nth root, for any natural number n. If n happens to be odd, one can do better: *every* number has an nth root. To prove this we just note that if the positive number α has the nth root x, i.e., if $x^n = \alpha$, then $(-x)^n = -\alpha$ (since n is odd), so $-\alpha$ has the nth root $-x$. The assertion, that for odd n any number α has an nth root, is equivalent to the statement that the equation

$$x^n - \alpha = 0$$

has a root if n is odd. Expressed in this way the result is susceptible of great generalization.

THEOREM 9 If n is odd, then any equation

$$x^n + a_{n-1}x^{n-1} + \cdots + a_0 = 0$$

has a root.

PROOF We obviously want to consider the function

$$f(x) = x^n + a_{n-1}x^{n-1} + \cdots + a_0;$$

we would like to prove that f is sometimes positive and sometimes negative. The intuitive idea is that for large $|x|$, the function is very much like $g(x) = x^n$ and, since n is odd, this function is positive for large positive x and negative for large negative x. A little algebra is all we need to make this intuitive idea work.

The proper analysis of the function f depends on writing

$$f(x) = x^n + a_{n-1}x^{n-1} + \cdots + a_0 = x^n\left(1 + \frac{a_{n-1}}{x} + \cdots + \frac{a_0}{x^n}\right).$$

Note that

$$\left|\frac{a_{n-1}}{x} + \frac{a_{n-2}}{x^2} + \cdots + \frac{a_0}{x^n}\right| \le \frac{|a_{n-1}|}{|x|} + \cdots + \frac{|a_0|}{|x^n|}.$$

Consequently, if we choose x satisfying

$$(*) \qquad |x| > 1, 2n|a_{n-1}|, \ldots, 2n|a_0|,$$

then $|x^k| > |x|$ and

$$\frac{|a_{n-k}|}{|x^k|} < \frac{|a_{n-k}|}{|x|} < \frac{|a_{n-k}|}{2n|a_{n-k}|} = \frac{1}{2n},$$

so

$$\left|\frac{a_{n-1}}{x}+\frac{a_{n-2}}{x^2}+\cdots+\frac{a_0}{x^n}\right| \le \underbrace{\frac{1}{2n}+\cdots+\frac{1}{2n}}_{n \text{ terms}} = \frac{1}{2}.$$

In other words,

$$-\frac{1}{2} \le \frac{a_{n-1}}{x}+\cdots+\frac{a_0}{x^n} \le \frac{1}{2},$$

which implies that

$$\frac{1}{2} \le 1+\frac{a_{n-1}}{x}+\cdots+\frac{a_0}{x^n}.$$

Therefore, if we choose an $x_1 > 0$ which satisfies $(*)$, then

$$\frac{(x_1)^n}{2} \le (x_1)^n\left(1+\frac{a_{n-1}}{x_1}+\cdots+\frac{a_0}{(x_1)^n}\right) = f(x_1),$$

so that $f(x_1) > 0$. On the other hand, if $x_2 < 0$ satisfies $(*)$, then $(x_2)^n < 0$ and

$$\frac{(x_2)^n}{2} \ge (x_2)^n\left(1+\frac{a_{n-1}}{x_2}+\cdots+\frac{a_0}{(x_2)^n}\right) = f(x_2),$$

so that $f(x_2) < 0$.

Now applying Theorem 1 to the interval $[x_2, x_1]$ we conclude that there is an x in $[x_2, x_1]$ such that $f(x) = 0$. ▮

Theorem 9 disposes of the problem of odd degree equations so happily that it would be frustrating to leave the problem of even degree equations completely undiscussed. At first sight, however, the problem seems insuperable. Some equations, like $x^2 - 1 = 0$, have a solution, and some, like $x^2 + 1 = 0$, do not—what more is there to say? If we are willing to consider a more general question, however, something interesting *can* be said. Instead of trying to solve the equation

$$x^n + a_{n-1}x^{n-1} + \cdots + a_0 = 0,$$

let us ask about the possibility of solving the equations

$$x^n + a_{n-1}x^{n-1} + \cdots + a_0 = c$$

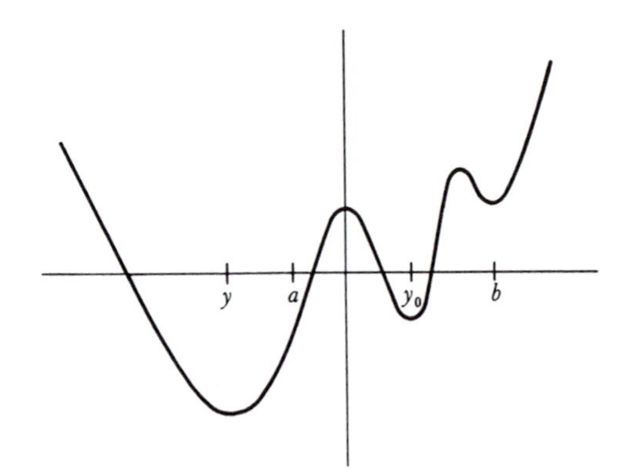

FIGURE 8

for all possible numbers c. This amount to allowing the constant term a_0 to vary. The information which can be given concerning the solution of these equations depends on a fact which is illustrated in Figure 8.

The graph of the function $f(x) = x^n + a_{n-1}x^{n-1} + \cdots + a_0$, with n even, contains, at least the way we have drawn it, a lowest point. In other words, there is a number y such that $f(y) \le f(x)$ for all numbers x—the function f takes on a minimum value, not just on each closed interval, but on the whole line. (Notice that this is false if n is odd.) The proof depends on Theorem 7, but a tricky application will be required. We can apply Theorem 7 to any interval $[a, b]$, and obtain a point y_0 such that $f(y_0)$ is the minimum value of f on $[a, b]$; but if $[a, b]$ happens to be the interval shown in Figure 8, for example, then the point y_0 will not be the place where f has its minimum value for the whole line. In the next theorem the entire point of the proof is to choose an interval $[a, b]$ in such a way that this cannot happen.

THEOREM 10 If n is even and $f(x) = x^n + a_{n-1}x^{n-1} + \cdots + a_0$, then there is a number y such that $f(y) \leq f(x)$ for all x.

PROOF As in the proof of Theorem 9, if

$$M = \max(1, 2n|a_{n-1}|, \ldots, 2n|a_0|),$$

then for all x with $|x| \geq M$, we have

$$\frac{1}{2} \leq 1 + \frac{a_{n-1}}{x} + \cdots + \frac{a_0}{x^n}.$$

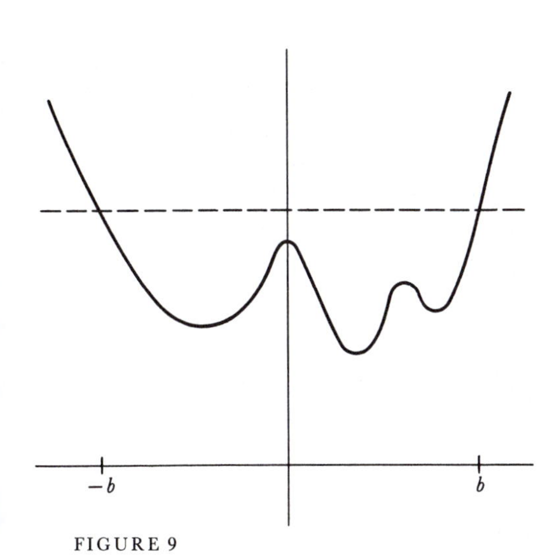

FIGURE 9

Since n is even, $x^n \geq 0$ for all x, so

$$\frac{x^n}{2} \leq x^n\left(1 + \frac{a_{n-1}}{x} + \cdots + \frac{a_0}{x^n}\right) = f(x),$$

provided that $|x| \geq M$. Now consider the number $f(0)$. Let $b > 0$ be a number such that $b^n \geq 2f(0)$ and also $b > M$. Then, if $x \geq b$, we have (Figure 9)

$$f(x) \geq \frac{x^n}{2} \geq \frac{b^n}{2} \geq f(0).$$

Similarly, if $x \leq -b$, then

$$f(x) \geq \frac{x^n}{2} \geq \frac{(-b)^n}{2} = \frac{b^n}{2} \geq f(0).$$

Summarizing:

$$\text{if } x \geq b \text{ or } x \leq -b, \text{ then } f(x) \geq f(0).$$

Now apply Theorem 7 to the function f on the interval $[-b, b]$. We conclude that there is a number y such that

$$(1) \qquad \text{if } -b \leq x \leq b, \text{ then } f(y) \leq f(x).$$

In particular, $f(y) \leq f(0)$. Thus

$$(2) \qquad \text{if } x \leq -b \text{ or } x \geq b, \text{ then } f(x) \geq f(0) \geq f(y).$$

Combining (1) and (2) we see that $f(y) \leq f(x)$ for all x. ▮

Theorem 10 now allows us to prove the following result.

THEOREM 11 Consider the equation

$$(*) \qquad x^n + a_{n-1}x^{n-1} + \cdots + a_0 = c,$$

and suppose n is even. Then there is a number m such that $(*)$ has a solution for $c \geq m$ and has no solution for $c < m$.

PROOF Let $f(x) = x^n + a_{n-1}x^{n-1} + \cdots + a_0$ (Figure 10).

According to Theorem 10 there is a number y such that $f(y) \leq f(x)$ for all x. Let $m = f(y)$. If $c < m$, then the equation $(*)$ obviously has no solution, since the left side always has a value $\geq m$. If $c = m$, then $(*)$ has y as a solution. Finally, suppose $c > m$. Let b be a number such that $b > y$ and $f(b) > c$. Then $f(y) = m < c < f(b)$. Consequently, by Theorem 4, there is some number x in $[y, b]$ such that $f(x) = c$, so x is a solution of $(*)$. ▮

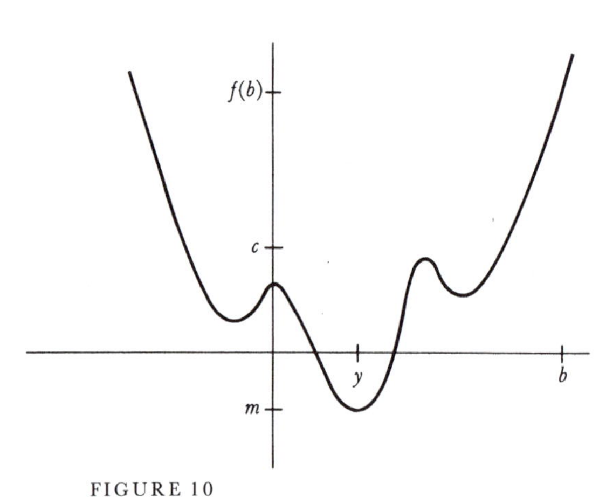

FIGURE 10

These consequences of Theorems 1, 2, and 3 are the only ones we will derive now (these theorems will play a fundamental role in everything we do later, however). Only one task remains—to prove Theorems 1, 2, and 3. Unfortunately, we cannot hope to do this—on the basis of our present knowledge about the real numbers (namely, P1–P12) a proof is *impossible.* There are several ways of convincing ourselves that this gloomy conclusion is actually the case. For example, the proof of Theorem 8 relies only on the proof of Theorem 1; if we could prove Theorem 1, then the proof of Theorem 8 would be complete, and we would have a proof that every positive number has a square root. As pointed out in Part I, it is impossible to prove this on the basis of P1–P12. Again, suppose we consider the function

$$f(x) = \frac{1}{x^2 - 2}.$$

If there were no number x with $x^2 = 2$, then f would be continuous, since the denominator would never $= 0$. But f is not bounded on $[0, 2]$. So Theorem 2 depends essentially on the existence of numbers other than rational numbers, and therefore on some property of the real numbers other than P1–P12.

Despite our inability to prove Theorems 1, 2, and 3, they are certainly results which we want to be true. If the pictures we have been drawing have any connection with the mathematics we are doing, if our notion of continuous function corresponds to any degree with our intuitive notion, Theorems 1, 2, and 3 have got to be true. Since a proof of any of these theorems must require some new property of **R** which has so far been overlooked, our present difficulties suggest a way to discover that property: let us try to construct a proof of Theorem 1, for example, and see what goes wrong.

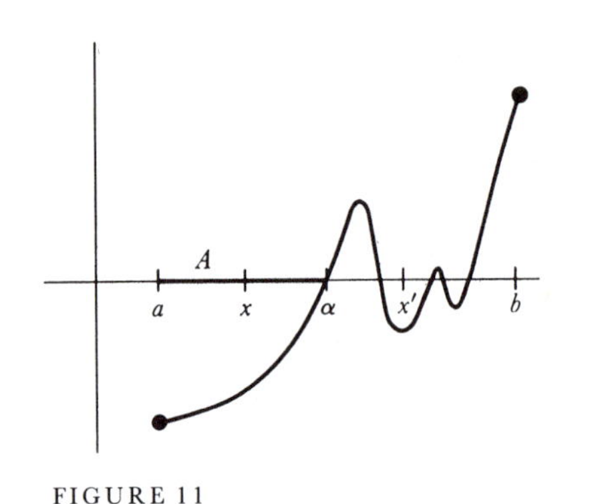

FIGURE 11

One idea which seems promising is to locate the first point where $f(x) = 0$, that is, the smallest x in $[a, b]$ such that $f(x) = 0$. To find this point, first consider the set A which contains all numbers x in $[a, b]$ such that f is negative on $[a, x]$. In Figure 11, x is such a point, while x' is not. The set A itself is indicated by a heavy line. Since f is negative at a, and positive at b, the set A contains some points greater than a, while all points sufficiently close to b are not in A. (We are here using the continuity of f on $[a, b]$, as well as Problem 6-15.)

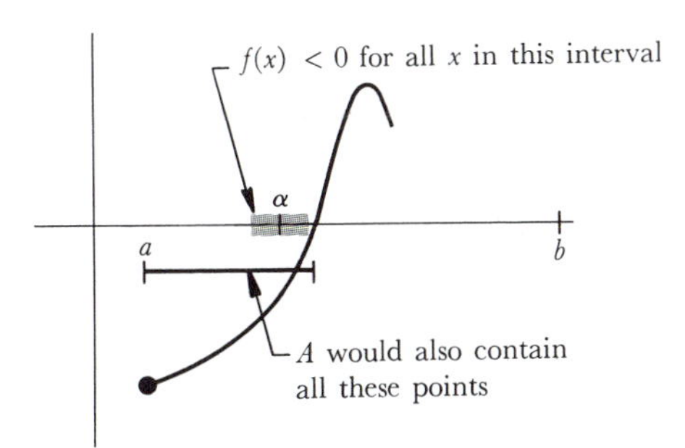

FIGURE 12

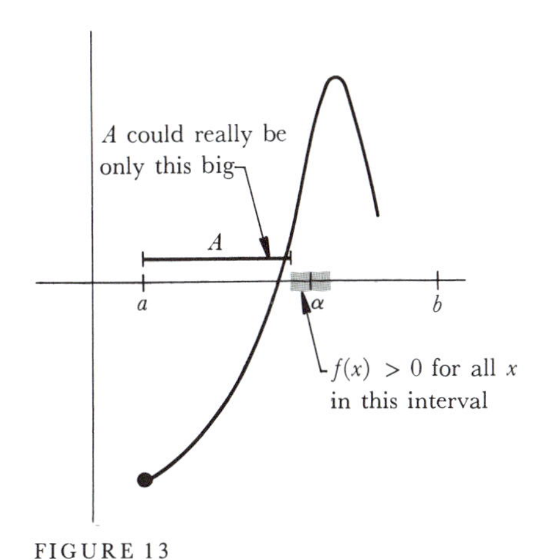

FIGURE 13

Now suppose α is the smallest number which is greater than all members of A; clearly $a < \alpha < b$. We claim that $f(\alpha) = 0$, and to prove this we only have to eliminate the possibilities $f(\alpha) < 0$ and $f(\alpha) > 0$.

Suppose first that $f(\alpha) < 0$. Then, by Theorem 6-3, $f(x)$ would be less than 0 for all x in a small interval containing α, in particular for some numbers bigger than α (Figure 12); but this contradicts the fact that α is bigger than every member of A, since the larger numbers would also be in A. Consequently, $f(\alpha) < 0$ is false.

On the other hand, suppose $f(\alpha) > 0$. Again applying Theorem 6-3, we see that $f(x)$ would be positive for all x in a small interval containing α, in particular for some numbers smaller than α (Figure 13). This means that these smaller numbers are all *not* in A. Consequently, one could have chosen an even smaller α which would be greater than all members of A. Once again we have a contradiction; $f(\alpha) > 0$ is also false. Hence $f(\alpha) = 0$ and, we are tempted to say, Q.E.D.

We know, however, that something must be wrong, since no new properties of $\mathbf{R}$ were ever used, and it does not require much scrutiny to find the dubious point. It is clear that we can choose a number α which is greater than all members of A (for example, we can choose $\alpha = b$), but it is not so clear that we can choose a *smallest* one. In fact, suppose A consists of all numbers $x \geq 0$ such that $x^2 < 2$. If the number $\sqrt{2}$ did not exist, there would not be a least number greater than all the members of A; for any $y > \sqrt{2}$ we chose, we could always choose a still smaller one.

Now that we have discovered the fallacy, it is almost obvious what additional property of the real numbers we need. All we must do is say it properly and use it. That is the business of the next chapter.

PROBLEMS

1. For each of the following functions, decide which are bounded above or below on the indicated interval, and which take on their maximum or minimum value. (Notice that f *might* have these properties even if f is not continuous, and even if the interval is not a closed interval.)

(i) $f(x) = x^2$ on $(-1, 1)$.

(ii) $f(x) = x^3$ on $(-1, 1)$.

(iii) $f(x) = x^2$ on $\mathbf{R}$.

(iv) $f(x) = x^2$ on $[0, \infty)$.

(v) $f(x) = \begin{cases} x^2, & x \leq a \\ a+2, & x > a \end{cases}$ on $(-a-1, a+1)$. (It will be necessary to consider several possibilities for a.)

(vi) $f(x) = \begin{cases} x^2, & x < a \\ a+2, & x \geq a \end{cases}$ on $[-a-1, a+1]$.

(vii) $f(x) = \begin{cases} 0, & x \text{ irrational} \\ 1/q & x = p/q \text{ in lowest terms} \end{cases}$ on $[0, 1]$.

(viii) $f(x) = \begin{cases} 1, & x \text{ irrational} \\ 1/q & x = p/q \text{ in lowest terms} \end{cases}$ on $[0, 1]$.

(ix) $f(x) = \begin{cases} 1, & x \text{ irrational} \\ -1/q & x = p/q \text{ in lowest terms} \end{cases}$ on $[0, 1]$.

(x) $f(x) = \begin{cases} x, & x \text{ rational} \\ 0 & x \text{ irrational} \end{cases}$ on $[0, a]$.

(xi) $f(x) = \sin^2(\cos x + \sqrt{a + a^2})$ on $[0, a^3]$.

(xii) $f(x) = [x]$ on $[0, a]$.

2. For each of the following polynomial functions f, find an integer n such that $f(x) = 0$ for some x between n and $n + 1$.

(i) $f(x) = x^3 - x + 3$.
(ii) $f(x) = x^5 + 5x^4 + 2x + 1$.
(iii) $f(x) = x^5 + x + 1$.
(iv) $f(x) = 4x^2 - 4x + 1$.

3. Prove that there is some number x such that

(i) $x^{179} + \dfrac{163}{1 + x^2 + \sin^2 x} = 119$.
(ii) $\sin x = x - 1$.

4. This problem is a continuation of Problem 3-7.

(a) If $n - k$ is even, and ≥ 0, find a polynomial function of degree n with exactly k roots.
(b) A root a of the polynomial function f is said to have **multiplicity** m if $f(x) = (x - a)^m g(x)$, where g is a polynomial function that does *not* have a as a root. Let f be a polynomial function of degree n. Suppose that f has k roots, counting multiplicities, i.e., suppose that k is the sum of the multiplicities of all the roots. Show that $n - k$ is even.

5. Suppose that f is continuous on $[a, b]$ and that $f(x)$ is always rational. What can be said about f?

6. Suppose that f is a *continuous* function on $[-1, 1]$ such that $x^2 + (f(x))^2 = 1$ for all x. (This means that $(x, f(x))$ always lies on the unit circle.) Show that either $f(x) = \sqrt{1 - x^2}$ for all x, or else $f(x) = -\sqrt{1 - x^2}$ for all x.

7. How many continuous functions f are there which satisfy $(f(x))^2 = x^2$ for all x?

8. Suppose that f and g are continuous, that $f^2 = g^2$, and that $f(x) \neq 0$ for all x. Prove that either $f(x) = g(x)$ for all x, or else $f(x) = -g(x)$ for all x.

9. (a) Suppose that f is continuous, that $f(x) = 0$ only for $x = a$, and that $f(x) > 0$ for some $x > a$ as well as for some $x < a$. What can be said about $f(x)$ for all $x \neq a$?

(b) Again assume that f is continuous and that $f(x) = 0$ only for $x = a$, but suppose, instead, that $f(x) > 0$ for some $x > a$ and $f(x) < 0$ for some $x < a$. Now what can be said about $f(x)$ for $x \neq a$?

*(c) Discuss the sign of $x^3 + x^2y + xy^2 + y^3$ when x and y are not both 0.

10. Suppose f and g are continuous on $[a, b]$ and that $f(a) < g(a)$, but $f(b) > g(b)$. Prove that $f(x) = g(x)$ for some x in $[a, b]$. (If your proof isn't very short, it's not the right one.)

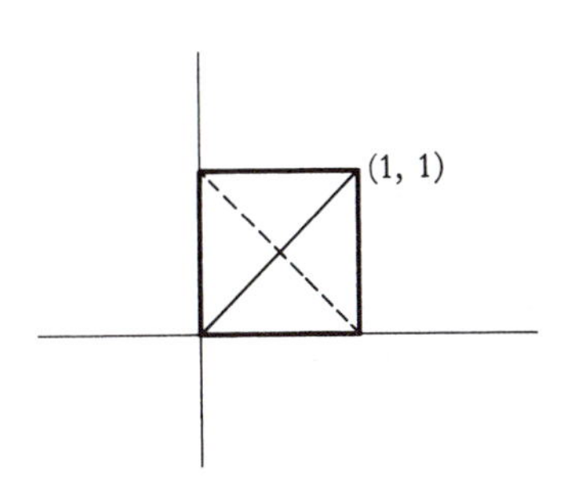

FIGURE 14

11. Suppose that f is a continuous function on $[0, 1]$ and that $f(x)$ is in $[0, 1]$ for each x (draw a picture). Prove that $f(x) = x$ for some number x.

12. (a) Problem 11 shows that f intersects the diagonal of the square in Figure 14 (solid line). Show that f must also intersect the other (dashed) diagonal.

(b) Prove the following more general fact: If g is continuous on $[0, 1]$ and $g(0) = 0$, $g(1) = 1$ or $g(0) = 1$, $g(1) = 0$, then $f(x) = g(x)$ for some x.

13. (a) Let $f(x) = \sin 1/x$ for $x \neq 0$ and let $f(0) = 0$. Is f continuous on $[-1, 1]$? Show that f satisfies the conclusion of the Intermediate Value Theorem on $[-1, 1]$; in other words, if f takes on two values somewhere on $[-1, 1]$, it also takes on every value in between.

*(b) Suppose that f satisfies the conclusion of the Intermediate Value Theorem, and that f takes on each value *only once*. Prove that f is continuous.

*(c) Generalize to the case where f takes on each value only finitely many times.

14. If f is a continuous function on $[0, 1]$, let $\|f\|$ be the maximum value of $|f|$ on $[0, 1]$.

(a) Prove that for any number c we have $\|cf\| = |c| \cdot \|f\|$.

*(b) Prove that $\|f + g\| \leq \|f\| + \|g\|$. Give an example where $\|f + g\| \neq \|f\| + \|g\|$.

(c) Prove that $\|h - f\| \leq \|h - g\| + \|g - f\|$.

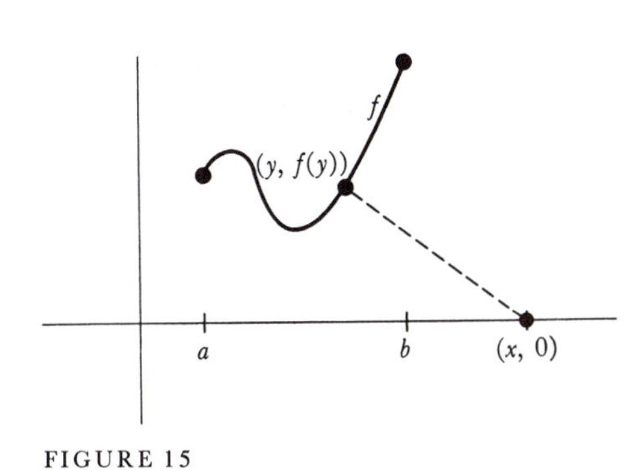

FIGURE 15

***15.** Suppose that ϕ is continuous and $\lim_{x\to\infty} \phi(x)/x^n = 0 = \lim_{x\to-\infty} \phi(x)/x^n$.

(a) Prove that if n is odd, then there is a number x such that $x^n + \phi(x) = 0$.

(b) Prove that if n is even, then there is a number y such that $y^n + \phi(y) \leq x^n + \phi(x)$ for all x.

Hint: Of which proofs does this problem test your understanding?

***16.** Let f be any polynomial function. Prove that there is some number y such that $|f(y)| \leq |f(x)|$ for all x.

***17.** Suppose that f is a continuous function with $f(x) > 0$ for all x, and $\lim_{x\to\infty} f(x) = 0 = \lim_{x\to-\infty} f(x)$. (Draw a picture.) Prove that there is some number y such that $f(y) \geq f(x)$ for all x.

***18.** (a) Suppose that f is continuous on $[a,b]$, and let x by any number. Prove that there is a point on the graph of f which is closest to $(x,0)$; in other words there is some y in $[a,b]$ such that the distance from $(x,0)$ to $(y, f(y))$ is $\le$ distance from $(x,0)$ to $(z, f(z))$ for all z in $[a,b]$. (See Figure 15.)

(b) Show that this same assertion is not necessarily true if $[a,b]$ is replaced by (a,b) throughout.

(c) Show that the assertion *is* true if $[a,b]$ is replaced by **R** throughout.

(d) In cases (a) and (c), let $g(x)$ be the minimum distance from $(x,0)$ to a point on the graph of f. Prove that $g(y) \le g(x) + |x-y|$, and conclude that g is continuous.

(e) Prove that there are numbers x_0 and x_1 in $[a,b]$ such that the distance from $(x_0,0)$ to $(x_1, f(x_1))$ is $\le$ the distance from $(x_0',0)$ to $(x_1', f(x_1'))$ for any x_0', x_1' in $[a,b]$.

****19.** (a) Suppose that f is continuous on $[0,1]$ and $f(0) = f(1)$. Let n be any natural number. Prove that there is some number x such that $f(x) = f(x+1/n)$, as shown in Figure 16 for $n = 4$. Hint: Consider the function $g(x) = f(x) - f(x+1/n)$; what would be true if $g(x) \ne 0$ for all x?

(b) Suppose $0 < a < 1$, but that a is not equal to $1/n$ for any natural number n. Find a function f which is continuous on $[0,1]$ and which satisfies $f(0) = f(1)$, but which does not satisfy $f(x) = f(x+a)$ for any x.

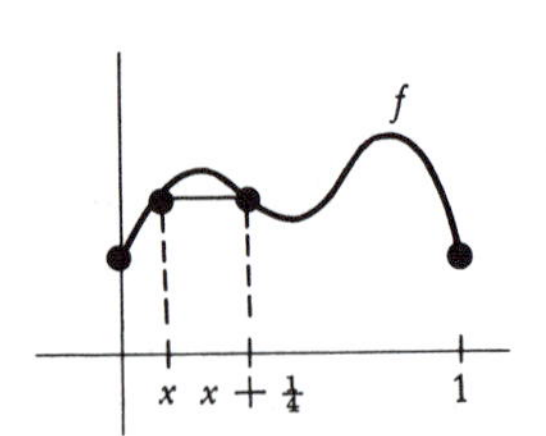

FIGURE 16

****20.** (a) Prove that there does not exist a continuous function f defined on **R** which takes on every value exactly twice. Hint: If $f(a) = f(b)$ for $a < b$, then either $f(x) > f(a)$ for all x in (a,b) or $f(x) < f(a)$ for all x in (a,b). Why? In the first case all values close to $f(a)$, but slightly larger than $f(a)$, are taken on somewhere in (a,b); this implies that $f(x) < f(a)$ for $x < a$ and $x > b$.

(b) Refine part (a) by proving that there is no continuous function f which takes on each value either 0 times or 2 times, i.e., which takes on exactly twice each value that it does take on. Hint: The previous hint implies that f has either a maximum or a minimum value (which must be taken on twice). What can be said about values close to the maximum value?

(c) Find a continuous function f which takes on every value exactly 3 times. More generally, find one which takes on every value exactly n times, if n is odd.

(d) Prove that if n is even, then there is no continuous f which takes on every value exactly n times. Hint: To treat the case $n = 4$, for example, let $f(x_1) = f(x_2) = f(x_3) = f(x_4)$. Then either $f(x) > 0$ for all x in two of the three intervals (x_1,x_2), (x_2,x_3), (x_3,x_4), or else $f(x) < 0$ for all x in two of these three intervals.

CHAPTER 8 LEAST UPPER BOUNDS

This chapter reveals the most important property of the real numbers. Nevertheless, it is merely a sequel to Chapter 7; the path which must be followed has already been indicated, and further discussion would be useless delay.

DEFINITION

A set A of real numbers is **bounded above** if there is a number x such that

$$x \geq a \quad \text{for every } a \text{ in } A.$$

Such a number x is called an **upper bound** for A.

Obviously A is bounded above if and only if there is a number x which is an upper bound for A (and in this case there will be lots of upper bounds for A); we often say, as a concession to idiomatic English, that "A has an upper bound" when we mean that there is a number which is an upper bound for A.

Notice that the term "bounded above" has now been used in two ways—first, in Chapter 7, in reference to functions, and now in reference to sets. This dual usage should cause no confusion, since it will always be clear whether we are talking about a set of numbers or a function. Moreover, the two definitions are closely connected: if A is the set $\{f(x) : a \leq x \leq b\}$, then the function f is bounded above on $[a, b]$ if and only if the set A is bounded above.

The entire collection **R** of real numbers, and the natural numbers **N**, are both examples of sets which are *not* bounded above. An example of a set which *is* bounded above is

$$A = \{x : 0 \leq x < 1\}.$$

To show that A is bounded above we need only name some upper bound for A, which is easy enough; for example, 138 is an upper bound for A, and so are 2, $1\frac{1}{2}$, $1\frac{1}{4}$, and 1. Clearly, 1 is the least upper bound of A; although the phrase just introduced is self-explanatory, in order to avoid any possible confusion (in particular, to ensure that we all know what the superlative of "less" means), we define this explicitly.

DEFINITION

A number x is a **least upper bound** of A if

(1) x is an upper bound of A,

and (2) if y is an upper bound of A, then $x \leq y$.

The use of the indefinite article "a" in this definition was merely a concession to temporary ignorance. Now that we have made a precise definition, it is easily seen that if x and y are both least upper bounds of A, then $x = y$. Indeed, in this case

$$x \leq y, \quad \text{since } y \text{ is an upper bound, and } x \text{ is a least upper bound,}$$
$$\text{and } y \leq x, \quad \text{since } x \text{ is an upper bound, and } y \text{ is a least upper bound;}$$

it follows that $x = y$. For this reason we speak of *the* least upper bound of A. The term **supremum** of A is synonymous and has one advantage. It abbreviates quite nicely to

$$\sup A \qquad \text{(pronounced "soup } A\text{")}$$

and saves us from the abbreviation

$$\text{lub } A$$

(which is nevertheless used by some authors).

There is a series of important definitions, analogous to those just given, which can now be treated more briefly. A set A of real numbers is **bounded below** if there is a number x such that

$$x \leq a \qquad \text{for every } a \text{ in } A.$$

Such a number x is called a **lower bound** for A. A number x is the **greatest lower bound** of A if

$$\text{(1)} \quad x \text{ is a lower bound of } A,$$
$$\text{and} \quad \text{(2)} \quad \text{if } y \text{ is a lower bound of } A, \text{ then } x \geq y.$$

The greatest lower bound of A is also called the **infimum** of A, abbreviated

$$\inf A;$$

some authors use the abbreviation

$$\text{glb } A.$$

One detail has been omitted from our discussion so far—the question of which sets have at least one, and hence exactly one, least upper bound or greatest lower bound. We will consider only least upper bounds, since the question for greatest lower bounds can then be answered easily (Problem 2).

If A is not bounded above, then A has no upper bound at all, so A certainly cannot be expected to have a least upper bound. It is tempting to say that A does have a least upper bound if it has *some* upper bound, but, like the principle of mathematical induction, this assertion can fail to be true in a rather special way. If $A = \emptyset$, then A is bounded above. Indeed, any number x is an upper bound for $\emptyset$:

$$x \geq y \quad \text{for every } y \text{ in } \emptyset$$

simply because there is no y in $\emptyset$. Since *every* number is an upper bound for $\emptyset$, there is surely no least upper bound for $\emptyset$. With this trivial exception however, our assertion is true—and very important, definitely important enough to warrant consideration of details. We are finally ready to state the last property of the real numbers which we need.

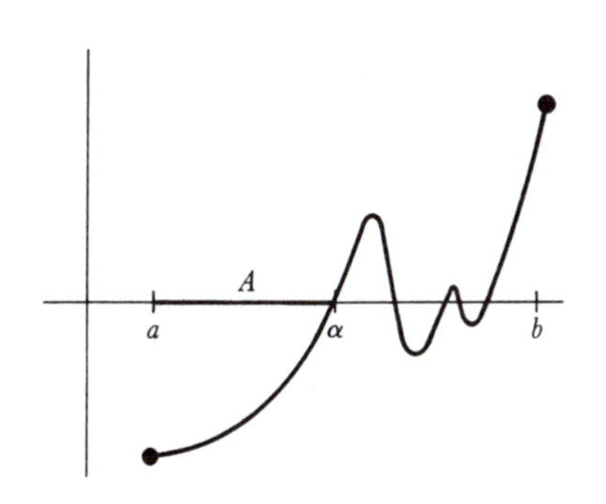

FIGURE 1

(P13) (The least upper bound property) If A is a set of real numbers, $A \neq \emptyset$, and A is bounded above, then A has a least upper bound.

Property P13 may strike you as anticlimactic, but that is actually one of its virtues. To complete our list of basic properties for the real numbers we require no particularly abstruse proposition, but only a property so simple that we might feel foolish for having overlooked it. Of course, the least upper bound property is not really so innocent as all that; after all, it does *not* hold for the rational numbers $\mathbf{Q}$. For example, if A is the set of all rational numbers x satisfying $x^2 < 2$, then there is no *rational* number y which is an upper bound for A and which is less than or equal to every other *rational* number which is an upper bond for A. It will become clear only gradually how significant P13 is, but we are already in a position to demonstrate its power, by supplying the proofs which were omitted in Chapter 7.

THEOREM 7-1 If f is continuous on $[a, b]$ and $f(a) < 0 < f(b)$, then there is some number x in $[a, b]$ such that $f(x) = 0$.

PROOF Our proof is merely a rigorous version of the outline developed at the end of Chapter 7—we will locate the smallest number x in $[a, b]$ with $f(x) = 0$.

Define the set A, shown in Figure 1, as follows:

$$A = \{x : a \leq x \leq b, \text{ and } f \text{ is negative on the interval } [a, x]\}.$$

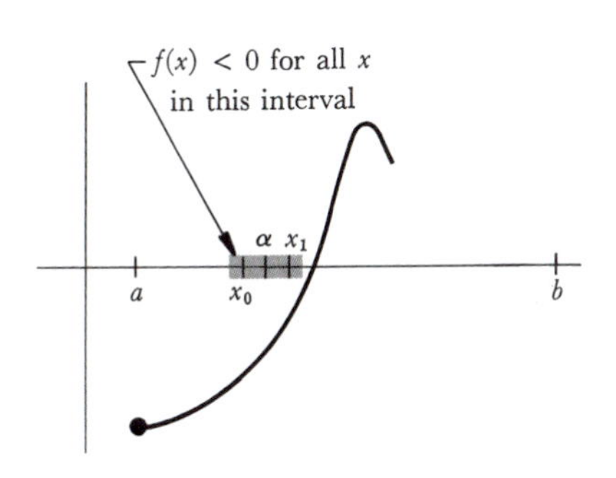

FIGURE 2

Clearly $A \neq \emptyset$, since a is in A; in fact, there is some $\delta > 0$ such that A contains all points x satisfying $a \leq x < a + \delta$; this follows from Problem 6-15, since f is continuous on $[a, b]$ and $f(a) < 0$. Similarly, b is an upper bound for A and, in fact, there is a $\delta > 0$ such that all points x satisfying $b - \delta < x \leq b$ are upper bounds for A; this also follows from Problem 6-15, since $f(b) > 0$.

From these remarks it follows that A has a least upper bound α and that $a < \alpha < b$. We now wish to show that $f(\alpha) = 0$, by eliminating the possibilities $f(\alpha) < 0$ and $f(\alpha) > 0$.

Suppose first that $f(\alpha) < 0$. By Theorem 6-3, there is a $\delta > 0$ such that $f(x) < 0$ for $\alpha - \delta < x < \alpha + \delta$ (Figure 2). Now there is some number x_0 in A which satisfies $\alpha - \delta < x_0 < \alpha$ (because otherwise α would not be the *least* upper bound of A). This means that f is negative on the whole interval $[a, x_0]$. But if x_1 is a number between α and $\alpha + \delta$, then f is also negative on the whole interval $[x_0, x_1]$. Therefore f is negative on the interval $[a, x_1]$, so x_1 is in A. But this contradicts the fact that α is an upper bound for A; our original assumption that $f(\alpha) < 0$ must be false.

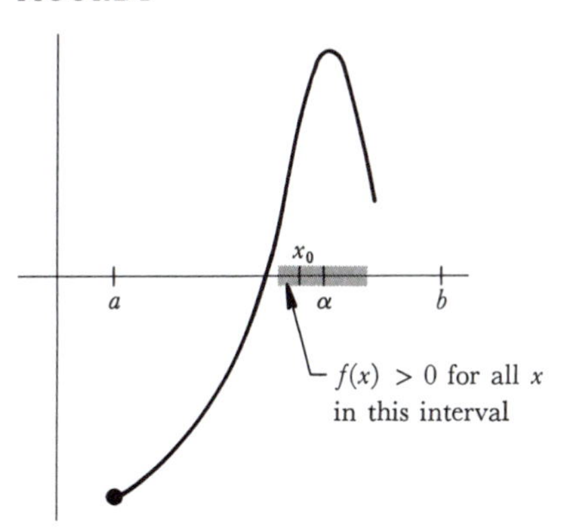

FIGURE 3

Suppose, on the other hand, that $f(\alpha) > 0$. Then there is a number $\delta > 0$ such that $f(x) > 0$ for $\alpha - \delta < x < \alpha + \delta$ (Figure 3). Once again we know that there is an x_0 in A satisfying $\alpha - \delta < x_0 < \alpha$; but this means that f is negative on $[a, x_0]$, which is impossible, since $f(x_0) > 0$. Thus the assumption $f(\alpha) > 0$ also leads to a contradiction, leaving $f(\alpha) = 0$ as the only possible alternative. ∎

The proofs of Theorems 2 and 3 of Chapter 7 require a simple preliminary result, which will play much the same role as Theorem 6-3 played in the previous proof.

THEOREM 1 If f is continuous at a, then there is a number $\delta > 0$ such that f is bounded above on the interval $(a - \delta, a + \delta)$ (see Figure 4).

PROOF Since $\lim_{x \to a} f(x) = f(a)$, there is, for every $\varepsilon > 0$, a $\delta > 0$ such that, for all x,

$$\text{if } |x - a| < \delta, \text{ then } |f(x) - f(a)| < \varepsilon.$$

It is only necessary to apply this statement to some particular ε (any one will do), for example, $\varepsilon = 1$. We conclude that there is a $\delta > 0$ such that, for all x,

$$\text{if } |x - a| < \delta, \text{ then } |f(x) - f(a)| < 1.$$

It follows, in particular, that if $|x - a| < \delta$, then $f(x) - f(a) < 1$. This completes the proof: on the interval $(a - \delta, a + \delta)$ the function f is bounded above by $f(a) + 1$. ∎

It should hardly be necessary to add that we can now also prove that f is bounded below on some interval $(a - \delta, a + \delta)$, and, finally, that f is bounded on some open interval containing a.

A more significant point is the observation that if $\lim_{x \to a^+} f(x) = f(a)$, then there

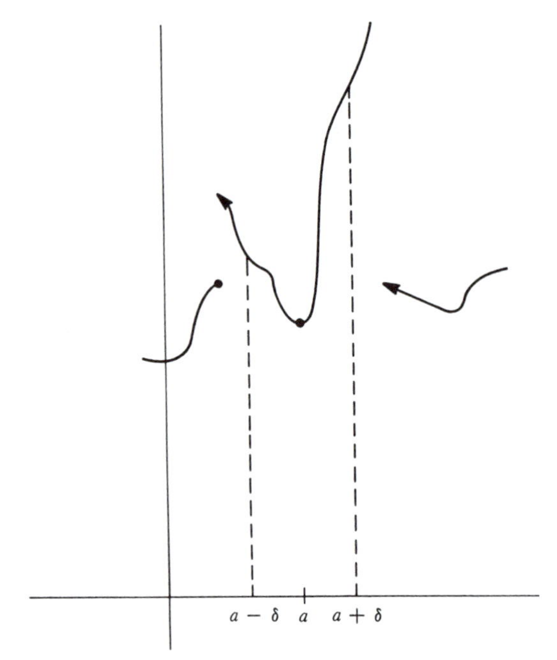

FIGURE 4

is a $\delta > 0$ such that f is bounded on the set $\{x : a \le x < a + \delta\}$, and a similar observation holds if $\lim_{x \to b^-} f(x) = f(b)$. Having made these observations (and assuming that you will supply the proofs), we tackle our second major theorem.

THEOREM 7-2 If f is continuous on $[a, b]$, then f is bounded above on $[a, b]$.

PROOF Let

$$A = \{x : a \le x \le b \text{ and } f \text{ is bounded above on } [a, x]\}.$$

Clearly $A \ne \emptyset$ (since a is in A), and A is bounded above (by b), so A has a least upper bound α. Notice that we are here applying the term "bounded above" both to the set A, which can be visualized as lying on the horizontal axis, and to f, i.e., to the sets $\{f(y) : a \le y \le x\}$, which can be visualized as lying on the vertical axis (Figure 5).

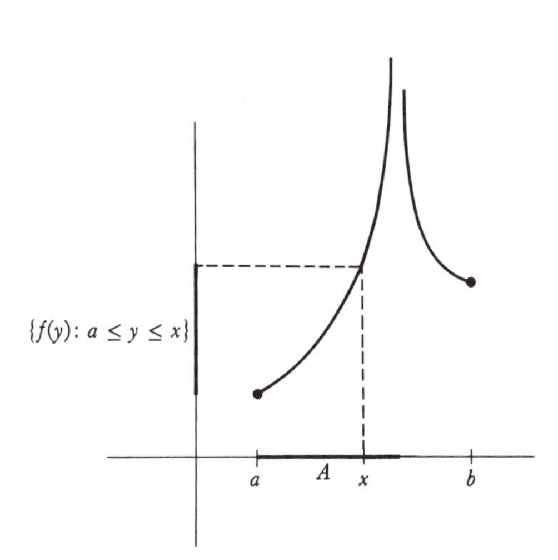

FIGURE 5

Our first step is to prove that we actually have $\alpha = b$. Suppose, instead, that $\alpha < b$. By Theorem 1 there is $\delta > 0$ such that f is bounded on $(\alpha - \delta, \alpha + \delta)$. Since α is the least upper bound of A there is some x_0 in A satisfying $\alpha - \delta < x_0 < \alpha$. This means that f is bounded on $[a, x_0]$. But if x_1 is any number with $\alpha < x_1 < \alpha + \delta$, then f is also bounded on $[x_0, x_1]$. Therefore f is bounded on $[a, x_1]$, so x_1 is in A, contradicting the fact that α is an upper bound for A. This contradiction shows that $\alpha = b$. One detail should be mentioned: this demonstration implicitly assumed that $a < \alpha$ [so that f would be defined on some interval $(\alpha - \delta, \alpha + \delta)$]; the possibility $a = \alpha$ can be ruled out similarly, using the existence of a $\delta > 0$ such that f is bounded on $\{x : a \le x < a + \delta\}$.

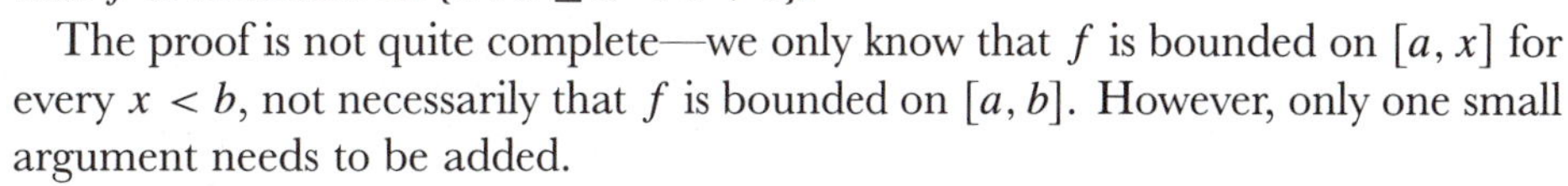

The proof is not quite complete—we only know that f is bounded on $[a, x]$ for every $x < b$, not necessarily that f is bounded on $[a, b]$. However, only one small argument needs to be added.

There is a $\delta > 0$ such that f is bounded on $\{x : b - \delta < x \le b\}$. There is x_0 in A such that $b - \delta < x_0 < b$. Thus f is bounded on $[a, x_0]$ and also on $[x_0, b]$, so f is bounded on $[a, b]$. ∎

To prove the third important theorem we resort to a trick.

THEOREM 7-3 If f is continuous on $[a, b]$, then there is a number y in $[a, b]$ such that $f(y) \ge f(x)$ for all x in $[a, b]$.

PROOF We already know that f is bounded on $[a, b]$, which means that the set

$$\{f(x) : x \text{ in } [a, b]\}$$

is bounded. This set is obviously not $\emptyset$, so it has a least upper bound α. Since $\alpha \ge f(x)$ for x in $[a, b]$ it suffices to show that $\alpha = f(y)$ for some y in $[a, b]$.

Suppose instead that $\alpha \ne f(y)$ for all y in $[a, b]$. Then the function g defined by

$$g(x) = \frac{1}{\alpha - f(x)}, \quad x \text{ in } [a, b]$$

is continuous on $[a, b]$, since the denominator of the right side is never 0. On the other hand, α is the least upper bound of $\{f(x) : x \text{ in } [a, b]\}$; this means that

$$\text{for every } \varepsilon > 0 \text{ there is } x \text{ in } [a, b] \text{ with } \alpha - f(x) < \varepsilon.$$

This, in turn, means that

$$\text{for every } \varepsilon > 0 \text{ there is } x \text{ in } [a, b] \text{ with } g(x) > 1/\varepsilon.$$

But *this* means that g is not bounded on $[a, b]$, contradicting the previous theorem. ▮

At the beginning of this chapter the set of natural numbers **N** was given as an example of an unbounded set. We are now going to *prove* that **N** is unbounded. After the difficult theorems proved in this chapter you may be startled to find such an "obvious" theorem winding up our proceedings. If so, you are, perhaps, allowing the geometrical picture of **R** to influence you too strongly. "Look," you may say, "the real numbers look like

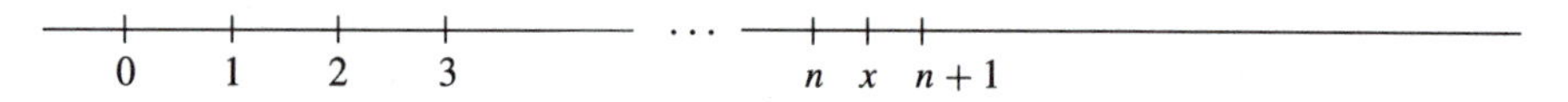

so every number x is between two integers n, $n + 1$ (unless x is itself an integer)." Basing the argument on a geometric picture is not a proof, however, and even the geometric picture contains an assumption: that if you place unit segments end-to-end you will eventually get a segment larger than any given segment. This axiom, often omitted from a first introduction to geometry, is usually attributed (not quite justly) to Archimedes, and the corresponding property for numbers, that **N** is not bounded, is called the *Archimedian property* of the real numbers. This property is *not* a consequence of P1–P12 (see reference [17] of the Suggested Reading), although it does hold for **Q**, of course. Once we have P13 however, there are no longer any problems.

THEOREM 2 **N** is not bounded above.

PROOF Suppose **N** were bounded above. Since $\mathbf{N} \neq \emptyset$, there would be a least upper bound α for **N**. Then

$$\alpha \geq n \qquad \text{for all } n \text{ in } \mathbf{N}.$$

Consequently,

$$\alpha \geq n + 1 \qquad \text{for all } n \text{ in } \mathbf{N},$$

since $n + 1$ is in **N** if n is in **N**. But this means that

$$\alpha - 1 \geq n \qquad \text{for all } n \text{ in } \mathbf{N},$$

and *this* means that $\alpha - 1$ is also an upper bound for **N**, contradicting the fact that α is the least upper bound. ▮

There is a consequence of Theorem 2 (actually an equivalent formulation) which we have very often assumed implicitly.

THEOREM 3 For any $\varepsilon > 0$ there is a natural number n with $1/n < \varepsilon$.

PROOF Suppose not; then $1/n \geq \varepsilon$ for all n in $\mathbf{N}$. Thus $n \leq 1/\varepsilon$ for all n in $\mathbf{N}$. But this means that $1/\varepsilon$ is an upper bound for $\mathbf{N}$, contradicting Theorem 2. ▮

A brief glance through Chapter 6 will show you that the result of Theorem 3 was used in the discussion of many examples. Of course, Theorem 3 was not available at the time, but the examples were so important that in order to give them some cheating was tolerated. As partial justification for this dishonesty we can claim that this result was never used in the proof of a *theorem*, but if your faith has been shaken, a review of all the proofs given so far is in order. Fortunately, such deception will not be necessary again. We have now stated every property of the real numbers that we will ever need. Henceforth, no more lies.

PROBLEMS

1. Find the least upper bound and the greatest lower bound (if they exist) of the following sets. Also decide which sets have greatest and least elements (i.e., decide when the least upper bound and greatest lower bound happens to belong to the set).

(i) $\left\{\frac{1}{n} : n \text{ in } \mathbf{N}\right\}$.

(ii) $\left\{\frac{1}{n} : n \text{ in } \mathbf{Z} \text{ and } n \neq 0.\right\}$.

(iii) $\{x : x = 0 \text{ or } x = 1/n \text{ for some } n \text{ in } \mathbf{N}\}$.

(iv) $\{x : 0 \leq x \leq \sqrt{2} \text{ and } x \text{ is rational}\}$.

(v) $\{x : x^2 + x + 1 \geq 0\}$.

(vi) $\{x : x^2 + x - 1 < 0\}$.

(vii) $\{x : x < 0 \text{ and } x^2 + x - 1 < 0\}$.

(viii) $\left\{\frac{1}{n} + (-1)^n : n \text{ in } \mathbf{N}\right\}$.

2. (a) Suppose $A \neq \emptyset$ is bounded below. Let $-A$ denote the set of all $-x$ for x in A. Prove that $-A \neq \emptyset$, that $-A$ is bounded above, and that $-\sup(-A)$ is the greatest lower bound of A.

(b) If $A \neq \emptyset$ is bounded below, let B be the set of all lower bounds of A. Show that $B \neq \emptyset$, that B is bounded above, and that $\sup B$ is the greatest lower bound of A.

3. Let f be a continuous function on $[a, b]$ with $f(a) < 0 < f(b)$.

(a) The proof of Theorem 1 showed that there is a smallest x in $[a, b]$ with $f(x) = 0$. Is there necessarily a second smallest x in $[a, b]$ with

$f(x) = 0$? Show that there is a largest x in $[a, b]$ with $f(x) = 0$. (Try to give an easy proof by considering a new function closely related to f.)

(b) The proof of Theorem 1 depended upon consideration of $A = \{x : a \le x \le b$ and f is negative on $[a, x]\}$. Give another proof of Theorem 1, which depends upon consideration of $B = \{x : a \le x \le b$ and $f(x) < 0\}$. Which point x in $[a, b]$ with $f(x) = 0$ will this proof locate? Give an example where the sets A and B are not the same.

***4.** (a) Suppose that f is continuous on $[a, b]$ and that $f(a) = f(b) = 0$. Suppose also that $f(x_0) > 0$ for some x_0 in $[a, b]$. Prove that there are numbers c and d with $a \le c < x_0 < d \le b$ such that $f(c) = f(d) = 0$, but $f(x) > 0$ for all x in (c, d). Hint: The previous problem can be used to good advantage.

(b) Suppose that f is continuous on $[a, b]$ and that $f(a) < f(b)$. Prove that there are numbers c and d with $a \le c < d \le b$ such that $f(c) = f(a)$ and $f(d) = f(b)$ and $f(a) < f(x) < f(d)$ for all x in (c, d).

5. (a) Suppose that $y - x > 1$. Prove that there is an integer k such that $x < k < y$. Hint: Let l by the largest integer satisfying $l \le x$, and consider $l + 1$.

(b) Suppose $x < y$. Prove that there is a rational number r such that $x < r < y$. Hint: If $1/n < y - x$, then $ny - nx > 1$. (Query: Why have parts (a) and (b) been postponed until this problem set?)

(c) Suppose that $r < s$ are rational numbers. Prove that there is an irrational number between r and s. Hint: As a start, you know that there is an irrational number between 0 and 1.

(d) Suppose that $x < y$. Prove that there is an irrational number between x and y. Hint: It is unnecessary to do any more work; this follows from (b) and (c).

***6.** A set A of real numbers is said to be **dense** if every open interval contains a point of A. For example, Problem 5 shows that the set of rational numbers and the set of irrational numbers are each dense.

(a) Prove that if f is continuous and $f(x) = 0$ for all numbers x in a dense set A, then $f(x) = 0$ for all x.

(b) Prove that if f and g are continuous and $f(x) = g(x)$ for all x in a dense set A, then $f(x) = g(x)$ for all x.

(c) If we assume instead that $f(x) \ge g(x)$ for all x in A, show that $f(x) \ge g(x)$ for all x. Can $\ge$ be replaced by $>$ throughout?

7. Prove that if f is continuous and $f(x + y) = f(x) + f(y)$ for all x and y, then there is a number c such that $f(x) = cx$ for all x. (This conclusion can be demonstrated simply by combining the results of two previous problems.) Point of information: There *do* exist *noncontinuous* functions f satisfying $f(x + y) = f(x) + f(y)$ for all x and y, but we cannot prove this now; in fact, this simple question involves ideas that are usually never mentioned in any undergraduate course. The Suggested Reading contains references.

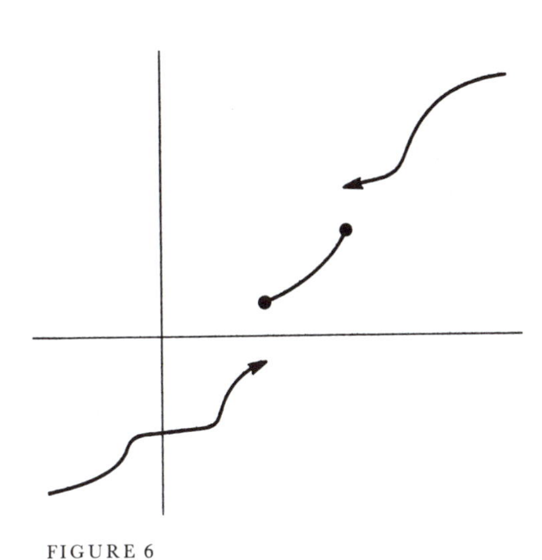

FIGURE 6

***8.** Suppose that f is a function such that $f(a) \le f(b)$ whenever $a < b$ (Figure 6).

(a) Prove that $\lim_{x \to a^-} f(x)$ and $\lim_{x \to a^+} f(x)$ both exist. Hint: Why is this problem in this chapter?

(b) Prove that f never has a removable discontinuity (this terminology comes from Problem 6-16).

(c) Prove that if f satisfies the conclusions of the Intermediate Value Theorem, then f is continuous.

***9.** If f is a bounded function on $[0, 1]$, let $|||f||| = \sup\{|f(x)| : x \text{ in } [0, 1]\}$. Prove analogues of the properties of $\|\ \|$ in Problem 7-14.

10. Suppose $\alpha > 0$. Prove that every number x can be written uniquely in the form $x = k\alpha + x'$, where k is an integer, and $0 \le x' < \alpha$.

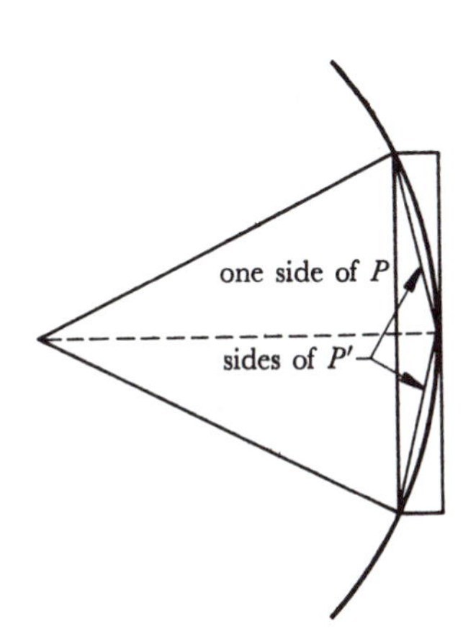

FIGURE 7

11. (a) Suppose that $a_1, a_2, a_3, \ldots$ is a sequence of positive numbers with $a_{n+1} \le a_n/2$. Prove that for any $\varepsilon > 0$ there is some n with $a_n < \varepsilon$.

(b) Suppose P is a regular polygon inscribed inside a circle. If P' is the inscribed regular polygon with twice as many sides, show that the difference between the area of the circle and the area of P' is less than half the difference between the area of the circle and the area of P (use Figure 7).

(c) Prove that there is a regular polygon P inscribed in a circle with area as close as desired to the area of the circle. In order to do part (c) you will need part (a). This was clear to the Greeks, who used part (a) as the basis for their entire treatment of proportion and area. By calculating the areas of polygons, this method ("the method of exhaustion") allows computations of π to any desired accuracy; Archimedes used it to show that $\frac{223}{71} < \pi < \frac{22}{7}$. But it has far greater theoretical importance:

*(d) Using the fact that the areas of two regular polygons with the same number of sides have the same ratio as the square of their sides, prove that the areas of two circles have the same ratios as the square of their radii. Hint: Deduce a contradiction from the assumption that the ratio of the areas is greater, or less, than the ratio of the square of the radii by inscribing appropriate polygons.

12. Suppose that A and B are two nonempty sets of numbers such that $x \le y$ for all x in A and all y in B.

(a) Prove that $\sup A \le y$ for all y in B.

(b) Prove that $\sup A \le \inf B$.

13. Let A and B be two nonempty sets of numbers which are bounded above, and let $A+B$ denote the set of all numbers $x+y$ with x in A and y in B. Prove that $\sup(A+B) = \sup A + \sup B$. Hint: The inequality $\sup(A+B) \le \sup A + \sup B$ is easy. Why? To prove that $\sup A + \sup B \le \sup(A+B)$ it suffices to prove that $\sup A + \sup B \le \sup(A+B) + \varepsilon$ for all $\varepsilon > 0$; begin by choosing x in A

and y in B with $\sup A - x < \varepsilon/2$ and $\sup B - y < \varepsilon/2$.

FIGURE 8

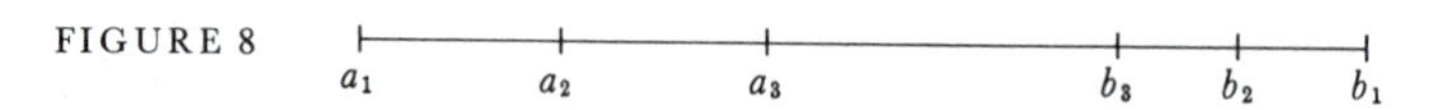

14. (a) Consider a sequence of closed intervals $I_1 = [a_1, b_1], I_2 = [a_2, b_2], \dots$. Suppose that $a_n \le a_{n+1}$ and $b_{n+1} \le b_n$ for all n (Figure 8). Prove that there is a point x which is in every I_n.

(b) Show that this conclusion is false if we consider open intervals instead of closed intervals.

The simple result of Problem 14(a) is called the "Nested Interval Theorem." It may be used to give alternative proofs of Theorems 1 and 2. The appropriate reasoning, outlined in the next two problems, illustrates a general method, called a "bisection argument."

***15.** Suppose f is continuous on $[a, b]$ and $f(a) < 0 < f(b)$. Then either $f((a+b)/2) = 0$, or f has different signs at the end points of the interval $[a, (a+b)/2]$, or f has different signs at the end points of $[(a+b)/2, b]$. Why? If $f((a+b)/2) \neq 0$, let I_1 be one of the two intervals on which f changes sign. Now bisect I_1. Either f is 0 at the midpoint, or f changes sign on one of the two intervals. Let I_2 be such an interval. Continue in this way, to define I_n for each n (unless f is 0 at some midpoint). Use the Nested Interval Theorem to find a point x where $f(x) = 0$.

***16.** Suppose f were continuous on $[a, b]$, but not bounded on $[a, b]$. Then f would be unbounded on either $[a, (a+b)/2]$ or $[(a+b)/2, b]$. Why? Let I_1 be one of these intervals on which f is unbounded. Proceed as in Problem 15 to obtain a contradiction.

17. (a) Let $A = \{x : x < \alpha\}$. Prove the following (they are all easy):

(i) If x is in A and $y < x$, then y is in A.
(ii) $A \neq \emptyset$.
(iii) $A \neq \mathbf{R}$.
(iv) If x is in A, then there is some number x' in A such that $x < x'$.

(b) Suppose, conversely, that A satisfies (i)–(iv). Prove that $A = \{x : x < \sup A\}$.

***18.** A number x is called an **almost upper bound** for A if there are only finitely many numbers y in A with $y \ge x$. An **almost lower bound** is defined similarly.

(a) Find all almost upper bounds and almost lower bounds of the sets in Problem 1.

(b) Suppose that A is a bounded infinite set. Prove that the set B of all almost upper bounds of A is nonempty, and bounded below.

(c) It follows from part (b) that inf B exists; this number is called the **limit superior** of A, and denoted by $\overline{\lim}\, A$ or $\limsup A$. Find $\overline{\lim}\, A$ for each set A in Problem 1.

(d) Define $\underline{\lim}\, A$, and find it for all A in Problem 1.

***19.** If A is a bounded infinite set prove

(a) $\underline{\lim}\, A \le \overline{\lim}\, A$.

(b) $\overline{\lim}\, A \le \sup A$.

(c) If $\overline{\lim}\, A < \sup A$, then A contains a largest element.

(d) The analogues of parts (b) and (c) for $\underline{\lim}$.

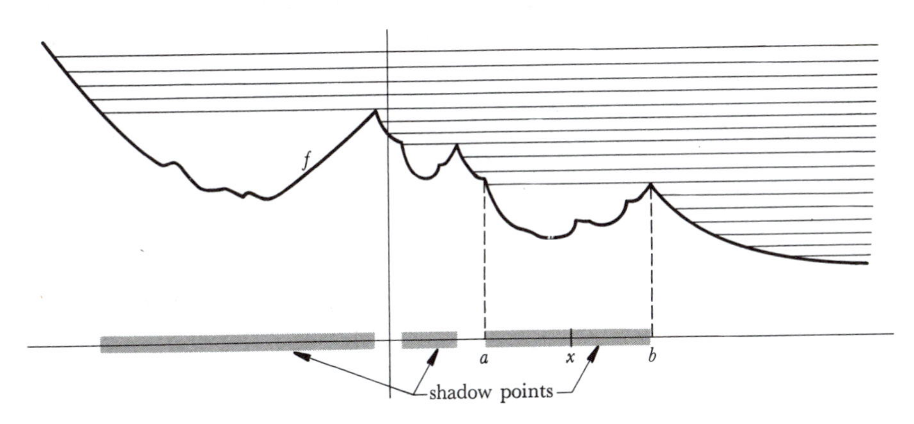

FIGURE 9

***20.** Let f be a continuous function on **R**. A point x is called a **shadow point** of f if there is a number $y > x$ with $f(y) > f(x)$. The rationale for this terminology is indicated in Figure 9; the parallel lines are the rays of the sun rising in the east (you are facing north). Suppose that all points of (a, b) are shadow points, but that a and b are not shadow points.

(a) For x in (a, b), prove that $f(x) \le f(b)$. Hint: Let $A = \{y : x \le y \le b$ and $f(x) \le f(y)\}$. If $\sup A$ were less than b, then $\sup A$ would be a shadow point. Use this fact to obtain a contradiction to the fact that b is not a shadow point.

(b) Now prove that $f(a) \le f(b)$. (This is a simple consequence of continuity.)

(c) Finally, using the fact that a is not a shadow point, prove that $f(a) = f(b)$.

This result is known as the Rising Sun Lemma. Aside from serving as a good illustration of the use of least upper bounds, it is instrumental in proving several beautiful theorems that do not appear in this book; see page 443.

APPENDIX. UNIFORM CONTINUITY

Now that we've come to the end of the "foundations," it might be appropriate to slip in one further fundamental concept. This notion is not used crucially in the rest of the book, but it can help clarify many points later on.

We know that the function $f(x) = x^2$ is continuous at a for all a. In other words,

> if a is any number, then for every $\varepsilon > 0$ there is some $\delta > 0$ such that, for all x, if $|x - a| < \delta$, then $|x^2 - a^2| < \varepsilon$.

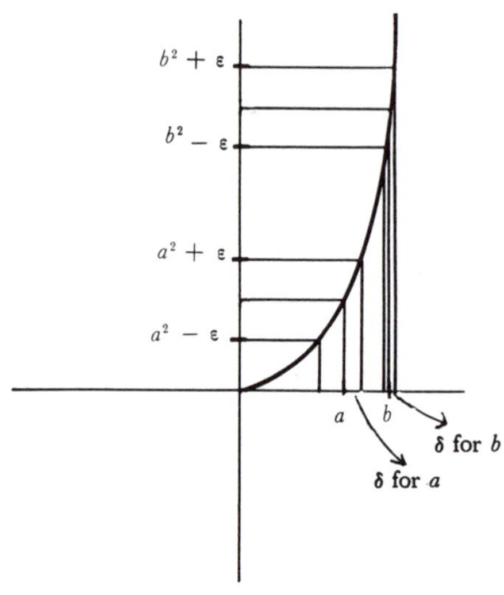

FIGURE 1

Of course, δ depends on ε. But δ *also depends* on a—the δ that works at a might not work at b (Figure 1). Indeed, it's clear that given $\varepsilon > 0$ there is no one $\delta > 0$ that works for all a, or even for all positive a. In fact, the number $a + \delta/2$ will certainly satisfy $|x - a| < \delta$, but if $a > 0$, then

$$\left|\left(a + \frac{\delta}{2}\right)^2 - a^2\right| = \left|a\delta + \frac{\delta^2}{4}\right| \geq a\delta,$$

and this won't be $< \varepsilon$ once $a > \varepsilon/\delta$. (This is just an admittedly confusing computational way of saying that f is growing faster and faster!)

On the other hand, for any $\varepsilon > 0$ there *will* be one $\delta > 0$ that works for all a in any interval $[-N, N]$. In fact, the δ which works at N or $-N$ will also work everywhere else in the interval.

As a final example, consider the function $f(x) = \sin 1/x$, or the function whose graph appears in Figure 18 on page 62. It is easy to see that, so long as $\varepsilon < 1$, there will not be one $\delta > 0$ that works for these functions at all points a in the open interval $(0, 1)$.

These examples illustrate important distinctions between the behavior of various continuous functions on certain intervals, and there is a special term to signal this distinction.

DEFINITION

> The function f is **uniformly continuous on an interval** A if for every $\varepsilon > 0$ there is some $\delta > 0$ such that, for all x and y in A,
>
> if $|x - y| < \delta$, then $|f(x) - f(y)| < \varepsilon$.

We've seen that a function can be continuous on the whole line, or on an open interval, without being uniformly continuous there. On the other hand, the function $f(x) = x^2$ did turn out to be uniformly continuous on any closed interval. This shouldn't be too surprising—it's the same sort of thing that occurs when we ask whether a function is bounded on an interval—and we would be led to suspect that any continuous function on a closed interval is also uniformly continuous on that interval. In order to prove this, we'll need to deal first with one subtle point.

Suppose that we have two intervals $[a, b]$ and $[b, c]$ with the common endpoint b, and a function f that is continuous on $[a, c]$. Let $\varepsilon > 0$ and suppose that the following two statements hold:

(i) if x and y are in $[a, b]$ and $|x - y| < \delta_1$, then $|f(x) - f(y)| < \varepsilon$,
(ii) if x and y are in $[b, c]$ and $|x - y| < \delta_2$, then $|f(x) - f(y)| < \varepsilon$.

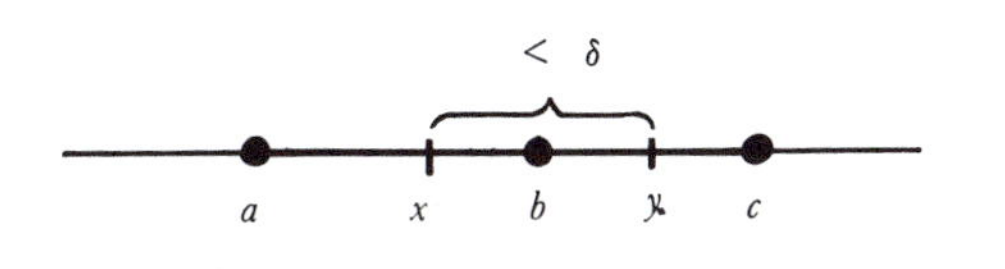

FIGURE 2

We'd like to know if there is some $\delta > 0$ such that $|f(x) - f(y)| < \varepsilon$ whenever x and y are points in $[a, c]$ with $|x - y| < \delta$. Our first inclination might be to choose δ as the minimum of δ_1 and δ_2. But it is easy to see what goes wrong (Figure 2): we might have x in $[a, b]$ and y in $[b, c]$, and then neither (i) nor (ii) tells us anything about $|f(x) - f(y)|$. So we have to be a little more cagey, and also use continuity of f at b.

LEMMA Let $a < b < c$ and let f be continuous on the interval $[a, c]$. Let $\varepsilon > 0$, and suppose that statements (i) and (ii) hold. Then there is a $\delta > 0$ such that,

$$\text{if } x \text{ and } y \text{ are in } [a, c] \text{ and } |x - y| < \delta, \text{ then } |f(x) - f(y)| < \varepsilon.$$

PROOF Since f is continuous at b, there is a $\delta_3 > 0$ such that,

$$\text{if } |x - b| < \delta_3, \text{ then } |f(x) - f(b)| < \frac{\varepsilon}{2}.$$

It follows that

$$\text{(iii)} \quad \text{if } |x - b| < \delta_3 \text{ and } |y - b| < \delta_3, \text{ then } |f(x) - f(y)| < \varepsilon.$$

Choose δ to be the minimum of δ_1, δ_2, and δ_3. We claim that this δ works. In fact, suppose that x and y are any two points in $[a, c]$ with $|x - y| < \delta$. If x and y are both in $[a, b]$, then $|f(x) - f(y)| < \varepsilon$ by (i); and if x and y are both in $[b, c]$, then $|f(x) - f(y)| < \varepsilon$ by (ii). The only other possibility is that

$$x < b < y \qquad \text{or} \qquad y < b < x.$$

In either case, since $|x - y| < \delta$, we also have $|x - b| < \delta$ and $|y - b| < \delta$. So $|f(x) - f(y)| < \varepsilon$ by (iii). ▮

THEOREM 1 If f is continuous on $[a, b]$, then f is uniformly continuous on $[a, b]$.

PROOF It's the usual trick, but we've got to be a little bit careful about the mechanism of the proof. For $\varepsilon > 0$ let's say that f is *ε-good* on $[a, b]$ if there is some $\delta > 0$ such that, for all y and z in $[a, b]$,

$$\text{if } |y - z| < \delta, \text{ then } |f(y) - f(z)| < \varepsilon.$$

Then we're trying to prove that f is ε-good on $[a, b]$ for all $\varepsilon > 0$.

Consider any particular $\varepsilon > 0$. Let

$$A = \{x : a \le x \le b \text{ and } f \text{ is } \varepsilon\text{-good on } [a, x]\}.$$

Then $A \neq \emptyset$ (since a is in A), and A is bounded above (by b), so A has a least upper bound α. We really should write α_ε, since A and α might depend on ε. But we won't since we intend to prove that $\alpha = b$, no matter what ε is.

Suppose that we had $\alpha < b$. Since f is continuous at α, there is some $\delta_0 > 0$ such that, if $|y - \alpha| < \delta_0$, then $|f(y) - f(\alpha)| < \varepsilon/2$. Consequently, if $|y - \alpha| < \delta_0$ and $|z - \alpha| < \delta_0$, then $|f(y) - f(z)| < \varepsilon$. So f is surely ε-good on the interval $[\alpha - \delta_0, \alpha + \delta_0]$. On the other hand, since α is the least upper bound of A, it is also clear that f is ε-good on $[a, \alpha - \delta_0]$. Then the Lemma implies that f is ε-good on $[a, a + \delta_0]$, so $a + \delta_0$ is in A, contradicting the fact that α is an upper bound.

To complete the proof we just have to show that $\alpha = b$ is actually in A. The argument for this is practically the same: Since f is continuous at b, there is some $\delta_0 > 0$ such that, if $|b - y| < \delta_0$, then $|f(y) - f(b)| < \varepsilon/2$. So f is ε-good on $[b - \delta_0, b]$. But f is also ε-good on $[a, b - \delta_0]$, so the Lemma implies that f is ε-good on $[a, b]$. ▮

PROBLEMS

1. (a) For which of the following values of α is the function $f(x) = x^\alpha$ uniformly continuous on $[0, \infty)$: $\alpha = 1/3, 1/2, 2, 3$?
(b) Find a function f that is continuous and bounded on $(0, 1]$, but not uniformly continuous on $(0, 1]$.
(c) Find a function f that is continuous and bounded on $[0, \infty)$ but which is not uniformly continuous on $[0, \infty)$.

2. (a) Prove that if f and g are uniformly continuous on A, then so is $f + g$.
(b) Prove that if f and g are uniformly continuous and bounded on A, then fg is uniformly continuous on A.
(c) Show that this conclusion does not hold if one of them isn't bounded.
(d) Suppose that f is uniformly continuous on A, that g is uniformly continuous on B, and that $f(x)$ is in B for all x in A. Prove that $g \circ f$ is uniformly continuous on A.

3. Use a "bisection argument" (page 140) to give another proof of Theorem 1.

4. Derive Theorem 7-2 as a consequence of Theorem 1.

PART 3
DERIVATIVES AND INTEGRALS

In 1604, at the height of
his scientific career, Galileo argued
that for a rectilinear motion
in which speed increases proportionally
to distance covered,
the law of motion should be
just that ($x = ct^2$)
which he had discovered
in the investigation of falling bodies.
Between 1695 and 1700
not a single one of the monthly issues
of Leipzig's Acta Eruditorum was published
without articles of Leibniz,
the Bernoulli brothers
or the Marquis de l'Hôpital treating,
with notation only slightly different from
that which we use today,
the most varied problems of
differential calculus, integral calculus
and the calculus of variations.
Thus in the space of almost precisely
one century
infinitesimal calculus or,
as we now call it in English,
The Calculus,
the calculating tool par excellence,
had been forged;
and nearly three centuries of
constant use have not completely dulled
this incomparable instrument.

NICHOLAS BOURBAKI

CHAPTER 9 DERIVATIVES

The derivative of a function is the first of the two major concepts of this section. Together with the integral, it constitutes the source from which calculus derives its particular flavor. While it is true that the concept of a function is fundamental, that you cannot do anything without limits or continuity, and that least upper bounds are essential, everything we have done until now has been preparation—if adequate, this section will be easier than the preceding ones—for the really exciting ideas to come, the powerful concepts that are truly characteristic of calculus.

Perhaps (some would say "certainly") the interest of the ideas to be introduced in this section stems from the intimate connection between the mathematical concepts and certain physical ideas. Many definitions, and even some theorems, may be described in terms of physical problems, often in a revealing way. In fact, the demands of physics were the original inspiration for these fundamental ideas of calculus, and we shall frequently mention the physical interpretations. But we shall always first define the ideas in precise mathematical form, and discuss their significance in terms of mathematical problems.

The collection of all functions exhibits such diversity that there is almost no hope of discovering any interesting general properties pertaining to all. Because continuous functions form such a restricted class, we might expect to find some nontrivial theorems pertaining to them, and the sudden abundance of theorems after Chapter 6 shows that this expectation is justified. But the most interesting and most powerful results about functions will be obtained only when we restrict our attention even further, to functions which have even greater claim to be called "reasonable," which are even better behaved than most continuous functions.

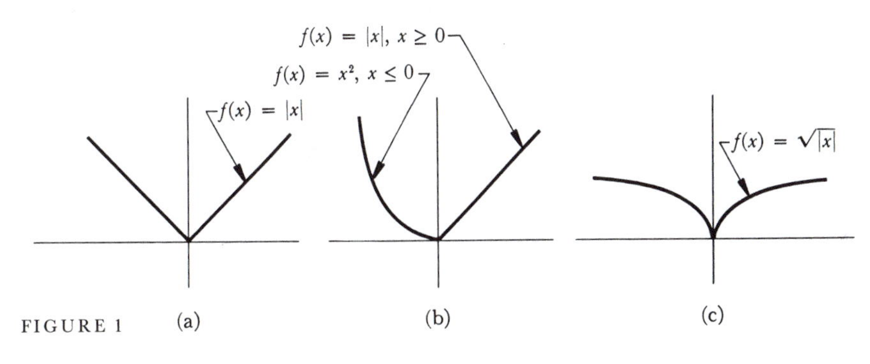

FIGURE 1

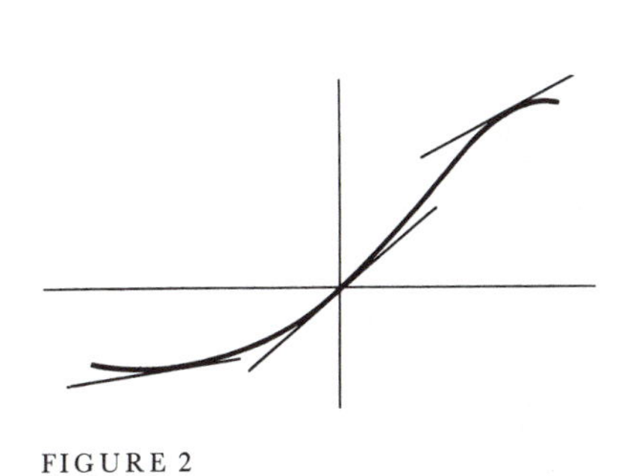

FIGURE 2

Figure 1 illustrates certain types of misbehavior which continuous functions can display. The graphs of these functions are "bent" at $(0, 0)$, unlike the graph of Figure 2, where it is possible to draw a "tangent line" at each point. The quotation marks have been used to avoid the suggestion that we have defined "bent" or

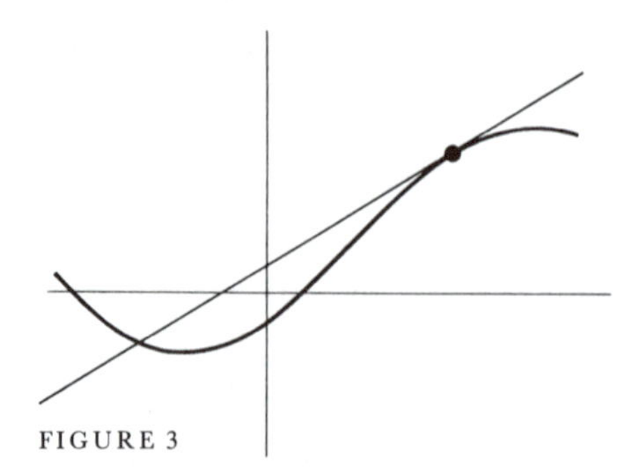

FIGURE 3

"tangent line," although we are suggesting that the graph might be "bent" at a point where a "tangent line" cannot be drawn. You have probably already noticed that a tangent line cannot be defined as a line which intersects the graph only once—such a definition would be both too restrictive and too permissive. With such a definition, the straight line shown in Figure 3 would not be a tangent line to the graph in that picture, while the parabola would have two tangent lines at each point (Figure 4), and the three functions in Figure 5 would have more than one tangent line at the points where they are "bent."

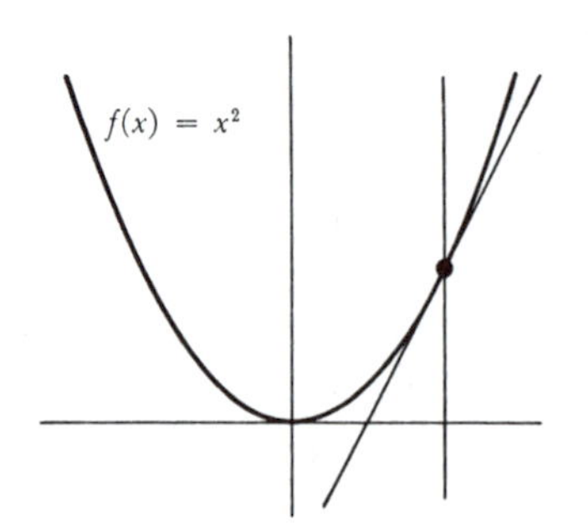

FIGURE 4

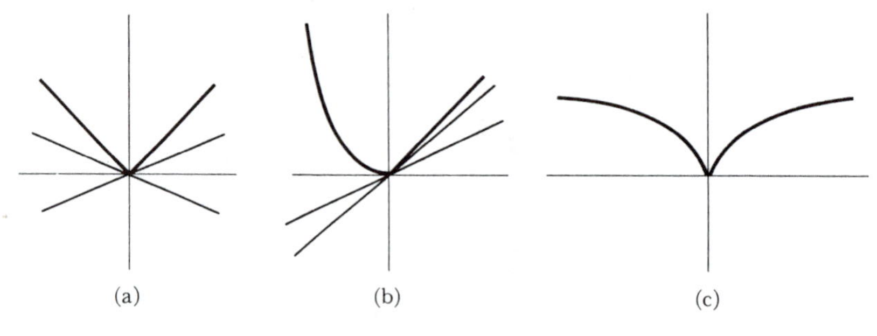

FIGURE 5

A more promising approach to the definition of a tangent line might start with "secant lines," and use the notion of limits. If $h \neq 0$, then the two distinct points $(a, f(a))$ and $(a+h, f(a+h))$ determine, as in Figure 6, a straight line whose slope is

$$\frac{f(a+h) - f(a)}{h}.$$

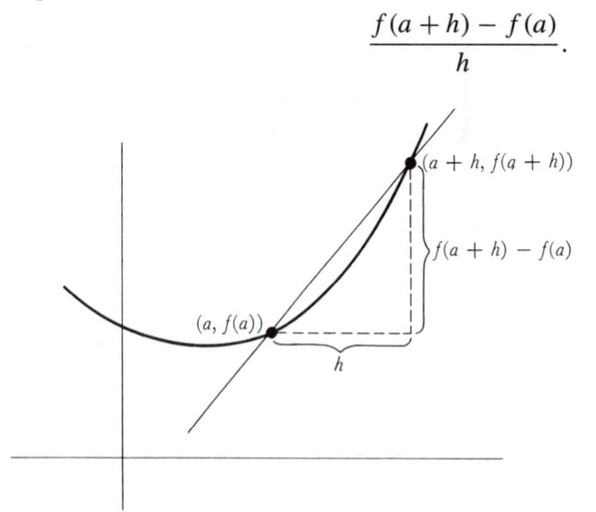

FIGURE 6

As Figure 7 illustrates, the "tangent line" at $(a, f(a))$ seems to be the limit, in some sense, of these "secant lines," as h approaches 0. We have never before talked about a "limit" of lines, but we *can* talk about the limit of their slopes: the slope of the tangent line through $(a, f(a))$ should be

$$\lim_{h \to 0} \frac{f(a+h) - f(a)}{h}.$$

We are ready for a definition, and some comments.

DEFINITION

The function f is **differentiable at a** if

$$\lim_{h \to 0} \frac{f(a+h) - f(a)}{h} \text{ exists.}$$

In this case the limit is denoted by $f'(a)$ and is called the **derivative of f at a.** (We also say that f is **differentiable** if f is differentiable at a for every a in the domain of f.)

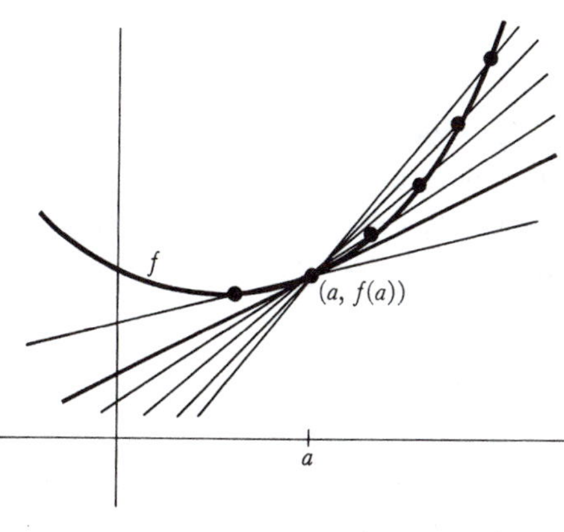

FIGURE 7

The first comment on our definition is really an addendum; we define the **tangent line** to the graph of f at $(a, f(a))$ to be the line through $(a, f(a))$ with slope $f'(a)$. This means that the tangent line at $(a, f(a))$ is defined only if f is differentiable at a.

The second comment refers to notation. The symbol $f'(a)$ is certainly reminiscent of functional notation. In fact, for any function f, we denote by f' the function whose domain is the set of all numbers a such that f is differentiable at a, and whose value at such a number a is

$$\lim_{h \to 0} \frac{f(a+h) - f(a)}{h}.$$

(To be very precise: f' is the collection of all pairs

$$\left(a, \lim_{h \to 0} \frac{f(a+h) - f(a)}{h}\right)$$

for which $\lim\limits_{h \to 0} [f(a+h) - f(a)]/h$ exists.) The function f' is called the **derivative** of f.

Our third comment, somewhat longer than the previous two, refers to the physical interpretation of the derivative. Consider a particle which is moving along a straight line (Figure 8(a)) on which we have chosen an "origin" point O, and a direction in which distances from O shall be written as positive numbers, the distance from O of points in the other direction being written as negative numbers. Let $s(t)$ denote the distance of the particle from O, at time t. The suggestive notation $s(t)$ has been chosen purposely; since a distance $s(t)$ is determined for each

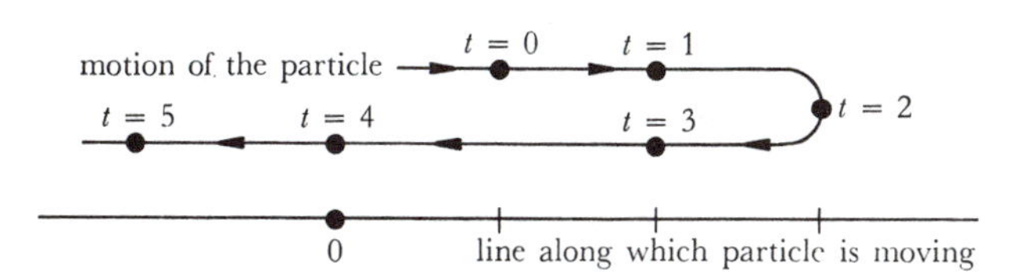

FIGURE 8(a)

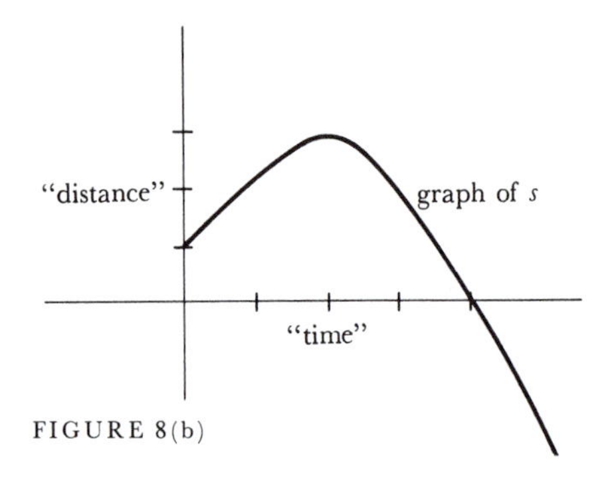

FIGURE 8(b)

number t, the physical situation automatically supplies us with a certain function s. The graph of s indicates the distance of the particle from O, on the vertical axis, in terms of the time, indicated on the horizontal axis (Figure 8(b)).

The quotient

$$\frac{s(a+h)-s(a)}{h}$$

has a natural physical interpretation. It is the "average velocity" of the particle during the time interval from a to $a+h$. For any particular a, this average speed depends on h, of course. On the other hand, the limit

$$\lim_{h\to 0}\frac{s(a+h)-s(a)}{h}$$

depends only on a (as well as the particular function s) and there are important physical reasons for considering this limit. We would like to speak of the "velocity of the particle at time a," but the usual definition of velocity is really a definition of average velocity; the only reasonable definition of "velocity at time a" (so-called "instantaneous velocity") is the limit

$$\lim_{h\to 0}\frac{s(a+h)-s(a)}{h}.$$

Thus we *define* the (**instantaneous**) **velocity** of the particle at a to be $s'(a)$. Notice that $s'(a)$ could easily be negative; the absolute value $|s'(a)|$ is sometimes called the (**instantaneous**) **speed**.

It is important to realize that instantaneous velocity is a theoretical concept, an abstraction which does not correspond precisely to any observable quantity. While it would not be fair to say that instantaneous velocity has nothing to do with average velocity, remember that $s'(t)$ is not

$$\frac{s(t+h)-s(t)}{h}$$

for any particular h, but merely the limit of these average velocities as h approaches 0. Thus, when velocities are measured in physics, what a physicist really measures is an average velocity over some (very small) time interval; such a procedure cannot be expected to give an exact answer, but this is really no defect, because physical measurements can never be exact anyway.

The velocity of a particle is often called the "rate of change of its position." This notion of the derivative, as a rate of change, applies to any other physical situation in which some quantity varies with time. For example, the "rate of change of mass" of a growing object means the derivative of the function m, where $m(t)$ is the mass at time t.

In order to become familiar with the basic definitions of this chapter, we will spend quite some time examining the derivatives of particular functions. Before proving the important theoretical results of Chapter 11, we want to have a good idea of what the derivative of a function looks like. The next chapter is devoted exclusively to one aspect of this problem—calculating the derivative of complicated functions. In this chapter we will emphasize the concepts, rather than the calculations, by considering a few simple examples. Simplest of all is a constant function, $f(x) = c$. In this case

$$\lim_{h\to 0} \frac{f(a+h) - f(a)}{h} = \lim_{h\to 0} \frac{c - c}{h} = 0.$$

Thus f is differentiable at a for every number a, and $f'(a) = 0$. This means that the tangent line to the graph of f always has slope 0, so the tangent line always coincides with the graph.

Constant functions are not the only ones whose graphs coincide with their tangent lines—this happens for any linear function $f(x) = cx + d$. Indeed

$$\begin{aligned} f'(a) &= \lim_{h\to 0} \frac{f(a+h) - f(a)}{h} \\ &= \lim_{h\to 0} \frac{c(a+h) + d - [ca + d]}{h} \\ &= \lim_{h\to 0} \frac{ch}{h} = c; \end{aligned}$$

the slope of the tangent line is c, the same as the slope of the graph of f.

A refreshing difference occurs for $f(x) = x^2$. Here

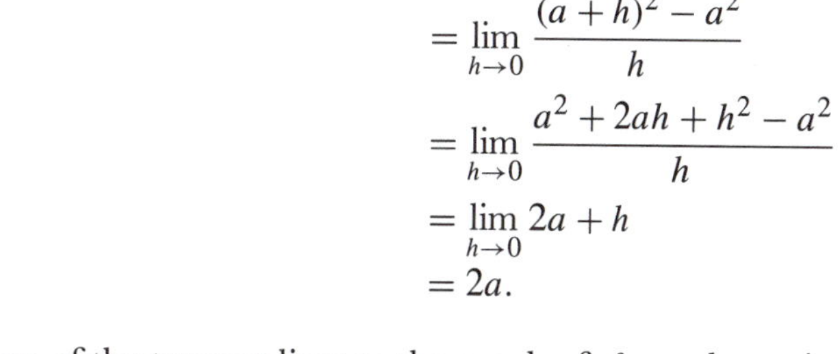

$$\begin{aligned} f'(a) &= \lim_{h\to 0} \frac{f(a+h) - f(a)}{h} \\ &= \lim_{h\to 0} \frac{(a+h)^2 - a^2}{h} \\ &= \lim_{h\to 0} \frac{a^2 + 2ah + h^2 - a^2}{h} \\ &= \lim_{h\to 0} 2a + h \\ &= 2a. \end{aligned}$$

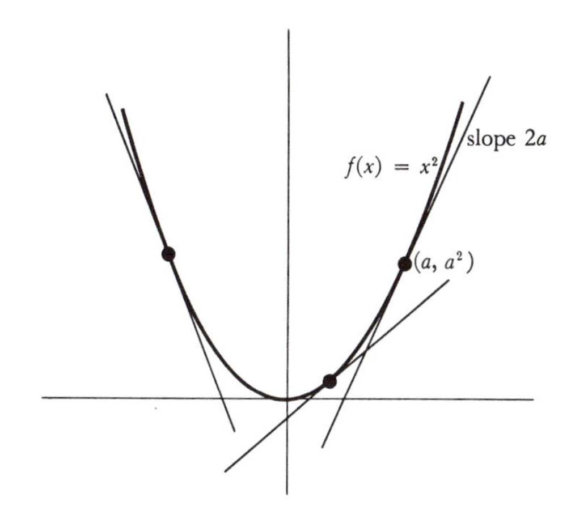

FIGURE 9

Some of the tangent lines to the graph of f are shown in Figure 9. In this picture each tangent line appears to intersect the graph only once, and this fact can be checked fairly easily: Since the tangent line through (a, a^2) has slope $2a$, it is the graph of the function

$$\begin{aligned} g(x) &= 2a(x - a) + a^2 \\ &= 2ax - a^2. \end{aligned}$$

Now, if the graphs of f and g intersect at a point $(x, f(x)) = (x, g(x))$, then

$$\begin{aligned} & x^2 = 2ax - a^2 \\ \text{or}\quad & x^2 - 2ax + a^2 = 0; \\ \text{so}\quad & (x - a)^2 = 0 \\ \text{or}\quad & x = a. \end{aligned}$$

In other words, (a, a^2) is the only point of intersection.

The function $f(x) = x^2$ happens to be quite special in this regard; usually a tangent line will intersect the graph more than once. Consider, for example, the function $f(x) = x^3$. In this case

$$\begin{aligned} f'(a) &= \lim_{h\to 0} \frac{f(a+h)-f(a)}{h} \\ &= \lim_{h\to 0} \frac{(a+h)^3 - a^3}{h} \\ &= \lim_{h\to 0} \frac{a^3 + 3a^2h + 3ah^2 + h^3 - a^3}{h} \\ &= \lim_{h\to 0} \frac{3a^2h + 3ah^2 + h^3}{h} \\ &= \lim_{h\to 0} 3a^2 + 3ah + h^2 \\ &= 3a^2. \end{aligned}$$

Thus the tangent line to the graph of f at (a, a^3) has slope $3a^2$. This means that the tangent line is the graph of

$$\begin{aligned} g(x) &= 3a^2(x-a) + a^3 \\ &= 3a^2x - 2a^3. \end{aligned}$$

The graphs of f and g intersect at the point $(x, f(x)) = (x, g(x))$ when

$$\begin{aligned} & x^3 = 3a^2x - 2a^3 \\ \text{or} \quad & x^3 - 3a^2x + 2a^3 = 0. \end{aligned}$$

This equation is easily solved if we remember that one solution of the equation has got to be $x = a$, so that $(x - a)$ is a factor of the left side; the other factor can then be found by dividing. We obtain

$$(x-a)(x^2 + ax - 2a^2) = 0.$$

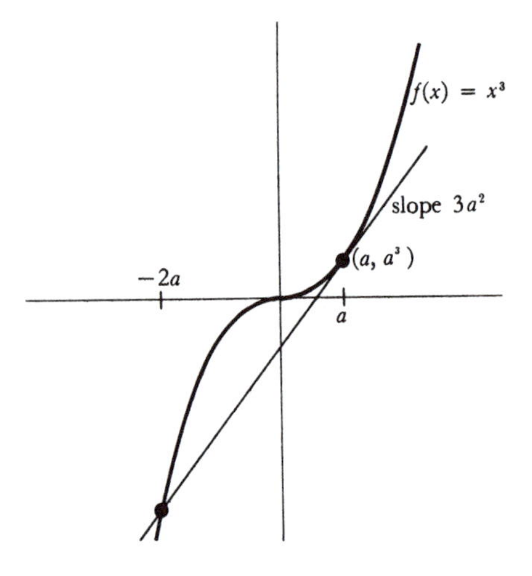

FIGURE 10

It so happens that $x^2 + ax - 2a^2$ also has $x - a$ as a factor; we obtain finally

$$(x-a)(x-a)(x+2a) = 0.$$

Thus, as illustrated in Figure 10, the tangent line through (a, a^3) also intersects the graph at the point $(-2a, -8a^3)$. These two points are always distinct, except when $a = 0$.

We have already found the derivative of sufficiently many functions to illustrate the classical, and still very popular, notation for derivatives. For a given function f, the derivative f' is often denoted by

$$\frac{df(x)}{dx}.$$

For example, the symbol

$$\frac{dx^2}{dx}$$

denotes the derivative of the function $f(x) = x^2$. Needless to say, the separate parts of the expression

$$\frac{df(x)}{dx}$$

are not supposed to have any sort of independent existence—the d's are *not* numbers, they *cannot* be canceled, and the entire expression is *not* the quotient of two other numbers "$df(x)$" and "dx." This notation is due to Leibniz (generally considered an independent co-discoverer of calculus, along with Newton), and is affectionately referred to as Leibnizian notation.* Although the notation $df(x)/dx$ seems very complicated, in concrete cases it may be shorter; after all, the symbol dx^2/dx is actually more concise than the phrase "the derivative of the function $f(x) = x^2$."

The following formulas state in standard Leibnizian notation all the information that we have found so far:

$$\begin{aligned} \frac{dc}{dx} &= 0, \\ \frac{d(ax+b)}{dx} &= a, \\ \frac{dx^2}{dx} &= 2x, \\ \frac{dx^3}{dx} &= 3x^2. \end{aligned}$$

Although the meaning of these formulas is clear enough, attempts at literal interpretation are hindered by the reasonable stricture that an equation should not contain a function on one side and a number on the other. For example, if the third equation is to be true, then either $df(x)/dx$ must denote $f'(x)$, rather than f', or else $2x$ must denote, not a number, but the function whose value at x is $2x$. It is really impossible to assert that one or the other of these alternatives is intended; in practice $df(x)/dx$ sometimes means f' and sometimes means $f'(x)$, while $2x$ may denote either a number or a function. Because of this ambiguity, most authors are reluctant to denote $f'(a)$ by

$$\frac{df(x)}{dx}(a);$$

instead $f'(a)$ is usually denoted by the barbaric, but unambiguous, symbol

$$\left.\frac{df(x)}{dx}\right|_{x=a}$$

*Leibniz was led to this symbol by his intuitive notion of the derivative, which he considered to be, not the limit of quotients $[f(x+h) - f(x)]/h$, but the "value" of this quotient when h is an "infinitely small" number. This "infinitely small" quantity was denoted by dx and the corresponding "infinitely small" difference $f(x+dx) - f(x)$ by $df(x)$. Although this point of view is impossible to reconcile with properties (P1)–(P13) of the real numbers, some people find this notion of the derivative congenial.

In addition to these difficulties, Leibnizian notation is associated with one more ambiguity. Although the notation dx^2/dx is absolutely standard, the notation $df(x)/dx$ is often replaced by df/dx. This, of course, is in conformity with the practice of confusing a function with its value at x. So strong is this tendency that functions are often indicated by a phrase like the following: "consider the function $y = x^2$." We will sometimes follow classical practice to the extent of using y as the name of a function, but we will nevertheless carefully distinguish between the function and its values—thus we will always say something like "consider the function (defined by) $y(x) = x^2$."

Despite the many ambiguities of Leibnizian notation, it is used almost exclusively in older mathematical writing, and is still used very frequently today. The staunchest opponents of Leibnizian notation admit that it will be around for quite some time, while its most ardent admirers would say that it will be around forever, and a good thing too! In any case, Leibnizian notation cannot be ignored completely.

The policy adopted in this book is to disallow Leibnizian notation within the text, but to include it in the Problems; several chapters contain a few (immediately recognizable) problems which are expressly designed to illustrate the vagaries of Leibnizian notation. Trusting that these problems will provide ample practice in this notation, we return to our basic task of examining some simple examples of derivatives.

The few functions examined so far have all been differentiable. To fully appreciate the significance of the derivative it is equally important to know some examples of functions which are *not* differentiable. The obvious candidates are the three functions first discussed in this chapter, and illustrated in Figure 1; if they turn out to be differentiable at 0 something has clearly gone wrong.

Consider first $f(x) = |x|$. In this case

$$\frac{f(0+h) - f(0)}{h} = \frac{|h|}{h}.$$

Now $|h|/h = 1$ for $h > 0$, and $|h|/h = -1$ for $h < 0$. This shows that

$$\lim_{h \to 0} \frac{f(h) - f(0)}{h} \text{ does not exist.}$$

In fact,

$$\lim_{h \to 0^+} \frac{f(h) - f(0)}{h} = 1$$

$$\text{and} \quad \lim_{h \to 0^-} \frac{f(h) - f(0)}{h} = -1.$$

(These two limits are sometimes called the **right-hand derivative** and the **left-hand derivative**, respectively, of f at 0.)

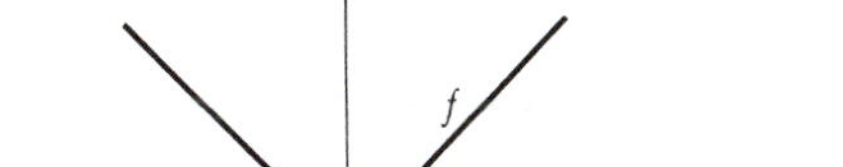

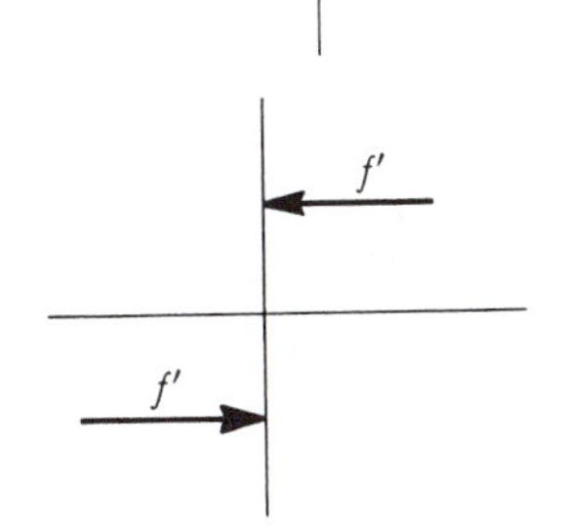

FIGURE 11

If $a \neq 0$, then $f'(a)$ does exist. In fact,

$$\begin{aligned} f'(x) &= 1 \qquad \text{if } x > 0, \\ f'(x) &= -1 \qquad \text{if } x < 0. \end{aligned}$$

The proof of this fact is left to you (it is easy if you remember the derivative of a linear function). The graphs of f and of f' are shown in Figure 11.

For the function

$$f(x) = \begin{cases} x^2, & x \leq 0 \\ x, & x \geq 0, \end{cases}$$

a similar difficulty arises in connection with $f'(0)$. We have

$$\frac{f(h) - f(0)}{h} = \begin{cases} \dfrac{h^2}{h} = h, & h < 0 \\ \dfrac{h}{h} = 1, & h > 0. \end{cases}$$

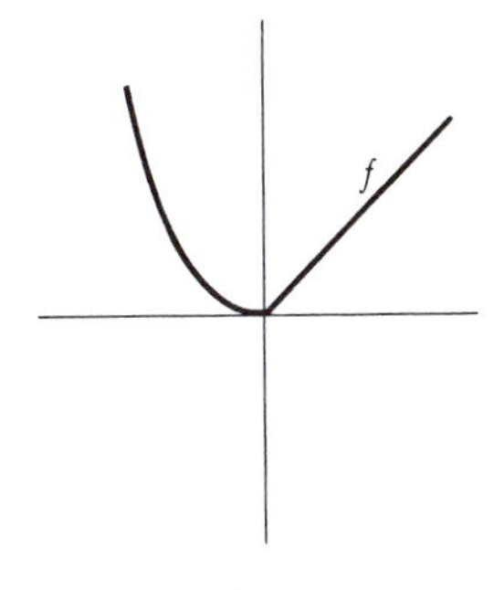

Therefore,

$$\lim_{h \to 0^-} \frac{f(h) - f(0)}{h} = 0,$$

$$\text{but} \quad \lim_{h \to 0^+} \frac{f(h) - f(0)}{h} = 1.$$

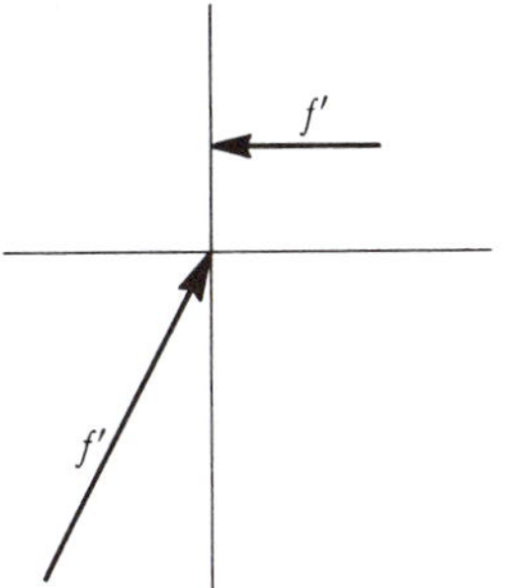

FIGURE 12

Thus $f'(0)$ does not exist; f is not differentiable at 0. Once again, however, $f'(x)$ exists for $x \neq 0$—it is easy to see that

$$f'(x) = \begin{cases} 2x, & x < 0 \\ 1, & x > 0. \end{cases}$$

The graphs of f and f' are shown in Figure 12.

Even worse things happen for $f(x) = \sqrt{|x|}$. For this function

$$\frac{f(h) - f(0)}{h} = \begin{cases} \dfrac{\sqrt{h}}{h} = \dfrac{1}{\sqrt{h}}, & h > 0 \\ \dfrac{\sqrt{-h}}{h} = -\dfrac{1}{\sqrt{-h}}, & h < 0. \end{cases}$$

In this case the right-hand limit

$$\lim_{h \to 0^+} \frac{f(h) - f(0)}{h} = \lim_{h \to 0^+} \frac{1}{\sqrt{h}}$$

does not exist; instead $1/\sqrt{h}$ becomes arbitrarily large as h approaches 0. And, what's more, $-1/\sqrt{-h}$ becomes arbitrarily large in absolute value, but *negative* (Figure 13).

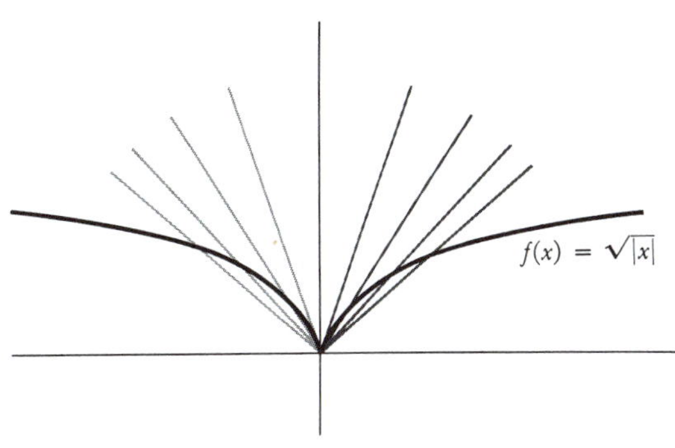

FIGURE 13

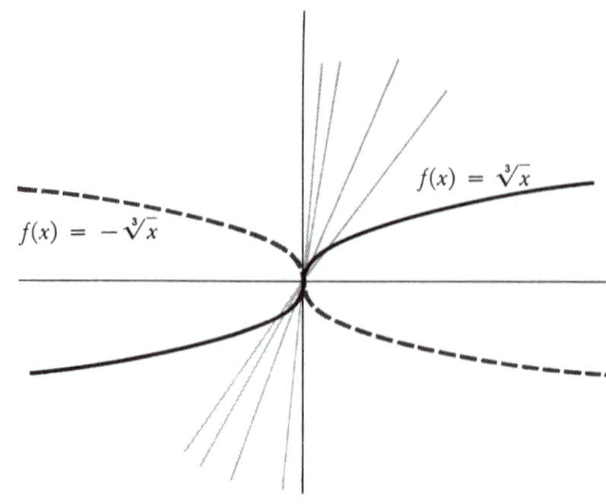

FIGURE 14

The function $f(x) = \sqrt[3]{x}$, although not differentiable at 0, is at least a little better behaved than this. The quotient

$$\frac{f(h)-f(0)}{h} = \frac{\sqrt[3]{h}}{h} = \frac{h^{1/3}}{h} = \frac{1}{h^{2/3}} = \frac{1}{(\sqrt[3]{h})^2}$$

simply becomes arbitrarily large as h goes to 0. Sometimes one says that f has an "infinite" derivative at 0. Geometrically this means that the graph of f has a "tangent line" which is parallel to the vertical axis (Figure 14). Of course, $f(x) = -\sqrt[3]{x}$ has the same geometric property, but one would say that f has a derivative of "negative infinity" at 0.

Remember that differentiability is supposed to be an improvement over mere continuity. This idea is supported by the many examples of functions which are continuous, but not differentiable; however, one important point remains to be noted:

THEOREM 1 If f is differentiable at a, then f is continuous at a.

PROOF

$$\begin{aligned}\lim_{h\to 0} f(a+h)-f(a) &= \lim_{h\to 0}\frac{f(a+h)-f(a)}{h}\cdot h\\ &= \lim_{h\to 0}\frac{f(a+h)-f(a)}{h}\cdot \lim_{h\to 0} h\\ &= f'(a)\cdot 0\\ &= 0.\end{aligned}$$

As we pointed out in Chapter 5, the equation $\lim_{h\to 0} f(a+h)-f(a) = 0$ is equivalent to $\lim_{x\to a} f(x) = f(a)$; thus f is continuous at a. ▮

It is very important to remember Theorem 1, and just as important to remember that the converse is not true. A differentiable function is continuous, but a continuous function need not be differentiable (keep in mind the function $f(x) = |x|$, and you will never forget which statement is true and which false).

The continuous functions examined so far have been differentiable at all points with at most one exception, but it is easy to give examples of continuous functions which are not differentiable at several points, even an infinite number (Figure 15). Actually, one can do much worse than this. There is a function which is *continuous*

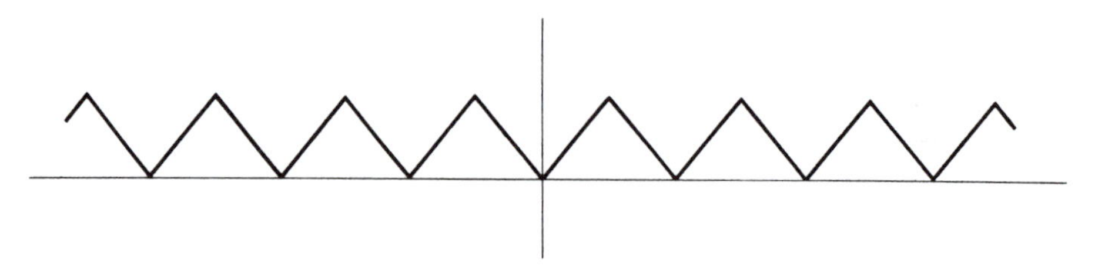
FIGURE 15

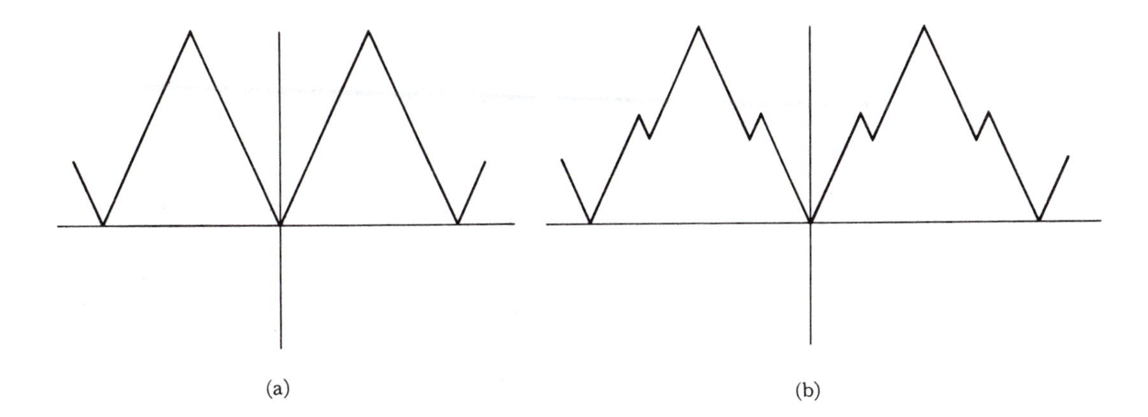

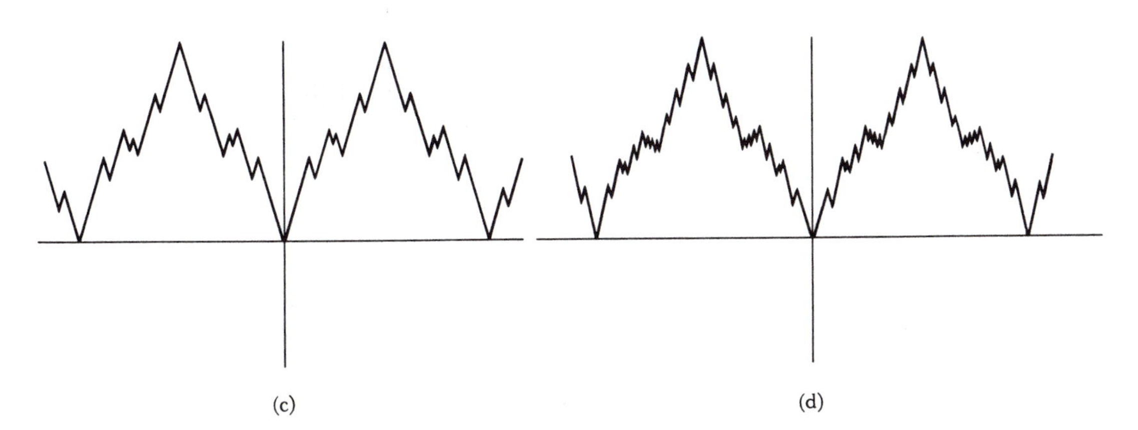

FIGURE 16

everywhere and *differentiable nowhere!* Unfortunately, the definition of this function will be inaccessible to us until Chapter 24, and I have been unable to persuade the artist to draw it (consider carefully what the graph should look like and you will sympathize with her point of view). It is possible to draw some rough approximations to the graph, however; several successively better approximations are shown in Figure 16.

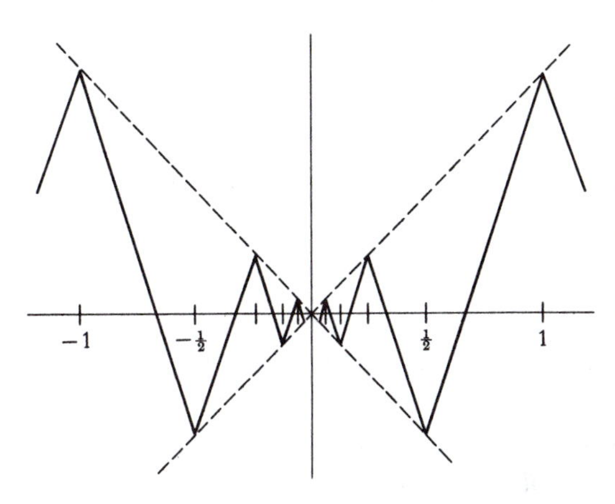

FIGURE 17

Although such spectacular examples of nondifferentiability must be postponed, we can, with a little ingenuity, find a continuous function which is not differentiable at infinitely many points, *all of which are in* $[0, 1]$. One such function is illustrated in Figure 17. The reader is given the problem of defining it precisely; it is a straight line version of the function

$$f(x) = \begin{cases} x \sin \dfrac{1}{x}, & x \neq 0 \\ 0, & x = 0. \end{cases}$$

This particular function f is itself quite sensitive to the question of differentiability. Indeed, for $h \neq 0$ we have

$$\frac{f(h)-f(0)}{h}=\frac{h\sin\dfrac{1}{h}-0}{h}=\sin\frac{1}{h}.$$

Long ago we proved that $\lim_{h\to 0}\sin 1/h$ does not exist, so f is not differentiable at 0. Geometrically, one can see that a tangent line cannot exist, by noting that the secant line through $(0,0)$ and $(h,f(h))$ in Figure 18 can have any slope between -1 and 1, no matter how small we require h to be.

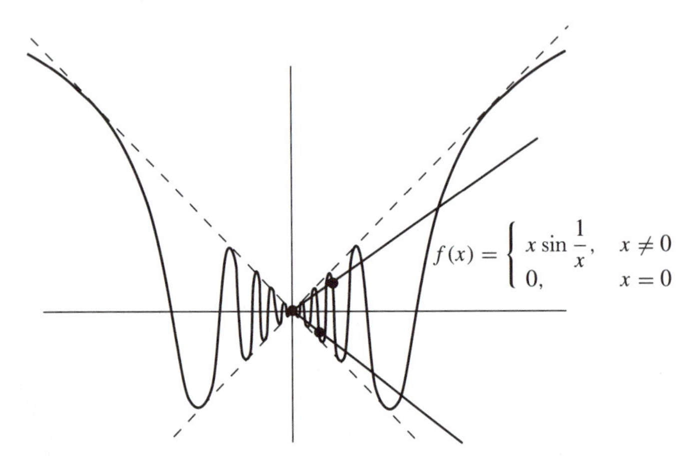

FIGURE 18

This finding represents something of a triumph; although continuous, the function f seems somehow quite unreasonable, and we can now enunciate one mathematically undesirable feature of this function—it is not differentiable at 0. Nevertheless, one should not become too enthusiastic about the criterion of differentiability. For example, the function

$$g(x)=\begin{cases} x^2\sin\dfrac{1}{x}, & x\neq 0\\ 0, & x=0\end{cases}$$

is differentiable at 0; in fact $g'(0)=0$:

$$\begin{aligned}\lim_{h\to 0}\frac{g(h)-g(0)}{h}&=\lim_{h\to 0}\frac{h^2\sin\dfrac{1}{h}}{h}\\ &=\lim_{h\to 0}h\sin\frac{1}{h}\\ &=0.\end{aligned}$$

The tangent line to the graph of g at $(0,0)$ is therefore the horizontal axis (Figure 19).

This example suggests that we should seek even more restrictive conditions on a function than mere differentiability. We can actually use the derivative to formulate such conditions if we introduce another set of definitions, the last of this chapter.

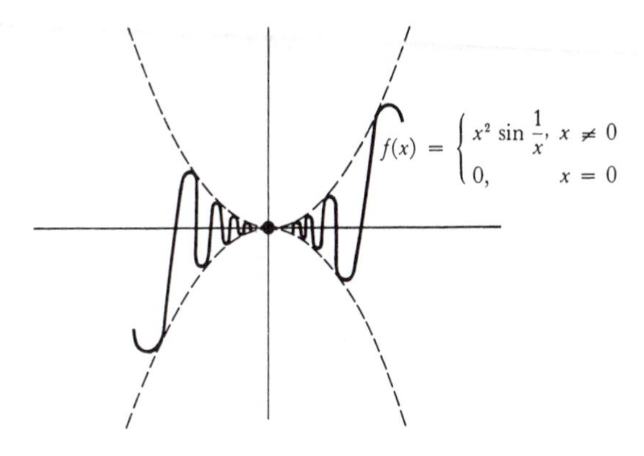

FIGURE 19

For any function f, we obtain, by taking the derivative, a new function f' (whose domain may be considerably smaller than that of f). The notion of differentiability can be applied to the function f', of course, yielding another function $(f')'$, whose domain consists of all points a such that f' is differentiable at a. The function $(f')'$ is usually written simply f'' and is called the **second derivative** of f. If $f''(a)$ exists, then f is said to be 2-times differentiable at a, and the number $f''(a)$ is called the **second derivative of f at a.**

In physics the second derivative is particularly important. If $s(t)$ is the position at time t of a particle moving along a straight line, then $s''(t)$ is called the **acceleration** at time t. Acceleration plays a special role in physics, because, as stated in Newton's laws of motion, the force on a particle is the product of its mass and its acceleration. Consequently you can feel the second derivative when you sit in an accelerating car.

There is no reason to stop at the second derivative—we can define $f''' = (f'')'$, $f'''' = (f''')'$, etc. This notation rapidly becomes unwieldy, so the following abbreviation is usually adopted (it is really a recursive definition):

$$\begin{aligned} f^{(1)} &= f', \\ f^{(k+1)} &= (f^{(k)})'. \end{aligned}$$

Thus

$$\begin{aligned} f^{(1)} &= f' \\ f^{(2)} &= f'' = (f')', \\ f^{(3)} &= f''' = (f'')', \\ f^{(4)} &= f'''' = (f''')', \\ &\text{etc.} \end{aligned}$$

The various functions $f^{(k)}$, for $k \geq 2$, are sometimes called **higher-order derivatives** of f.

Usually, we resort to the notation $f^{(k)}$ only for $k \geq 4$, but it is convenient to have $f^{(k)}$ defined for smaller k also. In fact, a reasonable definition can be made for $f^{(0)}$, namely,

$$f^{(0)} = f.$$

Leibnizian notation for higher-order derivatives should also be mentioned. The natural Leibnizian symbol for $f''(x)$, namely,

$$\frac{d\left(\dfrac{df(x)}{dx}\right)}{dx},$$

is abbreviated to

$$\frac{d^2f(x)}{(dx)^2}, \qquad \text{or more frequently to} \qquad \frac{d^2f(x)}{dx^2}$$

Similar notation is used for $f^{(k)}(x)$.

The following example illustrates the notation $f^{(k)}$, and also shows, in one very simple case, how various higher-order derivatives are related to the original function. Let $f(x) = x^2$. Then, as we have already checked,

$$\begin{aligned} f'(x) &= 2x, \\ f''(x) &= 2, \\ f'''(x) &= 0, \\ f^{(k)}(x) &= 0, \qquad \text{if } k \geq 3. \end{aligned}$$

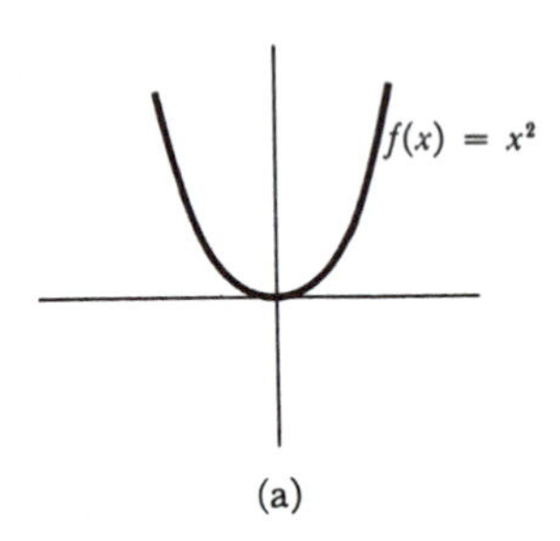

(a)

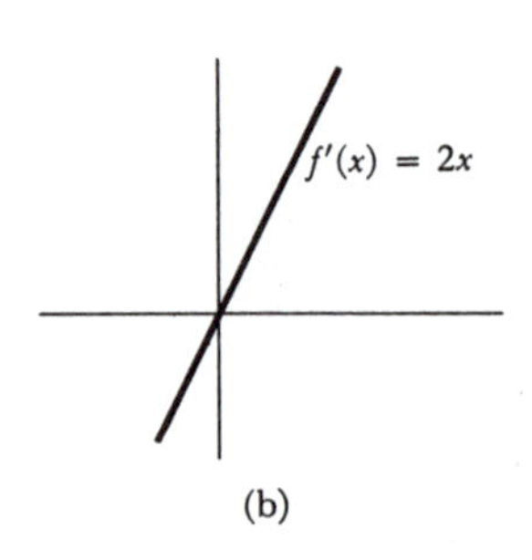

(b)

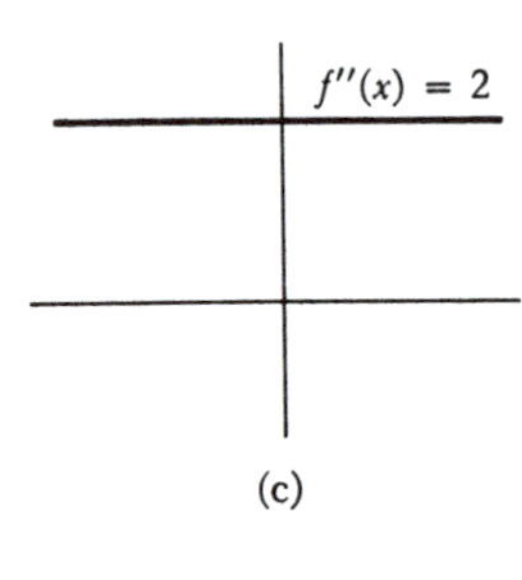

(c)

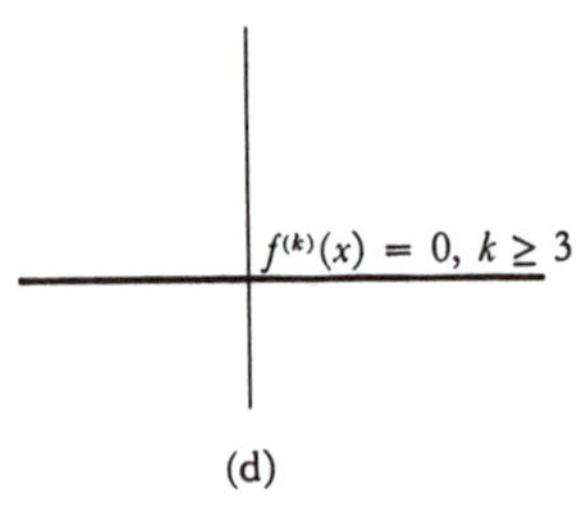

(d)

FIGURE 20

Figure 20 shows the function f, together with its various derivatives.

A rather more illuminating example is presented by the following function, whose graph is shown in Figure 21(a):

$$f(x) = \begin{cases} x^2, & x \geq 0 \\ -x^2, & x \leq 0. \end{cases}$$

It is easy to see that

$$\begin{aligned} f'(a) &= 2a \qquad \text{if } a > 0, \\ f'(a) &= -2a \qquad \text{if } a < 0. \end{aligned}$$

Moreover,

$$\begin{aligned} f'(0) &= \lim_{h\to 0} \frac{f(h) - f(0)}{h} \\ &= \lim_{h\to 0} \frac{f(h)}{h}. \end{aligned}$$

Now

$$\lim_{h\to 0^+} \frac{f(h)}{h} = \lim_{h\to 0^+} \frac{h^2}{h} = 0$$

and

$$\lim_{h\to 0^-} \frac{f(h)}{h} = \lim_{h\to 0^-} \frac{-h^2}{h} = 0,$$

so

$$f'(0) = \lim_{h\to 0} \frac{f(h)}{h} = 0.$$

This information can all be summarized as follows:

$$f'(x) = 2|x|.$$

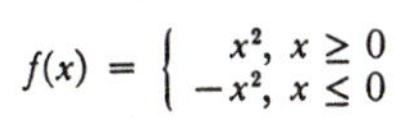

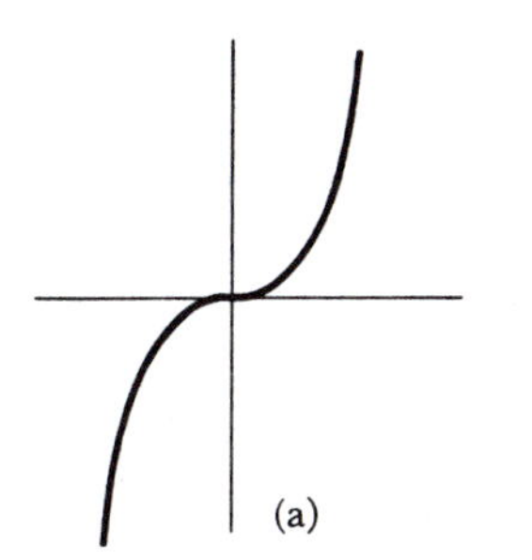

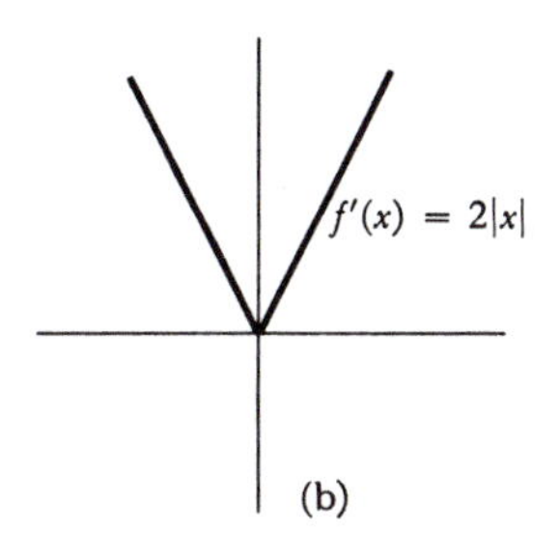

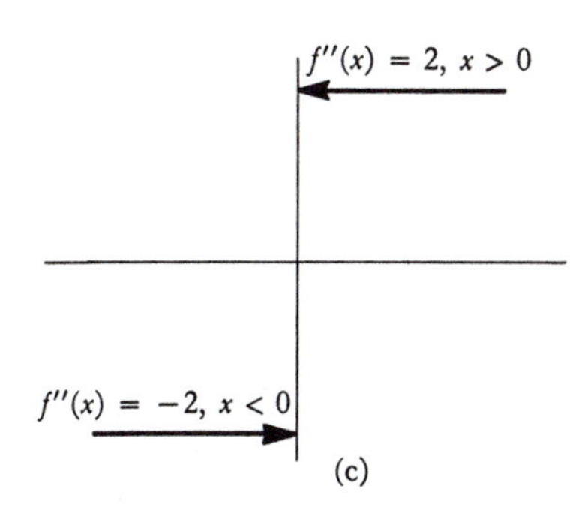

FIGURE 21

It follows that $f''(0)$ does not exist! Existence of the second derivative is thus a rather strong criterion for a function to satisfy. Even a "smooth looking" function like f reveals some irregularity when examined with the second derivative. This suggests that the irregular behavior of the function

$$g(x) = \begin{cases} x^2 \sin \dfrac{1}{x}, & x \neq 0 \\ 0, & x = 0 \end{cases}$$

might also be revealed by the second derivative. At the moment we know that $g'(0) = 0$, but we do not know $g'(a)$ for any $a \neq 0$, so it is hopeless to begin computing $g''(0)$. We will return to this question at the end of the next chapter, after we have perfected the technique of finding derivatives.

PROBLEMS

1. (a) Prove, working directly from the definition, that if $f(x) = 1/x$, then $f'(a) = -1/a^2$, for $a \neq 0$.
(b) Prove that the tangent line to the graph of f at $(a, 1/a)$ does not intersect the graph of f, except at $(a, 1/a)$.

2. (a) Prove that if $f(x) = 1/x^2$, then $f'(a) = -2/a^3$ for $a \neq 0$.
(b) Prove that the tangent line to f at $(a, 1/a^2)$ intersects f at one other point, which lies on the opposite side of the vertical axis.

3. Prove that if $f(x) = \sqrt{x}$, then $f'(a) = 1/(2\sqrt{a})$, for $a > 0$. (The expression you obtain for $[f(a+h) - f(a)]/h$ will require some algebraic face lifting, but the answer should suggest the right trick.)

4. For each natural number n, let $S_n(x) = x^n$. Remembering that $S_1{}'(x) = 1$, $S_2{}'(x) = 2x$, and $S_3{}'(x) = 3x^2$, conjecture a formula for $S_n{}'(x)$. Prove your conjecture. (The expression $(x + h)^n$ may be expanded by the binomial theorem.)

5. Find f' if $f(x) = [x]$.

6. Prove, starting from the definition (and drawing a picture to illustrate):

(a) if $g(x) = f(x) + c$, then $g'(x) = f'(x)$;
(b) if $g(x) = cf(x)$, then $g'(x) = cf'(x)$.

7. Suppose that $f(x) = x^3$.

(a) What is $f'(9)$, $f'(25)$, $f'(36)$?
(b) What is $f'(3^2)$, $f'(5^2)$, $f'(6^2)$?
(c) What is $f'(a^2)$, $f'(x^2)$?

If you do not find this problem silly, you are missing a very important point: $f'(x^2)$ means the derivative of f at the number which we happen to be calling x^2; it is *not* the derivative at x of the function $g(x) = f(x^2)$. Just to drive the point home:

(d) For $f(x) = x^3$, compare $f'(x^2)$ and $g'(x)$ where $g(x) = f(x^2)$.

8. (a) Suppose $g(x) = f(x+c)$. Prove (starting from the definition) that $g'(x) = f'(x+c)$. Draw a picture to illustrate this. To do this problem you must write out the definitions of $g'(x)$ and $f'(x+c)$ correctly. The purpose of Problem 7 was to convince you that although this problem is easy, it is not an utter triviality, and there is something to prove: you cannot simply put prime marks into the equation $g(x) = f(x+c)$. To emphasize this point:

(b) Prove that if $g(x) = f(cx)$, then $g'(x) = c \cdot f'(cx)$. Try to see pictorially why this should be true, also.

(c) Suppose that f is differentiable and periodic, with period a (i.e., $f(x+a) = f(x)$ for all x). Prove that f' is also periodic.

9. Find $f'(x)$ and also $f'(x+3)$ in the following cases. Be very methodical, or you will surely slip up somewhere. Consult the answers (after you do the problem, naturally).

(i) $f(x) = (x+3)^5$.
(ii) $f(x+3) = x^5$.
(iii) $f(x+3) = (x+5)^7$.

10. Find $f'(x)$ if $f(x) = g(t+x)$, and if $f(t) = g(t+x)$. The answers will *not* be the same.

11. (a) Prove that Galileo was wrong: if a body falls a distance $s(t)$ in t seconds, and s' is proportional to s, then s cannot be a function of the form $s(t) = ct^2$.

(b) Prove that the following facts are true about s if $s(t) = (a/2)t^2$ (the first fact will show why we switched from c to $a/2$):

(i) $s''(t) = a$ (the acceleration is constant).
(ii) $[s'(t)]^2 = 2as(t)$.

(c) If s is measured in feet, the value of a is 32. How many seconds do you have to get out of the way of a chandelier which falls from a 400-foot ceiling? If you don't make it, how fast will the chandelier be going when it hits you? Where was the chandelier when it was moving with half that speed?

12. Imagine a road on which the speed limit is specified at every single point. In other words, there is a certain function L such that the speed limit x miles from the beginning of the road is $L(x)$. Two cars, A and B, are driving along this road; car A's position at time t is $a(t)$, and car B's is $b(t)$.

(a) What equation expresses the fact that car A always travels at the speed limit? (The answer is *not* $a'(t) = L(t)$.)

(b) Suppose that A always goes at the speed limit, and that B's position at time t is A's position at time $t-1$. Show that B is also going at the speed limit at all times.

(c) Suppose B always stays a constant distance behind A. Under what conditions will B still always travel at the speed limit?

13. Suppose that $f(a) = g(a)$ and that the left-hand derivative of f at a equals the right-hand derivative of g at a. Define $h(x) = f(x)$ for $x \leq a$, and $h(x) = g(x)$ for $x \geq a$. Prove that h is differentiable at a.

14. Let $f(x) = x^2$ if x is rational, and $f(x) = 0$ if x is irrational. Prove that f is differentiable at 0. (Don't be scared by this function. Just write out the definition of $f'(0)$.)

***15.** (a) Let f be a function such that $|f(x)| \leq x^2$ for all x. Prove that f is differentiable at 0. (If you have done Problem 14 you should be able to do this.)

(b) This result can be generalized if x^2 is replaced by $|g(x)|$, where g has what property?

16. Let $\alpha > 1$. If f satisfies $|f(x)| \leq |x|^\alpha$, prove that f is differentiable at 0.

17. Let $0 < \beta < 1$. Prove that if f satisfies $|f(x)| \geq |x|^\beta$ and $f(0) = 0$, then f is not differentiable at 0.

***18.** Let $f(x) = 0$ for irrational x, and $1/q$ for $x = p/q$ in lowest terms. Prove that f is not differentiable at a for any a. Hint: It obviously suffices to prove this for irrational a. Why? If $a = m.a_1a_2a_3\ldots$ is the decimal expansion of a, consider $[f(a+h) - f(a)]/h$ for h rational, and also for

$$h = -0.00\ldots0a_{n+1}a_{n+2}\ldots.$$

19. (a) Suppose that $f(a) = g(a) = h(a)$, that $f(x) \leq g(x) \leq h(x)$ for all x, and that $f'(a) = h'(a)$. Prove that g is differentiable at a, and that $f'(a) = g'(a) = h'(a)$. (Begin with the definition of $g'(a)$.)

(b) Show that the conclusion does not follow if we omit the hypothesis $f(a) = g(a) = h(a)$.

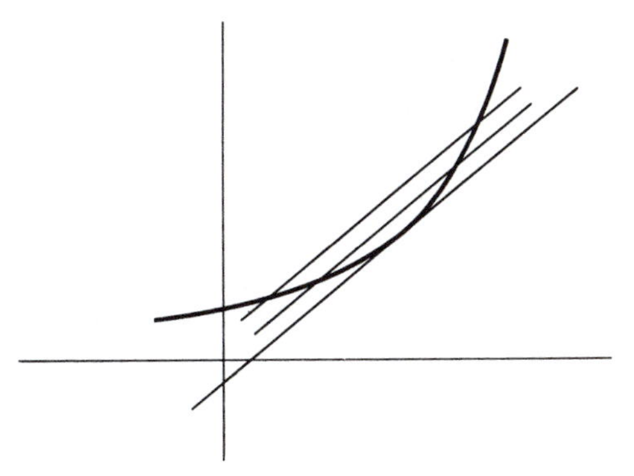
FIGURE 22

20. Let f be any polynomial function; we will see in the next chapter that f is differentiable. The tangent line to f at $(a, f(a))$ is the graph of $g(x) = f'(a)(x-a) + f(a)$. Thus $f(x) - g(x)$ is the polynomial function $d(x) = f(x) - f'(a)(x-a) - f(a)$. We have already seen that if $f(x) = x^2$, then $d(x) = (x-a)^2$, and if $f(x) = x^3$, then $d(x) = (x-a)^2(x+2a)$.

(a) Find $d(x)$ when $f(x) = x^4$, and show that it is divisible by $(x-a)^2$.

(b) There certainly seems to be some evidence that $d(x)$ is always divisible by $(x-a)^2$. Figure 22 provides an intuitive argument: usually, lines parallel to the tangent line will intersect the graph at two points; the tangent line intersects the graph only once near the point, so the intersection should be a "double intersection." To give a rigorous proof, first note that

$$\frac{d(x)}{x-a} = \frac{f(x) - f(a)}{x-a} - f'(a).$$

Now answer the following questions. Why is $f(x) - f(a)$ divisible by $(x-a)$? Why is there a polynomial function h such that $h(x) =$

$d(x)/(x-a)$ for $x \neq a$? Why is $\lim_{x \to a} h(x) = 0$? Why is $h(a) = 0$? Why does this solve the problem?

21. (a) Show that $f'(a) = \lim_{x \to a} [f(x) - f(a)]/(x - a)$. (Nothing deep here.)
(b) Show that derivatives are a "local property": if $f(x) = g(x)$ for all x in some open interval containing a, then $f'(a) = g'(a)$. (This means that in computing $f'(a)$, you can ignore $f(x)$ for any particular $x \neq a$. Of course you can't ignore $f(x)$ for all such x at once!)

***22.** (a) Suppose that f is differentiable at x. Prove that

$$f'(x) = \lim_{h \to 0} \frac{f(x+h) - f(x-h)}{2h}.$$

Hint: Remember an old algebraic trick—a number is not changed if the same quantity is added to and then subtracted from it.

**(b) Prove, more generally, that

$$f'(x) = \lim_{h,k \to 0^+} \frac{f(x+h) - f(x-k)}{h+k}.$$

***23.** Prove that if f is even, then $f'(x) = -f'(-x)$. (In order to minimize confusion, let $g(x) = f(-x)$; find $g'(x)$ and *then* remember what other thing g is.) Draw a picture!

***24.** Prove that if f is odd, then $f'(x) = f'(-x)$. Once again, draw a picture.

25. Problems 23 and 24 say that f' is even if f is odd, and odd if f is even. What can therefore be said about $f^{(k)}$?

26. Find $f''(x)$ if

(i) $f(x) = x^3$.
(ii) $f(x) = x^5$.
(iii) $f'(x) = x^4$.
(iv) $f(x+3) = x^5$.

27. If $S_n(x) = x^n$, and $0 \leq k \leq n$, prove that

$$\begin{aligned} S_n{}^{(k)}(x) &= \frac{n!}{(n-k)!} x^{n-k} \\ &= k! \binom{n}{k} x^{n-k}. \end{aligned}$$

***28.** (a) Find $f'(x)$ if $f(x) = |x|^3$. Find $f''(x)$. Does $f'''(x)$ exist for all x?
(b) Analyze f similarly if $f(x) = x^4$ for $x \geq 0$ and $f(x) = -x^4$ for $x \leq 0$.

***29.** Let $f(x) = x^n$ for $x \geq 0$ and let $f(x) = 0$ for $x \leq 0$. Prove that $f^{(n-1)}$ exists (and find a formula for it), but that $f^{(n)}(0)$ does not exist.

30. Interpret the following specimens of Leibnizian notation; each is a restatement of some fact occurring in a previous problem.

(i) $\dfrac{dx^n}{dx} = nx^{n-1}$

(ii) $\dfrac{dz}{dy} = -\dfrac{1}{y^2}$ if $z = \dfrac{1}{y}$.

(iii) $\dfrac{d[f(x)+c]}{dx} = \dfrac{df(x)}{dx}$.

(iv) $\dfrac{d[cf(x)]}{dx} = c\dfrac{df(x)}{dx}$.

(v) $\dfrac{dz}{dx} = \dfrac{dy}{dx}$ if $z = y + c$.

(vi) $\left.\dfrac{dx^3}{dx}\right|_{x=a^2} = 3a^4$.

(vii) $\left.\dfrac{df(x+a)}{dx}\right|_{x=b} = \left.\dfrac{df(x)}{dx}\right|_{x=b+a}$.

(viii) $\left.\dfrac{df(cx)}{dx}\right|_{x=b} = c \cdot \left.\dfrac{df(x)}{dx}\right|_{x=cb}$.

(ix) $\dfrac{df(cx)}{dx} = c \cdot \left.\dfrac{df(y)}{dy}\right|_{y=cx}$.

(x) $\dfrac{d^k x^n}{dx^k} = k!\dbinom{n}{k}x^{n-k}$.

CHAPTER 10 DIFFERENTIATION

The process of finding the derivative of a function is called *differentiation*. From the previous chapter you may have the impression that this process is usually laborious, requires recourse to the definition of the derivative, and depends upon successfully recognizing some limit. It is true that such a procedure is often the only possible approach—if you forget the definition of the derivative you are likely to be lost. Nevertheless, in this chapter we will learn to differentiate a large number of functions, without the necessity of even recalling the definition. A few theorems will provide a mechanical process for differentiating a large class of functions, which are formed from a few simple functions by the process of addition, multiplication, division, and composition. This description should suggest what theorems will be proved. We will first find the derivative of a few simple functions, and then prove theorems about the sum, products, quotients, and compositions of differentiable functions. The first theorem is merely a formal recognition of a computation carried out in the previous chapter.

THEOREM 1 If f is a constant function, $f(x) = c$, then

$$f'(a) = 0 \quad \text{for all numbers } a.$$

PROOF

$$f'(a) = \lim_{h \to 0} \frac{f(a+h) - f(a)}{h} = \lim_{h \to 0} \frac{c - c}{h} = 0. \blacksquare$$

The second theorem is also a special case of a computation in the last chapter.

THEOREM 2 If f is the identity function, $f(x) = x$, then

$$f'(a) = 1 \quad \text{for all numbers } a.$$

PROOF

$$\begin{aligned} f'(a) &= \lim_{h \to 0} \frac{f(a+h) - f(a)}{h} \\ &= \lim_{h \to 0} \frac{a + h - a}{h} \\ &= \lim_{h \to 0} \frac{h}{h} = 1. \blacksquare \end{aligned}$$

The derivative of the sum of two functions is just what one would hope—the sum of the derivatives.

THEOREM 3 If f and g are differentiable at a, then $f+g$ is also differentiable at a, and

$$(f+g)'(a) = f'(a) + g'(a).$$

PROOF

$$\begin{aligned}
(f+g)'(a) &= \lim_{h\to 0} \frac{(f+g)(a+h)-(f+g)(a)}{h} \\
&= \lim_{h\to 0} \frac{f(a+h)+g(a+h)-[f(a)+g(a)]}{h} \\
&= \lim_{h\to 0} \left[\frac{f(a+h)-f(a)}{h} + \frac{g(a+h)-g(a)}{h}\right] \\
&= \lim_{h\to 0} \frac{f(a+h)-f(a)}{h} + \lim_{h\to 0} \frac{g(a+h)-g(a)}{h} \\
&= f'(a) + g'(a). \quad \blacksquare
\end{aligned}$$

The formula for the derivative of a product is not as simple as one might wish, but it is nevertheless pleasantly symmetric, and the proof requires only a simple algebraic trick, which we have found useful before—a number is not changed if the same quantity is added to and subtracted from it.

THEOREM 4 If f and g are differentiable at a, then

$$(f\cdot g)'(a) = f'(a)\cdot g(a) + f(a)\cdot g'(a).$$

PROOF

$$\begin{aligned}
(f\cdot g)'(a) &= \lim_{h\to 0} \frac{(f\cdot g)(a+h)-(f\cdot g)(a)}{h} \\
&= \lim_{h\to 0} \frac{f(a+h)g(a+h)-f(a)g(a)}{h} \\
&= \lim_{h\to 0} \left[\frac{f(a+h)[g(a+h)-g(a)]}{h} + \frac{[f(a+h)-f(a)]g(a)}{h}\right] \\
&= \lim_{h\to 0} f(a+h)\cdot \lim_{h\to 0} \frac{g(a+h)-g(a)}{h} + \lim_{h\to 0} \frac{f(a+h)-f(a)}{h} \cdot \lim_{h\to 0} g(a) \\
&= f(a)\cdot g'(a) + f'(a)\cdot g(a).
\end{aligned}$$

(Notice that we have used Theorem 9-1 to conclude that $\lim_{h\to 0} f(a+h) = f(a)$.) $\blacksquare$

In one special case Theorem 4 simplifies considerably:

THEOREM 5 If $g(x) = cf(x)$ and f is differentiable at a, then g is differentiable at a, and

$$g'(a) = c\cdot f'(a).$$

PROOF If $h(x) = c$, so that $g = h\cdot f$, then by Theorem 4,

$$\begin{aligned}
g'(a) &= (h\cdot f)'(a) \\
&= h(a)\cdot f'(a) + h'(a)\cdot f(a) \\
&= c\cdot f'(a) + 0\cdot f(a) \\
&= c\cdot f'(a). \quad \blacksquare
\end{aligned}$$

Notice, in particular, that $(-f)'(a) = -f'(a)$, and consequently $(f - g)'(a) = (f + [-g])'(a) = f'(a) - g'(a)$.

To demonstrate what we have already achieved, we will compute the derivative of some more special functions.

THEOREM 6 If $f(x) = x^n$ for some natural number n, then

$$f'(a) = na^{n-1} \qquad \text{for all } a.$$

PROOF The proof will be by induction on n. For $n = 1$ this is simply Theorem 2. Now assume that the theorem is true for n, so that if $f(x) = x^n$, then

$$f'(a) = na^{n-1} \qquad \text{for all } a.$$

Let $g(x) = x^{n+1}$. If $I(x) = x$, the equation $x^{n+1} = x^n \cdot x$ can be written

$$g(x) = f(x) \cdot I(x) \qquad \text{for all } x;$$

thus $g = f \cdot I$. It follows from Theorem 4 that

$$\begin{aligned} g'(a) = (f \cdot I)'(a) &= f'(a) \cdot I(a) + f(a) \cdot I'(a) \\ &= na^{n-1} \cdot a + a^n \cdot 1 \\ &= na^n + a^n \\ &= (n+1)a^n, \qquad \text{for all } a. \end{aligned}$$

This is precisely the case $n + 1$ which we wished to prove. ▮

Putting together the theorems proved so far we can now find f' for f of the form

$$f(x) = a_n x^n + a_{n-1}x^{n-1} + \cdots + a_2 x^2 + a_1 x + a_0.$$

We obtain

$$f'(x) = na_n x^{n-1} + (n-1)a_{n-1}x^{n-2} + \cdots + 2a_2 x + a_1.$$

We can also find f'':

$$f''(x) = n(n-1)a_n x^{n-2} + (n-1)(n-2)a_{n-1}x^{n-3} + \cdots + 2a_2.$$

This process can be continued easily. Each differentiation reduces the highest power of x by 1, and eliminates one more a_i. It is a good idea to work out the derivatives f''', $f^{(4)}$, and perhaps $f^{(5)}$, until the pattern becomes quite clear. The last interesting derivative is

$$f^{(n)}(x) = n!a_n;$$

for $k > n$ we have

$$f^{(k)}(x) = 0.$$

Clearly, the next step in our program is to find the derivative of a quotient f/g. It is quite a bit simpler, and, because of Theorem 4, obviously sufficient to find the derivative of $1/g$.

THEOREM 7 If g is differentiable at a, and $g(a) \neq 0$, then $1/g$ is differentiable at a, and

$$\left(\frac{1}{g}\right)'(a) = \frac{-g'(a)}{[g(a)]^2}.$$

PROOF Before we even write

$$\frac{\left(\frac{1}{g}\right)(a+h) - \left(\frac{1}{g}\right)(a)}{h}$$

we must be sure that this expression makes sense—it is necessary to check that $(1/g)(a+h)$ is defined for sufficiently small h. This requires only two observations. Since g is, by hypothesis, differentiable at a, it follows from Theorem 9-1 that g is continuous at a. Since $g(a) \neq 0$, it follows from Theorem 6-3 that there is some $\delta > 0$ such that $g(a+h) \neq 0$ for $|h| < \delta$. Therefore $(1/g)(a+h)$ *does* make sense for small enough h, and we can write

$$\begin{aligned}
\lim_{h \to 0} \frac{\left(\frac{1}{g}\right)(a+h) - \left(\frac{1}{g}\right)(a)}{h} &= \lim_{h \to 0} \frac{\frac{1}{g(a+h)} - \frac{1}{g(a)}}{h} \\
&= \lim_{h \to 0} \frac{g(a) - g(a+h)}{h[g(a) \cdot g(a+h)]} \\
&= \lim_{h \to 0} \frac{-[g(a+h) - g(a)]}{h} \cdot \frac{1}{g(a)g(a+h)} \\
&= \lim_{h \to 0} \frac{-[g(a+h) - g(a)]}{h} \cdot \lim_{h \to 0} \frac{1}{g(a) \cdot g(a+h)} \\
&= -g'(a) \cdot \frac{1}{[g(a)]^2}.
\end{aligned}$$

(Notice that we have used continuity of g at a once again.) ▮

The general formula for the derivative of a quotient is now easy to derive. Though not particularly appealing, it is important, and must simply be memorized (I always use the incantation: "bottom times derivative of top, minus top times derivative of bottom, over bottom squared.")

THEOREM 8 If f and g are differentiable at a and $g(a) \neq 0$, then f/g is differentiable at a, and

$$\left(\frac{f}{g}\right)'(a) = \frac{g(a) \cdot f'(a) - f(a) \cdot g'(a)}{[g(a)]^2}.$$

PROOF Since $f/g = f \cdot (1/g)$ we have

$$\begin{aligned} \left(\frac{f}{g}\right)'(a) &= \left(f \cdot \frac{1}{g}\right)'(a) \\ &= f'(a) \cdot \left(\frac{1}{g}\right)(a) + f(a) \cdot \left(\frac{1}{g}\right)'(a) \\ &= \frac{f'(a)}{g(a)} + \frac{f(a)(-g'(a))}{[g(a)]^2} \\ &= \frac{f'(a) \cdot g(a) - f(a) \cdot g'(a)}{[g(a)]^2}. \ \blacksquare \end{aligned}$$

We can now differentiate a few more functions. For example,

$$\text{if } f(x) = \frac{x^2-1}{x^2+1}, \text{ then } f'(x) = \frac{(x^2+1)(2x) - (x^2-1)(2x)}{(x^2+1)^2} = \frac{4x}{(x^2+1)^2};$$

$$\text{if } f(x) = \frac{x}{x^2+1}, \text{ then } f'(x) = \frac{(x^2+1) - x(2x)}{(x^2+1)^2} = \frac{1-x^2}{(x^2+1)^2};$$

$$\text{if } f(x) = \frac{1}{x}, \quad \text{then } f'(x) = -\frac{1}{x^2} = (-1)x^{-2}.$$

Notice that the last example can be generalized: if

$$f(x) = x^{-n} = \frac{1}{x^n}, \qquad \text{for some natural number } n,$$

then

$$f'(x) = \frac{-nx^{n-1}}{x^{2n}} = (-n)x^{-n-1};$$

thus Theorem 6 actually holds both for positive and negative integers. If we interpret $f(x) = x^0$ to mean $f(x) = 1$, and $f'(x) = 0 \cdot x^{-1}$ to mean $f'(x) = 0$, then Theorem 6 is true for $n = 0$ also. (The word "interpret" is necessary because it is not clear how 0^0 should be defined and, in any case, $0 \cdot 0^{-1}$ is meaningless.)

Further progress in differentiation requires the knowledge of the derivatives of certain special functions to be studied later. One of these is the sine function. For the moment we shall divulge, and use, the following information, without proof:

$$\begin{aligned} \sin'(a) &= \cos a \qquad \text{for all } a, \\ \cos'(a) &= -\sin a \qquad \text{for all } a, \end{aligned}$$

This information allows us to differentiate many other functions. For example, if

$$f(x) = x \sin x,$$

then

$$\begin{aligned} f'(x) &= x \cos x + \sin x, \\ f''(x) &= -x \sin x + \cos x + \cos x \\ &= -x \sin x + 2 \cos x; \end{aligned}$$

if

$$g(x) = \sin^2 x = \sin x \cdot \sin x,$$

then

$$\begin{aligned} g'(x) &= \sin x \cos x + \cos x \sin x \\ &= 2 \sin x \cos x, \\ g''(x) &= 2[(\sin x)(-\sin x) + \cos x \cos x] \\ &= 2[\cos^2 x - \sin^2 x]; \end{aligned}$$

if

$$h(x) = \cos^2 x = \cos x \cdot x,$$

then

$$\begin{aligned} h'(x) &= (\cos x)(-\sin x) + (-\sin x) \cos x \\ &= -2 \sin x \cos x, \\ h''(x) &= -2[\cos^2 x - \sin^2 x]. \end{aligned}$$

Notice that

$$g'(x) + h'(x) = 0,$$

hardly surprising, since $(g + h)(x) = \sin^2 x + \cos^2 x = 1$. As we would expect, we also have $g''(x) + h''(x) = 0$.

The examples above involved only products of two functions. A function involving triple products can be handled by Theorem 4 also; in fact it can be handled in two ways. Remember that $f \cdot g \cdot h$ is an abbreviation for

$$(f \cdot g) \cdot h \quad \text{or} \quad f \cdot (g \cdot h).$$

Choosing the first of these, for example, we have

$$\begin{aligned} (f \cdot g \cdot h)'(x) &= (f \cdot g)'(x) \cdot h(x) + (f \cdot g)(x)h'(x) \\ &= [f'(x)g(x) + f(x)g'(x)]h(x) + f(x)g(x)h'(x) \\ &= f'(x)g(x)h(x) + f(x)g'(x)h(x) + f(x)g(x)h'(x). \end{aligned}$$

The choice of $f \cdot (g \cdot h)$ would, of course, have given the same result, with a different intermediate step. The final answer is completely symmetric and easily remembered:

> $(f \cdot g \cdot h)'$ is the sum of the three terms obtained by differentiating each of f, g, and h and multiplying by the other two.

For example, if

$$f(x) = x^3 \sin x \cos x,$$

then

$$f'(x) = 3x^2 \sin x \cos x + x^3 \cos x \cos x + x^3(\sin x)(-\sin x).$$

Products of more than 3 functions can be handled similarly. For example, you should have little difficulty deriving the formula

$$\begin{aligned} (f \cdot g \cdot h \cdot k)'(x) = f'(x)g(x)h(x)k(x) &+ f(x)g'(x)h(x)k(x) \\ &+ f(x)g(x)h'(x)k(x) + f(x)g(x)h(x)k'(x). \end{aligned}$$

You might even try to prove (by induction) the general formula:

$$(f_1 \cdot \ldots \cdot f_n)'(x) = \sum_{i=1}^{n} f_1(x) \cdot \ldots \cdot f_{i-1}(x) f_i{}'(x) f_{i+1}(x) \cdot \ldots \cdot f_n(x).$$

Differentiating the most interesting functions obviously requires a formula for $(f \circ g)'(x)$ in terms of f' and g'. To ensure that $f \circ g$ be differentiable at a, one reasonable hypothesis would seem to be that g be differentiable at a. Since the behavior of $f \circ g$ near a depends on the behavior of f near $g(a)$ (not near a), it also seems reasonable to assume that f is differentiable at $g(a)$. Indeed we shall prove that if g is differentiable at a and f is differentiable at $g(a)$, then $f \circ g$ is differentiable at a, and

$$(f \circ g)'(a) = f'(g(a)) \cdot g'(a).$$

This extremely important formula is called the *Chain Rule*, presumable because a composition of functions might be called a "chain" of functions. Notice that $(f \circ g)'$ is practically the product of f' and g', but not quite: f' must be evaluated at $g(a)$ and g' at a. Before attempting to prove this theorem we will try a few applications. Suppose

$$f(x) = \sin x^2.$$

Let us, temporarily, use S to denote the ("squaring") function $S(x) = x^2$. Then

$$f = \sin \circ S.$$

Therefore we have

$$\begin{aligned} f'(x) &= \sin'(S(x)) \cdot S'(x) \\ &= \cos x^2 \cdot 2x. \end{aligned}$$

Quite a different result is obtained if

$$f(x) = \sin^2 x.$$

In this case

$$f = S \circ \sin,$$

so

$$\begin{aligned} f'(x) &= S'(\sin x) \cdot \sin'(x) \\ &= 2 \sin x \cdot \cos x. \end{aligned}$$

Notice that this agrees (as it should) with the result obtained by writing $f = \sin \cdot \sin$ and using the product formula.

Although we have invented a special symbol, S, to name the "squaring" function, it does not take much practice to do problems like this without bothering to write down special symbols for functions, and without even bothering to write down the particular composition which f is—one soon becomes accustomed to taking f apart in one's head. The following differentiations may be used as practice for such mental gymnastics—if you find it necessary to work a few out on paper, by all means do so, but try to develop the knack of writing f' immediately after seeing the definition of f; problems of this sort are so simple that, if you just remember the Chain Rule, there is no thought necessary.

$$\begin{aligned}
&\text{if } f(x) = \sin x^3 && \text{then } f'(x) = \cos x^3 \cdot 3x^2 \\
&f(x) = \sin^3 x && f'(x) = 3\sin^2 x \cdot \cos x \\
&f(x) = \sin \frac{1}{x} && f'(x) = \cos \frac{1}{x} \cdot \left(\frac{-1}{x^2}\right) \\
&f(x) = \sin(\sin x) && f'(x) = \cos(\sin x) \cdot \cos x \\
&f(x) = \sin(x^3 + 3x^2) && f'(x) = \cos(x^3 + 3x^2) \cdot (3x^2 + 6x) \\
&f(x) = (x^3 + 3x^2)^{53} && f'(x) = 53(x^3 + 3x^2)^{52} \cdot (3x^2 + 6x).
\end{aligned}$$

A function like

$$f(x) = \sin^2 x^2 = [\sin x^2]^2,$$

which is the composition of three functions,

$$f = S \circ \sin \circ S,$$

can also be differentiated by the Chain Rule. It is only necessary to remember that a triple composition $f \circ g \circ h$ means $(f \circ g) \circ h$ or $f \circ (g \circ h)$. Thus if

$$f(x) = \sin^2 x^2$$

we can write

$$\begin{aligned}
f &= (S \circ \sin) \circ S, \\
f &= S \circ (\sin \circ S).
\end{aligned}$$

The derivative of either expression can be found by applying the Chain Rule twice; the only doubtful point is whether the two expressions lead to equally simple calculations. As a matter of fact, as any experienced differentiator knows, it is much better to use the second:

$$\boxed{f = S \circ (\sin \circ S).}$$

We can now write down $f'(x)$ in one fell swoop. To begin with, note that the first function to be differentiated is S, so the formula for $f'(x)$ begins

$$f'(x) = 2(\qquad) \cdot \rule{6em}{1em}\,.$$

Inside the parentheses we must put $\sin x^2$, the value at x of the second function, $\sin \circ S$. Thus we begin by writing

$$f'(x) = 2\sin x^2 \cdot \rule{6em}{1em}$$

(the parentheses weren't really necessary, after all). We must now multiply this much of the answer by the derivative of $\sin \circ S$ at x; this part is easy—it involves a composition of two functions, which we already know how to handle. We obtain, for the final answer,

$$f'(x) = 2\sin x^2 \cdot \cos x^2 \cdot 2x.$$

The following example is handled similarly. Suppose

$$\boxed{f(x) = \sin(\sin x^2).}$$

Without even bothering to write down f as a composition $g \circ h \circ k$ of three functions, we can see that the left-most one will be sin, so our expression for $f'(x)$ begins

$$f'(x) = \cos(\quad) \cdot \blacksquare.$$

Inside the parentheses we must put the value of $h \circ k(x)$; this is simply $\sin x^2$ (what you get from $\sin(\sin x^2)$ by deleting the first sin). So our expression for $f'(x)$ begins

$$f'(x) = \cos(\sin x^2) \cdot \blacksquare.$$

We can now forget about the first sin in $\sin(\sin x^2)$; we have to multiply what we have so far by the derivative of the function whose value at x is $\sin x^2$—which is again a problem we already know how to solve:

$$f'(x) = \cos(\sin x^2) \cdot \cos x^2 \cdot 2x.$$

Finally, here are the derivatives of some other functions which are the composition of sin and S, as well as some other triple compositions. You can probably just "see" that the answers are correct—if not, try writing out f as a composition:

$$\begin{aligned}
\text{if } f(x) &= \sin((\sin x)^2) & \text{then } f'(x) &= \cos((\sin x)^2) \cdot 2\sin x \cdot \cos x \\
f(x) &= [\sin(\sin x)]^2 & f'(x) &= 2\sin(\sin x) \cdot \cos(\sin x) \cdot \cos x \\
f(x) &= \sin(\sin(\sin x)) & f'(x) &= \cos(\sin(\sin x)) \cdot \cos(\sin x) \cdot \cos x \\
f(x) &= \sin^2(x \sin x) & f'(x) &= 2\sin(x \sin x) \cdot \cos(x \sin x) \\
& & &\qquad \cdot [\sin x + x \cos x] \\
f(x) &= \sin(\sin(x^2 \sin x)) & f'(x) &= \cos(\sin(x^2 \sin x)) \cdot \cos(x^2 \sin x) \\
& & &\qquad \cdot [2x \sin x + x^2 \cos x].
\end{aligned}$$

The rule for treating compositions of four (or even more) functions is easy—always (mentally) put in parentheses starting from the right,

$$f \circ (g \circ (h \circ k)),$$

and start reducing the calculation to the derivative of a composition of a smaller number of functions:

$$f'(g(h(k(x)))) \cdot \blacksquare.$$

For example, if

$$f(x) = \sin^2(\sin^2(x)) \qquad \begin{aligned}[t] [f &= S \circ \sin \circ S \circ \sin \\ &= S \circ (\sin \circ (S \circ \sin))] \end{aligned}$$

then

$$f'(x) = 2\sin(\sin^2 x) \cdot \cos(\sin^2 x) \cdot 2\sin x \cdot \cos x;$$

if

$$f(x) = \sin((\sin x^2)^2) \qquad [f = \sin \circ S \circ \sin \circ S = \sin \circ (S \circ (\sin \circ S))]$$

then

$$f'(x) = \cos((\sin x^2)^2) \cdot 2 \sin x^2 \cdot \cos x^2 \cdot 2x;$$

if

$$f(x) = \sin^2(\sin(\sin x)) \qquad [\text{fill in yourself, if necessary}]$$

then

$$f'(x) = 2\sin(\sin(\sin x)) \cdot \cos(\sin(\sin x)) \cdot \cos(\sin x) \cdot \cos x.$$

With these examples as reference, you require only one thing to become a master differentiator—practice. You can be safely turned loose on the exercises at the end of the chapter, and it is now high time that we proved the Chain Rule.

The following argument, while not a proof, indicates some of the tricks one might try, as well as some of the difficulties encountered. We begin, of course, with the definition—

$$\begin{aligned}(f \circ g)'(a) &= \lim_{h \to 0} \frac{(f \circ g)(a+h) - (f \circ g)(a)}{h} \\ &= \lim_{h \to 0} \frac{f(g(a+h)) - f(g(a))}{h}.\end{aligned}$$

Somewhere in here we would like the expression for $g'(a)$. One approach is to put it in by fiat:

$$\lim_{h \to 0} \frac{f(g(a+h)) - f(g(a))}{h} = \lim_{h \to 0} \frac{f(g(a+h)) - f(g(a))}{g(a+h) - g(a)} \cdot \frac{g(a+h) - g(a)}{h}.$$

This does not look bad, and it looks even better if we write

$$\begin{aligned}&\lim_{h \to 0} \frac{(f \circ g)(a+h) - (f \circ g)(a)}{h} \\ &\qquad = \lim_{h \to 0} \frac{f(g(a) + [g(a+h) - g(a)]) - f(g(a))}{g(a+h) - g(a)} \cdot \lim_{h \to 0} \frac{g(a+h) - g(a)}{h}.\end{aligned}$$

The second limit is the factor $g'(a)$ which we want. If we let $g(a+h) - g(a) = k$ (to be precise we should write $k(h)$), then the first limit is

$$\lim_{h \to 0} \frac{f(g(a) + k) - f(g(a))}{k}.$$

It looks as if this limit should be $f'(g(a))$, since continuity of g at a implies that k goes to 0 as h does. In fact, one can, and we soon will, make this sort of reasoning precise. There is already a problem, however, which you will have noticed if you are the kind of person who does not divide blindly. Even for $h \neq 0$ we might have $g(a+h) - g(a) = 0$, making the division and multiplication by $g(a+h) - g(a)$ meaningless. True, we only care about small h, but $g(a+h) - g(a)$ could be 0 for arbitrarily small h. The easiest way this can happen is for g to be a constant function, $g(x) = c$. Then $g(a+h) - g(a) = 0$ for all h. In this case, $f \circ g$ is also a constant function, $(f \circ g)(x) = f(c)$, so the Chain Rule does indeed hold:

$$(f \circ g)'(a) = 0 = f'(g(a)) \cdot g'(a).$$

However, there are also nonconstant functions g for which $g(a+h) - g(a) = 0$ for arbitrarily small h. For example, if $a = 0$, the function g might be

$$g(x) = \begin{cases} x^2 \sin \dfrac{1}{x}, & x \neq 0 \\ 0, & x = 0. \end{cases}$$

In this case, $g'(0) = 0$, as we showed in Chapter 9. If the Chain Rule is correct, we must have $(f \circ g)'(0) = 0$ for any differentiable f, and this is not exactly obvious. A proof of the Chain Rule can be found by considering such recalcitrant functions separately, but it is easier simply to abandon this approach, and use a trick.

THEOREM 9 (THE CHAIN RULE) If g is differentiable at a, and f is differentiable at $g(a)$, then $f \circ g$ is differentiable at a, and

$$(f \circ g)'(a) = f'(g(a)) \cdot g'(a).$$

PROOF Define a function ϕ as follows:

$$\phi(h) = \begin{cases} \dfrac{f(g(a+h)) - f(g(a))}{g(a+h) - g(a)}, & \text{if } g(a+h) - g(a) \neq 0 \\ f'(g(a)), & \text{if } g(a+h) - g(a) = 0. \end{cases}$$

It should be intuitively clear that ϕ is continuous at 0: When h is small, $g(a+h) - g(a)$ is also small, so if $g(a+h) - g(a)$ is not zero, then $\phi(h)$ will be close to $f'(g(a))$; and if it is zero, then $\phi(h)$ actually equals $f'(g(a))$, which is even better. Since the continuity of ϕ is the crux of the whole proof we will provide a careful translation of this intuitive argument.

We know that f is differentiable at $g(a)$. This means that

$$\lim_{k \to 0} \frac{f(g(a)+k) - f(g(a))}{k} = f'(g(a)).$$

Thus, if $\varepsilon > 0$ there is some number $\delta' > 0$ such that, for all k,

$$\text{(1)} \quad \text{if } 0 < |k| < \delta', \text{ then } \left| \frac{f(g(a)+k) - f(g(a))}{k} - f'(g(a)) \right| < \varepsilon.$$

Now g is differentiable at a, hence continuous at a, so there is a $\delta > 0$ such that, for all h,

$$\text{(2)} \quad \text{if } |h| < \delta, \text{ then } |g(a+h) - g(a)| < \delta'.$$

Consider now any h with $|h| < \delta$. If $k = g(a+h) - g(a) \neq 0$, then

$$\phi(h) = \frac{f(g(a+h)) - f(g(a))}{g(a+h) - g(a)} = \frac{f(g(a)+k) - f(g(a))}{k};$$

it follows from (2) that $|k| < \delta'$, and hence from (1) that

$$|\phi(h) - f'(g(a))| < \varepsilon.$$

On the other hand, if $g(a+h) - g(a) = 0$, then $\phi(h) = f'(g(a))$, so it is surely true that

$$|\phi(h) - f'(g(a))| < \varepsilon.$$

We have therefore proved that

$$\lim_{h\to 0} \phi(h) = f'(g(a)),$$

so ϕ is continuous at 0. The rest of the proof is easy. If $h \neq 0$, then we have

$$\frac{f(g(a+h) - f(g(a))}{h} = \phi(h) \cdot \frac{g(a+h) - g(a)}{h}$$

even if $g(a+h) - g(a) = 0$ (because in that case both sides are 0). Therefore

$$\begin{aligned}(f \circ g)'(a) &= \lim_{h\to 0} \frac{f(g(a+h)) - f(g(a))}{h} = \lim_{h\to 0} \phi(h) \cdot \lim_{h\to 0} \frac{g(a+h) - g(a)}{h} \\ &= f'(g(a)) \cdot g'(a). \ \blacksquare\end{aligned}$$

Now that we can differentiate so many functions so easily we can take another look at the function

$$f(x) = \begin{cases} x^2 \sin \dfrac{1}{x}, & x \neq 0 \\ 0, & x = 0. \end{cases}$$

In Chapter 9 we showed that $f'(0) = 0$, working straight from the definition (the only possible way). For $x \neq 0$ we can use the methods of this chapter. We have

$$f'(x) = 2x \sin \frac{1}{x} + x^2 \cos \frac{1}{x} \cdot \left(-\frac{1}{x^2}\right);$$

Thus

$$f'(x) = \begin{cases} 2x \sin \dfrac{1}{x} - \cos \dfrac{1}{x}, & x \neq 0 \\ 0, & x = 0. \end{cases}$$

As this formula reveals, the first derivative f' is indeed badly behaved at 0—it is not even continuous there. If we consider instead

$$f(x) = \begin{cases} x^3 \sin \dfrac{1}{x}, & x \neq 0 \\ 0, & x = 0, \end{cases}$$

then

$$f'(x) = \begin{cases} 3x^2 \sin \dfrac{1}{x} - x \cos \dfrac{1}{x}, & x \neq 0 \\ 0, & x = 0. \end{cases}$$

In this case f' is continuous at 0, but $f''(0)$ does not exist (because the expression $3x^2 \sin 1/x$ defines a function which is differentiable at 0 but the expression $-x \cos 1/x$ does not).

As you may suspect, increasing the power of x yet again produces another improvement. If

$$f(x) = \begin{cases} x^4 \sin \dfrac{1}{x}, & x \neq 0 \\ 0, & x = 0, \end{cases}$$

then

$$f'(x) = \begin{cases} 4x^3 \sin \dfrac{1}{x} - x^2 \cos \dfrac{1}{x}, & x \neq 0 \\ 0, & x = 0. \end{cases}$$

It is easy to compute, right from the definition, that $(f')'(0) = 0$, and $f''(x)$ is easy to find for $x \neq 0$:

$$f''(x) = \begin{cases} 12x^2 \sin \dfrac{1}{x} - 4x \cos \dfrac{1}{x} - 2x \cos \dfrac{1}{x} - \sin \dfrac{1}{x}, & x \neq 0 \\ 0, & x = 0. \end{cases}$$

In this case, the *second* derivative f'' is not continuous at 0. By now you may have guessed the pattern, which two of the problems ask you to establish: if

$$f(x) = \begin{cases} x^{2n} \sin \dfrac{1}{x}, & x \neq 0 \\ 0, & x = 0, \end{cases}$$

then $f'(0), \ldots, f^{(n)}(0)$ exist, but $f^{(n)}$ is not continuous at 0; if

$$f(x) = \begin{cases} x^{2n+1} \sin \dfrac{1}{x}, & x \neq 0 \\ 0, & x = 0, \end{cases}$$

then $f'(0), \ldots, f^{(n)}(0)$ exist, and $f^{(n)}$ is continuous at 0, but $f^{(n)}$ is not differentiable at 0. These examples may suggest that "reasonable" functions can be characterized by the possession of higher-order derivatives—no matter how hard we try to mask the infinite oscillation of $f(x) = \sin 1/x$, a derivative of sufficiently high order seems able to reveal the underlying irregularity. Unfortunately, we will see later that much worse things can happen.

After all these involved calculations, we will bring this chapter to a close with a minor remark. It is often tempting, and seems more elegant, to write some of the theorems in this chapter as equations about functions, rather than about their values. Thus Theorem 3 might be written

$$(f + g)' = f' + g',$$

Theorem 4 might be written as

$$(f \cdot g)' = f \cdot g' + f' \cdot g,$$

and Theorem 9 often appears in the form

$$(f \circ g)' = (f' \circ g) \cdot g'.$$

Strictly speaking, these equations may be false, because the functions on the left-hand side might have a larger domain than those on the right. Nevertheless, this is hardly worth worrying about. If f and g are differentiable everywhere in their domains, then these equations, and others like them, *are* true, and this is the only case any one cares about.

PROBLEMS

1. As a warm up exercise, find $f'(x)$ for each of the following f. (Don't worry about the domain of f or f'; just get a formula for $f'(x)$ that gives the right answer when it makes sense.)

(i) $f(x) = \sin(x + x^2)$.
(ii) $f(x) = \sin x + \sin x^2$.
(iii) $f(x) = \sin(\cos x)$.
(iv) $f(x) = \sin(\sin x)$.
(v) $f(x) = \sin\left(\frac{\cos x}{x}\right)$.
(vi) $f(x) = \frac{\sin(\cos x)}{x}$.
(vii) $f(x) = \sin(x + \sin x)$.
(viii) $f(x) = \sin(\cos(\sin x))$.

2. Find $f'(x)$ for each of the following functions f. (It took the author 20 minutes to compute the derivatives for the answer section, and it should not take you much longer. Although rapid calculation is not the goal of mathematics, if you hope to treat theoretical applications of the Chain Rule with aplomb, these concrete applications should be child's play—mathematicians like to pretend that they can't even add, but most of them can when they have to.)

(i) $f(x) = \sin((x + 1)^2(x + 2))$.
(ii) $f(x) = \sin^3(x^2 + \sin x)$.
(iii) $f(x) = \sin^2((x + \sin x)^2)$.
(iv) $f(x) = \sin\left(\frac{x^3}{\cos x^3}\right)$.
(v) $f(x) = \sin(x \sin x) + \sin(\sin x^2)$.
(vi) $f(x) = (\cos x)^{31^2}$.
(vii) $f(x) = \sin^2 x \sin x^2 \sin^2 x^2$.
(viii) $f(x) = \sin^3(\sin^2(\sin x))$.
(ix) $f(x) = (x + \sin^5 x)^6$.
(x) $f(x) = \sin(\sin(\sin(\sin(\sin x))))$.
(xi) $f(x) = \sin((\sin^7 x^7 + 1)^7)$.
(xii) $f(x) = (((x^2 + x)^3 + x)^4 + x)^5$.
(xiii) $f(x) = \sin(x^2 + \sin(x^2 + \sin x^2))$.
(xiv) $f(x) = \sin(6\cos(6\sin(6\cos 6x)))$.

(xv) $f(x) = \dfrac{\sin x^2 \sin^2 x}{1 + \sin x}$.

(xvi) $f(x) = \dfrac{1}{x - \dfrac{2}{x + \sin x}}$.

(xvii) $f(x) = \sin\left(\dfrac{x^3}{\sin\left(\dfrac{x^3}{\sin x}\right)}\right)$.

(xviii) $f(x) = \sin\left(\dfrac{x}{x - \sin\left(\dfrac{x}{x - \sin x}\right)}\right)$.

3. Find the derivatives of the functions tan, cotan, sec, cosec. (You don't have to memorize these formulas, although they will be needed once in a while; if you express your answers in the right way, they will be simple and somewhat symmetrical.)

4. For each of the following functions f, find $f'(f(x))$ (*not* $(f \circ f)'(x)$).

(i) $f(x) = \dfrac{1}{1+x}$.
(ii) $f(x) = \sin x$.
(iii) $f(x) = x^2$.
(iv) $f(x) = 17$.

5. For each of the following functions f, find $f(f'(x))$.

(i) $f(x) = \dfrac{1}{x}$.
(ii) $f(x) = x^2$.
(iii) $f(x) = 17$.
(iv) $f(x) = 17x$.

6. Find f' in terms of g' if

(i) $f(x) = g(x + g(a))$.
(ii) $f(x) = g(x \cdot g(a))$.
(iii) $f(x) = g(x + g(x))$.
(iv) $f(x) = g(x)(x - a)$.
(v) $f(x) = g(a)(x - a)$.
(vi) $f(x + 3) = g(x^2)$.

7. (a) A circular object is increasing in size in some unspecified manner, but it is known that when the radius is 6, the rate of change of the radius is 4. Find the rate of change of the area when the radius is 6. (If $r(t)$ and $A(t)$ represent the radius and the area at time t, then the functions r and A satisfy $A = \pi r^2$; a straightforward use of the Chain Rule is called for.)

(b) Suppose that we are now informed that the circular object we have been watching is really the cross section of a spherical object. Find the rate of change of the *volume* when the radius is 6. (You will clearly need to know a formula for the volume of a sphere; in case you have forgotten, the volume is $\frac{4}{3}\pi$ times the cube of the radius.)

(c) Now suppose that the rate of change of the area of the circular cross section is 5 when the radius is 3. Find the rate of change of the volume when the radius is 3. You should be able to do this problem in two ways: first, by using the formulas for the area and volume in terms of the radius; and then by expressing the volume in terms of the area (to use this method you will need Problem 9-3).

8. The area between two varying concentric circles is at all times 9π in^2. The rate of change of the area of the larger circle is 10π in^2/sec. How fast is the circumference of the smaller circle changing when it has area 16π in^2?

9. Particle A moves along the positive horizontal axis, and particle B along the graph of $f(x) = -\sqrt{3}x$, $x \le 0$. At a certain time, A is at the point (5,0) and moving with speed 3 units/sec; and B is at a distance of 3 units from the origin and moving with speed 4 units/sec. At what rate is the distance between A and B changing?

10. Let $f(x) = x^2 \sin 1/x$ for $x \ne 0$, and let $f(0) = 0$. Suppose also that h and k are two functions such that

$$\begin{aligned} h'(x) &= \sin^2(\sin(x+1)) \qquad & k'(x) &= f(x+1) \\ h(0) &= 3 & k(0) &= 0. \end{aligned}$$

Find

(i) $(f \circ h)'(0)$.
(ii) $(k \circ f)'(0)$.
(iii) $\alpha'(x^2)$, where $\alpha(x) = h(x^2)$. Exercise great care.

11. Find $f'(0)$ if

$$f(x) = \begin{cases} g(x) \sin \dfrac{1}{x}, & x \ne 0 \\ 0, & x = 0, \end{cases}$$

and

$$g(0) = g'(0) = 0.$$

12. Using the derivative of $f(x) = 1/x$, as found in Problem 9-1, find $(1/g)'(x)$ by the Chain Rule.

13. (a) Using Problem 9-3, find $f'(x)$ for $-1 < x < 1$, if $f(x) = \sqrt{1 - x^2}$.

(b) Prove that the tangent line to the graph of f at $(a, \sqrt{1 - a^2})$ intersects the graph only at that point (and thus show that the elementary geometry definition of the tangent line coincides with ours).

14. Prove similarly that the tangent lines to an ellipse or hyperbola intersect these sets only once.

15. If $f+g$ is differentiable at a, are f and g necessarily differentiable at a? If $f \cdot g$ and f are differentiable at a, what conditions on f imply that g is differentiable at a?

16. (a) Prove that if f is differentiable at a, then $|f|$ is also differentiable at a, provided that $f(a) \neq 0$.
(b) Give a counterexample if $f(a) = 0$.
(c) Prove that if f and g are differentiable at a, then the functions $\max(f, g)$ and $\min(f, g)$ are differentiable at a, provided that $f(a) \neq g(a)$.
(d) Give a counterexample if $f(a) = g(a)$.

17. If f is three times differentiable and $f'(x) \neq 0$, the *Schwarzian derivative* of f at x is defined to be

$$\mathscr{D}f(x) = \frac{f'''(x)}{f'(x)} - \frac{3}{2}\left(\frac{f''(x)}{f'(x)}\right)^2.$$

(a) Show that

$$\mathscr{D}(f \circ g) = [\mathscr{D}f \circ g] \cdot g'^2 + \mathscr{D}g.$$

(b) Show that if $f(x) = \dfrac{ax+b}{cx+d}$, with $ad - bc \neq 0$, then $\mathscr{D}f = 0$. Consequently, $\mathscr{D}(f \circ g) = \mathscr{D}g$.

18. Suppose that $f^{(n)}(a)$ and $g^{(n)}(a)$ exist. Prove *Leibniz's formula*:

$$(f \cdot g)^{(n)}(a) = \sum_{k=0}^{n} \binom{n}{k} f^{(k)}(a) \cdot g^{n-k}(a).$$

***19.** Prove that if $f^{(n)}(g(a))$ and $g^{(n)}(a)$ both exist, then $(f \circ g)^{(n)}(a)$ exists. A little experimentation should convince you that it is unwise to seek a formula for $(f \circ g)^{(n)}(a)$. In order to prove that $(f \circ g)^{(n)}(a)$ exists you will therefore have to devise a reasonable assertion about $(f \circ g)^{(n)}(a)$ which can be proved by induction. Try something like: "$(f \circ g)^{(n)}(a)$ exists and is a sum of terms each of which is a product of terms of the form . . . ."

20. (a) If $f(x) = a_n x^n + a_{n-1}x^{n-1} + \cdots + a_0$, find a function g such that $g' = f$. Find another.
(b) If

$$f(x) = \frac{b_2}{x^2} + \frac{b_3}{x^3} + \cdots + \frac{b_m}{x^m},$$

find a function g with $g' = f$.
(c) Is there a function

$$f(x) = a_n x^n + \cdots + a_0 + \frac{b_1}{x} + \cdots + \frac{b_m}{x^m}$$

such that $f'(x) = 1/x$?

21. Show that there is a polynomial function f of degree n such that

(a) $f'(x) = 0$ for precisely $n - 1$ numbers x.
(b) $f'(x) = 0$ for no x, if n is odd.
(c) $f'(x) = 0$ for exactly one x, if n is even.
(d) $f'(x) = 0$ for exactly k numbers x, if $n - k$ is odd.

22. (a) The number a is called a **double root** of the polynomial function f if $f(x) = (x - a)^2 g(x)$ for some polynomial function g. Prove that a is a double root of f if and only if a is a root of both f and f'.
(b) When does $f(x) = ax^2 + bx + c$ ($a \neq 0$) have a double root? What does the condition say geometrically?

23. If f is differentiable at a, let $d(x) = f(x) - f'(a)(x - a) - f(a)$. Find $d'(a)$. In connection with Problem 22, this gives another solution for Problem 9-20.

***24.** This problem is a companion to Problem 3-6. Let $a_1, \ldots, a_n$ and $b_1, \ldots, b_n$ be given numbers.

(a) If $x_1, \ldots, x_n$ are distinct numbers, prove that there is a polynomial function f of degree $2n - 1$, such that $f(x_j) = f'(x_j) = 0$ for $j \neq i$, and $f(x_i) = a_i$ and $f'(x_i) = b_i$. Hint: Remember Problem 22.
(b) Prove that there is a polynomial function f of degree $2n - 1$ with $f(x_i) = a_i$ and $f'(x_i) = b_i$ for all i.

***25.** Suppose that a and b are two consecutive roots of a polynomial function f, but that a and b are not double roots, so that we can write $f(x) = (x - a)(x - b)g(x)$ where $g(a) \neq 0$ and $g(b) \neq 0$.

(a) Prove that $g(a)$ and $g(b)$ have the same sign. (Remember that a and b are consecutive roots.)
(b) Prove that there is some number x with $a < x < b$ and $f'(x) = 0$. (Also draw a picture to illustrate this fact.) Hint: Compare the sign of $f'(a)$ and $f'(b)$.
(c) Now prove the same fact, even if a and b are multiple roots. Hint: If $f(x) = (x - a)^m (x - b)^n g(x)$ where $g(a) \neq 0$ and $g(b) \neq 0$, consider the polynomial function $h(x) = f'(x)/(x - a)^{m-1}(x - b)^{n-1}$.

This theorem was proved by the French mathematician Rolle, in connection with the problem of approximating roots of polynomials, but the result was not originally stated in terms of derivatives. In fact, Rolle was one of the mathematicians who never accepted the new notions of calculus. This was not such a pigheaded attitude, in view of the fact that for one hundred years no one could define limits in terms that did not verge on the mystic, but on the whole history has been particularly kind to Rolle; his name has become attached to a much more general result, to appear in the next chapter, which forms the basis for the most important theoretical results of calculus.

26. Suppose that $f(x) = xg(x)$ for some function g which is continuous at 0. Prove that f is differentiable at 0, and find $f'(0)$ in terms of g.

***27.** Suppose f is differentiable at 0, and that $f(0) = 0$. Prove that $f(x) = xg(x)$ for some function g which is continuous at 0. Hint: What happens if you try to write $g(x) = f(x)/x$?

28. If $f(x) = x^{-n}$ for n in $\mathbf{N}$, prove that

$$\begin{aligned} f^{(k)}(x) &= (-1)^k \frac{(n+k-1)!}{(n-1)!} x^{-n-k} \\ &= (-1)^k k! \binom{n+k-1}{k-1} x^{-n-k}, \qquad \text{for } x \neq 0. \end{aligned}$$

***29.** Prove that it is impossible to write $x = f(x)g(x)$ where f and g are differentiable and $f(0) = g(0) = 0$. Hint: Differentiate.

30. What is $f^{(k)}(x)$ if

(a) $f(x) = 1/(x-a)^n$?
*(b) $f(x) = 1/(x^2-1)$?

***31.** Let $f(x) = x^{2n} \sin 1/x$ if $x \neq 0$, and let $f(0) = 0$. Prove that $f'(0), \dots, f^{(n)}(0)$ exist, and that $f^{(n)}$ is not continuous at 0. (You will encounter the same basic difficulty as that in Problem 19.)

***32.** Let $f(x) = x^{2n+1} \sin 1/x$ if $x \neq 0$, and let $f(0) = 0$. Prove that $f'(0), \dots, f^{(n)}(0)$ exist, that $f^{(n)}$ is continuous at 0, and that $f^{(n)}$ is not differentiable at 0.

33. In Leibnizian notation the Chain Rule ought to read:

$$\frac{df(g(x))}{dx} = \left.\frac{df(y)}{dy}\right|_{y=g(x)} \cdot \frac{dg(x)}{dx}.$$

Instead, one usually finds the following statement: "Let $y = g(x)$ and $z = f(y)$. Then

$$\frac{dz}{dx} = \frac{dz}{dy} \cdot \frac{dy}{dx}."$$

Notice that the z in dz/dx denotes the composite function $f \circ g$, while the z in dz/dy denotes the function f; it is also understood that dz/dy will be "an expression involving y," and that in the final answer $g(x)$ must be substituted for y. In each of the following cases, find dz/dx by using this formula; then compare with Problem 1.

(i) $z = \sin y, \quad y = x + x^2.$
(ii) $z = \sin y, \quad y = \cos x.$
(iii) $z = \cos u, \quad u = \sin x.$
(iv) $z = \sin v, \quad v = \cos u, \quad u = \sin x.$

CHAPTER 11 SIGNIFICANCE OF THE DERIVATIVE

One aim in this chapter is to justify the time we have spent learning to find the derivative of a function. As we shall see, knowing just a little about f' tells us a lot about f. Extracting information about f from information about f' requires some difficult work, however, and we shall begin with the one theorem which is really easy.

This theorem is concerned with the maximum value of a function on an interval. Although we have used this term informally in Chapter 7, it is worthwhile to be precise, and also more general.

DEFINITION

Let f be a function and A a set of numbers contained in the domain of f. A point x in A is a **maximum point** for f on A if

$$f(x) \geq f(y) \quad \text{for every } y \text{ in } A.$$

The number $f(x)$ itself is called the **maximum value** of f on A (and we also say that f "has its maximum value on A at x").

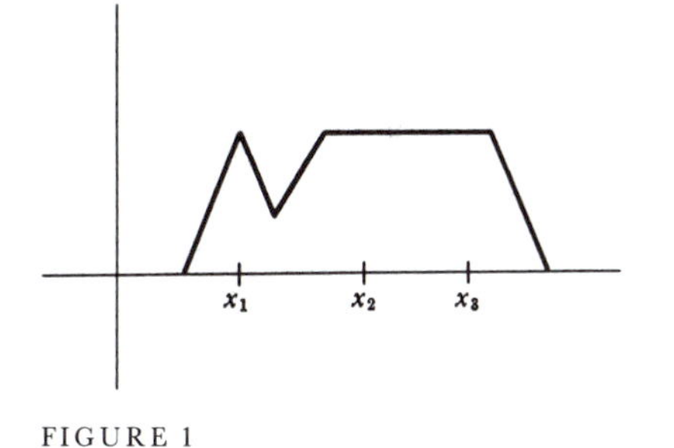

FIGURE 1

Notice that the maximum value of f on A could be $f(x)$ for several different x (Figure 1); in other words, a function f can have several different maximum points on A, although it can have at most one maximum value. Usually we shall be interested in the case where A is a closed interval $[a, b]$; if f is continuous, then Theorem 7-3 guarantees that f does indeed have a maximum value on $[a, b]$.

The definition of a minimum of f on A will be left to you. (One possible definition is the following: f has a minimum on A at x, if $-f$ has a maximum on A at x.)

We are now ready for a theorem which does not even depend upon the existence of least upper bounds.

THEOREM 1 Let f be any function defined on (a, b). If x is a maximum (or a minimum) point for f on (a, b), and f is differentiable at x, then $f'(x) = 0$.
(Notice that we do not assume differentiability, or even continuity, of f at other points.)

PROOF Consider the case where f has a maximum at x. Figure 2 illustrates the simple idea behind the whole argument—secants drawn through points to the left of $(x, f(x))$ have slopes ≥ 0, and secants drawn through points to the right of $(x, f(x))$ have slopes ≤ 0. Analytically, this argument proceeds as follows.

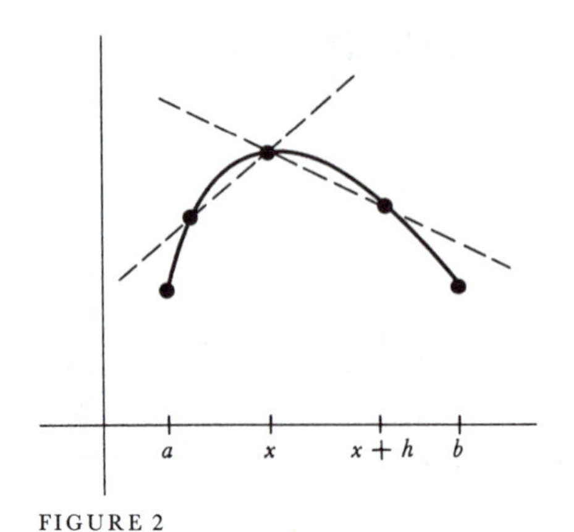

FIGURE 2

If h is any number such that $x+h$ is in (a,b), then

$$f(x) \geq f(x+h),$$

since f has a maximum on (a,b) at x. This means that

$$f(x+h) - f(x) \leq 0.$$

Thus, if $h > 0$ we have

$$\frac{f(x+h)-f(x)}{h} \leq 0.$$

and consequently

$$\lim_{h \to 0^+} \frac{f(x+h)-f(x)}{h} \leq 0.$$

On the other hand, if $h < 0$, we have

$$\frac{f(x+h)-f(x)}{h} \geq 0,$$

so

$$\lim_{h \to 0^-} \frac{f(x+h)-f(x)}{h} \geq 0.$$

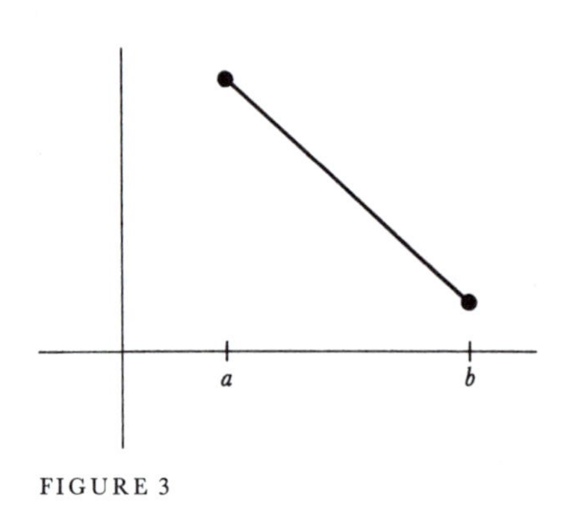

FIGURE 3

By hypothesis, f is differentiable at x, so these two limits must be equal, in fact equal to $f'(x)$. This means that

$$f'(x) \leq 0 \quad \text{and} \quad f'(x) \geq 0,$$

from which it follows that $f'(x) = 0$.

The case where f has a minimum at x is left to you (give a one-line proof). ▌

Notice (Figure 3) that we cannot replace (a,b) by $[a,b]$ in the statement of the theorem (unless we add to the hypothesis the condition that x is in (a,b).)

Since $f'(x)$ depends only on the values of f near x, it is almost obvious how to get a stronger version of Theorem 1. We begin with a definition which is illustrated in Figure 4.

DEFINITION Let f be a function, and A a set of numbers contained in the domain of f. A point x in A is a **local maximum [minimum] point** for f on A if there is some $\delta > 0$ such that x is a maximum [minimum] point for f on $A \cap (x-\delta, x+\delta)$.

THEOREM 2 If f is defined on (a,b) and has a local maximum (or minimum) at x, and f is differentiable at x, then $f'(x) = 0$.

PROOF You should see why this is an easy application of Theorem 1. ▌

The converse of Theorem 2 is definitely not true—it is possible for $f'(x)$ to be 0 even if x is not a local maximum or minimum point for f. The simplest example is provided by the function $f(x) = x^3$; in this case $f'(0) = 0$, but f has no local maximum or minimum anywhere.

Probably the most widespread misconceptions about calculus are concerned with the behavior of a function f near x when $f'(x) = 0$. The point made in the previous paragraph is so quickly forgotten by those who want the world to be simpler than it is, that we will repeat it: the converse of Theorem 2 is *not* true—the condition $f'(x) = 0$ does *not* imply that x is a local maximum or minimum point of f. Precisely for this reason, special terminology has been adopted to describe numbers x which satisfy the condition $f'(x) = 0$.

DEFINITION

A **critical point** of a function f is a number x such that

$$f'(x) = 0.$$

The number $f(x)$ itself is called a **critical value** of f.

The critical values of f, together with a few other numbers, turn out to be the ones which must be considered in order to find the maximum and minimum of a given function f. To the uninitiated, finding the maximum and minimum value of a function represents one of the most intriguing aspects of calculus, and there is no denying that problems of this sort are fun (until you have done your first hundred or so).

Let us consider first the problem of finding the maximum or minimum of f on a closed interval $[a, b]$. (Then, if f is continuous, we can at least be sure that a maximum and minimum value exist.) In order to locate the maximum and minimum of f three kinds of points must be considered:

(1) The critical points of f in $[a, b]$.
(2) The end points a and b.
(3) Points x in $[a, b]$ such that f is not differentiable at x.

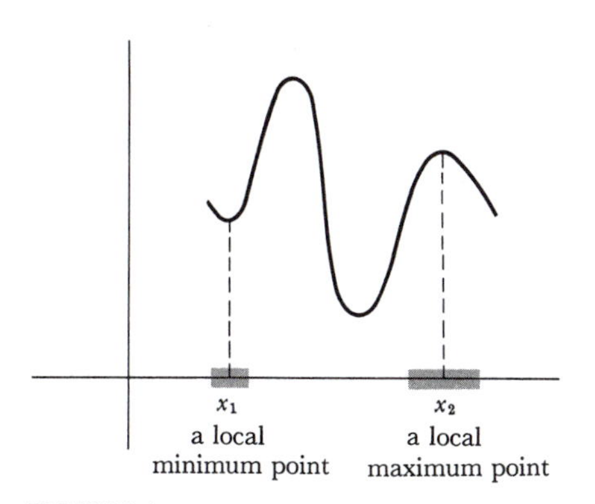

FIGURE 4

If x is a maximum point or a minimum point for f on $[a, b]$, then x must be in one of the three classes listed above: for if x is not in the second or third group, then x *is* in (a, b) and f *is* differentiable at x; consequently $f'(x) = 0$, by Theorem 1, and this means that x is in the first group.

If there are many points in these three categories, finding the maximum and minimum of f may still be a hopeless proposition, but when there are only a few critical points, and only a few points where f is not differentiable, the procedure is fairly straightforward: one simply finds $f(x)$ for each x satisfying $f'(x) = 0$, and $f(x)$ for each x such that f is not differentiable at x and, finally, $f(a)$ and $f(b)$. The biggest of these will be the maximum value of f, and the smallest will be the minimum. A simple example follows.

Suppose we wish to find the maximum and minimum value of the function

$$f(x) = x^3 - x$$

on the interval $[-1, 2]$. To begin with, we have

$$f'(x) = 3x^2 - 1,$$

so $f'(x) = 0$ when $3x^2 - 1 = 0$, that is, when

$$x = \sqrt{1/3} \quad \text{or} \quad -\sqrt{1/3}.$$

The numbers $\sqrt{1/3}$ and $-\sqrt{1/3}$ both lie in $[-1, 2]$, so the first group of candidates for the location of the maximum and the minimum is

$$(1) \quad \sqrt{1/3},\ -\sqrt{1/3}.$$

The second group contains the end points of the interval,

$$(2) \quad -1,\ 2.$$

The third group is empty, since f is differentiable everywhere. The final step is to compute

$$\begin{aligned} f(\sqrt{1/3}) &= (\sqrt{1/3}\,)^3 - \sqrt{1/3} = \tfrac{1}{3}\sqrt{1/3} - \sqrt{1/3} = -\tfrac{2}{3}\sqrt{1/3}, \\ f(-\sqrt{1/3}) &= (-\sqrt{1/3}\,)^3 - (-\sqrt{1/3}\,) = -\tfrac{1}{3}\sqrt{1/3} + \sqrt{1/3} = \tfrac{2}{3}\sqrt{1/3}, \\ f(-1) &= 0, \\ f(2) &= 6. \end{aligned}$$

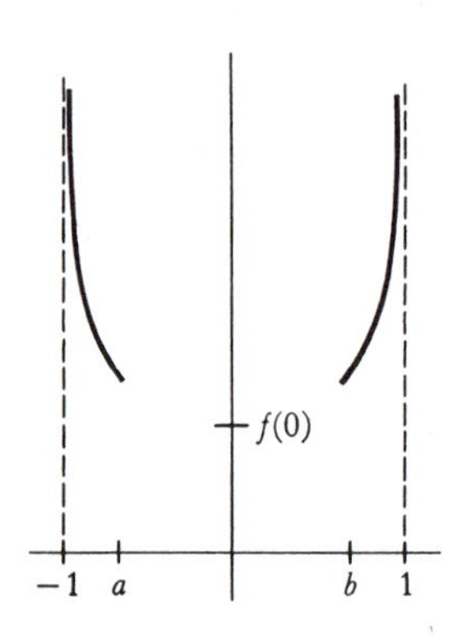

FIGURE 5

Clearly the minimum value is $-\frac{2}{3}\sqrt{1/3}$, occurring at $\sqrt{1/3}$, and the maximum value is 6, occurring at 2.

This sort of procedure, if feasible, will always locate the maximum and minimum value of a continuous function on a closed interval. If the function we are dealing with is not continuous, however, or if we are seeking the maximum or minimum on an open interval or the whole line, then we cannot even be sure beforehand that the maximum and minimum values exist, so all the information obtained by this procedure may say nothing. Nevertheless, a little ingenuity will often reveal the nature of things. In Chapter 7 we solved just such a problem when we showed that if n is even, then the function

$$f(x) = x^n + a_{n-1}x^{n-1} + \cdots + a_0$$

has a minimum value on the whole line. This proves that the minimum value must occur at some number x satisfying

$$0 = f'(x) = nx^{n-1} + (n-1)a_{n-1}x^{n-2} + \cdots + a_1.$$

If we can solve this equation, and compare the values of $f(x)$ for such x, we can actually find the minimum of f. One more example may be helpful. Suppose we wish to find the maximum and minimum, if they exist, of the function

$$f(x) = \frac{1}{1 - x^2}$$

on the open interval $(-1, 1)$. We have

$$f'(x) = \frac{2x}{(1-x^2)^2}$$

so $f'(x) = 0$ only for $x = 0$. We can see immediately that for x close to 1 or -1 the values of $f(x)$ become arbitrarily large, so f certainly does not have a maximum. This observation also makes it easy to show that f has a minimum at 0. We just note (Figure 5) that there will be numbers a and b, with

$$-1 < a < 0 \quad \text{and} \quad 0 < b < 1,$$

such that $f(x) > f(0)$ for

$$-1 < x \le a \quad \text{and} \quad b \le x < 1.$$

This means that the minimum of f on $[a, b]$ is the minimum of f on all of $(-1, 1)$. Now on $[a, b]$ the minimum occurs either at 0 (the only place where $f' = 0$), or at a or b, and a and b have already been ruled out, so the minimum value is $f(0) = 1$.

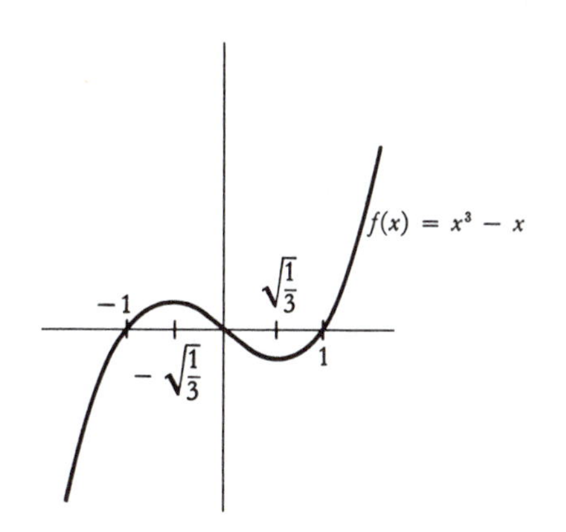

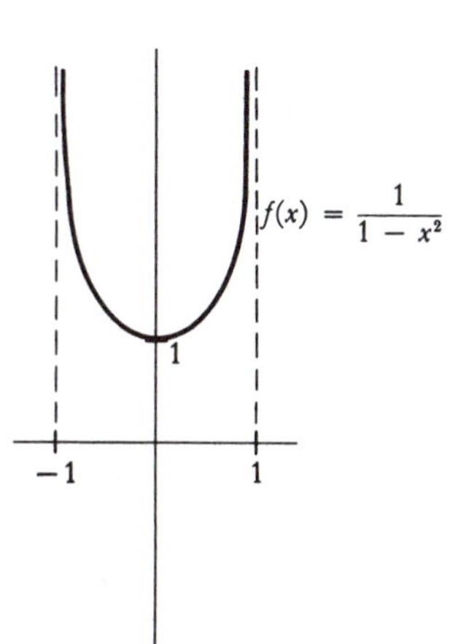

FIGURE 6

In solving these problems we purposely did not draw the graphs of $f(x) = x^3 - x$ and $f(x) = 1/(1 - x^2)$, but it is not cheating to draw the graph (Figure 6) as long as you do not rely solely on your picture to prove anything. As a matter of fact, we are now going to discuss a method of sketching the graph of a function that really gives enough information to be used in discussing maxima and minima—in fact we will be able to locate even *local* maxima and minima. This method involves consideration of the sign of $f'(x)$, and relies on some deep theorems.

The theorems about derivatives which have been proved so far, always yield information about f' in terms of information about f. This is true even of Theorem 1, although this theorem can sometimes be used to determine certain information about f, namely, the location of maxima and minima. When the derivative was first introduced, we emphasized that $f'(x)$ is not $[f(x+h) - f(x)]/h$ for any particular h, but only a limit of these numbers as h approaches 0; this fact becomes painfully relevant when one tries to extract information about f from information about f'. The simplest and most frustrating illustration of the difficulties encountered is afforded by the following question: If $f'(x) = 0$ for all x, must f be a constant function? It is impossible to imagine how f could be anything else, and this conviction is strengthened by considering the physical interpretation—if the velocity of a particle is always 0, surely the particle must be standing still! Nevertheless it is difficult even to begin a proof that only the constant functions satisfy $f'(x) = 0$ for all x. The hypothesis $f'(x) = 0$ only means that

$$\lim_{h \to 0} \frac{f(x+h) - f(x)}{h} = 0,$$

and it is not at all obvious how one can use the information about the limit to derive information about the function.

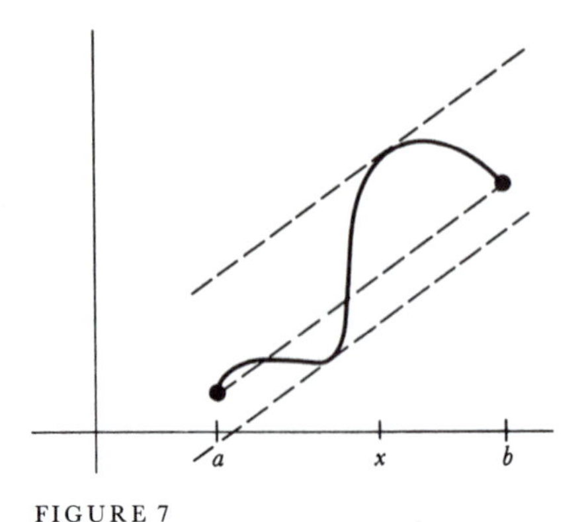

FIGURE 7

The fact that f is a constant function if $f'(x) = 0$ for all x, and many other facts of the same sort, can all be derived from a fundamental theorem, called the Mean Value Theorem, which states much stronger results. Figure 7 makes it plausible that if f is differentiable on $[a, b]$, then there is some x in (a, b) such that

$$f'(x) = \frac{f(b) - f(a)}{b - a}.$$

Geometrically this means that some tangent line is parallel to the line between $(a, f(a))$ and $(b, f(b))$. The Mean Value Theorem asserts that this is true—there is some x in (a, b) such that $f'(x)$, the instantaneous rate of change of f at x, is exactly equal to the average or "mean" change of f on $[a, b]$, this average change being $[f(b) - f(a)]/[b - a]$. (For example, if you travel 60 miles in one hour, then at some time you must have been traveling exactly 60 miles per hour.) This theorem is one of the most important theoretical tools of calculus—probably the deepest result about derivatives. From this statement you might conclude that the proof is difficult, but there you would be wrong—the hard theorems in this book have occurred long ago, in Chapter 7. It is true that if you try to prove the Mean Value Theorem yourself you will probably fail, but this is neither evidence that the theorem is hard, nor something to be ashamed of. The first proof of the theorem was an achievement, but today we can supply a proof which is quite simple. It helps to begin with a very special case.

FIGURE 8

THEOREM 3 (ROLLE'S THEOREM) If f is continuous on $[a, b]$ and differentiable on (a, b), and $f(a) = f(b)$, then there is a number x in (a, b) such that $f'(x) = 0$.

PROOF If follows from the continuity of f on $[a, b]$ that f has a maximum and a minimum value on $[a, b]$.

Suppose first that the maximum value occurs at a point x in (a, b). Then $f'(x) = 0$ by Theorem 1, and we are done (Figure 8).

Suppose next that the minimum value of f occurs at some point x in (a, b). Then, again, $f'(x) = 0$ by Theorem 1 (Figure 9).

Finally, suppose the maximum and minimum values both occur at the end points. Since $f(a) = f(b)$, the maximum and minimum values of f are equal, so f is a constant function (Figure 10), and for a constant function we can choose any x in (a, b). ▮

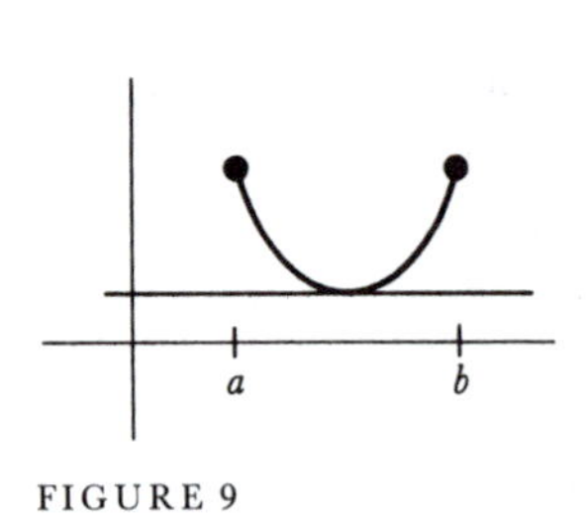

FIGURE 9

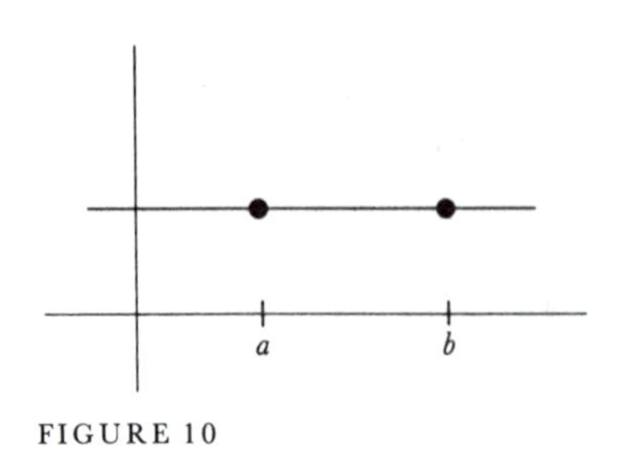

FIGURE 10

Notice that we really needed the hypothesis that f is differentiable everywhere on (a, b) in order to apply Theorem 1. Without this assumption the theorem is false (Figure 11).

You may wonder why a special name should be attached to a theorem as easily proved as Rolle's Theorem. The reason is, that although Rolle's Theorem is a special case of the Mean Value Theorem, it also yields a simple proof of the Mean Value Theorem. In order to prove the Mean Value Theorem we will apply Rolle's Theorem to the function which gives the length of the vertical segment shown in

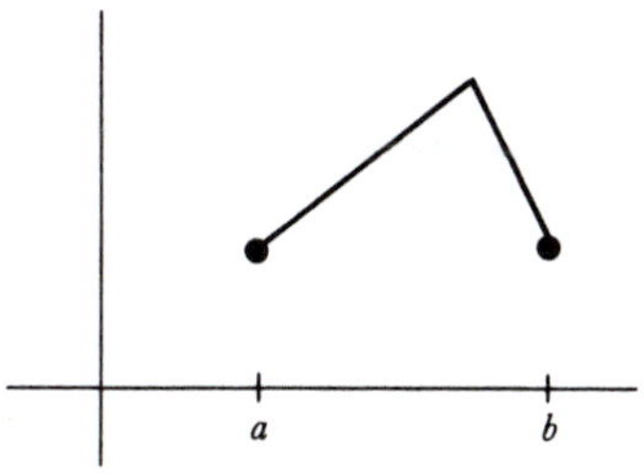

FIGURE 11

Figure 12; this is the difference between $f(x)$, and the height at x of the line L between $(a, f(a))$ and $(b, f(b))$. Since L is the graph of

$$g(x) = \left[\frac{f(b) - f(a)}{b - a}\right](x - a) + f(a),$$

we want to look at

$$f(x) - \left[\frac{f(b) - f(a)}{b - a}\right](x - a) - f(a).$$

As it turns out, the constant $f(a)$ is irrelevant.

THEOREM 4 (THE MEAN VALUE THEOREM) If f is continuous on $[a, b]$ and differentiable on (a, b), then there is a number x in (a, b) such that

$$f'(x) = \frac{f(b) - f(a)}{b - a}.$$

PROOF Let

$$h(x) = f(x) - \left[\frac{f(b) - f(a)}{b - a}\right](x - a).$$

Clearly, h is continuous on $[a, b]$ and differentiable on (a, b), and

$$\begin{aligned} h(a) &= f(a), \\ h(b) &= f(b) - \left[\frac{f(b) - f(a)}{b - a}\right](x - a) \\ &= f(a). \end{aligned}$$

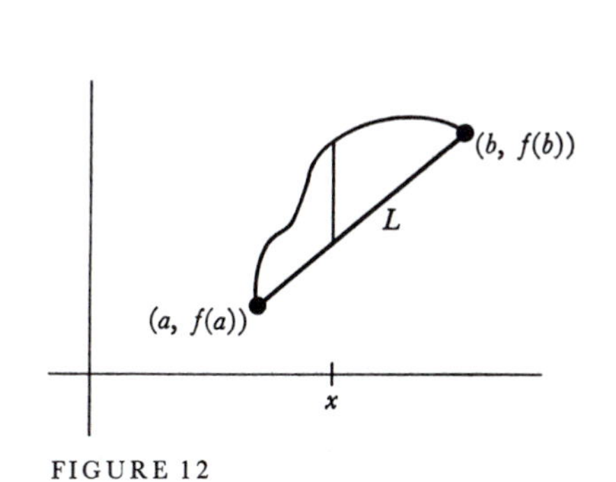

FIGURE 12

Consequently, we may apply Rolle's Theorem to h and conclude that there is some x in (a, b) such that

$$0 = h'(x) = f'(x) - \frac{f(b) - f(a)}{b - a},$$

so that

$$f'(x) = \frac{f(b) - f(a)}{b - a}. \blacksquare$$

Notice that the Mean Value Theorem still fits into the pattern exhibited by previous theorems—information about f yields information about f'. This information is so strong, however, that we can now go in the other direction.

COROLLARY 1 If f is defined on an interval and $f'(x) = 0$ for all x in the interval, then f is constant on the interval.

PROOF Let a and b be any two points in the interval with $a \neq b$. Then there is some x in

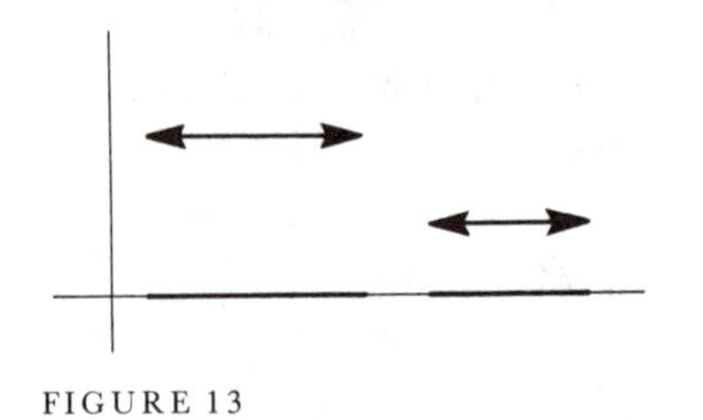

FIGURE 13

(a, b) such that

$$f'(x) = \frac{f(b) - f(a)}{b - a}.$$

But $f'(x) = 0$ for all x in the interval, so

$$0 = \frac{f(b) - f(a)}{b - a},$$

and consequently $f(a) = f(b)$. Thus the value of f at any two points in the interval is the same, i.e., f is constant on the interval. ▮

Naturally, Corollary 1 does not hold for functions defined on two or more intervals (Figure 13).

COROLLARY 2 If f and g are defined on the same interval, and $f'(x) = g'(x)$ for all x in the interval, then there is some number c such that $f = g + c$.

PROOF For all x in the interval we have $(f - g)'(x) = f'(x) - g'(x) = 0$ so, by Corollary 1, there is a number c such that $f - g = c$. ▮

The statement of the next corollary requires some terminology, which is illustrated in Figure 14.

DEFINITION

A function is **increasing** on an interval if $f(a) < f(b)$ whenever a and b are two numbers in the interval with $a < b$. The function f is **decreasing** on an interval if $f(a) > f(b)$ for all a and b in the interval with $a < b$. (We often say simply that f is increasing or decreasing, in which case the interval is understood to be the domain of f.)

COROLLARY 3 If $f'(x) > 0$ for all x in an interval, then f is increasing on the interval; if $f'(x) < 0$ for all x in the interval, then f is decreasing on the interval.

PROOF Consider the case where $f'(x) > 0$. Let a and b be two points in the interval with $a < b$. Then there is some x in (a, b) with

$$f'(x) = \frac{f(b) - f(a)}{b - a}.$$

But $f'(x) > 0$ for all x in (a, b), so

$$\frac{f(b) - f(a)}{b - a} > 0.$$

Since $b - a > 0$ it follows that $f(b) > f(a)$.

The proof when $f'(x) < 0$ for all x is left to you. ▮

Notice that although the converses of Corollary 1 and Corollary 2 are true (and obvious), the converse of Corollary 3 is not true. If f is increasing, it is easy to see that $f'(x) \geq 0$ for all x, but the equality sign might hold for some x (consider $f(x) = x^3$).

Corollary 3 provides enough information to get a good idea of the graph of a function with a minimal amount of point plotting. Consider, once more, the function $f(x) = x^3 - x$. We have

$$f'(x) = 3x^2 - 1.$$

We have already noted that $f'(x) = 0$ for $x = \sqrt{1/3}$ and $x = -\sqrt{1/3}$, and it is also possible to determine the sign of $f'(x)$ for all other x. Note that $3x^2 - 1 > 0$ precisely when

$$\begin{aligned} 3x^2 &> 1 \\ x^2 &> \tfrac{1}{3}, \end{aligned}$$
$$x > \sqrt{1/3} \quad \text{or} \quad x < -\sqrt{1/3};$$

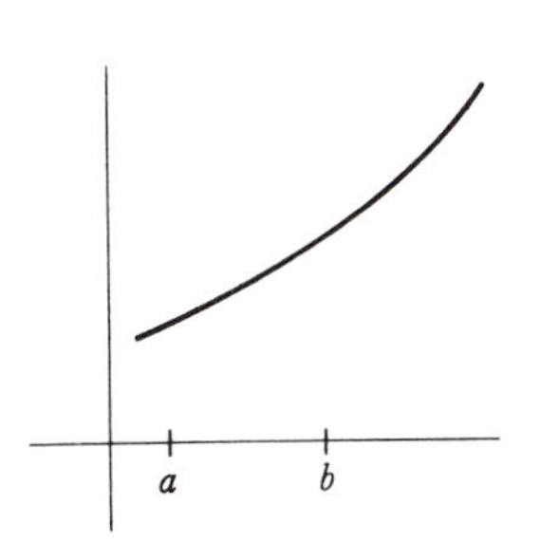

(a) an increasing function

thus $3x^2 - 1 < 0$ precisely when

$$-\sqrt{1/3} < x < \sqrt{1/3}.$$

Thus f is increasing for $x < -\sqrt{1/3}$, decreasing between $-\sqrt{1/3}$ and $\sqrt{1/3}$, and once again increasing for $x > \sqrt{1/3}$. Combining this information with the following facts

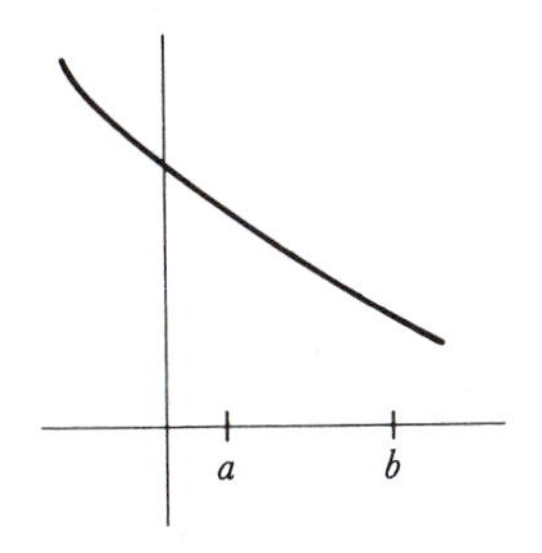

(b) a decreasing function

FIGURE 14

(1) $f(-\sqrt{1/3}) = \frac{2}{3}\sqrt{1/3}$,
$f(\sqrt{1/3}) = -\frac{2}{3}\sqrt{1/3}$,
(2) $f(x) = 0$ for $x = -1, 0, 1$,
(3) $f(x)$ gets large as x gets large, and large negative as x gets large negative,

it is possible to sketch a pretty respectable approximation to the graph (Figure 15).

By the way, notice that the intervals on which f increases and decreases could have been found without even bothering to examine the sign of f'. For example, since f' is continuous, and vanishes only at $-\sqrt{1/3}$ and $\sqrt{1/3}$, we know that f' always has the same sign on the interval $(-\sqrt{1/3}, \sqrt{1/3})$. Since $f(-\sqrt{1/3}) > f(\sqrt{1/3})$, it follows that f decreases on this interval. Similarly, f' always has the same sign on $(\sqrt{1/3}, \infty)$ and $f(x)$ is large for large x, so f must be increasing on $(\sqrt{1/3}, \infty)$. Another point worth noting: If f' is continuous, then the sign of f' on the interval between two adjacent critical points can be determined simply by finding the sign of $f'(x)$ for any *one* x in this interval.

Our sketch of the graph of $f(x) = x^3 - x$ contains sufficient information to allow us to say with confidence that $-\sqrt{1/3}$ is a local maximum point, and $\sqrt{1/3}$ a local minimum point. In fact, we can give a general scheme for decid-

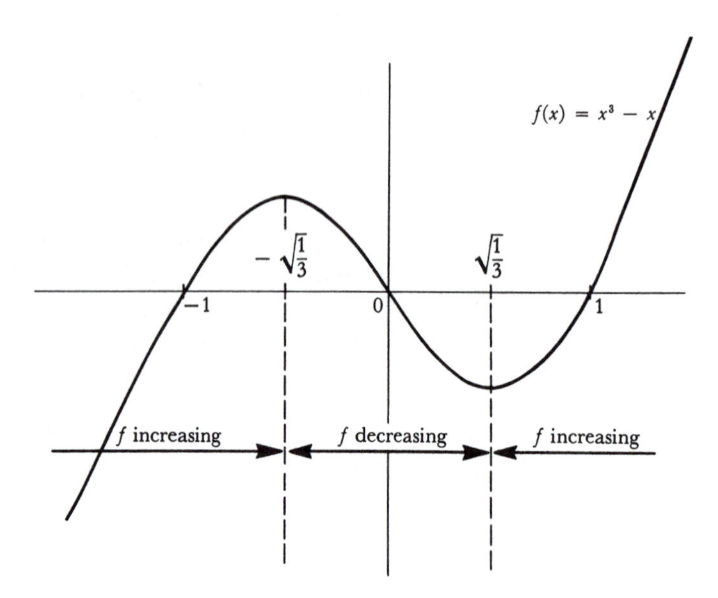

FIGURE 15

ing whether a critical point is a local maximum point, a local minimum point, or neither (Figure 16):

(1) if $f' > 0$ in some interval to the left of x and $f' < 0$ in some interval to the right of x, then x is a local maximum point.
(2) if $f' < 0$ in some interval to the left of x and $f' > 0$ in some interval to the right of x, then x is a local minimum point.
(3) if f' has the same sign in some interval to the left of x as it has in some interval to the right, then x is neither a local maximum nor a local minimum point.

(There is no point in memorizing these rules—you can always draw the pictures yourself.)

The polynomial functions can all be analyzed in this way, and it is even possible to describe the general form of the graph of such functions. To begin, we need a

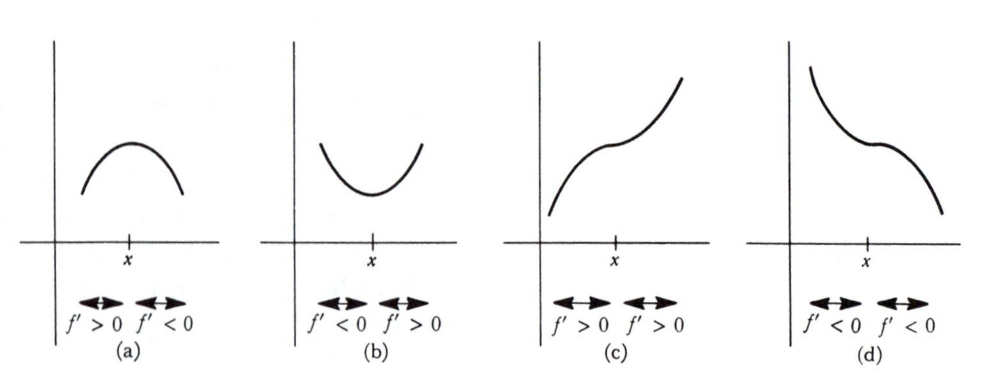

FIGURE 16

result already mentioned in Problem 3-7: If

$$f(x) = a_n x^n + a_{n-1}x^{n-1} + \cdots + a_0,$$

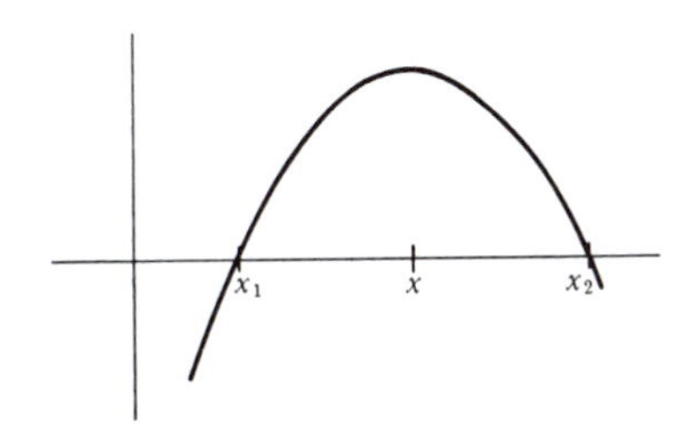

FIGURE 17

then f has at most n "roots," i.e., there are at most n numbers x such that $f(x) = 0$. Although this is really an algebraic theorem, calculus can be used to give an easy proof. Notice that if x_1 and x_2 are roots of f (Figure 17), so that $f(x_1) = f(x_2) = 0$, then by Rolle's Theorem there is a number x between x_1 and x_2 such that $f'(x) = 0$. This means that if f has k different roots $x_1 < x_2 < \cdots < x_k$, then f' has at least $k - 1$ different roots: one between x_1 and x_2, one between x_2 and x_3, etc. It is now easy to prove by induction that a polynomial function

$$f(x) = a_n x^n + a_{n-1}x^{n-1} + \cdots + a_0$$

has at most n roots: The statement is surely true for $n = 1$, and if we assume that it is true for n, then the polynomial

$$g(x) = b_{n+1}x^{n+1} + b_n x^n + \cdots + b_0$$

could not have more than $n + 1$ roots, since if it did, g' would have more than n roots.

With this information it is not hard to describe the graph of

$$f(x) = a_n x^n + a_{n-1}x^{n-1} + \cdots + a_0.$$

The derivative, being a polynomial function of degree $n - 1$, has at most $n - 1$ roots. Therefore f has at most $n - 1$ critical points. Of course, a critical point is not necessarily a local maximum or minimum point, but at any rate, if a and b are adjacent critical points of f, then f' will remain either positive or negative on (a, b), since f' is continuous; consequently, f will be either increasing or decreasing on (a, b). Thus f has at most n regions of decrease or increase.

As a specific example, consider the function

$$f(x) = x^4 - 2x^2.$$

Since

$$f'(x) = 4x^3 - 4x = 4x(x - 1)(x + 1),$$

the critical points of f are -1, 0, and 1, and

$$\begin{aligned} f(-1) &= -1, \\ f(0) &= 0, \\ f(1) &= -1. \end{aligned}$$

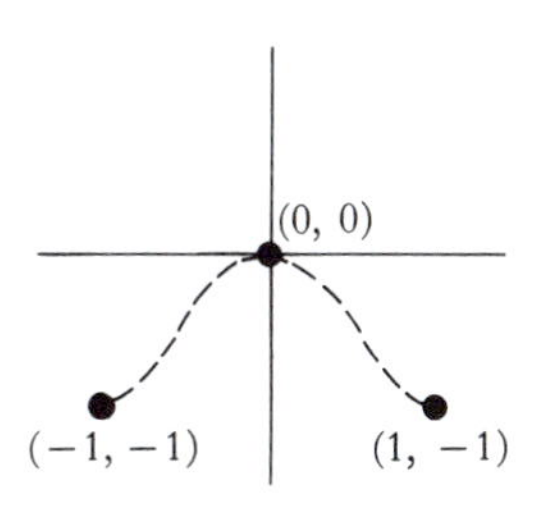

FIGURE 18

The behavior of f on the intervals between the critical points can be determined by one of the methods mentioned before. In particular, we could determine the sign of f' on these intervals simply be examining the formula for $f'(x)$. On the other hand, from the three critical values alone we can see (Figure 18) that f increases on $(-1, 0)$ and decreases on $(0, 1)$. To determine the sign of f' on

$(-\infty, -1)$ and $(1, \infty)$ we can compute

$$f'(-2) = 4 \cdot (-2)^3 - 4 \cdot (-2) = -24,$$
$$f'(2) = 4 \cdot 2^3 - 4 \cdot 2 = 24,$$

and conclude that f is decreasing on $(-\infty, -1)$ and increasing on $(1, \infty)$. These conclusions also follow from the fact that $f(x)$ is large for large x and for large negative x.

We can already produce a good sketch of the graph; two other pieces of information provide the finishing touches (Figure 19). First, it is easy to determine that $f(x) = 0$ for $x = 0, \pm\sqrt{2}$; second, it is clear that f is even, $f(x) = f(-x)$, so the graph is symmetric with respect to the vertical axis. The function $f(x) = x^3 - x$, already sketched in Figure 15, is odd, $f(x) = -f(-x)$, and is consequently symmetric with respect to the origin. Half the work of graph sketching may be saved by noticing these things in the beginning.

Several problems in this and succeeding chapters ask you to sketch the graphs of functions. In each case you should determine

(1) the critical points of f,
(2) the value of f at the critical points,
(3) the sign of f' in the regions between critical points (if this is not already clear),
(4) the numbers x such that $f(x) = 0$ (if possible),
(5) the behavior of $f(x)$ as x becomes large or large negative (if possible).

Finally, bear in mind that a quick check, to see whether the function is odd or even, may save a lot of work.

This sort of analysis, if performed with care, will usually reveal the basic shape of the graph, but sometimes there are special features which require a little more thought. It is impossible to anticipate all of these, but one piece of information is often very important. If f is not defined at certain points (for example, if f is a rational function whose denominator vanishes at some points), then the behavior of f near these points should be determined.

For example, consider the function

$$f(x) = \frac{x^2 - 2x + 2}{x - 1},$$

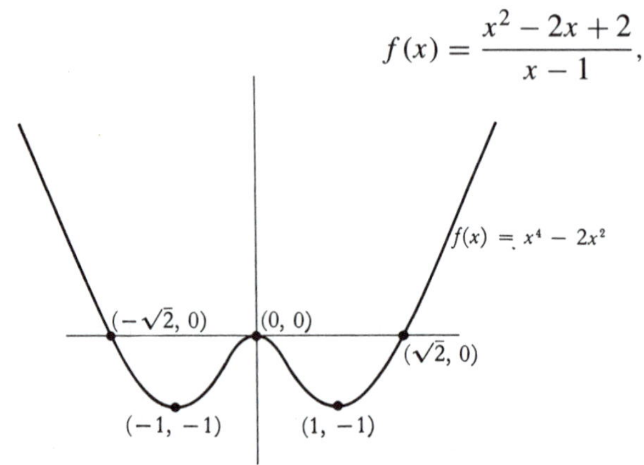

FIGURE 19

which is not defined at 1. We have

$$\begin{aligned} f'(x) &= \frac{(x-1)(2x-2)-(x^2-2x+2)}{(x-1)^2} \\ &= \frac{x(x-2)}{(x-1)^2}. \end{aligned}$$

Thus

(1) the critical points of f are 0, 2.

Moreover,

$$\begin{aligned} (2)\ f(0) &= -2, \\ f(2) &= 2. \end{aligned}$$

Because f is not defined on the whole interval $(0, 2)$, the sign of f' must be determined separately on the intervals $(0, 1)$ and $(1, 2)$, as well as on the intervals $(-\infty, 0)$ and $(2, \infty)$. We can do this by picking particular points in each of these intervals, or simply by staring hard at the formula for f'. Either way we find that

$$\begin{aligned} (3)\ f'(x) &> 0 &&\text{if} && x < 0, \\ f'(x) &< 0 &&\text{if} && 0 < x < 1, \\ f'(x) &< 0 &&\text{if} && 1 < x < 2, \\ f'(x) &> 0 &&\text{if} && 2 < x. \end{aligned}$$

Finally, we must determine the behavior of $f(x)$ as x becomes large or large negative, as well as when x approaches 1 (this information will also give us another way to determine the regions on which f increases and decreases). To examine the behavior as x becomes large we write

$$\frac{x^2-2x+2}{x-1} = x - 1 + \frac{1}{x-1};$$

clearly $f(x)$ is close to $x - 1$ (and slightly larger) when x is large, and $f(x)$ is close to $x - 1$ (but slightly smaller) when x is large negative. The behavior of f near 1 is also easy to determine; since

$$\lim_{x \to 1}(x^2 - 2x + 2) = 1 \neq 0,$$

the fraction

$$\frac{x^2-2x+2}{x-1}$$

becomes large as x approaches 1 from above and large negative as x approaches 1 from below.

All this information may seem a bit overwhelming, but there is only one way that it can be pieced together (Figure 20); be sure that you can account for each feature of the graph.

When this sketch has been completed, we might note that it looks like the graph

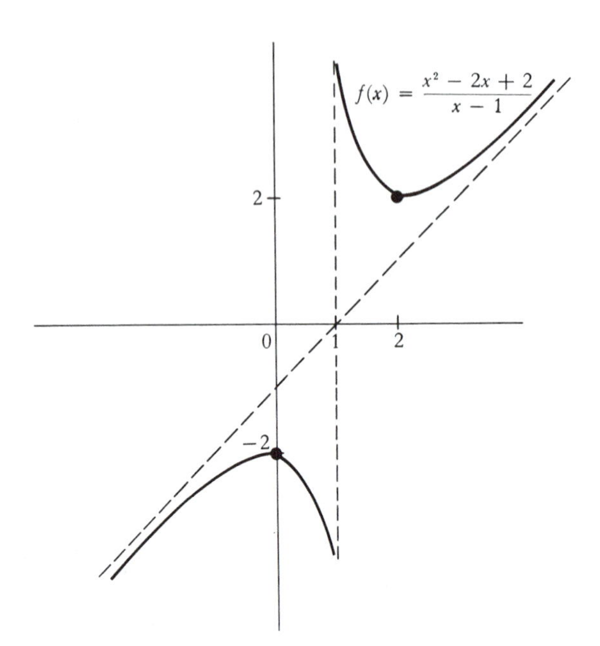

FIGURE 20

of an odd function shoved over 1 unit, and the expression

$$\frac{x^2 - 2x + 2}{x - 1} = \frac{(x-1)^2 + 1}{x - 1}$$

shows that this is indeed the case. However, this is one of those special features which should be investigated only after you have used the other information to get a good idea of the appearance of the graph.

Although the location of local maxima and minima of a function is always revealed by a detailed sketch of its graph, it is usually unnecessary to do so much work. There is a popular test for local maxima and minima which depends on the behavior of the function only at its critical points.

THEOREM 5 Suppose $f'(a) = 0$. If $f''(a) > 0$, then f has a local minimum at a; if $f''(a) < 0$, then f has a local maximum at a.

PROOF By definition,

$$f''(a) = \lim_{h \to 0} \frac{f'(a+h) - f'(a)}{h}.$$

Since $f'(a) = 0$, this can be written

$$f''(a) = \lim_{h \to 0} \frac{f'(a+h)}{h}.$$

Suppose now that $f''(a) > 0$. Then $f'(a+h)/h$ must be positive for sufficiently small h. Therefore:

$$f'(a+h) \text{ must be positive for sufficiently small } h > 0$$
$$\text{and } f'(a+h) \text{ must be negative for sufficiently small } h < 0.$$

This means (Corollary 3) that f is increasing in some interval to the right of a and f is decreasing in some interval to the left of a. Consequently, f has a local minimum at a.

The proof for the case $f''(a) < 0$ is similar. ▌

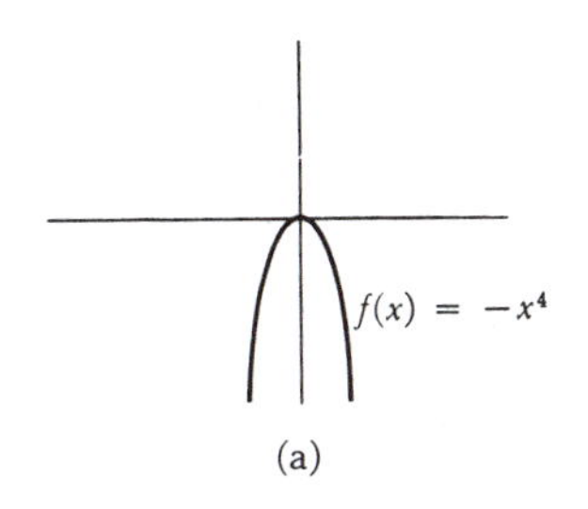

(a)

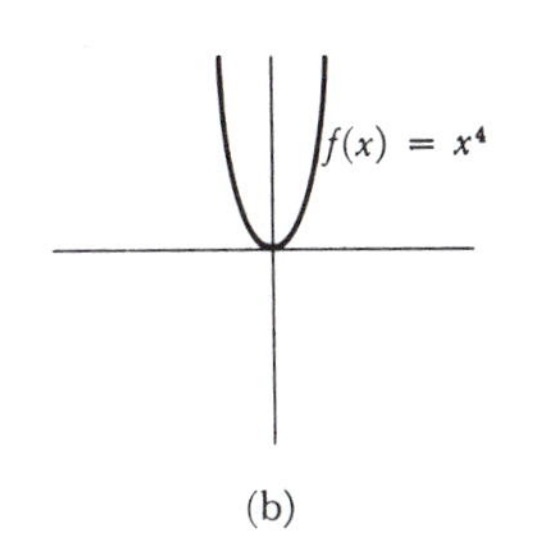

(b)

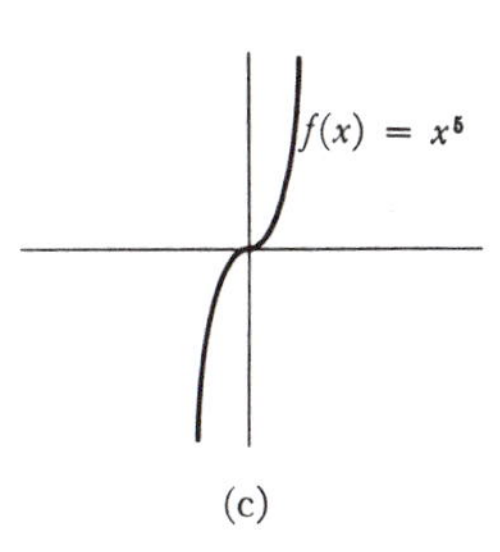

(c)

FIGURE 21

Theorem 5 may be applied to the function $f(x) = x^3 - x$, which has already been considered. We have

$$f'(x) = 3x^2 - 1$$
$$f''(x) = 6x.$$

At the critical points, $-\sqrt{1/3}$ and $\sqrt{1/3}$, we have

$$f''(-\sqrt{1/3}) = -6\sqrt{1/3} < 0,$$
$$f''(\sqrt{1/3}) = 6\sqrt{1/3} > 0.$$

Consequently, $-\sqrt{1/3}$ is a local maximum point and $\sqrt{1/3}$ is a local minimum point.

Although Theorem 5 will be found quite useful for polynomial functions, for many functions the second derivative is so complicated that it is easier to consider the sign of the first derivative. Moreover, if a is a critical point of f it may happen that $f''(a) = 0$. In this case, Theorem 5 provides no information: it is possible that a is a local maximum point, a local minimum point, or neither, as shown (Figure 21) by the functions

$$f(x) = -x^4, \quad f(x) = x^4, \quad f(x) = x^5;$$

in each case $f'(0) = f''(0) = 0$, but 0 is a local maximum point for the first, a local minimum point for the second, and neither a local maximum nor minimum point for the third. This point will be pursued further in Part IV.

It is interesting to note that Theorem 5 automatically proves a partial converse of itself.

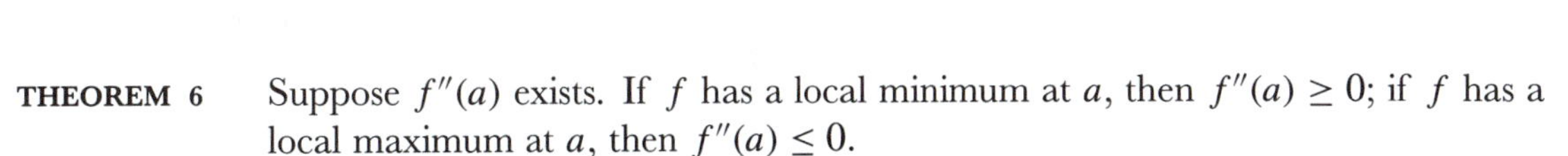

THEOREM 6 Suppose $f''(a)$ exists. If f has a local minimum at a, then $f''(a) \geq 0$; if f has a local maximum at a, then $f''(a) \leq 0$.

PROOF Suppose f has local minimum at a. If $f''(a) < 0$, then f would also have a local maximum at a, by Theorem 5. Thus f would be constant in some interval containing a, so that $f''(a) = 0$, a contradiction. Thus we must have $f''(a) \geq 0$.

The case of a local maximum is handled similarly. ▌

(This partial converse to Theorem 5 is the best we can hope for: the $\geq$ and $\leq$ signs cannot be replaced by $>$ and $<$, as shown by the functions $f(x) = x^4$ and $f(x) = -x^4$.)

The remainder of this chapter deals, not with graph sketching, or maxima and minima, but with three consequences of the Mean Value Theorem. The first is a simple, but very beautiful, theorem which plays an important role in Chapter 15, and which also sheds light on many examples which have occurred in previous chapters.

THEOREM 7 Suppose that f is continuous at a, and that $f'(x)$ exists for all x in some interval containing a, except perhaps for $x = a$. Suppose, moreover, that $\lim_{x \to a} f'(x)$ exists. Then $f'(a)$ also exists, and

$$f'(a) = \lim_{x \to a} f'(x).$$

PROOF By definition,

$$f'(a) = \lim_{h \to 0} \frac{f(a+h) - f(a)}{h}.$$

For sufficiently small $h > 0$ the function f will be continuous on $[a, a+h]$ and differentiable on $(a, a+h)$ (a similar assertion holds for sufficiently small $h < 0$). By the Mean Value Theorem there is a number α_h in $(a, a+h)$ such that

$$\frac{f(a+h) - f(a)}{h} = f'(\alpha_h).$$

Now α_h approaches a as h approaches 0, because α_h is in $(a, a+h)$; since $\lim_{x \to a} f'(x)$ exists, it follows that

$$f'(a) = \lim_{h \to 0} \frac{f(a+h) - f(a)}{h} = \lim_{h \to 0} f'(\alpha_h) = \lim_{x \to a} f'(x).$$

(It is a good idea to supply a rigorous ε-δ argument for this final step, which we have treated somewhat informally.) ▮

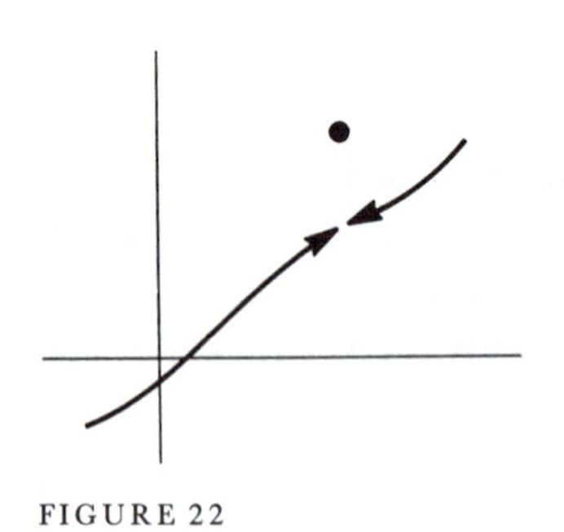

FIGURE 22

Even if f is an everywhere differentiable function, it is still possible for f' to be discontinuous. This happens, for example, if

$$f(x) = \begin{cases} x^2 \sin \dfrac{1}{x}, & x \neq 0 \\ 0, & x = 0. \end{cases}$$

According to Theorem 7, however, the graph of f' can never exhibit a discontinuity of the type shown in Figure 22. Problem 55 outlines the proof of another beautiful theorem which gives further information about the function f', and Problem 56 uses this result to strengthen Theorem 7.

The next theorem, a generalization of the Mean Value Theorem, is of interest mainly because of its applications.

THEOREM 8 (THE CAUCHY MEAN VALUE THEOREM) If f and g are continuous on $[a, b]$ and differentiable on (a, b), then there is a number x in (a, b) such that

$$[f(b) - f(a)]g'(x) = [g(b) - g(a)]f'(x).$$

(If $g(b) \neq g(a)$, and $g'(x) \neq 0$, this equation can be written

$$\frac{f(b) - f(a)}{g(b) - g(a)} = \frac{f'(x)}{g'(x)}.$$

Notice that if $g(x) = x$ for all x, then $g'(x) = 1$, and we obtain the Mean Value Theorem. On the other hand, applying the Mean Value Theorem to f and g separately, we find that there are x and y in (a, b) with

$$\frac{f(b) - f(a)}{g(b) - g(a)} = \frac{f'(x)}{g'(y)};$$

but there is no guarantee that the x and y found in this way will be equal. These remarks may suggest that the Cauchy Mean Value Theorem will be quite difficult to prove, but actually the simplest of tricks suffices.)

PROOF Let

$$h(x) = f(x)[g(b) - g(a)] - g(x)[f(b) - f(a)].$$

Then h is continuous on $[a, b]$, differentiable on (a, b), and

$$h(a) = f(a)g(b) - g(a)f(b) = h(b).$$

It follows from Rolle's Theorem that $h'(x) = 0$ for some x in (a, b), which means that

$$0 = f'(x)[g(b) - g(a)] - g'(x)[f(b) - f(a)]. \blacksquare$$

The Cauchy Mean Value Theorem is the basic tool needed to prove a theorem which facilitates evaluation of limits of the form

$$\lim_{x \to a} \frac{f(x)}{g(x)},$$

when

$$\lim_{x \to a} f(x) = 0 \quad \text{and} \quad \lim_{x \to a} g(x) = 0.$$

In this case, Theorem 5-2 is of no use. Every derivative is a limit of this form, and computing derivatives frequently requires a great deal of work. If some derivatives are known, however, many limits of this form can now be evaluated easily.

THEOREM 9 (L'HÔPITAL'S RULE) Suppose that

$$\lim_{x \to a} f(x) = 0 \quad \text{and} \quad \lim_{x \to a} g(x) = 0,$$

and suppose also that $\lim_{x \to a} f'(x)/g'(x)$ exists. Then $\lim_{x \to a} f(x)/g(x)$ exists, and

$$\lim_{x \to a} \frac{f(x)}{g(x)} = \lim_{x \to a} \frac{f'(x)}{g'(x)}.$$

(Notice that Theorem 7 is a special case.)

PROOF The hypothesis that $\lim_{x \to a} f'(x)/g'(x)$ exists contains two implicit assumptions:

(1) there is an interval $(a - \delta, a + \delta)$ such that $f'(x)$ and $g'(x)$ exist for all x in $(a - \delta, a + \delta)$ except, perhaps, for $x = a$,
(2) in this interval $g'(x) \neq 0$ with, once again, the possible exception of $x = a$.

On the other hand, f and g are not even assumed to be defined at a. If we define $f(a) = g(a) = 0$ (changing the previous values of $f(a)$ and $g(a)$, if necessary), then f and g are continuous at a. If $a < x < a + \delta$, then the Mean Value Theorem and the Cauchy Mean Value Theorem apply to f and g on the interval $[a, x]$ (and a similar statement holds for $a - \delta < x < a$). First applying the Mean Value Theorem to g, we see that $g(x) \neq 0$, for if $g(x) = 0$ there would be some x_1 in (a, x) with $g'(x_1) = 0$, contradicting (2). Now applying the Cauchy Mean Value Theorem to f and g, we see that there is a number α_x in (a, x) such that

$$[f(x) - 0]g'(\alpha_x) = [g(x) - 0]f'(\alpha_x)$$

or

$$\frac{f(x)}{g(x)} = \frac{f'(\alpha_x)}{g'(\alpha_x)}.$$

Now α_x approaches a as x approaches a, because α_x is in (a, x); since $\lim_{y \to a} f'(y)/g'(y)$ exists, it follows that

$$\lim_{x \to a} \frac{f(x)}{g(x)} = \lim_{x \to a} \frac{f'(\alpha_x)}{g'(\alpha_x)} = \lim_{y \to a} \frac{f'(y)}{g'(y)}.$$

(Once again, the reader is invited to supply the details of this part of the argument.) ▮

PROBLEMS

1. For each of the following functions, find the maximum and minimum values on the indicated intervals, by finding the points in the interval where the derivative is 0, and comparing the values at these points with the values at the end points.

(i) $f(x) = x^3 - x^2 - 8x + 1$ on $[-2, 2]$.
(ii) $f(x) = x^5 + x + 1$ on $[-1, 1]$.
(iii) $f(x) = 3x^4 - 8x^3 + 6x^2$ on $[-\frac{1}{2}, \frac{1}{2}]$.
(iv) $f(x) = \dfrac{1}{x^5 + x + 1}$ on $[-\frac{1}{2}, 1]$.
(v) $f(x) = \dfrac{x + 1}{x^2 + 1}$ on $[-1, \frac{1}{2}]$.
(vi) $f(x) = \dfrac{x}{x^2 - 1}$ on $[0, 5]$.

2. Now sketch the graph of each of the functions in Problem 1, and find all local maximum and minimum points.

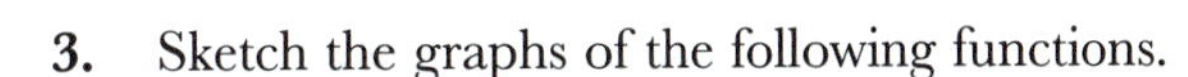

3. Sketch the graphs of the following functions.

(i) $f(x) = x + \dfrac{1}{x}$.

(ii) $f(x) = x + \dfrac{3}{x^2}$.

(iii) $f(x) = \dfrac{x^2}{x^2 - 1}$.

(iv) $f(x) = \dfrac{1}{1 + x^2}$.

4. (a) If $a_1 < \cdots < a_n$, find the minimum value of $f(x) = \sum_{i=1}^{n} (x - a_i)^2$.

*(b) Now find the minimum value of $f(x) = \sum_{i=1}^{n} |x - a_i|$. This is a problem where calculus won't help at all: on the intervals between the a_i's the function f is linear, so that the minimum clearly occurs at one of the a_i, and these are precisely the points where f is not differentiable. However, the answer is easy to find if you consider how $f(x)$ changes as you pass from one such interval to another.

*(c) Let $a > 0$. Show that the maximum value of

$$f(x) = \frac{1}{1 + |x|} + \frac{1}{1 + |x - a|}$$

is $(2 + a)/(1 + a)$. (The derivative can be found on each of the intervals $(-\infty, 0)$, $(0, a)$, and (a, ∞) separately.)

5. For each of the following functions, find all local maximum and minimum points.

(i) $f(x) = \begin{cases} x, & x \neq 3, 5, 7, 9 \\ 5, & x = 3 \\ -3, & x = 5 \\ 9, & x = 7 \\ 7, & x = 9. \end{cases}$

(ii) $f(x) = \begin{cases} 0, & x \text{ irrational} \\ 1/q, & x = p/q \text{ in lowest terms.} \end{cases}$

(iii) $f(x) = \begin{cases} x, & x \text{ rational} \\ 0, & x \text{ irrational.} \end{cases}$

(iv) $f(x) = \begin{cases} 1, & x = 1/n \text{ for some } n \text{ in } \mathbf{N} \\ 0, & \text{otherwise.} \end{cases}$

(v) $f(x) = \begin{cases} 1, & \text{if the decimal expansion of } x \text{ contains a } 5 \\ 0, & \text{otherwise.} \end{cases}$

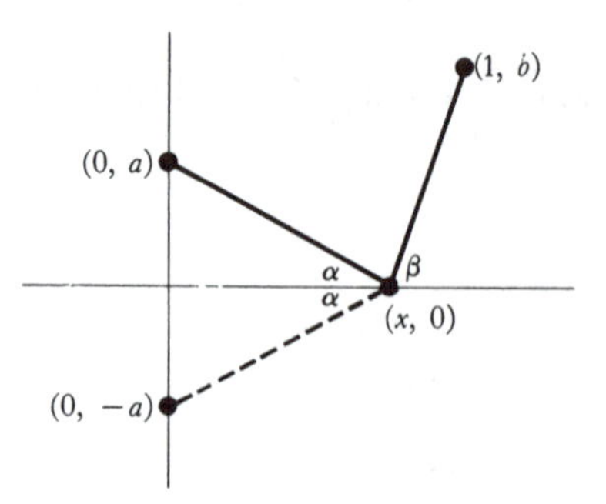

FIGURE 23

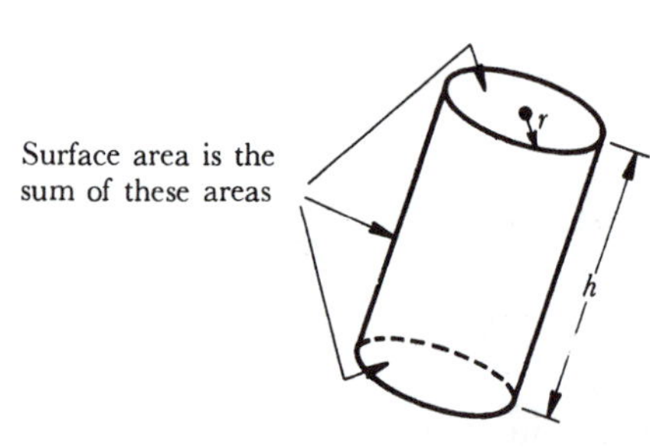

FIGURE 24

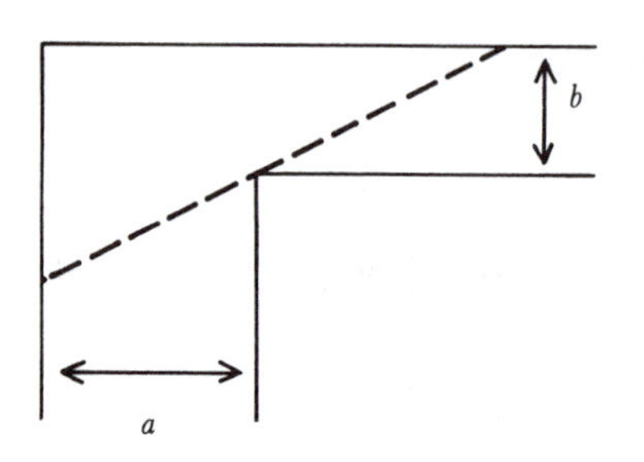

FIGURE 25

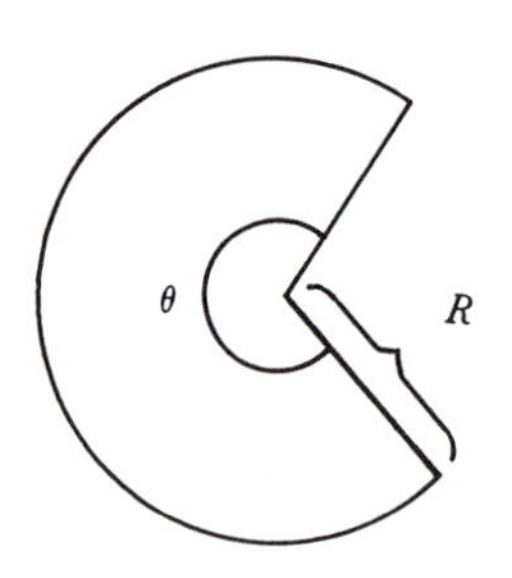

FIGURE 26

6. (a) Let (x_0, y_0) be a point of the plane, and let L be the graph of the function $f(x) = mx + b$. Find the point $\bar{x}$ such that the distance from (x_0, y_0) to $(\bar{x}, f(\bar{x}))$ is smallest. [Notice that minimizing this distance is the same as minimizing its square. This may simplify the computations somewhat.]

(b) Also find $\bar{x}$ by noting that the line from (x_0, y_0) to $(\bar{x}, f(\bar{x}))$ is perpendicular to L.

(c) Find the distance from (x_0, y_0) to L, i.e., the distance from (x_0, y_0) to $(\bar{x}, f(\bar{x}))$. [It will make the computations easier if you first assume that $b = 0$; then apply the result to the graph of $f(x) = mx$ and the point $(x_0, y_0 - b)$.] Compare with Problem 4-22.

(d) Consider a straight line described by the equation $Ax + By + C = 0$ (Problem 4-7). Show that the distance from (x_0, y_0) to this line is $(Ax_0 + By_0 + C)/\sqrt{A^2 + B^2}$.

7. The previous Problem suggests the following question: What is the relationship between the critical points of f and those of f^2?

8. A straight line is drawn from the point $(0, a)$ to the horizontal axis, and then back to $(1, b)$, as in Figure 23. Prove that the total length is shortest when the angles α and β are equal. (Naturally you must bring a function into the picture: express the length in terms of x, where $(x, 0)$ is the point on the horizontal axis. The dashed line in Figure 23 suggests an alternative geometric proof; in either case the problem can be solved without actually finding the point $(x, 0)$.)

9. Prove that of all rectangles with given perimeter, the square has the greatest area.

10. Find, among all right circular cylinders of fixed volume V, the one with smallest surface area (counting the areas of the faces at top and bottom, as in Figure 24).

11. A right triangle with hypotenuse of length a is rotated about one of its legs to generate a right circular cone. Find the greatest possible volume of such a cone.

12. Two hallways, of widths a and b, meet at right angles (Figure 25). What is the greatest possible length of a ladder which can be carried horizontally around the corner?

13. A garden is to be designed in the shape of a circular sector (Figure 26), with radius R and central angle θ. The garden is to have a fixed area A. For what value of R and θ (in radians) will the length of the fencing around the perimeter be minimized?

14. Show that the sum of a positive number and its reciprocal is at least 2.

15. Find the trapezoid of largest area that can be inscribed in a semicircle of radius a, with one base lying along the diameter.

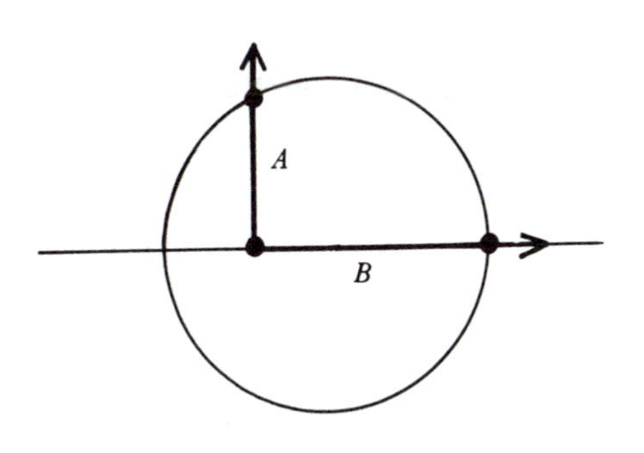

FIGURE 27

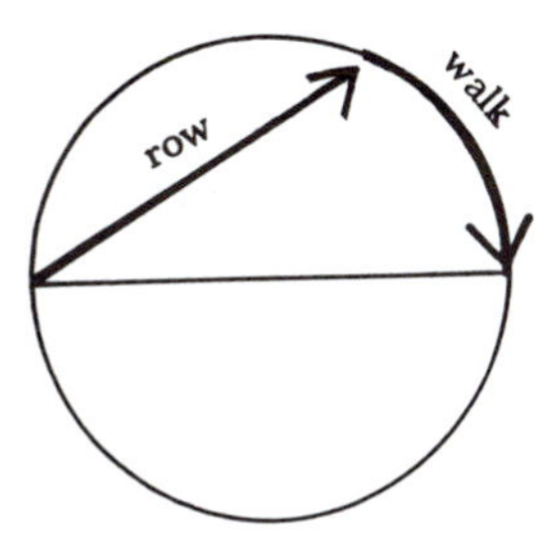

FIGURE 28

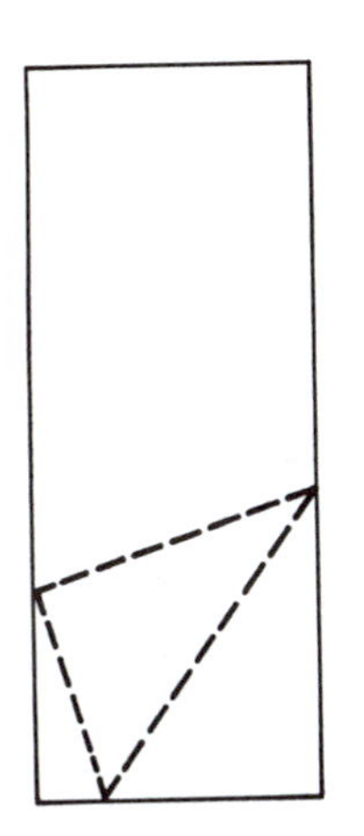

FIGURE 29

16. A right angle is moved along the diameter of a circle of radius a, as shown in Figure 27. What is the greatest possible length $(A + B)$ intercepted on it by the circle?

17. Ecological Ed must cross a circular lake of radius 1 mile. He can row across at 2 mph or walk around at 4 mph, or he can row part way and walk the rest (Figure 28). What route should he take so as to

(i) see as much scenery as possible?
(ii) cross as quickly as possible?

18. The lower right-hand corner of a page is folded over so that it just touches the left edge of the paper, as in Figure 29. If the width of the paper is α and the page is very long, show that the minimum length of the crease is $3\sqrt{3}\alpha/4$.

19. Figure 30 shows the graph of the *derivative* of f. Find all local maximum and minimum points of f.

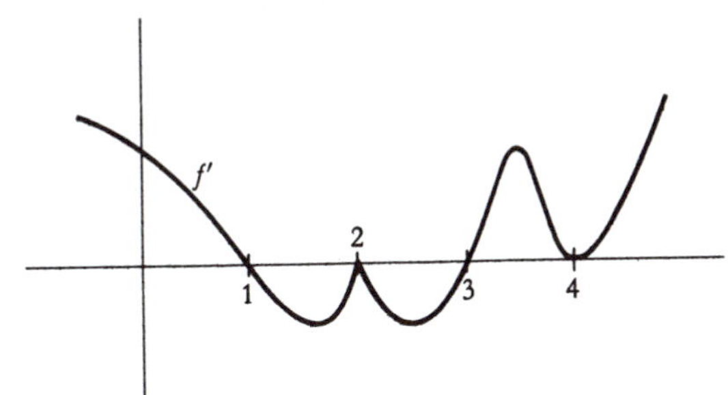

FIGURE 30

***20.** Suppose that f is a polynomial function, $f(x) = x^n + a_{n-1}x^{n-1} + \cdots + a_0$, with critical points -1, 1, 2, 3, 4, and corresponding critical values 6, 1, 2, 4, 3. Sketch the graph of f, distinguishing the cases n even and n odd.

***21.** (a) Suppose that the critical points of the polynomial function $f(x) = x^n + a_{n-1}x^{n-1} + \cdots + a_0$ are $-1, 1, 2, 3$, and $f''(-1) = 0$, $f''(1) > 0$, $f''(2) < 0$, $f''(3) = 0$. Sketch the graph of f as accurately as possible on the basis of this information.

(b) Does there exist a polynomial function with the above properties, except that 3 is not a critical point?

22. Describe the graph of a rational function (in very general terms, similar to the text's description of the graph of a polynomial function).

23. (a) Prove that two polynomial functions of degree m and n, respectively, intersect in at most $\max(m, n)$ points.

(b) For each m and n exhibit two polynomial functions of degree m and n which intersect $\max(m, n)$ times.

***24.** (a) Suppose that the polynomial function $f(x) = x^n + a_{n-1}x^{n-1} + \cdots + a_0$ has exactly k critical points and $f''(x) \neq 0$ for all critical points x. Show that $n - k$ is odd.

(b) For each n, show that there is a polynomial function f of degree n with k critical points if $n-k$ is odd.

(c) Suppose that the polynomial function $f(x) = x^n + a_{n-1}x^{n-1} + \cdots + a_0$ has k_1 local maximum points and k_2 local minimum points. Show that $k_2 = k_1 + 1$ if n is even, and $k_2 = k_1$ if n is odd.

(d) Let n, k_1, k_2 be three integers with $k_2 = k_1 + 1$ if n is even, and $k_2 = k_1$ if n is odd, and $k_1 + k_2 < n$. Show that there is a polynomial function f of degree n, with k_1 local maximum points and k_2 local minimum points.

Hint: Pick $a_1 < a_2 < \cdots < a_{k_1+k_2}$ and try $f'(x) = \prod_{i=1}^{k_1+k_2}(x-a_i)\cdot(1+x^2)^l$ for an appropriate number l.

25. (a) Prove that if $f'(x) \ge M$ for all x in $[a,b]$, then $f(b) \ge f(a) + M(b-a)$.

(b) Prove that if $f'(x) \le M$ for all x in $[a,b]$, then $f(b) \le f(a) + M(b-a)$.

(c) Formulate a similar theorem when $|f'(x)| \le M$ for all x in $[a,b]$.

***26.** Suppose that $f'(x) \ge M > 0$ for all x in $[0,1]$. Show that there is an interval of length $\frac{1}{4}$ on which $|f| \ge M/4$.

27. (a) Suppose that $f'(x) > g'(x)$ for all x, and that $f(a) = g(a)$. Show that $f(x) > g(x)$ for $x > a$ and $f(x) < g(x)$ for $x < a$.

(b) Show by an example that these conclusions do not follow without the hypothesis $f(a) = g(a)$.

28. Find all functions f such that

(a) $f'(x) = \sin x$.

(b) $f''(x) = x^3$.

(c) $f'''(x) = x + x^2$.

29. Although it is true that a weight dropped from rest will fall $s(t) = 16t^2$ feet after t seconds, this experimental fact does not mention the behavior of weights which are thrown upwards or downwards. On the other hand, the law $s''(t) = 32$ is always true and has just enough ambiguity to account for the behavior of a weight released from any height, with any initial velocity. For simplicity let us agree to measure heights upwards from ground level; in this case velocities are positive for rising bodies and negative for falling bodies, and all bodies fall according to the law $s''(t) = -32$.

(a) Show that s is of the form $s(t) = -16t^2 + \alpha t + \beta$.

(b) By setting $t = 0$ in the formula for s, and then in the formula for s', show that $s(t) = -16t^2 + v_0 t + s_0$, where s_0 is the height from which the body is released at time 0, and v_0 is the velocity with which it is released.

(c) A weight is thrown upwards with velocity v feet per second, at ground level. How high will it go? ("How high" means "what is the maximum height for all times".) What is its velocity at the moment it achieves its greatest height? What is its acceleration at that moment? When will it hit the ground again? What will its velocity be when it hits the ground again?

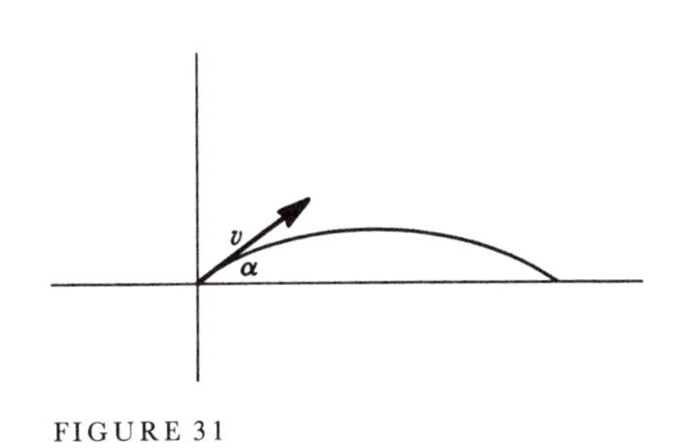

FIGURE 31

30. A cannon ball is shot from the ground with velocity v at an angle α (Figure 31) so that it has a vertical component of velocity $v \sin\alpha$ and a horizontal component $v \cos\alpha$. Its distance $s(t)$ above the ground obeys the law $s(t) = -16t^2 + (v \sin\alpha)t$, while its horizontal velocity remains constantly $v \cos\alpha$.

(a) Show that the path of the cannon ball is a parabola (find the position at each time t, and show that these points lie on a parabola).

(b) Find the angle α which will maximize the horizontal distance traveled by the cannon ball before striking the ground.

31. (a) Give an example of a function f for which $\lim_{x\to\infty} f(x)$ exists, but $\lim_{x\to\infty} f'(x)$ does not exist.

(b) Prove that if $\lim_{x\to\infty} f(x)$ and $\lim_{x\to\infty} f'(x)$ both exist, then $\lim_{x\to\infty} f'(x) = 0$.

(c) Prove that if $\lim_{x\to\infty} f(x)$ exists and $\lim_{x\to\infty} f''(x)$ exists, then $\lim_{x\to\infty} f''(x) = 0$. (See also Problem 20-15.)

32. Suppose that f and g are two differentiable functions which satisfy $fg' - f'g = 0$. Prove that if a and b are adjacent zeros of f, and $g(a)$ and $g(b)$ are not both 0, then $g(x) = 0$ for some x between a and b. (Naturally the same result holds with f and g interchanged; thus, the zeros of f and g separate each other.) Hint: Derive a contradiction from the assumption that $g(x) \neq 0$ for all x between a and b: if a number is not 0, there is a natural thing to do with it.

33. Suppose that $|f(x) - f(y)| \le |x - y|^n$ for $n > 1$. Prove that f is constant by considering f'. Compare with Problem 3-20.

34. A function f is *Lipschitz of order* α at x if there is a constant C such that

$$(*) \qquad |f(x) - f(y)| \le C|x - y|^\alpha$$

for all y in an interval around x. The function f is *Lipschitz of order* α *on an interval* if $(*)$ holds for all x and y in the interval.

(a) If f is Lipschitz of order $\alpha > 0$ at x, then f is continuous at x.

(b) If f is Lipschitz of order $\alpha > 0$ on an interval, then f is uniformly continuous on this interval (see Chapter 8, Appendix).

(c) If f is differentiable at x, then f is Lipschitz of order 1 at x. Is the converse true?

(d) If f is differentiable on $[a, b]$, is f Lipschitz of order 1 on $[a, b]$?

(e) If f is Lipschitz of order $\alpha > 1$ on $[a, b]$, then f is constant on $[a, b]$.

35. Prove that if

$$\frac{a_0}{1} + \frac{a_1}{2} + \cdots + \frac{a_n}{n+1} = 0,$$

then

$$a_0 + a_1 x + \cdots + a_n x^n = 0$$

for some x in $[0, 1]$.

36. Prove that the polynomial function $f_m(x) = x^3 - 3x + m$ never has two roots in $[0, 1]$, no matter what m may be. (This is an easy consequence of Rolle's Theorem. It is instructive, after giving an analytic proof, to graph f_0 and f_2, and consider where the graph of f_m lies in relation to them.)

37. Suppose that f is continuous and differentiable on $[0, 1]$, that $f(x)$ is in $[0, 1]$ for each x, and that $f'(x) \neq 1$ for all x in $[0, 1]$. Show that there is exactly one number x in $[0, 1]$ such that $f(x) = x$. (Half of this problem has been done already, in Problem 7-11.)

38. (a) Prove that the function $f(x) = x^2 - \cos x$ satisfies $f(x) = 0$ for precisely two numbers x.

(b) Prove the same for the function $f(x) = 2x^2 - x \sin x - \cos^2 x$. (Some preliminary estimates will be useful to restrict the possible location of the zeros of f.)

***39.** (a) Prove that if f is a twice differentiable function with $f(0) = 0$ and $f(1) = 1$ and $f'(0) = f'(1) = 0$, then $|f''(x)| \geq 4$ for some x in $[0, 1]$. In more picturesque terms: A particle which travels a unit distance in a unit time, and starts and ends with velocity 0, has at some time an acceleration ≥ 4. Hint: Prove that either $f''(x) > 4$ for some x in $[0, \frac{1}{2}]$, or else $f''(x) < -4$ for some x in $[\frac{1}{2}, 1]$.

(b) Show that in fact we must have $|f''(x)| > 4$ for some x in $[0, 1]$.

40. Suppose that f is a function such that $f'(x) = 1/x$ for all $x > 0$ and $f(1) = 0$. Prove that $f(xy) = f(x) + f(y)$ for all $x, y > 0$. Hint: Find $g'(x)$ when $g(x) = f(xy)$.

***41.** Suppose that f satisfies

$$f''(x) + f'(x)g(x) - f(x) = 0$$

for some function g. Prove that if f is 0 at two points, then f is 0 on the interval between them. Hint: Use Theorem 6.

42. Suppose that f is n-times differentiable and that $f(x) = 0$ for $n + 1$ different x. Prove that $f^{(n)}(x) = 0$ for some x.

43. Let $a_1, \ldots, a_{n+1}$ be arbitrary points in $[a, b]$, and let

$$Q(x) = \prod_{i=1}^{n+1} (x - x_i).$$

Suppose that f is $(n+1)$-times differentiable and that P is a polynomial function of degree $\le n$ such that $P(x_i) = f(x_i)$ for $i = 1, \dots, n+1$ (see page 49). Show that for each x in $[a,b]$ there is a number c in (a,b) such that

$$f(x) - P(x) = Q(x) \cdot \frac{f^{(n+1)}(c)}{(n+1)!}.$$

Hint: Consider the function

$$F(t) = Q(x)[f(t) - P(t)] - Q(t)[f(x) - P(x)].$$

Show that F is zero at $n+2$ different points in $[a,b]$, and use Problem 42.

44. Prove that

$$\tfrac{1}{9} < \sqrt{66} - 8 < \tfrac{1}{8}$$

(without computing $\sqrt{66}$ to 2 decimal places!).

45. Prove the following slight generalization of the Mean Value Theorem: If f is continuous and differentiable on (a,b) and $\lim_{y \to a^+} f(y)$ and $\lim_{y \to b^-} f(y)$ exist, then there is some x in (a,b) such that

$$f'(x) = \frac{\lim_{y \to b^-} f(y) - \lim_{y \to a^+} f(y)}{b-a}.$$

(Your proof should begin: "This is a trivial consequence of the Mean Value Theorem because ... ".)

46. Prove that the conclusion of the Cauchy Mean Value Theorem can be written in the form

$$\frac{f(b) - f(a)}{g(b) - g(a)} = \frac{f'(x)}{g'(x)},$$

under the additional assumptions that $g(b) \ne g(a)$ and that $f'(x)$ and $g'(x)$ are never simultaneously 0 on (a,b).

***47.** Prove that if f and g are continuous on $[a,b]$ and differentiable on (a,b), and $g'(x) \ne 0$ for x in (a,b), then there is some x in (a,b) with

$$\frac{f'(x)}{g'(x)} = \frac{f(x) - f(a)}{g(b) - g(x)}.$$

Hint: Multiply out first, to see what this really says.

48. What is wrong with the following use of l'Hôpital's Rule:

$$\lim_{x \to 1} \frac{x^3 + x - 2}{x^2 - 3x + 2} = \lim_{x \to 1} \frac{3x^2 + 1}{2x - 3} = \lim_{x \to 1} \frac{6x}{2} = 3.$$

(The limit is actually -4.)

49. Find the following limits:

(i) $\lim_{x \to 0} \frac{x}{\tan x}$.

(ii) $\lim_{x \to 0} \frac{\cos^2 x - 1}{x^2}$.

50. Find $f'(0)$ if

$$f(x) = \begin{cases} \frac{g(x)}{x}, & x \neq 0 \\ 0, & x = 0, \end{cases}$$

and $g(0) = g'(0) = 0$ and $g''(0) = 17$.

51. Prove the following forms of l'Hôpital's Rule (none requiring any essentially new reasoning).

(a) If $\lim_{x \to a^+} f(x) = \lim_{x \to a^+} g(x) = 0$, and $\lim_{x \to a^+} f'(x)/g'(x) = l$, then $\lim_{x \to a^+} f(x)/g(x) = l$ (and similarly for limits from below).

(b) If $\lim_{x \to a} f(x) = \lim_{x \to a} g(x) = 0$, and $\lim_{x \to a} f'(x)/g'(x) = \infty$, then $\lim_{x \to a} f(x)/g(x) = \infty$ (and similarly for $-\infty$, or if $x \to a$ is replaced by $x \to a^+$ or $x \to a^-$).

(c) If $\lim_{x \to \infty} f(x) = \lim_{x \to \infty} g(x) = 0$, and $\lim_{x \to \infty} f'(x)/g'(x) = l$, then $\lim_{x \to \infty} f(x)/g(x) = l$ (and similarly for $-\infty$). Hint: Consider $\lim_{x \to 0^+} f(1/x)/g(1/x)$.

(d) If $\lim_{x \to \infty} f(x) = \lim_{x \to \infty} g(x) = 0$, and $\lim_{x \to \infty} f'(x)/g'(x) = \infty$, then $\lim_{x \to \infty} f(x)/g(x) = \infty$.

52. There is another form of l'Hôpital's Rule which requires more than algebraic manipulations: If $\lim_{x \to \infty} f(x) = \lim_{x \to \infty} g(x) = \infty$, and $\lim_{x \to \infty} f'(x)/g'(x) = l$, then $\lim_{x \to \infty} f(x)/g(x) = l$. Prove this as follows.

(a) For every $\varepsilon > 0$ there is a number a such that

$$\left| \frac{f'(x)}{g'(x)} - l \right| < \varepsilon \quad \text{for } x > a.$$

Apply the Cauchy Mean Value Theorem to f and g on $[a, x]$ to show that

$$\left| \frac{f(x) - f(a)}{g(x) - g(a)} - l \right| < \varepsilon \quad \text{for } x > a.$$

(Why can we assume $g(x) - g(a) \neq 0$?)

(b) Now write

$$\frac{f(x)}{g(x)} = \frac{f(x) - f(a)}{g(x) - g(a)} \cdot \frac{f(x)}{f(x) - f(a)} \cdot \frac{g(x) - g(a)}{g(x)}$$

(why can we assume that $f(x) - f(a) \neq 0$ for large x?) and conclude that

$$\left| \frac{f(x)}{g(x)} - l \right| < 2\varepsilon \quad \text{for sufficiently large } x.$$

53. To complete the orgy of variations on l'Hôpital's Rule, use Problem 52 to prove a few more cases of the following general statement (there are so many possibilities that you should select just a few, if any, that interest you):

If $\lim_{x \to [\ ]} f(x) = \lim_{x \to [\ ]} g(x) = \{\quad\}$ and $\lim_{x \to [\ ]} f'(x)/g'(x) = (\quad)$, then $\lim_{x \to [\ ]} f(x)/g(x) = (\quad)$. Here [ ] can be a or a^+ or a^- or ∞ or $-\infty$, and { } can be 0 or ∞ or $-\infty$, and ( ) can be l or ∞ or $-\infty$.

***54.** (a) Suppose that f is differentiable on $[a, b]$. Prove that if the minimum of f on $[a, b]$ is at a, then $f'(a) \geq 0$, and if it is at b, then $f'(b) \leq 0$. (One half of the proof of Theorem 1 will go through.)

(b) Suppose that $f'(a) < 0$ and $f'(b) > 0$. Show that $f'(x) = 0$ for some x in (a, b). Hint: Consider the minimum of f on $[a, b]$; why must it be somewhere in (a, b)?

(c) Prove that if $f'(a) < c < f'(b)$, then $f'(x) = c$ for some x in (a, b). (This result is known as Darboux's Theorem.) Hint: Cook up an appropriate function to which part (b) may be applied.

55. Suppose that f is differentiable in some interval containing a, but that f' is discontinuous at a.

(a) The one-sided limits $\lim_{x \to a^+} f'(x)$ and $\lim_{x \to a^-} f'(x)$ cannot both exist. (This is just a minor variation on Theorem 7.)

(b) Neither of these one-sided limits can exist even in the sense of being $+\infty$ or $-\infty$. Hint: Use Darboux's Theorem (Problem 54).

***56.** It is easy to find a function f such that $|f|$ is differentiable but f is not. For example, we can choose $f(x) = 1$ for x rational and $f(x) = -1$ for x irrational. In this example f is not even continuous, nor is this a mere coincidence: Prove that if $|f|$ is differentiable at a, and f is continuous at a, then f is also differentiable at a. Hint: It suffices to consider only a with $f(a) = 0$. Why? In this case, what must $|f|'(a)$ be?

***57.** (a) Let $y \neq 0$ and let n be even. Prove that $x^n + y^n = (x + y)^n$ only when $x = 0$. Hint: If ${x_0}^n + y^n = (x_0 + y)^n$, apply Rolle's Theorem to $f(x) = x^n + y^n - (x + y)^n$ on $[0, x_0]$.

(b) Prove that if $y \neq 0$ and n is odd, then $x^n + y^n = (x+y)^n$ only if $x = 0$ or $x = -y$.

****58.** Use the method of Problem 57 to prove that if n is even and $f(x) = x^n$, then every tangent line to f intersects f only once.

****59.** Prove even more generally that if f' is increasing, then every tangent line intersects f only once.

***60.** Suppose that $f(0) = 0$ and f' is increasing. Prove that the function $g(x) = f(x)/x$ is increasing on $(0, \infty)$. Hint: Obviously you should look at $g'(x)$. Prove that it is positive by applying the Mean Value Theorem to f on the right interval (it will help to remember that the hypothesis $f(0) = 0$ is essential, as shown by the function $f(x) = 1 + x^2$).

***61.** Use derivatives to prove that if $n \geq 1$, then

$$(1+x)^n > 1 + nx \quad \text{for } -1 < x < 0 \text{ and } 0 < x$$

(notice that equality holds for $x = 0$).

62. Let $f(x) = x^4 \sin^2 1/x$ for $x \neq 0$, and let $f(0) = 0$ (Figure 32).

(a) Prove that 0 is a local minimum point for f.
(b) Prove that $f'(0) = f''(0) = 0$.

This function thus provides another example to show that Theorem 6 cannot be improved. It also illustrates a subtlety about maxima and minima that often goes unnoticed: a function may not be increasing in any interval to the right of a local minimum point, nor decreasing in any interval to the left.

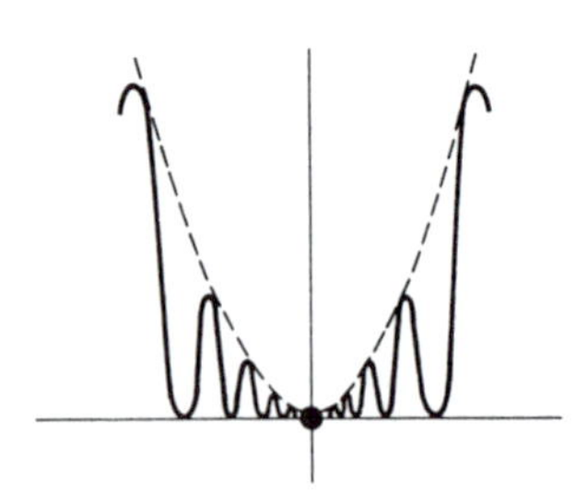

FIGURE 32

***63.** (a) Prove that if $f'(a) > 0$ and f' is continuous at a, then f is increasing in some interval containing a.

The next two parts of this problem show that continuity of f' is essential.

(b) If $g(x) = x^2 \sin 1/x$, show that there are numbers x arbitrarily close to 0 with $g'(x) = 1$ and also with $g'(x) = -1$.

(c) Suppose $0 < \alpha < 1$. Let $f(x) = \alpha x + x^2 \sin 1/x$ for $x \neq 0$, and let $f(0) = 0$ (see Figure 33). Show that f is not increasing in any open interval containing 0, by showing that in any interval there are points x with $f'(x) > 0$ and also points x with $f'(x) < 0$.

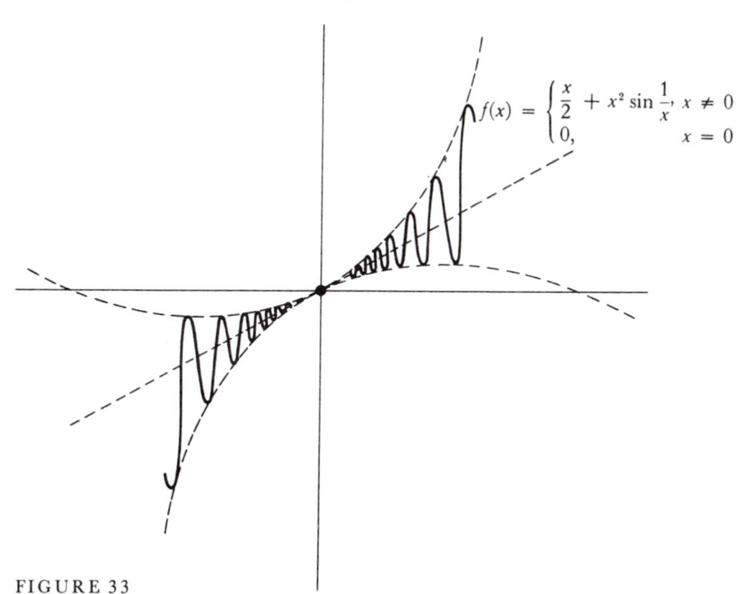

FIGURE 33

The behavior of f for $\alpha \geq 1$, which is much more difficult to analyze, is discussed in the next problem.

****64.** Let $f(x) = \alpha x + x^2 \sin 1/x$ for $x \neq 0$, and let $f(0) = 0$. In order to find the sign of $f'(x)$ when $\alpha \geq 1$ it is necessary to decide if $2x \sin 1/x - \cos 1/x$ is < -1 for any numbers x close to 0. It is a little more convenient to consider the function $g(y) = 2(\sin y)/y - \cos y$ for $y \neq 0$; we want to know if $g(y) < -1$ for large y. This question is quite delicate; the most significant part of $g(y)$ is $-\cos y$, which does reach the value -1, but this happens only when $\sin y = 0$, and it is not at all clear whether g itself can have values < -1. The obvious approach to this problem is to find the local minimum values of g. Unfortunately, it is impossible to solve the equation $g'(y) = 0$ explicitly, so more ingenuity is required.

(a) Show that if $g'(y) = 0$, then

$$\cos y = (\sin y)\left(\frac{2 - y^2}{2y}\right),$$

and conclude that

$$g(y) = (\sin y)\left(\frac{2 + y^2}{2y}\right).$$

(b) Now show that if $g'(y) = 0$, then

$$\sin^2 y = \frac{4y^2}{4 + y^4},$$

and conclude that

$$|g(y)| = \frac{2 + y^2}{\sqrt{4 + y^4}}.$$

(c) Using the fact that $(2 + y^2)/\sqrt{4 + y^4} > 1$, show that if $\alpha = 1$, then f is not increasing in any interval around 0.

(d) Using the fact that $\lim_{y\to\infty} (2 + y^2)/\sqrt{4 + y^4} = 1$, show that if $\alpha > 1$, then f *is* increasing in some interval around 0.

****65.** A function f is **increasing at** $\boldsymbol{a}$ if there is some number $\delta > 0$ such that

$$f(x) > f(a) \quad \text{if} \quad a < x < a + \delta$$

and

$$f(x) < f(a) \quad \text{if} \quad a - \delta < x < a.$$

Notice that this does *not* mean that f is increasing in the interval $(a - \delta, a + \delta)$; for example, the function shown in Figure 33 is increasing at 0, but is not an increasing function in any open interval containing 0.

(a) Suppose that f is continuous on $[0, 1]$ and that f is increasing at a for every a in $[0, 1]$. Prove that f is increasing on $[0, 1]$. (First convince yourself that there is something to be proved.) Hint: For $0 < b < 1$, prove that the minimum of f on $[b, 1]$ must be at b.

(b) Prove part (a) without the assumption that f is continuous, by considering for each b in $[0, 1]$ the set $S_b = \{x : f(y) \geq f(b) \text{ for all } y \text{ in } [b, x]\}$. (This part of the problem is not necessary for the other parts.) Hint: Prove that $S_b = \{x : b \leq x \leq 1\}$ by considering $\sup S_b$.

(c) If f is increasing at a and f is differentiable at a, prove that $f'(a) \geq 0$ (this is easy).

(d) If $f'(a) > 0$, prove that f is increasing at a (go right back to the definition of $f'(a)$).

(e) Use parts (a) and (d) to show, without using the Mean Value Theorem, that if f is continuous on $[0, 1]$ and $f'(a) > 0$ for all a in $[0, 1]$, then f is increasing on $[0, 1]$.

(f) Suppose that f is continuous on $[0, 1]$ and $f'(a) = 0$ for all a in $(0, 1)$. Apply part (e) to the function $g(x) = f(x) + \varepsilon x$ to show that $f(1) - f(0) > -\varepsilon$. Similarly, show that $f(1) - f(0) < \varepsilon$ by considering $h(x) = \varepsilon x - f(x)$. Conclude that $f(0) = f(1)$.

This particular proof that a function with zero derivative must be constant has many points in common with a proof of H. A. Schwarz, which may be the first rigorous proof ever given. Its discoverer, at least, seemed to think it was. See his exuberant letter in reference [40] of the Suggested Reading.

****66.** (a) If f is a constant function, then every point is a local maximum point for f. It is quite possible for this to happen even if f is not a constant function: for example, if $f(x) = 0$ for $x < 0$ and $f(x) = 1$ for $x \geq 0$. But prove, using Problem 8-4, that if f is continuous on $[a, b]$ and every point of $[a, b]$ is a local maximum point, then f is a constant function. The same result holds, of course, if every point of $[a, b]$ is a local minimum point.

(b) Suppose now that every point is either a local maximum or a local minimum point for f (but we don't preclude the possibility that some points are local maxima while others are local minima). Prove that f is constant, as follows. Suppose that $f(a_0) < f(b_0)$. We can assume that $f(a_0) < f(x) < f(b_0)$ for $a_0 < x < b_0$. (Why?) Using Theorem 1 of the Appendix to Chapter 8, partition $[a_0, b_0]$ into intervals on which $\sup f - \inf f < (f(b_0) - f(a_0))/2$; also choose the lengths of these intervals to be less than $(b_0 - a_0)/2$. Then there is one such interval $[a_1, b_1]$ with $a_0 < a_1 < b_1 < b_0$ and $f(a_1) < f(b_1)$. (Why?) Continue inductively and use the Nested Interval Theorem (Problem 8-14) to find a point x that cannot be a local maximum or minimum.

****67.** (a) A point x is called a **strict maximum point** for f on A if $f(x) > f(y)$ for all y in A with $y \neq x$ (compare with the definition of an ordinary maximum point). A **local strict maximum point** is defined in the obvious way. Find all local strict maximum points of the function

$$f(x) = \begin{cases} 0, & x \text{ irrational} \\ \dfrac{1}{q}, & x = \dfrac{p}{q} \text{ in lowest terms.} \end{cases}$$

It seems quite unlikely that a function can have a local strict maximum at *every* point (although the above example might give one pause for thought). Prove this as follows.

(b) Suppose that every point is a local strict maximum point for f. Let x_1 be any number and choose $a_1 < x_1 < b_1$ with $b_1 - a_1 < 1$ such that $f(x_1) > f(x)$ for all x in $[a_1, b_1]$. Let $x_2 \neq x_1$ be any point in (a_1, b_1) and choose $a_1 \leq a_2 < x_2 < b_2 \leq b_1$ with $b_2 - a_2 < \frac{1}{2}$ such that $f(x_2) > f(x)$ for all x in $[a_2, b_2]$. Continue in this way, and use the Nested Interval Theorem (Problem 8-14) to obtain a contradiction.

APPENDIX. CONVEXITY AND CONCAVITY

Although the graph of a function can be sketched quite accurately on the basis of the information provided by the derivative, some subtle aspects of the graph are revealed only by examining the second derivative. These details were purposely omitted previously because graph sketching is complicated enough without worrying about them, and the additional information obtained is often not worth the effort. Also, correct proofs of the relevant facts are sufficiently difficult to be placed in an appendix. Despite these discouraging remarks, the information presented here is well worth assimilating, because the notions of convexity and concavity are far more important than as mere aids to graph sketching. Moreover, the proofs have a pleasantly geometric flavor not often found in calculus theorems. Indeed, the basic definition is geometric in nature (see Figure 1).

DEFINITION 1

> A function f is **convex** on an interval, if for all a and b in the interval, the line segment joining $(a, f(a))$ and $(b, f(b))$ lies above the graph of f.

The geometric condition appearing in this definition can be expressed in an analytic way that is sometimes more useful in proofs. The straight line between $(a, f(a))$ and $(b, f(b))$ is the graph of the function g defined by

$$g(x) = \frac{f(b) - f(a)}{b - a}(x - a) + f(a).$$

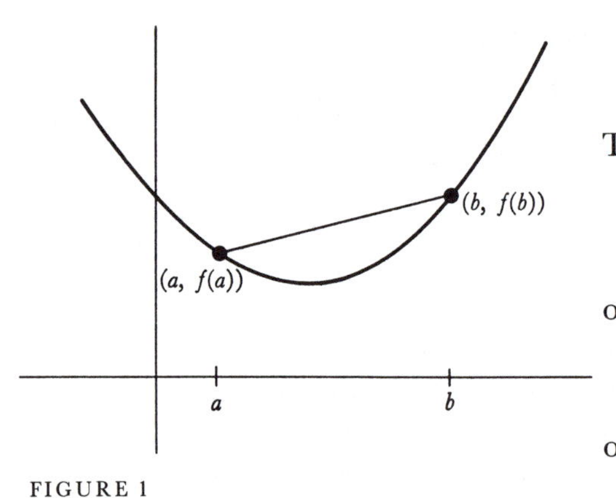

FIGURE 1

This line lies above the graph of f at x if $g(x) > f(x)$, that is, if

$$\frac{f(b) - f(a)}{b - a}(x - a) + f(a) > f(x)$$

or

$$\frac{f(b) - f(a)}{b - a}(x - a) > f(x) - f(a)$$

or

$$\frac{f(b) - f(a)}{b - a} > \frac{f(x) - f(a)}{x - a}.$$

We therefore have an equivalent definition of convexity.

DEFINITION 2

> A function f is **convex** on an interval if for a, x, and b in the interval with $a < x < b$ we have
>
> $$\frac{f(x) - f(a)}{x - a} < \frac{f(b) - f(a)}{b - a}.$$

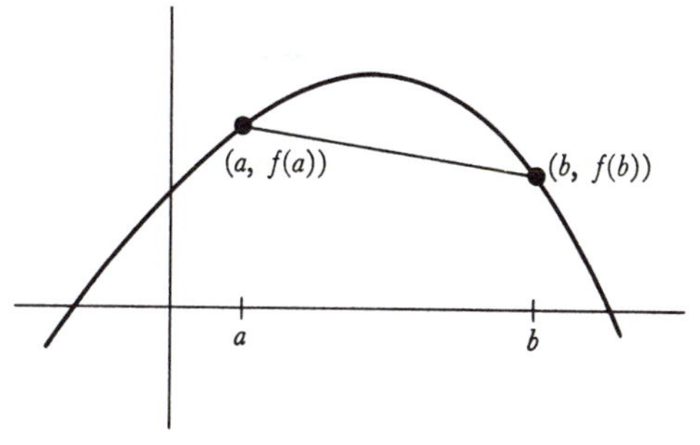

FIGURE 2

If the word "over" in Definition 1 is replaced by "under" or, equivalently, if the inequality in Definition 2 is replaced by

$$\frac{f(x) - f(a)}{x - a} > \frac{f(b) - f(a)}{b - a},$$

we obtain the definition of a **concave** function (Figure 2). It is not hard to see that the concave functions are precisely the ones of the form $-f$, where f is convex. For this reason, the next three theorems about convex functions have immediate corollaries about concave functions, so simple that we will not even bother to state them.

Figure 3 shows some tangent lines of a convex function. Two things seem to be true:

(1) The graph of f lies above the tangent line at $(a, f(a))$ except at the point $(a, f(a))$ itself (this point is called the **point of contact** of the tangent line).
(2) If $a < b$, then the slope of the tangent line at $(a, f(a))$ is less than the slope of the tangent line at $(b, f(b))$; that is, f' is increasing.

As a matter of fact these observations are true, and the proofs are not difficult.

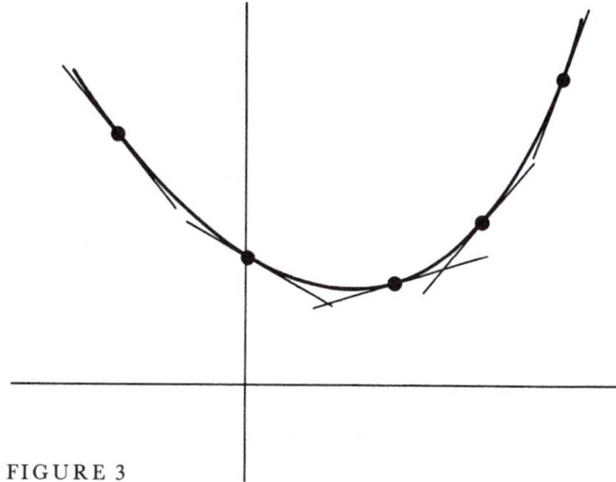

FIGURE 3

THEOREM 1 Let f be convex. If f is differentiable at a, then the graph of f lies above the tangent line through $(a, f(a))$, except at $(a, f(a))$ itself. If $a < b$ and f is differentiable at a and b, then $f'(a) < f'(b)$.

PROOF If $0 < h_1 < h_2$, then as Figure 4 indicates,

$$(1) \qquad \frac{f(a + h_1) - f(a)}{h_1} < \frac{f(a + h_2) - f(a)}{h_2}.$$

A nonpictorial proof can be derived immediately from Definition 2 applied to $a < a + h_1 < a + h_2$. Inequality (1) shows that the values of

$$\frac{f(a + h) - f(a)}{h}$$

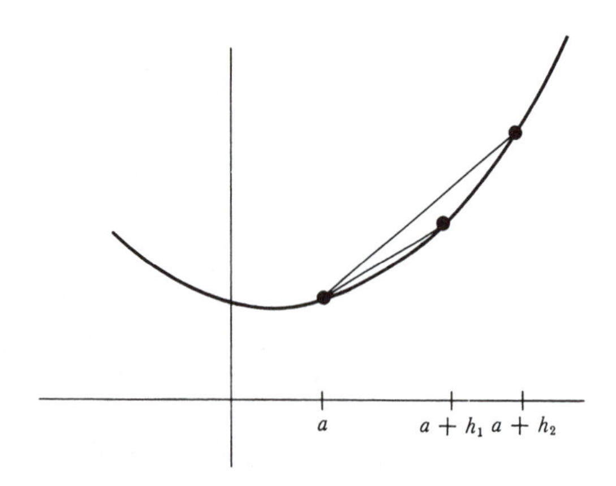

FIGURE 4

decrease as $h \to 0^+$. Consequently,

$$f'(a) < \frac{f(a+h) - f(a)}{h} \quad \text{for } h > 0$$

(in fact $f'(a)$ is the greatest lower bound of all these numbers). But this means that for $h > 0$ the secant line through $(a, f(a))$ and $(a+h, f(a+h))$ has larger slope than the tangent line, which implies that $(a+h, f(a+h))$ lies above the tangent line (an analytic translation of this argument is easily supplied).

For negative h there is a similar situation (Figure 5): if $h_2 < h_1 < 0$, then

$$\frac{f(a+h_1) - f(a)}{h_1} > \frac{f(a+h_2) - f(a)}{h_2}.$$

This shows that the slope of the tangent line is greater than

$$\frac{f(a+h) - f(a)}{h} \quad \text{for } h < 0$$

(in fact $f'(a)$ is the least upper bound of all these numbers), so that $f(a+h)$ lies above the tangent line if $h < 0$. This proves the first part of the theorem.

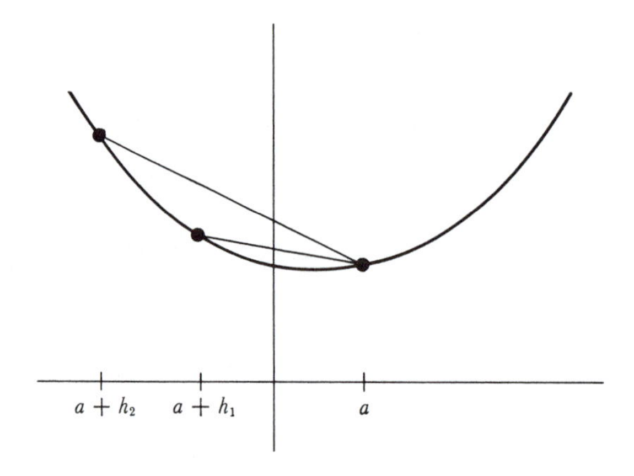

FIGURE 5

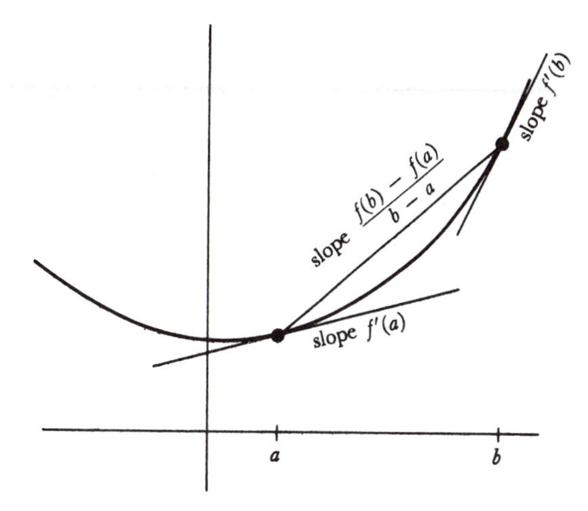

FIGURE 6

Now suppose that $a < b$. Then, as we have already seen (Figure 6),

$$f'(a) < \frac{f(a + (b - a)) - f(a)}{b - a} \quad \text{since } b - a > 0$$
$$= \frac{f(b) - f(a)}{b - a}$$

and

$$f'(b) > \frac{f(b + (a - b)) - f(b)}{a - b} \quad \text{since } a - b < 0$$
$$= \frac{f(a) - f(b)}{a - b} = \frac{f(b) - f(a)}{b - a}.$$

Combining these inequalities, we obtain $f'(a) < f'(b)$. ▮

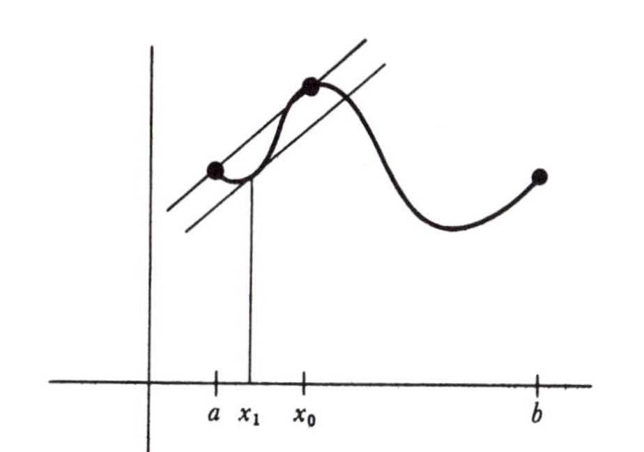

FIGURE 7

Theorem 1 has two converses. Here the proofs will be a little more difficult. We begin with a lemma that plays the same role in the next theorem that Rolle's Theorem plays in the proof of the Mean Value Theorem. It states that if f' is increasing, then the graph of f lies below any secant line *which happens to be horizontal.*

LEMMA Suppose f is differentiable and f' is increasing. If $a < b$ and $f(a) = f(b)$, then $f(x) < f(a) = f(b)$ for $a < x < b$.

PROOF Suppose first that $f(x) > f(a) = f(b)$ for some x in (a, b). Then the maximum of f on $[a, b]$ occurs at some point x_0 in (a, b) with $f(x_0) > f(a)$ and, of course, $f'(x_0) = 0$ (Figure 7). On the other hand, applying the Mean Value Theorem to the interval $[a, x_0]$, we find that there is x_1 with $a < x_1 < x_0$ and

$$f'(x_1) = \frac{f(x_0) - f(a)}{x_0 - a} > 0,$$

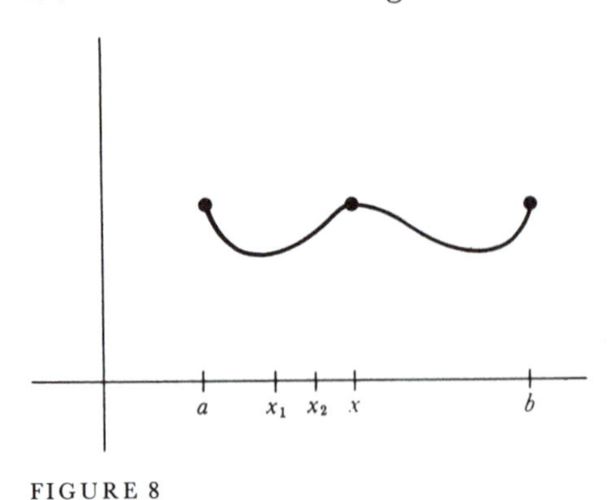

FIGURE 8

contradicting the fact that f' is increasing. This proves that $f(x) \leq f(a) = f(b)$ for $a < x < b$, and it only remains to prove that $f(x) = f(a)$ is also impossible for x in (a, b).

Suppose $f(x) = f(a)$ for some x in (a, b). We know that f is not constant on $[a, x]$ (if it were, f' would not be increasing on $[a, x]$) so there is (Figure 8) some x_1 with $a < x_1 < x$ and $f(x_1) < f(a)$. Applying the Mean Value Theorem to $[x_1, x]$ we conclude that there is x_2 with $x_1 < x_2 < x$ and

$$f'(x_2) = \frac{f(x) - f(x_1)}{x - x_1} > 0.$$

On the other hand, $f'(x) = 0$, since a local maximum occurs at x. Again this contradicts the hypothesis that f' is increasing. ▮

We now attack the general case by the same sort of algebraic machinations that we used in the proof of the Mean Value Theorem.

THEOREM 2 If f is differentiable and f' is increasing, then f is convex.

PROOF Let $a < b$. Define g by

$$g(x) = f(x) - \frac{f(b) - f(a)}{b - a}(x - a).$$

It is easy to see that g' is also increasing; moreover, $g(a) = g(b) = f(a)$. Applying the lemma to g we conclude that

$$g(x) < f(a) \quad \text{if} \quad a < x < b.$$

In other words, if $a < x < b$, then

$$f(x) - \frac{f(b) - f(a)}{b - a}(x - a) < f(a)$$

or

$$\frac{f(x) - f(a)}{x - a} < \frac{f(b) - f(a)}{b - a}.$$

Hence f is convex. ▮

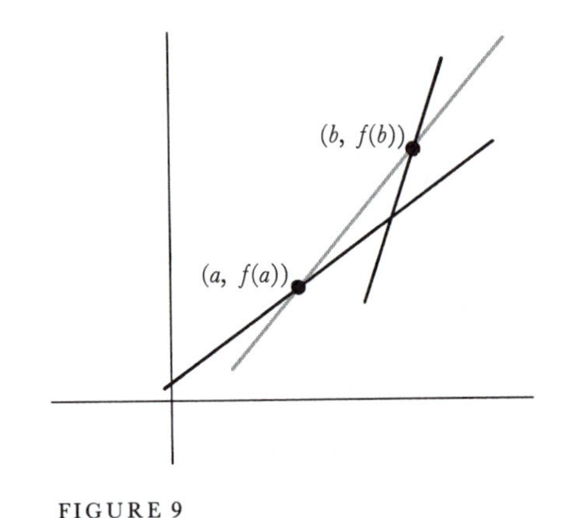

FIGURE 9

THEOREM 3 If f is differentiable and the graph of f lies above each tangent line except at the point of contact, then f is convex.

PROOF Let $a < b$. It is clear from Figure 9 that if $(b, f(b))$ lies above the tangent line at $(a, f(a))$, and $(a, f(a))$ lies above the tangent line at $(b, f(b))$, then the slope of the tangent line at $(b, f(b))$ must be larger than the slope of the tangent line at $(a, f(a))$. The following argument just says this with equations.

Since the tangent line at $(a, f(a))$ is the graph of the function

$$g(x) = f'(a)(x - a) + f(a),$$

and since $(b, f(b))$ lies above the tangent line, we have

$$(1) \quad f(b) > f'(a)(b-a) + f(a).$$

Similarly, since the tangent line at $(b, f(b))$ is the graph of

$$h(x) = f'(b)(x-b) + f(b),$$

and $(a, f(a))$ lies above the tangent line at $(b, f(b))$, we have

$$(2) \quad f(a) > f'(b)(a-b) + f(b).$$

It follows from (1) and (2) that $f'(a) < f'(b)$.

It now follows from Theorem 2 that f is convex. ▌

If a function f has a reasonable second derivative, the information given in these theorems can be used to discover the regions in which f is convex or concave. Consider, for example, the function

$$f(x) = \frac{1}{1+x^2}.$$

For this function,

$$f'(x) = \frac{-2x}{(1+x^2)^2}.$$

Thus $f'(x) = 0$ only for $x = 0$, and $f(0) = 1$, while

$$\begin{aligned} f'(x) > 0 \quad &\text{if} \quad x < 0, \\ f'(x) < 0 \quad &\text{if} \quad x > 0. \end{aligned}$$

Moreover,

$$\begin{aligned} &f(x) > 0 \quad \text{for all } x, \\ &f(x) \to 0 \quad \text{as } x \to \infty \text{ or } -\infty, \\ &f \text{ is even.} \end{aligned}$$

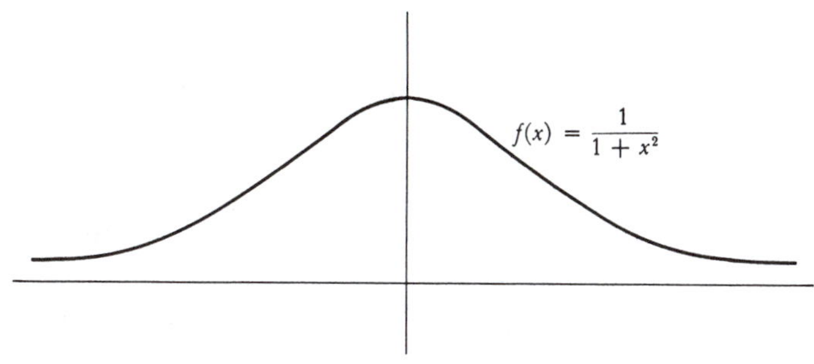

FIGURE 10

The graph of f therefore looks something like Figure 10. We now compute

$$\begin{aligned} f''(x) &= \frac{(1+x^2)^2(-2)+2x\cdot[2(1+x^2)\cdot 2x]}{(1+x^2)^4} \\ &= \frac{2(3x^2-1)}{(1+x^2)^3}. \end{aligned}$$

It is not hard to determine the sign of $f''(x)$. Note first that $f''(x)=0$ only when $x=\sqrt{1/3}$ or $-\sqrt{1/3}$. Since f'' is clearly continuous, it must keep the same sign on each of the sets

$$\begin{gathered} (-\infty,-\sqrt{1/3}\,), \\ (-\sqrt{1/3},\sqrt{1/3}\,), \\ (\sqrt{1/3},\infty). \end{gathered}$$

Since we easily compute, for example, that

$$\begin{aligned} f''(-1) &= \tfrac{1}{2} > 0, \\ f''(0) &= -2 < 0, \\ f''(1) &= \tfrac{1}{2} > 0, \end{aligned}$$

we conclude that

$$\begin{aligned} f'' &> 0 \text{ on } (-\infty,-\sqrt{1/3}\,) \text{ and } (\sqrt{1/3},\infty), \\ f'' &< 0 \text{ on } (-\sqrt{1/3},\sqrt{1/3}\,). \end{aligned}$$

Since $f''>0$ means f' is increasing, it follows from Theorem 2 that f is convex on $(-\infty,-\sqrt{1/3}\,)$ and $(\sqrt{1/3},\infty)$, while on $(-\sqrt{1/3},\sqrt{1/3}\,)$ f is concave (Figure 11).

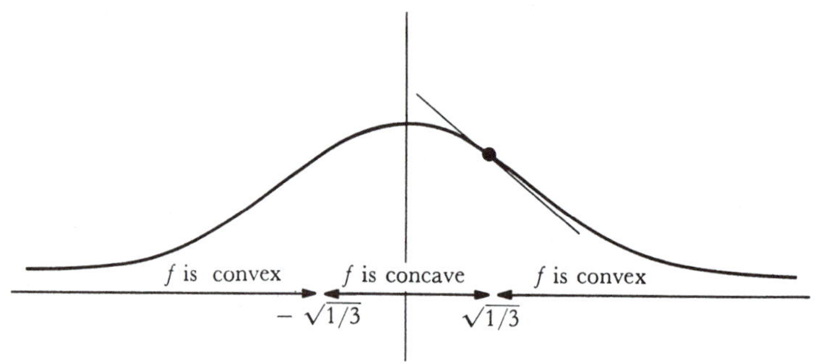

FIGURE 11

Notice that at $(\sqrt{1/3},\frac{3}{4})$ the tangent line lies below the part of the graph to the right, since f is convex on $(\sqrt{1/3},\infty)$, and above the part of the graph to the left, since f is concave on $(-\sqrt{1/3},\sqrt{1/3}\,)$; thus the tangent line crosses the graph. In general, a number a is called an **inflection point** of f if the tangent line to the graph of f at $(a,f(a))$ crosses the graph; thus $\sqrt{1/3}$ and $-\sqrt{1/3}$ are inflection points of $f(x)=1/(1+x^2)$. Note that the condition $f''(a)=0$ does *not* ensure that a is an inflection point of f; for example, if $f(x)=x^4$, then $f''(0)=0$, but f is convex, so the tangent line at $(0,0)$ certainly doesn't cross the graph of f. In order for a to be an inflection point of a function f, it is necessary that f'' should have different signs to the left and right of a.

This example illustrates the procedure which may be used to analyze any function f. After the graph has been sketched, using the information provided by f', the zeros of f'' are computed and the sign of f'' is determined on the intervals between consecutive zeros. On intervals where $f'' > 0$ the function is convex; on intervals where $f'' < 0$ the function is concave. Knowledge of the regions of convexity and concavity of f can often prevent absurd misinterpretation of other data about f. Several functions, which can be analyzed in this way, are given in the problems, which also contain some theoretical questions.

To round out our discussion of convexity and concavity, we will prove one further result that you may already have begun to suspect. We have seen that convex and concave functions have the property that every tangent line intersects the graph just once; a few drawings will probably convince you that no other functions have this property. The proof of this assertion is rather tricky; it is closely related to the proof of Theorem 2 of the next chapter, and is probably best deferred until after that proof has been read.

THEOREM 4 If f is differentiable on an interval and intersects each of its tangent lines just once, then f is either convex or concave on that interval.

PROOF There are two parts to the proof.

(1) First we claim that no straight line can intersect the graph of f in *three* different points. Suppose, on the contrary, that some straight line did intersect the graph of f at $(a, f(a))$, $(b, f(b))$ and $(c, f(c))$, with $a < b < c$ (Figure 12). Then we would have

$$\frac{f(b)-f(a)}{b-a} = \frac{f(c)-f(a)}{c-a}. \tag{1}$$

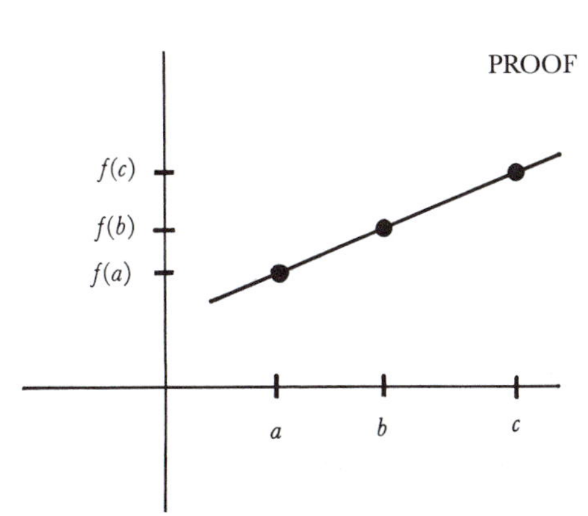

FIGURE 12

Consider the function

$$g(x) = \frac{f(x)-f(a)}{x-a} \qquad \text{for } x \text{ in } [b, c].$$

Equation (1) says that $g(b) = g(c)$. So by Rolle's Theorem, there is some number x in (b, c) where $0 = g'(x)$, and thus

$$0 = (x-a)f'(x) - [f(x) - f(a)]$$

or

$$f'(x) = \frac{f(x)-f(a)}{x-a}.$$

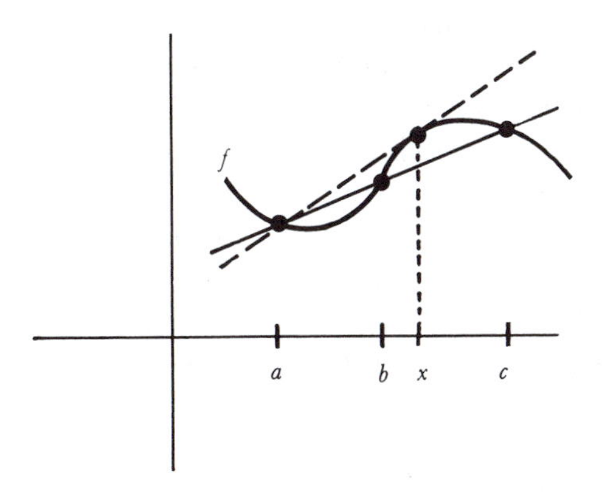

FIGURE 13

But this says (Figure 13) that the tangent line at $(x, f(x))$ passes through $(a, f(a))$, contradicting the hypotheses.

(2) Suppose that $a_0 < b_0 < c_0$ and $a_1 < b_1 < c_1$ are points in the interval. Let

$$\begin{aligned} x_t &= (1-t)a_0 + ta_1 \\ y_t &= (1-t)b_0 + tb_1 \qquad 0 \le t \le 1. \\ z_t &= (1-t)c_0 + tc_1 \end{aligned}$$

Then $x_0 = a_0$ and $x_1 = a_1$ and (Problem 4-2) the points x_t all lie between a_0 and a_1, with analogous statements for y_t and z_t. Moreover,

$$x_t < y_t < z_t \qquad \text{for} \qquad 0 \le t \le 1.$$

Now consider the function

$$g(t) = \frac{f(y_t) - f(x_t)}{y_t - x_t} - \frac{f(z_t) - f(x_t)}{z_t - x_t} \qquad \text{for } 0 \le t \le 1.$$

By step (1), $g(t) \neq 0$ for all t in $[0, 1]$. So either $g(t) > 0$ for all t in $[0, 1]$ or $g(t) < 0$ for all t in $[0, 1]$. Thus, either f is convex or f is concave (compare pages 231–232). ▮

PROBLEMS

1. Sketch, indicating regions of convexity and concavity and points of inflection, the functions in Problem 11-1 (consider (iv) as double starred).

2. Figure 30 in Chapter 11 shows the graph of f'. Sketch the graph of f.

3. Find two convex functions f and g such that $f(x) = g(x)$ if and only if x is an integer.

4. Show that f is convex on an interval if and only if for all x and y in the interval we have

$$f(tx + (1-t)y) < tf(x) + (1-t)f(y), \qquad \text{for } 0 < t < 1.$$

(This is just a restatement of the definition, but a useful one.)

5. (a) Prove that if f and g are convex and f is increasing, then $f \circ g$ is convex. (It will be easiest to use Problem 4.)
(b) Give an example where $g \circ f$ is not convex.
(c) Suppose that f and g are twice differentiable. Give another proof of the result of part (a) by considering second derivatives.

6. (a) Suppose that f is differentiable and convex on an interval. Show that either f is increasing, or else f is decreasing, or else there is a number c such that f is decreasing to the left of c and increasing to the right of c.
(b) Use this fact to give another proof of the result in Problem 5(a) when f and g are (one-time) differentiable. (You will have to be a little careful when comparing $f'(g(x))$ and $f'(g(y))$ for $x < y$.)
(c) Prove the result in part (a) without assuming f differentiable. You will have to keep track of several different cases, but no particularly clever ideas are needed. Begin by showing that if $a < b$ and $f(a) < f(b)$, then f is increasing to the right of b; and if $f(a) > f(b)$, then f is decreasing to the left of a.

***7.** Let f be a twice-differentiable function with the following properties: $f(x) > 0$ for $x \ge 0$, and f is decreasing, and $f'(0) = 0$. Prove that

$f''(x) = 0$ for some $x > 0$ (so that in reasonable cases f will have an inflection point at x—an example is given by $f(x) = 1/(1+x^2)$). Every hypothesis in this theorem is essential, as shown by $f(x) = 1 - x^2$, which is not positive for all x; by $f(x) = x^2$, which is not decreasing; and by $f(x) = 1/(x+1)$, which does not satisfy $f'(0) = 0$. Hint: Choose $x_0 > 0$ with $f'(x_0) < 0$. We cannot have $f'(y) \le f'(x_0)$ for all $y > x_0$. Why not? So $f'(x_1) > f'(x_0)$ for some $x_1 > x_0$. Consider f' on $[0, x_1]$.

***8.** (a) Prove that if f is convex, then $f([x+y]/2) < [f(x) + f(y)]/2$.
(b) Suppose that f satisfies this condition. Show that $f(kx + (1-k)y) < kf(x) + (1-k)f(y)$ whenever k is a rational number, between 0 and 1, of the form $m/2^n$. Hint: Part (a) is the special case $n = 1$. Use induction, employing part (a) at each step.
(c) Suppose that f satisfies the condition in part (a) and f is continuous. Show that f is convex.

***9.** Let $p_1, \dots, p_n$ by *positive* numbers with $\sum_{i=1}^{n} p_i = 1$.

(a) For any numbers $x_1, \dots, x_n$ show that $\sum_{i=1}^{n} p_i x_i$ lies between the smallest and the largest x_i.
(b) Show the same for $(1/t)\sum_{i=1}^{n-1} p_i x_i$, where $t = \sum_{i=1}^{n-1} p_i$.
(c) Prove *Jensen's inequality:* If f is convex, then $f\left(\sum_{i=1}^{n} p_i x_i\right) \le \sum_{i=1}^{n} p_i f(x_i)$.
Hint: Use Problem 4, noting that $p_n = 1 - t$. (Part (b) is needed to show that $(1/t)\sum_{i=1}^{n-1} p_i x_i$ is in the domain of f if $x_1, \dots, x_n$ are.)

***10.** (a) For any function f, the right-hand derivative, $\lim_{h \to 0^+} [f(a+h) - f(a)]/h$, is denoted by $f_+{}'(a)$, and the left-hand derivative is denoted by $f_-{}'(a)$. The proof of Theorem 1 actually shows that $f_+{}'$ and $f_-{}'$ always exist if f is convex. Check this assertion, and also show that $f_+{}'$ and $f_-{}'$ are increasing, and that $f_-{}'(a) \le f_+{}'(a)$.
**(b) Show that if f is convex, then $f_+{}'(a) = f_-{}'(a)$ if and only if $f_+{}'$ is continuous at a. (Thus f is differentiable precisely when $f_+{}'$ is continuous.) Hint: $[f(b) - f(a)]/(b-a)$ is close to $f_-{}'(a)$ for $b < a$ close to a, and $f_+{}'(b)$ is less than this quotient.

***11.** (a) Prove that a convex function on **R**, or on any open interval, must be continuous.
(b) Give an example of a convex function on a closed interval that is *not* continuous, and explain exactly what kinds of discontinuities are possible.

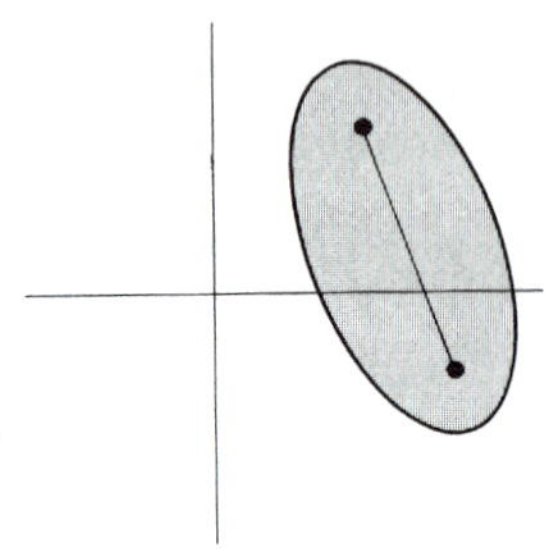

(a) a convex subset of the plane

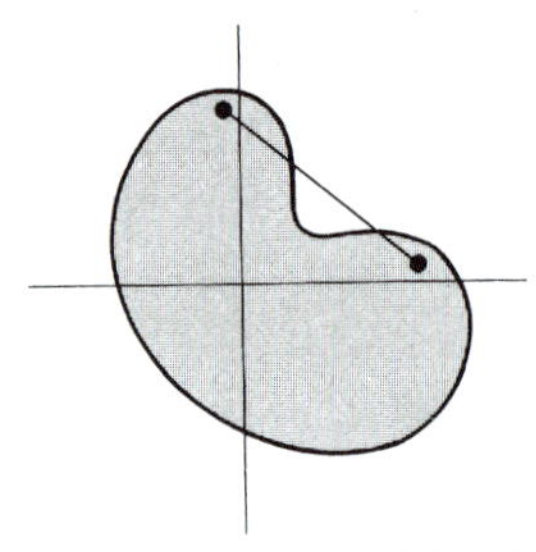

(b) a nonconvex subset of the plane

FIGURE 14

12. Call a function f *weakly convex* on an interval if for $a < b < c$ in this interval we have

$$\frac{f(x) - f(a)}{x - a} \le \frac{f(b) - f(a)}{b - a}.$$

(a) Show that a weakly convex function is convex if and only if its graph contains no straight line segments. (Sometimes a weakly convex function is simply called "convex," while convex functions in our sense are called "strictly convex".)

(b) Reformulate the theorems of this section for weakly convex functions.

13. A set A of points in the plane is called *convex* if A contains the line segment joining any two points in it (Figure 14). For a function f, let A_f be the set of points (x, y) with $y \ge f(x)$, so that A_f is the set of points on or above the graph of f. Show that A is convex if and only if f is weakly convex, in the terminology of the previous problem. Further information on convex sets will be found in reference [19] of the Suggested Reading.

CHAPTER 12 INVERSE FUNCTIONS

We now have at our disposal quite powerful methods for investigating functions; what we lack is an adequate supply of functions to which these methods may be applied. We have studied various ways of forming new functions from old—addition, multiplication, division, and composition—but using these alone, we can produce only the rational functions (even the sine function, although frequently used for examples, has never been defined). In the next few chapters we will begin to construct new functions in quite sophisticated ways, but there is one important method which will practically double the usefulness of any other method we discover.

If we recall that a function is a collection of pairs of numbers, we might hit upon the bright idea of simply reversing all the pairs. Thus from the function

$$f = \{ (1, 2), (3, 4), (5, 9), (13, 8) \},$$

we obtain

$$g = \{ (2, 1), (4, 3), (9, 5), (8, 13) \}.$$

While $f(1) = 2$ and $f(3) = 4$, we have $g(2) = 1$ and $g(4) = 3$.

Unfortunately, this bright idea does not always work. If

$$f = \{ (1, 2), (3, 4), (5, 9), (13, 4) \},$$

then the collection

$$\{ (2, 1), (4, 3), (9, 5), (4, 13) \}$$

is not a function at all, since it contains both $(4, 3)$ and $(4, 13)$. It is clear where the trouble lies: $f(3) = f(13)$, even though $3 \neq 13$. This is the only sort of thing that can go wrong, and it is worthwhile giving a name to the functions for which this does not happen.

DEFINITION

A function f is **one-one** (read "one-to-one") if $f(a) \neq f(b)$ whenever $a \neq b$.

The identity function I is obviously one-one, and so is the following modification:

$$g(x) = \begin{cases} x, & x \neq 3, 5 \\ 3, & x = 5 \\ 5, & x = 3. \end{cases}$$

The function $f(x) = x^2$ is not one-one, since $f(-1) = f(1)$, but if we define

$$g(x) = x^2, \qquad x \geq 0$$

(and leave g undefined for $x < 0$), then g is one-one, because g is increasing (since $g'(x) = 2x > 0$, for $x > 0$). This observation is easily generalized: If n is a natural number and

$$f(x) = x^n, \qquad x \geq 0,$$

then f is one-one. If n is odd, one can do better: the function

$$f(x) = x^n \qquad \text{for all } x$$

is one-one (since $f'(x) = nx^{n-1} > 0$, for all $x \neq 0$).

It is particularly easy to decide from the graph of f whether f is one-one: the condition $f(a) \neq f(b)$ for $a \neq b$ means that no *horizontal* line intersects the graph of f twice (Figure 1).

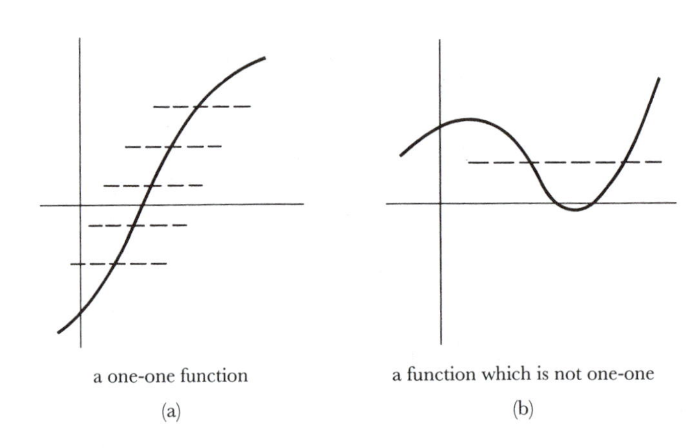

FIGURE 1

If we reverse all the pairs in (a not necessarily one-one function) f we obtain, in any case, some collection of pairs. It is popular to abstain from this procedure unless f is one-one, but there is no particular reason to do so—instead of a definition with restrictive conditions we obtain a definition and a theorem.

DEFINITION

For any function f, the **inverse** of f, denoted by $\boldsymbol{f^{-1}}$, is the set of all pairs (a, b) for which the pair (b, a) is in f.

THEOREM 1

f^{-1} is a function if and only if f is one-one.

PROOF

Suppose first that f is one-one. Let (a, b) and (a, c) be two pairs in f^{-1}. Then (b, a) and (c, a) are in f, so $a = f(b)$ and $a = f(c)$; since f is one-one this implies that $b = c$. Thus f^{-1} is a function.

Conversely, suppose that f^{-1} is a function. If $f(b) = f(c)$, then f contains the pairs $(b, f(b))$ and $(c, f(c)) = (c, f(b))$, so $(f(b), b)$ and $(f(b), c)$ are in f^{-1}. Since f^{-1} is a function this implies that $b = c$. Thus f is one-one. ▮

The graphs of f and f^{-1} are so closely related that it is possible to use the graph of f to visualize the graph of f^{-1}. Since the graph of f^{-1} consists of all pairs (a, b) with (b, a) in the graph of f, one obtains the graph of f^{-1} from the graph of f by interchanging the horizontal and vertical axes. If f has the graph shown in Figure 2(a),

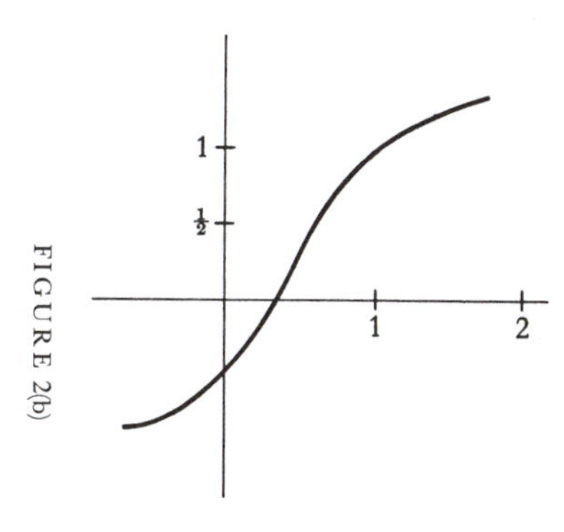

FIGURE 2(a)

FIGURE 2(b)

and you rotate this page counter clockwise through a right angle, then the graph of f^{-1} appears on your left (Figure 2(b)). The only trouble is that the numbering on the horizontal axis goes in the wrong direction, so you must flip this picture over to get the usual picture of f^{-1}, which appears on your right (Figure 3).

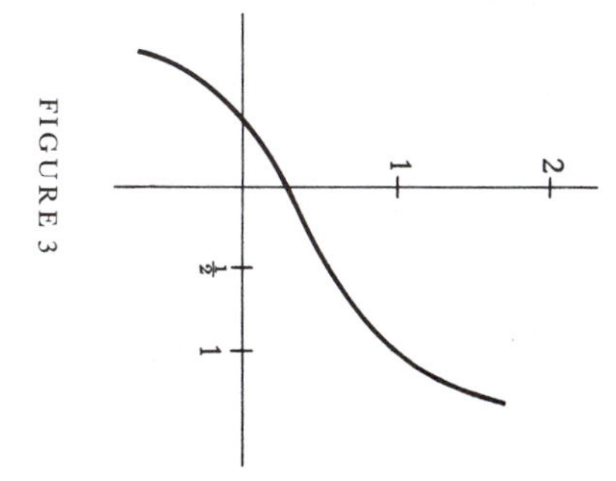

FIGURE 3

This procedure is awkward with books and impossible with blackboards, so it is fortunate that there is another way of constructing the graph of f^{-1}. The points

(a, b) and (b, a) are reflections of each other through the graph of $I(x) = x$, which is called the **diagonal** (Figure 4). To obtain the graph of f^{-1} we merely reflect the graph of f through this line (Figure 5).

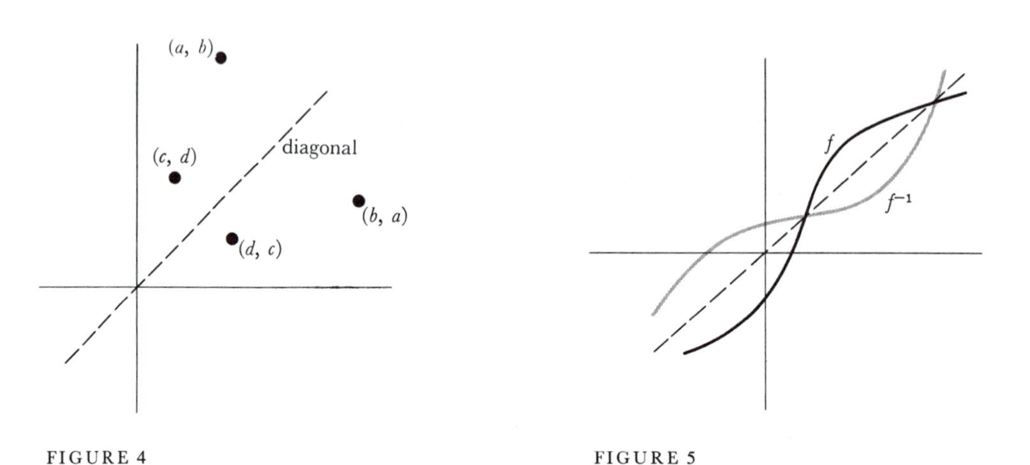

FIGURE 4

FIGURE 5

Reflecting through the diagonal twice will clearly leave us right back where we started; this means that $(f^{-1})^{-1} = f$, which is also clear from the definition. In conjunction with Theorem 1, this equation has a significant consequence: if f is a one-one function, then the function f^{-1} is also one-one (since $(f^{-1})^{-1}$ is a function).

There are a few other simple manipulations with inverse functions of which you should be aware. Since (a, b) is in f precisely when (b, a) is in f^{-1}, it follows that

$$b = f(a) \qquad \text{means the same as} \qquad a = f^{-1}(b).$$

Thus $f^{-1}(b)$ is the (unique) number a such that $f(a) = b$; for example, if $f(x) = x^3$, then $f^{-1}(b)$ is the unique number a such that $a^3 = b$, and this number is, by definition, $\sqrt[3]{b}$.

The fact that $f^{-1}(x)$ is the number y such that $f(y) = x$ can be restated in a much more compact form:

$$f(f^{-1}(x)) = x, \qquad \text{for all } x \text{ in the domain of } f^{-1}.$$

Moreover,

$$f^{-1}(f(x)) = x, \qquad \text{for all } x \text{ in the domain of } f;$$

this follows from the previous equation upon replacing f by f^{-1}. These two important equations can be written

$$f \circ f^{-1} = I,$$
$$f^{-1} \circ f = I$$

(except that the right side will have a bigger domain if the domain of f or f^{-1} is not all of **R**).

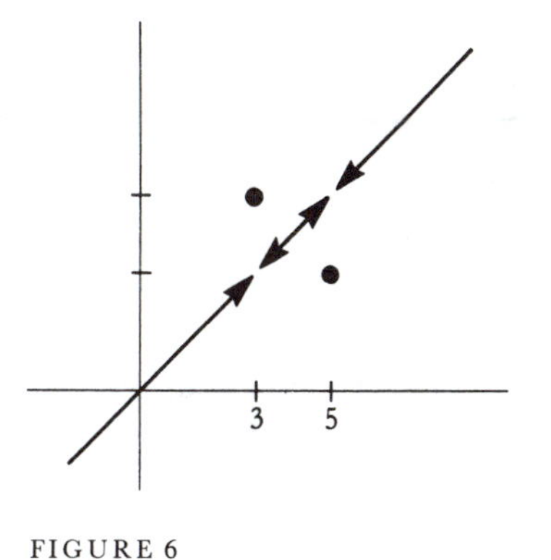

FIGURE 6

Since many standard functions will be defined as the inverses of other functions, it is quite important that we be able to tell which functions are one-one. We have already hinted which class of functions are most easily dealt with—increasing and decreasing functions are obviously one-one. Moreover, if f is increasing, then f^{-1} is also increasing, and if f is decreasing, then f^{-1} is decreasing (the proof is left to you). In addition, f is increasing if and only if $-f$ is decreasing, a very useful fact to remember.

It is certainly not true that every one-one function is either increasing or decreasing. One example has already been mentioned, and is now graphed in Figure 6:

$$g(x) = \begin{cases} x, & x \neq 3, 5 \\ 3, & x = 5 \\ 5, & x = 3. \end{cases}$$

Figure 7 shows that there are even continuous one-one functions which are neither increasing nor decreasing. But if you try drawing a few pictures you will soon agree that every one-one continuous function defined on an interval is either increasing or decreasing. It's possible to give a straightforward, but cumbersome, proof of this fact that involves keeping track of a lot of cases (very much like Problem 6(c) in the previous Appendix). The following proof dispenses with all these unpleasant details, although it is rather tricky.

THEOREM 2 If f is continuous and one-one of an interval, then f is either increasing or decreasing on that interval.

PROOF Let $a_0 < b_0$ be two numbers in the interval. Since f is one-one, we know that

$$\begin{array}{lll} \text{either} & \text{(i)} & f(b_0) - f(a_0) > 0 \\ \text{or} & \text{(ii)} & f(b_0) - f(a_0) < 0. \end{array}$$

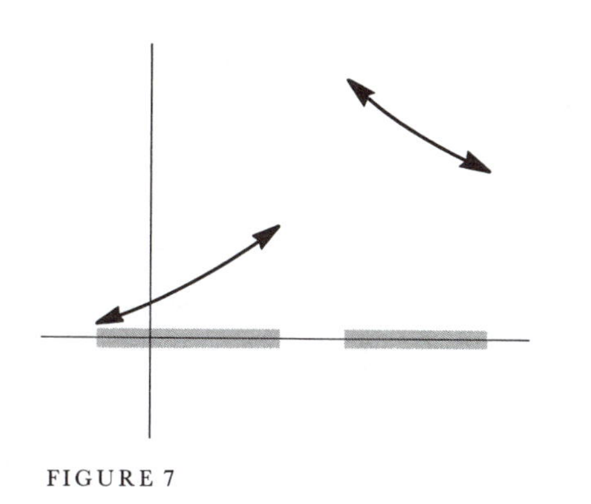

FIGURE 7

We will assume that (i) is true, and show that the same inequality holds for any $a_1 < b_1$ in the interval, so that f in increasing. (A similar argument shows that if (ii) is true, then f is decreasing.)

Let

$$\left.\begin{aligned} x_t &= (1-t)a_0 + ta_1 \\ y_t &= (1-t)b_0 + tb_1 \end{aligned}\right\} \quad \text{for } 0 \le t \le 1.$$

Then $x_0 = a_0$ and $x_1 = a_1$ and the points x_t all lie between a_0 and a_1 (Problem 4-2). An analogous statement holds for y_t. So x_t and y_t are all in the domain of f. Moreover, since $a_0 < b_0$ and $a_1 < b_1$, we also have

$$x_t < y_t \qquad \text{for } 0 \le t \le 1.$$

Now consider the function

$$g(t) = f(y_t) - f(x_t) \qquad \text{for } 0 \le t \le 1.$$

Using Theorem 6-2, it is easy to see that g is continuous on $[0, 1]$. Moreover, $g(t)$ is never 0, since $x_t < y_t$ and f is one-one. Consequently, $g(t)$ is either positive for all t in $[0, 1]$ or negative for all t in $[0, 1]$ (otherwise, by the Intermediate Value

Theorem it would also by 0 somewhere in $[0, 1]$). But $g(0) > 0$ by (i). So also $g(1) > 0$, which means that (i) also holds for a_1, b_1. ▌

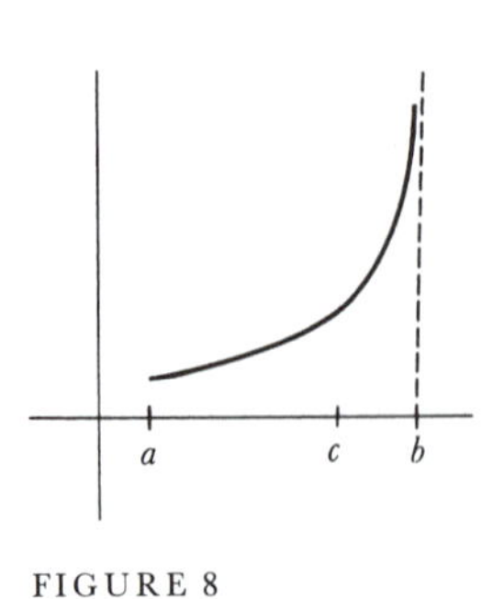

FIGURE 8

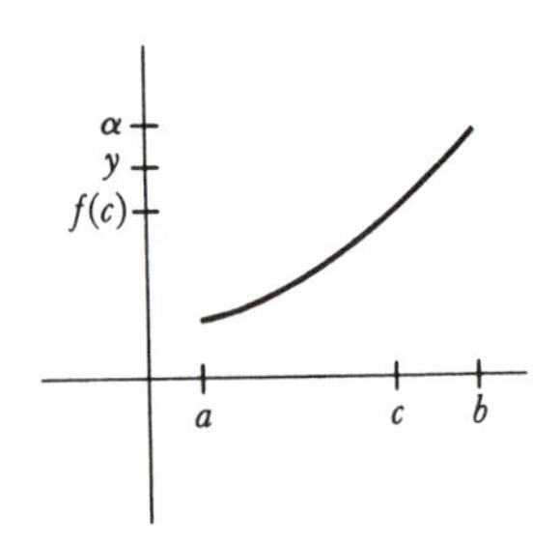

FIGURE 9

Henceforth we shall be concerned almost exclusively with continuous increasing or decreasing functions which are defined on an interval. If f is such a function, it is possible to say quite precisely what the domain of f^{-1} will be like.

Suppose first that f is a continuous increasing function on the closed interval $[a, b]$. Then, by the Intermediate Value Theorem, f takes on every value between $f(a)$ and $f(b)$. Therefore, the domain of f^{-1} is the closed interval $[f(a), f(b)]$. Similarly, if f is continuous and decreasing on $[a, b]$, then the domain of f^{-1} is $[f(b), f(a)]$.

If f is a continuous increasing function on an *open* interval (a, b) the analysis becomes a bit more difficult. To begin with, let us choose some point c in (a, b). We will first decide which values $> f(c)$ are taken on by f. One possibility is that f takes on arbitrarily large values (Figure 8). In this case f takes on *all* values $> f(c)$, by the Intermediate Value Theorem. If, on the other hand, f does not take on arbitrarily large values, then $A = \{f(x) : c \le x < b\}$ is bounded above, so A has a least upper bound α (Figure 9). Now suppose y is any number with $f(c) < y < \alpha$. Then f takes on some value $f(x) > y$ (because α is the least upper bound of A). By the Intermediate Value Theorem, f actually takes on the value y. Notice that f cannot take on the value α itself; for if $\alpha = f(x)$ for $a < x < b$ and we choose t with $x < t < b$, then $f(t) > \alpha$, which is impossible.

Precisely the same arguments work for values less than $f(c)$: either f takes on all values less than $f(c)$ or there is a number $\beta < f(c)$ such that f takes on all values between β and $f(c)$, but not β itself.

This entire argument can be repeated if f is decreasing, and if the domain of f is $\mathbf{R}$ or (a, ∞) or $(-\infty, a)$. Summarizing: if f is a continuous increasing, or decreasing, function whose domain is an interval having one of the forms

$$(a, b),\ (-\infty, b),\ (a, \infty),\ \text{or } \mathbf{R},$$

then the domain of f^{-1} is also an interval which has one of these four forms.

Now that we have completed this preliminary analysis of continuous one-one functions, it is possible to begin asking which important properties of a one-one function are inherited by its inverse. For continuity there is no problem.

THEOREM 3 If f is continuous and one-one on an interval, then f^{-1} is also continuous.

PROOF We know by Theorem 2 that f is either increasing or decreasing. We might as well assume that f is increasing, since we can then take care of the other case by applying the usual trick of considering $-f$.

We must show that

$$\lim_{x \to b} f^{-1}(x) = f^{-1}(b)$$

for each b in the domain of f^{-1}. Such a number b is of the form $f(a)$ for some a in the domain of f. For any $\varepsilon > 0$, we want to find a $\delta > 0$ such that, for all x,

$$\text{if } f(a) - \delta < x < f(a) + \delta, \text{ then } a - \varepsilon < f^{-1}(x) < a + \varepsilon.$$

Figure 10 suggests the way of finding δ (remember that by looking sideways you see the graph of f^{-1}): since

$$a - \varepsilon < a < a + \varepsilon,$$

it follows that

$$f(a - \varepsilon) < f(a) < f(a + \varepsilon);$$

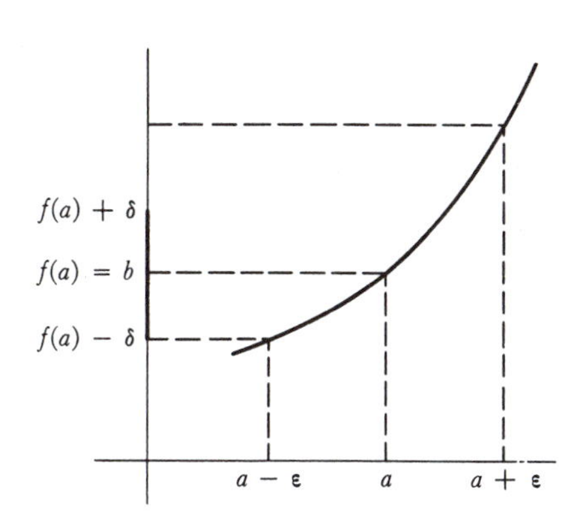

FIGURE 10

we let δ be the smaller of $f(a+\varepsilon) - f(a)$ and $f(a) - f(a-\varepsilon)$. Figure 10 contains the entire proof that this δ works, and what follows is simply a verbal account of the information contained in this picture.

Our choice of δ ensures that

$$f(a - \varepsilon) \le f(a) - \delta \text{ and } f(a) + \delta \le f(a + \varepsilon).$$

Consequently, if

$$f(a) - \delta < x < f(a) + \delta,$$

then

$$f(a - \varepsilon) < x < f(a + \varepsilon).$$

Since f is increasing, f^{-1} is also increasing, and we obtain

$$f^{-1}(f(a - \varepsilon)) < f^{-1}(x) < f^{-1}(f(a + \varepsilon)),$$

i.e.,

$$a - \varepsilon < f^{-1}(x) < a + \varepsilon,$$

which is precisely what we want. ▌

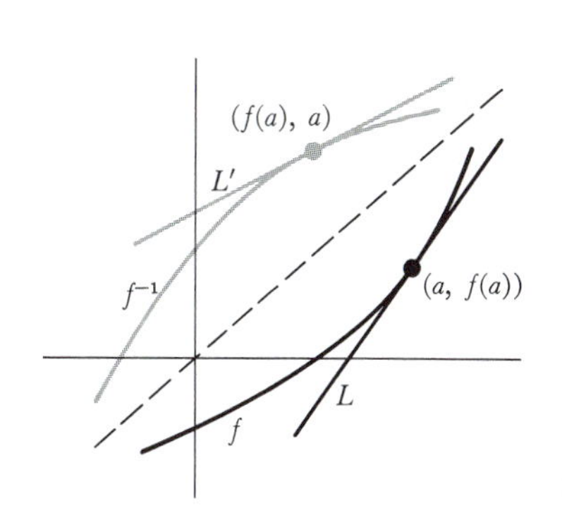

FIGURE 11

Having successfully investigated continuity of f^{-1}, it is only reasonable to tackle differentiability. Again, a picture indicates just what result ought to be true. Figure 11 shows the graph of a one-one function f with a tangent line L through $(a, f(a))$. If this entire picture is reflected through the diagonal, it shows the graph of f^{-1} and the tangent line L' through $(f(a), a)$. The slope of L' is the reciprocal of the slope of L. In other words, it appears that

$$(f^{-1})'(f(a)) = \frac{1}{f'(a)}.$$

This formula can equally well be written in a way which expresses $(f^{-1})'(b)$ directly, for each b in the domain of f^{-1}:

$$(f^{-1})'(b) = \frac{1}{f'(f^{-1}(b))}.$$

Unlike the argument for continuity, this pictorial "proof" becomes somewhat involved when formulated analytically. There is another approach which might be tried. Since we know that

$$f(f^{-1}(x)) = x,$$

it is tempting to prove the desired formula by applying the Chain Rule:

$$f'(f^{-1}(x)) \cdot (f^{-1})'(x) = 1,$$

so

$$(f^{-1})'(x) = \frac{1}{f'(f^{-1}(x))}.$$

Unfortunately, this is not a proof that f^{-1} is differentiable, since the Chain Rule cannot be applied unless f^{-1} is already known to be differentiable. But this argument does show what $(f^{-1})'(x)$ will have to be if f^{-1} *is* differentiable, and it can also be used to obtain some important preliminary information.

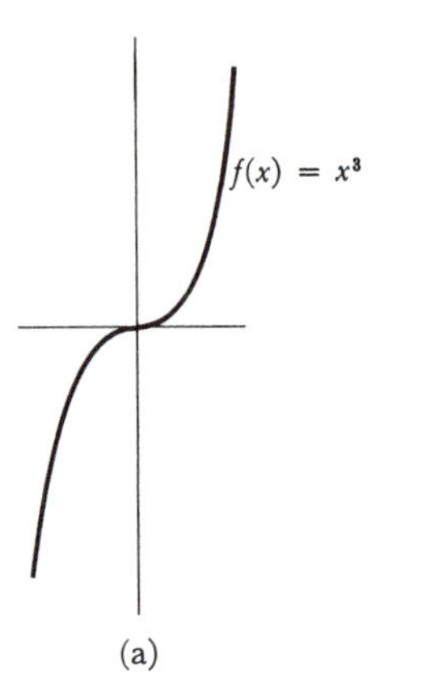

(a)

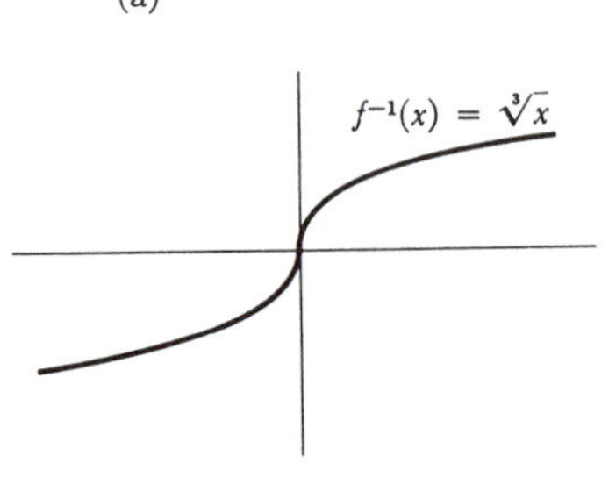

(b)

FIGURE 12

THEOREM 4 If f is a continuous one-one function defined on an interval and $f'(f^{-1}(a)) = 0$, then f^{-1} is *not* differentiable at a.

PROOF We have

$$f(f^{-1}(x)) = x.$$

If f^{-1} *were* differentiable at a, the Chain Rule would imply that

$$f'(f^{-1}(a)) \cdot (f^{-1})'(a) = 1,$$

hence

$$0 \cdot (f^{-1})'(a) = 1,$$

which is absurd. ▮

A simple example to which Theorem 4 applies is the function $f(x) = x^3$. Since $f'(0) = 0$ and $0 = f^{-1}(0)$, the function f^{-1} is not differentiable at 0 (Figure 12).

Having decided where an inverse function cannot be differentiable, we are now ready for the rigorous proof that in all other cases the derivative is given by the formula which we have already "derived" in two different ways. Notice that the following argument uses *continuity* of f^{-1}, which we have already proved.

THEOREM 5 Let f be a continuous one-one function defined on an interval, and suppose that f is differentiable at $f^{-1}(b)$, with derivative $f'(f^{-1}(b)) \neq 0$. Then f^{-1} is differentiable at b, and

$$(f^{-1})'(b) = \frac{1}{f'(f^{-1}(b))}.$$

PROOF Let $b = f(a)$. Then

$$\lim_{h\to 0} \frac{f^{-1}(b+h) - f^{-1}(b)}{h}$$

$$= \lim_{h\to 0} \frac{f^{-1}(b+h) - a}{h}$$

Now every number $b+h$ in the domain of f^{-1} can be written in the form

$$b+h=f(a+k)$$

for a unique k (we should really write $k(h)$, but we will stick with k for simplicity). Then

$$\begin{aligned}\lim_{h\to 0}\frac{f^{-1}(b+h)-a}{h}\\ &=\lim_{h\to 0}\frac{f^{-1}(f(a+k))-a}{f(a+k)-b}\\ &=\lim_{h\to 0}\frac{k}{f(a+k)-f(a)}.\end{aligned}$$

We are clearly on the right track! It is not hard to get an explicit expression for k; since

$$b+h=f(a+k)$$

we have

$$f^{-1}(b+h)=a+k$$

or

$$k=f^{-1}(b+h)-f^{-1}(b).$$

Now by Theorem 3, the function f^{-1} is continuous at b. This means that k approaches 0 as h approaches 0. Since

$$\lim_{k\to 0}\frac{f(a+k)-f(a)}{k}=f'(a)=f'(f^{-1}(b))\neq 0,$$

this implies that

$$(f^{-1})'(b)=\frac{1}{f'(f^{-1}(b))}. \blacksquare$$

The work we have done on inverse functions will be amply repaid later, but here is an immediate dividend. For n odd, let

$$f_n(x)=x^n \qquad \text{for all } x;$$

for n even, let

$$f_n(x)=x^n, \qquad x\geq 0.$$

Then f_n is a continuous one-one function, whose inverse function is

$$g_n(x)=\sqrt[n]{x}=x^{1/n}.$$

By Theorem 5 we have, for $x \neq 0$,

$$\begin{aligned} g_n{}'(x) &= \frac{1}{f_n{}'(f_n{}^{-1}(x))} \\ &= \frac{1}{n(f_n{}^{-1}(x))^{n-1}} \\ &= \frac{1}{n(x^{1/n})^{n-1}} \\ &= \frac{1}{n} \cdot \frac{1}{x^{1-(1/n)}} \\ &= \frac{1}{n} \cdot x^{(1/n)-1}. \end{aligned}$$

Thus, if $f(x) = x^a$, and a is an integer or the reciprocal of a natural number, then $f'(x) = ax^{a-1}$. It is now easy to check that this formula is true if a is any rational number: Let $a = m/n$, where m is an integer, and n is a natural number; if

$$f(x) = x^{m/n} = \left(x^{1/n}\right)^m,$$

then, by the Chain Rule,

$$\begin{aligned} f'(x) &= m\left(x^{1/n}\right)^{m-1} \cdot \frac{1}{n} \cdot x^{(1/n)-1} \\ &= \frac{m}{n} \cdot x^{[(m/n)-(1/n)]+[(1/n)-1]} \\ &= \frac{m}{n} x^{(m/n)-1}. \end{aligned}$$

Although we now have a formula for $f'(x)$ when $f(x) = x^a$ and a is rational, the treatment of the function $f(x) = x^a$ for irrational a will have to be saved for later—at the moment we do not even know the *meaning* of a symbol like $x^{\sqrt{2}}$. Actually, inverse functions will be involved crucially in the definition of x^a for irrational a. Indeed, in the next few chapters several important functions will be defined in terms of their inverse functions.

PROBLEMS

1. Find f^{-1} for each of the following f.

(i) $f(x) = x^3 + 1$.

(ii) $f(x) = (x-1)^3$.

(iii) $f(x) = \begin{cases} x, & x \text{ rational} \\ -x, & x \text{ irrational}. \end{cases}$

(iv) $f(x) = \begin{cases} -x^2 & x \geq 0 \\ 1 - x^3, & x < 0. \end{cases}$

(v) $f(x) = \begin{cases} x, & x \neq a_1, \ldots, a_n \\ a_{i+1} & x = a_i, \quad i = 1, \ldots, n-1 \\ a_1, & x = a_n. \end{cases}$

(vi) $f(x) = x + [x]$.

(vii) $f(0.a_1a_2a_3\dots) = 0.a_2a_1a_3\dots$. (Decimal representation is being used.)

(viii) $f(x) = \dfrac{x}{1-x^2}$, $-1 < x < 1$.

2. Describe the graph of f^{-1} when

(i) f is increasing and always positive.
(ii) f is increasing and always negative.
(iii) f is decreasing and always positive.
(iv) f is decreasing and always negative.

3. Prove that if f is increasing, then so is f^{-1}, and similarly for decreasing functions.

4. If f and g are increasing, is $f+g$? Or $f \cdot g$? Or $f \circ g$?

5. (a) Prove that if f and g are one-one, then $f \circ g$ is also one-one. Find $(f \circ g)^{-1}$ in terms of f^{-1} and g^{-1}. Hint: The answer is *not* $f^{-1} \circ g^{-1}$.
(b) Find g^{-1} in terms of f^{-1} if $g(x) = 1 + f(x)$.

6. Show that $f(x) = \dfrac{ax+b}{cx+d}$ is one-one if and only if $ad - bc \neq 0$, and find f^{-1} in this case.

7. On which intervals $[a, b]$ will the following functions be one-one?

(i) $f(x) = x^3 - 3x^2$.
(ii) $f(x) = x^5 + x$.
(iii) $f(x) = (1 + x^2)^{-1}$.
(iv) $f(x) = \dfrac{x+1}{x^2+1}$.

8. Suppose that f is differentiable with derivative $f'(x) = (1 + x^3)^{-1/2}$. Show that $g = f^{-1}$ satisfies $g''(x) = \frac{3}{2}g(x)^2$.

9. Suppose that f is a one-one function and that f^{-1} has a derivative which is nowhere 0. Prove that f is differentiable. Hint: There is a one-step proof.

10. The Schwarzian derivative $\mathcal{D}f$ was defined in Problem 10-17.

(a) Prove that if $\mathcal{D}f(x)$ exists for all x, then $\mathcal{D}f^{-1}(x)$ also exists for all x in the domain of f^{-1}.
(b) Find a formula for $\mathcal{D}f^{-1}(x)$.

***11.** (a) Prove that there is a differentiable function f such that $[f(x)]^5 + f(x) + x = 0$ for all x. Hint: Show that f can be expressed as an inverse function. The easiest way to do this is to find f^{-1}. And the easiest way to do *this* is to set $x = f^{-1}(y)$.
(b) Find f' in terms of f, using an appropriate theorem of this chapter.
(c) Find f' in another way, by simply differentiating the equation defining f.

The function in Problem 11 is often said to be **defined implicitly** by the equation $y^5+y+x=0$. The situation for this equation is quite special, however. As the next problem shows, an equation does not usually define a function implicitly on the whole line, and in some regions more than one function may be defined implicitly.

12. (a) What are the two differentiable functions f which are defined implicitly on $(-1,1)$ by the equation $x^2+y^2=1$, i.e., which satisfy $x^2+[f(x)]^2=1$ for all x in $(-1,1)$? Notice that there are no solutions defined outside $[-1,1]$.

(b) Which functions f satisfy $x^2+[f(x)]^2=-1$?

*(c) Which differentiable functions f satisfy $[f(x)]^3-3f(x)=x$? Hint: It will help to first draw the graph of the function $g(x)=x^3-3x$.

In general, determining on what intervals a differentiable function is defined implicitly by a particular equation may be a delicate affair, and is best discussed in the context of advanced calculus. If we *assume* that f is such a differentiable solution, however, then a formula for $f'(x)$ can be derived, exactly as in Problem 11(c), by differentiating both sides of the equation defining f (a process known as "implicit differentiation"):

13. (a) Apply this method to the equation $[f(x)]^2+x^2=1$. Notice that your answer will involve $f(x)$; this is only to be expected, since there is more than one function defined implicitly by the equation $y^2+x^2=1$.

(b) But check that your answer works for both of the functions f found in Problem 12(a).

(c) Apply this same method to $[f(x)]^3-3f(x)=x$.

14. (a) Use implicit differentiation to find $f'(x)$ and $f''(x)$ for the functions f defined implicitly by the equation $x^3+y^3=7$.

(b) One of these functions f satisfies $f(-1)=2$. Find $f'(-1)$ and $f''(-1)$ for this f.

15. The collection of all points (x,y) such that $3x^3+4x^2y-xy^2+2y^3=4$ forms a certain curve in the plane. Find the equation of the tangent line to this curve at the point $(-1,1)$.

16. Leibnizian notation is particularly convenient for implicit differentiation. Because y is so consistently used as an abbreviation for $f(x)$, the equation in x and y which defines f implicitly will automatically stand for the equation which f is supposed to satisfy. How would the following computation be written in our notation?

$$y^4+y^3+xy=1,$$

$$4y^3\frac{dy}{dx}+3y^2\frac{dy}{dx}+y+x\frac{dy}{dx}=0,$$

$$\frac{dy}{dx}=\frac{-y}{4y^3+3y^2+x}.$$

17. As long as Leibnizian notation has entered the picture, the Leibnizian notation for derivatives of inverse functions should be mentioned. If dy/dx denotes the derivative of f, then the derivative of f^{-1} is denoted by dx/dy. Write out Theorem 5 in this notation. The resulting equation will show you another reason why Leibnizian notation has such a large following. It will also explain at which point $(f^{-1})'$ is to be calculated when using the dx/dy notation. What is the significance of the following computation?

$$x = y^n,$$
$$y = x^{1/n},$$
$$\frac{dx^{1/n}}{dx} = \frac{dy}{dx} = \frac{1}{\dfrac{dx}{dy}} = \frac{1}{ny^{n-1}}.$$

18. Suppose that f is a differentiable one-one function with a nowhere zero derivative and that $f = F'$. Let $G(x) = xf^{-1}(x) - F(f^{-1}(x))$. Prove that $G'(x) = f^{-1}(x)$. (Disregarding details, this problem tells us a very interesting fact: if we know a function whose derivative is f, then we also know one whose derivative is f^{-1}. But how could anyone ever guess the function G? Two different ways are outlined in Problems 14-17 and 19-15.)

19. Suppose h is a function such that $h'(x) = \sin^2(\sin(x+1))$ and $h(0) = 3$. Find

(i) $(h^{-1})'(3)$.
(ii) $(\beta^{-1})'(3)$, where $\beta(x) = h(x+1)$.

20. Find a formula for $(f^{-1})''(x)$.

***21.** Prove that if $f^{(k)}(f^{-1}(x))$ exists, and is nonzero, then $(f^{-1})^{(k)}(x)$ exists.

22. (a) Prove that an increasing and a decreasing function intersect at most once.
(b) Find two continuous increasing functions f and g such that $f(x) = g(x)$ precisely when x is an integer.
(c) Find a continuous increasing function f and a continuous decreasing function g, defined on **R**, which do not intersect at all.

***23.** (a) If f is a continuous function on **R** and $f = f^{-1}$, prove that there is at least one x such that $f(x) = x$. (What does the condition $f = f^{-1}$ mean geometrically?)
(b) Give several examples of continuous f such that $f = f^{-1}$ and $f(x) = x$ for exactly one x. Hint: Try decreasing f, and remember the geometric interpretation. One possibility is $f(x) = -x$.
(c) Prove that if f is an increasing function such that $f = f^{-1}$, then $f(x) = x$ for all x. Hint: Although the geometric interpretation will be immediately convincing, the simplest proof (about 2 lines) is to rule out the possibilities $f(x) < x$ and $f(x) > x$.

***24.** Which functions have the property that the graph is still the graph of a function when reflected through the graph of $-I$ (the "antidiagonal")?

25. A function f is **nondecreasing** if $f(x) \leq f(y)$ whenever $x < y$. (To be more precise we should stipulate that the domain of f be an interval.) A **nonincreasing** function is defined similarly. Caution: Some writers use "increasing" instead of "nondecreasing," and "strictly increasing" for our "increasing."

(a) Prove that if f is nondecreasing, but not increasing, then f is constant on some interval. (Beware of unintentional puns: "not increasing" is not the same as "nonincreasing.")

(b) Prove that if f is differentiable and nondecreasing, then $f'(x) \geq 0$ for all x.

(c) Prove that if $f'(x) \geq 0$ for all x, then f is nondecreasing.

***26.** (a) Suppose that $f(x) > 0$ for all x, and that f is decreasing. Prove that there is a *continuous* decreasing function g such that $0 < g(x) \leq f(x)$ for all x.

(b) Show that we can even arrange that g will satisfy $\lim_{x \to \infty} g(x)/f(x) = 0$.

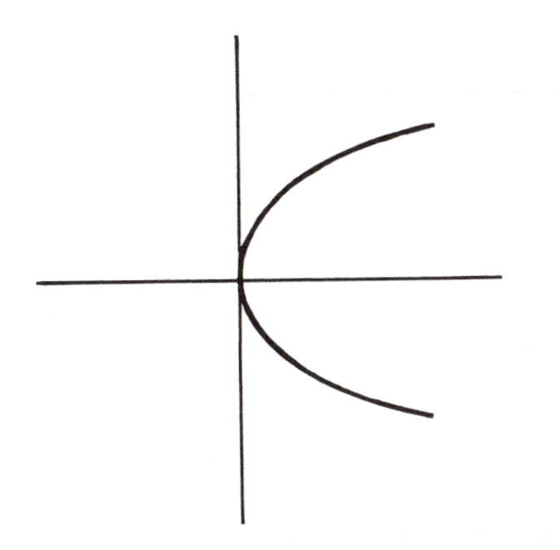

FIGURE 1

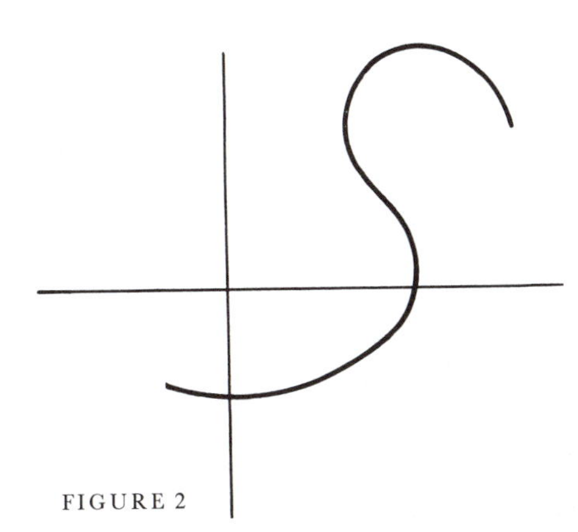

FIGURE 2

APPENDIX. PARAMETRIC REPRESENTATION OF CURVES

The material in this chapter serves to emphasize something that we noticed a long time ago—a perfectly nice looking curve need not be the graph of a function (Figure 1). In other words, we may not be able to describe it as the set of all points $(x, f(x))$. Of course, we might be able to describe the curve as the set of all points $(f(x), x)$; for example, the curve in Figure 1 is the set of all points (x^2, x). But even this trick doesn't work in most cases. It won't allow us to describe the circle, consisting of all points (x, y) with $x^2 + y^2 = 1$, or an ellipse, and it can't be used to describe a curve like the one in Figure 2.

The simplest way of describing curves in the plane in general harks back to the physical conception of a curve as the path of a particle moving in the plane. At each time t, the particle is at a certain point, which has two coordinates; to indicate the dependence of these coordinates on the time t, we can call them $u(t)$ and $v(t)$. Thus, we end up with *two* functions. Conversely, given two functions u and v, we can consider the curve consisting of all points $(u(t), v(t))$. This curve is said to be represented *parametrically* by u and v, and the pair of functions u, v is called a parametric representation of the curve. The curve represented parametrically by u and v thus consists of all pairs (x, y) with $x = u(t)$ and $y = v(t)$. It is often described briefly as "the curve $x = u(t)$, $y = v(t)$." Notice that the graph of a function f can always be described parametrically, as the curve $x = t$, $y = f(t)$.

Instead of considering a curve in the plane as defined by two functions, we can obtain a conceptually simpler picture if we broaden our original definition of function somewhat. Instead of considering a rule which associates a number with another number, we can consider a "function c from real numbers to the plane," i.e., a rule c that associates, to each number t, a *point in the plane*, which we can denote by $c(t)$. With this notion, a curve is just a function from some interval of real numbers to the plane.

Of course, these two different descriptions of a curve are essentially the same: A pair of (ordinary) functions u and v determines a single function c from the real numbers to the plane by the rule

$$c(t) = (u(t), v(t)),$$

and, conversely, given a function c from the real numbers to the plane, each $c(t)$ is a point in the plane, so it is a pair of numbers, which we can call $u(t)$ and $v(t)$, so that we have unique functions u and v satisfying this equation.

In Appendix 1 to Chapter 4, we used the term "vector" to describe a point in the plane. In conformity with this usage, a curve in the plane may also be called a "vector-valued function." The conventions of that Appendix would lead us to write $c(t) = (c_1(t), c_2(t))$, but in this Appendix we'll continue to use notation like $c(t) = (u(t), v(t))$ to minimize the use of subscripts.

A simple example of a vector-valued function that is quite useful is

$$\mathbf{e}(t) = (\cos t, \sin t),$$

which goes round and round the unit circle (Figure 3).

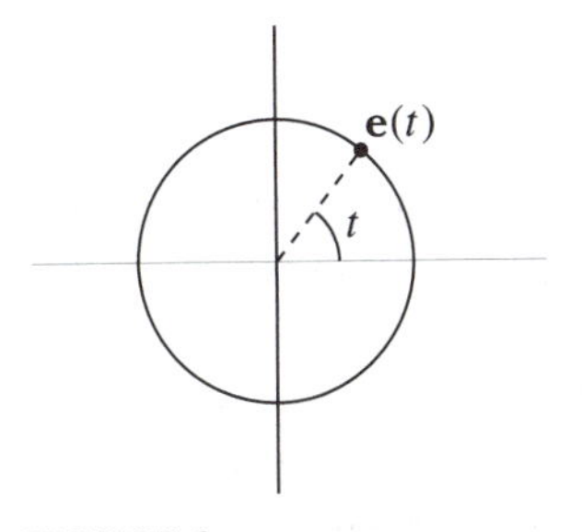

FIGURE 3

For two (ordinary) functions f and g, we defined new functions $f+g$ and $f\cdot g$ by the rules

$$(f+g)(x)=f(x)+g(x), \tag{1}$$

$$(f\cdot g)(x)=f(x)\cdot g(x). \tag{2}$$

Since we have defined a way of adding vectors, we can imitate the first of these definitions for vector-valued functions c and d: we define the vector-valued function $c+d$ by

$$(c+d)(t)=c(t)+d(t),$$

where the $+$ on the right-hand side is now *the sum of vectors*. This simply amounts to saying that if

$$c(t)=(u(t),v(t)),$$
$$d(t)=(w(t),z(t)),$$

then

$$(c+d)(t)=\big(u(t),v(t)\big)+\big(w(t),z(t)\big)=\big(u(t)+w(t),v(t)+z(t)\big).$$

Recall that we have also defined $a\cdot v$ for a number a and a vector v. To extend this to vector-valued functions, we want to consider an ordinary *function* α and a vector-valued function c, so that for each t we have a number $\alpha(t)$ and a vector $c(t)$. Then we can define a new vector-valued function $\alpha\cdot c$ by

$$(\alpha\cdot c)(t)=\alpha(t)\cdot c(t),$$

where the $\cdot$ on the right-hand side is the product of a number and a vector. This simply amounts to saying that

$$(\alpha\cdot c)(t)=\alpha(t)\cdot(u(t),v(t))=\big(\alpha(t)\cdot u(t),\ \alpha(t)\cdot v(t)\big).$$

Notice that the curve $\alpha\cdot\mathbf{e}$,

$$(\alpha\cdot\mathbf{e})(t)=(\alpha(t)\cos t,\ \alpha(t)\sin t),$$

is already quite general (Figure 4). In the notation of Appendix 3 to Chapter 4, the point $(\alpha\cdot\mathbf{e})(t)$ has polar coordinates $\alpha(t)$ and t, so that $(\alpha\cdot\mathbf{e})(t)$ is the "graph of α in polar coordinates."

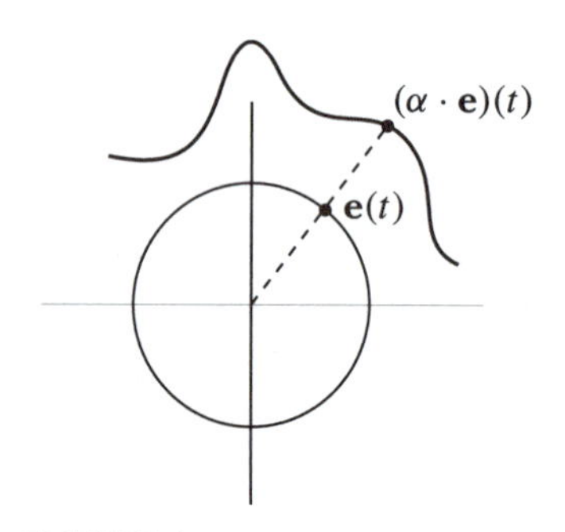

FIGURE 4

Even more generally, given any vector-valued function c, we can define new functions r and θ by

$$c(t)=r(t)\cdot\mathbf{e}(\theta(t)),$$

where $r(t)$ is just the distance from the origin to $c(t)$, and $\theta(t)$ is some choice of the angle of $c(t)$ (as usual, the function θ isn't defined unambiguously, so one has to be careful when using this way of writing an arbitrary curve c).

We aren't in a position to extend (2) to vector-valued functions in general, since we haven't defined the product of two vectors. However, Problems 2 and 4 of Appendix 1 to Chapter 4 define two *real-valued* products $v\bullet w$ and $\det(v,w)$. It should be clear, given vector-valued functions c and d, how we would define two ordinary (real-valued) functions

$$c \bullet d \qquad \text{and} \qquad \det(c, d).$$

Beyond imitating simple arithmetic operations on functions, we can consider more interesting problems, like limits. For $c(t) = (u(t), v(t))$, we can define

$$(*) \qquad \lim_{t\to a} c(t) = \lim_{t\to a}(u(t), v(t)) \quad \text{to be} \quad \left(\lim_{t\to a} u(t), \lim_{t\to a} v(t)\right).$$

Rules like

$$\lim_{t\to a} c + d = \lim_{t\to a} c + \lim_{t\to a} d,$$
$$\lim_{t\to a} \alpha \cdot c = \lim_{t\to a} \alpha(t) \cdot \lim_{t\to a} c$$

follow immediately. Problem 10 shows how to give an equivalent definition that imitates the basic definition of limits directly.

Limits lead us of course to derivatives. For

$$c(t) = (u(t), v(t))$$

we can define c' by the straightforward definition

$$c'(a) = \big(u'(a), v'(a)\big).$$

We could also try to imitate the basic definition:

$$c'(a) = \lim_{h\to 0} \frac{c(a+h) - c(a)}{h},$$

where the fraction on the right-hand side is understood to mean

$$\frac{1}{h} \cdot [c(a+h) - c(a)].$$

As a matter of fact, these two definitions are equivalent, because

$$\begin{aligned} \lim_{h\to 0} \frac{c(a+h) - c(a)}{h} &= \lim_{h\to 0} \left(\frac{u(a+h) - u(a)}{h}, \frac{v(a+h) - v(a)}{h}\right) \\ &= \left(\lim_{h\to 0} \frac{u(a+h) - u(a)}{h}, \lim_{h\to 0} \frac{v(a+h) - v(a)}{h}\right) \\ &\qquad \text{by our definition } (*) \text{ of limits} \\ &= \big(u'(a), v'(a)\big). \end{aligned}$$

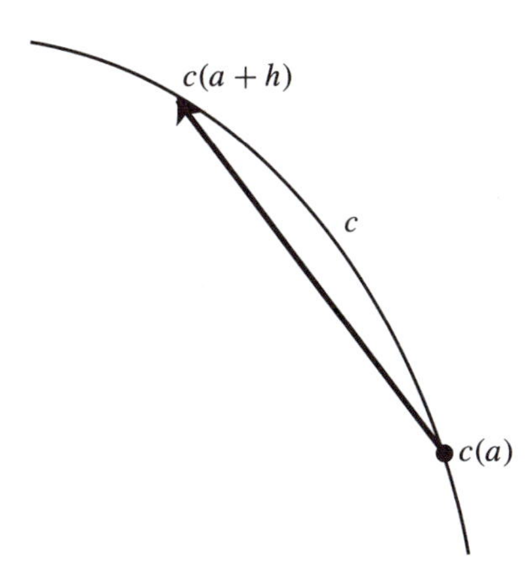

FIGURE 5

Figure 5 shows $c(a+h)$ and $c(a)$, as well as the arrow from $c(a)$ to $c(a+h)$; as we showed in Appendix 1 to Chapter 4, this arrow is $c(a+h) - c(a)$, except moved over so that it starts at $c(a)$. As $h \to 0$, this arrow would appear to move closer and closer to the tangent of our curve, so it seems reasonable to *define* the tangent line of c at $c(a)$ to be the straight line along $c'(a)$, when $c'(a)$ is moved over so that it starts at $c(a)$. In other words, we define the tangent line of c at $c(a)$ as the set of all points

$$c(a) + s \cdot c'(a);$$

for $s = 0$ we get the point $c(a)$ itself, for $s = 1$ we get $c(a) + c'(a)$, etc. (Note, however, that this definition does not make sense when $c'(a) = 0$.) Problem 1 shows that this definition agrees with the old one when our curve c is defined by

$$c(t) = (t, f(t)),$$

so that we simply have the graph of f.

Once again, various old formulas have analogues. For example,

$$\begin{aligned}(c+d)'(a) &= c'(a) + d'(a),\\ (\alpha \cdot c)'(a) &= \alpha'(a) \cdot c(a) + \alpha(a) \cdot c'(a),\end{aligned}$$

or, as equations involving functions,

$$\begin{aligned}(c+d)' &= c' + d',\\ (\alpha \cdot c)' &= \alpha' \cdot c + \alpha \cdot c'.\end{aligned}$$

These formulas can be derived immediately from the definition in terms of the component functions. They can also be derived from the definition as a limit, by imitating previous proofs; for the second, we would of course use the standard trick of writing

$$\alpha(a+h)c(a+h) - \alpha(a)c(a) = \\ \alpha(a+h) \cdot [c(a+h) - c(a)] + [\alpha(a+h) - \alpha(a)] \cdot c(a).$$

We can also consider the function

$$d(t) = c(p(t)) = (c \circ p)(t),$$

where p is now an ordinary function, from numbers to numbers. The new curve d passes through the same points as c, except at different times; thus p corresponds to a "reparameterization" of c. For

$$\begin{aligned}c &= (u, v),\\ d &= (u \circ p, v \circ p),\end{aligned}$$

we obtain

$$\begin{aligned}d'(a) &= \big((u \circ p)'(a), (v \circ p)'(a)\big)\\ &= \big(p'(a)u'(p(a)), p'(a)v'(p(a))\big)\\ &= p'(a) \cdot \big(u'(p(a)), v'(p(a))\big)\\ &= p'(a) \cdot c'(p(a)),\end{aligned}$$

or simply

$$d' = p' \cdot (c' \circ p).$$

Notice that if $p(a) = a$, so that d and c actually pass through the same point at time a, then $d'(a) = p'(a) \cdot c'(a)$, so that the tangent vector $d'(a)$ is just a multiple of $c'(a)$. This means that the tangent *line* to c at $c(a)$ is the same as the tangent line to the reparameterized curve d at $d(a) = c(a)$. The one exception occurs when $p'(a) = 0$, since the tangent line for d is then undefined, even though the tangent line for c may be defined. For example, $d(t) = c(t^3)$ won't have a tangent line defined at $t = 0$, even though it's merely a reparameterization of c.

Finally, since we can define real-valued functions

$$(c \cdot d)(t) = c(t) \cdot d(t),$$
$$\det(c, d)(t) = \det(c(t), d(t)),$$

we ought to have formulas for the derivatives of these new functions. As you might guess, the proper formulas are

$$(c \cdot d)'(a) = c(a) \cdot d'(a) + c'(a) \cdot d(a),$$
$$[\det(c, d)]'(a) = \det(c', d)(a) + \det(c, d')(a).$$

These can be derived by straightforward calculations from the definitions in terms of the component functions. But it is more elegant to imitate the proof of the ordinary product rule, using the simple formulas in Problems 2 and 4 of Appendix 1 to Chapter 4, and, of course, the "standard trick" referred to above.

PROBLEMS

1. (a) For a function f, the "point-slope form" (Problem 4-6) of the tangent line at $(a, f(a))$ can be written as $y - f(a) = (x - a)f'(a)$, so that the tangent line consists of all points of the form

$$\big(x, f(a) + (x - a)f'(a)\big).$$

Conclude that the tangent line consists of all points of the form

$$\big(a + s, f(a) + sf'(a)\big).$$

(b) If c is the curve $c(t) = (t, f(t))$, conclude that the tangent line of c at $(a, f(a))$ [using our new definition] is the same as the tangent line of f at $(a, f(a))$.

2. Let $c(t) = (f(t), t^2)$, where f is the function shown in Figure 21 of Chapter 9. Show that c lies along the graph of the non-differentiable function $h(x) = |x|$, but that $c'(0) = 0$. In other words, a reparameterization can "hide" a corner. For this reason, we are usually only interested in curves c with c' never 0.

3. Suppose that $x = u(t)$, $y = v(t)$ is a parametric representation of a curve, and that $u' \neq 0$ on some interval.

(a) Show that on this interval the curve lies along the graph of $f = v \circ u^{-1}$.

(b) Show that at the point $x = u(t)$ we have

$$f'(x) = \frac{v'(t)}{u'(t)}.$$

In Leibnizian notation this is often written suggestively as

$$\frac{dy}{dx} = \frac{\dfrac{dy}{dt}}{\dfrac{dx}{dt}}.$$

(c) We also have

$$f''(x) = \frac{u'(t)v''(t) - v'(t)u''(t)}{(u'(t))^3}.$$

4. Consider a function f defined implicitly by the equation $x^{2/3} + y^{2/3} = 1$. Compute $f'(x)$ in two ways:

(i) By implicit differentiation.
(ii) By considering the parametric representation $x = \cos^3 t$, $y = \sin^3 t$.

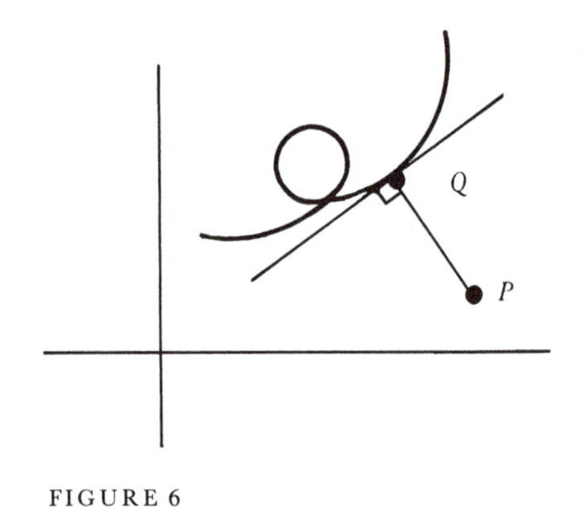

FIGURE 6

5. Let $x = u(t)$, $y = v(t)$ be the parametric representation of a curve, with u and v differentiable, and let $P = (x_0, y_0)$ be a point in the plane. Prove that if the point $Q = (u(\bar{t}), v(\bar{t}))$ on the curve is closest to (x_0, y_0), and $u'(\bar{t})$ and $v'(\bar{t})$ are not both 0, then the line from P to Q is perpendicular to the tangent line of the curve at Q (Figure 6). The same result holds if Q is furthest from (x_0, y_0).

We've seen that the "graph of f in polar coordinates" is the curve

$$(f \cdot \mathbf{e})(t) = (f(t)\cos t, f(t)\sin t);$$

in other words, the graph of f in polar coordinates is the curve with the parametric representation

$$x = f(\theta)\cos\theta, \qquad y = f(\theta)\sin\theta.$$

6. (a) Show that for the graph of f in polar coordinates the slope of the tangent line at the point with polar coordinates $(f(\theta), \theta)$ is

$$\frac{f(\theta)\cos\theta + f'(\theta)\sin\theta}{-f(\theta)\sin\theta + f'(\theta)\cos\theta}.$$

(b) Show that if $f(\theta) = 0$ and f is differentiable at 0, then the line through the origin making an angle of θ with the positive horizontal axis is a tangent line of the graph of f in polar coordinates. Use this result to add some details to the graph of the Archimedian spiral in Appendix 3 of Chapter 4, and to the graphs in Problems 3 and 10 of that Appendix as well.

(c) Suppose that the point with polar coordinates $(f(\theta), \theta)$ is further from the origin O than any other point on the graph of f. What can you say about the tangent line to the graph at this point? Compare with Problem 5.

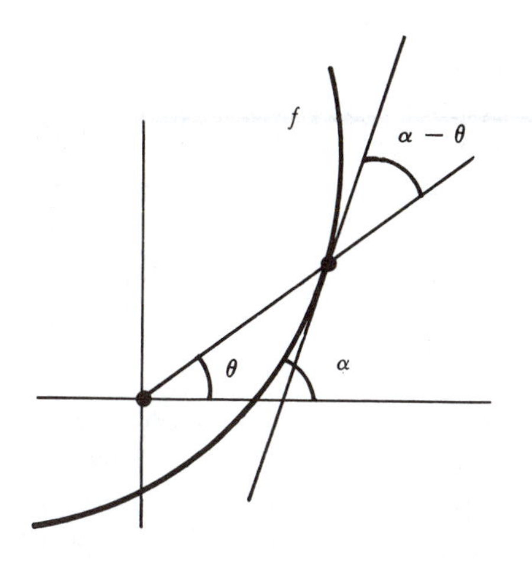

FIGURE 7

(d) Suppose that the tangent line to the graph of f at the point with polar coordinates $(f(\theta), \theta)$ makes an angle of α with the horizontal axis (Figure 7), so that $\alpha - \theta$ is the angle between the tangent line and the ray from O to the point. Show that

$$\tan(\alpha - \theta) = \frac{f(\theta)}{f'(\theta)}.$$

7. (a) In Problem 5 of Appendix 1 to Chapter 4 we found that the cardioid $r = 1 - \sin\theta$ is also described by the equation $(x^2 + y^2 + y)^2 = x^2 + y^2$. Find the slope of the tangent line at a point on the cardioid in two ways:

(i) By implicit differentiation.
(ii) By using the previous problem.

(b) Check that at the origin the tangent lines are vertical, as they appear to be in Figure 8.

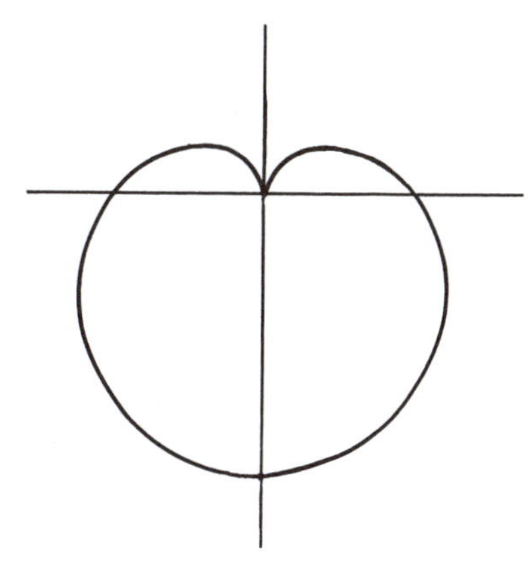

FIGURE 8

The next problem uses the material from Chapter 15, in particular, radian measure, and the inverse trigonometric functions and their properties.

8. A *cycloid* is defined as the path traced out by a point on the rim of a rolling wheel of radius a. You can see a beautiful cycloid by pasting a reflector on the edge of a bicycle wheel and having a friend ride slowly in front of the headlights of your car at night. Lacking a car, bicycle, or trusting friend, you can settle instead for Figure 9.

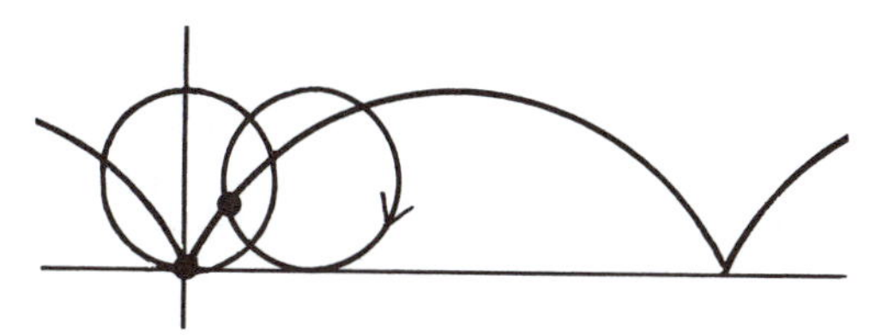

FIGURE 9

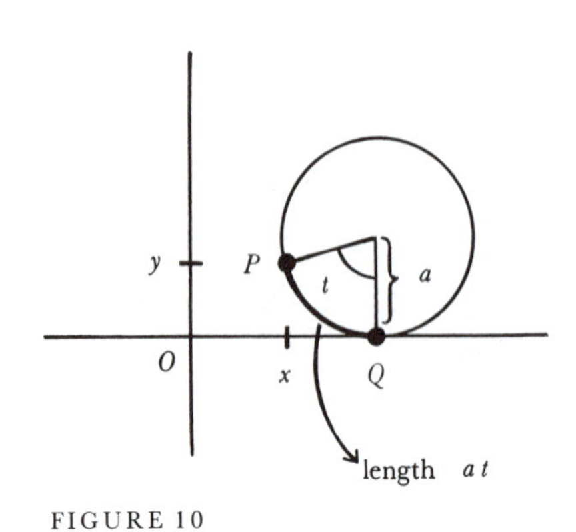

FIGURE 10

(a) Let $u(t)$ and $v(t)$ be the coordinates of the point on the rim after the wheel has rotated through an angle of t (radians). This means that the arc of the wheel rim from P to Q in Figure 10 has length at. Since the wheel is rolling, at is also the distance from O to Q. Show that we have the parametric representation of the cycloid

$$u(t) = a(t - \sin t)$$
$$v(t) = a(1 - \cos t).$$

Figure 11 shows the curves we obtain if the distance from the point to the center of the wheel is (a) less than the radius or (b) greater than the radius. In the latter case, the curve is not the graph of a function; at certain times the point is moving backwards, even though the wheel is moving forwards!

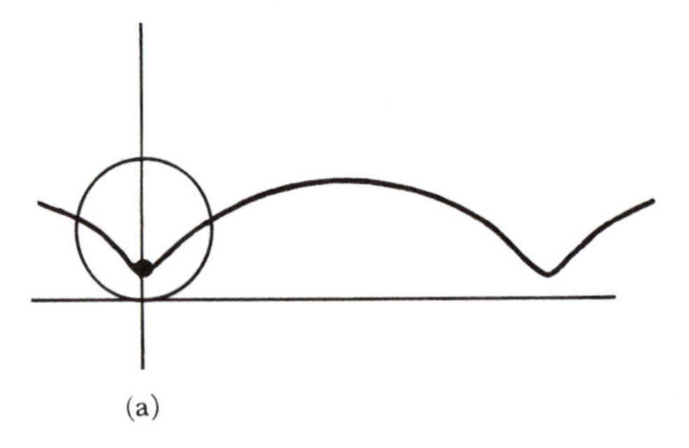

(a)

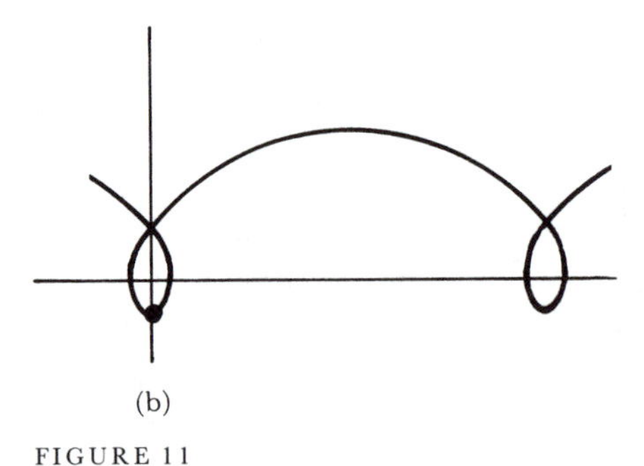

(b)

FIGURE 11

In Figure 9 we drew the cycloid as the graph of a function, but we really need to check that this is the case:

(b) Compute $u'(t)$ and conclude that u is increasing. Problem 3 then shows that the cycloid is the graph of $f = v \circ u^{-1}$, and allows us to compute $f'(t)$.

It isn't possible to get an explicit formula for f, but we can come close.

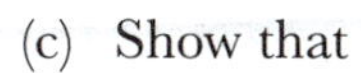

(c) Show that

$$u(t) = a \arccos \frac{a - v(t)}{a} \pm \sqrt{[2a - v(t)]v(t)}.$$

Hint: first solve for t in terms of $v(t)$.

(d) The first half of the first arch of the cycloid is the graph of g^{-1}, where

$$g(y) = a \arccos \frac{a - y}{y} - \sqrt{(2a - y)y}.$$

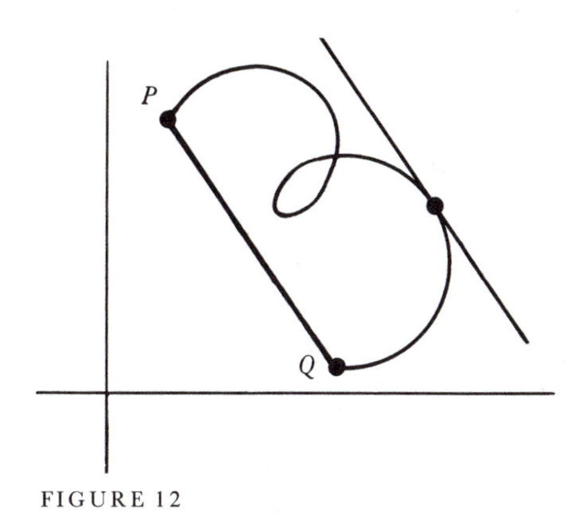

FIGURE 12

9. Let u and v be continuous on $[a, b]$ and differentiable on (a, b); then u and v give a parametric representation of a curve from $P = (u(a), u(b))$ to $Q = (v(a), v(b))$. Geometrically, it seems clear (Figure 12) that at some point on the curve the tangent line is parallel to the line segment from P to Q. Prove this analytically. Hint: This problem will give a geometric interpretation for one of the theorems in Chapter 11.

10. The following definition of a limit for a vector-valued function is the direct analogue of the definition for ordinary functions:

$\lim_{t \to a} c(t) = l$ means that for every $\varepsilon > 0$ there is some $\delta > 0$ such that, for all t, if $0 < |t - a| < \delta$, then $\|c(t) - l\| < \varepsilon$.

Here $\|\ \|$ is the *norm*, defined in Problem 2 of Appendix 1 to Chapter 4. If $l = (l_1, l_2)$, then

$$\|c(t) - l\|^2 = |u(t) - l_1|^2 + |v(t) - l_2|^2.$$

(a) Conclude that

$$|u(t) - l_1| \le \|c(t) - l\| \qquad \text{and} \qquad |v(t) - l_2| \le \|c(t) - l\|,$$

and show that if $\lim_{t \to a} c(t) = l$ according to the above definition, then we also have

$$\lim_{t \to a} u(t) = l_1 \qquad \text{and} \qquad \lim_{t \to a} v(t) = l_2,$$

so that $\lim_{t \to a} c(t) = l$ according to our definition $(*)$ in terms of component functions, on page 243.

(b) Conversely, show that if $\lim_{t \to a} c(t) = l$ according to the definition in terms of component functions, then also $\lim_{t \to a} c(t) = l$ according to the above definition.

CHAPTER 13 INTEGRALS

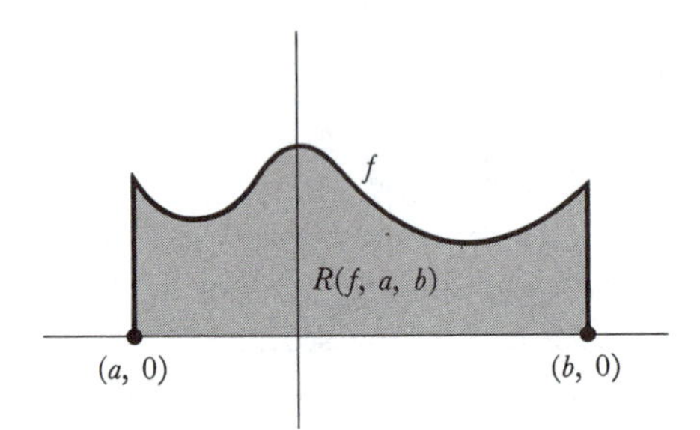

FIGURE 1

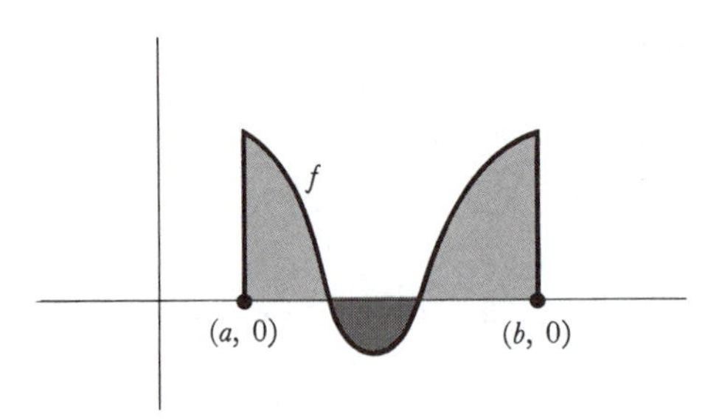

FIGURE 2

The derivative does not display its full strength until allied with the "integral," the second main concept of Part III. At first this topic may seem to be a complete digression—in this chapter derivatives do not appear even once! The study of integrals does require a long preparation, but once this preliminary work has been completed, integrals will be an invaluable tool for creating new functions, and the derivative will reappear in Chapter 14, more powerful than ever.

Although ultimately to be defined in a quite complicated way, the integral formalizes a simple, intuitive concept—that of area. By now it should come as no surprise to learn that the definition of an intuitive concept can present great difficulties—"area" is certainly no exception.

In elementary geometry, formulas are derived for the areas of many plane figures, but a little reflection shows that an acceptable definition of area is seldom given. The area of a region is sometimes defined as the number of squares, with sides of length 1, which fit in the region. But this definition is hopelessly inadequate for any but the simplest regions. For example, a circle of radius 1 supposedly has as area the irrational number π, but it is not at all clear what "π squares" means. Even if we consider a circle of radius $1/\sqrt{\pi}$, which supposedly has area 1, it is hard to say in what way a unit square fits in this circle, since it does not seem possible to divide the unit square into pieces which can be arranged to form a circle.

In this chapter we will only try to define the area of some very special regions (Figure 1)—those which are bounded by the horizontal axis, the vertical lines through $(a, 0)$ and $(b, 0)$, and the graph of a function f such that $f(x) \geq 0$ for all x in $[a, b]$. It is convenient to indicate this region by $R(f, a, b)$. Notice that these regions include rectangles and triangles, as well as many other important geometric figures.

The number which we will eventually assign as the area of $R(f, a, b)$ will be called the *integral* of f on $[a, b]$. Actually, the integral will be defined even for functions f which do not satisfy the condition $f(x) \geq 0$ for all x in $[a, b]$. If f is the function graphed in Figure 2, the integral will represent the difference of the area of the lightly shaded region and the area of the heavily shaded region (the "algebraic area" of $R(f, a, b)$).

The idea behind the prospective definition is indicated in Figure 3. The interval $[a, b]$ has been divided into four subintervals

$$[t_0, t_1] \quad [t_1, t_2] \quad [t_2, t_3] \quad [t_3, t_4]$$

by means of numbers t_0, t_1, t_2, t_3, t_4 with

$$a = t_0 < t_1 < t_2 < t_3 < t_4 = b$$

(the numbering of the subscripts begins with 0 so that the largest subscript will equal the number of subintervals).

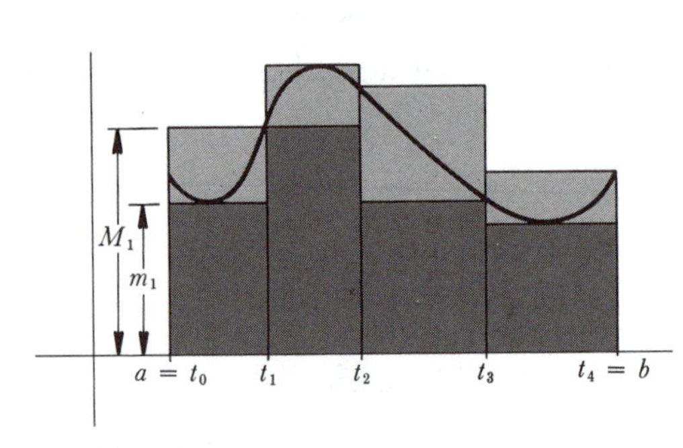

FIGURE 3

On the first interval $[t_0, t_1]$ the function f has the minimum value m_1 and the maximum value M_1; similarly, on the ith interval $[t_{i-1}, t_i]$ let the minimum value of f be m_i and let the maximum value be M_i. The sum

$$s = m_1(t_1 - t_0) + m_2(t_2 - t_1) + m_3(t_3 - t_2) + m_4(t_4 - t_3)$$

represents the total area of rectangles lying inside the region $R(f, a, b)$, while the sum

$$S = M_1(t_1 - t_0) + M_2(t_2 - t_1) + M_3(t_3 - t_2) + M_4(t_4 - t_3)$$

represents the total area of rectangles containing the region $R(f, a, b)$. The guiding principle of our attempt to define the area A of $R(f, a, b)$ is the observation that A should satisfy

$$s \leq A \quad \text{and} \quad A \leq S,$$

and that this should be true, *no matter how the interval* $[a, b]$ *is subdivided.* It is to be hoped that these requirements will determine A. The following definitions begin to formalize, and eliminate some of the implicit assumptions in, this discussion.

DEFINITION

Let $a < b$. A **partition** of the interval $[a, b]$ is a finite collection of points in $[a, b]$, one of which is a, and one of which is b.

The points in a partition can be numbered $t_0, \ldots, t_n$ so that

$$a = t_0 < t_1 < \cdots < t_{n-1} < t_n = b;$$

we shall always assume that such a numbering has been assigned.

DEFINITION

Suppose f is bounded on $[a, b]$ and $P = \{t_0, \ldots, t_n\}$ is a partition of $[a, b]$. Let

$$m_i = \inf\{f(x) : t_{i-1} \leq x \leq t_i\},$$
$$M_i = \sup\{f(x) : t_{i-1} \leq x \leq t_i\}.$$

The **lower sum** of f for P, denoted by $L(f, P)$, is defined as

$$L(f, P) = \sum_{i=1}^{n} m_i(t_i - t_{i-1}).$$

The **upper sum** of f for P, denoted by $U(f, P)$, is defined as

$$U(f, P) = \sum_{i=1}^{n} M_i(t_i - t_{i-1}).$$

The lower and upper sums correspond to the sums s and S in the previous example; they are supposed to represent the total areas of rectangles lying below and above the graph of f. Notice, however, that despite the geometric motivation, these sums have been defined precisely without any appeal to a concept of "area."

Two details of the definition deserve comment. The requirement that f be bounded on $[a, b]$ is essential in order that all the m_i and M_i be defined. Note, also, that it was necessary to define the numbers m_i and M_i as inf's and sup's, rather than as minima and maxima, since f was not assumed continuous.

One thing is clear about lower and upper sums: If P is any partition, then

$$L(f, P) \leq U(f, P),$$

because

$$L(f, P) = \sum_{i=1}^{n} m_i(t_i - t_{i-1}),$$

$$U(f, P) = \sum_{i=1}^{n} M_i(t_i - t_{i_1}),$$

and for each i we have

$$m_i(t_i - t_{i-1}) \leq M_i(t_i - t_{i-1}).$$

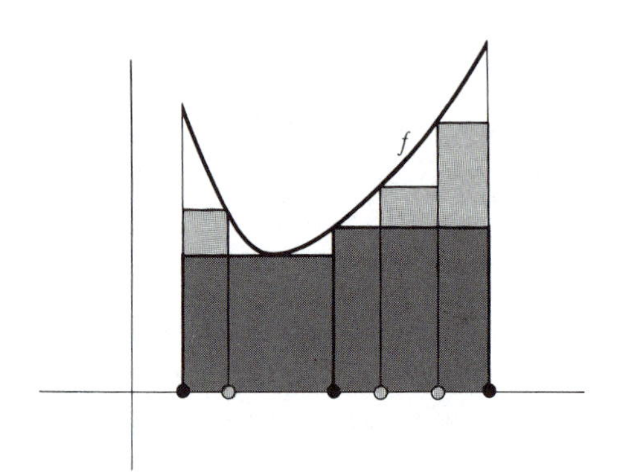

FIGURE 4

On the other hand, something less obvious *ought* to be true: If P_1 and P_2 are *any* two partitions of $[a, b]$, then it should be the case that

$$L(f, P_1) \leq U(f, P_2),$$

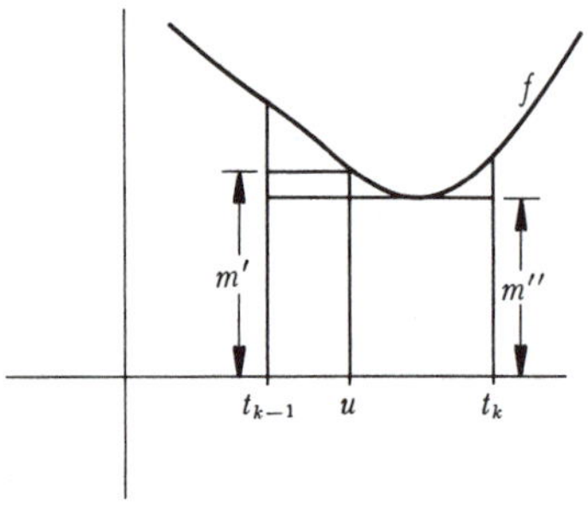

because $L(f, P_1)$ should be $\leq$ area $R(f, a, b)$, and $U(f, P_2)$ should be $\geq$ area $R(f, a, b)$. This remark proves nothing (since the "area of $R(f, a, b)$" has not even been defined yet), but it does indicate that if there is to be any hope of defining the area of $R(f, a, b)$, a proof that $L(f, P_1) \leq U(f, P_2)$ should come first. The proof which we are about to give depends upon a lemma which concerns the behavior of lower and upper sums when more points are included in a partition. In Figure 4 the partition P contains the points in black, and Q contains both the points in black and the points in grey. The picture indicates that the rectangles drawn for the partition Q are a better approximation to the region $R(f, a, b)$ than those for the original partition P. To be precise:

FIGURE 5

LEMMA If Q contains P (i.e., if all points of P are also in Q), then

$$L(f, P) \leq L(f, Q),$$
$$U(f, P) \geq U(f, Q).$$

PROOF Consider first the special case (Figure 5) in which Q contains just one more point than P:

$$P = \{t_0, \ldots, t_n\},$$
$$Q = \{t_0, \ldots, t_{k-1}, u, t_k, \ldots, t_n\},$$

where

$$a = t_0 < t_1 < \cdots < t_{k-1} < u < t_k < \cdots < t_n = b.$$

Let

$$m' = \inf\{f(x) : t_{k-1} \le x \le u\},$$
$$m'' = \inf\{f(x) : u \le x \le t_k\}.$$

Then

$$L(f, P) = \sum_{i=1}^{n} m_i(t_i - t_{i-1}),$$
$$L(f, Q) = \sum_{i=1}^{k-1} m_i(t_i - t_{i-1}) + m'(u - t_{k-1}) + m''(t_k - u) + \sum_{i=k+1}^{n} m_i(t_i - t_{i-1}).$$

To prove that $L(f, P) \le L(f, Q)$ it therefore suffices to show that

$$m_k(t_k - t_{k-1}) \le m'(u - t_{k-1}) + m''(t_k - u).$$

Now the set $\{f(x) : t_{k-1} \le x \le t_k\}$ contains all the numbers in $\{f(x) : t_{k-1} \le x \le u\}$, and possibly some smaller ones, so the greatest lower bound of the first set is *less than or equal to* the greatest lower bound of the second; thus

$$m_k \le m'.$$

Similarly,

$$m_k \le m''.$$

Therefore,

$$m_k(t_k - t_{k-1}) = m_k(u - t_{k-1}) + m_k(t_k - u) \le m'(u - t_{k-1}) + m''(t_k - u).$$

This proves, in this special case, that $L(f, P) \le L(f, Q)$. The proof that $U(f, P) \ge U(f, Q)$ is similar, and is left to you as an easy, but valuable, exercise.

The general case can now be deduced quite easily. The partition Q can be obtained from P by adding one point at a time; in other words, there is a sequence of partitions

$$P = P_1, P_2, \dots, P_\alpha = Q$$

such that P_{j+1} contains just one more point than P_j. Then

$$L(f, P) = L(f, P_1) \le L(f, P_2) \le \cdots \le L(f, P_\alpha) = L(f, Q),$$

and

$$U(f, P) = U(f, P_1) \ge U(f, P_2) \ge \cdots \ge U(f, P_\alpha) = U(f, Q).$$

The theorem we wish to prove is a simple consequence of this lemma.

THEOREM 1 Let P_1 and P_2 be partitions of $[a, b]$, and let f be a function which is bounded on $[a, b]$. Then

$$L(f, P_1) \le U(f, P_2).$$

PROOF There is a partition P which contains both P_1 and P_2 (let P consist of all points in both P_1 and P_2). According to the lemma,

$$L(f, P_1) \le L(f, P) \le U(f, P) \le U(f, P_2). \blacksquare$$

It follows from Theorem 1 that any upper sum $U(f, P')$ is an upper bound for the set of all lower sums $L(f, P)$. Consequently, any upper sum $U(f, P')$ is greater than or equal to the *least* upper bound of all lower sums:

$$\sup\{L(f, P) : P \text{ a partition of } [a, b]\} \le U(f, P'),$$

for every P'. This, in turn, means that $\sup\{L(f, P)\}$ is a lower bound for the set of all upper sums of f. Consequently,

$$\sup\{L(f, P)\} \le \inf\{U(f, P)\}.$$

It is clear that both of these numbers are between the lower sum and upper sum of f for *all* partitions:

$$L(f, P') \le \sup\{L(f, P)\} \le U(f, P'),$$
$$L(f, P') \le \inf\{U(f, P)\} \le U(f, P'),$$

for all partitions P'.

It may well happen that

$$\sup\{L(f, P)\} = \inf\{U(f, P\};$$

in this case, this is the *only* number between the lower sum and upper sum of f for all partitions, and this number is consequently an ideal candidate for the area of $R(f, a, b)$. On the other hand, if

$$\sup\{L(f, P)\} < \inf\{U(f, P)\},$$

then every number x between $\sup\{L(f, P)\}$ and $\inf\{U(f, P)\}$ will satisfy

$$L(f, P') \le x \le U(f, P')$$

for all partitions P'.

It is not at all clear just when such an embarrassment of riches will occur. The following two examples, although not as interesting as many which will soon appear, show that both phenomena are possible.

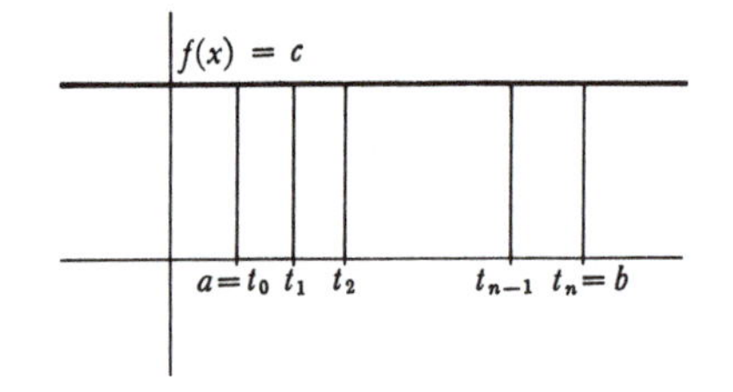

FIGURE 6

Suppose first that $f(x) = c$ for all x in $[a, b]$ (Figure 6). If $P = \{t_0, \dots, t_n\}$ is any partition of $[a, b]$, then

$$m_i = M_i = c,$$

so

$$L(f, P) = \sum_{i=1}^{n} c(t_i - t_{i-1}) = c(b - a),$$

$$U(f, P) = \sum_{i=1}^{n} c(t_i - t_{i-1}) = c(b - a).$$

In this case, all lower sums and upper sums are equal, and

$$\sup\{L(f,P)\} = \inf\{U(f,P)\} = c(b-a).$$

Now consider (Figure 7) the function f defined by

$$f(x) = \begin{cases} 0, & x \text{ irrational} \\ 1, & x \text{ rational.} \end{cases}$$

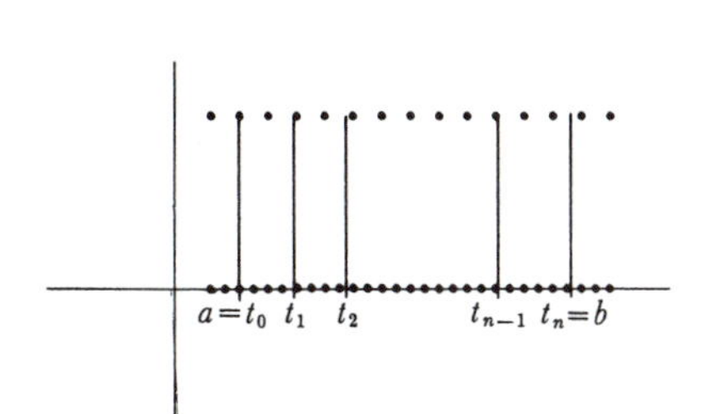

FIGURE 7

If $P = \{t_0, \dots, t_n\}$ is any partition, then

$$m_i = 0, \text{ since there is an irrational number in } [t_{i-1}, t_i],$$

and

$$M_i = 1, \text{ since there is a rational number in } [t_{i-1}, t_i].$$

Therefore,

$$L(f,P) = \sum_{i=1}^{n} 0 \cdot (t_i - t_{i-1}) = 0,$$
$$U(f,P) = \sum_{i=1}^{n} 1 \cdot (t_i - t_{i-1}) = b-a.$$

Thus, in this case it is certainly *not* true that $\sup\{L(f,P)\} = \inf\{U(f,P)\}$. The principle upon which the definition of area was to be based provides insufficient information to determine a specific area for $R(f,a,b)$—any number between 0 and $b-a$ seems equally good. On the other hand, the region $R(f,a,b)$ is so weird that we might with justice refuse to assign it any area at all. In fact, we can maintain, more generally, that whenever

$$\sup\{L(f,P)\} \neq \inf\{U(f,P)\},$$

the region $R(f,a,b)$ is too unreasonable to deserve having an area. As our appeal to the word "unreasonable" suggests, we are about to cloak our ignorance in terminology.

DEFINITION

A function f which is bounded on $[a,b]$ is **integrable** on $[a,b]$ if

$$\sup\{L(f,P) : P \text{ a partition of } [a,b]\} = \inf\{U(f,P) : P \text{ a partition of } [a,b]\}.$$

In this case, this common number is called the **integral** of f on $[a,b]$ and is denoted by

$$\int_a^b f.$$

(The symbol $\int$ is called an *integral sign* and was originally an elongated s, for "sum;" the numbers a and b are called the *lower* and *upper limits of integration.*) The integral $\int_a^b f$ is also called the **area** of $R(f,a,b)$ when $f(x) \geq 0$ for all x in $[a,b]$.

If f is integrable, then according to this definition,

$$L(f, P) \leq \int_a^b f \leq U(f, P) \quad \text{for all partitions } P \text{ of } [a, b].$$

Moreover, $\int_a^b f$ is the *unique* number with this property.

This definition merely pinpoints, and does not solve, the problem discussed before: we do not know which functions are integrable (nor do we know how to find the integral of f on $[a, b]$ when f *is* integrable). At present we know only two examples:

$$\text{(1)} \quad \text{if } f(x) = c, \text{ then } f \text{ is integrable on } [a, b] \text{ and } \int_a^b f = c \cdot (b - a).$$

(Notice that this integral assigns the expected area to a rectangle.)

$$\text{(2)} \quad \text{if } f(x) = \begin{cases} 0, & x \text{ irrational} \\ 1, & x \text{ rational,} \end{cases} \quad \text{then } f \text{ is not integrable on } [a, b].$$

Several more examples will be given before discussing these problems further. Even for these examples, however, it helps to have the following simple criterion for integrability stated explicitly.

THEOREM 2 If f is bounded on $[a, b]$, then f is integrable on $[a, b]$ if and only if for every $\varepsilon > 0$ there is a partition P of $[a, b]$ such that

$$U(f, P) - L(f, P) < \varepsilon.$$

PROOF Suppose first that for every $\varepsilon > 0$ there is a partition P with

$$U(f, P) - L(f, P) < \varepsilon.$$

Since

$$\begin{aligned} \inf\{U(f, P')\} &\leq U(f, P), \\ \sup\{L(f, P')\} &\geq L(f, P), \end{aligned}$$

it follows that

$$\inf\{U(f, P')\} - \sup\{L(f, P')\} < \varepsilon.$$

Since this is true for all $\varepsilon > 0$, it follows that

$$\sup\{L(f, P')\} = \inf\{U(f, P')\};$$

by definition, then, f is integrable. The proof of the converse assertion is similar: If f is integrable, then

$$\sup\{L(f, P)\} = \inf\{U(f, P)\}.$$

This means that for each $\varepsilon > 0$ there are partitions P', P'' with

$$U(f, P'') - L(f, P') < \varepsilon.$$

Let P be a partition which contains both P' and P''. Then, according to the lemma,

$$U(f, P) \leq U(f, P''),$$
$$L(f, P) \geq L(f, P');$$

consequently,

$$U(f, P) - L(f, P) \leq U(f, P'') - L(f, P') < \varepsilon. \blacksquare$$

Although the mechanics of the proof take up a little space, it should be clear that Theorem 2 amounts to nothing more than a restatement of the definition of integrability. Nevertheless, it is a very convenient restatement because there is no mention of sup's and inf's, which are often difficult to work with. The next example illustrates this point, and also serves as a good introduction to the type of reasoning which the complicated definition of the integral necessitates, even in very simple situations.

Let f be defined on $[0, 2]$ by

$$f(x) = \begin{cases} 0, & x \neq 1 \\ 1, & x = 1. \end{cases}$$

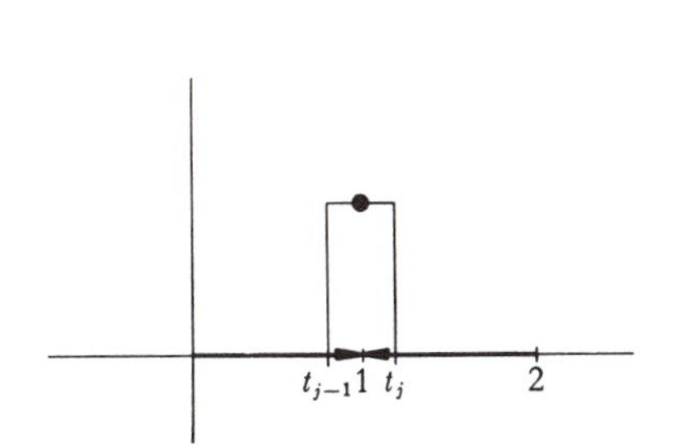

FIGURE 8

Suppose $P = \{t_0, \ldots, t_n\}$ is a partition of $[0, 2]$ with

$$t_{j-1} < 1 < t_j$$

(see Figure 8). Then

$$m_i = M_i = 0 \quad \text{if} \quad i \neq j,$$

but

$$m_j = 0 \qquad \text{and} \qquad M_j = 1.$$

Since

$$L(f, P) = \sum_{i=1}^{j-1} m_i(t_i - t_{i-1}) + m_j(t_j - t_{j-1}) + \sum_{i=j+1}^{n} m_i(t_i - t_{i-1}),$$
$$U(f, P) = \sum_{i=1}^{j-1} M_i(t_i - t_{i-1}) + M_j(t_j - t_{j-1}) + \sum_{i=j+1}^{n} M_i(t_i - t_{i-1}),$$

we have

$$U(f, P) - L(f, P) = t_j - t_{j-1}.$$

This certainly shows that f is integrable: to obtain a partition P with

$$U(f, P) - L(f, P) < \varepsilon,$$

it is only necessary to choose a partition with

$$t_{j-1} < 1 < t_j \quad \text{and} \quad t_j - t_{j-1} < \varepsilon.$$

Moreover, it is clear that

$$L(f, P) \leq 0 \leq U(f, P) \quad \text{for all partitions } P.$$

Since f is integrable, there is only *one* number between all lower and upper sums, namely, the integral of f, so

$$\int_0^2 f = 0.$$

Although the discontinuity of f was responsible for the difficulties in this example, even worse problems arise for very simple continuous functions. For example, let $f(x) = x$, and for simplicity consider an interval $[0, b]$, where $b > 0$. If $P = \{t_0, \dots, t_n\}$ is a partition of $[0, b]$, then (Figure 9)

$$m_i = t_{i-1} \qquad \text{and} \qquad M_i = t_i$$

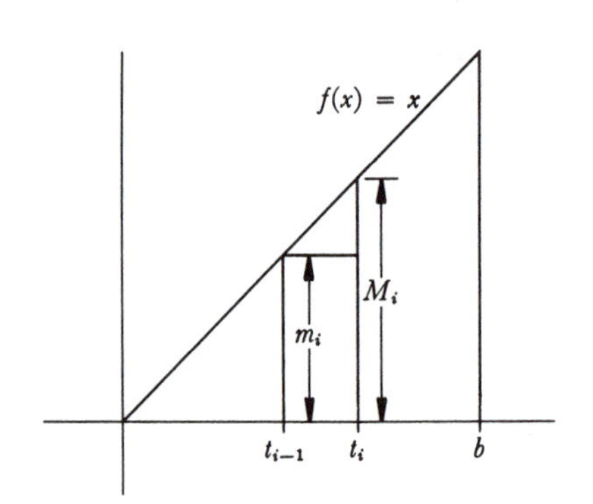

FIGURE 9

and therefore

$$\begin{aligned} L(f, P) &= \sum_{i=1}^{n} t_{i-1}(t_i - t_{i-1}) \\ &= t_0(t_1 - t_0) + t_1(t_2 - t_1) + \cdots + t_{n-1}(t_n - t_{n-1}), \\ U(f, P) &= \sum_{i=1}^{n} t_i(t_i - t_{i-1}) \\ &= t_1(t_1 - t_0) + t_2(t_2 - t_1) + \cdots + t_n(t_n - t_{n-1}). \end{aligned}$$

Neither of these formulas is particularly appealing, but both simplify considerably for partitions $P_n = \{t_0, \dots, t_n\}$ into n *equal* subintervals. In this case, the length $t_i - t_{i-1}$ of each subinterval is b/n, so

$$\begin{aligned} t_0 &= 0, \\ t_1 &= \frac{b}{n}, \\ t_2 &= \frac{2b}{n}, \text{ etc;} \end{aligned}$$

in general

$$t_i = \frac{ib}{n}.$$

Then

$$\begin{aligned} L(f, P_n) &= \sum_{i=1}^{n} t_{i-1}(t_i - t_{i-1}) \\ &= \sum_{i=1}^{n} \left\{ \frac{(i-1)b}{n} \right\} \cdot \frac{b}{n} \\ &= \left[\sum_{i=1}^{n} (i-1) \right] \frac{b^2}{n^2} \\ &= \left(\sum_{j=0}^{n-1} j \right) \frac{b^2}{n^2}. \end{aligned}$$

Remembering the formula

$$1 + \cdots + k = \frac{k(k+1)}{2},$$

this can be written

$$\begin{aligned} L(f, P_n) &= \frac{(n-1)(n)}{2} \cdot \frac{b^2}{n^2} \\ &= \frac{n-1}{n} \cdot \frac{b^2}{2}. \end{aligned}$$

Similarly,

$$\begin{aligned} U(f, P_n) &= \sum_{i=1}^{n} t_i(t_i - t_{i-1}) \\ &= \sum_{i=1}^{n} \frac{ib}{n} \cdot \frac{b}{n} \\ &= \frac{n(n+1)}{2} \cdot \frac{b^2}{n^2} \\ &= \frac{n+1}{n} \cdot \frac{b^2}{2}. \end{aligned}$$

If n is very large, both $L(f, P_n)$ and $U(f, P_n)$ are close to $b^2/2$, and this remark makes it easy to show that f is integrable. Notice first that

$$U(f, P_n) - L(f, P_n) = \frac{2}{n} \cdot \frac{b^2}{2}.$$

This shows that there are partitions P_n with $U(f, P_n) - L(f, P_n)$ as small as desired. By Theorem 2 the function f is integrable. Moreover, $\int_0^b f$ may now be found with only a little work. It is clear, first of all, that

$$L(f, P_n) \leq \frac{b^2}{2} \leq U(f, P_n) \quad \text{for all } n.$$

This inequality shows only that $b^2/2$ lies between certain special upper and lower sums, but we have just seen that $U(f, P_n) - L(f, P_n)$ can be made as small as desired, so there is *only one* number with this property. Since the integral certainly has this property, we can conclude that

$$\int_0^b f = \frac{b^2}{2}.$$

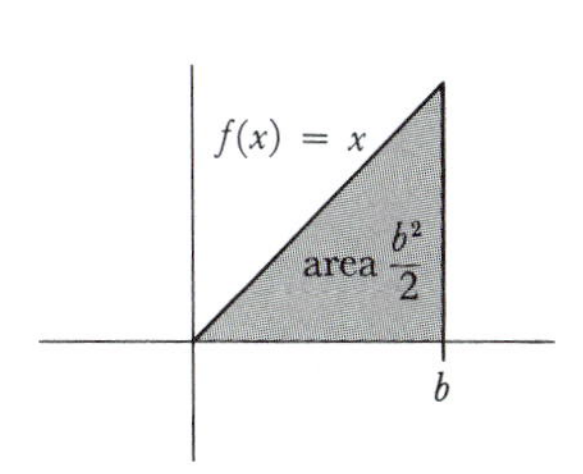

FIGURE 10

Notice that this equation assigns area $b^2/2$ to a right triangle with base and altitude b (Figure 10). Using more involved calculations, or appealing to Theorem 4, it can be shown that

$$\int_a^b f = \frac{b^2}{2} - \frac{a^2}{2}.$$

The function $f(x) = x^2$ presents even greater difficulties. In this case (Figure 11), if $P = \{t_0, \dots, t_n\}$ is a partition of $[0, b]$, then

$$m_i = f(t_{i-1}) = (t_{i-1})^2 \quad \text{and} \quad M_i = f(t_i) = {t_i}^2.$$

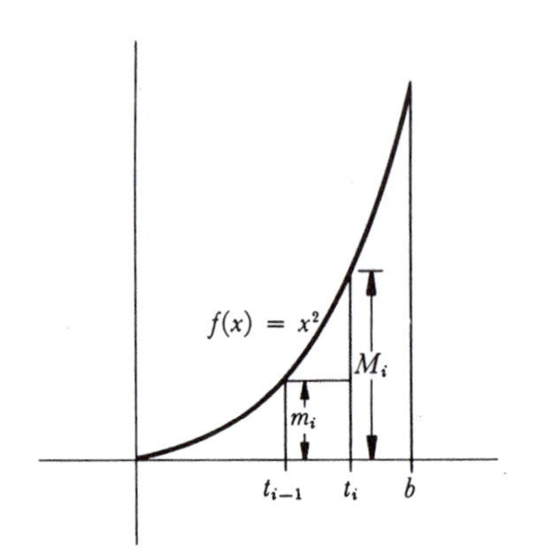

FIGURE 11

Choosing, once again, a partition $P_n = \{t_0, \dots, t_n\}$ into n equal parts, so that

$$t_i = \frac{i \cdot b}{n}$$

the lower and upper sums become

$$\begin{aligned}
L(f, P_n) &= \sum_{i=1}^{n} (t_{i-1})^2 \cdot (t_i - t_{i-1}) \\
&= \sum_{i=1}^{n} (i-1)^2 \frac{b^2}{n^2} \cdot \frac{b}{n} \\
&= \frac{b^3}{n^3} \cdot \sum_{j=0}^{n-1} j^2, \\
U(f, P_n) &= \sum_{i=1}^{n} {t_i}^2 \cdot (t_i - t_{i-1}) \\
&= \sum_{i=1}^{n} i^2 \frac{b^2}{n^2} \cdot \frac{b}{n} \\
&= \frac{b^3}{n^3} \sum_{j=1}^{n} j^2.
\end{aligned}$$

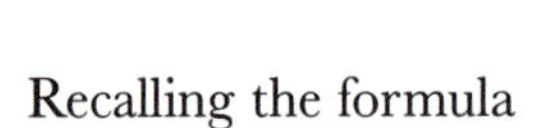

Recalling the formula

$$1^2 + \cdots + k^2 = \tfrac{1}{6} k(k+1)(2k+1)$$

from Problem 2-1, these sums can be written as

$$\begin{aligned}
L(f, P_n) &= \frac{b^3}{n^3} \cdot \frac{1}{6}(n-1)(n)(2n-1), \\
U(f, P_n) &= \frac{b^3}{n^3} \cdot \frac{1}{6}(n+1)(2n+1).
\end{aligned}$$

It is not too hard to show that

$$L(f, P_n) \le \frac{b^3}{3} \le U(f, P_n),$$

and that $U(f, P_n) - L(f, P_n)$ can be made as small as desired, by choosing n sufficiently large. The same sort of reasoning as before then shows that

$$\int_0^b f = \frac{b^3}{3}.$$

This calculation already represents a nontrivial result—the area of the region bounded by a parabola is not usually derived in elementary geometry. Nevertheless, the result was known to Archimedes, who derived it in essentially the same way. The only superiority we can claim is that in the next chapter we will discover a much simpler way to arrive at this result.

Some of our investigations can be summarized as follows:

$$\int_a^b f = c \cdot (b - a) \quad \text{if} \quad f(x) = c \text{ for all } x,$$
$$\int_a^b f = \frac{b^2}{2} - \frac{a^2}{2} \quad \text{if} \quad f(x) = x \text{ for all } x,$$
$$\int_a^b f = \frac{b^3}{3} - \frac{a^3}{3} \quad \text{if} \quad f(x) = x^2 \text{ for all } x.$$

This list already reveals that the notation $\int_a^b f$ suffers from the lack of a convenient notation for naming functions defined by formulas. For this reason an alternative notation,* analogous to the notation $\lim_{x \to a} f(x)$, is also useful:

$$\int_a^b f(x)\, dx \quad \text{means precisely the same as} \quad \int_a^b f.$$

Thus

$$\int_a^b c\, dx = c \cdot (b - a),$$
$$\int_a^b x\, dx = \frac{b^2}{2} - \frac{a^2}{2},$$
$$\int_a^b x^2\, dx = \frac{b^3}{3} - \frac{a^3}{3}.$$

Notice that, as in the notation $\lim_{x \to a} f(x)$, the symbol x can be replaced by any other letter (except f, a, or b, of course):

$$\int_a^b f(x)\, dx = \int_a^b f(t)\, dt = \int_a^b f(\alpha)\, d\alpha = \int_a^b f(y)\, dy = \int_a^b f(c)\, dc.$$

The symbol dx has no meaning in isolation, any more than the symbol $x \to$ has any meaning, except in the context $\lim_{x \to a} f(x)$. In the equation

$$\int_a^b x^2\, dx = \frac{b^3}{3} - \frac{a^3}{3},$$

*The notation $\int_a^b f(x)\, dx$ is actually the older, and was for many years the only, symbol for the integral. Leibniz used this symbol because he considered the integral to be the sum (denoted by $\int$) of infinitely many rectangles with height $f(x)$ and "infinitely small" width dx. Later writers used $x_0, \dots, x_n$ to denote the points of a partition, and abbreviated $x_i - x_{i-1}$ by Δx_i. The integral was defined as the limit as Δx_i approaches 0 of the sums $\sum_{i=1}^{n} f(x_i)\, \Delta x_i$ (analogous to lower and upper sums). The fact that the limit is obtained by changing $\sum$ to $\int$, $f(x_i)$ to $f(x)$, and Δx_i to dx, delights many people.

the *entire* symbol $x^2\,dx$ may be regarded as an abbreviation for:

the function f such that $f(x) = x^2$ for all x.

This notation for the integral is as flexible as the notation $\lim_{x\to a} f(x)$. Several examples may aid in the interpretation of various types of formulas which frequently appear; we have made use of Theorems 5 and 6.*

$$(1)\quad \int_a^b (x+y)\,dx = \int_a^b x\,dx + \int_a^b y\,dx = \frac{b^2}{2} - \frac{a^2}{2} + y(b-a).$$

$$(2)\quad \int_a^x (y+t)\,dy = \int_a^x y\,dy + \int_a^x t\,dy = \frac{x^2}{2} - \frac{a^2}{2} + t(x-a).$$

$$\begin{aligned}(3)\quad \int_a^b \left(\int_a^x (1+t)\,dz\right) dx &= \int_a^b (1+t)(x-a)\,dx \\ &= (1+t)\int_a^b (x-a)\,dx \\ &= (1+t)\left[\frac{b^2}{2} - \frac{a^2}{2} - a(b-a)\right].\end{aligned}$$

$$\begin{aligned}(4)\quad \int_a^b \left(\int_c^d (x+y)\,dy\right) dx &= \int_a^b \left[x(d-c) + \frac{d^2}{2} - \frac{c^2}{2}\right] dx \\ &= \left(\frac{d^2}{2} - \frac{c^2}{2}\right)(b-a) + (d-c)\int_a^b x\,dx \\ &= \left(\frac{d^2}{2} - \frac{c^2}{2}\right)(b-a) + (d-c)\left(\frac{b^2}{2} - \frac{a^2}{2}\right).\end{aligned}$$

The computations of $\int_a^b x\,dx$ and $\int_a^b x^2\,dx$ may suggest that evaluating integrals is generally difficult or impossible. As a matter of fact, the integrals of most functions *are* impossible to determine exactly (*although they may be computed to any degree of accuracy desired by calculating lower and upper sums*). Nevertheless, as we shall see in the next chapter, the integral of many functions can be computed very easily.

Even though most integrals cannot be computed exactly, it is important at least to know when a function f *is* integrable on $[a, b]$. Although it is possible to say precisely which functions are integrable, the criterion for integrability is a little too difficult to be stated here, and we will have to settle for partial results. The next Theorem gives the most useful result, but the proof given here uses material from the Appendix to Chapter 8. If you prefer, you can wait until the end of the next chapter, when a totally different proof will be given.

*Lest chaos overtake the reader when consulting other books, equation (1) requires an important qualification. This equation interprets $\int_a^b y\,dx$ to mean the integral of the function f such that each value $f(x)$ is the number y. But classical notation often uses y for $y(x)$, so $\int_a^b y\,dx$ might mean the integral of some arbitrary *function* y.

THEOREM 3 If f is continuous on $[a, b]$, then f is integrable on $[a, b]$.

PROOF Notice, first, that f is bounded on $[a, b]$, because it is continuous on $[a, b]$. To prove that f is integrable on $[a, b]$, we want to use Theorem 2, and show that for every $\varepsilon > 0$ there is a partition P of $[a, b]$ such that

$$U(f, P) - L(f, P) < \varepsilon.$$

Now we know, by Theorem 1 of the Appendix to Chapter 8, that f is uniformly continuous on $[a, b]$. So there is some $\delta > 0$ such that for all x and y in $[a, b]$,

$$\text{if } |x - y| < \delta, \text{ then } |f(x) - f(y)| < \frac{\varepsilon}{2(b-a)}.$$

The trick is simply to choose a partition $P = \{t_0, \dots, t_n\}$ such that each $|t_i - t_{i-1}| < \delta$. Then for each i we have

$$|f(x) - f(y)| < \frac{\varepsilon}{2(b-a)} \qquad \text{for all } x, y \text{ in } [t_{i-1}, t_i],$$

and it follows easily that

$$M_i - m_i \le \frac{\varepsilon}{2(b-a)} < \frac{\varepsilon}{b-a}.$$

Since this is true for all i, we then have

$$\begin{aligned} U(f, P) - L(f, P) &= \sum_{i=1}^{n} (M_i - m_i)(t_i - t_{i-1}) \\ &< \frac{\varepsilon}{b-a} \sum_{i=1}^{n} t_i - t_{i-1} \\ &= \frac{\varepsilon}{b-a} \cdot b - a \\ &= \varepsilon, \end{aligned}$$

which is what we wanted. ▮

Although this theorem will provide all the information necessary for the use of integrals in this book, it is more satisfying to have a somewhat larger supply of integrable functions. Several problems treat this question in detail. It will help to know the following three theorems, which show that f is integrable on $[a, b]$, if it is integrable on $[a, c]$ and $[c, b]$; that $f + g$ is integrable if f and g are; and that $c \cdot f$ is integrable if f is integrable and c is any number.

As a simple application of these theorems, recall that if f is 0 except at one point, where its value is 1, then f is integrable. Multiplying this function by c, it follows that the same is true if the value of f at the exceptional point is c. Adding such a function to an integrable function, we see that the value of an integrable function may be changed arbitrarily at one point without destroying integrability. By breaking up the interval into many subintervals, we see that the value can be changed at finitely many points.

The proofs of these theorems usually use the alternative criterion for integrability in Theorem 2; as some of our previous demonstrations illustrate, the details of the argument often conspire to obscure the point of the proof. It is a good idea to attempt proofs of your own, consulting those given here as a last resort, or as a check. This will probably clarify the proofs, and will certainly give good practice in the techniques used in some of the problems.

THEOREM 4 Let $a < c < b$. If f is integrable on $[a, b]$, then f is integrable on $[a, c]$ and on $[c, b]$. Conversely, if f is integrable on $[a, c]$ and on $[c, b]$, then f is integrable on $[a, b]$. Finally, if f is integrable on $[a, b]$, then

$$\int_a^b f = \int_a^c f + \int_c^b f.$$

PROOF Suppose f is integrable on $[a, b]$. If $\varepsilon > 0$, there is a partition $P = \{t_0, \ldots, t_n\}$ of $[a, b]$ such that

$$U(f, P) - L(f, P) < \varepsilon.$$

We might as well assume that $c = t_j$ for some j. (Otherwise, let Q be the partition which contains $t_0, \ldots, t_n$ and c; then Q contains P, so $U(f, Q) - L(f, Q) \leq U(f, P) - L(f, P) < \varepsilon$.)

Now $P' = \{t_0, \ldots, t_j\}$ is a partition of $[a, c]$ and $P'' = \{t_j, \ldots, t_n\}$ is a partition of $[c, b]$ (Figure 12). Since

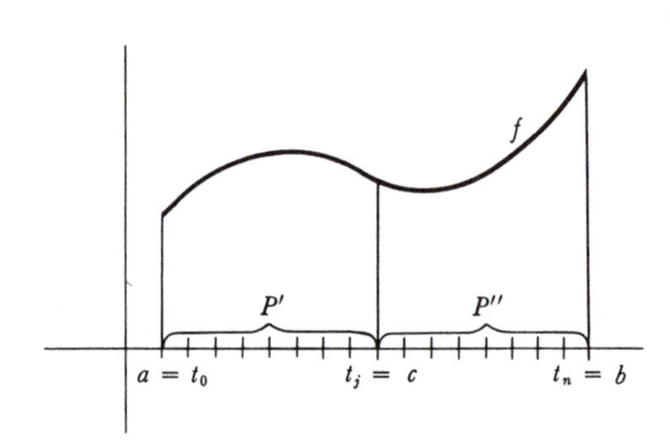

FIGURE 12

$$L(f, P) = L(f, P') + L(f, P''),$$
$$U(f, P) = U(f, P') + U(f, P''),$$

we have

$$[U(f, P') - L(f, P')] + [U(f, P'') - L(f, P'')] = U(f, P) - L(f, P) < \varepsilon.$$

Since each of the terms in brackets is nonnegative, each is less than ε. This shows that f is integrable on $[a, c]$ and $[c, b]$. Note also that

$$L(f, P') \leq \int_a^c f \leq U(f, P'),$$
$$L(f, P'') \leq \int_c^b f \leq U(f, P''),$$

so that

$$L(f, P) \leq \int_a^c f + \int_c^b f \leq U(f, P).$$

Since this is true for any P, this proves that

$$\int_a^c f + \int_c^b f = \int_a^b f.$$

Now suppose that f is integrable on $[a, c]$ and on $[c, b]$. If $\varepsilon > 0$, there is a partition P' of $[a, c]$ and a partition P'' of $[c, b]$ such that

$$U(f, P') - L(f, P') < \varepsilon/2,$$
$$U(f, P'') - L(f, P'') < \varepsilon/2.$$

If P is the partition of $[a, b]$ containing all the points of P' and P'', then

$$L(f, P) = L(f, P') + L(f, P''),$$
$$U(f, P) = U(f, P') + U(f, P'');$$

consequently,

$$U(f, P) - L(f, P) = [U(f, P') - L(f, P')] + [U(f, P'') - L(f, P'')] < \varepsilon. \blacksquare$$

Theorem 4 is the basis for some minor notational conventions. The integral $\int_a^b f$ was defined only for $a < b$. We now add the definitions

$$\int_a^a f = 0 \qquad \text{and} \qquad \int_a^b f = -\int_b^a f \quad \text{if } a > b.$$

With these definitions, the equation $\int_a^c f + \int_c^b f = \int_a^b f$ holds for all a, c, b even if $a < c < b$ is not true (the proof of this assertion is a rather tedious case-by-case check).

THEOREM 5 If f and g are integrable on $[a, b]$, then $f + g$ is integrable on $[a, b]$ and

$$\int_a^b (f + g) = \int_a^b f + \int_a^b g.$$

PROOF Let $P = \{t_0, \ldots, t_n\}$ be any partition of $[a, b]$. Let

$$m_i = \inf\{(f + g)(x) : t_{i-1} \le x \le t_i\},$$
$$m_i{}' = \inf\{f(x) : t_{i-1} \le x \le t_i\},$$
$$m_i{}'' = \inf\{g(x) : t_{i-1} \le x \le t_i\},$$

and define M_i, $M_i{}'$, $M_i{}''$ similarly. It is not necessarily true that

$$m_i = m_i{}' + m_i{}'',$$

but it is true (Problem 10) that

$$m_i \ge m_i{}' + m_i{}''.$$

Similarly,

$$M_i \le M_i{}' + M_i{}''.$$

Therefore,

$$L(f, P) + L(g, P) \le L(f + g, P)$$

and

$$U(f + g, P) \le U(f, P) + U(g, P).$$

Thus,

$$L(f, P) + L(g, P) \le L(f + g, P) \le U(f + g, P) \le U(f, P) + U(g, P).$$

Since f and g are integrable, there are partitions P', P'' with

$$U(f, P') - L(f, P') < \varepsilon/2,$$
$$U(g, P'') - L(g, P'') < \varepsilon/2.$$

If P contains both P' and P'', then

$$U(f, P) + U(g, P) - [L(f, P) + L(g, P)] < \varepsilon,$$

and consequently

$$U(f + g, P) - L(f + g, P) < \varepsilon.$$

This proves that $f + g$ is integrable on $[a, b]$. Moreover,

$$\begin{aligned} (1) \qquad L(f, P) + L(g, P) &\leq L(f + g, P) \\ &\leq \int_a^b (f + g) \\ &\leq U(f + g, P) \leq U(f, P) + U(g, P); \end{aligned}$$

and also

$$(2) \qquad L(f, P) + L(g, P) \leq \int_a^b f + \int_a^b g \leq U(f, P) + U(g, P).$$

Since $U(f, P) - L(f, P)$ and $U(g, p) - L(g, P)$ can both be made as small as desired, it follows that

$$U(f, P) + U(g, P) - [L(f, P) + L(g, P)]$$

can also be made as small as desired; it therefore follows from (1) and (2) that

$$\int_a^b (f + g) = \int_a^b f + \int_a^b g. \blacksquare$$

THEOREM 6 If f is integrable on $[a, b]$, then for any number c, the function cf is integrable on $[a, b]$ and

$$\int_a^b cf = c \cdot \int_a^b f.$$

PROOF The proof (which is much easier than that of Theorem 5) is left to you. It is a good idea to treat separately the cases $c \geq 0$ and $c \leq 0$. Why? $\blacksquare$

(Theorem 6 is just a special case of the more general theorem that $f \cdot g$ is integrable on $[a, b]$, if f and g are, but this result is quite hard to prove (see Problem 38).)

In this chapter we have acquired only one complicated definition, a few simply theorems with intricate proofs, and one theorem which required material from the Appendix to Chapter 8. This is not because integrals constitute a more difficult topic than derivatives, but because powerful tools developed in previous chapters have been allowed to remain dormant. The most significant discovery of calculus is the fact that the integral and the derivative are intimately related—once we learn the connection, the integral will become as useful as the derivative, and as easy to use. The connection between derivatives and integrals deserves a separate chapter, but the preparations which we will make in this chapter may serve as a hint. We first state a simple inequality concerning integrals, which plays a role in many important theorems.

THEOREM 7 Suppose f is integrable on $[a, b]$ and that

$$m \leq f(x) \leq M \quad \text{for all } x \text{ in } [a, b].$$

Then

$$m(b-a) \leq \int_a^b f \leq M(b-a).$$

PROOF It is clear that

$$m(b-a) \leq L(f, P) \quad \text{and} \quad U(f, P) \leq M(b-a)$$

for every partition P. Since $\int_a^b f = \sup\{L(f, P)\} = \inf\{U(f, P)\}$, the desired inequality follows immediately. ▮

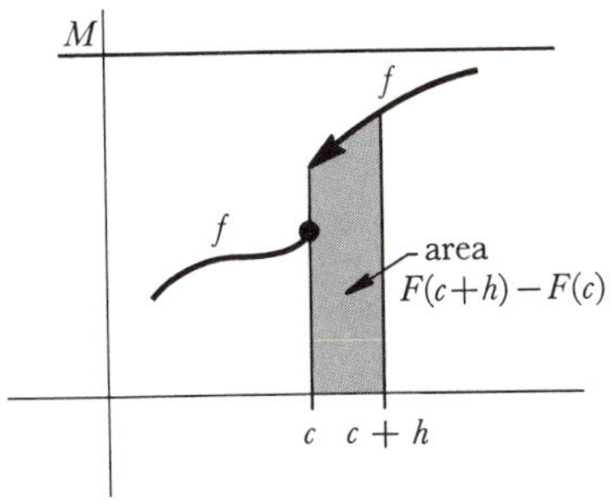

Suppose now that f is integrable on $[a, b]$. We can define a new function f on $[a, b]$ by

$$F(x) = \int_a^x f = \int_a^x f(t)\,dt.$$

(This depends on Theorem 4.) We have seen that f may be integrable even if it is not continuous, and the Problems give examples of integrable functions which are quite pathological. The behavior of F is therefore a very pleasant surprise.

FIGURE 13

THEOREM 8 If f is integrable on $[a, b]$ and F is defined on $[a, b]$ by

$$F(x) = \int_a^x f,$$

then F is continuous on $[a, b]$.

PROOF Suppose c is in $[a, b]$. Since f is integrable on $[a, b]$ it is, by definition, bounded on $[a, b]$; let M be a number such that

$$|f(x)| \leq M \quad \text{for all } x \text{ in } [a, b].$$

If $h > 0$, then (Figure 13)

$$F(c+h) - F(c) = \int_a^{c+h} f - \int_a^c f = \int_c^{c+h} f.$$

Since

$$-M \leq f(x) \leq M \quad \text{for all } x,$$

it follows from Theorem 7 that

$$-M \cdot h \leq \int_c^{c+h} f \leq Mh;$$

in other words,

$$(1) \qquad -M \cdot h \leq F(c+h) - F(c) \leq M \cdot h.$$

If $h < 0$, a similar inequality can be derived: Note that

$$F(c+h) - F(c) = \int_c^{c+h} f = -\int_{c+h}^c f.$$

Applying Theorem 7 to the interval $[c+h, c]$, of length $-h$, we obtain

$$Mh \leq \int_{c+h}^{c} f \leq -Mh;$$

multiplying by -1, which reverses all the inequalities, we have

$$(2) \qquad Mh \leq F(c+h) - F(c) \leq -Mh.$$

Inequalities (1) and (2) can be combined:

$$|F(c+h) - F(c)| \leq M \cdot |h|.$$

Therefore, if $\varepsilon > 0$, we have

$$|F(c+h) - F(c)| < \varepsilon,$$

provided that $|h| < \varepsilon/M$. This proves that

$$\lim_{h \to 0} F(c+h) = F(c);$$

in other words F is continuous at c. ▌

Figure 14 compares f and $F(x) = \int_a^x f$ for various functions f; it appears that F is always better behaved than f. In the next chapter we will see how true this is.

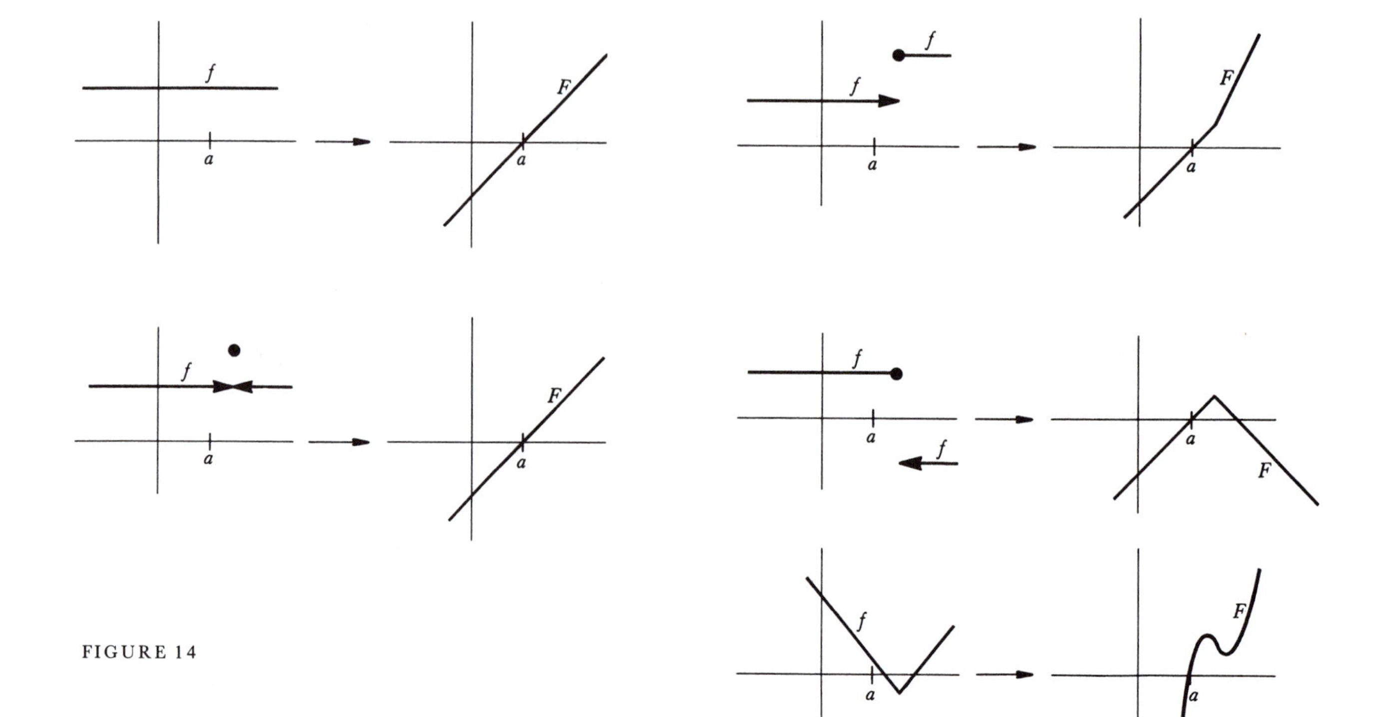

FIGURE 14

PROBLEMS

1. Prove that $\int_0^b x^3\,dx = b^4/4$, by considering partitions into n equal subintervals, using the formula for $\sum_{i=1}^{n} i^3$ which was found in Problem 2-6. This problem requires only a straightforward imitation of calculations in the text, but you should write it up as a formal proof to make certain that all the fine points of the argument are clear.

2. Prove, similarly, that $\int_0^b x^4\,dx = b^5/5$.

***3.** (a) Using Problem 2-7, show that the sum $\sum_{k=1}^{n} k^p/n^{p+1}$ can be made as close to $1/(p+1)$ as desired, by choosing n large enough.

(b) Prove that $\int_0^b x^p\,dx = b^{p+1}/(p+1)$.

***4.** This problem outlines a clever way to find $\displaystyle\int_a^b x^p\,dx$ for $0 < a < b$. (The result for $a = 0$ will then follow by continuity.) The trick is to use partitions $P = \{t_0, \dots, t_n\}$ for which all *ratios* $r = t_i/t_{i-1}$ are equal, instead of using partitions for which all differences $t_i - t_{i-1}$ are equal.

(a) Show that for such a partition P we have

$$t_i = a \cdot c^{i/n} \qquad \text{for } c = \frac{b}{a}.$$

(b) If $f(x) = x^p$, show, using the formula in Problem 2-5, that

$$\begin{aligned} U(f,P) &= a^{p+1}(1 - c^{-1/n}) \sum_{i=1}^{n} (c^{(p+1)/n})^i \\ &= (a^{p+1} - b^{p+1})c^{(p+1)/n} \frac{1 - c^{-1/n}}{1 - c^{(p+1)/n}} \\ &= (b^{p+1} - a^{p+1})c^{p/n} \cdot \frac{1}{1 + c^{1/n} + \cdots + c^{p/n}} \end{aligned}$$

and find a similar formula for $L(f,P)$.

(c) Conclude that

$$\int_a^b x^p\,dx = \frac{b^{p+1} - a^{p+1}}{p+1}.$$

5. Evaluate without doing any computations:

(i) $\displaystyle\int_{-1}^{1} x^3\sqrt{1 - x^2}\,dx.$

(ii) $\displaystyle\int_{-1}^{1} (x^5 + 3)\sqrt{1 - x^2}\,dx.$

6. Prove that

$$\int_0^x \frac{\sin t}{t+1}\,dt > 0$$

for all $x > 0$.

7. Decide which of the following functions are integrable on $[0, 2]$, and calculate the integral when you can.

(i) $f(x) = \begin{cases} x, & 0 \le x < 1 \\ x-2, & 1 \le x \le 2. \end{cases}$

(ii) $f(x) = \begin{cases} x, & 0 \le x \le 1 \\ x-2, & 1 < x \le 2. \end{cases}$

(iii) $f(x) = x + [x]$.

(iv) $f(x) = \begin{cases} x + [x], & x \text{ rational} \\ 0, & x \text{ irrational}. \end{cases}$

(v) $f(x) = \begin{cases} 1, & x \text{ of the form } a + b\sqrt{2} \text{ for rational } a \text{ and } b \\ 0, & x \text{ not of this form}. \end{cases}$

(vi) $f(x) = \begin{cases} \dfrac{1}{\left[\dfrac{1}{x}\right]}, & 0 < x \le 1 \\ 0, & x = 0 \text{ or } x > 1. \end{cases}$

(vii) f is the function shown in Figure 15.

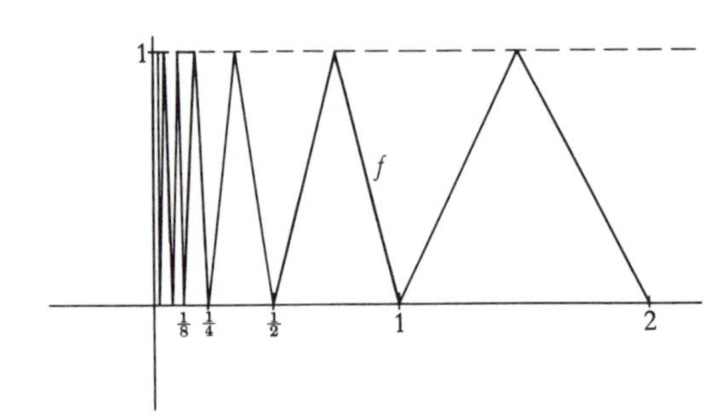

FIGURE 15

8. Find the areas of the regions bounded by

(i) the graphs of $f(x) = x^2$ and $g(x) = \dfrac{x^2}{2} + 2$.

(ii) the graphs of $f(x) = x^2$ and $g(x) = -x^2$ and the vertical lines through $(-1, 0)$ and $(1, 0)$.

(iii) the graphs of $f(x) = x^2$ and $g(x) = 1 - x^2$.

(iv) the graphs of $f(x) = x^2$ and $g(x) = 1 - x^2$ and $h(x) = 2$.

(v) the graphs of $f(x) = x^2$ and $g(x) = x^2 - 2x + 4$ and the vertical axis.

(vi) the graph of $f(x) = \sqrt{x}$, the horizontal axis, and the vertical line through $(2, 0)$. (Don't try to find $\int_0^2 \sqrt{x}\,dx$; you should see a way of guessing the answer, using only integrals that you already know how to evaluate. The questions that this example should suggest are considered in Problem 21.)

9. Find

$$\int_a^b \left(\int_c^d f(x)g(y)\,dy \right) dx$$

in terms of $\int_a^b f$ and $\int_c^d g$. (This problem is an exercise in notation, with a vengeance; it is crucial that you recognize a constant when it appears.)

10. Prove, using the notation of Theorem 5, that

$$m_i{}' + m_i{}'' = \inf\{f(x_1) + g(x_2) : t_{i-1} \le x_1, x_2 \le t_i\} \le m_i.$$

11. (a) Which functions have the property that every lower sum equals every upper sum?

(b) Which functions have the property that some upper sum equals some (other) lower sum?

(c) Which continuous functions have the property that all lower sums are equal?

*(d) Which integrable functions have the property that all lower sums are equal? (Bear in mind that one such function is $f(x) = 0$ for x irrational, $f(x) = 1/q$ for $x = p/q$ in lowest terms.) Hint: You will need the notion of a dense set, introduced in Problem 8-6, as well as the results of Problem 30.

12. If $a < b < c < d$ and f is integrable on $[a, d]$, prove that f is integrable on $[b, c]$. (Don't work hard.)

13. (a) Prove that if f is integrable on $[a, b]$ and $f(x) \ge 0$ for all x in $[a, b]$, then $\int_a^b f \ge 0$.

(b) Prove that if f and g are integrable on $[a, b]$ and $f(x) \ge g(x)$ for all x in $[a, b]$, then $\int_a^b f \ge \int_a^b g$. (By now it should be unnecessary to warn that if you work hard on part (b) you are wasting time.)

14. Prove that

$$\int_a^b f(x)\,dx = \int_{a+c}^{b+c} f(x - c)\,dx.$$

(The geometric interpretation should make this very plausible.) Hint: Every partition $P = \{t_0, \dots, t_n\}$ of $[a, b]$ gives rise to a partition $P' = \{t_0 + c, \dots, t_n + c\}$ of $[a + c, b + c]$, and conversely.

***15.** Prove that

$$\int_1^a \frac{1}{t}\,dt + \int_1^b \frac{1}{t}\,dt = \int_1^{ab} \frac{1}{t}\,dt.$$

Hint: This can be written $\int_1^a 1/t\,dt = \int_b^{ab} 1/t\,dt$. Every partition $P = \{t_0, \dots, t_n\}$ of $[1, a]$ gives rise to a partition $P' = \{bt_0, \dots, bt_n\}$ of $[b, ab]$, and conversely.

***16.** Prove that

$$\int_{ca}^{cb} f(t)\,dt = c \int_a^b f(ct)\,dt.$$

(Notice that Problem 15 is a special case.)

17. Given that the area enclosed by the unit circle, described by the equation $x^2 + y^2 = 1$, is π, use Problem 16 to show that the area enclosed by the ellipse described by the equation $x^2/a^2 + y^2/b^2 = 1$ is πab.

18. This problem outlines yet another way to compute $\int_a^b x^n\,dx$; it was used by Cavalieri, one of the mathematicians working just before the invention of calculus.

(a) Let $c_n = \int_0^1 x^n\,dx$. Use Problem 16 to show that $\int_0^a x^n\,dx = c_n a^{n+1}$.

(b) Problem 14 shows that

$$\int_0^{2a} x^n\,dx = \int_{-a}^{a} (x+a)^n\,dx.$$

Use this formula to prove that

$$2^{n+1} c_n a^{n+1} = 2a^{n+1} \sum_{k \text{ even}} \binom{n}{k} c_k.$$

(c) Now use Problem 2-3 to prove that $c_n = 1/(n+1)$.

19. Suppose that f is bounded on $[a, b]$ and that f is continuous at each point in $[a, b]$ with the exception of x_0 in (a, b). Prove that f is integrable on $[a, b]$. Hint: Imitate one of the examples in the text.

20. Suppose that f is nondecreasing on $[a, b]$. Notice that f is automatically bounded on $[a, b]$, because $f(a) \le f(x) \le f(b)$ for x in $[a, b]$.

(a) If $P = \{t_0, \dots, t_n\}$ is a partition of $[a, b]$, what is $L(f, P)$ and $U(f, P)$?

(b) Suppose that $t_i - t_{i-1} = \delta$ for each i. Prove that $U(f, P) - L(f, P) = \delta[f(b) - f(a)]$.

(c) Prove that f is integrable.

(d) Give an example of a nondecreasing function on $[0, 1]$ which is discontinuous at infinitely many points.

It might be of interest to compare this problem with the following extract from Newton's *Principia.**

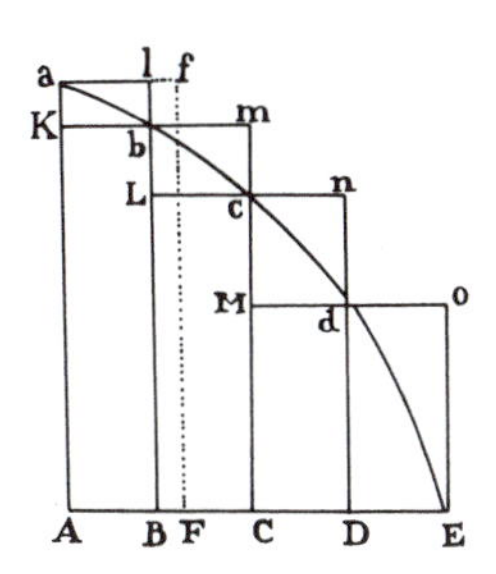

LEMMA II

If in any figure AacE, *terminated by the right lines* Aa, AE, *and the curve* acE, *there be inscribed any number of parallelograms* Ab, Bc, Cd, *&c., comprehended under equal bases* AB, BC, CD, *&c., and the sides,* Bb, Cc, Dd, *&c., parallel to one side* Aa *of the figure; and the parallelograms* aKbl, bLcm, cMdn, *&c., are completed: then if the breadth of those parallelograms be supposed to be diminished, and their number to be augmented* in infinitum, *I say, that the ultimate ratios which the inscribed figure* AKbLcMdD, *the circumscribed figure* AalbmcndoE, *and curvilinear figure* AabcdE, *will have to one another, are ratios of equality.*

For the difference of the inscribed and circumscribed figures is the sum of the parallelograms Kl, Lm, Mn, Do, that is (from the equality of all their bases), the rectangle under one of their bases Kb and the sum of their altitudes Aa, that is, the rectangle ABla. But this rectangle, because its breadth AB is supposed diminished *in infinitum*, becomes less than any given space. And therefore (by Lem. 1) the figures inscribed and circumscribed become ultimately equal one to the other; and much more will the intermediate curvilinear figure be ultimately equal to either. Q.E.D.

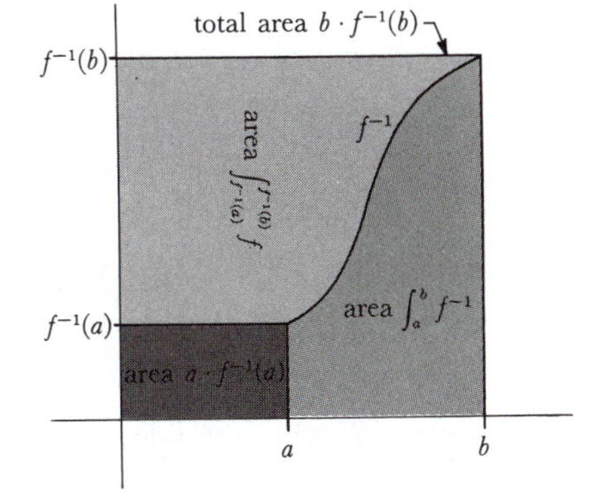

FIGURE 16

***21.** Suppose that f is increasing. Figure 16 suggests that

$$\int_a^b f^{-1} = bf^{-1}(b) - af^{-1}(a) - \int_{f^{-1}(a)}^{f^{-1}(b)} f.$$

(a) If $P = \{t_0, \dots, t_n\}$ is a partition of $[a, b]$, let $P' = \{f^{-1}(t_0), \dots, f^{-1}(t_n)\}$. Prove that, as suggested in Figure 17,

$$L(f^{-1}, P) + U(f, P') = bf^{-1}(b) - af^{-1}(a).$$

(b) Now prove the formula stated above.

(c) Find $\int_a^b \sqrt[n]{x}\, dx$ for $0 \le a < b$.

22. Suppose that f is a continuous increasing function with $f(0) = 0$. Prove that for $a, b > 0$ we have *Young's inequality*,

$$ab \le \int_0^a f(x)\, dx + \int_0^b f^{-1}(x)\, dx,$$

and that equality holds if and only if $b = f(a)$. Hint: Draw a picture like Figure 16!

*Newton's *Principia*, A Revision of Mott's Translation, by Florian Cajori. University of California Press, Berkeley, California, 1946.

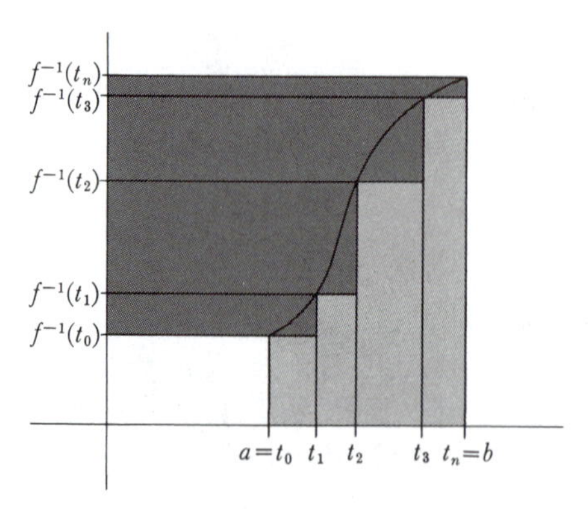

FIGURE 17

23. (a) Prove that if f is integrable on $[a, b]$ and $m \leq f(x) \leq M$ for all x in $[a, b]$, then

$$\int_a^b f(x)\,dx = (b-a)\mu$$

for some number μ with $m \leq \mu \leq M$.

(b) Prove that if f is continuous on $[a, b]$, then

$$\int_a^b f(x)\,dx = (b-a)f(\xi)$$

for some ξ in $[a, b]$; and show by an example that continuity is essential.

(c) More generally, suppose that f is continuous on $[a, b]$ and that g is integrable and nonnegative on $[a, b]$. Prove that

$$\int_a^b f(x)g(x)\,dx = f(\xi)\int_a^b g(x)\,dx$$

for some ξ in $[a, b]$. This result is called the Mean Value Theorem for Integrals.

(d) Deduce the same result if g is integrable and nonpositive on $[a, b]$.

(e) Show that one of these two hypotheses for g is essential.

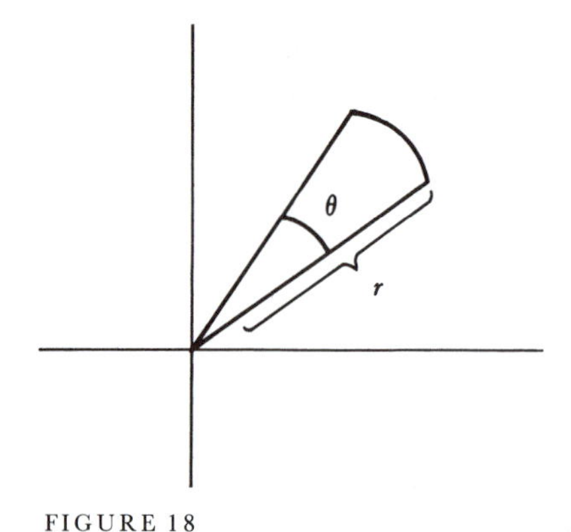

FIGURE 18

24. In this problem we consider the graph of a function in polar coordinates (Chapter 4, Appendix 3). Figure 18 shows a sector of a circle, with central angle θ. When θ is measured in radians (Chapter 15), the area of this sector is $r^2 \cdot \dfrac{\theta}{2}$. Now consider the region A shown in Figure 19, where the curve is the graph in polar coordinates of the continuous function f. Show that

$$\text{area } A = \frac{1}{2}\int_{\theta_0}^{\theta_1} f(\theta)^2 d\theta.$$

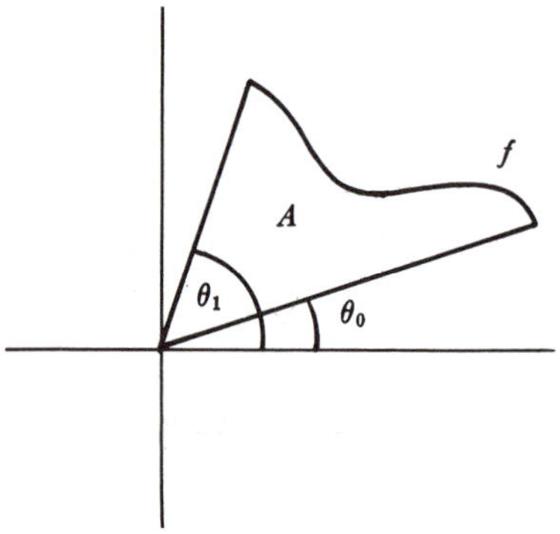

FIGURE 19

***25.** Let f be a continuous function on $[a, b]$. If $P = \{t_0, \ldots, t_n\}$ is a partition of $[a, b]$, define

$$\ell(f, P) = \sum_{i=1}^{n} \sqrt{(t_i - t_{i-1})^2 + [f(t_i) - f(t_{i-1})]^2}.$$

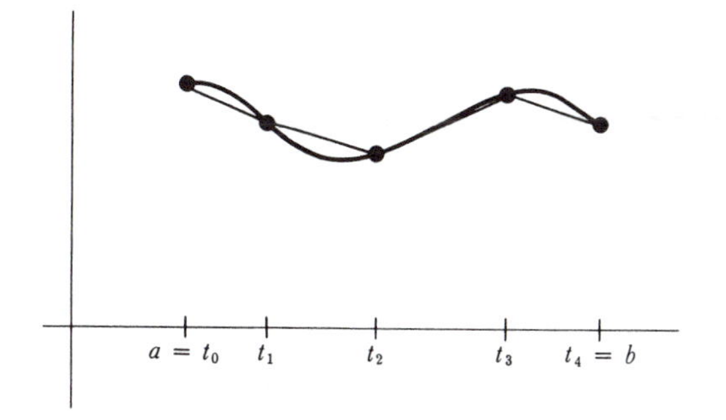

FIGURE 20

The number $\ell(f, P)$ represents the length of a polygonal curve inscribed in the graph of f (see Figure 20). We define the **length** of f on $[a, b]$ to be the least upper bound of all $\ell(f, P)$ for all partitions P (provided that the set of all such $\ell(f, P)$ is bounded above).

(a) If f is a linear function on $[a, b]$, prove that the length of f is the distance from $(a, f(a))$ to $(b, f(b))$.

(b) If f is not linear, prove that there is a partition $P = \{a, t, b\}$ of $[a, b]$ such that $\ell(f, P)$ is greater than the distance from $(a, f(a))$ to $(b, f(b))$. (You will need Problem 4-9.)

(c) Conclude that of all functions f on $[a, b]$ with $f(a) = c$ and $f(b) = d$, the length of the linear function is less than the length of any other. (Or, in conventional but hopelessly muddled terminology: "A straight line is the shortest distance between two points".)

(d) Suppose that f' is bounded on $[a, b]$. If P is any partition of $[a, b]$ show that

$$L\left(\sqrt{1 + (f')^2}, P\right) \leq \ell(f, P) \leq U\left(\sqrt{1 + (f')^2}, P\right).$$

Hint: Use the Mean Value Theorem.

(e) Why is $\sup\left\{L(\sqrt{1 + (f')^2}, P)\right\} \leq \sup\{\ell(f, P)\}$? (This is easy.)

(f) Now show that $\sup\{\ell(f, P)\} \leq \inf\left\{U(\sqrt{1 + (f')^2}, P)\right\}$, thereby proving that the length of f on $[a, b]$ is $\displaystyle\int_a^b \sqrt{1 + (f')^2}$, if $\sqrt{1 + (f')^2}$ is integrable on $[a, b]$. Hint: It suffices to show that if P' and P'' are any two partitions, then $\ell(f, P') \leq U\left(\sqrt{1 + (f')^2}, P''\right)$. If P contains the points of both P' and P'', how does $\ell(f, P')$ compare to $\ell(f, P)$?

(g) Let $\mathscr{L}(x)$ be the length of the graph of f on $[a, x]$, and let $d(x)$ be the length of the straight line segment from $(a, f(a))$ to $(x, f(x))$. Show that

$$\lim_{x \to a} \frac{\mathscr{L}(x)}{d(x)} = 1.$$

Hint: It will help to use a couple of Mean Value Theorems.

26. A function s defined on $[a, b]$ is called a **step function** if there is a partition $P = \{t_0, \dots, t_n\}$ of $[a, b]$ such that s is a constant on each (t_{i-1}, t_i) (the values of s at t_i may be arbitrary).

(a) Prove that if f is integrable on $[a, b]$, then for any $\varepsilon > 0$ there is a step function $s_1 \leq f$ with $\displaystyle\int_a^b f - \int_a^b s_1 < \varepsilon$, and also a step function $s_2 \geq f$ with $\displaystyle\int_a^b s_2 - \int_a^b f < \varepsilon$.

(b) Suppose that for all $\varepsilon > 0$ there are step functions $s_1 \leq f$ and $s_2 \geq f$ such that $\displaystyle\int_a^b s_2 - \int_a^b s_1 < \varepsilon$. Prove that f is integrable.

(c) Find a function f which is not a step function, but which satisfies $\int_a^b f = L(f, P)$ for some partition P of $[a, b]$.

***27.** Prove that if f is integrable on $[a, b]$, then for any $\varepsilon > 0$ there are continuous functions $g \le f \le h$ with $\int_a^b h - \int_a^b g < \varepsilon$. Hint: First get step functions with this property, and then continuous ones. A picture will help immensely.

28. (a) Show that if s_1 and s_2 are step functions on $[a, b]$, then $s_1 + s_2$ is also.

(b) Prove, without using Theorem 5, that $\int_a^b (s_1 + s_2) = \int_a^b s_1 + \int_a^b s_2$.

(c) Use part (b) (and Problem 26) to give an alternative proof of Theorem 5.

29. Suppose that f is integrable on $[a, b]$. Prove that there is a number x in $[a, b]$ such that $\int_a^x f = \int_x^b f$. Show by example that it is *not* always possible to choose x to be in (a, b).

***30.** The purpose of this problem is to show that if f is integrable on $[a, b]$, then f must be continuous at many points in $[a, b]$.

(a) Let $P = \{t_0, \dots, t_n\}$ be a partition of $[a, b]$ with $U(f, P) - L(f, P) < b - a$. Prove that for some i we have $M_i - m_i < 1$.

(b) Prove that there are numbers a_1 and b_1 with $a < a_1 < b_1 < b$ and $\sup\{f(x) : a_1 \le x \le b_1\} - \inf\{f(x) : a_1 \le x \le b_1\} < 1$. (You can choose $[a_1, b_1] = [t_{i-1}, t_i]$ from part (a) unless $i = 1$ or n; and in these two cases a very simple device solves the problem.)

(c) Prove that there are numbers a_2 and b_2 with $a_1 < a_2 < b_2 < b_1$ and $\sup\{f(x) : a_2 \le x \le b_2\} - \inf\{f(x) : a_2 \le x \le b_2\} < \frac{1}{2}$.

(d) Continue in this way to find a sequence of intervals $I_n = [a_n, b_n]$ such that $\sup\{f(x) : x \text{ in } I_n\} - \inf\{f(x) : x \text{ in } I_n\} < 1/n$. Apply the Nested Intervals Theorem (Problem 8-14) to find a point x at which f is continuous.

(e) Prove that f is continuous at infinitely many points in $[a, b]$.

***31.** Recall, from Problem 13, that $\int_a^b f \ge 0$ if $f(x) \ge 0$ for all x in $[a, b]$.

(a) Give an example where $f(x) \ge 0$ for all x, and $f(x) > 0$ for some x in $[a, b]$, and $\int_a^b f = 0$.

(b) Suppose $f(x) \ge 0$ for all x in $[a, b]$ and f is continuous at x_0 in $[a, b]$ and $f(x_0) > 0$. Prove that $\int_a^b f > 0$. Hint: It suffices to find one lower sum $L(f, P)$ which is positive.

(c) Suppose f is integrable on $[a, b]$ and $f(x) > 0$ for all x in $[a, b]$. Prove that $\int_a^b f > 0$. Hint: You will need Problem 30; indeed that was one reason for including Problem 30.

***32.** (a) Suppose that f is continuous on $[a,b]$ and $\int_a^b fg = 0$ for all continuous functions g on $[a,b]$. Prove that $f = 0$. (This is easy; there is an obvious g to choose.)

(b) Suppose f is continuous on $[a,b]$ and that $\int_a^b fg = 0$ for those continuous functions g on $[a,b]$ which satisfy the extra conditions $g(a) = g(b) = 0$. Prove that $f = 0$. (This innocent looking fact is an important lemma in the calculus of variations; see the Suggested Reading for references.) Hint: Derive a contradiction from the assumption $f(x_0) > 0$ or $f(x_0) < 0$; the g you pick will depend on the behavior of f near x_0.

33. Let $f(x) = x$ for x rational and $f(x) = 0$ for x irrational.

(a) Compute $L(f, P)$ for all partitions P of $[0,1]$.
(b) Find $\inf\{U(f, P) : P \text{ a partition of } [0, 1]\}$.

***34.** Let $f(x) = 0$ for irrational x, and $1/q$ if $x = p/q$ in lowest terms. Show that f is integrable on $[0, 1]$ and that $\int_0^1 f = 0$. (Every lower sum is clearly 0; you must figure out how to make upper sums small.)

***35.** Find two functions f and g which are integrable, but whose composition $g \circ f$ is not. Hint: Problem 34 is relevant.

***36.** Let f be a bounded function on $[a,b]$ and let P be a partition of $[a,b]$. Let M_i and m_i have their usual meanings, and let $M_i{}'$ and $m_i{}'$ have the corresponding meanings for the function $|f|$.

(a) Prove that $M_i{}' - m_i{}' \le M_i - m_i$.
(b) Prove that if f is integrable on $[a,b]$, then so is $|f|$.
(c) Prove that if f and g are integrable on $[a,b]$, then so are $\max(f, g)$ and $\min(f, g)$.
(d) Prove that f is integrable on $[a,b]$ if and only if its "positive part" $\max(f, 0)$ and its "negative part" $\min(f, 0)$ are integrable on $[a,b]$.

37. Prove that if f is integrable on $[a,b]$, then

$$\left|\int_a^b f(t)\,dt\right| \le \int_a^b |f(t)|\,dt.$$

Hint: This follows easily from a certain string of inequalities; Problem 1-14 is relevant.

***38.** Suppose f and g are integrable on $[a,b]$ and $f(x), g(x) \ge 0$ for all x in $[a,b]$. Let P be a partition of $[a,b]$. Let $M_i{}'$ and $m_i{}'$ denote the appropriate sup's and inf's for f, define $M_i{}''$ and $m_i{}''$ similarly for g, and define M_i and m_i similarly for fg.

(a) Prove that $M_i \le M_i{}'M_i{}''$ and $m_i \ge m_i{}'m_i{}''$.

(b) Show that

$$U(fg, P) - L(fg, P) \leq \sum_{i=1}^{n} [M_i' M_i'' - m_i' m_i''](t_i - t_{i-1}).$$

(c) Using the fact that f and g are bounded, so that $|f(x)|, |g(x)| \leq M$ for x in $[a, b]$, show that

$$U(fg, P) - L(fg, P)$$
$$\leq M \left\{ \sum_{i=1}^{n} [M_i' - m_i'](t_i - t_{i-1}) + \sum_{i=1}^{n} [M_i'' - m_i''](t_i - t_{i-1}) \right\}.$$

(d) Prove that fg is integrable.

(e) Now eliminate the restriction that $f(x), g(x) \geq 0$ for x in $[a, b]$.

39. Suppose that f and g are integrable on $[a, b]$. The *Cauchy-Schwarz inequality* states that

$$\left(\int_a^b fg \right)^2 \leq \left(\int_a^b f^2 \right) \left(\int_a^b g^2 \right).$$

(a) Show that the Schwarz inequality is a special case of the Cauchy-Schwarz inequality.

(b) Give three proofs of the Cauchy-Schwarz inequality by imitating the proofs of the Schwarz inequality in Problem 2-21. (The last one will take some imagination.)

(c) If equality holds, is it necessarily true that $f = \lambda g$ for some λ? What if f and g are continuous?

(d) Prove that $\left(\int_0^1 f \right)^2 \leq \left(\int_0^1 f^2 \right)$. Is this result true if 0 and 1 are replaced by a and b?

***40.** Suppose that f is continuous and $\lim_{x \to \infty} f(x) = a$. Prove that

$$\lim_{x \to \infty} \frac{1}{x} \int_0^x f(t)\, dt = a.$$

Hint: The condition $\lim_{x \to \infty} f(x) = a$ implies that $f(t)$ is close to a for $t \geq$ some N. This means that $\int_N^{N+M} f(t)\, dt$ is close to Ma. If M is large in comparison to N, then $Ma/(N + M)$ is close to a.

APPENDIX. RIEMANN SUMS

Suppose that $P = \{t_0, \dots, t_n\}$ is a partition of $[a, b]$, and that for each i we choose some point x_i in $[t_{i-1}, t_i]$. Then we clearly have

$$L(f, P) \le \sum_{i=1}^{n} f(x_i)(t_i - t_{i-1}) \le U(f, P).$$

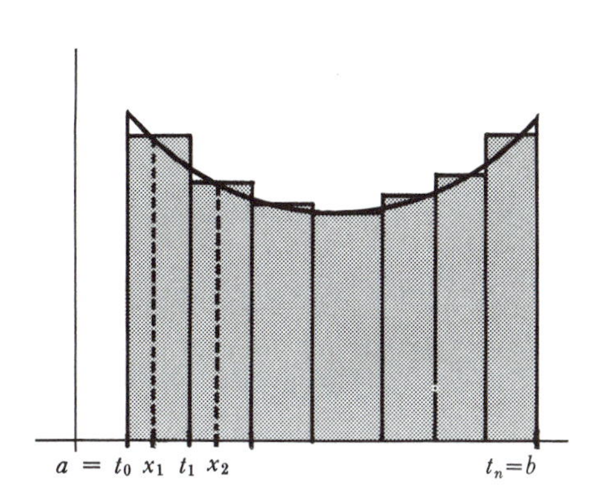

FIGURE 1

Any sum $\sum_{i=1}^{n} f(x_i)(t_i - t_{i-1})$ is called a *Riemann sum* of f for P. Figure 1 shows the geometric interpretation of a Riemann sum; it is the total area of n rectangles that lie partly below the graph of f and partly above it. Because of the arbitrary way in which the heights of the rectangles have been picked, we can't say for sure whether a particular Riemann sum is less than or greater than the integral $\int_a^b f(x)\,dx$. But it does seem that the overlaps shouldn't matter too much; if the bases of all the rectangles are narrow enough, then the Riemann sum ought to be close to the integral. The following theorem states this precisely.

THEOREM 1 Suppose that f is integrable on $[a, b]$. Then for every $\varepsilon > 0$ there is some $\delta > 0$ such that, if $P = \{t_0, \dots, t_n\}$ is any partition of $[a, b]$ with all lengths $t_i - t_{i-1} < \delta$, then

$$\left| \sum_{i=1}^{n} f(x_i)(t_i - t_{i-1}) - \int_a^b f(x)\,dx \right| < \varepsilon,$$

for any Riemann sum formed by choosing x_i in $[t_{i-1}, t_i]$.

PROOF First we will prove the theorem when f is continuous. As in the proof that a continuous function is integrable (Theorem 13-3), we will use Theorem 1 from the Appendix to Chapter 8, so you might want to skip it. But if you've already read the proof of Theorem 13-3, this part of the proof will be a snap—in fact, it's practically the same.

Given $\varepsilon > 0$, choose $\delta > 0$ so that for all x and y in $[a, b]$

$$\text{if } |x - y| < \delta, \text{ then } |f(x) - f(y)| < \frac{\varepsilon}{2(b-a)}.$$

Now consider any partition $P = \{t_0, \dots, t_n\}$ with each $t_i - t_{i-1} < \delta$, and any x_i in $[t_{i-1}, t_i]$. Then, as we saw in the proof of Theorem 13-3, we have

$$(1) \qquad U(f, P) - L(f, P) < \varepsilon.$$

But we also have

$$(2) \qquad L(f, P) \le \sum_{i=1}^{n} f(x_i)(t_i - t_{i-1}) \le U(f, P)$$

and

$$(3) \qquad L(f, P) \le \int_a^b f(x)\,dx \le U(f, P).$$

The desired inequality, for our continuous function f, follows immediately from (1), (2) and (3).

The argument in the general case is simple (though perhaps a bit messy), using Problem 13-27, which says that there are continuous functions $g \le f \le h$ satisfying

$$\int_a^b g \le \int_a^b f \le \int_a^b h, \tag{4}$$

with

$$\int_a^b h - \int_a^b g < \varepsilon.$$

We have

$$\sum_{i=1}^n g(x_i)(t_i - t_{i-1}) \le \sum_{i=1}^n f(x_i)(t_i - t_{i-1}) \le \sum_{i=1}^n h(x_i)(t_i - t_{i-1}),$$

and since the theorem holds for continuous functions, we know that for $t_i - t_{i-1} < \delta$, the left- and right-hand sides of this inequality are close to the left- and right-hand sides of (4). This implies that the two middle terms,

$$\int_a^b f \qquad \text{and} \qquad \sum_{i=1}^n f(x_i)(t_i - t_{i-1}),$$

must be close to $\int_a^b h - \int_a^b f$, which is small. Detailed inequalities are left to the skeptical reader. ▮

The moral of this tale is that anything which looks like a good approximation to an integral really is, provided that all the lengths $t_i - t_{i-1}$ of the intervals in the partition are small enough. Some of the following problems should bring home this message with even greater force.

PROBLEMS

1. Suppose that f and g are continuous functions on $[a, b]$. For a partition $P = \{t_0, \dots, t_n\}$ of $[a, b]$ choose a set of points x_i in $[t_{i-1}, t_i]$ and another set of points u_i in $[t_{i-1}, t_i]$. Consider the sum

$$\sum_{i=1}^n f(x_i)g(u_i)(t_i - t_{i-1}).$$

Notice that this is *not* a Riemann sum of fg for P. Nevertheless, show that all such sums will be within ε of $\int_a^b fg$ provided that the partition P has all lengths $t_i - t_{i-1}$ small enough. Hint: Estimate the difference between such a sum and a Riemann sum; you will need to use uniform continuity.

2. This problem is similar to, but somewhat harder than, the previous one. Suppose that f and g are continuous nonnegative functions on $[a, b]$. For a partition P, consider sums

$$\sum_{i=1}^{n} \sqrt{f(x_i) + g(u_i)}\,(t_i - t_{i-1}).$$

Show that these sums will be within ε of $\int_a^b \sqrt{f + g}$ if all $t_i - t_{i-1}$ are small enough. Hint: Use the fact that the square-root function is uniformly continuous on a closed interval $[0, M]$.

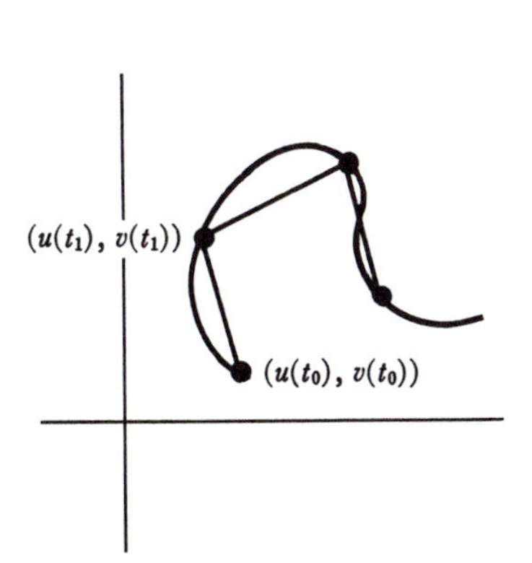

FIGURE 2

3. Finally, we're ready to tackle something big! (Compare Problem 13-25.) Consider a curve c given parametrically by two functions u and v on $[a, b]$. For a partition $P = \{t_0, \dots, t_n\}$ of $[a, b]$ we define

$$\ell(c, P) = \sum_{i=1}^{n} \sqrt{[u(t_i) - u(t_{i-1})]^2 + [v(t_i) - v(t_{i-1})]^2};$$

this represents the length of an inscribed polygonal curve (Figure 2). We define the length of c to be the least upper bound of all $\ell(f, P)$, if it exists. Prove that if u' and v' are continuous on $[a, b]$, then the length of c is

$$\int_a^b \sqrt{u'^2 + v'^2}.$$

4. Let f' be continuous on the interval $[\theta_0, \theta_1]$. Show that the graph of f in polar coordinates on this interval has the length

$$\int_{\theta_0}^{\theta_1} \sqrt{f^2 + f'^2}.$$

5. Using Theorem 1, show that the Cauchy-Schwarz inequality (Problem 13-39) is a consequence of the Schwarz inequality.

CHAPTER 14 THE FUNDAMENTAL THEOREM OF CALCULUS

From the hints given in the previous chapter you may have already guessed the first theorem of this chapter. We know that if f is integrable, then $F(x) = \int_a^x f$ is continuous; it is only fitting that we ask what happens when the original function f is continuous. It turns out that F is differentiable (and its derivative is especially simple).

THEOREM 1 (THE FIRST FUNDAMENTAL THEOREM OF CALCULUS) Let f be integrable on $[a, b]$, and define F on $[a, b]$ by

$$F(x) = \int_a^x f.$$

If f is continuous at c in $[a, b]$, then F is differentiable at c, and

$$F'(c) = f(c).$$

(If $c = a$ or b, then $F'(c)$ is understood to mean the right- or left-hand derivative of F.)

PROOF We will assume that c is in (a, b); the easy modifications for $c = a$ or b may be supplied by the reader. By definition,

$$F'(c) = \lim_{h \to 0} \frac{F(c+h) - F(c)}{h}.$$

Suppose first that $h > 0$. Then

$$F(c+h) - F(c) = \int_c^{c+h} f.$$

Define m_h and M_h as follows (Figure 1):

$$\begin{aligned} m_h &= \inf\{f(x) : c \le x \le c+h\}, \\ M_h &= \sup\{f(x) : c \le x \le c+h\}. \end{aligned}$$

It follows from Theorem 13-7 that

$$m_h \cdot h \le \int_c^{c+h} f \le M_h \cdot h.$$

Therefore

$$m_h \le \frac{F(c+h) - F(c)}{h} \le M_h.$$

If $h < 0$, only a few details of the argument have to be changed. Let

$$\begin{aligned} m_h &= \inf\{f(x) : c+h \le x \le c\}, \\ M_h &= \sup\{f(x) : c+h \le x \le c\}. \end{aligned}$$

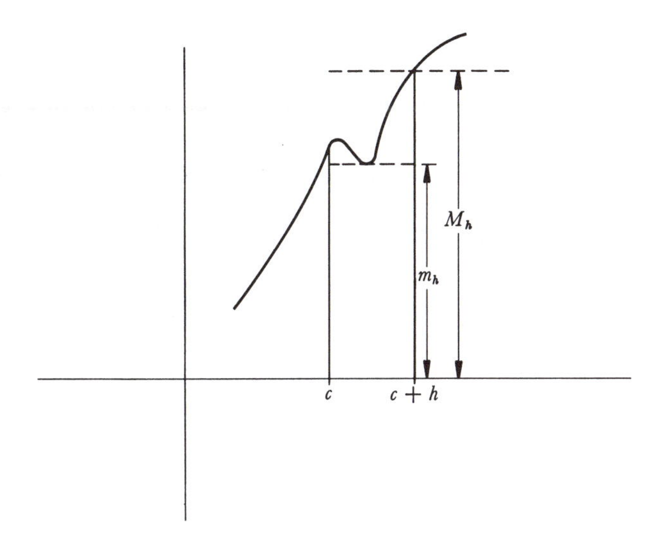

FIGURE 1

Then

$$m_h \cdot (-h) \leq \int_{c+h}^{c} f \leq M_h \cdot (-h).$$

Since

$$F(c+h) - F(c) = \int_{c}^{c+h} f = -\int_{c+h}^{c} f$$

this yields

$$m_h \cdot h \geq F(c+h) - F(c) \geq M_h \cdot h.$$

Since $h < 0$, dividing by h reverses the inequality again, yielding the same result as before:

$$m_h \leq \frac{F(c+h) - F(c)}{h} \leq M_h.$$

This inequality is true for any integrable function, continuous or not. Since f is continuous at c, however,

$$\lim_{h \to 0} m_h = \lim_{h \to 0} M_h = f(c),$$

and this proves that

$$F'(c) = \lim_{h \to 0} \frac{F(c+h) - F(c)}{h} = f(c). \blacksquare$$

Although Theorem 1 deals only with the function obtained by varying the upper limit of integration, a simple trick shows what happens when the lower limit is varied. If G is defined by

$$G(x) = \int_{x}^{b} f,$$

then

$$G(x) = \int_a^b f - \int_a^x f.$$

Consequently, if f is continuous at c, then

$$G'(c) = -f(c).$$

The minus sign appearing here is very fortunate, and allows us to extend Theorem 1 to the situation where the function

$$F(x) = \int_a^x f$$

is defined even for $x < a$. In this case we can write

$$F(x) = -\int_x^a f,$$

so if $c < a$ we have

$$F'(c) = -(-f(c)) = f(c),$$

exactly as before.

Notice that in either case, differentiability of F at c is ensured by continuity of f at c alone. Nevertheless, Theorem 1 is most interesting when f is continuous at all points in $[a, b]$. In this case F is differentiable at all points in $[a, b]$ and

$$F' = f.$$

In general, it is extremely difficult to decide whether a given function f is the derivative of some other function; for this reason Theorem 11-7 and Problems 11-54 and 11-55 are particularly interesting, since they reveal certain properties which f must have. If f is continuous, however, there is no problem at all—according to Theorem 1, f *is* the derivative of some function, namely the function

$$F(x) = \int_a^x f.$$

Theorem 1 has a simple corollary which frequently reduces computations of integrals to a triviality.

COROLLARY If f is continuous on $[a, b]$ and $f = g'$ for some function g, then

$$\int_a^b f = g(b) - g(a).$$

PROOF Let

$$F(x) = \int_a^x f.$$

Then $F' = f = g'$ on $[a, b]$. Consequently, there is a number c such that

$$F = g + c.$$

The number c can be evaluated easily: note that

$$0 = F(a) = g(a) + c,$$

so $c = -g(a)$; thus

$$F(x) = g(x) - g(a).$$

This is true, in particular, for $x = b$. Thus

$$\int_a^b f = F(b) = g(b) - g(a). \blacksquare$$

The proof of this corollary tends, at first sight, to make the corollary seem useless: after all, what good is it to know that

$$\int_a^b f = g(b) - b(a)$$

if g is, for example, $g(x) = \int_a^x f$? The point, of course, is that one might happen to know a quite different function g with this property. For example, if

$$g(x) = \frac{x^3}{3} \quad \text{and} \quad f(x) = x^2,$$

then $g'(x) = f(x)$ so we obtain, without ever computing lower and upper sums:

$$\int_a^b x^2\, dx = \frac{b^3}{3} - \frac{a^3}{3}.$$

One can treat other powers similarly; if n is a natural number and $g(x) = x^{n+1}/(n+1)$, then $g'(x) = x^n$, so

$$\int_a^b x^n\, dx = \frac{b^{n+1}}{n+1} - \frac{a^{n+1}}{n+1}.$$

For any natural number n, the function $f(x) = x^{-n}$ is not bounded on any interval containing 0, but if a and b are both positive or both negative, then

$$\int_a^b x^{-n}\, dx = \frac{b^{-n+1}}{-n+1} - \frac{a^{-n+1}}{-n+1}.$$

Naturally this formula is only true for $n \neq -1$. *We do not know a simple expression for*

$$\int_a^b \frac{1}{x}\, dx.$$

The problem of computing this integral is discussed later, but it provides a good opportunity to warn against a serious error. The conclusion of Corollary 1 is often confused with the definition of integrals—many students think that $\int_a^b f$ is defined as: "$g(b) - g(a)$, where g is a function whose derivative is f." This "definition" is not only wrong—it is useless. One reason is that a function f may be integrable without being the derivative of another function. For example, if $f(x) = 0$ for $x \neq 1$ and $f(1) = 1$, then f is integrable, but f cannot be a derivative (why not?). There is also another reason that is much more important: If f is continuous, then we know that $f = g'$ for some function g; but we know this *only because of Theorem 1*. The function $f(x) = 1/x$ provides an excellent illustration: if $x > 0$, then $f(x) = g'(x)$, where

$$g(x) = \int_1^x \frac{1}{t}\, dt,$$

and we know of no simpler function g with this property.

The corollary to Theorem 1 is so useful that it is frequently called the Second Fundamental Theorem of Calculus. In this book, that name is reserved for a somewhat stronger result (which in practice, however, is not much more useful). As we have just mentioned, a function f might be of the form g' even if f is not continuous. If f is integrable, then it is still true that

$$\int_a^b f = g(b) - g(a).$$

The proof, however, must be entirely different—we cannot use Theorem 1, so we must return to the definition of integrals.

THEOREM 2 (THE SECOND FUNDAMENTAL THEOREM OF CALCULUS) If f is integrable on $[a, b]$ and $f = g'$ for some function g, then

$$\int_a^b f = g(b) - g(a).$$

PROOF Let $P = \{t_0, \dots, t_n\}$ be any partition of $[a, b]$. By the Mean Value Theorem there is a point x_i in $[t_{i-1}, t_i]$ such that

$$\begin{aligned} g(t_i) - g(t_{i-1}) &= g'(x_i)(t_i - t_{i-1}) \\ &= f(x_i)(t_i - t_{i-1}). \end{aligned}$$

If

$$\begin{aligned} m_i &= \inf\{f(x) : t_{i-1} \le x \le t_i\}, \\ M_i &= \sup\{f(x) : t_{i-1} \le x \le t_i\}, \end{aligned}$$

then clearly

$$m_i(t_i - t_{i-1}) \le f(x_i)(t_i - t_{i-1}) \le M_i(t_i - t_{i-1}),$$

that is,

$$m_i(t_i - t_{i-1}) \le g(t_i) - g(t_{i-1}) \le M_i(t_i - t_{i-1}).$$

Adding these equations for $i = 1, \dots, n$ we obtain

$$\sum_{i=1}^n m_i(t_i - t_{i-1}) \le g(b) - g(a) \le \sum_{i=1}^n M_i(t_i - t_{i-1})$$

so that

$$L(f, P) \le g(b) - g(a) \le U(f, P)$$

for every partition P. But this means that

$$g(b) - g(a) = \int_a^b f. \blacksquare$$

We have already used the corollary to Theorem 1 (or, equivalently, Theorem 2) to find the integrals of a few elementary functions:

$$\int_a^b x^n\,dx = \frac{b^{n+1}}{n+1} - \frac{a^{n+1}}{n+1}, \quad n \neq -1. \qquad \begin{array}{l}(a \text{ and } b \text{ both positive or}\\ \text{both negative if } n > 0).\end{array}$$

As we pointed out in Chapter 13, this integral does not always represent the area bounded by the graph of the function, the horizontal axis, and the vertical lines through $(a, 0)$ and $(b, 0)$. For example, if $a < 0 < b$, then

$$\int_a^b x^3\,dx$$

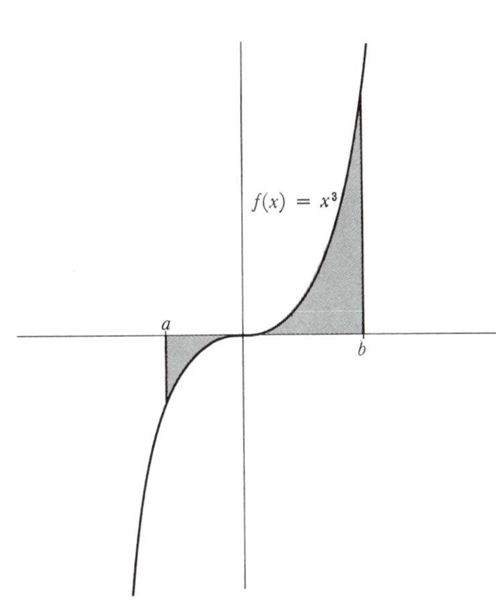

FIGURE 2

does not represent the area of the region shown in Figure 2, which is given instead by

$$-\left(\int_a^0 x^3\,dx\right) + \int_0^b x^3\,dx = -\left(\frac{0^4}{4} - \frac{a^4}{4}\right) + \left(\frac{b^4}{4} - \frac{0^4}{4}\right)$$
$$= \frac{a^4}{4} + \frac{b^4}{4}.$$

Similar care must be exercised in finding the areas of regions which are bounded by the graphs of more than one function—a problem which may frequently involve considerable ingenuity in any case. Suppose, to take a simple example first, that we wish to find the area of the region, shown in Figure 3, between the graphs of the functions

$$f(x) = x^2 \quad \text{and} \quad g(x) = x^3$$

on the interval $[0, 1]$. If $0 \le x \le 1$, then $0 \le x^3 \le x^2$, so that the graph of g lies below that of f. The area of the region of interest to us is therefore

$$\text{area } R(f, 0, 1) - \text{ area } R(g, 0, 1),$$

which is

$$\int_0^1 x^2\,dx - \int_0^1 x^3\,dx = \tfrac{1}{3} - \tfrac{1}{4} = \tfrac{1}{12}.$$

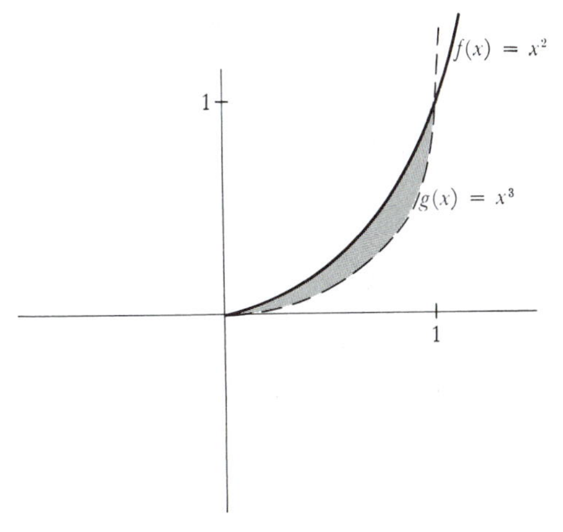

FIGURE 3

This area could have been expressed as

$$\int_a^b (f - g).$$

If $g(x) \le f(x)$ for all x in $[a, b]$, then this integral always gives the area bounded by f and g, *even if f and g are sometimes negative.* The easiest way to see this is shown in Figure 4. If c is a number such that $f + c$ and $g + c$ are nonnegative on $[a, b]$, then the region R_1, bounded by f and g, has the same area as the region R_2,

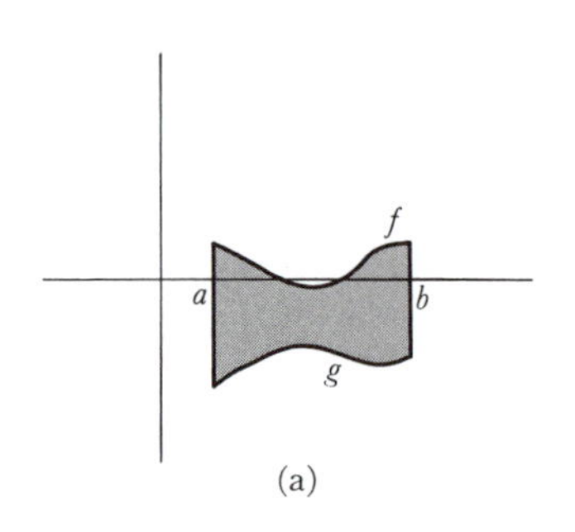

(a)

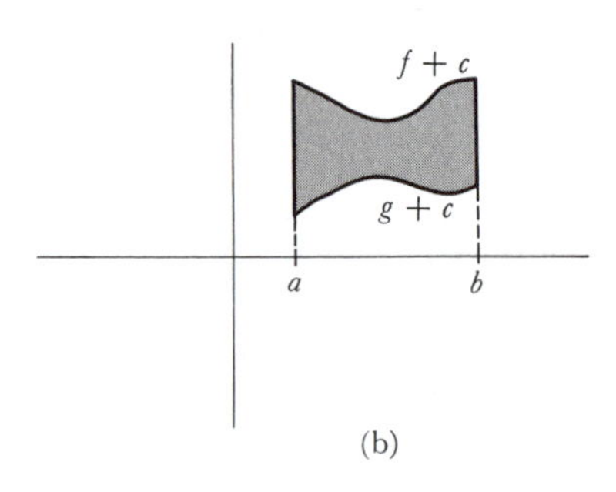

(b)

FIGURE 4

bounded by $f + c$ and $g + c$. Consequently,

$$\begin{aligned}\text{area } R_1 = \text{area } R_2 &= \int_a^b (f+c) - \int_a^b (g+c)\\ &= \int_a^b [(f+c)-(g+c)]\\ &= \int_a^b (f-g).\end{aligned}$$

This observation is useful in the following problem: Find the area of the region bounded by the graphs of

$$f(x) = x^3 - x \quad \text{and} \quad g(x) = x^2.$$

The first necessity is to determine this region more precisely. The graphs of f and g intersect when

$$\begin{aligned} & x^3 - x = x^2,\\ \text{or}\quad & x^3 - x^2 - x = 0,\\ \text{or}\quad & x(x^2 - x - 1) = 0,\\ \text{or}\quad & x = 0, \frac{1+\sqrt{5}}{2}, \frac{1-\sqrt{5}}{2}.\end{aligned}$$

On the interval $([1-\sqrt{5}]/2, 0)$ we have $x^3 - x \geq x^2$ and on the interval $(0, [1+\sqrt{5}]/2)$ we have $x^2 \geq x^3 - x$. These assertions are apparent from the graphs (Figure 5), but they can also be checked easily, as follows. Since $f(x) = g(x)$ only if $x = 0$, $[1+\sqrt{5}]/2$, or $[1-\sqrt{5}]/2$, the function $f - g$ does not change sign on the intervals $([1-\sqrt{5}]/2, 0)$ and $(0, [1+\sqrt{5}]/2)$; it is therefore only necessary to observe, for example, that

$$\begin{aligned}(-\tfrac{1}{2})^3 - (-\tfrac{1}{2}) - (-\tfrac{1}{2})^2 &= \tfrac{1}{8} > 0,\\ 1^3 - 1 - 1^2 &= -1 < 0,\end{aligned}$$

to conclude that

$$\begin{aligned} f - g \geq 0 &\quad \text{on } ([1-\sqrt{5}]/2, 0),\\ f - g \leq 0 &\quad \text{on } (0, [1+\sqrt{5}]/2).\end{aligned}$$

The area of the region in question is thus

$$\int_{\frac{1-\sqrt{5}}{2}}^{0} (x^3 - x - x^2)\,dx + \int_0^{\frac{1+\sqrt{5}}{2}} [x^2 - (x^3 - x)]\,dx.$$

As this example reveals, one of the major problems involved in finding the areas of a region may be the exact determination of the region. There are, however, more substantial problems of a logical nature—we have thus far defined the areas of some very special regions only, which do not even include some of the regions whose areas have just been computed! We have simply assumed that area made

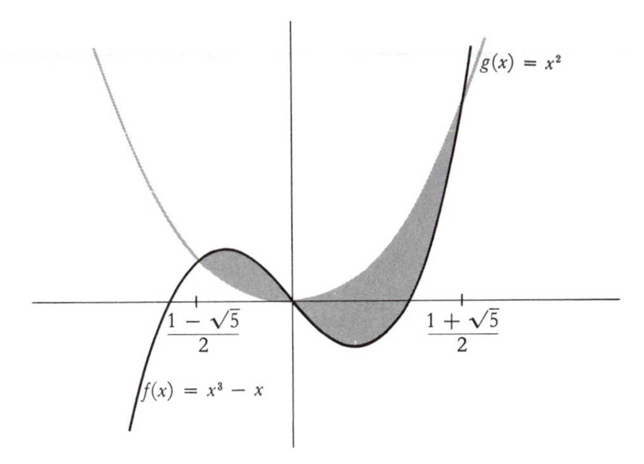

FIGURE 5

sense for these regions, and that certain reasonable properties of "area" do hold. These remarks are not meant to suggest that you should regard exercising ingenuity to compute areas as beneath you, but are meant to indicate that a better approach to the definition of area is available, although its proper place is somewhere in advanced calculus. The desire to define area was the motivation, both in this book and historically, for the definition of the integral, but the integral does not really provide the best method of *defining* areas, although it is frequently the proper tool for *computing* them.

It may be discouraging to learn that integrals are not suitable for the very purpose for which they were invented, but we will soon see how essential they are for other purposes. The most important use of integrals has already been emphasized: if f is continuous, the integral provides a function y such that

$$y'(x) = f(x).$$

This equation is the simplest example of a "differential equation" (an equation for a function y which involves derivatives of y). The Fundamental Theorem of Calculus says that this differential equation has a solution, if f is continuous. In succeeding chapters, and in various problems, we will solve more complicated equations, but the solution almost always depends somehow on the integral; in order to solve a differential equation it is necessary to construct a new function, and the integral is one of the best ways of doing this.

Since the differentiable functions provided by the Fundamental Theorem of Calculus will play such a prominent role in later work, it is very important to realize that these functions may be combined, like less esoteric functions, to yield still more functions, whose derivatives can be found by the Chain Rule.

Suppose, for example, that

$$f(x) = \int_a^{x^3} \frac{1}{1+\sin^2 t}\, dt.$$

Although the notation tends to disguise the fact somewhat, f is the composition of the functions

$$C(x) = x^3 \quad \text{and} \quad F(x) = \int_a^x \frac{1}{1+\sin^2 t}\,dt.$$

In fact, $f(x) = F(C(x))$; in other words, $f = F \circ C$. Therefore, by the Chain Rule,

$$\begin{aligned} f'(x) &= F'(C(x)) \cdot C'(x) \\ &= F'(x^3) \cdot 3x^2 \\ &= \frac{1}{1+\sin^2 x^3} \cdot 3x^2. \end{aligned}$$

If f is defined, instead, as

$$f(x) = \int_{x^3}^a \frac{1}{1+\sin^2 t}\,dt,$$

then

$$f'(x) = -\frac{1}{1+\sin^2 x^3} \cdot 3x^2.$$

If f is defined as the *reverse* composition,

$$f(x) = \left(\int_a^x \frac{1}{1+\sin^2 t}\,dt\right)^3,$$

then

$$\begin{aligned} f'(x) &= C'(F(x)) \cdot F'(x) \\ &= 3\left(\int_a^x \frac{1}{1+\sin^2 t}\,dt\right)^2 \cdot \frac{1}{1+\sin^2 x}. \end{aligned}$$

Similarly, if

$$\begin{aligned} f(x) &= \int_a^{\sin x} \frac{1}{1+\sin^2 t}\,dt, \\ g(x) &= \int_{\sin x}^a \frac{1}{1+\sin^2 t}\,dt, \\ h(x) &= \sin\left(\int_a^x \frac{1}{1+\sin^2 t}\,dt\right), \end{aligned}$$

then

$$\begin{aligned} f'(x) &= \frac{1}{1+\sin^2(\sin x)} \cdot \cos x, \\ g'(x) &= \frac{-1}{1+\sin^2(\sin x)} \cdot \cos x, \\ h'(x) &= \cos\left(\int_a^x \frac{1}{1+\sin^2 t}\,dt\right) \cdot \frac{1}{1+\sin^2 x}. \end{aligned}$$

The formidable appearing function

$$f(x) = \int_a^{\left(\int_a^x \frac{1}{1+\sin^2 t}\,dt\right)} \frac{1}{1+\sin^2 t}\,dt$$

is also a composition; in fact, $f = F \circ F$. Therefore

$$\begin{aligned} f'(x) &= F'(F(x)) \cdot F'(x) \\ &= \frac{1}{1+\sin^2\left(\int_a^x \frac{1}{1+\sin^2 t}\,dt\right)} \cdot \frac{1}{1+\sin^2 x}. \end{aligned}$$

As these examples reveal, the expression occurring above (or below) the integral sign indicates the function which will appear on the *right* when f is written as a composition. As a final example, consider the triple compositions

$$f(x) = \int_a^{\left(\int_a^{x^3} \frac{1}{1+\sin^2 t}\,dt\right)} \frac{1}{1+\sin^2 t}\,dt, \qquad g(x) = \int_a^{\left[\int_a^{\left(\int_a^x \frac{1}{1+\sin^2 t}\,dt\right)} \frac{1}{1+\sin^2 t}\,dt\right]} \frac{1}{1+\sin^2 t}\,dt,$$

which can be written

$$f = F \circ F \circ C \quad \text{and} \quad g = F \circ F \circ F.$$

Omitting the intermediate steps (which you may supply, if you still feel insecure), we obtain

$$f'(x) = \frac{1}{1+\sin^2\left(\int_a^{x^3} \frac{1}{1+\sin^2 t}\,dt\right)} \cdot \frac{1}{1+\sin^2 x^3} \cdot 3x^2,$$

$$g'(x) = \frac{1}{1+\sin^2\left[\int_a^{\left(\int_a^x \frac{1}{1+\sin^2 t}\,dt\right)} \frac{1}{1+\sin^2 t}\,dt\right]} \cdot \frac{1}{1+\sin^2\left(\int_a^x \frac{1}{1+\sin^2 t}\,dt\right)} \cdot \frac{1}{1+\sin^2 x}.$$

Like the simpler differentiations of Chapter 10, these manipulations should become much easier after the practice provided by some of the problems, and, like the problems of Chapter 10, these differentiations are simply a test of your understanding of the Chain Rule, in the somewhat unfamiliar context provided by the Fundamental Theorem of Calculus.

The powerful uses to which the integral will be put in the following chapters all depend on the Fundamental Theorem of Calculus, yet the proof of that theorem was quite easy—it seems that all the real work went into the definition of the integral. Actually, this is not quite true. In order to apply Theorem 1 to a continuous function we need to know that if f is continuous on $[a, b]$, then f is integrable on $[a, b]$. Although we've already offered one proof of this result, there is a more elementary argument that you might prefer. Like most "elementary" arguments, it's quite tricky, but it has the virtue that it will force a review of the proof of Theorem 1.

If f is any bounded function on $[a, b]$, then

$$\sup\{L(f, P)\} \quad \text{and} \quad \inf\{U(f, P)\}$$

will both exist, even if f is not integrable. These numbers are called the **lower integral** of f on $[a, b]$ and the **upper integral** of f on $[a, b]$, respectively, and will be denoted by

$$\mathbf{L}\int_a^b f \quad \text{and} \quad \mathbf{U}\int_a^b f.$$

The lower and upper integrals both have several properties which the integral possesses. In particular, if $a < c < b$, then

$$\mathbf{L}\int_a^b f = \mathbf{L}\int_a^c f + \mathbf{L}\int_c^b f \quad \text{and} \quad \mathbf{U}\int_a^b f = \mathbf{U}\int_a^c f + \mathbf{U}\int_c^b f,$$

and if $m \le f(x) \le M$ for all x in $[a, b]$, then

$$m(b-a) \le \mathbf{L}\int_a^b f \le \mathbf{U}\int_a^b f \le M(b-a).$$

The proofs of these facts are left as an exercise, since they are quite similar to the corresponding proofs for integrals. The results for integrals are actually a corollary of the results for upper and lower integrals, because f is integrable precisely when

$$\mathbf{L}\int_a^b f = \mathbf{U}\int_a^b f.$$

We will prove that a continuous function f is integrable by showing that this equality always holds for continuous functions. It is actually easier to show that

$$\mathbf{L}\int_a^x f = \mathbf{U}\int_a^x f$$

for all x in $[a, b]$; the trick is to note that most of the proof of Theorem 1 didn't even depend on the fact that f was integrable!

THEOREM 13-3 If f is continuous on $[a, b]$, then f is integrable on $[a, b]$.

PROOF Define functions L and U on $[a, b]$ by

$$L(x) = \mathbf{L}\int_a^x f \quad \text{and} \quad U(x) = \mathbf{U}\int_a^x f.$$

Let x be in (a, b). If $h > 0$ and

$$m_h = \inf\{f(t) : x \le t \le x + h\},$$
$$M_h = \sup\{f(t) : x \le t \le x + h\},$$

then

$$m_h \cdot h \le \mathbf{L}\int_x^{x+h} f \le \mathbf{U}\int_x^{x+h} f \le M_h \cdot h,$$

so

$$m_h \cdot h \le L(x+h) - L(x) \le U(x+h) - U(x) \le M_h \cdot h$$

or

$$m_h \le \frac{L(x+h) - L(x)}{h} \le \frac{U(x+h) - U(x)}{h} \le M_h.$$

If $h < 0$ and

$$\begin{aligned} m_h &= \inf\{f(t) : x+h \le t \le x\}, \\ M_h &= \sup\{f(t) : x+h \le t \le x\}, \end{aligned}$$

one obtains the same inequality, precisely as in the proof of Theorem 1.

Since f is continuous at x, we have

$$\lim_{h\to 0} m_h = \lim_{h\to 0} M_h = f(x),$$

and this proves that

$$L'(x) = U'(x) = f(x) \quad \text{for } x \text{ in } (a,b).$$

This means that there is a number c such that

$$U(x) = L(x) + c \quad \text{for all } x \text{ in } [a,b].$$

Since

$$U(a) = L(a) = 0,$$

the number c must equal 0, so

$$U(x) = L(x) \quad \text{for all } x \text{ in } [a,b].$$

In particular,

$$\mathbf{U}\int_a^b f = U(b) = L(b) = \mathbf{L}\int_a^b f,$$

and this means that f is integrable on $[a,b]$. ▮

PROBLEMS

1. Find the derivatives of each of the following functions.

(i) $F(x) = \displaystyle\int_a^{x^3} \sin^3 t\, dt.$

(ii) $F(x) = \displaystyle\int_3^{\left(\int_1^x \sin^3 t\, dt\right)} \frac{1}{1+\sin^6 t + t^2}\, dt$

(iii) $F(x) = \displaystyle\int_{15}^{x} \left(\int_8^y \frac{1}{1+t^2+\sin^2 t}\, dt \right) dy.$

(iv) $F(x) = \displaystyle\int_x^b \frac{1}{1+t^2+\sin^2 t}\, dt.$

(v) $F(x) = \displaystyle\int_a^b \frac{x}{1+t^2+\sin^2 t}\,dt.$

(vi) $F(x) = \sin\left(\displaystyle\int_0^x \sin\left(\int_0^y \sin^3 t\,dt\right) dy\right).$

(vii) F^{-1}, where $F(x) = \displaystyle\int_1^x \frac{1}{t}\,dt.$

(viii) F^{-1}, where $F(x) = \displaystyle\int_0^x \frac{1}{\sqrt{1-t^2}}\,dt.$

(Find $(F^{-1})'(x)$ in terms of $F^{-1}(x)$.)

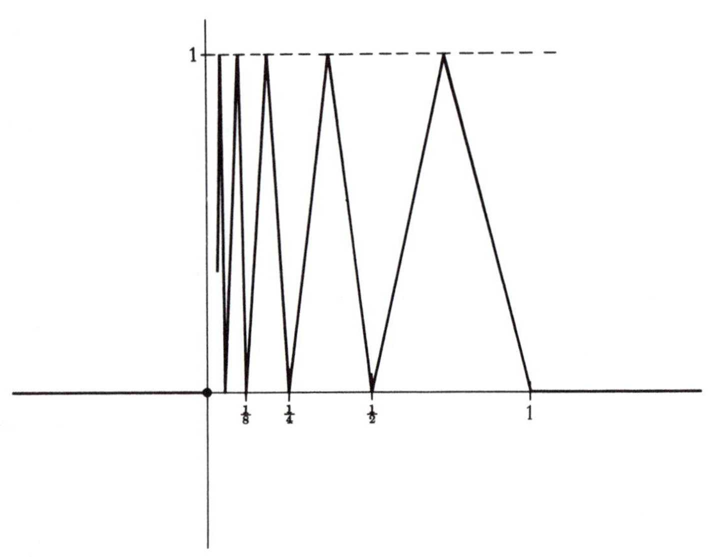

FIGURE 6

2. For each of the following f, if $F(x) = \int_0^x f$, at which points x is $F'(x) = f(x)$? (Caution: it might happen that $F'(x) = f(x)$, even if f is not continuous at x.)

(i) $f(x) = 0$ if $x \le 1$, $f(x) = 1$ if $x > 1$.
(ii) $f(x) = 0$ if $x < 1$, $f(x) = 1$ if $x \ge 1$.
(iii) $f(x) = 0$ if $x \ne 1$, $f(x) = 1$ if $x = 1$.
(iv) $f(x) = 0$ if x is irrational, $f(x) = 1/q$ if $x = p/q$ in lowest terms.
(v) $f(x) = 0$ if $x \le 0$, $f(x) = x$ if $x \ge 0$.
(vi) $f(x) = 0$ if $x \le 0$ or $x > 1$, $f(x) = 1/[1/x]$ if $0 < x \le 1$.
(vii) f is the function shown in Figure 6.
(viii) $f(x) = 1$ if $x = 1/n$ for some n in $\mathbf{N}$, $f(x) = 0$ otherwise.

3. Let f be integrable on $[a, b]$, let c be in (a, b), and let

$$F(x) = \int_a^x f, \quad a \le x \le b.$$

For each of the following statements, give either a proof or a counterexample.

(a) If f is differentiable at c, then F is differentiable at c.

(b) If f is differentiable at c, then F' is continuous at c.
(c) If f' is continuous at c, then F' is continuous at c.

4. Show that the values of the following expressions do not depend on x:

(i) $$\int_0^x \frac{1}{1+t^2}\,dt + \int_0^{1/x} \frac{1}{1+t^2}\,dt.$$

(ii) $$\int_{-\cos x}^{\sin x} \frac{1}{\sqrt{1-t^2}}\,dt, \qquad x \in [0, \pi/2].$$

5. Find $(f^{-1})'(0)$ if

(i) $$f(x) = \int_0^x 1 + \sin(\sin t)\,dt.$$

(ii) $$f(x) = \int_1^x \cos(\cos t)\,dt.$$

(Don't try to evaluate f explicitly.)

6. Find a function g such that

(i) $$\int_0^x tg(t)\,dt = x + x^2.$$

(ii) $$\int_0^{x^2} tg(t)\,dt = x + x^2.$$

(Notice that g is not assumed continuous at 0.)

7. Find all continuous functions f satisfying

$$\int_0^x f = (f(x))^2 + C.$$

for some constant C.

***8.** Suppose that f is a differentiable function with $f(0) = 0$ and $0 < f' \le 1$. Prove that for all $x \ge 0$ we have

$$\int_0^x f^3 \le \left(\int_0^x f\right)^2.$$

***9.** Let

$$f(x) = \begin{cases} \cos\dfrac{1}{x}, & x \neq 0 \\ 0, & x = 0. \end{cases}$$

Is the function $F(x) = \int_0^x f$ differentiable at 0? Hint: Stare at page 177.

10. Use Problem 13-23 to prove that

(i) $$\frac{1}{7\sqrt{2}} \le \int_0^1 \frac{x^6}{\sqrt{1+x^2}}\,dx \le \frac{1}{7}.$$

(ii) $$\frac{3}{8} \le \int_0^{1/2} \sqrt{\frac{1-x}{1+x}}\,dx \le \frac{\sqrt{3}}{4}.$$

11. Find $F'(x)$ if $F(x) = \int_0^x xf(t)\,dt$. (The answer is *not* $xf(x)$; you should perform an obvious manipulation on the integral before trying to find F'.)

12. Prove that if f is continuous, then

$$\int_0^x f(u)(x-u)\,du = \int_0^x \left(\int_0^u f(t)\,dt\right) du.$$

Hint: Differentiate both sides, making use of Problem 11.

***13.** Use Problem 12 to prove that

$$\int_0^x f(u)(x-u)^2\,du = 2\int_0^x \left(\int_0^{u_2} \left(\int_0^{u_1} f(t)\,dt\right) du_1\right) du_2.$$

14. Find a function f such that $f'''(x) = 1 \,/\, \sqrt{1+\sin^2 x}$. (This problem is supposed to be easy; don't misinterpret the word "find.")

***15.** A function f is **periodic**, with **period** a, if $f(x+a) = f(x)$ for all x.

(a) If f is periodic with period a and integrable on $[0, a]$, show that

$$\int_0^a f = \int_b^{b+a} f \quad \text{for all } b.$$

(b) Find a function f such that f is not periodic, but f' is. Hint: Choose a periodic g for which it can be guaranteed that $f(x) = \int_0^x g$ is not periodic.

(c) Suppose that f' is periodic with period a. Prove that f is periodic if and only if $f(a) = f(0)$.

16. Find $\int_0^b \sqrt[n]{x}\,dx$, by simply guessing a function f with $f'(x) = \sqrt[n]{x}$, and using the Second Fundamental Theorem of Calculus. Then check with Problem 13-21.

***17.** Use the Fundamental Theorem of Calculus and Problem 13-21 to derive the result stated in Problem 12-18.

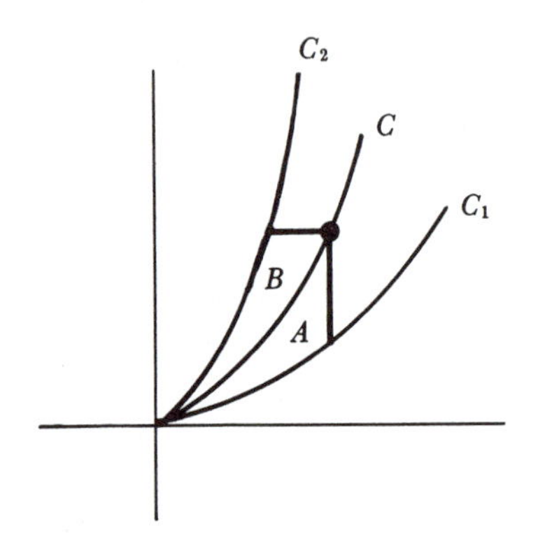

FIGURE 7

18. Let C_1, C and C_2 be curves passing through the origin, as shown in Figure 7. Each point on C can be joined to a point of C_1 with a vertical line segment and to a point of C_2 with a horizontal line segment. We will say that C *bisects* C_1 and C_2 if the regions A and B have equal areas for every point on C.

(a) If C_1 is the graph of $f(x) = x^2$, $x \geq 0$ and C is the graph of $f(x) = 2x^2$, $x \geq 0$, find C_2 so that C bisects C_1 and C_2.

(b) More generally, find C_2 if C_1 is the graph of $f(x) = x^m$, and C is the graph of $f(x) = cx^m$ for some $c > 1$.

19. (a) Find the derivatives of $F(x) = \int_1^x 1/t\,dt$ and $G(x) = \int_b^{bx} 1/t\,dt$.
(b) Now give a new proof for Problem 13-15.

***20.** Use the Fundamental Theorem of Calculus and Darboux's Theorem (Problem 11-54) to give another proof of the Intermediate Value Theorem.

21. Prove that if h is continuous, f and g are differentiable, and

$$F(x) = \int_{f(x)}^{g(x)} h(t)\,dt,$$

then $F'(x) = h(g(x)) \cdot g'(x) - h(f(x)) \cdot f'(x)$. Hint: Try to reduce this to the two cases you can already handle, with a constant either as the lower or the upper limit of integration.

22. Suppose that f' is integrable on $[0, 1]$ and $f(0) = 0$. Prove that for all x in $[0, 1]$ we have

$$|f(x)| \le \sqrt{\int_0^1 |f'|^2}.$$

Show also that the hypothesis $f(0) = 0$ is needed. Hint: Problem 13-39.

***23.** (a) Suppose $G' = g$ and $F' = f$. Prove that if the function y satisfies the differential equation

$$(*) \qquad g(y(x)) \cdot y'(x) = f(x) \quad \text{for all } x \text{ in some interval,}$$

then there is a number c such that

$$(**) \qquad G(y(x)) = F(x) + c \quad \text{for all } x \text{ in this interval.}$$

(b) Show, conversely, that if y satisfies $(**)$, then y is a solution of $(*)$.

(c) Find what condition y must satisfy if

$$y'(x) = \frac{1 + x^2}{1 + y(x)}.$$

(In this case $g(t) = 1 + t$ and $f(t) = 1 + t^2$.) Then "solve" the resulting equations to find all possible solutions y (no solution will have $\mathbf{R}$ as its domain).

(d) Find what condition y must satisfy if

$$y'(x) = \frac{-1}{1 + 5[y(x)]^4}.$$

(An appeal to Problem 12-11 will show that there *are* functions satisfying the resulting equation.)

(e) Find all functions y satisfying

$$y(x)y'(x) = -x.$$

Find the solution y satisfying $y(0) = -1$.

24. In Problem 10-17 we found that the Schwarzian derivative

$$\frac{f'''(x)}{f'(x)} - \frac{3}{2}\left(\frac{f''(x)}{f'(x)}\right)^2$$

was 0 for $f(x) = (ax + b)/(cx + d)$. Now suppose that f is any function whose Schwarzian derivative is 0.

(a) f''^2/f'^3 is a constant function.
(b) f is the form $f(x) = (ax+b)/(cx+d)$. Hint: Consider $u = f'$ and apply the previous problem.

***25.** The limit $\lim_{N\to\infty} \int_a^N f$, if it exists, is denoted by $\int_a^\infty f$ (or $\int_a^\infty f(x)\,dx$), and called an "improper integral."

(a) Determine $\int_1^\infty x^r\,dx$, if $r < -1$.
(b) Use Problem 13-15 to show that $\int_1^\infty 1/x\,dx$ does not exist. Hint: What can you say about $\int_1^{2^n} 1/x\,dx$?
(c) Suppose that $f(x) \ge 0$ for $x \ge 0$ and that $\int_0^\infty f$ exists. Prove that if $0 \le g(x) \le f(x)$ for all $x \ge 0$, and g is integrable on each interval $[0, N]$, then $\int_0^\infty g$ also exists.
(d) Explain why $\int_0^\infty 1/(1+x^2)\,dx$ exists. Hint: Split this integral up at 1.

26. Decide whether or not the following improper integrals exist.

(i) $$\int_0^\infty \frac{1}{\sqrt{1+x^3}}\,dx.$$
(ii) $$\int_0^\infty \frac{x}{1+x^{3/2}}\,dx.$$
(iii) $$\int_0^\infty \frac{1}{x\sqrt{1+x}}\,dx.$$

***27.** The improper integral $\int_{-\infty}^a f$ is defined in the obvious way, as $\lim_{N\to-\infty} \int_N^a f$. But another kind of improper integral $\int_{-\infty}^\infty f$ is defined in a nonobvious way: it is $\int_0^\infty f + \int_{-\infty}^0 f$, provided these improper integrals both exist.

(a) Explain why $\int_{-\infty}^\infty 1/(1+x^2)\,dx$ exists.
(b) Explain why $\int_{-\infty}^\infty x\,dx$ does not exist. (But notice that $\lim_{N\to\infty} \int_{-N}^N x\,dx$ *does* exist.)
(c) Prove that if $\int_{-\infty}^\infty f$ exists, then $\lim_{N\to\infty} \int_{-N}^N f$ exists and equals $\int_{-\infty}^\infty f$. Show moreover, that $\lim_{N\to\infty} \int_{-N}^{N+1} f$ and $\lim_{N\to\infty} \int_{-N^2}^N f$ both exist and equal $\int_{-\infty}^\infty f$. Can you state a reasonable generalization of these facts? (If you can't, you will have a miserable time trying to do these special cases!)

***28.** There is another kind of "improper integral" in which the interval is bounded, but the *function* is unbounded:

(a) If $a > 0$, find $\lim_{\varepsilon\to 0^+} \int_\varepsilon^a 1/\sqrt{x}\,dx$. This limit is denoted by $\int_0^a 1/\sqrt{x}\,dx$, even though the function $f(x) = 1/\sqrt{x}$ is not bounded on $[0, a]$, no matter how we define $f(0)$.
(b) Find $\int_0^a x^r\,dx$ if $-1 < r < 0$.
(c) Use Problem 13-15 to show that $\int_0^a x^{-1}\,dx$ does not make sense, even as a limit.

(d) Invent a reasonable definition of $\int_a^0 |x|^r\,dx$ for $a < 0$ and compute it for $-1 < r < 0$.

(e) Invent a reasonable definition of $\int_{-1}^1 (1-x^2)^{-1/2}\,dx$, as a sum of two limits, and show that the limits exist. Hint: Why does $\int_{-1}^0 (1+x)^{-1/2}\,dx$ exist? How does $(1+x)^{-1/2}$ compare with $(1-x^2)^{-1/2}$ for $-1 < x < 0$?

29. (a) If f is continuous on $[0, 1]$, compute $\displaystyle\lim_{x\to 0^+} x\int_x^1 \frac{f(t)}{t}\,dt$.

(b) If f is integrable on $[0, 1]$ and $\displaystyle\lim_{x\to 0} f(x)$ exists, compute $\displaystyle\lim_{x\to 0^+} x\int_x^1 \frac{f(t)}{t^2}\,dt$.

***30.** It is possible, finally, to combine the two possible extensions of the notion of the integral.

(a) If $f(x) = 1/\sqrt{x}$ for $0 \le x \le 1$ and $f(x) = 1/x^2$ for $x \ge 1$, find $\displaystyle\int_0^\infty f(x)\,dx$ (after deciding what this should mean).

(b) Show that $\displaystyle\int_0^\infty x^r\,dx$ never makes sense. (Distinguish the cases $-1 < r < 0$ and $r < -1$. In one case things go wrong at 0, in the other case at ∞; for $r = -1$ things go wrong at both places.)

CHAPTER 15 THE TRIGONOMETRIC FUNCTIONS

The definitions of the functions sin and cos are considerably more subtle than one might suspect. For this reason, this chapter begins with some informal and intuitive definitions, which should not be scrutinized too carefully, as they shall soon be replaced by the formal definitions which we really intend to use.

In elementary geometry an angle is simply the union of two half-lines with a common initial point (Figure 1).

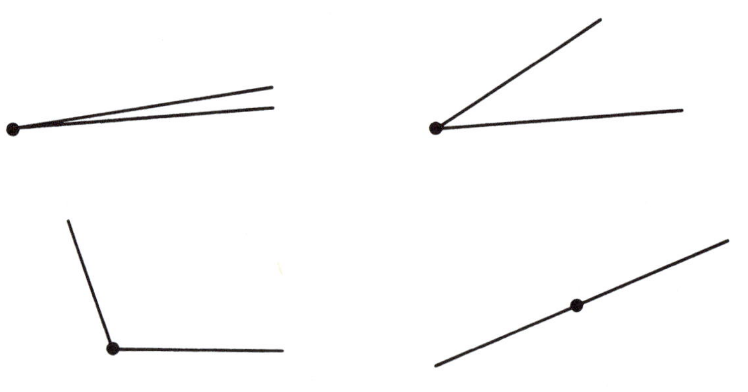

FIGURE 1

More useful for trigonometry are "directed angles," which may be regarded as pairs (l_1, l_2) of half-lines with the same initial point, visualized as in Figure 2.

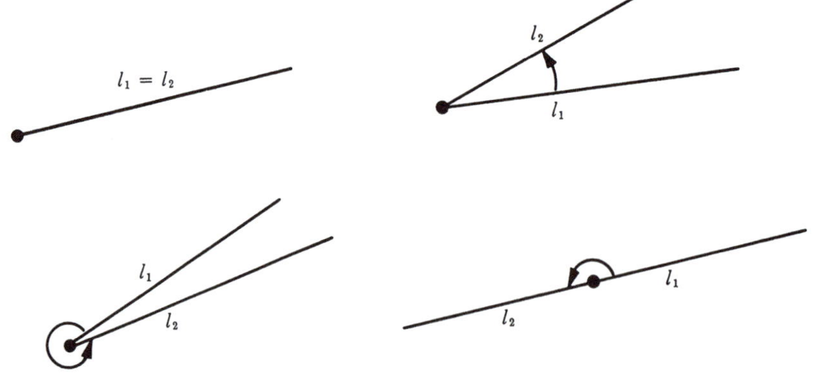

FIGURE 2

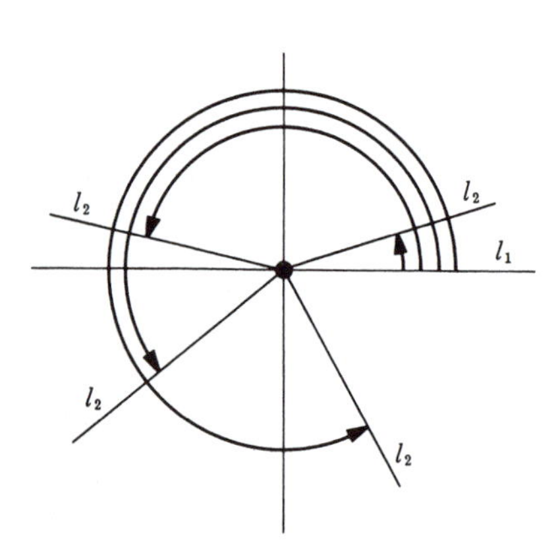

FIGURE 3

If for l_1 we always choose the positive half of the horizontal axis, a directed angle is described completely by the second half-line (Figure 3).

Since each half-line intersects the unit circle precisely once, a directed angle is described, even more simply, by a point on the unit circle (Figure 4), that is, by a point (x, y) with $x^2 + y^2 = 1$.

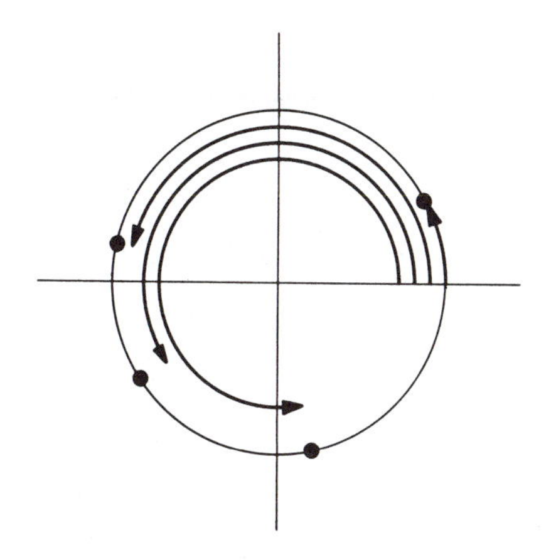

FIGURE 4

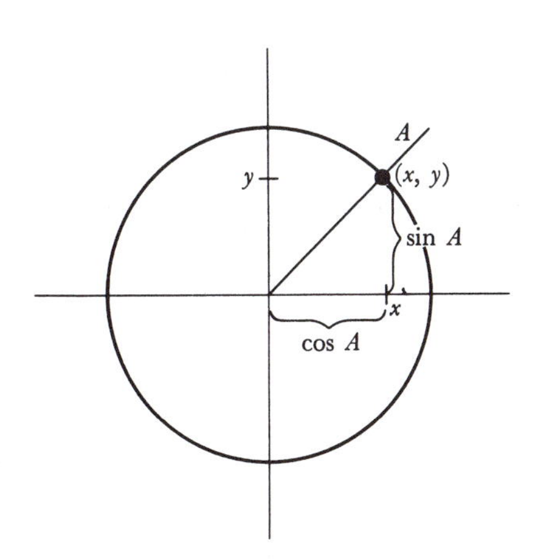

FIGURE 5

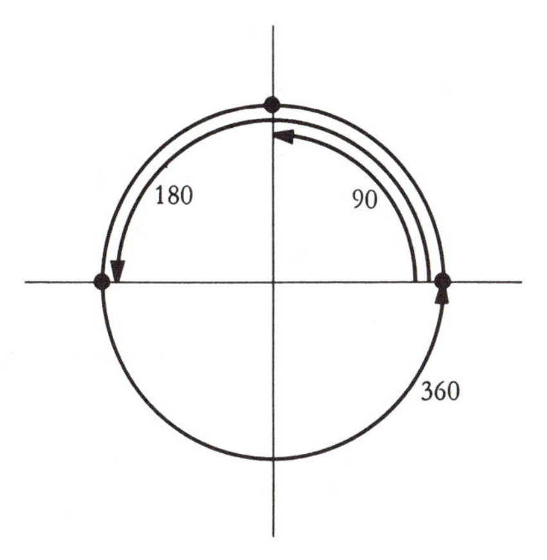

FIGURE 6

The sine and cosine of a directed angle can now be defined as follows (Figure 5): a directed angle is determined by a point (x, y) with $x^2 + y^2 = 1$; the sine of the angle is defined as y, and the cosine as x.

Despite the aura of precision surrounding the previous paragraph, we are not yet finished with the definitions of sin and cos. Indeed, we have barely begun. What we have defined is the sine and cosine of a directed angle; what we *want* to define is $\sin x$ and $\cos x$ for each *number* x. The usual procedure for doing this depends on associating an angle to every number. The oldest method is to "measure angles in degrees." An angle "all the way around" is associated to 360, an angle "half-way around" is associated to 180, an angle "a quarter way around" to 90, etc. (Figure 6). The angle associated, in this manner, to the number x, is called "the angle of x degrees." The angle of 0 degrees is the same as the angle of 360 degrees, and this ambiguity is purposely extended further, so that an angle of 90 degrees is also an angle of $360 + 90$ degrees, etc. One can now define a function, which we will denote by $\sin^\circ$, as follows:

$$\sin^\circ(x) = \text{sine of the angle of } x \text{ degrees.}$$

There are two difficulties with this approach. Although it may be clear what we mean by an angle of 90 or 45 degrees, it is not quite clear what an angle of $\sqrt{2}$ degrees is, for example. Even if this difficulty could be circumvented, it is unlikely that this system, depending as it does on the arbitrary choice of 360, will lead to elegant results—it would be sheer luck if the function $\sin^\circ$ had mathematically pleasing properties.

"Radian measure" appears to offer a remedy for both these defects. Given any number x, choose a point P on the unit circle such that x is the length of the arc of the circle beginning at $(1, 0)$ and running counterclockwise to P (Figure 7). The directed angle determined by P is called "the angle of x radians." Since the length of the whole circle is 2π, the angle of x radians and the angle of $2\pi + x$ radians are identical. A function $\sin^r$ can now be defined as follows:

$$\sin^r(x) = \text{sine of the angle of } x \text{ radians.}$$

This same method can easily be adopted to define $\sin^\circ$; since we want to have $\sin^\circ 360 = \sin^r 2\pi$, we can define

$$\sin^\circ x = \sin^r \frac{2\pi x}{360} = \sin^r \frac{\pi x}{180}.$$

We shall soon drop the superscript r in $\sin^r$, since $\sin^r$ (and not $\sin^\circ$) is the only function which will interest us; before we do, a few words of warning are advisable.

The expressions $\sin^\circ x$ and $\sin^r x$ are sometimes written

$$\sin x^\circ$$
$$\sin x \text{ radians,}$$

but this notation is quite misleading; a number x is simply a number—it does not carry a banner indicating that it is "in degrees" or "in radians." If the meaning

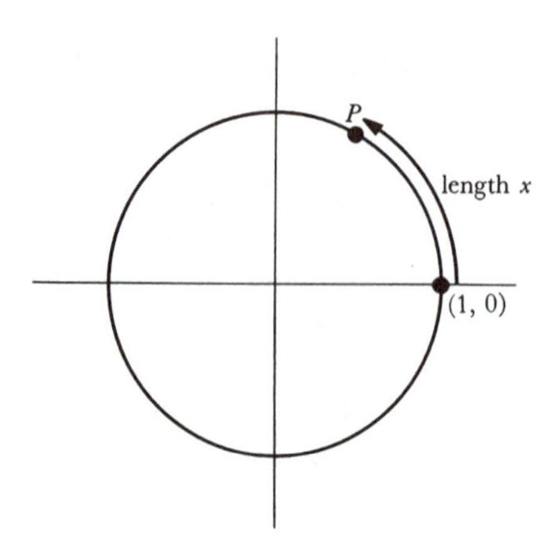

FIGURE 7

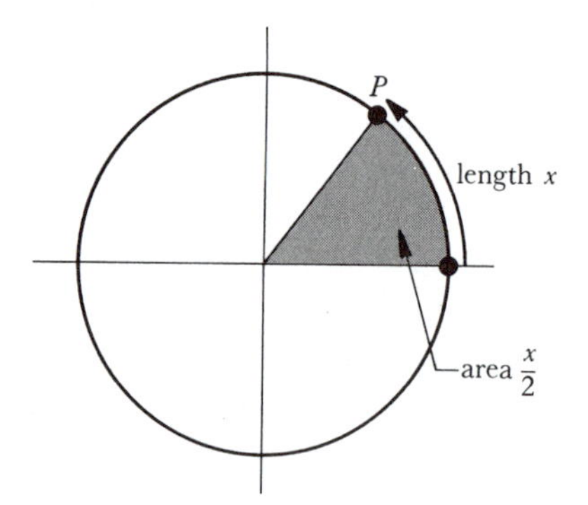

FIGURE 8

of the notation "$\sin x$" is in doubt one usually asks:

"Is x in degrees or radians?"

but what one means is:

"Do you mean '$\sin^{\circ}$' or '$\sin^{r}$'?"

Even for mathematicians, addicted to precision, these remarks might be dispensable, were it not for the fact that failure to take them into account will lead to incorrect answers to certain problems (an example is given in Problem 19).

Although the function $\sin^{r}$ is the function which we wish to denote simply by sin (and use exclusively henceforth), there is a difficulty involved even in the definition of $\sin^{r}$. Our proposed definition depends on the concept of the length of a curve. Although the length of a curve has been defined in several problems, it is also easy to reformulate the definition in terms of areas. (A treatment in terms of length is outlined in Problem 28.)

Suppose that x is the length of the arc of the unit circle from $(1, 0)$ to P; this arc thus contains $x/2\pi$ of the total length 2π of the circumference of the unit circle. Let S denote the "sector" shown in Figure 8; S is bounded by the unit circle, the horizontal axis, and the half-line through $(0, 0)$ and P. The area of S should be $x/2\pi$ times the area inside the unit circle, which we expect to be π; thus S should have area

$$\frac{x}{2\pi} \cdot \pi = \frac{x}{2}.$$

We can therefore define $\cos x$ and $\sin x$ as the coordinates of the point P which determines a sector of area $x/2$.

With these remarks as background, the rigorous definition of the functions sin and cos now begins. The first definition identifies π as the area of the unit circle—more precisely, as twice the area of a semicircle (Figure 9).

DEFINITION

$$\pi = 2 \cdot \int_{-1}^{1} \sqrt{1 - x^2}\, dx.$$

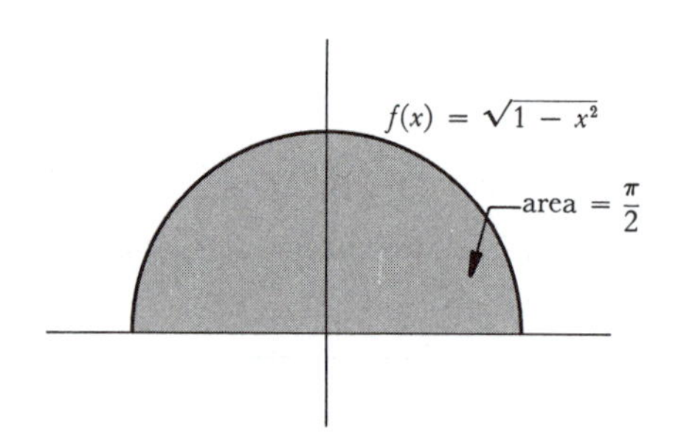

FIGURE 9

(This definition is not offered simply as an embellishment; to define the trigonometric functions it will be necessary to first define $\sin x$ and $\cos x$ only for $0 \le x \le \pi$.)

The second definition is meant to describe, for $-1 \le x \le 1$, the area $A(x)$ of the sector bounded by the unit circle, the horizontal axis, and the half-line through $(x, \sqrt{1 - x^2})$. If $0 \le x \le 1$, this area can be expressed (Figure 10) as the sum of the area of a triangle and the area of a region under the unit circle:

$$\frac{x\sqrt{1 - x^2}}{2} + \int_{x}^{1} \sqrt{1 - t^2}\, dt.$$

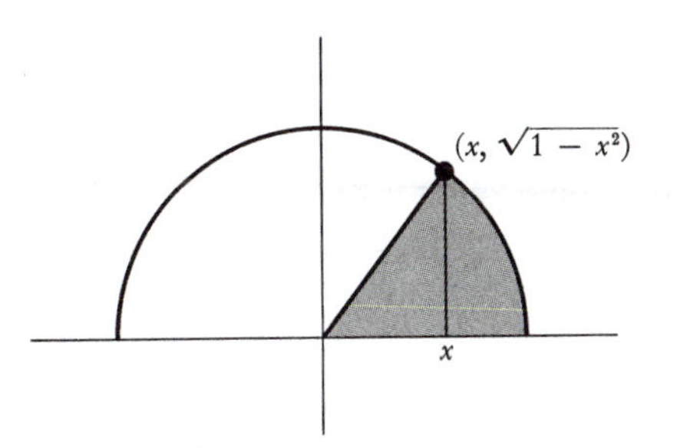

FIGURE 10

This same formula happens to work for $-1 \le x \le 0$ also. In this case (Figure 11), the term

$$\frac{x\sqrt{1-x^2}}{2}$$

is negative, and represents the area of the triangle which must be subtracted from the term

$$\int_x^1 \sqrt{1-t^2}\,dt.$$

DEFINITION

> If $-1 \le x \le 1$, then
>
> $$\boldsymbol{A(x)} = \frac{x\sqrt{1-x^2}}{2} + \int_x^1 \sqrt{1-t^2}\,dt.$$

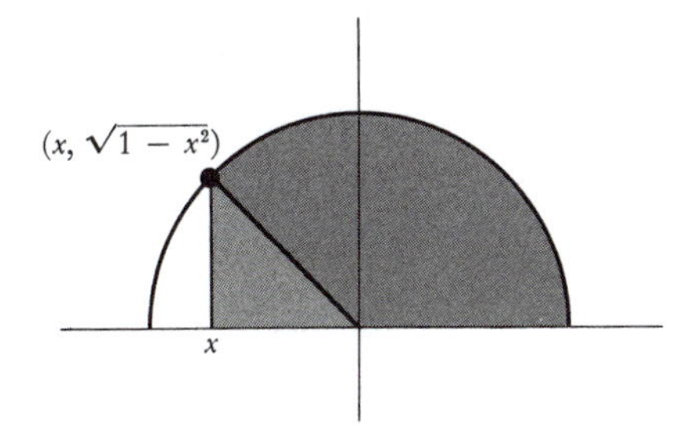

FIGURE 11

Notice that if $-1 < x < 1$, then A is differentiable at x and (using the Fundamental Theorem of Calculus),

$$\begin{aligned}
A'(x) &= \frac{1}{2}\left[x \cdot \frac{-2x}{2\sqrt{1-x^2}} + \sqrt{1-x^2}\right] - \sqrt{1-x^2} \\
&= \frac{1}{2}\left[\frac{-x^2 + (1-x^2)}{\sqrt{1-x^2}}\right] - \sqrt{1-x^2} \\
&= \frac{1-2x^2}{2\sqrt{1-x^2}} - \sqrt{1-x^2} \\
&= \frac{1-2x^2-2(1-x^2)}{2\sqrt{1-x^2}} \\
&= \frac{-1}{2\sqrt{1-x^2}}.
\end{aligned}$$

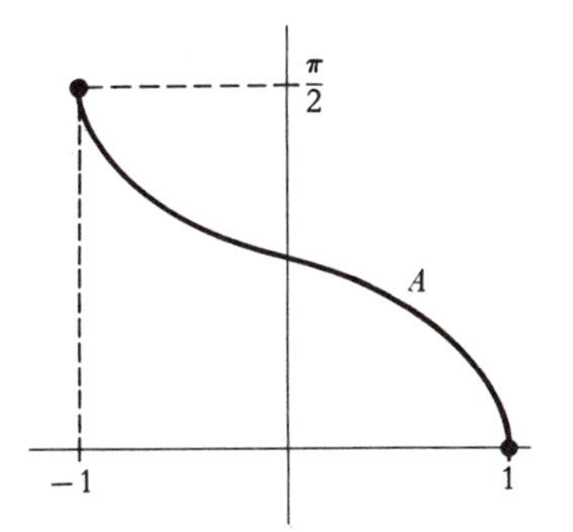

FIGURE 12

Notice also (Figure 12) that on the interval $[-1, 1]$ the function A decreases from

$$A(-1) = 0 + \int_{-1}^1 \sqrt{1-t^2}\,dt = \frac{\pi}{2}$$

to $A(1) = 0$. This follows directly from the definition of A, and also from the fact that its derivative is negative on $(-1, 1)$.

For $0 \le x \le \pi$ we wish to define $\cos x$ and $\sin x$ as the coordinates of a point $P = (\cos x, \sin x)$ on the unit circle which determines a sector whose area is $x/2$ (Figure 13). In other words:

DEFINITION

> If $0 \le x \le \pi$, then **cos *x*** is the unique number in $[-1, 1]$ such that
>
> $$A(\cos x) = \frac{x}{2};$$
>
> and
>
> $$\boldsymbol{\sin x} = \sqrt{1-(\cos x)^2}.$$

This definition actually requires a few words of justification. In order to know that there *is* a number y satisfying $A(y) = x/2$, we use the fact that A is continuous, and that A takes on the values 0 and $\pi/2$. This tacit appeal to the Intermediate Value Theorem is crucial, if we want to make our preliminary definition precise. Having made, and justified, our definition, we can now proceed quite rapidly.

THEOREM 1 If $0 < x < \pi$, then

$$\cos'(x) = -\sin x,$$
$$\sin'(x) = \cos x.$$

PROOF If $B = 2A$, then the definition $A(\cos x) = x/2$ can be written

$$B(\cos x) = x;$$

in other words, cos is just the inverse of B. We have already computed that

$$A'(x) = -\frac{1}{2\sqrt{1-x^2}},$$

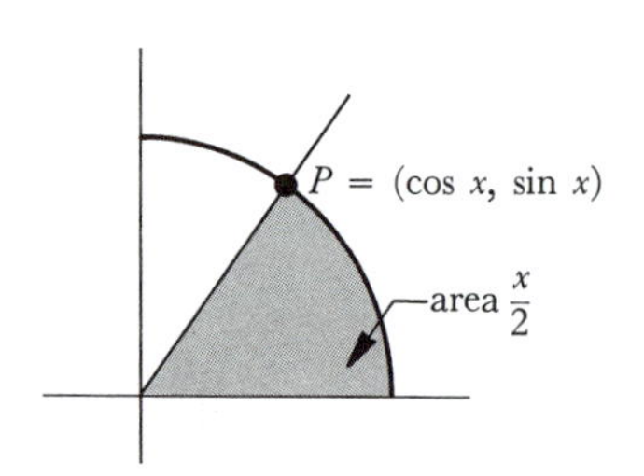

FIGURE 13

from which we conclude that

$$B'(x) = -\frac{1}{\sqrt{1-x^2}}.$$

Consequently,

$$\begin{aligned}\cos'(x) &= (B^{-1})'(x)\\ &= \frac{1}{B'(B^{-1}(x))}\\ &= \frac{1}{-\dfrac{1}{\sqrt{1-[B^{-1}(x)]^2}}}\\ &= -\sqrt{1-(\cos x)^2}\\ &= -\sin x.\end{aligned}$$

Since

$$\sin x = \sqrt{1-(\cos x)^2},$$

we also obtain

$$\begin{aligned}\sin'(x) &= \frac{1}{2}\cdot\frac{-2\cos x\cdot\cos'(x)}{\sqrt{1-(\cos x)^2}}\\ &= \frac{\cos x\sin x}{\sin x}\\ &= \cos x. \blacksquare\end{aligned}$$

The information contained in Theorem 1 can be used to sketch the graphs of

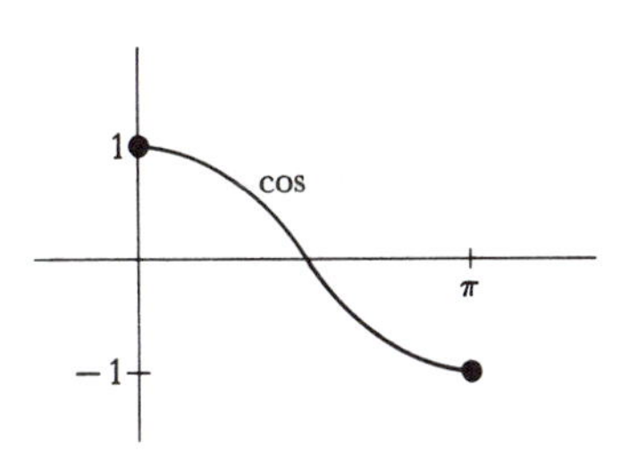

FIGURE 14

sin and cos on the interval $[0, \pi]$. Since

$$\cos'(x) = -\sin x < 0, \quad 0 < x < \pi,$$

the function cos decreases from $\cos 0 = 1$ to $\cos \pi = -1$ (Figure 14). Consequently, $\cos y = 0$ for a unique y in $[0, \pi]$. To find y, we note that the definition of cos,

$$A(\cos x) = \frac{x}{2},$$

means that

$$A(0) = \frac{y}{2},$$

so

$$y = 2\int_0^1 \sqrt{1-t^2}\,dt.$$

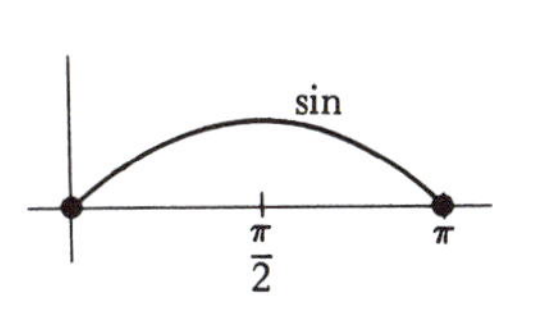

FIGURE 15

It is easy to see that

$$\int_{-1}^{0} \sqrt{1-t^2}\,dt = \int_0^1 \sqrt{1-t^2}\,dt$$

so we can also write

$$y = \int_{-1}^{1} \sqrt{1-t^2}\,dt = \frac{\pi}{2}.$$

Now we have

$$\sin'(x) = \cos x \begin{cases} > 0, & 0 < x < \pi/2 \\ < 0, & \pi/2 < x < \pi, \end{cases}$$

so sin increases on $[0, \pi/2]$ from $\sin 0 = 0$ to $\sin \pi/2 = 1$, and then decreases on $[\pi/2, \pi]$ to $\sin \pi = 0$ (Figure 15).

The values of $\sin x$ and $\cos x$ for x not in $[0, \pi]$ are most easily defined by a two-step piecing together process:

(1) If $\pi \le x \le 2\pi$, then

$$\sin x = -\sin(2\pi - x),$$
$$\cos x = \cos(2\pi - x).$$

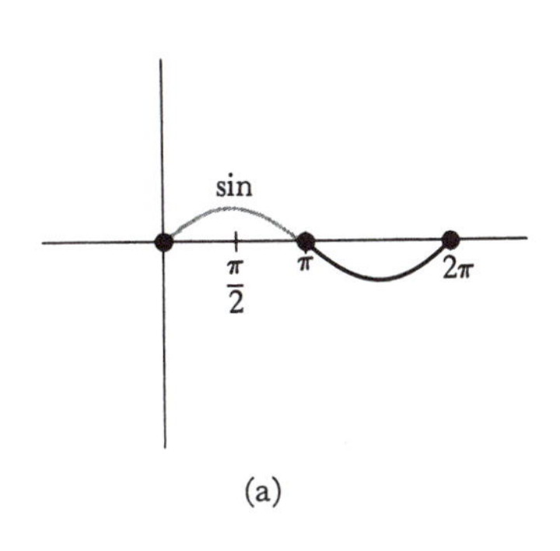

(a)

Figure 16 shows the graphs of sin and cos on $[0, 2\pi]$.

(2) If $x = 2\pi k + x'$ for some integer k, and some x' in $[0, 2\pi]$, then

$$\sin x = \sin x',$$
$$\cos x = \cos x'.$$

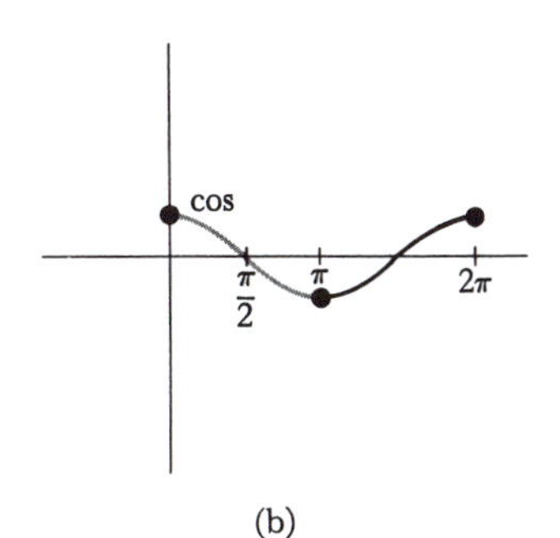

(b)

FIGURE 16

Figure 17 shows the graphs of sin and cos, now defined on all of **R**.

Having extended the functions sin and cos to **R**, we must now check that the basic properties of these functions continue to hold. In most cases this is easy. For example, it is clear that the equation

$$\sin^2 x + \cos^2 x = 1$$

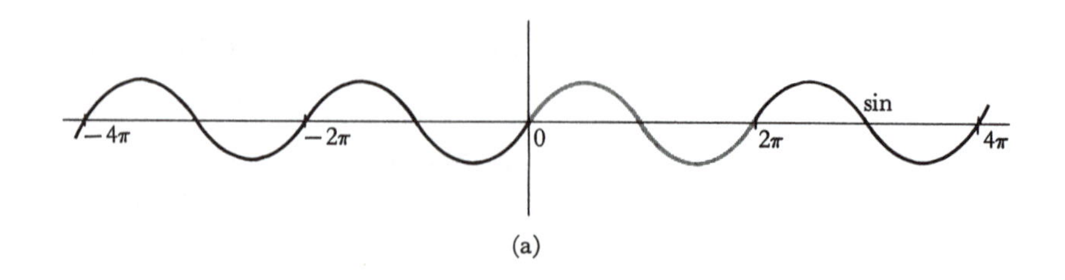

(a)

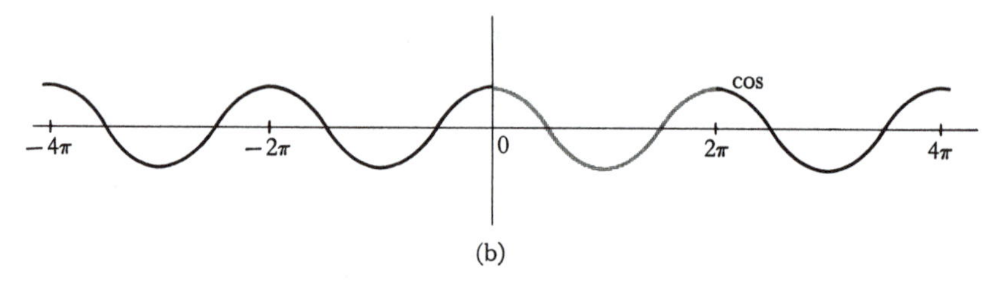

(b)

FIGURE 17

holds for all x. It is also not hard to prove that

$$\begin{aligned} \sin'(x) &= \cos x, \\ \cos'(x) &= -\sin x, \end{aligned}$$

if x is not a multiple of π. For example, if $\pi < x < 2\pi$, then

$$\sin x = -\sin(2\pi - x),$$

so

$$\begin{aligned} \sin'(x) &= -\sin'(2\pi - x) \cdot (-1) \\ &= \cos(2\pi - x) \\ &= \cos x. \end{aligned}$$

If x is a multiple of π we resort to a trick; it is only necessary to apply Theorem 11-7 to conclude that the same formulas are true in this case also.

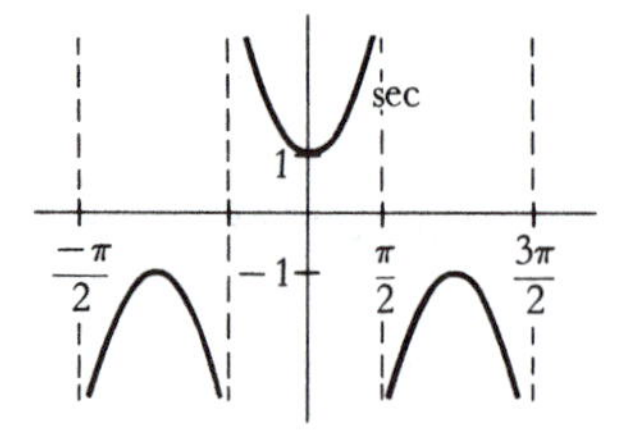

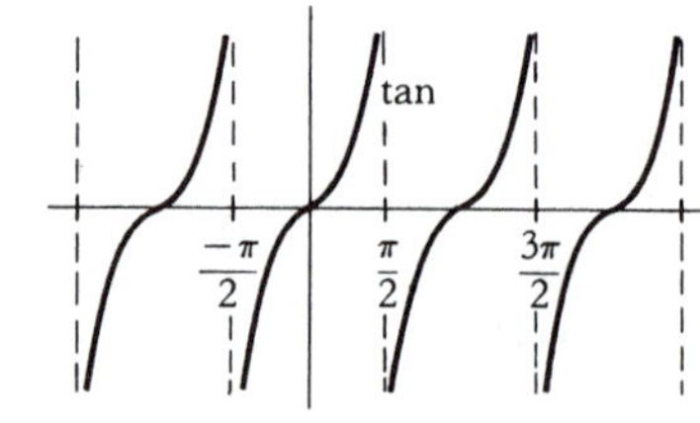

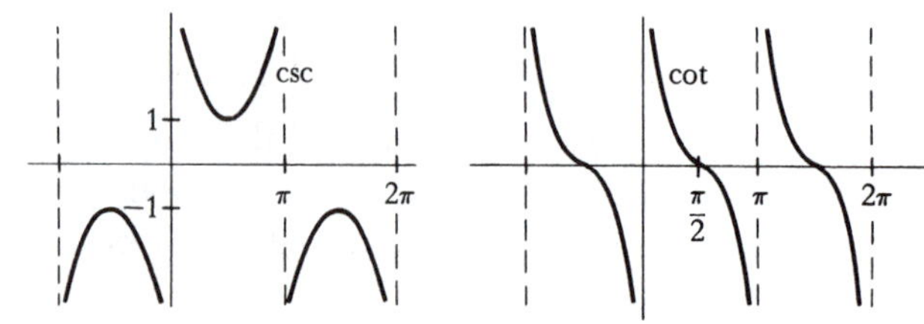

FIGURE 18

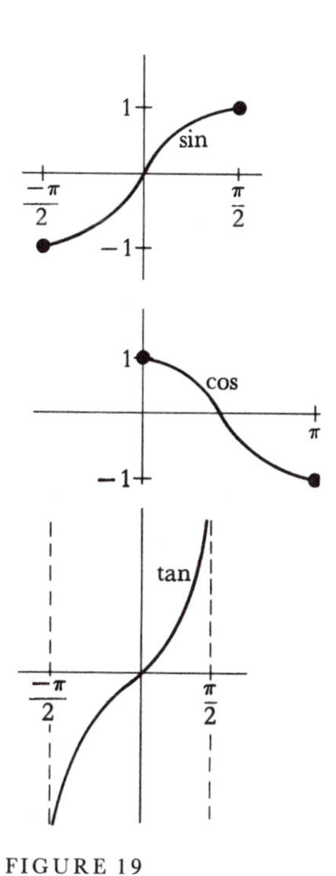

FIGURE 19

The other standard trigonometric functions present no difficulty at all. We define

$$\left.\begin{aligned}\sec x &= \frac{1}{\cos x}\\ \tan x &= \frac{\sin x}{\cos x}\end{aligned}\right\}\quad x \neq k\pi + \pi/2,$$

$$\left.\begin{aligned}\csc x &= \frac{1}{\sin x}\\ \cot x &= \frac{\cos x}{\sin x}\end{aligned}\right\}\quad x \neq k\pi.$$

The graphs are sketched in Figure 18. It is a good idea to convince yourself that the general features of these graphs can be predicted from the derivatives of these functions, which are listed in the next theorem (there is no need to memorize the statement of the theorem, since the results can be rederived whenever needed.)

THEOREM 2 If $x \neq k\pi + \pi/2$, then

$$\begin{aligned}\sec'(x) &= \sec x \tan x,\\ \tan'(x) &= \sec^2 x.\end{aligned}$$

If $x \neq k\pi$, then

$$\begin{aligned}\csc'(x) &= -\csc x \cot x,\\ \cot'(x) &= -\csc^2 x.\end{aligned}$$

PROOF Left to you (a straightforward computation). ▌

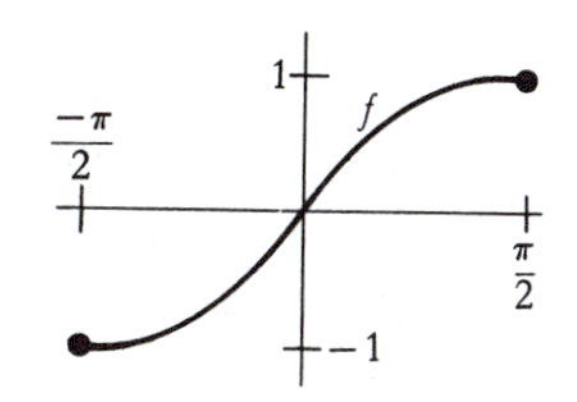

(a)

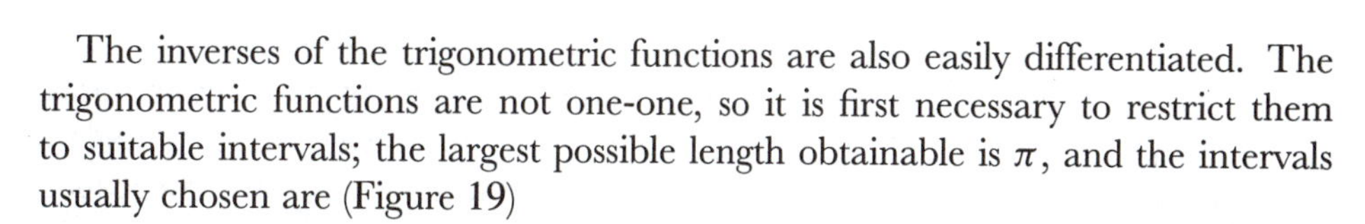

The inverses of the trigonometric functions are also easily differentiated. The trigonometric functions are not one-one, so it is first necessary to restrict them to suitable intervals; the largest possible length obtainable is π, and the intervals usually chosen are (Figure 19)

$$\begin{aligned}&[-\pi/2, \pi/2] &&\text{for sin,}\\ &[0, \pi] &&\text{for cos,}\\ &(-\pi/2, \pi/2) &&\text{for tan.}\end{aligned}$$

(The inverses of the other trigonometric functions are so rarely used that they will not even be discussed here.)

The inverse of the function

$$f(x) = \sin x, \qquad -\pi/2 \leq x \leq \pi/2$$

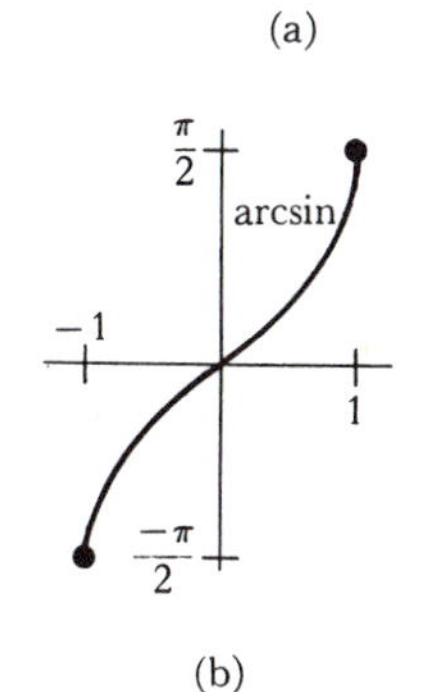

(b)

FIGURE 20

is denoted by **arcsin** (Figure 20); the domain of arcsin is $[-1, 1]$. The notation $\sin^{-1}$ has been avoided because arcsin is not the inverse of sin (which is not one-one), but of the restricted function f; sometimes this function f is denoted by Sin, and arcsin by Sin^{-1}.

The inverse of the function

$$g(x) = \cos x, \qquad 0 \le x \le \pi$$

is denoted by **arccos** (Figure 21); the domain of arccos is $[-1, 1]$. Sometimes g is denoted by Cos, and arccos by Cos^{-1}.

The inverse of the function

$$h(x) = \tan x, \qquad -\pi/2 < x < \pi/2$$

is denoted by **arctan** (Figure 22); arctan is one of the simplest examples of a differentiable function which is bounded even though it is one-one on all of **R**. Sometimes the function h is denoted by Tan, and arctan by Tan^{-1}.

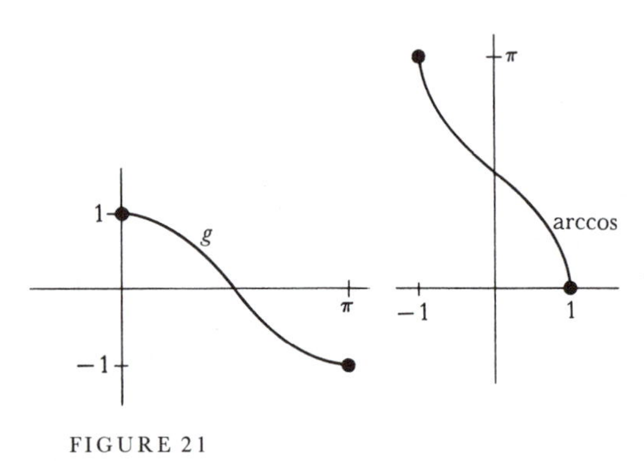

FIGURE 21

The derivatives of the inverse trigonometric functions are surprisingly simple, and do not involve trigonometric functions at all. Finding the derivatives is a simple matter, but to express them in a suitable form we will have to simplify expressions like

$$\cos(\arcsin x), \qquad \sec(\arctan x).$$

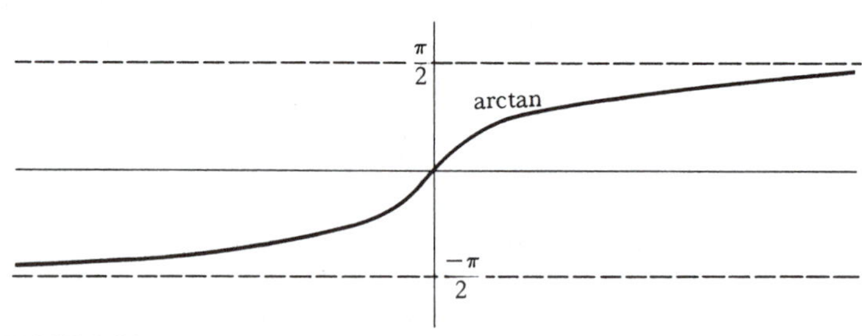

FIGURE 22

A little picture is the best way to remember the correct simplifications. For example, Figure 23 shows a directed angle whose sine is x—the angle shown is thus an angle of $(\arcsin x)$ radians; consequently $\cos(\arcsin x)$ is the length of the other side, namely, $\sqrt{1 - x^2}$. However, in the proof of the next theorem we will not resort to such pictures.

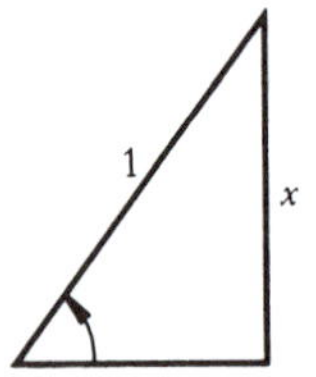

FIGURE 23

THEOREM 3 If $-1 < x < 1$, then

$$\arcsin'(x) = \frac{1}{\sqrt{1 - x^2}},$$

$$\arccos'(x) = \frac{-1}{\sqrt{1 - x^2}}.$$

Moreover, for all x we have

$$\arctan'(x) = \frac{1}{1 + x^2}.$$

PROOF

$$\begin{aligned}\arcsin'(x) &= (f^{-1})'(x)\\ &= \frac{1}{f'(f^{-1}(x))}\\ &= \frac{1}{\sin'(\arcsin x)}\\ &= \frac{1}{\cos(\arcsin x)}.\end{aligned}$$

Now

$$[\sin(\arcsin x)]^2 + [\cos(\arcsin x)]^2 = 1,$$

that is,

$$x^2 + [\cos(\arcsin x)]^2 = 1;$$

therefore,

$$\cos(\arcsin x) = \sqrt{1 - x^2}.$$

(The positive square root is to be taken because $\arcsin x$ is in $(-\pi/2, \pi/2)$, so $\cos(\arcsin x) > 0$.) This proves the first formula.

The second formula has already been established (in the proof of Theorem 1). It is also possible to imitate the proof for the first formula, a valuable exercise if that proof presented any difficulties. The third formula is proved as follows.

$$\begin{aligned}\arctan'(x) &= (h^{-1})'(x)\\ &= \frac{1}{h'(h^{-1}(x))}\\ &= \frac{1}{\tan'(\arctan x)}\\ &= \frac{1}{\sec^2(\arctan x)}\end{aligned}$$

Dividing both sides of the identity

$$\sin^2 a + \cos^2 a = 1$$

by $\cos^2 a$ yields

$$\tan^2 a + 1 = \sec^2 a.$$

It follows that

$$[\tan(\arctan x)]^2 + 1 = \sec^2(\arctan x),$$

or

$$x^2 + 1 = \sec^2(\arctan x),$$

which proves the third formula. ▮

The traditional proof of the formula $\sin'(x) = \cos x$ (quite different from the one given here) is outlined in Problem 27. This proof depends upon first establishing the limit

$$\lim_{h\to 0} \frac{\sin h}{h} = 1,$$

and the "addition formula"

$$\sin(x+y) = \sin x \cos y + \cos x \sin y.$$

Both of these formulas can be derived easily now that the derivative of sin and cos are known. The first is just the special case $\sin'(0) = \cos 0$. The second depends on a beautiful characterization of the functions sin and cos. In order to derive this result we need a lemma whose proof involves a clever trick; a more straightforward proof will be supplied in Part IV.

LEMMA Suppose f has a second derivative everywhere and that

$$\begin{aligned} f'' + f &= 0, \\ f(0) &= 0, \\ f'(0) &= 0. \end{aligned}$$

Then $f = 0$.

PROOF Multiplying both sides of the first equation by f' yields

$$f'f'' + ff' = 0.$$

Thus

$$[(f')^2 + f^2]' = 2(f'f'' + ff') = 0,$$

so $(f')^2 + f^2$ is a constant function. From $f(0) = 0$ and $f'(0) = 0$ it follows that the constant is 0; thus

$$[f'(x)]^2 + [f(x)]^2 = 0 \quad \text{for all } x.$$

This implies that

$$f(x) = 0 \quad \text{for all } x. \blacksquare$$

THEOREM 4 If f has a second derivative everywhere and

$$\begin{aligned} f'' + f &= 0, \\ f(0) &= a, \\ f'(0) &= b, \end{aligned}$$

then

$$f = b \cdot \sin + a \cdot \cos.$$

(In particular, if $f(0) = 0$ and $f'(0) = 1$, then $f = \sin$; if $f(0) = 1$ and $f'(0) = 0$, then $f = \cos$.)

PROOF Let

$$g(x) = f(x) - b\sin x - a\cos x.$$

Then

$$g'(x) = f'(x) - b\cos x + a\sin x,$$
$$g''(x) = f''(x) + b\sin x + a\cos x.$$

Consequently,

$$g'' + g = 0,$$
$$g(0) = 0,$$
$$g'(0) = 0,$$

which shows that

$$0 = g(x) = f(x) - b\sin x - a\cos x, \qquad \text{for all } x. \blacksquare$$

THEOREM 5 If x and y are any two numbers, then

$$\sin(x + y) = \sin x \cos y + \cos x \sin y,$$
$$\cos(x + y) = \cos x \cos y - \sin x \sin y.$$

PROOF For any particular number y we can define a function f by

$$f(x) = \sin(x + y).$$

Then

$$f'(x) = \cos(x + y)$$
$$f''(x) = -\sin(x + y).$$

Consequently,

$$f'' + f = 0,$$
$$f(0) = \sin y,$$
$$f'(0) = \cos y.$$

It follows from Theorem 4 that

$$f = (\cos y) \cdot \sin + (\sin y) \cdot \cos;$$

that is,

$$\sin(x + y) = \cos y \sin x + \sin y \cos x, \qquad \text{for all } x.$$

Since any number y could have been chosen to begin with, this proves the first formula for all x and y.

The second formula is proved similarly. ▌

As a conclusion to this chapter, and as a prelude to Chapter 18, we will mention an alternative approach to the definition of the function sin. Since

$$\arcsin'(x) = \frac{1}{\sqrt{1 - x^2}} \quad \text{for } -1 < x < 1,$$

it follows from the Second Fundamental Theorem of Calculus that

$$\arcsin x = \arcsin x - \arcsin 0 = \int_0^x \frac{1}{\sqrt{1-t^2}}\,dt.$$

This equation could have been taken as the *definition* of arcsin. It would follow immediately that

$$\arcsin'(x) = \frac{1}{\sqrt{1-x^2}};$$

the function sin could then be defined as $(\arcsin)^{-1}$ and the formula for the derivative of an inverse function would show that

$$\sin'(x) = \sqrt{1-\sin^2 x},$$

which could be defined as $\cos x$. Eventually, one could show that $A(\cos x) = x/2$, recovering at the very end of the development the definition with which we started. While much of this presentation would proceed more rapidly, the definition would be utterly unmotivated; the reasonableness of the definitions would be known to the author, but not to the student, for whom it was intended! Nevertheless, as we shall see in Chapter 18, an approach of this sort is sometimes very reasonable indeed.

PROBLEMS

1. Differentiate each of the following functions.

(i) $f(x) = \arctan(\arctan(\arctan x))$.
(ii) $f(x) = \arcsin(\arctan(\arccos x))$.
(iii) $f(x) = \arctan(\tan x \arctan x)$.
(iv) $f(x) = \arcsin\left(\dfrac{1}{\sqrt{1+x^2}}\right)$.

2. Find the following limits by l'Hôpital's Rule.

(i) $\displaystyle\lim_{x\to 0} \frac{\sin x - x + x^3/6}{x^3}$.

(ii) $\displaystyle\lim_{x\to 0} \frac{\sin x - x + x^3/6}{x^4}$.

(iii) $\displaystyle\lim_{x\to 0} \frac{\cos x - 1 + x^2/2}{x^2}$.

(iv) $\displaystyle\lim_{x\to 0} \frac{\cos x - 1 + x^2/2}{x^4}$.

(v) $\displaystyle\lim_{x\to 0} \frac{\arctan x - x + x^3/3}{x^3}$.

(vi) $\displaystyle\lim_{x\to 0} \left(\frac{1}{x} - \frac{1}{\sin x}\right)$.

3. Let $f(x) = \begin{cases} \dfrac{\sin x}{x}, & x \neq 0 \\ 1, & x = 0. \end{cases}$

(a) Find $f'(0)$.
(b) Find $f''(0)$.

At this point, you will almost certainly have to use l'Hôpital's Rule, but in Chapter 24 we will be able to find $f^{(k)}(0)$ for all k, with almost no work at all.

4. Graph the following functions.

(a) $f(x) = \sin 2x$.
(b) $f(x) = \sin(x^2)$. (A pretty respectable sketch of this graph can be obtained using only a picture of the graph of sin. Indeed, pure thought is your only hope in this problem, because determining the sign of the derivative $f'(x) = \cos(x^2)\cdot 2x$ is no easier than determining the behavior of f directly. The formula for $f'(x)$ does indicate one important fact, however—$f'(0) = 0$, which must be true since f is even, and which should be clear in your graph.)
(c) $f(x) = \sin x + \sin 2x$. (It will probably be instructive to first draw the graphs of $g(x) = \sin x$ and $h(x) = \sin 2x$ carefully on the same set of axes, from 0 to 2π, and guess what the sum will look like. You can easily find out how many critical points f has on $[0, 2\pi]$ by considering the derivative of f. You can then determine the nature of these critical points by finding out the sign of f at each point; your sketch will probably suggest the answer.)
(d) $f(x) = \tan x - x$. (First determine the behavior of f in $(-\pi/2, \pi/2)$; in the intervals $(k\pi - \pi/2, k\pi + \pi/2)$ the graph of f will look exactly the same, except moved up a certain amount. Why?)
(e) $f(x) = \sin x - x$. (The material in the Appendix to Chapter 11 will be particularly helpful for this function.)
(f) $f(x) = \begin{cases} \dfrac{\sin x}{x}, & x \neq 0 \\ 1, & x = 0. \end{cases}$

(Part (d) should enable you to determine approximately where the zeros of f' are located. Notice that f is even and continuous at 0; also consider the size of f for large x.)
(g) $f(x) = x \sin x$.

***5.** The *hyperbolic spiral* is the graph of the function $f(\theta) = a/\theta$ in polar coordinates (Chapter 4, Appendix 3). Sketch this curve, paying particular attention to its behavior for θ close to 0.

6. Prove the addition formula for cos.

7. (a) From the addition formula for sin and cos derive formulas for $\sin 2x$, $\cos 2x$, $\sin 3x$, and $\cos 3x$.

(b) Use these formulas to find the following values of the trigonometric functions (usually deduced by geometric arguments in elementary trigonometry):

$$\sin\frac{\pi}{4} = \cos\frac{\pi}{4} = \frac{\sqrt{2}}{2},$$
$$\tan\frac{\pi}{4} = 1,$$
$$\sin\frac{\pi}{6} = \frac{1}{2},$$
$$\cos\frac{\pi}{6} = \frac{\sqrt{3}}{2}.$$

8. (a) Show that $A\sin(x+B)$ can be written as $a\sin x + b\cos x$ for suitable a and b. (One of the theorems in this chapter provides a one-line proof. You should also be able to figure out what a and b are.)

(b) Conversely, given a and b, find numbers A and B such that $a\sin x + b\cos x = A\sin(x+B)$ for all x.

(c) Use part (b) to graph $f(x) = \sqrt{3}\sin x + \cos x$.

9. (a) Prove that

$$\tan(x+y) = \frac{\tan x + \tan y}{1 - \tan x \tan y}$$

provided that x, y, and $x+y$ are not of the form $k\pi + \pi/2$. (Use the addition formulas for sin and cos.)

(b) Prove that

$$\arctan x + \arctan y = \arctan\left(\frac{x+y}{1-xy}\right),$$

indicating any necessary restrictions on x and y. Hint: Replace x by $\arctan x$ and y by $\arctan y$ in part (a).

10. Prove that

$$\arcsin\alpha + \arcsin\beta = \arcsin\left(\alpha\sqrt{1-\beta^2} + \beta\sqrt{1-\alpha^2}\right),$$

indicating any restrictions on α and β.

11.

Prove that if m and n are any numbers, then

$$\sin mx \sin nx = \tfrac{1}{2}[\cos(m-n)x - \cos(m+n)x],$$
$$\sin mx \cos nx = \tfrac{1}{2}[\sin(m+n)x + \sin(m-n)x],$$
$$\cos mx \cos nx = \tfrac{1}{2}[\cos(m+n)x + \cos(m-n)x].$$

12. Prove that if m and n are natural numbers, then

$$\int_{-\pi}^{\pi} \sin mx \sin nx \, dx = \begin{cases} 0, & m \neq n \\ \pi, & m = n, \end{cases}$$

$$\int_{-\pi}^{\pi} \cos mx \cos nx \, dx = \begin{cases} 0, & m \neq n \\ \pi, & m = n, \end{cases}$$

$$\int_{-\pi}^{\pi} \sin mx \cos nx \, dx = 0.$$

These relations are particularly important in the theory of Fourier series. Although this topic will receive serious attention only in the Suggested Reading, the next problem provides a hint as to their importance.

13. (a) If f is integrable on $[-\pi, \pi]$, show that the minimum value of

$$\int_{-\pi}^{\pi} (f(x) - a\cos nx)^2 \, dx$$

occurs when

$$a = \frac{1}{\pi}\int_{-\pi}^{\pi} f(x)\cos nx \, dx,$$

and the minimum value of

$$\int_{-\pi}^{\pi} (f(x) - a\sin nx)^2 \, dx$$

when

$$a = \frac{1}{\pi}\int_{-\pi}^{\pi} f(x)\sin nx \, dx.$$

(In each case, bring a outside the integral sign, obtaining a quadratic expression in a.)

(b) Define

$$a_n = \frac{1}{\pi}\int_{-\pi}^{\pi} f(x)\cos nx \, dx, \qquad n = 0, 1, 2, \dots,$$

$$b_n = \frac{1}{\pi}\int_{-\pi}^{\pi} f(x)\sin nx \, dx, \qquad n = 1, 2, 3, \dots.$$

Show that if c_i and d_i are any numbers, then

$$\int_{-\pi}^{\pi} \left(f(x) - \left[\frac{c_0}{2} + \sum_{n=1}^{N} c_n \cos nx + d_n \sin nx \right] \right)^2 dx$$

$$= \int_{-\pi}^{\pi} [f(x)]^2 \, dx - 2\pi \left(\frac{a_0 c_0}{2} + \sum_{n=1}^{N} a_n c_n + b_n d_n \right) + \pi \left(\frac{{c_0}^2}{2} + \sum_{n=1}^{N} {c_n}^2 + {d_n}^2 \right)$$

$$= \int_{-\pi}^{\pi} [f(x)]^2 \, dx - \pi \left(\frac{{a_0}^2}{2} + \sum_{n=1}^{N} {a_n}^2 + {b_n}^2 \right)$$

$$+ \pi \left(\left(\frac{c_0}{\sqrt{2}} - \frac{a_0}{\sqrt{2}} \right)^2 + \sum_{n=1}^{N} (c_n - a_n)^2 + (d_n - b_n)^2 \right),$$

thus showing that the first integral is smallest when $a_i = c_i$ and $b_i = d_i$. In other words, among all "linear combinations" of the functions $s_n(x) = \sin nx$ and $c_n(x) = \cos nx$ for $1 \le n \le N$, the particular function

$$g(x) = \frac{a_0}{2} + \sum_{n=1}^{N} a_n \cos nx + b_n \sin nx$$

has the "closest fit" to f on $[-\pi, \pi]$.

14. (a) Find a formula for $\sin x + \sin y$. (Notice that this also gives a formula for $\sin x - \sin y$.) Hint: First find a formula for $\sin(a+b) + \sin(a-b)$. What good does that do?
(b) Also find a formula for $\cos x + \cos y$ and $\cos x - \cos y$.

15. (a) Starting from the formula for $\cos 2x$, derive formulas for $\sin^2 x$ and $\cos^2 x$ in terms of $\cos 2x$.
(b) Prove that

$$\cos\frac{x}{2} = \sqrt{\frac{1+\cos x}{2}} \quad \text{and} \quad \sin\frac{x}{2} = \sqrt{\frac{1-\cos x}{2}}$$

for $0 \le x \le \pi/2$.
(c) Use part (a) to find $\int_a^b \sin^2 x\, dx$ and $\int_a^b \cos^2 x\, dx$.
(d) Graph $f(x) = \sin^2 x$.

16. Find $\sin(\arctan x)$ and $\cos(\arctan x)$ as expressions not involving trigonometric functions. Hint: $y = \arctan x$ means that $x = \tan y = \sin y / \cos y = \sin y / \sqrt{1 - \sin^2 y}$.

17. If $x = \tan u/2$, express $\sin u$ and $\cos u$ in terms of x. (Use Problem 16; the answers should be very simple expressions.)

18. (a) Prove that $\sin(x + \pi/2) = \cos x$. (All along we have been drawing the graphs of sin and cos as if this were the case.)
(b) What is $\arcsin(\cos x)$ and $\arccos(\sin x)$?

19. (a) Find $\displaystyle\int_0^1 \frac{1}{1+t^2}\, dt$. Hint: The answer is not 45.
(b) Find $\displaystyle\int_0^\infty \frac{1}{1+t^2}\, dt$.

20. Find $\displaystyle\lim_{x\to\infty} x \sin\frac{1}{x}$.

21. (a) Define functions $\sin^\circ$ and $\cos^\circ$ by $\sin^\circ(x) = \sin(\pi x/180)$ and $\cos^\circ(x) = \cos(\pi x/180)$. Find $(\sin^\circ)'$ and $(\cos^\circ)'$ in terms of these same functions.
(b) Find $\displaystyle\lim_{x\to 0} \frac{\sin^\circ x}{x}$ and $\displaystyle\lim_{x\to\infty} x \sin^\circ \frac{1}{x}$.

22. Prove that every point on the unit circle is of the form $(\cos\theta, \sin\theta)$ for at least one (and hence for infinitely many) numbers θ.

23. (a) Prove that π is the maximum possible length of an interval on which sin is one-one, and that such an interval must be of the form $[2k\pi - \pi/2, 2k\pi + \pi/2]$ or $[2k\pi + \pi/2, 2(k+1)\pi - \pi/2]$.

(b) Suppose we let $g(x) = \sin x$ for x in $(2k\pi - \pi/2, 2k\pi + \pi/2)$. What is $(g^{-1})'$?

24. Let $f(x) = \sec x$ for $0 \le x \le \pi$. Find the domain of f^{-1} and sketch its graph.

25. Prove that $|\sin x - \sin y| < |x - y|$ for all numbers $x \ne y$. Hint: The same statement, with $<$ replaced by $\le$, is a very straightforward consequence of a well-known theorem; simple supplementary considerations then allow $\le$ to be improved to $<$.

***26.** It is an excellent test of intuition to predict the value of

$$\lim_{\lambda\to\infty} \int_a^b f(x)\sin \lambda x\, dx.$$

Continuous functions should be most accessible to intuition, but once you get the right idea for a proof the limit can easily be established for any integrable f.

(a) Show that $\lim_{\lambda\to\infty} \int_c^d \sin\lambda x\, dx = 0$, by computing the integral explicitly.

(b) Show that if s is a step function on $[a, b]$ (terminology from Problem 13-26), then $\lim_{\lambda\to\infty} \int_a^b s(x)\sin\lambda x\, dx = 0$.

(c) Finally, use Problem 13-26 to show that $\lim_{\lambda\to\infty} \int_a^b f(x)\sin\lambda x\, dx = 0$ for any function f which is integrable on $[a, b]$. This result, like Problem 12, plays an important role in the theory of Fourier series; it is known as the Riemann-Lebesgue Lemma.

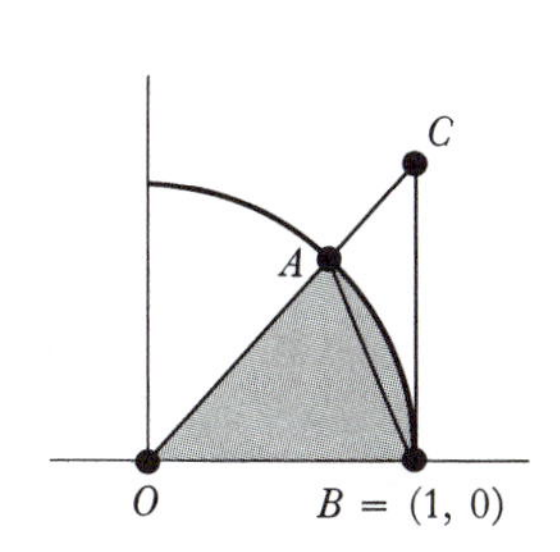

FIGURE 24

27. This problem outlines the classical approach to the trigonometric functions. The shaded sector in Figure 24 has area $x/2$.

(a) By considering the triangles OAB and OCB prove that if $0 < x < \pi/4$, then

$$\frac{\sin x}{2} < \frac{x}{2} < \frac{\sin x}{2\cos x}.$$

(b) Conclude that

$$\cos x < \frac{\sin x}{x} < 1,$$

and prove that

$$\lim_{x\to 0} \frac{\sin x}{x} = 1.$$

(c) Use this limit to find

$$\lim_{x\to 0} \frac{1 - \cos x}{x}.$$

(d) Using parts (b) and (c), and the addition formula for sin, find $\sin'(x)$, starting from the definition of the derivative.

***28.** This problem gives a treatment of the trigonometric functions in terms of length, and uses Problem 13-25. Let $f(x) = \sqrt{1-x^2}$ for $-1 \le x \le 1$. Define $\mathscr{L}(x)$ to be the length of f on $[x, 1]$.

(a) Show that

$$\mathscr{L}(x) = \int_x^1 \frac{1}{\sqrt{1-t^2}}\,dt.$$

(This is actually an improper integral, as defined in Problem 14-28.)

(b) Show that

$$\mathscr{L}'(x) = -\frac{1}{\sqrt{1-x^2}} \qquad \text{for } -1 < x < 1.$$

(c) Define π as $\mathscr{L}(-1)$. For $0 \le x \le \pi$, define $\cos x$ by $\mathscr{L}(\cos x) = x$, and define $\sin x = \sqrt{1-\cos^2 x}$. Prove that $\cos'(x) = -\sin x$ and $\sin'(x) = \cos x$ for $0 < x < \pi$.

***29.** Yet another development of the trigonometric functions was briefly mentioned in the text—starting with inverse functions defined by integrals. It is convenient to begin with arctan, since this function is defined for all x. To do this problem, pretend that you have never heard of the trigonometric functions.

(a) Let $\alpha(x) = \int_0^x (1+t^2)^{-1}\,dt$. Prove that α is odd and increasing, and that $\lim_{x\to\infty} \alpha(x)$ and $\lim_{x\to-\infty} \alpha(x)$ both exist, and are negatives of each other. If we define $\pi = 2\lim_{x\to\infty} \alpha(x)$, then α^{-1} is defined on $(-\pi/2, \pi/2)$.

(b) Show that $(\alpha^{-1})'(x) = 1 + [\alpha^{-1}(x)]^2$.

(c) For $x = k\pi + x'$ with $x' \ne \pi/2$ or $-\pi/2$, define $\tan x = \alpha^{-1}(x')$. Then define $\cos x = 1/\sqrt{1+\tan^2 x}$, for x not of the form $k\pi + \pi/2$ or $k\pi - \pi/2$, and $\cos(k\pi \pm \pi/2) = 0$. Prove first that $\cos'(x) = -\tan x \cos x$, and then that $\cos''(x) = -\cos x$ for all x.

***30.** If we are willing to assume that certain differential equations have solutions, another approach to the trigonometric functions is possible. Suppose, in particular, that there is some function y_0 which is not always 0 and which satisfies $y_0'' + y_0 = 0$.

(a) Prove that $y_0{}^2 + (y_0{}')^2$ is constant, and conclude that either $y_0(0) \ne 0$ or $y_0{}'(0) \ne 0$.

(b) Prove that there is a function s satisfying $s'' + s = 0$ and $s(0) = 0$ and $s'(0) = 1$. Hint: Try s of the form $ay_0 + by_0{}'$.

If we define $\sin = s$ and $\cos = s'$, then almost all facts about trigonometric functions become trivial. There is one point which requires work, however—producing the number π. This is most easily done using an exercise from the Appendix to Chapter 11:

(c) Use Problem 7 of the Appendix to Chapter 11 to prove that $\cos x$ cannot be positive for all $x > 0$. It follows that there is a smallest $x_0 > 0$ with $\cos x_0 = 0$, and we can define $\pi = 2x_0$.

(d) Prove that $\sin \pi/2 = 1$. (Since $\sin^2 + \cos^2 = 1$, we have $\sin \pi/2 = \pm 1$; the problem is to decide why $\sin \pi/2$ is positive.)

(e) Find $\cos \pi$, $\sin \pi$, $\cos 2\pi$, and $\sin 2\pi$. (Naturally you may use any addition formulas, since these can be derived once we know that $\sin' = \cos$ and $\cos' = -\sin$.)

(f) Prove that cos and sin are periodic with period 2π.

31. (a) After all the work involved in the definition of sin, it would be disconcerting to find that sin is actually a rational function. Prove that it isn't. (There is a simple property of sin which a rational function cannot possibly have.)

(b) Prove that sin isn't even defined implicitly by an algebraic equation; that is, there do not exist rational functions $f_0, \dots, f_{n-1}$ such that

$$(\sin x)^n + f_{n-1}(x)(\sin x)^{n-1} + \cdots + f_0(x) = 0 \quad \text{for all } x.$$

Hint: Prove that $f_0 = 0$, so that $\sin x$ can be factored out. The remaining factor is 0 except perhaps at multiples of 2π. But this implies that it is 0 for all x. (Why?) You are now set up for a proof by induction.

***32.** Suppose that ϕ_1 and ϕ_2 satisfy

$$\phi_1'' + g_1\phi_1 = 0,$$
$$\phi_2'' + g_2\phi_2 = 0,$$

and that $g_2 > g_1$.

(a) Show that

$$\phi_1''\phi_2 - \phi_2''\phi_1 - (g_2 - g_1)\phi_1\phi_2 = 0.$$

(b) Show that if $\phi_1(x) > 0$ and $\phi_2(x) > 0$ for all x in (a, b), then

$$\int_a^b [\phi_1''\phi_2 - \phi_2''\phi_1] > 0,$$

and conclude that

$$[\phi_1'(b)\phi_2(b) - \phi_1'(a)\phi_2(a)] + [\phi_1(b)\phi_2'(b) - \phi_1(a)\phi_2'(a)] > 0.$$

(c) Show that in this case we cannot have $\phi_1(a) = \phi_1(b) = 0$. Hint: Consider the sign of $\phi_1'(a)$ and $\phi_1'(b)$.

(d) Show that the equations $\phi_1(a) = \phi_1(b) = 0$ are also impossible if $\phi_1 > 0$, $\phi_2 < 0$ or $\phi_1 < 0$, $\phi_2 > 0$, or $\phi_1 < 0$, $\phi_2 < 0$ on (a, b). (You should be able to do this with almost no extra work.)

The net result of this problem may be stated as follows: if a and b are consecutive zeros of ϕ_1, then ϕ_2 must have a zero somewhere between a and b. This result, in a slightly more general form, is known as the Sturm Comparison Theorem. As a particular example, any solution of the differential equation

$$y'' + (x+1)y = 0$$

must have zeros on the positive horizontal axis which are within π of each other.

33. (a) Using the formula for $\sin x - \sin y$ derived in Problem 14, show that

$$\sin(k+\tfrac{1}{2})x - \sin(k-\tfrac{1}{2})x = 2\sin\frac{x}{2}\cos kx.$$

(b) Conclude that

$$\frac{1}{2} + \cos x + \cos 2x + \cdots + \cos nx = \frac{\sin(n+\frac{1}{2})x}{2\sin\dfrac{x}{2}}.$$

Like two other results in this problem set, this equation is very important in the study of Fourier series, and we also make use of it in Problems 19-42 and 23-19.

(c) Similarly, derive the formula

$$\sin x + \sin 2x + \cdots + \sin nx = \frac{\sin\left(\dfrac{n+1}{2}x\right)\sin\left(\dfrac{n}{2}x\right)}{\sin\dfrac{x}{2}}.$$

(A more natural derivation of these formulas will be given in Problem 27-14.)

(d) Use parts (b) and (c) to find $\int_0^b \sin x\,dx$ and $\int_0^b \cos x\,dx$ directly from the definition of the integral.

*CHAPTER 16 π IS IRRATIONAL

This short chapter, diverging from the main stream of the book, is included to demonstrate that we are already in a position to do some sophisticated mathematics. This entire chapter is devoted to an elementary proof that π is irrational. Like many "elementary" proofs of deep theorems, the motivation for many steps in our proof cannot be supplied; nevertheless, it is still quite possible to follow the proof step-by-step.

Two observations must be made before the proof. The first concerns the function

$$f_n(x) = \frac{x^n(1-x)^n}{n!},$$

which clearly satisfies

$$0 < f_n(x) < \frac{1}{n!} \quad \text{for } 0 < x < 1.$$

An important property of the function f_n is revealed by considering the expression obtained by actually multiplying out $x^n(1-x)^n$. The lowest power of x appearing will be n and the highest power will be $2n$. Thus f_n can be written in the form

$$f_n(x) = \frac{1}{n!}\sum_{i=n}^{2n} c_i x^i,$$

where the numbers c_i are integers. It is clear from this expression that

$$f_n{}^{(k)}(0) = 0 \quad \text{if } k < n \text{ or } k > 2n.$$

Moreover,

$$\begin{aligned}
f_n{}^{(n)}(x) &= \frac{1}{n!}[n!\,c_n + \text{ terms involving } x] \\
f_n{}^{(n+1)}(x) &= \frac{1}{n!}[(n+1)!\,c_{n+1} + \text{ terms involving } x] \\
&\;\;\vdots \\
f_n{}^{(2n)}(x) &= \frac{1}{n!}[(2n)!\,c_{2n}].
\end{aligned}$$

This means that

$$\begin{aligned} f_n{}^{(n)}(0) &= c_n, \\ f_n{}^{(n+1)}(0) &= (n+1)c_{n+1} \\ &\vdots \\ f_n{}^{(2n)}(0) &= (2n)(2n-1)\cdot\ldots\cdot(n+1)c_{2n}, \end{aligned}$$

where the numbers on the right are all integers. Thus

$$f_n{}^{(k)}(0) \text{ is an } \textit{integer} \text{ for all } k.$$

The relation

$$f_n(x) = f_n(1-x)$$

implies that

$$f_n{}^{(k)}(x) = (-1)^k f_n{}^{(k)}(1-x);$$

therefore,

$$f_n{}^{(k)}(1) \text{ is also an } \textit{integer} \text{ for all } k.$$

The proof that π is irrational requires one further observation: if a is any number, and $\varepsilon > 0$, then for sufficiently large n we will have

$$\frac{a^n}{n!} < \varepsilon.$$

To prove this, notice that if $n \geq 2a$, then

$$\frac{a^{n+1}}{(n+1)!} = \frac{a}{n+1} \cdot \frac{a^n}{n!} < \frac{1}{2} \cdot \frac{a^n}{n!}.$$

Now let n_0 be any natural number with $n_0 \geq 2a$. Then, whatever value

$$\frac{a^{n_0}}{(n_0)!}$$

may have, the succeeding values satisfy

$$\begin{aligned} \frac{a^{(n_0+1)}}{(n_0+1)!} &< \frac{1}{2} \cdot \frac{a^{n_0}}{(n_0)!} \\ \frac{a^{(n_0+2)}}{(n_0+2)!} &< \frac{1}{2} \cdot \frac{a^{(n_0+1)}}{(n_0+1)!} < \frac{1}{2} \cdot \frac{1}{2} \cdot \frac{a^{n_0}}{(n_0)!} \\ &\vdots \\ \frac{a^{(n_0+k)}}{(n_0+k)!} &< \frac{1}{2^k} \cdot \frac{a^{n_0}}{(n_0)!}. \end{aligned}$$

If k is so large that $\dfrac{a^{n_0}}{(n_0)!\,\varepsilon} < 2^k$, then

$$\frac{a^{(n_0+k)}}{(n_0+k)!} < \varepsilon,$$

which is the desired result. Having made these observations, we are ready for the one theorem in this chapter.

THEOREM 1 The number π is irrational; in fact, π^2 is irrational. (Notice that the irrationality of π^2 implies the irrationality of π, for if π were rational, then π^2 certainly would be.)

PROOF Suppose π^2 were rational, so that

$$\pi^2 = \frac{a}{b}$$

for some positive integers a and b. Let

$$(1) \quad G(x) = b^n[\pi^{2n} f_n(x) - \pi^{2n-2} f_n{}''(x) + \pi^{2n-4} f_n{}^{(4)}(x) - \cdots + (-1)^n f_n{}^{(2n)}(x)].$$

Notice that each of the factors

$$b^n \pi^{2n-2k} = b^n(\pi^2)^{n-k} = b^n\left(\frac{a}{b}\right)^{n-k} = a^{n-k}b^k$$

is an integer. Since $f_n{}^{(k)}(0)$ and $f_n{}^{(k)}(1)$ are integers, this shows that

$$G(0) \text{ and } G(1) \text{ are integers.}$$

Differentiating G twice yields

$$(2) \qquad G''(x) = b^n[\pi^{2n} f_n{}''(x) - \pi^{2n-2} f_n{}^{(4)}(x) + \cdots + (-1)^n f_n{}^{(2n+2)}(x)].$$

The last term, $(-1)^n f_n{}^{(2n+2)}(x)$, is zero. Thus, adding (1) and (2) gives

$$(3) \qquad G''(x) + \pi^2 G(x) = b^n \pi^{2n+2} f_n(x) = \pi^2 a^n f_n(x).$$

Now let

$$H(x) = G'(x)\sin \pi x - \pi G(x) \cos \pi x.$$

Then

$$\begin{aligned} H'(x) &= \pi G'(x)\cos \pi x + G''(x)\sin \pi x - \pi G'(x)\cos \pi x + \pi^2 G(x)\sin \pi x \\ &= [G''(x) + \pi^2 G(x)]\sin \pi x \\ &= \pi^2 a^n f_n(x)\sin \pi x, \text{ by (3).} \end{aligned}$$

By the Second Fundamental Theorem of Calculus,

$$\begin{aligned} \pi^2 \int_0^1 a^n f_n(x)\sin \pi x\,dx &= H(1) - H(0) \\ &= G'(1)\sin \pi - \pi G(1)\cos \pi - G'(0)\sin 0 + \pi G(0)\cos 0 \\ &= \pi[G(1) + G(0)]. \end{aligned}$$

Thus

$$\pi \int_0^1 a^n f_n(x)\sin \pi x\,dx \quad \text{is an } \textit{integer.}$$

On the other hand, $0 < f_n(x) < 1/n!$ for $0 < x < 1$, so

$$0 < \pi a^n f_n(x) \sin \pi x < \frac{\pi a^n}{n!} \quad \text{for } 0 < x < 1.$$

Consequently,

$$0 < \pi \int_0^1 a^n f_n(x) \sin \pi x \, dx < \frac{\pi a^n}{n!}.$$

This reasoning was completely independent of the value of n. Now if n is large enough, then

$$0 < \pi \int_0^1 a^n f_n(x) \sin \pi x \, dx < \frac{\pi a^n}{n!} < 1.$$

But this is absurd, because the integral is an integer, and there is no integer between 0 and 1. Thus our original assumption must have been incorrect: π^2 is irrational. ▮

This proof is admittedly mysterious; perhaps most mysterious of all is the way that π enters the proof—it almost looks as if we have proved π irrational without ever mentioning a definition of π. A close reexamination of the proof will show that precisely one property of π is essential—

$$\sin(\pi) = 0.$$

The proof really depends on the properties of the function sin, and proves the irrationality of the smallest positive number x with $\sin x = 0$. In fact, very few properties of sin are required, namely,

$$\begin{aligned} \sin' &= \cos, \\ \cos' &= -\sin, \\ \sin(0) &= 0, \\ \cos(0) &= 1. \end{aligned}$$

Even this list could be shortened; as far as the proof is concerned, cos might just as well be defined as $\sin'$. The properties of sin required in the proof may then be written

$$\begin{aligned} \sin'' + \sin &= 0, \\ \sin(0) &= 0, \\ \sin'(0) &= 1. \end{aligned}$$

Of course, this is not really very surprising at all, since, as we have seen in the previous chapter, these properties characterize the function sin completely.

PROBLEMS

1. (a) Prove that the areas of triangles OAB and OAC in Figure 1 are related by the equation

$$\text{area } OAC = \frac{1}{2}\sqrt{\frac{1-\sqrt{1-16(\text{area } OAB)^2}}{2}}.$$

Hint: Solve the equations $xy = 2(\text{area } OAB)$, $x^2 + y^2 = 1$, for y.

(b) Let P_m be the regular polygon of m sides inscribed in the unit circle. If A_m is the area of P_m show that

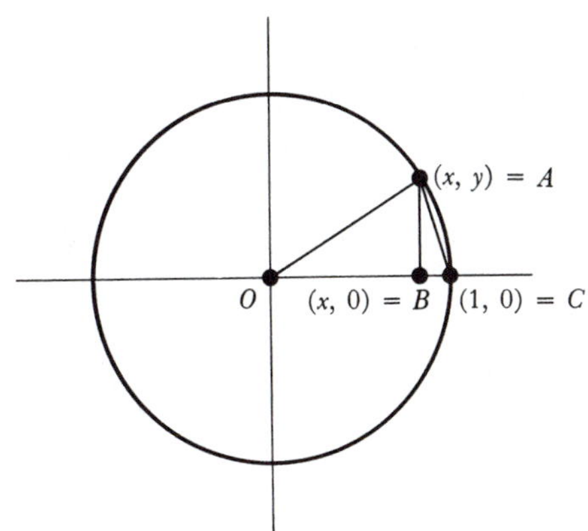

FIGURE 1

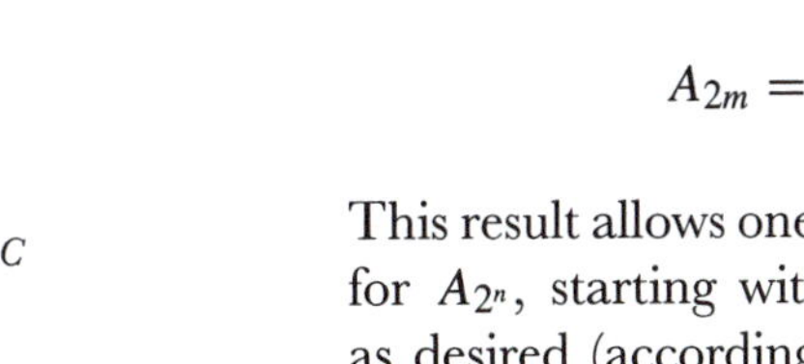

$$A_{2m} = \frac{m}{2}\sqrt{2 - 2\sqrt{1-(2A_m/m)^2}}.$$

This result allows one to obtain (more and more complicated) expressions for A_{2^n}, starting with $A_4 = 2$, and thus to compute π as accurately as desired (according to Problem 8-11). Although better methods will appear in Chapter 20, a slight variant of this approach yields a very interesting expression for π:

2. (a) Using the fact that

$$\frac{\text{area}(OAB)}{\text{area}(OAC)} = OB,$$

show that if α_m is the distance from O to one side of P_m, then

$$\frac{A_m}{A_{2m}} = \alpha_m.$$

(b) Show that

$$\frac{2}{A_{2^k}} = \alpha_4 \cdot \alpha_8 \cdot \ldots \cdot \alpha_{2^{k-1}}.$$

(c) Using the fact that

$$\alpha_m = \cos\frac{\pi}{m},$$

and the formula $\cos x/2 = \sqrt{\dfrac{1+\cos x}{2}}$ (Problem 15-15), prove that

$$\alpha_4 = \sqrt{\frac{1}{2}}$$

$$\alpha_8 = \sqrt{\frac{1}{2} + \frac{1}{2}\sqrt{\frac{1}{2}}},$$

$$\alpha_{16} = \sqrt{\frac{1}{2} + \frac{1}{2}\sqrt{\frac{1}{2} + \frac{1}{2}\sqrt{\frac{1}{2}}}},$$

etc.

Together with part (b), this shows that $2/\pi$ can be written as an "infinite product"

$$\frac{2}{\pi} = \sqrt{\frac{1}{2}} \cdot \sqrt{\frac{1}{2} + \frac{1}{2}\sqrt{\frac{1}{2}}} \cdot \sqrt{\frac{1}{2} + \frac{1}{2}\sqrt{\frac{1}{2} + \frac{1}{2}\sqrt{\frac{1}{2}}}} \cdot \ldots ;$$

to be precise, this equation means that the product of the first n factors can be made as close to $2/\pi$ as desired, by choosing n sufficiently large. This product was discovered by François Viète in 1579, and is only one of many fascinating expressions for π, some of which are mentioned later.

*CHAPTER 17 PLANETARY MOTION

Nature and Nature's Laws lay hid in night
God said "Let Newton be," and all was light.

Alexander Pope

Unlike Chapter 16, a short chapter diverging from the main stream of the book, this long chapter diverges from the main stream of the book to demonstrate that we are already in a position to do some real physics.

In 1609 Kepler published his first two laws of planetary motion. The first law describes the shape of planetary orbits:

The planets move in ellipses, with the sun at one focus.

The second law involves the area swept out by the segment from the sun to the planet (the 'radius vector from the sun to the planet') in various time intervals (Figure 1):

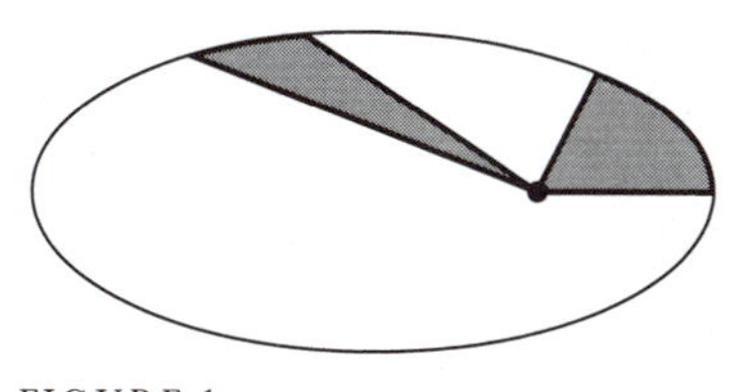

FIGURE 1

Equal areas are swept out by the radius vector in equal times. (Equivalently, the area swept out in time t is proportional to t.)

Kepler's third law, published in 1619, relates the motions of different planets. If a is the major axis of a planet's elliptical orbit and T is its period, the time it takes the planet to return to a given position, then:

The ratio a^3/T^2 is the same for all planets.

Newton's great accomplishment was to show (using his general law that the force on a body is its mass times its acceleration) that Kepler's laws follow from the assumption that the planets are attracted to the sun by a force (the gravitational force of the sun) always directed toward the sun, proportional to the mass of the planet, and satisfying an inverse square law; that is, by a force directed toward the sun whose magnitude varies inversely with the square of the distance from the sun to the planet and directly with the mass of the planet. Since force is mass times acceleration, this is equivalent simply to saying that the magnitude of the acceleration is a constant divided by the square of the distance from the sun.

Newton's analysis actually established three results that correlate with Kepler's individual laws. The first of Newton's results concerns Kepler's second law (which was actually discovered first, nicely preserving the symmetry of the situation):

Kepler's second law is true precisely for 'central forces', i.e., if and only the force between the sun and the planet always lies along the line between the sun and the planet.

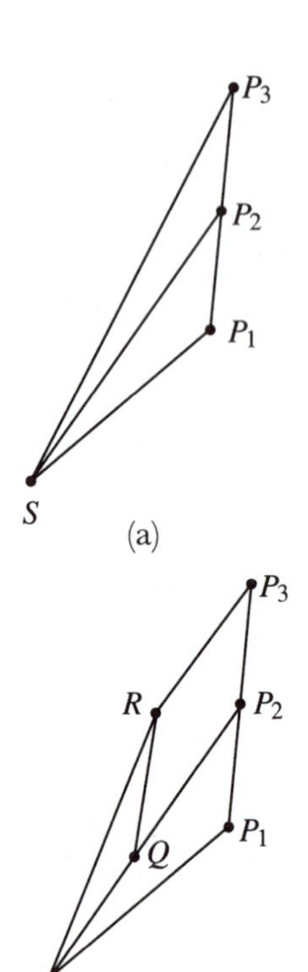

FIGURE 2

Although Newton is revered as the discoverer of calculus, and indeed invented calculus precisely in order to treat such problems, his derivation hardly seems to use calculus at all. Instead of considering a force that varies continuously as the planet moves, Newton first considers short equal time intervals and assumes that a momentary force is exerted at the ends of each of these intervals.

To be specific, let us imagine that during the first time interval the planet moves along the line P_1P_2, with uniform velocity (Figure 2a). If, during the next equal time interval, the planet continued to move along this line, it would end up at P_3, where the length of P_1P_2 equals the length of P_2P_3. This would imply that the triangle SP_1P_2 has the same are as the triangle SP_2P_3 (since they have equal bases, and the same height)—this just says that Kepler's law holds in the special case where the force is 0.

Now suppose (Figure 2b) that at the moment the planet arrives at P_2 it experiences a force exerted *along the line from S to P_2*, which by itself would cause the planet to move to the point Q. Combined with the motion that the planet already has, this causes the planet to move to R, the vertex opposite P_2 in the parallelogram whose sides are P_2P3 and P_2Q.

Thus, the area swept out in the second time interval is actually the triangle SP_2R. But the area of triangle SP_2R is equal to the area of triangle SP_3P_2, since they have the same base SP_2, and the same heights (since RP_3 is parallel to SP_2). Hence, finally, the area of triangle SP_2R is the same as the area of the original triangle SP_1P_2! Conversely, if the triangle SRP_2 has the same area as SP_1P_2, and hence the same area as SP_3P_2, then RP_3 must be parallel to SP_2, and this implies that Q must lie along SP_2.

Of course, this isn't quite the sort of argument one would expect to find in a modern book, but in its own charming way it shows physically just *why* the result should be true.

To analyze planetary motion we will be using the material in the Appendix to Chapter 12, and the "determinant" det defined in Problem 4 of Appendix 1 to Chapter 4. We describe the motion of the planet by the parameterized curve

$$c(t) = r(t)(\cos\theta(t), \sin\theta(t)),$$

so that r always gives the length of the line from the sun to the planet, while θ gives the angle. It will be convenient to write this also as

$$(1) \qquad c(t) = r(t) \cdot \mathbf{e}(\theta(t)),$$

where

$$\mathbf{e}(t) = (\cos t, \sin t)$$

is just the parameterized curve that runs along the unit circle. Note that

$$\mathbf{e}'(t) = (-\sin t, \cos t)$$

is also a vector of unit length, but perpendicular to $\mathbf{e}(t)$, and that we also have

$$\det\big(\mathbf{e}(t), \mathbf{e}'(t)\big) = 1. \tag{2}$$

Differentiating (1), using the formulas on page 244, we obtain

$$c'(t) = r'(t) \cdot \mathbf{e}(\theta(t)) + r(t)\theta'(t) \cdot \mathbf{e}'(\theta(t)), \tag{3}$$

and combining with (1), together with the formulas in Problem 6 of Appendix 1 to Chapter 4, we get

$$\begin{aligned}\det\big(c(t), c'(t)\big) &= r(t)r'(t)\det\big(\mathbf{e}(\theta(t)), \mathbf{e}(\theta(t))\big) + r(t)^2\theta'(t)\det\big(\mathbf{e}(\theta(t)), \mathbf{e}'(\theta(t))\big)\\ &= r(t)^2\theta'(t)\det\big(\mathbf{e}(\theta(t)), \mathbf{e}'(\theta(t))\big),\end{aligned}$$

since $\det(v, v)$ is always 0. Using (2) we then get

$$\det(c, c') = r^2\theta'. \tag{4}$$

As we will see, $r^2\theta'$ turns out to have another important interpretation.

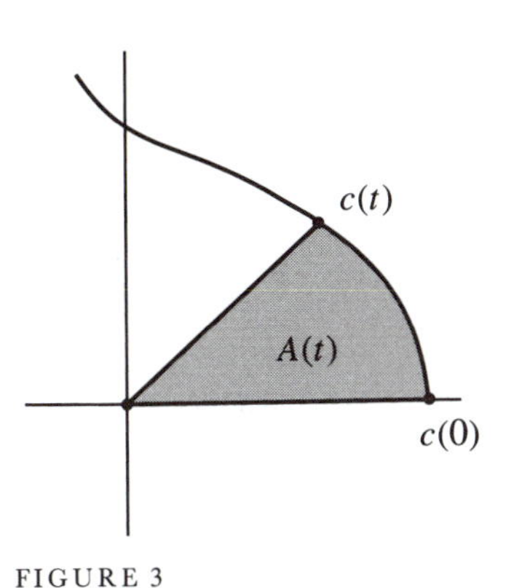

FIGURE 3

Suppose that $A(t)$ is the area swept out from time 0 to t (Figure 3). We want to get a formula for $A'(t)$, and, in the spirit of Newton, we'll begin by making an educated guess. Figure 4 shows $A(t+h) - A(t)$, together with a straight line segment between $c(t)$ and $c(t+h)$. It is easy to write down a formula for the area of the triangle $\Delta(h)$ with vertices O, $c(t)$, and $c(t+h)$: according to Problems 4 and 5 of Appendix 1 to Chapter 4, the area is

$$\text{area}(\Delta(h)) = \tfrac{1}{2}\det\big(c(t), c(t+h) - c(t)\big).$$

Since the triangle $\Delta(h)$ has practically the same area as the region $A(t+h) - A(t)$, this shows (or practically shows) that

$$\begin{aligned}A'(t) &= \lim_{h\to 0}\frac{A(t+h) - A(t)}{h}\\ &= \lim_{h\to 0}\frac{\text{area }\Delta(h)}{h}\\ &= \tfrac{1}{2}\det\left(c(t), \lim_{h\to 0}\frac{c(t+h) - c(t)}{h}\right)\\ &= \tfrac{1}{2}\det\big(c(t), c'(t)\big).\end{aligned}$$

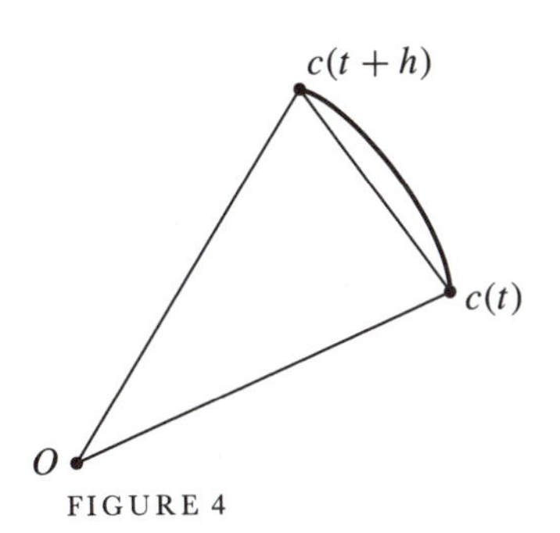

FIGURE 4

A rigorous derivation, establishing more in the process, can be made using Problem 13-24, which gives a formula for the area of a region determined by the graph of a function in polar coordinates. According to this Problem, we can write

$$A(t) = \frac{1}{2}\int_0^{\theta(t)} \rho(\phi)^2\, d\phi \tag{$*$}$$

if our parameterized curve $c(t) = r(t) \cdot \mathbf{e}(\theta(t))$ is the graph of the function ρ in polar coordinates (here we've used ϕ for the angular polar coordinate, to avoid confusion with the function θ used to describe the curve c).

Now the function ρ is just

$$\rho = r \circ \theta^{-1}$$

[for any particular angle ϕ, $\theta^{-1}(\phi)$ is the time at which the curve c has angular polar coordinate ϕ, so $r(\theta^{-1}(t))$ is the radius coordinate corresponding to ϕ]. Although the presence of the inverse function might look a bit forbidding, it's actually quite innocent: Applying the First Fundamental Theorem of Calculus and the Chain Rule to (∗) we immediately get

$$\begin{aligned} A'(t) &= \tfrac{1}{2}\rho(\theta(t))^2 \cdot \theta'(t) \\ &= \tfrac{1}{2}r(t)^2\theta'(t), \qquad \text{since } \rho = r \circ \theta^{-1}. \end{aligned}$$

Briefly,

$$A' = \tfrac{1}{2}r^2\theta'.$$

Combining with (4), we thus have

$$\text{(5)} \qquad \boxed{A' = \tfrac{1}{2}\det(c, c') = \tfrac{1}{2}r^2\theta'.}$$

Now we're ready to consider Kepler's second law. Notice that *Kepler's second law is equivalent to saying that A' is constant*, and thus it is equivalent to $A'' = 0$. But

$$\begin{aligned} A'' = \tfrac{1}{2}\big[\det(c, c')\big]' &= \tfrac{1}{2}\det(c', c') + \tfrac{1}{2}\det(c, c'') \qquad \text{(see page 245)} \\ &= \tfrac{1}{2}\det(c, c''). \end{aligned}$$

So

$$\boxed{\text{Kepler's second law is equivalent to } \det(c, c'') = 0.}$$

Putting this all together we have:

THEOREM 1 Kepler's second law is true if and only if the force is central, and in this case each planetary path $c(t) = r(t) \cdot \mathbf{e}(\theta(t))$ satisfies the equation

$$(K_2) \qquad r^2\theta' = \det(c, c') = \text{constant}.$$

PROOF Saying that the force is central just means that it always points along $c(t)$. Since $c''(t)$ is in the direction of the force, that is equivalent to saying that $c''(t)$ always points along $c(t)$. And this is equivalent to saying that we always have

$$\det(c, c'') = 0.$$

We've just seen that this is equivalent to Kepler's second law.

Moreover, this equation implies that $\big[\det(c, c')\big]' = 0$, which by (5) gives (K_2). ▌

Newton next showed that if the gravitational force of the sun is a central force and also satisfies an inverse square law, then the path of any object in it will be a conic section having the sun at one focus. Planets, of course, correspond to the case where the conic section is an ellipse, and this is also true for comets that visit the sun periodically; parabolas and hyperbolas represent objects that come from outside the solar system, and eventually continue on their merry way back outside the system.

THEOREM 2 If the gravitational force of the sun is a central force that satisfies an inverse square law, then the path of any body in it will be a conic section having the sun at one focus.

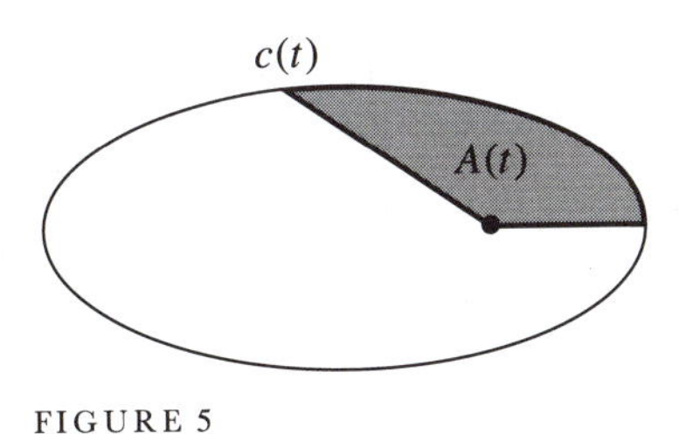

FIGURE 5

PROOF Notice that out conclusion specifies the shape of the path, not a particular parameterization. But this parameterization is essentially determined by Theorem 1: the hypothesis of a central force implies that the area $A(t)$ (Figure 5) is proportional to t, so determining $c(t)$ is essentially equivalent to determining A for arbitrary points on the ellipse. Unfortunately, the areas of such segments cannot be determined explicitly.[1] This means that we have to determine the *shape* of the path $c = r(t) \cdot \mathbf{e}(\theta(t))$ without finding its parameterization! Since it is the function $r \circ \theta^{-1}$ which actually describes the shape of the path in polar coordinates, we shouldn't be surprised to find θ^{-1} entering into the proof.

By Theorem 1, the hypothesis of a central force implies that

$$(K_2) \qquad r^2\theta' = \det(c, c') = M$$

for some constant M. The hypothesis of an inverse square law can be written

$$(*) \qquad c''(t) = -\frac{H}{r(t)^2}\,\mathbf{e}(\theta(t))$$

for some constant H. Using (K_2), this can be written

$$\frac{c''(t)}{\theta'(t)} = -\frac{H}{M}\,\mathbf{e}(\theta(t)).$$

Notice that the left-hand side of this equation is

$$[c' \circ \theta^{-1}]'(\theta(t)).$$

So if we let

$$D = c' \circ \theta^{-1}$$

(this is the main trick—"we consider c' as a function of θ"), then the equation can be written as

$$D'(\theta(t)) = -\frac{H}{M}\,\mathbf{e}(\theta(t)) = -\frac{H}{M}\big(\cos\theta(t), \sin\theta(t)\big),$$

[1] More precisely, we can't write down a solution in terms of familiar "standard functions," like sin, arcsin, etc.

and we can write this simply as

$$D'(u) = -\frac{H}{M}(\cos u, \sin u) = \left(-\frac{H}{M}\cos u, -\frac{H}{M}\sin u\right)$$

[for all u of the form $\theta(t)$ for some t, which happens to be all u], completely eliminating θ.

The equation that we have just obtained is simply a pair of equations, for the components of D, each of which we can easily solve individually; we thus find that

$$D(u) = \left(\frac{H \cdot \sin u}{-M} + A, \frac{H \cdot \cos u}{M} + B\right)$$

for two constants A and B. Letting $u = \theta(t)$ again we thus have an explicit formula for c':

$$c' = \left(\frac{H \cdot \sin \theta}{-M} + A, \frac{H \cdot \cos \theta}{M} + B\right).$$

[Here $\sin \theta$ really stands for $\sin \circ \theta$, etc., abbreviations that we will use throughout.]

Although we can't get an explicit formula for c itself, if we substitute this equation, together with $c = r(\cos \theta, \sin \theta)$, into the equation

$$\det(c, c') = M \qquad \text{(equation } (K_2)),$$

we get

$$r\left[\frac{H}{M}\cos^2 \theta + B \cos \theta + \frac{H}{M}\sin^2 \theta - A \sin \theta\right] = M,$$

which simplifies to

$$r\left[\frac{H}{M^2} + \frac{B}{M}\cos \theta - \frac{A}{M}\sin \theta\right] = 1.$$

Problem 15-8 shows that this can be written in the form

$$r(t)\left[\frac{H}{M^2} + C \cos(\theta(t) + D)\right] = 1,$$

for some constants C and D. We can let $D = 0$, since this simply amounts to rotating our polar coordinate system (choosing which ray corresponds to $\theta = 0$), so we can write, finally,

$$r[1 + \varepsilon \cos \theta] = \frac{M^2}{H} = \Lambda.$$

But this is the formula for a conic section derived in Appendix 3 of Chapter 4. ▮

In terms of the constant M in the equation

$$r^2\theta' = M$$

and the constant Λ in the equation of the orbit

$$r[1 + \varepsilon \cos \theta] = \Lambda$$

the last equation in our proof shows that we can rewrite (*) as

$$(**) \qquad c''(t) = -\frac{M^2}{\Lambda} \cdot \frac{1}{r^2}\,\mathbf{e}(\theta(t)).$$

Recall (page 87) that the major axis a of the ellipse is given by

$$(a) \qquad a = \frac{\Lambda}{1-\varepsilon^2},$$

while the minor axis b is given by

$$(b) \qquad b = \frac{\Lambda}{\sqrt{1-\varepsilon^2}}.$$

Consequently,

$$(c) \qquad \frac{b^2}{\Lambda} = a.$$

Remember that equation (5) gives

$$A'(t) = \tfrac{1}{2}r^2\theta' = \tfrac{1}{2}M,$$

and thus

$$A(t) = \tfrac{1}{2}Mt.$$

We can therefore interpret M in terms of the period T of the orbit. This period T is, by definition, the value of t for which we have $\theta(t) = 2\pi$, so that we obtain the complete ellipse. Hence

$$\text{area of the ellipse} = A(T) = \tfrac{1}{2}MT,$$

or

$$M = \frac{2(\text{area of the ellipse})}{T} = \frac{2\pi ab}{T} \qquad \text{by Problem 13-17.}$$

Hence the constant M^2/Λ in (**) is

$$\begin{aligned}\frac{M^2}{\Lambda} &= \frac{4\pi^2a^2b^2}{T^2\Lambda}\\ &= \frac{4\pi^2a^3}{T^2}, \qquad \text{using (c).}\end{aligned}$$

This completes the final step of Newton's analysis:

THEOREM 3 Kepler's third law is true if and only if the acceleration $c''(t)$ of any planet, moving on an ellipse, satisfies

$$c''(t) = -G \cdot \frac{1}{r^2}\,\mathbf{e}(\theta(t))$$

for a constant G that does not depend on the planet.

It should be mentioned that the converse of Theorem 2 is also true. To prove this, we first want to establish one further consequence of Kepler's second law. Recall that for

$$\mathbf{e}(t) = (\cos t, \sin t)$$

we have

$$\mathbf{e}'(t) = (-\sin t, \cos t).$$

Consequently,

$$\mathbf{e}''(t) = (-\cos t, -\sin t) = -\mathbf{e}(t).$$

Now differentiating (3) gives

$$\begin{aligned} c''(t) = r''(t) \cdot \mathbf{e}(\theta(t)) + r'(t)\theta'(t) \cdot \mathbf{e}'(\theta(t)) \\ + r'(t)\theta'(t) \cdot \mathbf{e}'(\theta(t)) + r(t)\theta''(t) \cdot \mathbf{e}'(\theta(t)) + r(t)\theta'(t)\theta'(t) \cdot \mathbf{e}''(\theta(t)). \end{aligned}$$

Using $\mathbf{e}''(t) = -\mathbf{e}(t)$ we get

$$c''(t) = \left[r''(t) - r(t)\theta'(t)^2\right] \cdot \mathbf{e}(\theta(t)) + \left[2r'(t)\theta'(t) + r(t)\theta''(t)\right] \cdot \mathbf{e}'(\theta(t)).$$

Since Kepler's second law implies central forces, hence that $c''(t)$ is always a multiple of $c(t)$, and thus always a multiple of $\mathbf{e}(\theta(t))$, the coefficient of $\mathbf{e}'(\theta(t))$ must be 0 [as a matter of fact, we can see this directly by taking the derivative of formula (K_2)]. Thus Kepler's second law implies that

$$(6) \qquad c''(t) = \left[r''(t) - r(t)\theta'(t)^2\right] \cdot \mathbf{e}(\theta(t)).$$

THEOREM 4 If the path of a planet moving under a central gravitational force is an ellipse with the sun as focus, then the force must satisfies an inverse square law.

PROOF As in Theorem 2, notice that the hypothesis on the shape of the path, together with the hypothesis of a central force, which is equivalent to Kepler's second law, essentially determines the parameterization. But we can't write down an explicit solution, so we have to obtain information about the acceleration without actually knowing what it is.

Once again, the hypothesis of a central force implies that

$$(K_2) \qquad r^2\theta' = M,$$

for some constant M, and the hypothesis that the path is an ellipse with the sun as focus implies that it satisfies the equation

$$(\mathrm{A}) \qquad r[1 + \varepsilon\cos\theta] = \Lambda,$$

for some ε and Λ. For our (not especially illuminating) proof, we will keep differentiating and substituting from these two equations.

First we differentiate (A) to obtain

$$r'[1 + \varepsilon\cos\theta] - \varepsilon r\theta'\sin\theta = 0.$$

Multiplying by r this becomes

$$rr'[1 + \varepsilon\cos\theta] - \varepsilon r^2\theta'\sin\theta = 0.$$

Using both (A) and (K_2), this becomes

$$\Lambda r' - \varepsilon M \sin\theta = 0.$$

Differentiating again, we get

$$\Lambda r'' - \varepsilon M \theta' \cos\theta = 0.$$

Using (K_2) we get

$$\Lambda r'' - \frac{\varepsilon M^2}{r^2} \cos\theta = 0,$$

and then using (A) we get

$$\Lambda r'' - \frac{M^2}{r^2}\left[\frac{\Lambda}{r} - 1\right] = 0.$$

Substituting from (K_2) yet again, we get

$$\Lambda[r'' - r(\theta')^2] + \frac{M^2}{r^2} = 0,$$

or

$$r'' - r(\theta')^2 = -\frac{M^2}{\Lambda r^2}.$$

Comparing with (6), we obtain

$$c''(t) = -\frac{M^2}{\Lambda r^2}\,\mathbf{e}(\theta(t)),$$

which is precisely what we wanted to show: the force is inversely proportional to the square of the distance from the sun to the planet. ▮

CHAPTER 18 THE LOGARITHM AND EXPONENTIAL FUNCTIONS

In Chapter 15 the integral provided a rigorous formulation for a preliminary definition of the functions sin and cos. In this chapter the integral plays a more essential role. For certain functions even a preliminary definition presents difficulties. For example, consider the function

$$f(x) = 10^x.$$

This function is assumed to be defined for all x and to have an inverse function, defined for positive x, which is the "logarithm to the base 10,"

$$f^{-1}(x) = \log_{10} x.$$

In algebra, 10^x is usually defined only for *rational* x, while the definition for irrational x is quietly ignored. A brief review of the definition for rational x will not only explain this omission, but also recall an important principle behind the definition of 10^x.

The symbol 10^n is first defined for natural numbers n. This notation turns out to be extremely convenient, especially for multiplying very large numbers, because

$$10^n \cdot 10^m = 10^{n+m}.$$

The extension of the definition of 10^x to rational x is motivated by the desire to preserve this equation; this requirement actually forces upon us the customary definition. Since we want the equation

$$10^0 \cdot 10^n = 10^{0+n} = 10^n$$

to be true, we must define $10^0 = 1$; since we want the equation

$$10^{-n} \cdot 10^n = 10^0 = 1$$

to be true, we must define $10^{-n} = 1/10^n$; since we want the equation

$$\underbrace{10^{1/n} \cdot \ldots \cdot 10^{1/n}}_{n \text{ times}} = 10^{\overbrace{1/n + \cdots + 1/n}^{}} = 10^1 = 10$$

$$\underbrace{10^{1/n} \cdot \ldots \cdot 10^{1/n}}_{n \text{ times}} = 10^{\underbrace{1/n+\cdots+1/n}_{n \text{ times}}} = 10^1 = 10$$

to be true, we must define $10^{1/n} = \sqrt[n]{10}$; and since we want the equation

$$\underbrace{10^{1/n} \cdot \ldots \cdot 10^{1/n}}_{m \text{ times}} = 10^{\underbrace{1/n+\cdots+1/n}_{m \text{ times}}} = 10^{m/n}$$

to be true, we must define $10^{m/n} = (\sqrt[n]{10})^m$.

Unfortunately, at this point the program comes to a dead halt. We have been guided by the principle that 10^x should be defined so as to ensure that $10^{x+y} = 10^x 10^y$; but this principle does not suggest any simple algebraic way of defining

10^x for irrational x. For this reason we will try some more sophisticated ways of finding a function f such that

$$(*) \qquad f(x+y) = f(x) \cdot f(y) \quad \text{for all } x \text{ and } y.$$

Of course, we are interested in a function which is not always zero, so we might add the condition $f(1) \neq 0$. If we add the more specific condition $f(1) = 10$, then $(*)$ will imply that $f(x) = 10^x$ for rational x, and 10^x could be *defined* as $f(x)$ for other x; in general $f(x)$ will equal $[f(1)]^x$ for rational x.

One way to find such a function is suggested if we try to solve an apparently more difficult problem: find a *differentiable* function f such that

$$\begin{gathered} f(x+y) = f(x) \cdot f(y) \quad \text{for all } x \text{ and } y, \\ f(1) = 10. \end{gathered}$$

Assuming that such a function exists, we can try to find f'—knowing the derivative of f might provide a clue to the definition of f itself. Now

$$\begin{aligned} f'(x) &= \lim_{h \to 0} \frac{f(x+h) - f(x)}{h} \\ &= \lim_{h \to 0} \frac{f(x) \cdot f(h) - f(x)}{h} \\ &= f(x) \cdot \lim_{h \to 0} \frac{f(h) - 1}{h}. \end{aligned}$$

The answer thus depends on

$$f'(0) = \lim_{h \to 0} \frac{f(h) - 1}{h};$$

for the moment assume this limit exists, and denote it by α. Then

$$f'(x) = \alpha \cdot f(x) \quad \text{for all } x.$$

Even if α could be computed, this approach seems self-defeating. The derivative of f has been expressed in terms of f again.

If we examine the inverse function $f^{-1} = \log_{10}$, the whole situation appears in a new light:

$$\begin{aligned} \log_{10}{}'(x) &= \frac{1}{f'(f^{-1}(x))} \\ &= \frac{1}{\alpha \cdot f(f^{-1}(x))} = \frac{1}{\alpha x}. \end{aligned}$$

The derivative of f^{-1} is about as simple as one could ask! And, what is even more interesting, of all the integrals $\int_a^b x^n \, dx$ examined previously, the integral $\int_a^b x^{-1} \, dx$ is the only one which we cannot evaluate. Since $\log_{10} 1 = 0$ we should have

$$\frac{1}{\alpha} \int_1^x \frac{1}{t} \, dt = \log_{10} x - \log_{10} 1 = \log_{10} x.$$

This suggests that we define $\log_{10} x$ as $(1/\alpha)\int_1^x t^{-1}\,dt$. The difficulty is that α is unknown. One way of evading this difficulty is to define

$$\log x = \int_1^x \frac{1}{t}\,dt,$$

and hope that this integral will be the logarithm to *some* base, which might be determined later. In any case, the function defined in this way is surely more reasonable, from a mathematical point of view, than $\log_{10}$. The usefulness of $\log_{10}$ depends on the important role of the number 10 in arabic notation (and thus ultimately on the fact that we have ten fingers), while the function log provides a notation for an extremely simple integral which cannot be evaluated in terms of any functions already known to us.

DEFINITION If $x > 0$, then

$$\mathbf{log}\ x = \int_1^x \frac{1}{t}\,dt.$$

The graph of log is shown in Figure 1. Notice that if $x > 1$, then $\log x > 0$, and if $0 < x < 1$, then $\log x < 0$, since, by our conventions,

$$\int_1^x \frac{1}{t}\,dt = -\int_x^1 \frac{1}{t}\,dt < 0.$$

For $x \le 0$, a number $\log x$ cannot be defined in this way, because $f(t) = 1/t$ is not bounded on $[x, 1]$.

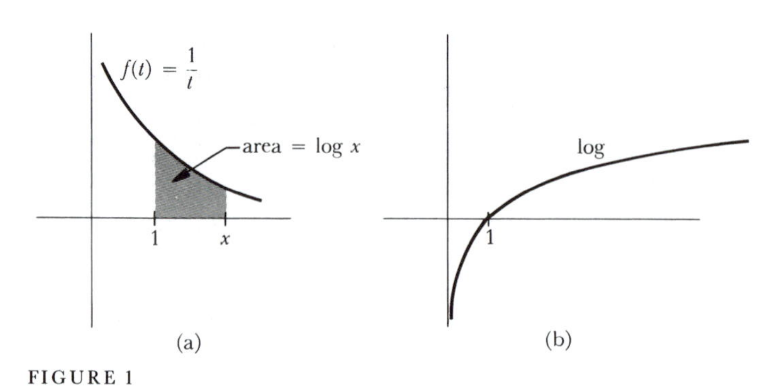

FIGURE 1

The justification for the notation "log" comes from the following theorem.

THEOREM 1 If $x, y > 0$, then

$$\log(xy) = \log x + \log y.$$

PROOF Notice first that $\log'(x) = 1/x$, by the Fundamental Theorem of Calculus. Now choose a number $y > 0$ and let

$$f(x) = \log(xy).$$

Then

$$f'(x) = \log'(xy) \cdot y = \frac{1}{xy} \cdot y = \frac{1}{x}.$$

Thus $f' = \log'$. This means that there is a number c such that

$$f(x) = \log x + c \quad \text{for all } x > 0,$$

that is,

$$\log(xy) = \log x + c \quad \text{for all } x > 0.$$

The number c can be evaluated by noting that when $x = 1$ we obtain

$$\begin{aligned}\log(1 \cdot y) &= \log 1 + c \\ &= c.\end{aligned}$$

Thus

$$\log(xy) = \log x + \log y \qquad \text{for all } x.$$

Since this is true for all $y > 0$, the theorem is proved. ▮

COROLLARY 1 If n is a natural number and $x > 0$, then

$$\log(x^n) = n \log x.$$

PROOF Let to you (use induction). ▮

COROLLARY 2 If $x, y > 0$, then

$$\log\left(\frac{x}{y}\right) = \log x - \log y.$$

PROOF This follows from the equations

$$\log x = \log\left(\frac{x}{y} \cdot y\right) = \log\left(\frac{x}{y}\right) + \log y. \ ▮$$

Theorem 1 provides some important information about the graph of log. The function log is clearly increasing, but since $\log'(x) = 1/x$, the derivative becomes very small as x becomes large, and log consequently grows more and more slowly. It is not immediately clear whether log is bounded or unbounded on **R**. Observe, however, that for a natural number n,

$$\log(2^n) = n \log 2 \quad (\text{and } \log 2 > 0);$$

it follows that log is, in fact, not bounded above. Similarly,

$$\log\left(\frac{1}{2^n}\right) = \log 1 - \log 2^n = -n \log 2;$$

therefore log is not bounded below on $(0, 1)$. Since log is continuous, it actually takes on all values. Therefore $\mathbf{R}$ is the domain of the function $\log^{-1}$. This important function has a special name, whose appropriateness will soon become clear.

DEFINITION The "exponential function," **exp**, is defined as $\log^{-1}$.

The graph of exp is shown in Figure 2. Since $\log x$ is defined only for $x > 0$, we always have $\exp(x) > 0$. The derivative of the function exp is easy to determine.

THEOREM 2 For all numbers x,

$$\exp'(x) = \exp(x).$$

PROOF

$$\begin{aligned}\exp'(x) = (\log^{-1})'(x) &= \frac{1}{\log'(\log^{-1}(x))} \\ &= \frac{1}{\dfrac{1}{\log^{-1}(x)}} \\ &= \log^{-1}(x) = \exp(x). \blacksquare\end{aligned}$$

A second important property of exp is an easy consequence of Theorem 1.

THEOREM 3 If x and y are any two numbers, then

$$\exp(x + y) = \exp(x) \cdot \exp(y).$$

PROOF Let $x' = \exp(x)$ and $y' = \exp(y)$, so that

$$\begin{aligned} x &= \log x', \\ y &= \log y'. \end{aligned}$$

Then

$$x + y = \log x' + \log y' = \log(x'y').$$

This means that

$$\exp(x + y) = x'y' = \exp(x) \cdot \exp(y). \blacksquare$$

FIGURE 2

This theorem, and the discussion at the beginning of this chapter, suggest that $\exp(1)$ is particularly important. There is, in fact, a special symbol for this number.

DEFINITION

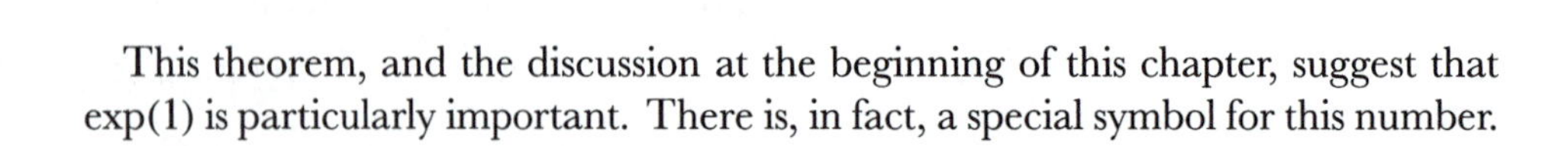

$$\boldsymbol{e} = \exp(1).$$

This definition is equivalent to the equation

$$1 = \log e = \int_1^e \frac{1}{t}\,dt.$$

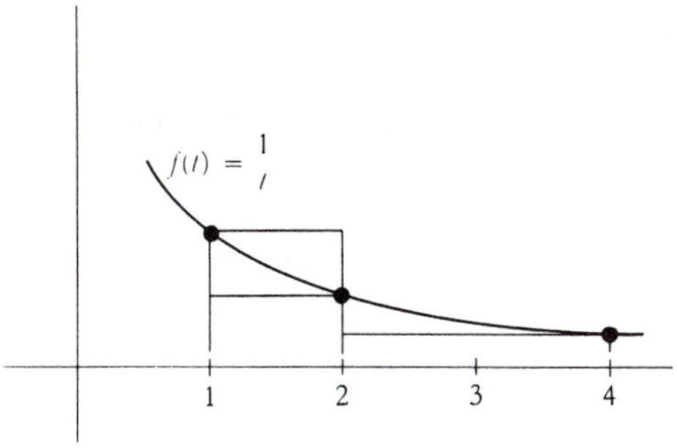

As illustrated in Figure 3,

$$\int_1^2 \frac{1}{t}\,dt < 1, \quad \text{since } 1 \cdot (2-1) \text{ is an upper sum for } f(t) = 1/t \text{ on } [1,2],$$

and

$$\int_1^4 \frac{1}{t}\,dt > 1, \quad \text{since } \tfrac{1}{2} \cdot (2-1) + \tfrac{1}{4} \cdot (4-2) = 1 \text{ is a lower sum for } f(t) = 1/t \text{ on } [1,4].$$

FIGURE 3

Thus

$$\int_1^2 \frac{1}{t}\,dt < \int_1^e \frac{1}{t}\,dt < \int_1^4 \frac{1}{t}\,dt,$$

which shows that

$$2 < e < 4.$$

In Chapter 20 we will find much better approximations for e, and also prove that e is irrational (the proof is much easier than the proof that π is irrational!).

As we remarked at the beginning of the chapter, the equation

$$\exp(x + y) = \exp(x) \cdot \exp(y)$$

implies that

$$\begin{aligned} \exp(x) &= [\exp(1)]^x \\ &= e^x, \quad \text{for all } \textit{rational } x. \end{aligned}$$

Since exp is defined for all x and $\exp(x) = e^x$ for rational x, it is consistent with our earlier use of the exponential notation to *define* e^x as $\exp(x)$ for all x.

DEFINITION

For any number x,

$$e^x = \exp(x).$$

The terminology "exponential function" should now be clear. We have succeeded in defining e^x for an arbitrary (even irrational) exponent x. We have not yet defined a^x, if $a \neq e$, but there is a reasonable principle to guide us in the attempt. If x is *rational*, then

$$a^x = (e^{\log a})^x = e^{x \log a}.$$

But the last expression is defined for *all* x, so we can use it to define a^x.

DEFINITION

> If $a > 0$, then, for any real number x,
> $$a^x = e^{x \log a}.$$
> (If $a = e$ this definition clearly agrees with the previous one.)

The requirement $a > 0$ is necessary, in order that $\log a$ be defined. This is not unduly restrictive since, for example, we would not even expect

$$(-1)^{1/2} \stackrel{?}{=} \sqrt{-1}$$

to be defined. (Of course, for certain rational x, the symbol a^x will make sense, according to the old definition; for example,

$$(-1)^{1/3} = \sqrt[3]{-1} = -1.)$$

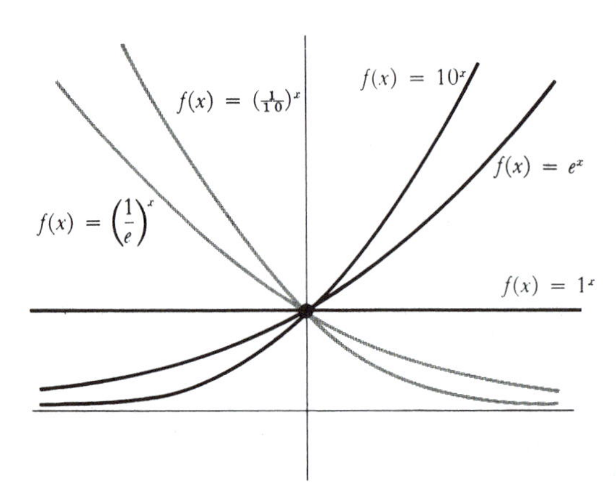

FIGURE 4

Our definition of a^x was designed to ensure that

$$(e^x)^y = e^{xy} \quad \text{for all } x \text{ and } y.$$

As we would hope, this equation turns out to be true when e is replaced by any number $a > 0$. The proof is a moderately involved unraveling of terminology. At the same time we will prove the other important properties of a^x.

THEOREM 4 If $a > 0$, then

$$(1) \quad (a^b)^c = a^{bc} \quad \text{for all } b, c.$$

(Notice that a^b will automatically be positive, so $(a^b)^c$ will be defined);

$$(2) \quad a^1 = a \text{ and } a^{x+y} = a^x \cdot a^y \quad \text{for all } x, y.$$

(Notice that (2) implies that this definition of a^x agrees with the old one for all rational x.)

PROOF

$$(1) \quad (a^b)^c = e^{c \log a^b} = e^{c \log(e^{b \log a})} = e^{c(b \log a)} = e^{cb \log a} = a^{bc}.$$

(Each of the steps in this string of equalities depends upon our last definition, or the fact that $\exp = \log^{-1}$.)

$$(2) \quad \begin{aligned} a^1 &= e^{1 \log a} = e^{\log a} = a, \\ a^{x+y} &= e^{(x+y)\log a} = e^{x \log a + y \log a} = e^{x \log a} \cdot e^{y \log a} = a^x \cdot a^y. \ \blacksquare \end{aligned}$$

Figure 4 shows the graphs of $f(x) = a^x$ for several different a. The behavior of the function depends on whether $a < 1$, $a = 1$, or $a > 1$. If $a = 1$, then

$f(x) = 1^x = 1$. Suppose $a > 1$. In this case $\log a > 0$. Thus,

$$\begin{array}{ll} \text{if} & x < y, \\ \text{then} & x \log a < y \log a, \\ \text{so} & e^{x \log a} < e^{y \log a}, \\ \text{i.e.,} & a^x < a^y. \end{array}$$

Thus the function $f(x) = a^x$ is increasing. On the other hand, if $0 < a < 1$, so that $\log a < 0$, the same sort of reasoning shows that the function $f(x) = a^x$ is decreasing. In either case, if $a > 0$ and $a \neq 1$, then $f(x) = a^x$ is one-one. Since exp takes on every positive value it is also easy to see that a^x takes on every positive value. Thus the inverse function is defined for all positive numbers, and takes on all values. If $f(x) = a^x$, then f^{-1} is the function usually denoted by $\log_a$ (Figure 5).

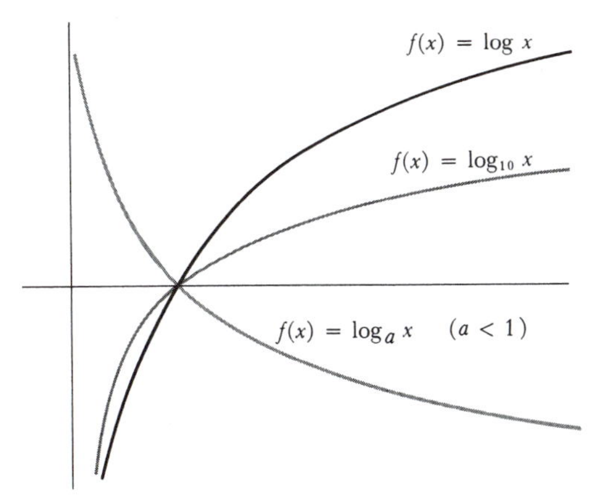

FIGURE 5

Just as a^x can be expressed in terms of exp, so $\log_a$ can be expressed in terms of log. Indeed,

$$\begin{array}{ll} \text{if} & y = \log_a x, \\ \text{then} & x = a^y = e^{y \log a}, \\ \text{so} & \log x = y \log a, \\ \text{or} & y = \dfrac{\log x}{\log a}. \end{array}$$

In other words,

$$\log_a x = \frac{\log x}{\log a}.$$

The derivatives of $f(x) = a^x$ and $g(x) = \log_a x$ are both easy to find:

$$\begin{aligned} f(x) &= e^{x \log a}, \quad \text{so } f'(x) = \log a \cdot e^{x \log a} = \log a \cdot a^x, \\ g(x) &= \frac{\log x}{\log a}, \quad \text{so } g'(x) = \frac{1}{x \log a}. \end{aligned}$$

A more complicated function like

$$f(x) = g(x)^{h(x)}$$

is also easy to differentiate, if you remember that, *by definition*,

$$f(x) = e^{h(x) \log g(x)};$$

it follows from the Chain Rule that

$$\begin{aligned} f'(x) &= e^{h(x) \log g(x)} \cdot \left[h'(x) \log g(x) + h(x) \frac{g'(x)}{g(x)} \right] \\ &= g(x)^{h(x)} \cdot \left[h'(x) \log g(x) + h(x) \frac{g'(x)}{g(x)} \right]. \end{aligned}$$

There is no point in remembering this formula—simply apply the principle behind it in any specific case that arises; it does help, however, to remember that the first factor in the derivative will be $g(x)^{h(x)}$.

There is one special case of the above formula which *is* worth remembering. The function $f(x) = x^a$ was previously defined only for rational a. We can now define and find the derivative of the function $f(x) = x^a$ for any number a; the result is just what we would expect:

$$f(x) = x^a = e^{a \log x}$$

so

$$f'(x) = \frac{a}{x} \cdot e^{a \log x} = \frac{a}{x} \cdot x^a = ax^{a-1}.$$

Algebraic manipulations with the exponential functions will become second nature after a little practice—just remember that all the rules which ought to work actually do. The basic properties of exp are still those stated in Theorems 2 and 3:

$$\exp'(x) = \exp(x),$$
$$\exp(x + y) = \exp(x) \cdot \exp(y).$$

In fact, each of these properties comes close to characterizing the function exp. Naturally, exp is not the only function f satisfying $f' = f$, for if $f = ce^x$, then $f'(x) = ce^x = f(x)$; these functions are the only ones with this property, however.

THEOREM 5 If f is differentiable and

$$f'(x) = f(x) \quad \text{for all } x,$$

then there is a number c such that

$$f(x) = ce^x \quad \text{for all } x.$$

PROOF Let

$$g(x) = \frac{f(x)}{e^x}.$$

(This is permissible, since $e^x \neq 0$ for all x.) Then

$$g'(x) = \frac{e^x f'(x) - f(x)e^x}{(e^x)^2} = 0.$$

Therefore there is a number c such that

$$g(x) = \frac{f(x)}{e^x} = c \quad \text{for all } x. \blacksquare$$

The second basic property of exp requires a more involved discussion. The function exp is clearly not the only function f which satisfies

$$f(x + y) = f(x) \cdot f(y).$$

In fact, $f(x) = 0$ or any function of the form $f(x) = a^x$ also satisfies this equation. But the true story is much more complex than this—there are infinitely many other functions which satisfy this property, but it is impossible, without appealing to more advanced mathematics, to prove that there is even one function other than those already mentioned! It is for this reason that the definition of 10^x is so difficult: there are infinitely many functions f which satisfy

$$f(x+y) = f(x) \cdot f(y),$$
$$f(1) = 10,$$

but which are *not* the function $f(x) = 10^x$! One thing is true however—any *continuous* function f satisfying

$$f(x+y) = f(x) \cdot f(y)$$

must be of the form $f(x) = a^x$ or $f(x) = 0$. (Problem 38 indicates the way to prove this, and also has a few words to say about discontinuous functions with this property.)

In addition to the two basic properties stated in Theorems 2 and 3, the function exp has one further property which is very important—exp "grows faster than any polynomial." In other words,

THEOREM 6 For any natural number n,

$$\lim_{x\to\infty} \frac{e^x}{x^n} = \infty.$$

PROOF The proof consists of several steps.

Step 1. $e^x > x$ for all x, and consequently $\lim_{x\to\infty} e^x = \infty$ (this may be considered to be the case $n = 0$).

To prove this statement (which is clear for $x \le 0$) it suffices to show that

$$x > \log x \quad \text{for all } x > 0.$$

If $x < 1$ this is clearly true, since $\log x < 0$. If $x > 1$, then (Figure 6) $x - 1$ is an upper sum for $f(t) = 1/t$ on $[1, x]$, so $\log x < x - 1 < x$.

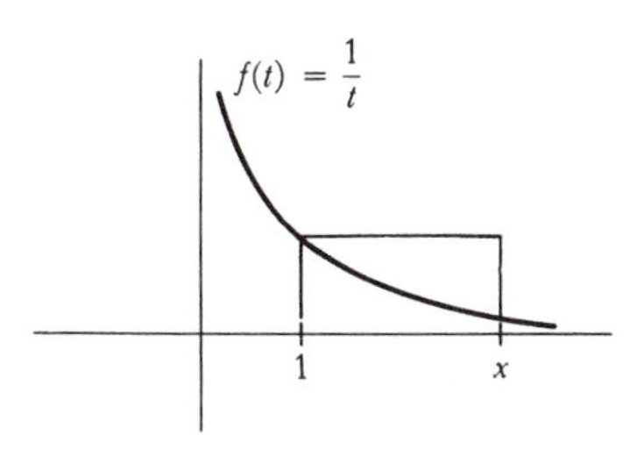

FIGURE 6

Step 2. $\lim_{x\to\infty} \frac{e^x}{x} = \infty.$

To prove this, note that

$$\frac{e^x}{x} = \frac{e^{x/2} \cdot e^{x/2}}{\frac{x}{2} \cdot 2} = \frac{1}{2}\left(\frac{e^{x/2}}{\frac{x}{2}}\right) \cdot e^{x/2}.$$

By Step 1, the expression in parentheses is greater than 1, and $\lim_{x\to\infty} e^{x/2} = \infty$; this shows that $\lim_{x\to\infty} e^x/x = \infty$.

Step 3. $\lim_{x\to\infty} \frac{e^x}{x^n} = \infty.$

Note that

$$\frac{e^x}{x^n} = \frac{(e^{x/n})^n}{\left(\frac{x}{n}\right)^n \cdot n^n} = \frac{1}{n^n} \cdot \left(\frac{e^{x/n}}{\frac{x}{n}}\right)^n.$$

The expression in parentheses becomes arbitrarily large, by Step 2, so the nth power certainly becomes arbitrarily large. ▮

It is now possible to examine carefully the following very interesting function: $f(x) = e^{-1/x^2}$, $x \neq 0$. We have

$$f'(x) = e^{-1/x^2} \cdot \frac{2}{x^3}.$$

Therefore,

$$\begin{aligned} f'(x) < 0 \quad & \text{for } x < 0, \\ f'(x) > 0 \quad & \text{for } x > 0, \end{aligned}$$

so f is decreasing for negative x and increasing for positive x. Moreover, if $|x|$ is large, then x^2 is large, so $-1/x^2$ is close to 0, so e^{-1/x^2} is close to 1 (Figure 7).

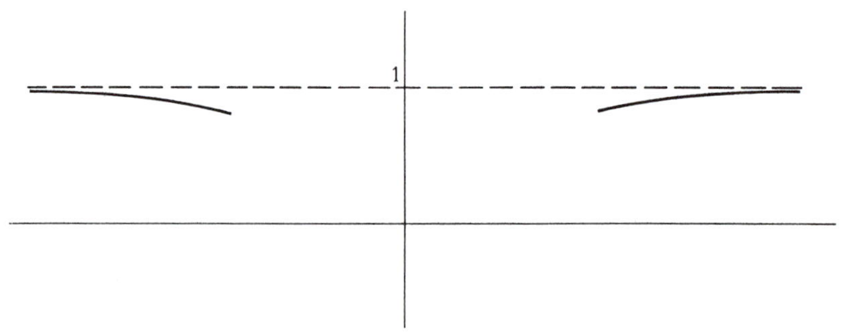

FIGURE 7

The behavior of f near 0 is more interesting. If x is small, then $1/x^2$ is large, so e^{1/x^2} is large, so $e^{-1/x^2} = 1/(e^{1/x^2})$ is small. This argument, suitably stated with ε's and δ's, shows that

$$\lim_{x \to 0} e^{-1/x^2} = 0.$$

Therefore, if we define

$$f(x) = \begin{cases} e^{-1/x^2}, & x \neq 0 \\ 0, & x = 0, \end{cases}$$

then the function f is continuous (Figure 8). In fact, f is actually differentiable

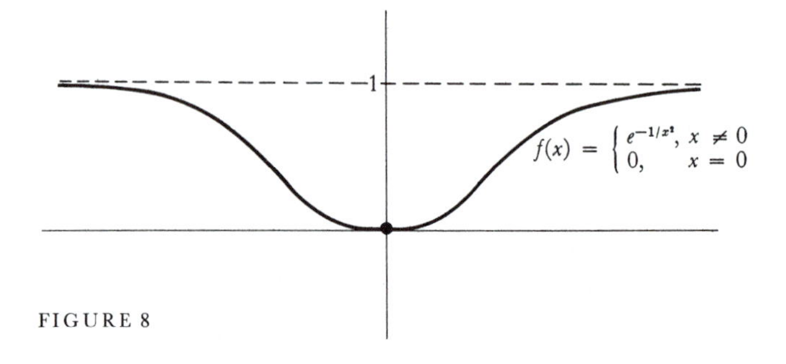

FIGURE 8

at 0: Indeed

$$f'(0) = \lim_{h \to 0} \frac{e^{-1/h^2}}{h} = \lim_{h \to 0} \frac{1/h}{e^{(1/h)^2}},$$

and

$$\lim_{h \to 0^+} \frac{1/h}{e^{(1/h)^2}} = \lim_{x \to \infty} \frac{x}{e^{(x^2)}}, \qquad \text{while} \qquad \lim_{h \to 0^-} \frac{1/h}{e^{(1/h)^2}} = -\lim_{x \to \infty} \frac{x}{e^{(x^2)}}.$$

We already know that

$$\lim_{x \to \infty} \frac{e^x}{x} = \infty;$$

it is all the more true that

$$\lim_{x \to \infty} \frac{e^{(x^2)}}{x} = \infty,$$

and this means that

$$\lim_{x \to \infty} \frac{x}{e^{(x^2)}} = 0.$$

Thus

$$f'(x) = \begin{cases} e^{-1/x^2} \cdot \dfrac{2}{x^3}, & x \neq 0 \\ 0, & x = 0. \end{cases}$$

We can now compute that

$$\begin{aligned} f''(0) &= \lim_{h \to 0} \frac{f'(h) - f'(0)}{h} \\ &= \lim_{h \to 0} \frac{e^{-1/h^2} \cdot \dfrac{2}{h^3}}{h} \\ &= \lim_{h \to 0} \frac{2 \cdot e^{-1/h^2}}{h^4} = \lim_{h \to 0} \frac{2 \cdot \dfrac{1}{h^4}}{e^{1/h^2}} = \lim_{x \to \infty} \frac{2x^4}{e^{(x^2)}}; \end{aligned}$$

an argument similar to the one above shows that $f''(0) = 0$. Thus

$$f''(x) = \begin{cases} e^{-1/x^2} \cdot \dfrac{-6}{x^4} + e^{-1/x^2} \cdot \dfrac{4}{x^6}, & x \neq 0 \\ 0, & x = 0. \end{cases}$$

This argument can be continued. In fact, using induction it can be shown (Problem 40) that $f^{(k)}(0) = 0$ for *every* k. The function f is *extremely* flat at 0, and approaches 0 so quickly that it can mask many irregularities of other functions. For example (Figure 9), suppose that

$$f(x) = \begin{cases} e^{-1/x^2} \cdot \sin \dfrac{1}{x}, & x \neq 0 \\ 0, & x = 0. \end{cases}$$

It can be shown (Problem 41) that for this function it is also true that $f^{(k)}(0) = 0$ for all k. This example shows, perhaps more strikingly than any other, just how bad a function can be, and still be infinitely differentiable. In Part IV we will investigate even more restrictive conditions on a function, which will finally rule out behavior of this sort.

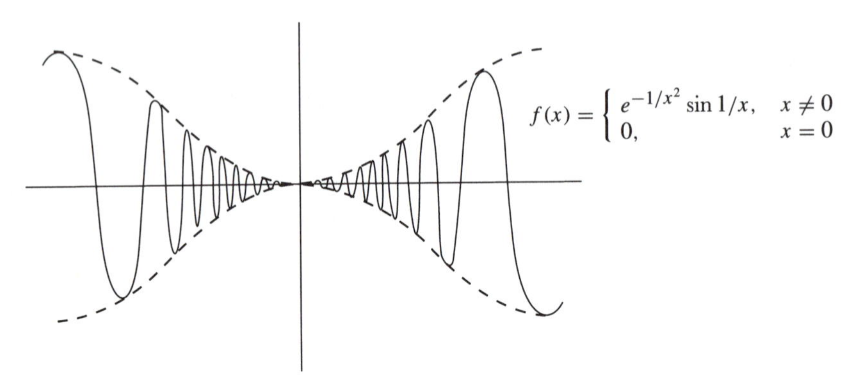

FIGURE 9

PROBLEMS

1. Differentiate each of the following functions (remember that a^{b^c} always denotes $a^{(b^c)}$).

(i) $f(x) = e^{e^{e^{e^x}}}$.

(ii) $f(x) = \log(1 + \log(1 + \log(1 + e^{1+e^{1+x}})))$.

(iii) $f(x) = (\sin x)^{\sin(\sin x)}$.

(iv) $f(x) = e^{\left(\int_0^x e^{-t^2}\,dt\right)}$.

(v) $f(x) = \sin x^{\sin x^{\sin x}}$.

(vi) $f(x) = \log_{(e^x)} \sin x$.

(vii) $f(x) = \left[\arcsin\left(\dfrac{x}{\sin x}\right)\right]^{\log(\sin e^x)}$.

(viii) $f(x) = (\log(3 + e^4))e^{4x} + (\arcsin x)^{\log 3}$.

(ix) $f(x) = (\log x)^{\log x}$.

(x) $f(x) = x^x$.

2. (a) The derivative of $\log \circ f$ is f'/f.

This expression is called the *logarithmic derivative* of f. It is often easier to compute than f', since products and powers in the expression for f become sums and products in the expression for $\log \circ f$. The derivative f' can then be recovered simply by multiplying by f; this process is called *logarithmic differentiation.*

(b) Use logarithmic differentiation to find $f'(x)$ for each of the following.

(i) $f(x) = (1 + x)(1 + e^{x^2})$.

(ii) $f(x) = \dfrac{(3-x)^{1/3}x^2}{(1-x)(3+x)^{2/3}}$.

(iii) $f(x) = (\sin x)^{\cos x} + (\cos x)^{\sin x}$.

(iv) $f(x) = \dfrac{e^x - e^{-x}}{e^{2x}(1+x^3)}$.

3. Find

$$\int_a^b \frac{f'(t)}{f(t)}\,dt$$

(for $f > 0$ on $[a, b]$).

4. Graph each of the following functions.

(a) $f(x) = e^{x+1}$.

(b) $f(x) = e^{\sin x}$.

(c) $f(x) = e^x + e^{-x}$.

(d) $f(x) = e^x - e^{-x}$.

(Compare the graph with the graphs of exp and 1/exp.) [applies to (c) and (d)]

(e) $f(x) = \dfrac{e^x - e^{-x}}{e^x + e^{-x}} = \dfrac{e^{2x}-1}{e^{2x}+1} = 1 - \dfrac{2}{e^{2x}+1}$.

5. Find the following limits by l'Hôpital's Rule.

(i) $\displaystyle\lim_{x\to 0} \frac{e^x - 1 - x - x^2/2}{x^2}$.

(ii) $\displaystyle\lim_{x\to 0} \frac{e^x - 1 - x - x^2/2 - x^3/6}{x^3}$.

(iii) $\displaystyle\lim_{x\to 0} \frac{e^x - 1 - x - x^2/2}{x^3}$.

(iv) $\displaystyle\lim_{x\to 0} \frac{\log(1+x) - x + x^2/2}{x^2}$.

(v) $\displaystyle\lim_{x\to 0} \frac{\log(1+x) - x + x^2/2}{x^3}$.

(vi) $\displaystyle\lim_{x\to 0} \frac{\log(1+x) - x + x^2/2 - x^3/3}{x^3}$.

6. The functions

$$\sinh x = \frac{e^x - e^{-x}}{2},$$

$$\cosh x = \frac{e^x + e^{-x}}{2},$$

$$\tanh x = \frac{e^x - e^{-x}}{e^x + e^{-x}} = 1 - \frac{2}{e^{2x}+1},$$

are called the **hyperbolic sine**, **hyperbolic cosine**, and **hyperbolic tangent**, respectively (but usually read 'sinch,' 'cosh,' and 'tanch'). There are many analogies between these functions and their ordinary trigonometric counterparts. One analogy is illustrated in Figure 10; a proof that the region

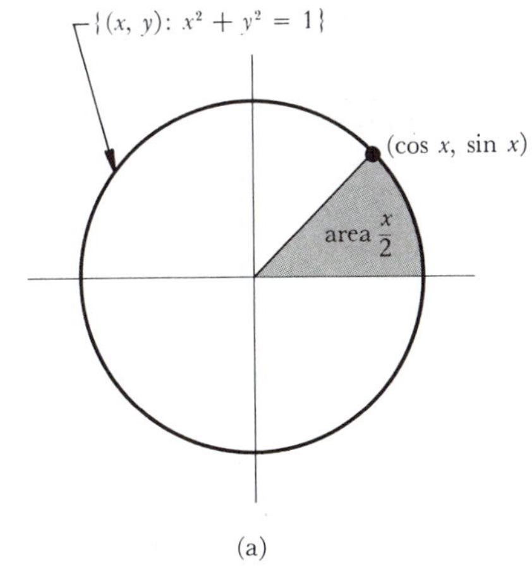

(a)

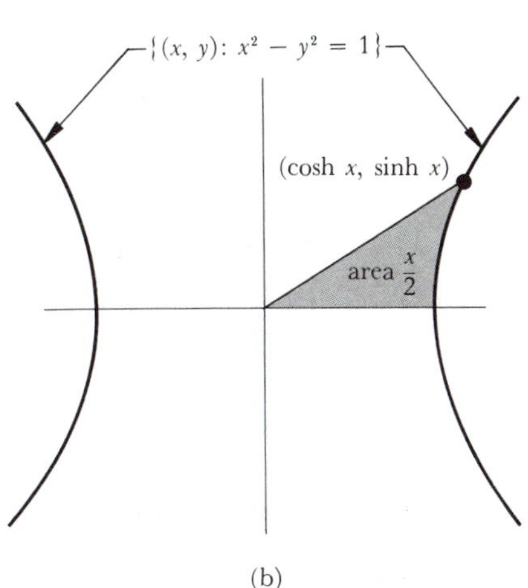

(b)

FIGURE 10

shown in Figure 10(b) really has area $x/2$ is best deferred until the next chapter, when we will develop methods of computing integrals. Other analogies are discussed in the following three problems, but the deepest analogies must wait until Chapter 27. If you have not already done Problem 4, graph the functions sinh, cosh, and tanh.

7. Prove that

(a) $\cosh^2 - \sinh^2 = 1$.
(b) $\tanh^2 + 1/\cosh^2 = 1$.
(c) $\sinh(x+y) = \sinh x \cosh x + \cosh x \sinh y$.
(d) $\cosh(x+y) = \cosh x \cosh y + \sinh x \sinh y$.
(e) $\sinh' = \cosh$.
(f) $\cosh' = \sinh$.
(g) $\tanh' = \dfrac{1}{\cosh^2}$.

8. The functions sinh and tanh are one-one; their inverses $\sinh^{-1}$ and $\tanh^{-1}$, are defined on **R** and $(-1, 1)$, respectively. These inverse functions are sometimes denoted by arg sinh and arg tanh (the "argument" of the hyperbolic sine and tangent). If cosh is restricted to $[0, \infty)$ it has an inverse, denoted by arg cosh, or simply $\cosh^{-1}$, which is defined on $[1, \infty)$. Prove, using the information in Problem 7, that

(a) $\sinh(\cosh^{-1} x) = \sqrt{x^2 - 1}$.
(b) $\cosh(\sinh^{-1} x) = \sqrt{1 + x^2}$.
(c) $(\sinh^{-1})'(x) = \dfrac{1}{\sqrt{1+x^2}}$.
(d) $(\cosh^{-1})'(x) = \dfrac{1}{\sqrt{x^2-1}}$ for $x > 1$.
(e) $(\tanh^{-1})'(x) = \dfrac{1}{1-x^2}$ for $|x| < 1$.

9. (a) Find an explicit formula for $\sinh^{-1}$, $\cosh^{-1}$, and $\tanh^{-1}$ (by solving the equation $y = \sinh^{-1} x$ for x in terms of y, etc.).
(b) Find

$$\int_a^b \frac{1}{\sqrt{1+x^2}}\,dx,$$

$$\int_a^b \frac{1}{\sqrt{x^2-1}}\,dx \quad \text{for } a, b > 1 \text{ or } a, b < 1,$$

$$\int_a^b \frac{1}{1-x^2}\,dx \quad \text{for } |a|, |b| < 1.$$

Compare your answer for the third integral with that obtained by writing

$$\frac{1}{1-x^2} = \frac{1}{2}\left[\frac{1}{1-x} + \frac{1}{1+x}\right].$$

10. Show that

$$F(x) = \int_2^x \frac{1}{\log t}\,dt$$

is not bounded on $[2, \infty)$.

11. Let f be a nondecreasing function on $[1, \infty)$, and define

$$F(x) = \int_1^x \frac{f(t)}{t}\,dt, \qquad x \geq 1.$$

Prove that f is bounded on $[1, \infty)$ if and only if $F/\log$ is bounded on $[1, \infty)$.

12. Find

(a) $\lim_{x\to\infty} a^x$ for $0 < a < 1$. (Remember the definition!)

(b) $\lim_{x\to\infty} \dfrac{x}{(\log x)^n}$.

(c) $\lim_{x\to\infty} \dfrac{(\log x)^n}{x}$.

(d) $\lim_{x\to 0^+} x(\log x)^n$. Hint: $x(\log x)^n = \dfrac{(-1)^n\left(\log \dfrac{1}{x}\right)^n}{\dfrac{1}{x}}$.

(e) $\lim_{x\to 0^+} x^x$.

13. Graph $f(x) = x^x$ for $x > 0$. (Use Problem 12(e).)

14. (a) Find the minimum value of $f(x) = e^x/x^n$ for $x > 0$, and conclude that $f(x) > e^n/n^n$ for $x > n$.
(b) Using the expression $f'(x) = e^x(x - n)/x^{n+1}$, prove that $f'(x) > e^{n+1}/(n+1)^{n+1}$ for $x > n + 1$, and thus obtain another proof that $\lim_{x\to\infty} f(x) = \infty$.

15. Graph $f(x) = e^x/x^n$.

16. (a) Find $\lim_{y\to 0} \log(1 + y)/y$. (You can use l'Hôpital's Rule, but that would be silly.)
(b) Find $\lim_{x\to\infty} x\log(1 + 1/x)$.
(c) Prove that $e = \lim_{x\to\infty} (1 + 1/x)^x$.
(d) Prove that $e^a = \lim_{x\to\infty} (1 + a/x)^x$. (It is possible to derive this from part (c) with just a little algebraic fiddling.)
*(e) Prove that $\log b = \lim_{x\to\infty} x(b^{1/x} - 1)$.

17. Graph $f(x) = (1 + 1/x)^x$ for $x > 0$. (Use Problem 16(c).)

18. If a bank gives a percent interest per annum, then an initial investment I yields $I(1 + a/100)$ after 1 year. If the bank compounds the interest (counts the accrued interest as part of the capital for computing interest the next year), then the initial investment grows to $I(1+a/100)^n$ after n years. Now suppose that interest is given twice a year. The final amount after n years is, alas, not $I(1+a/100)^{2n}$, but merely $I(1+a/200)^{2n}$—although interest is awarded twice as often, the interest must be halved in each calculation, since the interest is $a/2$ per half year. This amount is larger than $I(1+a/100)^n$, but not that much larger. Suppose that the bank now compounds the interest continuously, i.e., the bank considers what the investment would yield when compounding k times a year, and then takes the least upper bound of all these numbers. How much will an initial investment of 1 dollar yield after 1 year?

19. (a) Let $f(x) = \log|x|$ for $x \neq 0$. Prove that $f'(x) = 1/x$ for $x \neq 0$.
(b) If $f(x) \neq 0$ for all x, prove that $(\log|f|)' = f'/f$.

20. Suppose that on some interval the function f satisfies $f' = cf$ for some number c.

(a) Assuming that f is never 0, use Problem 19(b) to prove that $|f(x)| = le^{cx}$ for some number l (> 0). It follows that $f(x) = ke^{cx}$ for some k.
(b) Show that this result holds without the added assumption that f is never 0. Hint: Show that f can't be 0 at the endpoint of an open interval on which it is nowhere 0.
(c) Give a simpler proof that $f(x) = ke^{cx}$ for some k by considering the function $g(x) = f(x)/e^{cx}$.
(d) Suppose that $f' = fg'$ for some g. Show that $f(x) = ke^{g(x)}$ for some k.

***21.** A radioactive substance diminishes at a rate proportional to the amount present (since all atoms have equal probability of disintegrating, the total disintegration is proportional to the number of atoms remaining). If $A(t)$ is the amount at time t, this means that $A'(t) = cA(t)$ for some c (which represents the probability that an atom will disintegrate).

(a) Find $A(t)$ in terms of the amount $A_0 = A(0)$ present at time 0.
(b) Show that there is a number τ (the "half-life" of the radioactive element) with the property that $A(t+\tau) = A(t)/2$.

***22.** *Newton's law of cooling* states that an object cools at a rate proportional to the difference of its temperature and the temperature of the surrounding medium. Find the temperature $T(t)$ of the object at time t, in terms of its temperature T_0 at time 0, assuming that the temperature of the surrounding medium is kept at a constant, M. Hint: To solve the differential equation expressing Newton's law, remember that $T' = (T - M)'$.

***23.** Prove that if $f(x) = \int_0^x f(t)\,dt$, then $f = 0$.

24. Find all continuous functions f satisfying

(i) $\displaystyle\int_0^x f = e^x$.

(ii) $\displaystyle\int_0^{x^2} f = 1 - e^{2x^2}.$

25. Given a differentiable function f, find all continuous functions g satisfying

$$\int_0^{f(x)} fg = g(f(x)) - 1.$$

***26.** Find all functions f satisfying $f'(t) = f(t) + \displaystyle\int_0^1 f(t)\,dt.$

27. Find all continuous functions f which satisfy the equation

$$(f(x))^2 = \int_0^x f(t)\frac{t}{1+t^2}\,dt.$$

28. (a) Let f and g be continuous nonnegative functions on $[a, b]$, and let $C > 0$. Suppose that

$$f(x) \leq C + \int_a^x fg \qquad a \leq x \leq b.$$

Prove *Gronwall's inequality*:

$$f(x) \leq Ce^{\int_a^x g}.$$

Hint: Consider the derivative of the function $h(x) = C + \int_a^x fg$.

(b) Use a limiting argument to show that this result holds even for $C = 0$.

(c) Suppose that $f'(x) = g(x)f(x)$ for some continuous function g, and that $f(0) = 0$. Then $f = 0$. (Compare Problem 20.)

29. (a) Prove that

$$1 + x + \frac{x^2}{2!} + \frac{x^3}{3!} + \cdots + \frac{x^n}{n!} \leq e^x \quad \text{for } x \geq 0.$$

Hint: Use induction on n, and compare derivatives.

(b) Give a new proof that $\lim_{x\to\infty} e^x/x^n = \infty$.

30. Give yet another proof of this fact, using the appropriate form of l'Hôpital's Rule. (See Problem 11-53.)

31. (a) Evaluate $\displaystyle\lim_{x\to\infty} e^{-x^2}\int_0^x e^{t^2}\,dt$. (You should be able to make an educated guess before doing any calculations.)

(b) Evaluate the following limits.

(i) $$\lim_{x\to\infty} e^{-x^2}\int_x^{x+(1/x)} e^{t^2}\,dt.$$

(ii) $$\lim_{x\to\infty} e^{-x^2}\int_x^{x+(\log x/x)} e^{t^2}\,dt.$$

(iii) $$\lim_{x\to\infty} e^{-x^2}\int_x^{x+(\log x/2x)} e^{t^2}\,dt.$$

32. This problem outlines the classical approach to logarithms and exponentials. To begin with, we will simply assume that the function $f(x) = a^x$, defined in an elementary way for rational x, can somehow be extended to a continuous one-one function, obeying the same algebraic rules, on the whole line. (See Problem 22-29 for a direct proof of this.) The inverse of f will then be denoted by $\log_a$.

(a) Show, directly from the definition, that

$$\begin{aligned}\log_a{}'(x) &= \lim_{h\to 0} \log_a \left(1 + \frac{h}{x}\right)^{1/h} \\ &= \frac{1}{x} \cdot \log_a \left(\lim_{k\to 0} (1+k)^{1/k} \right).\end{aligned}$$

Thus, the whole problem has been reduced to the determination of $\lim_{h\to 0}(1+h)^{1/h}$. If we can show that this has a limit e, then $\log_e{}'(x) = \frac{1}{x} \cdot \log_e e = \frac{1}{x}$, and consequently $\exp = \log_e^{-1}$ has derivative $\exp'(x) = \exp(x)$.

(b) Let $a_n = \left(1 + \frac{1}{n}\right)^n$ for natural numbers n. Using the binomial theorem, show that

$$a_n = 2 + \sum_{k=2}^{n} \frac{1}{k!}\left(1 - \frac{1}{n}\right)\left(1 - \frac{2}{n}\right) \cdot \ldots \cdot \left(1 - \frac{k-1}{n}\right).$$

Conclude that $a_n < a_{n+1}$.

(c) Using the fact that $1/k! \le 1/2^{k-1}$ for $k \ge 2$, show that all $a_n < 3$. Thus, the set of numbers $\{a_1, a_2, a_3, \ldots\}$ is bounded, and therefore has a least upper bound e. Show that for any $\varepsilon > 0$ we have $e - a_n < \varepsilon$ for large enough n.

(d) If $n \le x \le n+1$, then

$$\left(1 + \frac{1}{n+1}\right)^n \le \left(1 + \frac{1}{x}\right)^x \le \left(1 + \frac{1}{n+1}\right)^{n+1}.$$

Conclude that $\lim_{x\to\infty}\left(1 + \frac{1}{x}\right)^x = e$. Also show that $\lim_{x\to -\infty}\left(1 + \frac{1}{x}\right)^x = e$, and conclude that $\lim_{h\to 0}(1+h)^{1/h} = e$.

***33.** A point P is moving along a line segment AB of length 10^7 while another point Q moves along an infinite ray (Figure 11). The velocity of P is always equal to the distance from P to B (in other words, if $P(t)$ is the position of P at time t, then $P'(t) = 10^7 - P(t)$), while Q moves with constant velocity $Q'(t) = 10^7$. The distance traveled by Q after time t is defined to be the *Napierian logarithm* of the distance from P to B at time t. Thus

$$10^7 t = \text{Nap log}[10^7 - P(t)].$$

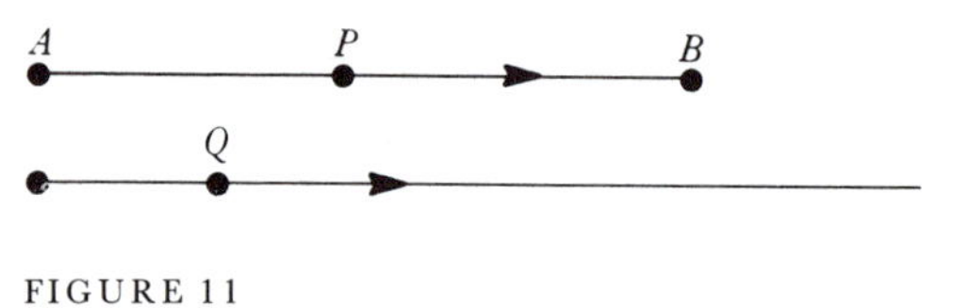

FIGURE 11

This was the definition of logarithms given by Napier (1550–1617) in his publication of 1614, *Mirifici logarithmonum canonis description* (A Description of the Wonderful Law of Logarithms); work which was done *before* the use of exponents was invented! The number 10^7 was chosen because Napier's tables (intended for astronomical and navigational calculations), listed the logarithms of sines of angles, for which the best possible available tables extended to seven decimal places, and Napier wanted to avoid fractions. Prove that

$$\text{Nap}\log x = 10^7 \log \frac{10^7}{x}.$$

Hint: Use the same trick as in Problem 22 to solve the equation for P.

***34.** (a) Sketch the graph of $f(x) = (\log x)/x$ (paying particular attention to the behavior near 0 and ∞).

(b) Which is larger, e^π or π^e?

(c) Prove that if $0 < x \le 1$, or $x = e$, then the only number y satisfying $x^y = y^x$ is $y = x$; but if $x > 1$, $x \ne e$, then there is precisely one number $y \ne x$ satisfying $x^y = y^x$; moreover, if $x < e$, then $y > e$, and if $x > e$, then $y < e$. (Interpret these statements in terms of the graph in part (a)!)

(d) Prove that if x and y are natural numbers and $x^y = y^x$, then $x = y$ or $x = 2$, $y = 4$, or $x = 4$, $y = 2$.

(e) Show that the set of all pairs (x, y) with $x^y = y^x$ consists of a curve and a straight line which intersect; find the intersection and draw a rough sketch.

**(f) For $1 < x < e$ let $g(x)$ be the unique number $> e$ with $x^{g(x)} = g(x)^x$. Prove that g is differentiable. (It is a good idea to consider separate functions,

$$f_1(x) = \frac{\log x}{x}, \quad 0 < x < e$$

$$f_2(x) = \frac{\log x}{x}, \quad e < x$$

and write g in terms of f_1 and f_2. You should be able to show that

$$g'(x) = \frac{[g(x)]^2}{1 - \log g(x)} \cdot \frac{1 - \log x}{x^2}$$

if you do this part properly.)

***35.** This problem uses the material from the Appendix to Chapter 11.

(a) Prove that exp is convex and log is concave.

(b) Prove that if $\sum_{i=1}^{n} p_i = 1$ and all $p_i > 0$, then

$$z_1{}^{p_1} \cdot \ldots \cdot z_n{}^{p_n} < p_1 z_1 + \cdots + p_n z_n.$$

(Use Problem 9 from the Appendix to Chapter 11.)

(c) Deduce another proof that $G_n \le A_n$ (Problem 2-22).

36. (a) Let f be a positive function on $[a, b]$, and let P_n be the partition of $[a, b]$ into n equal intervals. Use Problem 2-22 to show that

$$\frac{1}{b-a}L(\log f, P_n) \le \log\left(\frac{1}{b-a}L(f, P_n)\right).$$

(b) Use the Appendix to Chapter 13 to conclude that for all integrable $f > 0$ we have

$$\frac{1}{b-a}\int_a^b \log f \le \log\left(\frac{1}{b-a}\int_a^b f\right).$$

A more direct approach is illustrated in the next part:

(c) In Problem 35, Problem 2-22 was deduced as a special case of the inequality

$$g\left(\sum_{i=1}^n p_i x_i\right) \le \sum_{i=1}^n p_i g(x_i)$$

for $p_i > 0$, $\sum_{i=1}^n p_i = 1$ and g convex. For g concave we have the reverse inequality

$$\sum_{i=1}^n p_i g(x_i) \le g\left(\sum_{i=1}^n p_i x_i\right).$$

Apply this with $g = \log$ to prove the result of part (b) directly for any integrable f.

(d) State a general theorem of which part (b) is just a special case.

37. Suppose f satisfies $f' = f$ and $f(x+y) = f(x)f(y)$ for all x and y. Prove that $f = \exp$ or $f = 0$.

***38.** Prove that if f is continuous and $f(x+y) = f(x)f(y)$ for all x and y, then either $f = 0$ or $f(x) = [f(1)]^x$ for all x. Hint: Show that $f(x) = [f(1)]^x$ for rational x, and then use Problem 8-6. This problem is closely related to Problem 8-7, and the information mentioned at the end of Problem 8-7 can be used to show that there are discontinuous functions f satisfying $f(x+y) = f(x)f(y)$.

***39.** Prove that if f is a continuous function defined on the positive real numbers, and $f(xy) = f(x) + f(y)$ for all positive x and y, then $f = 0$ or $f(x) = f(e)\log x$ for all $x > 0$. Hint: Consider $g(x) = f(e^x)$.

***40.** Prove that if $f(x) = e^{-1/x^2}$ for $x \ne 0$, and $f(0) = 0$, then $f^{(k)}(0) = 0$ for all k.

***41.** Prove that if $f(x) = e^{-1/x^2}\sin 1/x$ for $x \ne 0$, and $f(0) = 0$, then $f^{(k)}(0) = 0$ for all k.

42. (a) Prove that if α is a root of the equation

$$(*) \quad a_n x^n + a_{n-1}x^{n-1} + \cdots + a_1 x + a_0 = 0,$$

then the function $y(x) = e^{\alpha x}$ satisfies the differential equation

$$(**) \quad a_n y^{(n)} + a_{n-1} y^{(n-1)} + \cdots + a_1 y' + a_0 y = 0.$$

*(b) Prove that if α is a double root of $(*)$, then $y(x) = xe^{\alpha x}$ also satisfies $(**)$. Hint: Remember that if α is a double root of a polynomial equation $f(x) = 0$, then $f'(\alpha) = 0$.

*(c) Prove that if α is a root of $(*)$ of order r, then $y(x) = x^k e^{\alpha x}$ is a solution for $0 \le k \le r - 1$.

If $(*)$ has n real numbers as roots (counting multiplicities), part (c) gives n solutions $y_1, \dots, y_n$ of $(**)$.

(d) Prove that in this case the function $c_1 y_1 + \cdots + c_n y_n$ also satisfies $(**)$.

It is a theorem that in this case these are the only solutions of $(**)$. Problem 20 and the next two problems prove special cases of this theorem, and the general case is considered in Problem 20-19. In Chapter 27 we will see what to do when $(*)$ does not have n real numbers as roots.

***43.** Suppose that f satisfies $f'' - f = 0$ and $f(0) = f'(0) = 0$. Prove that $f = 0$ as follows.

(a) Show that $f^2 - (f')^2 = 0$.

(b) Suppose that $f(x) \ne 0$ for all x in some interval (a, b). Show that either $f(x) = ce^x$ or else $f(x) = ce^{-x}$ for all x in (a, b), for some constant c.

**(c) If $f(x_0) \ne 0$ for $x_0 > 0$, say, then there would be a number a such that $0 \le a < x_0$ and $f(a) = 0$, while $f(x) \ne 0$ for $a < x < x_0$. Why? Use this fact and part (b) to deduce a contradiction.

***44.** (a) Show that if f satisfies $f'' - f = 0$, then $f(x) = ae^x + be^{-x}$ for some a and b. (First figure out what a and b should be in terms of $f(0)$ and $f'(0)$, and then use Problem 43.)

(b) Show also that $f = a \sinh + b \cosh$ for some (other) a and b.

45. Find all functions f satisfying

(a) $f^{(n)} = f^{(n-1)}$.

(b) $f^{(n)} = f^{(n-2)}$.

***46.** This problem, a companion to Problem 15-30, outlines a treatment of the exponential function starting from the assumption that the differential equation $f' = f$ has a nonzero solution.

(a) Suppose there is a function $f \ne 0$ with $f' = f$. Prove that $f(x) \ne 0$ for each x by considering the function $g(x) = f(x_0 + x)f(x - x_0)$, where $f(x_0) \ne 0$.

(b) Show that there is a function f satisfying $f' = f$ and $f(0) = 1$.
(c) For this f show that $f(x+y) = f(x) \cdot f(y)$ by considering the function $g(x) = f(x+y)/f(x)$.
(d) Prove that f is one-one and that $(f^{-1})'(x) = 1/x$.

47. Let f and g be continuous functions such that $\lim_{x\to\infty} f(x) = \lim_{x\to\infty} g(x) = \infty$. We say that f *grows faster than* g ($f \gg g$) if

$$\lim_{x\to\infty} \frac{f(x)}{g(x)} = \infty,$$

and we say that f and g *grow at the same rate* ($f \sim g$) if

$$\lim_{x\to\infty} \frac{f(x)}{g(x)} \text{ exists and is } \neq 0, \infty.$$

For example, $\exp \gg P$ for any polynomial function P, and $P \gg \log^n$ for any positive integer n.

(a) Given f and g, with $\lim_{x\to\infty} f(x) = \lim_{x\to\infty} g(x) = \infty$, is it necessarily true that one of the three conditions $f \gg g$ or $g \gg f$ or $f \sim g$ holds?
(b) If $f \gg g$, then $f + g \sim f$.
(c) If

$$\frac{\log f}{\log g} \geq c > 1$$

for sufficiently large x, then $f \gg g$.
(d) If $f \gg g$ and $F(x) = \int_0^x f$, $G(x) = \int_0^x g$, does it necessarily follow that $F \gg G$?
(e) Arrange each of the following sets of functions in increasing order of growth (for convenience, we indicate each function simply by giving its value at x):

(i) x^3, e^x, $x^3 + \log(x^3)$, $\log 4x$, $(\log x)^x$, x^x, $x + e^{-5x}$, $x^3 \log x$.
(ii) $x \log^2 x$, e^{5x}, $\log(x^x)$, e^{x^2}, x^x, $x^{\log x}$, $(\log x)^x$.
(iii) e^x, x^e, x^x, e^{x^2}, 2^x, $e^{x/2}$, $(\log x)^{2x}$.

48. Suppose that $g_1, g_2, g_3, \ldots$ are continuous functions. Show that there is a continuous function f which grows faster than each g_i.

49. Prove that $\log_{10} 2$ is irrational.

CHAPTER 19 INTEGRATION IN ELEMENTARY TERMS

Every computation of a derivative yields, according to the Second Fundamental Theorem of Calculus, a formula about integrals. For example,

$$\text{if } F(x) = x(\log x) - x \quad \text{then } F'(x) = \log x;$$

consequently,

$$\int_a^b \log x\,dx = F(b) - F(a) = b(\log b) - b - [a(\log a) - a], \quad 0 < a, b.$$

Formulas of this sort are simplified considerably if we adopt the notation

$$F(x)\Big|_a^b = F(b) - F(a).$$

We may then write

$$\int_a^b \log x\,dx = x(\log x) - x\Big|_a^b.$$

This evaluation of $\int_a^b \log x\,dx$ depended on the lucky guess that log is the derivative of the function $F(x) = x(\log x) - x$. In general, a function F satisfying $F' = f$ is called a **primitive** of f. Of course, **a continuous function f always has a primitive**, namely,

$$F(x) = \int_a^x f,$$

but in this chapter we will try to find a primitive which can be written in terms of familiar functions like sin, log, etc. A function which can be written in this way is called an elementary function. To be precise,* an **elementary function** is one which can be obtained by addition, multiplication, division, and composition from the rational functions, the trigonometric functions and their inverses, and the functions log and exp.

It should be stated at the very outset that elementary primitives usually cannot be found. For example, there is no *elementary* function F such that

$$F'(x) = e^{-x^2} \quad \text{for all } x$$

(this is not merely a report on the present state of mathematical ignorance; it is a (difficult) theorem that no such function exists). And, what is even worse, you

*The definition which we will give is precise, but not really accurate, or at least not quite standard. Usually the elementary functions are defined to include "algebraic" functions, that is, functions g satisfying an equation

$$(g(x))^n + f_{n-1}(x)(g(x))^{n-1} + \cdots + f_0(x) = 0,$$

where the f_i are rational functions. But for our purposes these functions can be ignored.

will have no way of knowing whether or not an elementary primitive *can* be found (you will just have to hope that the problems for this chapter contain no misprints). Because the search for elementary primitives is so uncertain, finding one is often peculiarly satisfying. If we observe that the function

$$F(x) = x \arctan x - \frac{\log(1+x^2)}{2}$$

satisfies

$$F'(x) = \arctan x$$

(just how we would ever be led to such an observation is quite another matter), so that

$$\int_a^b \arctan x \, dx = x \arctan x - \left.\frac{\log(1+x^2)}{2}\right|_a^b,$$

then we may feel that we have "really" evaluated $\int_a^b \arctan x \, dx$.

This chapter consists of little more than methods for finding elementary primitives of given elementary functions (a process known simply as "integration"), together with some notation, abbreviations, and conventions designed to facilitate this procedure. This preoccupation with elementary functions can be justified by three considerations:

(1) Integration is a standard topic in calculus, and everyone should know about it.
(2) Every once in a while you might actually need to evaluate an integral, under conditions which do not allow you to consult any of the standard integral tables (for example, you might take a (physics) course in which you are expected to be able to integrate).
(3) The most useful "methods" of integration are actually very important theorems (that apply to all functions, not just elementary ones).

Naturally, the last reason is the crucial one. Even if you intend to forget how to integrate (and you probably will forget some details the first time through), you must never forget the basic methods.

These basic methods are theorems which allow us to express primitives of one function in terms of primitives of other functions. To begin integrating we will therefore need a list of primitives for *some* functions; such a list can be obtained simply by differentiating various well-known functions. The list given below makes use of a standard symbol which requires some explanation. The symbol

$$\int f \quad \text{or} \quad \int f(x)\, dx$$

means "a primitive of f" or, more precisely, "the collection of all primitives of f." The symbol $\int f$ will often by used in stating theorems, while $\int f(x)\, dx$ is most useful in formulas like the following:

$$\int x^3\, dx = \frac{x^4}{4}.$$

This "equation" means that the function $F(x) = x^4/4$ satisfies $F'(x) = x^3$. It cannot be interpreted literally because the right side is a number, not a function, but in this one context we will allow such discrepancies; our aim is to make the integration process as mechanical as possible, and we will resort to any possible device. Another feature of the equation deserves mention. Most people write

$$\int x^3\,dx = \frac{x^4}{4} + C$$

to emphasize that the primitives of $f(x) = x^3$ are precisely the functions of the form $F(x) = x^4/4 + C$ for some number C. Although it is possible (Problem 13) to obtain contradictions if this point is disregarded, in practice such difficulties do not arise, and concern for this constant is merely an annoyance.

There is one important convention accompanying this notation: the letter appearing on the right side of the equation should match with the letter appearing after the "d" on the left side—thus

$$\int u^3\,du = \frac{u^4}{4},$$

$$\int tx\,dx = \frac{tx^2}{2},$$

$$\int tx\,dt = \frac{xt^2}{2}.$$

A function in $\int f(x)\,dx$, i.e., a primitive of f, is often called an "indefinite integral" of f, while $\int_a^b f(x)\,dx$ is called, by way of contrast, a "definite integral." This suggestive notation works out quite well in practice, but it is important not to be led astray. At the risk of boring you, the following fact is emphasized once again: the integral $\int_a^b f(x)\,dx$ is *not* defined as "$F(b) - F(a)$, where F is an indefinite integral of f" (if you do not find this statement repetitious, it is time to reread Chapter 13).

We can verify the formulas in the following short table of indefinite integrals simply by differentiating the functions indicated on the right side.

$$\int a\,dx = ax$$

$$\int x^n\,dx = \frac{x^{n+1}}{n+1}, \qquad n \neq -1$$

$$\int \frac{1}{x}\,dx = \log x$$ ($\int \frac{1}{x}\,dx$ is often written $\int \frac{dx}{x}$ for convenience; similar abbreviations are used in the last two examples of this table.)

$$\int e^x\,dx = e^x$$

$$\int \sin x\,dx = -\cos x$$

$$\int \cos x \, dx = \sin x$$

$$\int \sec^2 x \, dx = \tan x$$

$$\int \sec x \tan x \, dx = \sec x$$

$$\int \frac{dx}{1+x^2} = \arctan x$$

$$\int \frac{dx}{\sqrt{1-x^2}} = \arcsin x$$

Two general formulas of the same nature are consequences of theorems about differentiation:

$$\int [f(x) + g(x)] \, dx = \int f(x) \, dx + \int g(x) \, dx,$$

$$\int c \cdot f(x) \, dx = c \cdot \int f(x) \, dx.$$

These equations should be interpreted as meaning that a primitive of $f + g$ can be obtained by adding a primitive of f to a primitive of g, while a primitive of $c \cdot f$ can be obtained by multiplying a primitive of f by c.

Notice the consequences of these formulas for definite integrals: If f and g are continuous, then

$$\int_a^b [f(x) + g(x)] \, dx = \int_a^b f(x) \, dx + \int_a^b g(x) \, dx,$$

$$\int_a^b c \cdot f(x) \, dx = c \cdot \int_a^b f(x) \, dx.$$

These follow from the previous formulas, since each definite integral may be written as the difference of the values at a and b of a corresponding primitive. Continuity is required in order to know that these primitives exist. (Of course, the formulas are also true when f and g are merely integrable, but recall how much more difficult the proofs are in this case.)

The product formula for the derivative yields a more interesting theorem, which will be written in several different ways.

THEOREM 1 (INTEGRATION BY PARTS) If f' and g' are continuous, then

$$\int fg' = fg - \int f'g,$$

$$\int f(x)g'(x) \, dx = f(x)g(x) - \int f'(x)g(x) \, dx,$$

$$\int_a^b f(x)g'(x) \, dx = f(x)g(x) \Big|_a^b - \int_a^b f'(x)g(x) \, dx.$$

(Notice that in the second equation $f(x)g(x)$ denotes the *function* $f \cdot g$.)

PROOF The formula

$$(fg)' = f'g + fg'$$

can be written

$$fg' = (fg)' - f'g.$$

Thus

$$\int fg' = \int (fg)' - \int f'g,$$

and fg can be chosen as one of the functions denoted by $\int (fg)'$. This proves the first formula.

The second formula is merely a restatement of the first, and the third formula follows immediately from either of the first two. ∎

As the following examples illustrate, integration by parts is useful when the function to be integrated can be considered as a product of a function f, whose derivative is simpler than f, and another function which is obviously of the form g'.

$$\int \underset{f\ g'}{x e^x}\, dx = \underset{f\ g}{x e^x} - \int \underset{f'\ g}{1 \cdot e^x}\, dx = xe^x - e^x$$

$$\int \underset{f\ \ g'}{x \sin x}\, dx = \underset{f}{x} \cdot \underset{g}{(-\cos x)} - \int \underset{f'}{1} \cdot \underset{g}{(-\cos x)}\, dx = -x \cos x + \sin x$$

There are two special tricks which often work with integration by parts. The first is to consider the function g' to be the factor 1, which can always be written in.

$$\int \log x\, dx = \int \underset{g'}{1} \cdot \underset{f}{\log x}\, dx = \underset{g}{x} \underset{f}{\log x} - \int \underset{g}{x} \cdot \underset{f'}{(1/x)}\, dx = x(\log x) - x.$$

The second trick is to use integration by parts to find $\int h$ in terms of $\int h$ again, and then solve for $\int h$. A simple example is the calculation

$$\int \underset{g'}{(1/x)} \cdot \underset{f}{\log x}\, dx = \underset{g}{\log x} \cdot \underset{f}{\log x} - \int \underset{f'}{(1/x)} \cdot \underset{g}{\log x}\, dx,$$

which implies that

$$2 \int \frac{1}{x} \log x\, dx = (\log x)^2$$

or

$$\int \frac{1}{x} \log x \, dx = \frac{(\log x)^2}{2}.$$

A more complicated calculation is often required:

$$\begin{aligned}
\int \underset{\substack{\downarrow \\ f}}{e^x} \underset{\substack{\downarrow \\ g'}}{\sin x} \, dx &= \underset{\substack{\downarrow \\ f}}{e^x} \cdot \underset{\substack{\downarrow \\ g}}{(-\cos x)} - \int \underset{\substack{\downarrow \\ f'}}{e^x} \cdot \underset{\substack{\downarrow \\ g}}{(-\cos x)} \, dx \\
&= -e^x \cos x + \int \underset{\substack{\downarrow \\ u}}{e^x} \underset{\substack{\downarrow \\ v'}}{\cos x} \, dx \\
&= -e^x \cos x + [\underset{\substack{\downarrow \\ u}}{e^x} \cdot \underset{\substack{\downarrow \\ v}}{(\sin x)} - \int \underset{\substack{\downarrow \\ u'}}{e^x} \underset{\substack{\downarrow \\ v}}{(\sin x)} \, dx];
\end{aligned}$$

therefore,

$$2 \int e^x \sin x \, dx = e^x (\sin x - \cos x)$$

or

$$\int e^x \sin x \, dx = \frac{e^x (\sin x - \cos x)}{2}.$$

Since integration by parts depends upon recognizing that a function is of the form g', the more functions you can already integrate, the greater your chances for success. It is frequently reasonable to do a preliminary integration before tackling the main problem. For example, we can use parts to integrate

$$\int (\log x)^2 \, dx = \int \underset{\substack{\downarrow \\ f}}{(\log x)} \underset{\substack{\downarrow \\ g'}}{(\log x)} \, dx$$

if we recall that $\int \log x \, dx = x(\log x) - x$ (this formula was itself derived by integration by parts); we have

$$\begin{aligned}
\int \underset{\substack{\downarrow \\ f}}{(\log x)} \underset{\substack{\downarrow \\ g'}}{(\log x)} \, dx &= \underset{\substack{\downarrow \\ f}}{(\log x)} \underset{\substack{\downarrow \\ g}}{[x(\log x) - x]} - \int \underset{\substack{\downarrow \\ f'}}{(1/x)} \underset{\substack{\downarrow \\ g}}{[x(\log x) - x]} \, dx \\
&= (\log x)[x(\log x) - x] - \int [\log x - 1] \, dx \\
&= (\log x)[x(\log x) - x] - \int \log x \, dx + \int 1 \, dx \\
&= (\log x)[x(\log x) - x] - [x(\log x) - x] + x \\
&= x(\log x)^2 - 2x(\log x) + 2x.
\end{aligned}$$

The most important method of integration is a consequence of the Chain Rule. The use of this method requires considerably more ingenuity than integrating by parts, and even the explanation of the method is more difficult. We will therefore develop this method in stages, stating the theorem for definite integrals first, and saving the treatment of indefinite integrals for later.

THEOREM 2 (THE SUBSTITUTION FORMULA) If f and g' are continuous, then

$$\int_{g(a)}^{g(b)} f = \int_a^b (f \circ g) \cdot g'$$

$$\int_{g(a)}^{g(b)} f(u)\,du = \int_a^b f(g(x)) \cdot g'(x)\,dx.$$

PROOF If F is a primitive of f, then the left side is $F(g(b)) - F(g(a))$. On the other hand,

$$(F \circ g)' = (F' \circ g) \cdot g' = (f \circ g) \cdot g',$$

so $F \circ g$ is a primitive of $(f \circ g) \cdot g'$ and the right side is

$$(F \circ g)(b) - (F \circ g)(a) = F(g(b)) - F(g(a)). \blacksquare$$

The simplest uses of the substitution formula depend upon recognizing that a given function is of the form $(f \circ g) \cdot g'$. For example, the integration of

$$\int_a^b \sin^5 x \cos x\,dx \quad \left(= \int_a^b (\sin x)^5 \cos x\,dx \right)$$

is facilitated by the appearance of the factor $\cos x$, which will be the factor $g'(x)$ for $g(x) = \sin x$; the remaining expression, $(\sin x)^5$, can be written as $(g(x))^5 = f(g(x))$, for $f(u) = u^5$. Thus

$$\int_a^b \sin^5 x \cos x\,dx \qquad \begin{bmatrix} g(x) = \sin x \\ f(u) = u^5 \end{bmatrix}$$

$$= \int_a^b f(g(x))g'(x)\,dx = \int_{g(a)}^{g(b)} f(u)\,du$$

$$= \int_{\sin a}^{\sin b} u^5\,du = \frac{\sin^6 b}{6} - \frac{\sin^6 a}{6}.$$

The integration of $\int_a^b \tan x\,dx$ can be treated similarly if we write

$$\int_a^b \tan x\,dx = -\int_a^b \frac{-\sin x}{\cos x}\,dx.$$

In this case the factor $-\sin x$ is $g'(x)$, where $g(x) = \cos x$; the remaining factor $1/\cos x$ can then be written $f(\cos x)$ for $f(u) = 1/u$. Hence

$$\int_a^b \tan x\,dx \qquad \begin{bmatrix} g(x) = \cos x \\ f(u) = \dfrac{1}{u} \end{bmatrix}$$

$$= -\int_a^b f(g(x))g'(x)\,dx = -\int_{g(a)}^{g(b)} f(u)\,du$$

$$= -\int_{\cos a}^{\cos b} \frac{1}{u}\,du = \log(\cos a) - \log(\cos b).$$

Finally, to find

$$\int_a^b \frac{1}{x\log x}\,dx,$$

notice that $1/x = g'(x)$ where $g(x) = \log x$, and that $1/\log x = f(g(x))$ for $f(u) = 1/u$. Thus

$$\int_a^b \frac{1}{x\log x}\,dx \qquad \left[\begin{aligned} g(x) &= \log x \\ f(u) &= \frac{1}{u}\end{aligned}\right]$$

$$= \int_a^b f(g(x))g'(x)\,dx = \int_{g(a)}^{g(b)} f(u)\,du$$

$$= \int_{\log a}^{\log b} \frac{1}{u}\,du = \log(\log b) - \log(\log a).$$

Fortunately, these uses of the substitution formula can be shortened considerably. The intermediate steps, which involve writing

$$\int_a^b f(g(x))g'(x)\,dx = \int_{g(a)}^{g(b)} f(u)\,du,$$

can easily be eliminated by noticing the following: To go from the left side to the right side,

$$\text{substitute} \begin{cases} u \text{ for } g(x) \\ du \text{ for } g'(x)\,dx \end{cases}$$

(and change the limits of integration);

the substitutions can be performed directly on the original function (accounting for the name of this theorem). For example,

$$\int_a^b \sin^5 x \cos x\,dx \left[\text{substitute} \quad \begin{matrix} u \text{ for } \sin x \\ du \text{ for } \cos x\,dx \end{matrix}\right] = \int_{\sin a}^{\sin b} u^5\,du,$$

and similarly

$$\int_a^b \frac{-\sin x}{\cos x}\,dx \left[\text{substitute} \quad \begin{matrix} u \text{ for } \cos x \\ du \text{ for } -\sin x\,dx \end{matrix}\right] = \int_{\cos a}^{\cos b} \frac{1}{u}\,du.$$

Usually we abbreviate this method even more, and say simply:

$$\text{“Let } \begin{aligned} u &= g(x) \\ du &= g'(x)\,dx.\text{”} \end{aligned}$$

Thus

$$\int_a^b \frac{1}{x\log x}\,dx \left[\begin{aligned} \text{let } u &= \log x \\ du &= \frac{1}{x}\,dx \end{aligned}\right] = \int_{\log a}^{\log b} \frac{1}{u}\,du.$$

In this chapter we are usually interested in primitives rather than definite integrals, but if we can find $\int_a^b f(x)\,dx$ for all a and b, then we can certainly find

$\int f(x)\,dx$. For example, since

$$\int_a^b \sin^5 x \cos x\,dx = \frac{\sin^6 b}{6} - \frac{\sin^6 a}{6},$$

it follows that

$$\int \sin^5 x \cos x\,dx = \frac{\sin^6 x}{6}.$$

Similarly,

$$\int \tan x\,dx = -\log\cos x,$$

$$\int \frac{1}{x\log x}\,dx = \log(\log x).$$

It is quite uneconomical to obtain primitives from the substitution formula by first finding definite integrals. Instead, the two steps can be combined, to yield the following procedure:

> (1) Let
>
> $$u = g(x),$$
> $$du = g'(x)\,dx;$$
>
> (after this manipulation only the letter u should appear, *not* the letter x).
>
> (2) Find a primitive (as an expression involving u).
>
> (3) Substitute $g(x)$ back for u.

Thus, to find

$$\int \sin^5 x \cos x\,dx,$$

(1) let

$$u = \sin x,$$
$$du = \cos x\,dx$$

so that we obtain

$$\int u^5\,du;$$

(2) evaluate

$$\int u^5\,du = \frac{u^6}{6};$$

(3) remember to substitute $\sin x$ back for u, so that

$$\int \sin^5 x \cos x\,dx = \frac{\sin^6 x}{6}.$$

Similarly, if

$$u = \log x,$$
$$du = \frac{1}{x}\,dx,$$

then

$$\int \frac{1}{x \log x}\,dx \quad \text{becomes} \quad \int \frac{1}{u}\,du = \log u,$$

so that

$$\int \frac{1}{x \log x}\,dx = \log(\log x).$$

To evaluate

$$\int \frac{x}{1+x^2}\,dx,$$

let

$$u = 1 + x^2,$$
$$du = 2x\,dx;$$

the factor 2 which has just popped up causes no problem—the integral becomes

$$\frac{1}{2}\int \frac{1}{u}\,du = \frac{1}{2}\log u,$$

so

$$\int \frac{x}{1+x^2}\,dx = \frac{1}{2}\log(1+x^2).$$

(This result may be combined with integration by parts to yield

$$\begin{aligned}\int 1 \cdot \arctan x\,dx &= x \arctan x - \int \frac{x}{1+x^2}\,dx \\ &= x \arctan x - \tfrac{1}{2}\log(1+x^2),\end{aligned}$$

a formula that has already been mentioned.)

These applications of the substitution formula* illustrate the most straightforward and least interesting types—once the suitable factor $g'(x)$ is recognized, the whole problem may even become simple enough to do mentally. The following three problems require only the information provided by the short table of indefinite integrals at the beginning of the chapter and, of course, the right substitution

*The substitution formula is often written in the form

$$\int f(u)\,du = \int f(g(x))g'(x)\,dx, \qquad u = g(x).$$

This formula cannot be taken literally (after all, $\int f(u)\,du$ should mean a primitive of f and the symbol $\int f(g(x))g'(x)\,dx$ should mean a primitive of $(f \circ g) \cdot g'$; these are certainly not equal). However, it may be regarded as a symbolic summary of the procedure which we have developed. If we use Leibniz's notation, and a little fudging, the formula reads particularly well:

$$\int f(u)\,du = \int f(u)\frac{du}{dx}\,dx.$$

(the third problem has been disguised a little by some algebraic chicanery).

$$\int \sec^2 x \tan^5 x \, dx,$$

$$\int (\cos x) e^{\sin x} \, dx,$$

$$\int \frac{e^x}{\sqrt{1 - e^{2x}}} \, dx.$$

If you have not succeeded in finding the right substitutions, you should be able to guess them from the answers, which are $(\tan^6 x)/6$, $e^{\sin x}$, and $\arcsin e^x$. At first you may find these problems too hard to do in your head, but at least when g is of the very simple form $g(x) = ax + b$ you should not have to waste time writing out the substitution. The following integrations should all be clear. (The only worrisome detail is the proper positioning of the constant—should the answer to the second be $e^{3x}/3$ or $3e^{3x}$? I always take care of these problems as follows. Clearly $\int e^{3x} \, dx = e^{3x} \cdot$ (something). Now if I differentiate $F(x) = e^{3x}$, I get $F'(x) = 3e^{3x}$, so the "something" must be $\frac{1}{3}$, to cancel the 3.)

$$\int \frac{dx}{x+3} = \log(x+3),$$

$$\int e^{3x} \, dx = \frac{e^{3x}}{3},$$

$$\int \cos 4x \, dx = \frac{\sin 4x}{4},$$

$$\int \sin(2x+1) \, dx = \frac{-\cos(2x+1)}{2},$$

$$\int \frac{dx}{1+4x^2} = \frac{\arctan 2x}{2}.$$

More interesting uses of the substitution formula occur when the factor $g'(x)$ does *not* appear. There are two main types of substitutions where this happens. Consider first

$$\int \frac{1+e^x}{1-e^x} \, dx.$$

The prominent appearance of the expression e^x suggests the simplifying substitution

$$u = e^x,$$
$$du = e^x \, dx.$$

Although the expression $e^x \, dx$ does not appear, it can always be put in:

$$\int \frac{1+e^x}{1-e^x} \, dx = \int \frac{1+e^x}{1-e^x} \cdot \frac{1}{e^x} \cdot e^x \, dx.$$

We therefore obtain

$$\int \frac{1+u}{1-u} \cdot \frac{1}{u} \, du,$$

which can be evaluated by the algebraic trick

$$\int \frac{1+u}{1-u} \cdot \frac{1}{u}\,du = \int \frac{2}{1-u} + \frac{1}{u}\,du = -2\log(1-u) + \log u,$$

so that

$$\int \frac{1+e^x}{1-e^x}\,dx = -2\log(1-e^x) + \log e^x = -2\log(1-e^x) + x.$$

There is an alternative and preferable way of handling this problem, which does not require multiplying and dividing by e^x. If we write

$$u = e^x, \qquad x = \log u,$$
$$dx = \frac{1}{u}\,dx,$$

then

$$\int \frac{1+e^x}{1-e^x}\,dx \quad \text{immediately becomes} \quad \int \frac{1+u}{1-u} \cdot \frac{1}{u}\,du.$$

Most substitution problems are much easier if one resorts to this trick of expressing x in terms of u, and dx in terms of du, instead of vice versa. It is not hard to see why this trick always works (as long as the function expressing u in terms of x is one-one for all x under consideration): If we apply the substitution

$$u = g(x), \qquad x = g^{-1}(u)$$
$$dx = (g^{-1})'(u)\,du$$

to the integral

$$\int f(g(x))\,dx,$$

we obtain

$$(1) \qquad \int f(u)(g^{-1})'(u)\,du.$$

On the other hand, if we apply the straightforward substitution

$$u = g(x)$$
$$du = g'(x)\,dx$$

to the same integral,

$$\int f(g(x))\,dx = \int f(g(x)) \cdot \frac{1}{g'(x)} \cdot g'(x)\,dx,$$

we obtain

$$(2) \qquad \int f(u) \cdot \frac{1}{g'(g^{-1})(u))}\,du.$$

The integrals (1) and (2) are identical, since $(g^{-1})'(u) = 1/g'(g^{-1}(u))$.

As another concrete example, consider

$$\int \frac{e^{2x}}{\sqrt{e^x+1}}\,dx.$$

In this case we will go the whole hog and replace the entire expression $\sqrt{e^x+1}$ by one letter. Thus we choose the substitution

$$\begin{aligned} u &= \sqrt{e^x+1}, \\ u^2 &= e^x+1, \\ u^2-1 &= e^x, \quad x = \log(u^2-1), \\ dx &= \frac{2u}{u^2-1}\,du. \end{aligned}$$

The integral then becomes

$$\int \frac{(u^2-1)^2}{u} \cdot \frac{2u}{u^2-1}\,du = 2\int u^2-1\,du = \frac{2u^3}{3} - 2u.$$

Thus

$$\int \frac{e^{2x}}{\sqrt{e^x+1}}\,dx = \frac{2}{3}(e^x+1)^{3/2} - 2(e^x+1)^{1/2}.$$

Another example, which illustrates the second main type of substitution that can occur, is the integral

$$\int \sqrt{1-x^2}\,dx.$$

In this case, instead of replacing a complicated expression by a simpler one, we will replace x by $\sin u$, because $\sqrt{1-\sin^2 u} = \cos u$. This really means that we are using the substitution $u = \arcsin x$, but it is the expression for x in terms of u which helps us find the expression to be substituted for dx. Thus,

$$\begin{aligned} \text{let} \quad x &= \sin u, \quad [u = \arcsin x] \\ dx &= \cos u\,du; \end{aligned}$$

then the integral becomes

$$\int \sqrt{1-\sin^2 u}\,\cos u\,du = \int \cos^2 u\,du.$$

The evaluation of this integral depends on the equation

$$\cos^2 u = \frac{1+\cos 2u}{2}$$

(see the discussion of trigonometric functions below) so that

$$\int \cos^2 u\,du = \int \frac{1+\cos 2u}{2}\,du = \frac{u}{2} + \frac{\sin 2u}{4},$$

and

$$\begin{aligned} \int \sqrt{1-x^2}\,dx &= \frac{\arcsin x}{2} + \frac{\sin(2\arcsin x)}{4} \\ &= \frac{\arcsin x}{2} + \frac{1}{2}\sin(\arcsin x)\cdot\cos(\arcsin x) \\ &= \frac{\arcsin x}{2} + \frac{1}{2}x\sqrt{1-x^2}. \end{aligned}$$

Substitution and integration by parts are the only fundamental methods which you have to learn; with their aid primitives can be found for a large number of functions. Nevertheless, as some of our examples reveal, success often depends upon some additional tricks. The most important are listed below. Using these you should be able to integrate all the functions in Problems 1 to 9 (a few other interesting tricks are explained in some of the remaining problems).

1. TRIGONOMETRIC FUNCTIONS

Since

$$\sin^2 x + \cos^2 x = 1$$

and

$$\cos 2x = \cos^2 x - \sin^2 x,$$

we obtain

$$\cos 2x = \cos^2 x - (1 - \cos^2 x) = 2\cos^2 x - 1,$$
$$\cos 2x = (1 - \sin^2 x) - \sin^2 x = 1 - 2\sin^2 x,$$

or

$$\sin^2 x = \frac{1 - \cos 2x}{2},$$
$$\cos^2 x = \frac{1 + \cos 2x}{2}.$$

These formulas may be used to integrate

$$\int \sin^n x\, dx,$$
$$\int \cos^n x\, dx,$$

if n is even. Substituting

$$\frac{(1 - \cos 2x)}{2} \quad \text{or} \quad \frac{(1 + \cos 2x)}{2}$$

for $\sin^2 x$ or $\cos^2 x$ yields a sum of terms involving lower powers of cos. For example,

$$\int \sin^4 x\, dx = \int \left(\frac{1 - \cos 2x}{2}\right)^2 dx = \int \frac{1}{4}\, dx - \frac{1}{2}\int \cos 2x\, dx + \frac{1}{4}\int \cos^2 2x\, dx$$

and

$$\int \cos^2 2x\, dx = \int \frac{1 + \cos 4x}{2}\, dx.$$

If n is odd, $n = 2k + 1$, then

$$\int \sin^n x\, dx = \int \sin x (1 - \cos^2 x)^k\, dx;$$

the latter expression, multiplied out, involves terms of the form $\sin x \cos^l x$, all of which can be integrated easily. The integral for $\cos^n x$ is treated similarly. An integral

$$\int \sin^n x \cos^m x \, dx$$

is handled the same way if n or m is odd. If n and m are both even, use the formulas for $\sin^2 x$ and $\cos^2 x$.

A final important trigonometric integral is

$$\int \frac{1}{\cos x} \, dx = \int \sec x \, dx = \log(\sec x + \tan x).$$

Although there are several ways of "deriving" this result, by means of the methods already at our disposal (Problem 12), it is simplest to check this formula by differentiating the right side, and to memorize it.

2. REDUCTION FORMULAS

Integration by parts yields (Problem 20)

$$\int \sin^n x \, dx = -\frac{1}{n} \sin^{n-1} x \cos x + \frac{n-1}{n} \int \sin^{n-2} x \, dx,$$

$$\int \cos^n x \, dx = \frac{1}{n} \cos^{n-1} x \sin x + \frac{n-1}{n} \int \cos^{n-2} x \, dx,$$

$$\int \frac{1}{(x^2+1)^n} \, dx = \frac{1}{2n-2} \frac{x}{(x^2+1)^{n-1}} + \frac{2n-3}{2n-2} \int \frac{1}{(x^2+1)^{n-1}} \, dx$$

and many similar formulas. The first two, used repeatedly, give a different method for evaluating primitives of $\sin^n$ or $\cos^n$. The third is very important for integrating a large general class of functions, which will complete our discussion.

3. RATIONAL FUNCTIONS

Consider a rational function p/q where

$$p(x) = a_n x^n + a_{n-1} x^{n-1} + \cdots + a_0,$$
$$q(x) = b_m x^m + b_{m-1} x^{m-1} + \cdots + b_0.$$

We might as well assume that $a_n = b_m = 1$. Moreover, we can assume that $n < m$, for otherwise we may express p/q as a polynomial function plus a rational function which *is* of this form by dividing (the calculation

$$\frac{u^2}{u-1} = u + 1 + \frac{1}{u-1}$$

is a simple example). The integration of an arbitrary rational function depends on two facts; the first follows from the "Fundamental Theorem of Algebra" (see Chapter 26, Theorem 2 and Problem 26-3), but the second will not be proved in this book.

THEOREM Every polynomial function

$$q(x) = x^m + b_{m-1}x^{m-1} + \cdots + b_0$$

can be written as a product

$$q(x) = (x-\alpha_1)^{r_1} \cdot \ldots \cdot (x-\alpha_k)^{r_k}(x^2+\beta_1 x+\gamma_1)^{s_1} \cdot \ldots \cdot (x^2+\beta_l x+\gamma_l)^{s_l}$$

(where $r_1 + \cdots + r_k + 2(s_1 + \cdots + s_l) = m$).

(In this expression, identical factors have been collected together, so that all $x - \alpha_i$ and $x^2 + \beta_i x + \gamma_i$ may be assumed distinct. Moreover, we assume that each quadratic factor cannot be factored further. This means that

$$\beta_i{}^2 - 4\gamma_i < 0,$$

since otherwise we can factor

$$x^2 + \beta_i x + \gamma_i = \left[x - \left(\frac{-\beta_i + \sqrt{\beta_i{}^2 - 4\gamma_i}}{2}\right)\right] \cdot \left[x - \left(\frac{-\beta_i - \sqrt{\beta_i{}^2 - 4\gamma_i}}{2}\right)\right]$$

into linear factors.)

THEOREM If $n < m$ and

$$\begin{aligned} p(x) &= x^n + a_{n-1}x^{n-1} + \cdots + a_0, \\ q(x) &= x^m + b_{m-1}x^{m-1} + \cdots + b_0 \\ &= (x-\alpha_1)^{r_1} \cdot \ldots \cdot (x-\alpha_k)^{r_k}(x^2+\beta_1 x+\gamma_1)^{s_1} \cdot \ldots \cdot (x^2+\beta_l x+\gamma_l)^{s_l}, \end{aligned}$$

then $p(x)/q(x)$ can be written in the form

$$\begin{aligned} \frac{p(x)}{q(x)} = &\left[\frac{a_{1,1}}{(x-\alpha_1)} + \cdots + \frac{a_{1,r_1}}{(x-\alpha_1)^{r_1}}\right] + \cdots \\ &+ \left[\frac{\alpha_{k,1}}{(x-\alpha_k)} + \cdots + \frac{\alpha_{k,r_k}}{(x-\alpha_k)^{r_k}}\right] \\ &+ \left[\frac{b_{1,1}x + c_{1,1}}{(x^2+\beta_1 x+\gamma_1)} + \cdots + \frac{b_{1,s_1}x + c_{1,s_1}}{(x^2+\beta_1 x+\gamma_1)^{s_1}}\right] + \cdots \\ &+ \left[\frac{b_{l,1}x + c_{l,1}}{(x^2+\beta_l x+\gamma_l)} + \cdots + \frac{b_{l,s_l}x + c_{l,s_l}}{(x^2+\beta_l x+\gamma_l)^{s_l}}\right]. \end{aligned}$$

This expression, known as the "partial fraction decomposition" of $p(x)/q(x)$, is so complicated that it is simpler to examine the following example, which illustrates such an expression and shows how to find it. According to the theorem, it is possible to write

$$\begin{aligned} &\frac{2x^7 + 8x^6 + 13x^5 + 20x^4 + 15x^3 + 16x^2 + 7x + 10}{(x^2+x+1)^2(x^2+2x+2)(x-1)^2} \\ &\qquad = \frac{a}{x-1} + \frac{b}{(x-1)^2} + \frac{cx+d}{x^2+2x+2} + \frac{ex+f}{x^2+x+1} + \frac{gx+h}{(x^2+x+1)^2}. \end{aligned}$$

To find the numbers a, b, c, d, e, f, g, and h, write the right side as a polynomial over the common denominator $(x^2+x+1)^2(x^2+2x+3)(x-1)^2$; the numerator becomes

$$\begin{aligned} &a(x-1)(x^2+2x+2)(x^2+x+1)^2+b(x^2+2x+2)(x^2+x+1)^2 \\ &+(cx+d)(x-1)^2(x^2+x+1)^2+(ex+f)(x-1)^2(x^2+2x+2)(x^2+x+1) \\ &\qquad\qquad\qquad\qquad\qquad +(gx+h)(x-1)^2(x^2+2x+2). \end{aligned}$$

Actually multiplying this out (!) we obtain a polynomial of degree 8, whose coefficients are combinations of $a,\dots,h$. Equating these coefficients with the coefficients of $2x^7+8x^6+13x^5+20x^4+15x^3+16x^2+7x+10$ (the coefficient of x^8 is 0) we obtain 8 equations in the eight unknowns $a,\dots,h$. After heroic calculations these can be solved to give

$$\begin{array}{llll} a=1, & b=2, & c=1, & d=3, \\ e=0, & f=0, & g=0, & h=1. \end{array}$$

Thus

$$\begin{aligned} &\int \frac{2x^7+5x^6+13x^5+20x^4+17x^3+16x^2+7x+7}{(x^2+x+1)^2(x^2+2x+2)(x-1)^2}\,dx \\ &= \int \frac{1}{(x-1)}\,dx + \int \frac{2}{(x-1)^2}\,dx + \int \frac{1}{(x^2+x+1)^2}\,dx + \int \frac{x+3}{x^2+2x+2}\,dx. \end{aligned}$$

(In simpler cases the requisite calculations may actually be feasible. I obtained this particular example by *starting* with the partial fraction decomposition and converting it into one fraction.)

We are already in a position to find each of the integrals appearing in the above expression; the calculations will illustrate all the difficulties which arise in integrating rational functions.

The first two integrals are simple:

$$\begin{aligned} \int \frac{1}{x-1}\,dx &= \log(x-1), \\ \int \frac{2}{(x-1)^2}\,dx &= \frac{-2}{x-1}. \end{aligned}$$

The third integration depends on "completing the square":

$$\begin{aligned} x^2+x+1 &= (x+\tfrac{1}{2})^2+\tfrac{3}{4} \\ &= \frac{3}{4}\left[\left(\frac{x+\frac{1}{2}}{\sqrt{\frac{3}{4}}}\right)^2+1\right]. \end{aligned}$$

(If we had obtained $-\frac{3}{4}$ instead of $\frac{3}{4}$ we could not take the square root, but in this case our original quadratic factor could have been factored into linear factors.) We can now write

$$\int \frac{1}{(x^2+x+1)^2}\,dx = \frac{16}{9}\int \frac{1}{\left[\left(\dfrac{x+\frac{1}{2}}{\sqrt{\frac{3}{4}}}\right)+1\right]^2}\,dx.$$

The substitution

$$u = \frac{x+\frac{1}{2}}{\sqrt{\frac{3}{4}}},$$
$$du = \frac{1}{\sqrt{\frac{3}{4}}}\,dx,$$

changes this integral to

$$\frac{16}{9}\int \frac{\sqrt{\frac{3}{4}}}{(u^2+1)^2}\,du,$$

which can be computed using the third reduction formula given above.

Finally, to evaluate

$$\int \frac{x+3}{(x^2+2x+2)}\,dx$$

we write

$$\int \frac{x+3}{x^2+2x+2}\,dx = \frac{1}{2}\int \frac{2x+2}{x^2+2x+2}\,dx + \int \frac{2}{(x+1)^2+1}\,dx.$$

The first integral on the right side has been purposely constructed so that we can evaluate it by using the substitution

$$u = x^2+2x+2,$$
$$du = (2x+2)\,dx$$

The second integral on the right, which is just the difference of the other two, is simply $2\arctan(x+1)$. If the original integral were

$$\int \frac{x+3}{(x^2+2x+2)^n}\,dx = \frac{1}{2}\int \frac{2x+2}{(x^2+2x+2)^n}\,dx + \int \frac{2}{[(x+1)^2+1]^n}\,dx,$$

the first integral on the right would still be evaluated by the same substitution. The second integral would be evaluated by means of a reduction formula.

This example has probably convinced you that integration of rational functions is a theoretical curiosity only, especially since it is necessary to find the factorization of $q(x)$ before you can even begin. This is only partly true. We have already seen that simple rational functions sometimes arise, as in the integration

$$\int \frac{1+e^x}{1-e^x}\,dx;$$

another important example is the integral

$$\int \frac{1}{x^2-1}\,dx = \int \frac{\frac{1}{2}}{x-1} - \frac{\frac{1}{2}}{x+1}\,dx = \frac{1}{2}\log(x-1) - \frac{1}{2}\log(x+1).$$

Moreover, if a problem has been reduced to the integration of a rational function, it is then certain that an elementary primitive exists, even when the difficulty or impossibility of finding the factors of the denominator may preclude writing this primitive explicitly.

PROBLEMS

1. This problem contains some integrals which require little more than algebraic manipulation, and consequently test your ability to discover algebraic tricks, rather than your understanding of the integration processes. Nevertheless, any one of these tricks might be an important preliminary step in an honest integration problem. Moreover, you want to have some feel for which integrals are easy, so that you can see when the end of an integration process is in sight. The answer section, if you resort to it, will only reveal what algebra you should have used.

(i) $\displaystyle\int \frac{\sqrt[5]{x^3} + \sqrt[6]{x}}{\sqrt{x}}\,dx.$

(ii) $\displaystyle\int \frac{dx}{\sqrt{x-1}+\sqrt{x+1}}.$

(iii) $\displaystyle\int \frac{e^x + e^{2x} + e^{3x}}{e^{4x}}\,dx.$

(iv) $\displaystyle\int \frac{a^x}{b^x}\,dx.$

(v) $\displaystyle\int \tan^2 x\,dx.$ (Trigonometric integrals are always very touchy, because there are so many trigonometric identities that an easy problem can easily look hard.)

(vi) $\displaystyle\int \frac{dx}{a^2+x^2}.$

(vii) $\displaystyle\int \frac{dx}{\sqrt{a^2-x^2}}.$

(viii) $\displaystyle\int \frac{dx}{1+\sin x}.$

(ix) $\displaystyle\int \frac{8x^2+6x+4}{x+1}\,dx.$

(x) $\displaystyle\int \frac{1}{\sqrt{2x-x^2}}\,dx.$

2. The following integrations involve simple substitutions, most of which you should be able to do in your head.

(i) $\displaystyle\int e^x \sin e^x\,dx.$

(ii) $\displaystyle\int xe^{-x^2}\,dx.$

(iii) $\displaystyle\int \frac{\log x}{x}\,dx.$ (In the text this was done by parts.)

(iv) $\displaystyle\int \frac{e^x\,dx}{e^{2x}+2e^x+1}.$

(v) $\displaystyle\int e^{e^x}e^x\,dx.$

(vi) $\displaystyle\int \frac{x\,dx}{\sqrt{1-x^4}}.$

(vii) $\displaystyle\int \frac{e^{\sqrt{x}}}{\sqrt{x}}\,dx.$

(viii) $\displaystyle\int x\sqrt{1-x^2}\,dx.$

(ix) $\displaystyle\int \log(\cos x)\tan x\,dx.$

(x) $\displaystyle\int \frac{\log(\log x)}{x\log x}\,dx.$

3. Integration by parts.

(i) $\displaystyle\int x^2e^x\,dx.$

(ii) $\displaystyle\int x^3e^{x^2}\,dx.$

(iii) $\displaystyle\int e^{ax}\sin bx\,dx.$

(iv) $\displaystyle\int x^2\sin x\,dx.$

(v) $\displaystyle\int (\log x)^3\,dx.$

(vi) $\displaystyle\int \frac{\log(\log x)}{x}\,dx.$

(vii) $\displaystyle\int \sec^3 x\,dx.$ (This is a tricky and important integral that often comes up. If you do not succeed in evaluating it, be sure to consult the answers.)

(viii) $\displaystyle\int \cos(\log x)\,dx.$

(ix) $\displaystyle\int \sqrt{x}\log x\,dx.$

(x) $\displaystyle\int x(\log x)^2\,dx.$

4. The following integrations can all be done with substitutions of the form $x = \sin u$, $x = \cos u$, etc. To do some of these you will need to remember that

$$\int \sec x\, dx = \log(\sec x + \tan x)$$

as well as the following formula, which can also be checked by differentiation:

$$\int \csc x\, dx = -\log(\csc x + \cot x).$$

In addition, at this point the derivatives of all the trigonometric functions should be kept handy.

(i) $\displaystyle\int \frac{dx}{\sqrt{1-x^2}}$. (You already know this integral, but use the substitution $x = \sin u$ anyway, just to see how it works out.)

(ii) $\displaystyle\int \frac{dx}{\sqrt{1+x^2}}$. (Since $\tan^2 u + 1 = \sec^2 u$, you want to use the substitution $x = \tan u$.)

(iii) $\displaystyle\int \frac{dx}{\sqrt{x^2-1}}$.

(iv) $\displaystyle\int \frac{dx}{x\sqrt{x^2-1}}$. (The answer will be a certain inverse function that was given short shrift in the text.)

(v) $\displaystyle\int \frac{dx}{x\sqrt{1-x^2}}$.

(vi) $\displaystyle\int \frac{dx}{x\sqrt{1+x^2}}$.

(vii) $\displaystyle\int x^3\sqrt{1-x^2}\, dx$.

(viii) $\displaystyle\int \sqrt{1-x^2}\, dx$.

[(vii) and (viii):] You will need to remember the methods for integrating powers of sin and cos.

(ix) $\displaystyle\int \sqrt{1+x^2}\, dx$.

(x) $\displaystyle\int \sqrt{x^2-1}\, dx$.

5. The following integrations involve substitutions of various types. There is no substitute for cleverness, but there is a general rule to follow: substitute for an expression which appears frequently or prominently; if two different troublesome expressions appear, try to express them both in terms of some new expression. And don't forget that it usually helps to express x directly in terms of u, to find out the proper expression to substitute for dx.

(i) $\displaystyle\int \frac{dx}{1+\sqrt{x+1}}$.

(ii) $\displaystyle\int \frac{dx}{1+e^x}$.

(iii) $\displaystyle\int \frac{dx}{\sqrt{x}+\sqrt[3]{x}}$.

(iv) $\displaystyle\int \frac{dx}{\sqrt{1+e^x}}$. (The substitution $u = e^x$ leads to an integral requiring yet another substitution; this is all right, but both substitutions can be done at once.)

(v) $\displaystyle\int \frac{dx}{2+\tan x}$.

(vi) $\displaystyle\int \frac{dx}{\sqrt{\sqrt{x}+1}}$. (Another place where one substitution can be made to do the work of two.)

(vii) $\displaystyle\int \frac{4^x+1}{2^x+1}\,dx$.

(viii) $\displaystyle\int e^{\sqrt{x}}\,dx$.

(ix) $\displaystyle\int \frac{\sqrt{1-x}}{1-\sqrt{x}}\,dx$. (In this case two successive substitutions work out best; there are two obvious candidates for the first substitution, and either will work.)

*(x) $\displaystyle\int \sqrt{\frac{x-1}{x+1}}\cdot\frac{1}{x^2}\,dx$.

6. The previous problem provided gratis a haphazard selection of rational functions to be integrated. Here is a more systematic selection.

(i) $\displaystyle\int \frac{2x^2+7x-1}{x^3+x^2-x-1}\,dx$.

(ii) $\displaystyle\int \frac{2x+1}{x^3-3x^2+3x-1}\,dx$.

(iii) $\displaystyle\int \frac{x^3+7x^2-5x+5}{(x-1)^2(x+1)^3}\,dx$.

(iv) $\displaystyle\int \frac{2x^2+x+1}{(x+3)(x-1)^2}\,dx$.

(v) $\displaystyle\int \frac{x+4}{x^2+1}\,dx$.

(vi) $\displaystyle\int \frac{x^3+x+2}{x^4+2x^2+1}\,dx$.

(vii) $\displaystyle\int \frac{3x^2+3x+1}{x^3+2x^2+2x+1}\,dx$.

(viii) $\displaystyle\int \frac{dx}{x^4+1}$.

(ix) $\displaystyle\int \frac{2x}{(x^2+x+1)^2}\,dx$.

(x) $\displaystyle\int \frac{3x}{(x^2+x+1)^3}\,dx$.

***7.** Potpourri. (No holds barred.) The following integrations involve all the methods of the previous problems

(i) $\displaystyle\int \frac{\arctan x}{1+x^2}\,dx.$

(ii) $\displaystyle\int \frac{x\arctan x}{(1+x^2)^3}\,dx.$

(iii) $\displaystyle\int \log\sqrt{1+x^2}\,dx.$

(iv) $\displaystyle\int x\log\sqrt{1+x^2}\,dx.$

(v) $\displaystyle\int \frac{x^2-1}{x^2+1}\cdot\frac{1}{\sqrt{1+x^4}}\,dx.$

(vi) $\displaystyle\int \arcsin\sqrt{x}\,dx.$

(vii) $\displaystyle\int \frac{x}{1+\sin x}\,dx.$

(viii) $\displaystyle\int e^{\sin x}\cdot\frac{x\cos^3 x-\sin x}{\cos^2 x}\,dx.$

(ix) $\displaystyle\int \sqrt{\tan x}\,dx.$

(x) $\displaystyle\int \frac{dx}{x^6+1}.$ (To factor x^6+1, first factor y^3+1, using Problem 1-1.)

The following two problems provide still more practice at integration, if you need it (and can bear it). Problem 8 involves algebraic and trigonometric manipulations and integration by parts, while Problem 9 involves substitutions. (Of course, in many cases the resulting integrals will require still further manipulations.)

8. Find the following integrals.

(i) $\displaystyle\int \log(a^2+x^2)\,dx.$

(ii) $\displaystyle\int \frac{1+\cos x}{\sin^2 x}\,dx.$

(iii) $\displaystyle\int \frac{x+1}{\sqrt{4-x^2}}\,dx.$

(iv) $\displaystyle\int x\arctan x\,dx.$

(v) $\displaystyle\int \sin^3 x\,dx.$

(vi) $\displaystyle\int \frac{\sin^3 x}{\cos^2 x}\,dx.$

(vii) $\displaystyle\int x^2 \arctan x\,dx.$

(viii) $\displaystyle\int \frac{x\,dx}{\sqrt{x^2-2x+2}}.$

(ix) $\displaystyle\int \sec^3 x \tan x\,dx.$

(x) $\displaystyle\int x \tan^2 x\,dx.$

9. Find the following integrals.

(i) $\displaystyle\int \frac{dx}{(a^2+x^2)^2}.$

(ii) $\displaystyle\int \sqrt{1-\sin x}\,dx.$

(iii) $\displaystyle\int \arctan\sqrt{x}\,dx.$

(iv) $\displaystyle\int \sin\sqrt{x+1}\,dx.$

(v) $\displaystyle\int \frac{\sqrt{x^3-2}}{x}\,dx.$

(vi) $\displaystyle\int \log(x+\sqrt{x^2-1}\,)\,dx.$

(vii) $\displaystyle\int \log(x+\sqrt{x}\,)\,dx.$

(viii) $\displaystyle\int \frac{dx}{x-x^{3/5}}.$

(ix) $\displaystyle\int (\arcsin x)^2\,dx.$

(x) $\displaystyle\int x^5 \arctan(x^2)\,dx.$

10. If you have done Problem 18-9, the integrals (ii) and (iii) in Problem 4 will look very familiar. In general, the substitution $x = \cosh u$ often works for integrals involving $\sqrt{x^2-1}$, while $x = \sinh u$ is the thing to try for integrals involving $\sqrt{x^2+1}$. Try these substitutions on the other integrals in Problem 4. (The method is not really recommended; it is easier to stick with trigonometric substitutions.)

***11.** The world's sneakiest substitution is undoubtedly

$$t = \tan\frac{x}{2}, \qquad x = 2\arctan t,$$
$$dx = \frac{2}{1+t^2}\,dt.$$

As we found in Problem 15-17, this substitution leads to the expressions

$$\sin x = \frac{2t}{1+t^2}, \qquad \cos x = \frac{1-t^2}{1+t^2}.$$

This substitution thus transforms any integral which involves only sin and cos, combined by addition, multiplication, and division, into the integral of a rational function. Find

(i) $\displaystyle\int \frac{dx}{1+\sin x}$. (Compare your answer with Problem 1(viii).)

(ii) $\displaystyle\int \frac{dx}{1-\sin^2 x}$. (In this case it is better to let $t = \tan x$. Why?)

(iii) $\displaystyle\int \frac{dx}{a\sin x + b\cos x}$. (There is also another way to do this, using Problem 15-8.)

(iv) $\displaystyle\int \sin^2 x\, dx$. (An exercise to convince you that this substitution should be used only as a last resort.)

(v) $\displaystyle\int \frac{dx}{3+5\sin x}$. (A last resort.)

***12.** Derive the formula for $\int \sec x\, dx$ in the following two ways:

(a) By writing

$$\begin{aligned}\frac{1}{\cos x} &= \frac{\cos x}{\cos^2 x}\\ &= \frac{\cos x}{1-\sin^2 x}\\ &= \frac{1}{2}\left[\frac{\cos x}{1+\sin x} + \frac{\cos x}{1-\sin x}\right],\end{aligned}$$

an expression obviously inspired by partial fraction decompositions. Be sure to note that $\int \cos x/(1-\sin x)\, dx = -\log(1-\sin x)$; the minus sign is very important. And remember that $\frac{1}{2}\log\alpha = \log\sqrt{\alpha}$. From there on, keep doing algebra, and trust to luck.

(b) By using the substitution $t = \tan x/2$. One again, quite a bit of manipulation is required to put the answer in the desired form; the expression $\tan x/2$ can be attacked by using Problem 15-9, or both answers can be expressed in terms of t. There is another expression for $\int \sec x\, dx$, which is less cumbersome than $\log(\sec x + \tan x)$; using Problem 15-9, we obtain

$$\int \sec x\, dx = \log\left(\frac{1+\tan\dfrac{x}{2}}{1-\tan\dfrac{x}{2}}\right) = \log\left(\tan\left(\frac{x}{2}+\frac{\pi}{4}\right)\right).$$

This last expression was actually the one first discovered, and was due, not to any mathematician's cleverness, but to a curious historical accident: In 1599 Wright computed nautical tables that amounted to definite integrals of sec. When the first tables for the logarithms of tangents were produced, the correspondence between the two tables was immediately noticed (but remained unexplained until the invention of calculus).

13. The derivation of $\int e^x \sin x\, dx$ given in the text seems to prove that the only primitive of $f(x) = e^x \sin x$ is $F(x) = e^x(\sin x - \cos x)/2$, whereas $F(x) = e^x(\sin x - \cos x)/2 + C$ is also a primitive for any number C. Where does C come from? (What is the meaning of the equation

$$\int e^x \sin x\, dx = e^x \sin x - e^x \cos x - \int e^x \sin x\, dx?)$$

14. Suppose that f'' is continuous and that

$$\int_0^{\pi} [f(x) + f''(x)] \sin x\, dx = 2.$$

Given that $f(\pi) = 1$, compute $f(0)$.

15. (a) Find $\int \arcsin x\, dx$, using the same trick that worked for log and arctan.
*(b) Generalize this trick: Find $\int f^{-1}(x)\, dx$ in terms of $\int f(x)\, dx$. Compare with Problems 12-18 and 14-17.

16. (a) Find $\int \sin^4 x\, dx$ in two different ways: first using the reduction formula, and then using the formula for $\sin^2 x$.
(b) Combine your answers to obtain an impressive trigonometric identity.

17. Express $\int \log(\log x)\, dx$ in terms of $\int (\log x)^{-1}\, dx$. (Neither is expressible in terms of elementary functions.)

18. Express $\int x^2 e^{-x^2}\, dx$ in terms of $\int e^{-x^2}\, dx$.

19. Prove that the function $f(x) = e^x/(e^{5x} + e^x + 1)$ has an elementary primitive. (Do not try to find it!)

20. Prove the reduction formulas in the text. For the third one write

$$\int \frac{dx}{(1+x^2)^n} = \int \frac{dx}{(1+x^2)^{n-1}} - \int \frac{x^2\, dx}{(1+x^2)^n}$$

and work on the last integral. (Another possibility is to use the substitution $x = \tan u$.)

21. Find a reduction formula for

(a) $\displaystyle\int x^n e^x\, dx$

(b) $\displaystyle\int (\log x)^n\, dx$.

***22.** Prove that

$$\int_1^{\cosh x} \sqrt{t^2 - 1}\, dt = \frac{\cosh x \sinh x}{2} - \frac{x}{2}.$$

(See Problem 18-6 for the significance of this computation.)

23. Prove that

$$\int_a^b f(x)\,dx = \int_a^b f(a+b-x)\,dx.$$

(A geometric interpretation makes this clear, but it is also a good exercise in the handling of limits of integration during a substitution.)

24. Prove that the area of a circle of radius r is πr^2. (Naturally you must remember that π is defined as the area of the unit circle.)

25. Let ϕ be a nonnegative integrable function such that $\phi(x) = 0$ for $|x| \geq 1$ and such that $\int_{-1}^{1} \phi = 1$. For $h > 0$, let

$$\phi_h(x) = \frac{1}{h}\phi(x/h).$$

(a) Show that $\phi_h(x) = 0$ for $|x| \geq h$ and that $\int_{-h}^{h} \phi_h = 1$.

(b) Let f be integrable on $[-1, 1]$ and continuous at 0. Show that

$$\lim_{h\to 0^+} \int_{-1}^{1} \phi_h f = \lim_{h\to 0^+} \int_{-h}^{h} \phi_h f = f(0).$$

(c) Show that

$$\lim_{h\to 0^+} \int_{-1}^{1} \frac{h}{h^2+x^2}\,dx = \pi.$$

The final part of this problem might appear, at first sight, to be an exact analogue of part (b), but it actually requires more careful argument.

(d) Let f be integrable on $[-1, 1]$ and continuous at 0. Show that

$$\lim_{h\to 0^+} \int_{-1}^{1} \frac{h}{h^2+x^2} f(x)\,dx = \pi f(0).$$

Hint: If h is small, then $h/(h^2+x^2)$ will be small on most of $[-1, 1]$.

The next two problems use the formula

$$\frac{1}{2}\int_{\theta_0}^{\theta_1} f(\theta)^2\,d\theta,$$

derived in Problem 13-24, for the area of a region bounded by the graph of f in polar coordinates.

26. For each of the following functions, find the area bounded by the graphs in polar coordinates. (Be careful about the proper range for θ, or you will get nonsensical results!)

(i) $f(\theta) = a\sin\theta$.
(ii) $f(\theta) = 2 + \cos\theta$.
(iii) $f(\theta)^2 = 2a^2\cos 2\theta$.
(iv) $f(\theta) = a\cos 2\theta$.

FIGURE 1

27. Figure 1 shows the graph of f in polar coordinates; the region OAB thus has area $\frac{1}{2}\int_{\theta_0}^{\theta_1} f(\theta)^2\,d\theta$. Now suppose that this graph also happens to be the ordinary graph of some function g. Then the region OAB also has area

$$\text{area } \Delta Ox_1B + \int_{x_1}^{x_0} g - \text{ area } \Delta Ox_0A.$$

Prove analytically that these two numbers are indeed the same. Hint: The function g is determined by the equations

$$x = f(\theta)\cos\theta, \qquad g(x) = f(\theta)\sin\theta.$$

The next four problems use the formulas, derived in Problems 3 and 4 of the Appendix to Chapter 13, for the length of a curve represented parametrically (and, in particular, as the graph of a function in polar coordinates).

28. Let c be a curve represented parametrically by u and v on $[a,b]$, and let h be an increasing function with $h(\bar{a}) = a$ and $h(\bar{b}) = b$. Then on $[\bar{a},\bar{b}]$ the functions $\bar{u} = u \circ h$, $\bar{v} = v \circ h$ give a parametric representation of another curve $\bar{c}$; intuitively, $\bar{c}$ is just the same curve c traversed at a different rate.

(a) Show, directly from the definition of length, that the length of c on $[a,b]$ equals the length of $\bar{c}$ on $[\bar{a},\bar{b}]$.

(b) Assuming differentiability of any functions required, show that the lengths are equal by using the integral formula for length, and the appropriate substitution.

29. Find the length of the following curves, all described as the graphs of functions, except for (iii), which is represented parametrically.

(i) $f(x) = \frac{1}{3}(x^2+2)^{3/2}, \quad 0 \le x \le 1.$

(ii) $f(x) = x^3 + \frac{1}{12x}, \quad 1 \le x \le 2.$

(iii) $x = a^3\cos^3 t, \quad y = a^3\sin^3 t, \quad 0 \le t \le 2\pi.$

(iv) $f(x) = \log(\cos x), \quad 0 \le x \le \pi/6.$

(v) $f(x) = \log x, \quad 1 \le x \le e.$

(vi) $f(x) = \arcsin e^x, \quad -\log 2 \le x \le 0.$

30. For the following functions, find the length of the graph in polar coordinates.

(i) $f(\theta) = a\cos\theta.$

(ii) $f(\theta) = a(1-\cos\theta).$

(iii) $f(\theta) = a\sin^2\theta/2.$

(iv) $f(\theta) = \theta \quad 0 \le \theta \le 2\pi.$

(v) $f(\theta) = 3\sec\theta \quad 0 \le \theta \le \pi/3.$

31. In Problem 8 of the Appendix to Chapter 12 we described the cycloid, which has the parametric representation

$$x = u(t) = a(t - \sin t), \qquad y = v(t) = a(1 - \cos t).$$

(a) Find the length of one arch of the cycloid. [Answer: $8a$.]
(b) Recall that the cycloid is the graph of $v \circ u^{-1}$. Find the area under one arch of the cycloid by using the appropriate substitution in $\int f$ and evaluating the resultant integral. [Answer: $3\pi a^2$.]

32. Use induction and integration by parts to generalize Problem 14-13:

$$\int_0^x \frac{f(u)(x-u)^n}{n!}\,du = \int_0^x \left(\int_0^{u_n} \left(\cdots \left(\int_0^{u_1} f(t)\,dt \right) du_1 \right) \cdots \right) du_n.$$

33. If f' is continuous on $[a, b]$, use integration by parts to prove the Riemann-Lebesgue Lemma for f:

$$\lim_{\lambda \to \infty} \int_a^b f(t) \sin(\lambda t)\,dt = 0.$$

This result is just a special case of Problem 15-26, but it can be used to prove the general case (in much the same way that the Riemann-Lebesgue Lemma was derived in Problem 15-26 from the special case in which f is a step function).

34. The Mean Value Theorem for Integrals was introduced in Problem 13-23. The "Second Mean Value Theorem for Integrals" states the following. Suppose that f is integrable on $[a, b]$ and that ϕ is either nondecreasing or nonincreasing on $[a, b]$. Then there is a number ξ in $[a, b]$ such that

$$\int_a^b f(x)\phi(x)\,dx = \phi(a) \int_a^\xi f(x)\,dx + \phi(b) \int_\xi^b f(x)\,dx.$$

In this problem, we will assume that f is continuous and that ϕ is differentiable, with a continuous derivative ϕ'.

(a) Prove that if the result is true for nonincreasing ϕ, then it is also true for nondecreasing ϕ.
(b) Prove that if the result is true for nonincreasing ϕ satisfying $\phi(b) = 0$, then it is true for all nonincreasing ϕ.

Thus, we can assume that ϕ is nonincreasing and $\phi(b) = 0$. In this case, we have to prove that

$$\int_a^b f(x)\phi(x) = \phi(a) \int_a^\xi f(x)\,dx.$$

(c) Prove this by using integration by parts.
(d) Show that the hypothesis that ϕ is either nondecreasing or nonincreasing is needed.

From this special case of the Second Mean Value Theorem for Integrals, the general case could be derived by some approximation arguments, just as in the case of the Riemann-Lebesgue Lemma. But there is a more instructive way, outlined in the next problem.

35. (a) Given $a_1, \dots, a_n$ and $b_1, \dots, b_n$, let $s_k = a_1 + \dots + a_k$. Show that

$$(*) \quad a_1b_1 + \dots + a_nb_n = s_1(b_1 - b_2) + s_2(b_2 - b_3) + \dots + s_{n-1}(b_{n-1} - b_n) + s_nb_n$$

This disarmingly simple formula is sometimes called "Abel's formula for summation by parts." It may be regarded as an analogue for sums of the integration by parts formula

$$\int_a^b f'(x)g(x)\,dx = f(b)g(b) - f(a)g(a) - \int_a^b f(x)g'(x)\,dx,$$

especially if we use Riemann sums (Chapter 13, Appendix). In fact, for a partition $P = \{t_0, \dots, t_n\}$ of $[a, b]$, the left side is approximately

$$(1) \qquad \sum_{k=1}^{n} f'(t_k)g(t_{k-1})(t_k - t_{k-1}),$$

while the right side is approximately

$$f(b)g(b) - f(a)g(a) - \sum_{k=1}^{n} f(t_k)g'(t_k)(t_k - t_{k-1})$$

which is approximately

$$\begin{aligned} f(b)g(b) - f(a)g(a) &- \sum_{k=1}^{n} f(t_k)\frac{g(t_k) - g(t_{k-1})}{t_k - t_{k-1}}(t_k - t_{k-1}) \\ &= f(b)g(b) - f(a)g(a) + \sum_{k=1}^{n} f(t_k)[g(t_{k-1}) - g(t_k)] \\ &= f(b)g(b) - f(a)g(a) + \sum_{k=1}^{n}[f(t_k) - f(a)] \cdot [g(t_{k-1}) - g(t_k)] \\ &\qquad + f(a)\sum_{k=1}^{n} g(t_{k-1}) - g(t_k). \end{aligned}$$

Since the right-most sum is just $g(a) - g(b)$, this works out to be

$$(2) \qquad [f(b) - f(a)]g(b) + \sum_{k=1}^{n}[f(t_k) - f(a)] \cdot [g(t_{k-1}) - g(t_k)].$$

If we choose

$$a_k = f'(t_k)(t_k - t_{k-1}), \qquad b_k = g(t_{k-1})$$

then

$$(1) \quad \text{is} \quad \sum_{k=1}^{n} a_kb_k,$$

which is the left side of $(*)$, while

$$s_k = \sum_{i=1}^{k} f'(t_i)(t_i - t_{i-1}) \quad \text{is approximately} \quad \sum_{i=1}^{k} f(t_i) - f(t_{i-1}) = f(t_k) - f(a),$$

so

$$(2) \quad \text{is approximately} \quad s_nb_n + \sum_{k=1}^{n} s_k(b_k - b_{k-1}),$$

which is the right side of $(*)$.

This discussion is not meant to suggest that Abel's formula can actually be derived from the formula for integration by parts, or *vice versa.* But, as we shall see, Abel's formula can often be used as a substitute for integration by parts in situations where the functions in question aren't differentiable.

(b) Suppose that $\{b_n\}$ is nonincreasing, with $b_n \geq 0$ for each n, and that

$$m \leq a_1 + \cdots + a_n \leq M$$

for all n. Prove Abel's Lemma:

$$b_1 m \leq a_1 b_1 + \cdots + a_n b_n \leq b_1 M.$$

(And, moreover,

$$b_k m \leq a_k b_k + \cdots + a_n b_n \leq b_k M,$$

a formula which only looks more general, but really isn't.)

(c) Let f be integrable on $[a, b]$ and let ϕ be nonincreasing on $[a, b]$ with $\phi(b) = 0$. Let $P = \{t_0, \ldots, t_n\}$ be a partition of $[a, b]$. Show that the sum

$$\sum_{i=1}^{n} f(t_{i-1})\phi(t_{i-1})(t_i - t_{i-1})$$

lies between the smallest and the largest of the sums

$$\phi(a) \sum_{i=1}^{k} f(t_{i-1})(t_i - t_{i-1}).$$

Conclude that

$$\int_a^b f(x)\phi(x)\,dx$$

lies between the minimum and the maximum of

$$\phi(a) \int_a^x f(t)\,dt,$$

and that it therefore equals $\phi(a) \displaystyle\int_a^{\xi} f(t)\,dt$ for some ξ in $[a, b]$.

36. (a) Show that the following improper integrals both converge.

(i) $\displaystyle\int_0^1 \sin\left(x + \frac{1}{x}\right) dx.$

(ii) $\displaystyle\int_0^1 \sin^2\left(x + \frac{1}{x}\right) dx.$

(b) Decide which of the following improper integrals converge.

(i) $\displaystyle\int_1^{\infty} \sin\left(\frac{1}{x}\right) dx.$

(ii) $\displaystyle\int_1^{\infty} \sin^2\left(\frac{1}{x}\right) dx.$

37. (a) Compute the (improper) integral $\int_0^1 \log x\,dx$.

(b) Show that the improper integral $\int_0^\pi \log(\sin x)\,dx$ converges.

(c) Use the substitution $x = 2u$ to show that

$$\int_0^\pi \log(\sin x)\,dx = 2\int_0^{\pi/2} \log(\sin x)\,dx + 2\int_0^{\pi/2} \log(\cos x)\,dx + \pi \log 2.$$

(d) Compute $\int_0^{\pi/2} \log(\cos x)\,dx$.

(e) Using the relation $\cos x = \sin(\pi/2 - x)$, compute $\int_0^\pi \log(\sin x)\,dx$.

38. Prove the following version of integration by parts for improper integrals:

$$\int_a^\infty u'(x)v(x)\,dx = u(x)v(x)\Big|_a^\infty - \int_a^\infty u(x)v'(x)\,dx.$$

The first symbol on the right side means, of course,

$$\lim_{x\to\infty} u(x)v(x) - u(a)v(a).$$

***39.** One of the most important functions in analysis is the gamma function,

$$\Gamma(x) = \int_0^\infty e^{-t}t^{x-1}\,dt.$$

(a) Prove that the improper integral $\Gamma(x)$ is defined if $x > 0$.
(b) Use integration by parts (more precisely, the improper integral version in the previous problem) to prove that

$$\Gamma(x+1) = x\Gamma(x).$$

(c) Show that $\Gamma(1) = 1$, and conclude that $\Gamma(n) = (n-1)!$ for all natural numbers n.

The gamma function thus provides a simple example of a continuous function which "interpolates" the values of $n!$ for natural numbers n. Of course there are infinitely many continuous functions f with $f(n) = (n-1)!$; there are even infinitely many continuous functions f with $f(x+1) = xf(x)$ for all $x > 0$. However, the gamma function has the important additional property that $\log \circ \Gamma$ is convex, a condition which expresses the extreme smoothness of this function. A beautiful theorem due to Harold Bohr and Johannes Mollerup states that Γ is the only function f with $\log \circ f$ convex, $f(1) = 1$ and $f(x+1) = xf(x)$. See the Suggested Reading for a reference.

***40.** (a) Use the reduction formula for $\int \sin^n x\,dx$ to show that

$$\int_0^{\pi/2} \sin^n x\,dx = \frac{n-1}{n}\int_0^{\pi/2} \sin^{n-2} x\,dx.$$

(b) Now show that

$$\int_0^{\pi/2} \sin^{2n+1} x\,dx = \frac{2}{3}\cdot\frac{4}{5}\cdot\frac{6}{7}\cdot\ldots\cdot\frac{2n}{2n+1},$$

$$\int_0^{\pi/2} \sin^{2n} x\,dx = \frac{\pi}{2}\cdot\frac{1}{2}\cdot\frac{3}{4}\cdot\frac{5}{6}\cdot\ldots\cdot\frac{2n-1}{2n},$$

and conclude that

$$\frac{\pi}{2} = \frac{2}{1}\cdot\frac{2}{3}\cdot\frac{4}{3}\cdot\frac{4}{5}\cdot\frac{6}{5}\cdot\frac{6}{7}\cdot\ldots\cdot\frac{2n}{2n-1}\cdot\frac{2n}{2n+1}\cdot\frac{\displaystyle\int_0^{\pi/2}\sin^{2n} x\,dx}{\displaystyle\int_0^{\pi/2}\sin^{2n+1} x\,dx}.$$

(c) Show that the quotient of the two integrals in this expression is between 1 and $1 + 1/2n$, starting with the inequalities

$$0 < \sin^{2n+1} x \le \sin^{2n} x \le \sin^{2n-1} x \quad \text{for } 0 < x < \pi/2.$$

This result, which shows that the products

$$\frac{2}{1}\cdot\frac{2}{3}\cdot\frac{4}{3}\cdot\frac{4}{5}\cdot\frac{6}{5}\cdot\frac{6}{7}\cdot\ldots\cdot\frac{2n}{2n-1}\cdot\frac{2n}{2n+1}$$

can be made as close to $\pi/2$ as desired, is usually written as an infinite product, known as Wallis' product:

$$\frac{\pi}{2} = \frac{2}{1}\cdot\frac{2}{3}\cdot\frac{4}{3}\cdot\frac{4}{5}\cdot\frac{6}{5}\cdot\frac{6}{7}\cdot\ldots.$$

(d) Show also that the products

$$\frac{1}{\sqrt{n}}\,\frac{2\cdot 4\cdot 6\cdot\ldots\cdot 2n}{1\cdot 3\cdot 5\cdot\ldots\cdot(2n-1)}$$

can be made as close to $\sqrt{\pi}$ as desired. (This fact is used in the next problem and in Problem 27-19.)

**41. It is an astonishing fact that improper integrals $\int_0^\infty f(x)\,dx$ can often be computed in cases where ordinary integrals $\int_a^b f(x)\,dx$ cannot. There is no elementary formula for $\int_a^b e^{-x^2}\,dx$, but we can find the value of $\int_0^\infty e^{-x^2}\,dx$ precisely! There are many ways of evaluating this integral, but most require some advanced techniques; the following method involves a fair amount of work, but no facts that you do not already know.

(a) Show that

$$\int_0^1 (1-x^2)^n\,dx = \frac{2}{3}\cdot\frac{4}{5}\cdot\dots\cdot\frac{2n}{2n+1},$$

$$\int_0^\infty \frac{1}{(1+x^2)^n}\,dx = \frac{\pi}{2}\cdot\frac{1}{2}\cdot\frac{3}{4}\cdot\dots\cdot\frac{2n-3}{2n-2}.$$

(This can be done using reduction formulas, or by appropriate substitutions, combined with the previous problem.)

(b) Prove, using the derivative, that

$$1-x^2 \le e^{-x^2} \qquad \text{for } 0 \le x \le 1.$$

$$e^{-x^2} \le \frac{1}{1+x^2} \qquad \text{for } 0 \le x.$$

(c) Integrate the nth powers of these inequalities from 0 to 1 and from 0 to ∞, respectively. Then use the substitution $y = \sqrt{n}\,x$ to show that

$$\begin{aligned}\sqrt{n}\,\frac{2}{3}\cdot\frac{4}{5}\cdot\dots\cdot\frac{2n}{2n+1} \\ &\le \int_0^{\sqrt{n}} e^{-y^2}\,dy \le \int_0^\infty e^{-y^2}\,dy \\ &\le \frac{\pi}{2}\sqrt{n}\,\frac{1}{2}\cdot\frac{3}{4}\cdot\dots\cdot\frac{2n-3}{2n-2}.\end{aligned}$$

(d) Now use Problem 40(d) to show that

$$\int_0^\infty e^{-y^2}\,dy = \frac{\sqrt{\pi}}{2}.$$

****42.** (a) Use integration by parts to show that

$$\int_a^b \frac{\sin x}{x}\,dx = \frac{\cos a}{a} - \frac{\cos b}{b} - \int_a^b \frac{\cos x}{x^2}\,dx,$$

and conclude that $\int_0^\infty (\sin x)/x\,dx$ exists. (Use the left side to investigate the limit as $a \to 0^+$ and the right side for the limit as $b \to \infty$.)

(b) Use Problem 15-33 to show that

$$\int_0^\pi \frac{\sin(n+\frac{1}{2})t}{\sin\dfrac{t}{2}}\,dt = \pi$$

for any natural number n.

(c) Prove that

$$\lim_{\lambda\to\pi} \int_0^\pi \sin(\lambda+\tfrac{1}{2})t\left[\frac{2}{t} - \frac{1}{\sin\dfrac{t}{2}}\right]dt = 0.$$

Hint: The term in brackets is bounded by Problem 15-2(vi); the Riemann-Lebesgue Lemma then applies.

(d) Use the substitution $u = (\lambda + \frac{1}{2})t$ and part (b) to show that

$$\int_0^\infty \frac{\sin x}{x}\,dx = \frac{\pi}{2}.$$

43. Given the value of $\int_0^\infty (\sin x)/x\,dx$ from Problem 42, compute

$$\int_0^\infty \left(\frac{\sin x}{x}\right)^2 dx$$

by using integration by parts. (As in Problem 37, the formula for $\sin 2x$ will play an important role.)

***44.** (a) Use the substitution $u = t^x$ to show that

$$\Gamma(x) = \frac{1}{x}\int_0^\infty e^{-u^{1/x}}\,du.$$

(b) Find $\Gamma(\frac{1}{2})$.

***45.** (a) Suppose that $\dfrac{f(x)}{x}$ is integrable on every interval $[a, b]$ for $0 < a < b$, and that $\lim_{x\to 0} f(x) = A$ and $\lim_{x\to\infty} f(x) = B$. Prove that for all $\alpha, \beta > 0$ we have

$$\int_0^\infty \frac{f(\alpha x) - f(\beta x)}{x}\,dx = (A - B)\log\frac{\beta}{\alpha}.$$

Hint: To estimate $\displaystyle\int_\varepsilon^N \frac{f(\alpha x) - f(\beta x)}{x}\,dx$ use two different substitutions.

(b) Now suppose instead that $\displaystyle\int_a^\infty \frac{f(x)}{x}\,dx$ converges for all $a > 0$ and that $\lim_{x\to 0} f(x) = A$. Prove that

$$\int_0^\infty \frac{f(\alpha x) - f(\beta x)}{x}\,dx = A\log\frac{\beta}{\alpha}.$$

(c) Compute the following integrals:

(i) $\displaystyle\int_0^\infty \frac{e^{-\alpha x} - e^{-\beta x}}{x}\,dx.$

(ii) $\displaystyle\int_0^\infty \frac{\cos(\alpha x) - \cos(\beta x)}{x}\,dx.$

In Chapter 13 we said, rather blithely, that integrals may be computed to any degree of accuracy desired by calculating lower and upper sums. But an applied mathematician, who really has to do the calculation, rather than just talking about doing it, may not be overjoyed at the prospect of computing lower sums to evaluate an integral to three decimal places, say (a degree of accuracy that might easily be needed in certain circumstances). The next three problems show how more refined methods can make the calculations much more efficient.

We ought to mention at the outset that computing upper and lower sums might not even be practical, since it might not be possible to compute the quantities m_i and M_i for each interval $[t_{i-1}, t_i]$. It is far more reasonable simply to pick points x_i in $[t_{i-1}, t_i]$ and consider $\sum_{i=1}^{n} f(x_i) \cdot (t_i - t_{i-1})$. This represents the sum of the areas of certain rectangles which partially overlap the graph of f—see Figure 1 in the Appendix to Chapter 13. But we will get a much better result if we instead choose the trapezoids shown in Figure 2.

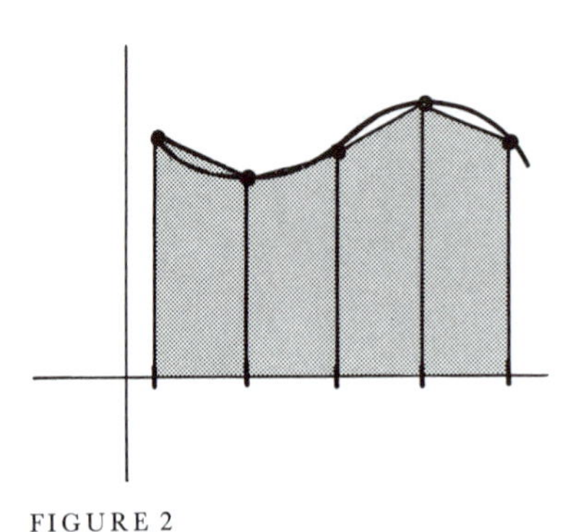

FIGURE 2

Suppose, in particular, that we divide $[a, b]$ into n equal intervals, by means of the points

$$t_i = a + i\left(\frac{b-a}{n}\right) = a + ih.$$

Then the trapezoid with base $[t_{i-1}, t_i]$ has area

$$\frac{f(t_{i-1}) + f(t_i)}{2} \cdot (t_i - t_{i-1})$$

and the sum of all these areas is simply

$$\begin{aligned}\Sigma_n &= h\left[\frac{f(t_1) + f(a)}{2} + \frac{f(t_2) + f(t_1)}{2} + \cdots + \frac{f(b) + f(t_{n-1})}{2}\right] \\ &= \frac{h}{2}\left[f(a) + 2\sum_{i=1}^{n-1} f(a + ih) + f(b)\right], \qquad h = \frac{b-a}{n}.\end{aligned}$$

This method of approximating an integral is called the *trapezoid rule.* Notice that to obtain Σ_{2n} from Σ_n it isn't necessary to recompute the old $f(t_i)$; their contribution to Σ_{2n} is just $\frac{1}{2}\Sigma_n$. So in practice it is best to compute Σ_2, Σ_4, Σ_8, ... to get approximations to $\int_a^b f$. In the next problem we will estimate $\int_a^b f - \Sigma_n$.

46. (a) Suppose that f'' is continuous. Let P_i be the linear function which agrees with f at t_{i-1} and t_i. Using Problem 11-43, show that if n_i and N_i are the minimum and maximum of f'' on $[t_{i-1}, t_i]$ and

$$I = \int_{t_{i-1}}^{t_i} (x - t_{i-1})(x - t_i)\,dx$$

then

$$\frac{n_i I}{2} \geq \int_{t_{i-1}}^{t_i} (f - P_i) \geq \frac{N_i I}{2}.$$

(b) Evaluate I to get

$$\frac{n_i h^3}{12} \leq \int_{t_{i-1}}^{t_i} (f - P_i) \leq \frac{N_i h^3}{12}.$$

(c) Conclude that there is some c in $[a, b]$ with

$$\int_a^b f = \Sigma_n - \frac{(b-a)^3}{12n^2} f''(c).$$

Notice that the "error term" $(b-a)^3 f''(c)/12n^2$ varies as $1/n^2$ (while the error obtained using ordinary sums varies as $1/n$).

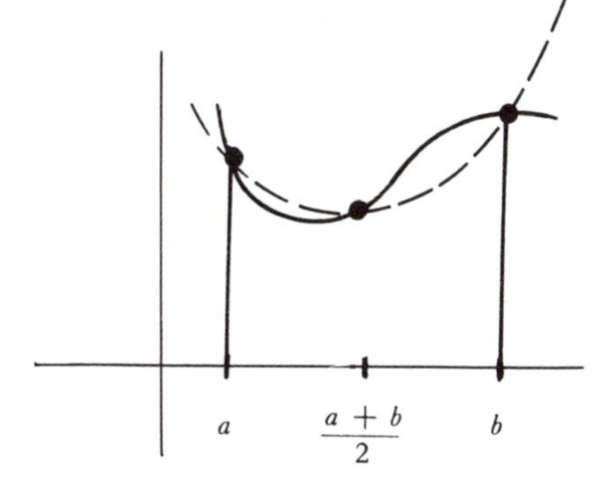

FIGURE 3

We can obtain still more accurate results if we approximate f by quadratic functions rather than by linear functions. We first consider what happens when the interval $[a, b]$ is divided into two equal intervals (Figure 3).

47. (a) Suppose first that $a = 0$ and $b = 2$. Let P be the second degree polynomial function which agrees with f at 0, 1, and 2 (Problem 3-6). Show that

$$\int_0^2 P = \frac{1}{3}[f(0) + 4f(1) + f(2)].$$

(b) Conclude that in the general case

$$\int_a^b P = \frac{b-a}{6}\left[f(a) + 4f\left(\frac{a+b}{2}\right) + f(b)\right].$$

(c) Naturally $\int_a^b P = \int_a^b f$ when f is a quadratic polynomial. But, remarkably enough, this same relation holds when f is a cubic polynomial! Prove this, using Problem 11-43; note that f''' is a constant.

The previous problem shows that we do not have to do any new calculations to compute $\int_a^b Q$ when Q is a *cubic* polynomial which agrees with f at a, b, and $\frac{a+b}{2}$: we still have

$$\int_a^b Q = \frac{b-a}{6}\left[f(a) + 4f\left(\frac{a+b}{2}\right) + f(b)\right].$$

But there is much more lee-way in choosing Q, which we can use to our advantage:

48. (a) Show that there is a cubic polynomial function Q satisfying

$$Q(a) = f(a), \qquad Q(b) = f(b), \qquad Q\left(\frac{a+b}{2}\right) = f\left(\frac{a+b}{2}\right)$$
$$Q'\left(\frac{a+b}{2}\right) = f'\left(\frac{a+b}{2}\right).$$

Hint: Clearly $Q(x) = P(x) + A(x-a)(x-b)\left(x - \frac{a+b}{2}\right)$ for some A.

(b) Prove that for every x we have

$$f(x) - Q(x) = (x-a)\left(x - \frac{a+b}{2}\right)^2 (x-b)\frac{f^{(4)}(\xi)}{4!}$$

for some ξ in $[a, b]$. Hint: Imitate the proof of Problem 11-43.

(c) Conclude that if $f^{(4)}$ is continuous, then

$$\int_a^b f = \frac{b-a}{6}\left[f(a) + 4f\left(\frac{a+b}{2}\right) + f(b)\right] - \frac{(b-a)^5}{2880} f^{(4)}(c)$$

for some c in $[a, b]$.

(d) Now divide $[a, b]$ into $2n$ intervals by means of the points

$$t_i = a + ih, \qquad h = \frac{b-a}{2n}.$$

Prove *Simpson's rule*:

$$\int_a^b f = \frac{b-a}{n}\left(f(a) + 4\sum_{i=1}^{n} f(t_{2i-1}) + 2\sum_{i=1}^{n-1} f(t_{2i}) + f(b)\right)$$
$$- \frac{(b-a)^5}{2880n^4} f^{(4)}(\bar{c})$$

for some $\bar{c}$ in $[a, b]$.

APPENDIX. THE COSMOPOLITAN INTEGRAL

We originally introduced integrals in order to find the area under the graph of a function, but the integral is considerably more versatile than that. For example, Problem 13-24 used the integral to express the area of a region of quite another sort. Moreover, Problem 13-25 showed that the integral can also be used to express the lengths of curves—though, as we've seen in Appendix to Chapter 13, a lot of work may be necessary to consider the general case! This result was probably a little more surprising, since the integral seems, at first blush, to be a very two-dimensional creature. Actually, the integral makes its appearance in quite a few geometric formulas, which we will present in this Appendix. To derive these formulas we will assume some results from elementary geometry (and allow a little fudging).

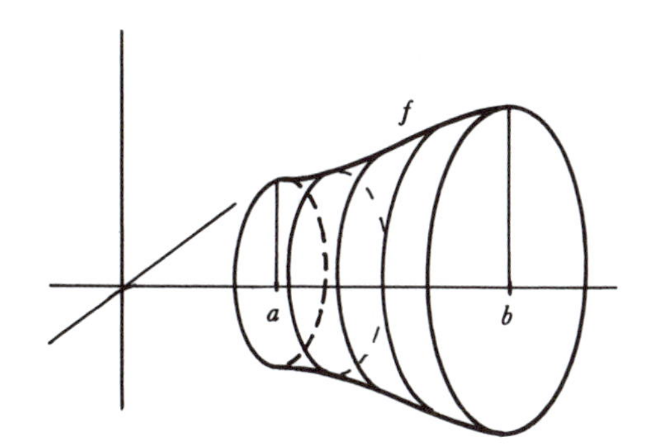

FIGURE 1

Instead of going down to one-dimensional objects, we'll begin by tackling some three-dimensional ones. There are some very special solids whose volumes can be expressed by integrals. The simplest such solid V is a "volume of revolution," obtained by revolving the region under the graph of $f \geq 0$ on $[a, b]$ around the horizontal axis, when we regard the plane as situated in space (Figure 1). If $P = \{t_0, \dots, t_n\}$ is any partition of $[a, b]$, and m_i and M_i have their usual meanings, then

$$\pi m_i{}^2(t_i - t_{i-1})$$

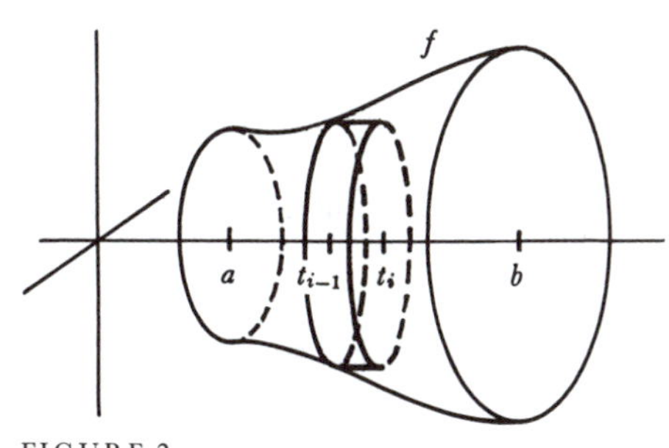

FIGURE 2

is the volume of a disc that lies inside the solid V (Figure 2). Similarly, $\pi M_i{}^2(t_i - t_{i-1})$ is the volume of a disc that contains the part of V between t_{i-1} and t_i. Consequently,

$$\pi \sum_{i=1}^{n} m_i{}^2(t_i - t_{i-1}) \leq \text{ volume } V \leq \pi \sum_{i=1}^{n} M_i{}^2(t_i - t_{i-1}).$$

But the sums on the ends of this inequality are just the lower and upper sums for f^2 on $[a, b]$:

$$\pi \cdot L(f^2, P) \leq \text{volume } V \leq \pi \cdot U(f^2, P).$$

Consequently, the volume of V must be given by

$$\text{volumn } V = \pi \int_a^b f(x)^2 \, dx.$$

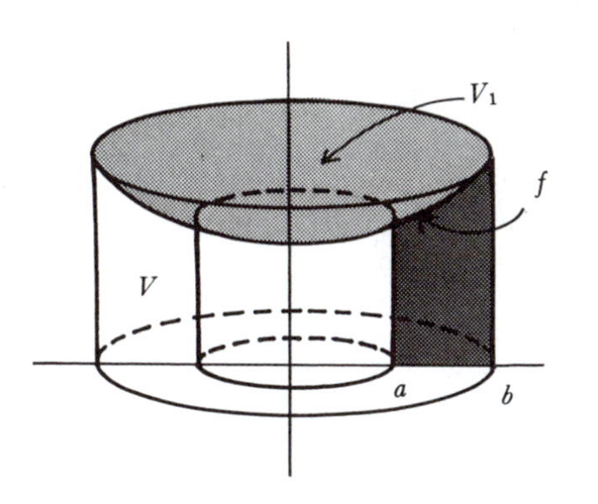

FIGURE 3

This method of finding volumes is affectionately referred to as the "disc method."

Figure 3 shows a more complicated solid V obtained by revolving the region under the graph of f around the *vertical* axis (V is the solid left over when we start with the big cylinder of radius b and take away both the small cylinder of radius a and the solid V_1 sitting right on top of it). In this case we assume $a \geq 0$ as well as $f \geq 0$. Figures 4 and 5 indicate some other possible shapes for V.

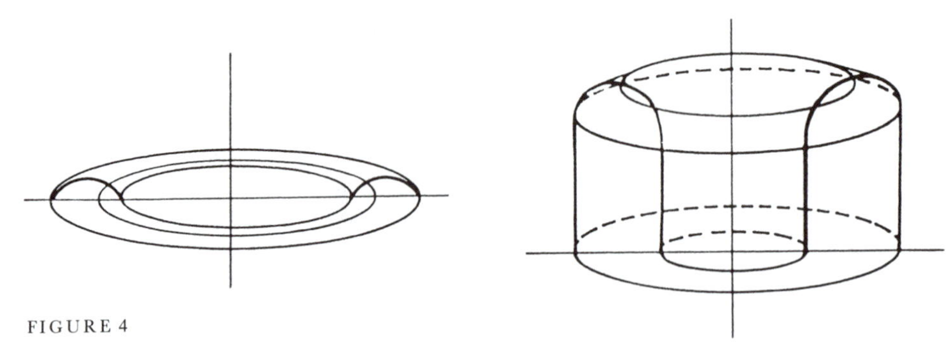

FIGURE 4

FIGURE 5

For a partition $P = \{t_0, \ldots, t_n\}$ we consider the "shells" obtained by rotating the rectangle with base $[t_{i-1}, t_i]$ and height m_i or M_i (Figure 6). Adding the volumes of these shells we obtain

$$\pi \sum_{i=1}^{n} m_i(t_i{}^2 - t_{i-1}{}^2) \leq \text{volume } V \leq \pi \sum_{i=1}^{n} M_i(t_i{}^2 - t_{i-1}{}^2),$$

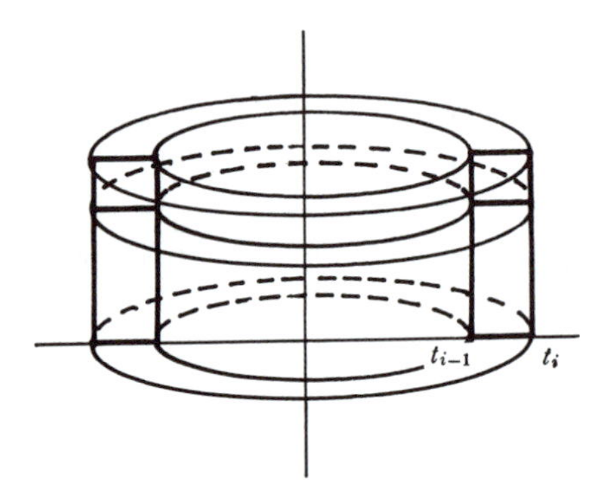

FIGURE 6

which we can write as

$$\pi \sum_{i=1}^{n} m_i(t_i + t_{i-1})(t_i - t_{i-1}) \leq \text{volume } V \leq \pi \sum_{i=1}^{n} M_i(t_i + t_{i-1})(t_i - t_{i-1}).$$

Now these sums are not lower or upper sums of anything. But Problem 1 of the Appendix to Chapter 13 shows that each sum

$$\sum_{i=1}^{n} m_i t_i(t_i - t_{i-1}) \qquad \text{and} \qquad \sum_{i=1}^{n} m_i t_{i-1}(t_i - t_{i-1})$$

can be made as close as desired to $\displaystyle\int_a^b xf(x)\,dx$ by choosing the lengths $t_i - t_{i-1}$ small enough. The same is true of the sums on the right, so we find that

$$\text{volume } V = 2\pi \int_a^b xf(x)\,dx;$$

this is the so-called "shell method" of finding volumes.

The surface area of certain curved regions can also be expressed in terms of integrals. Before we tackle complicated regions, a little review of elementary geometric formulas may be appreciated here.

Figure 7 shows a right pyramid made up of triangles with bases of length l and altitude s. The total surface area of the sides of the pyramid is thus

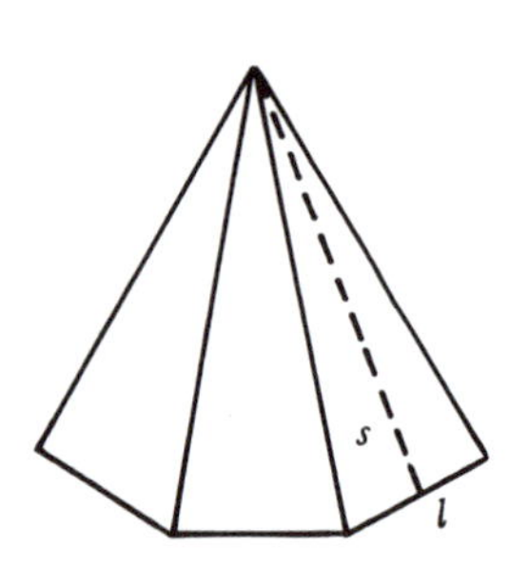

FIGURE 7

$$\frac{1}{2}ps,$$

where p is the perimeter of the base. By choosing the base to be a regular polygon

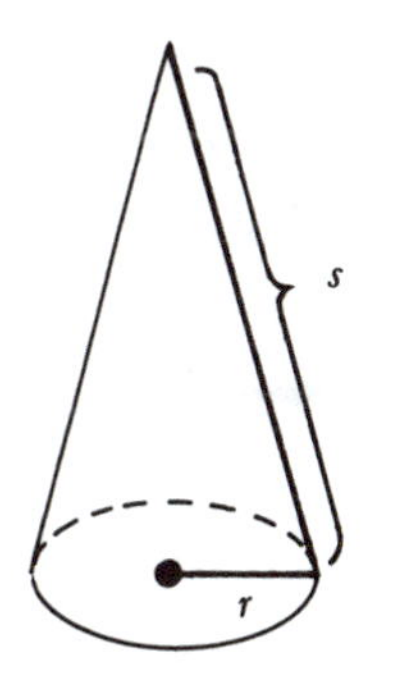

FIGURE 8

with a large number of sides we see that the area of a right circular cone (Figure 8) must be

$$\frac{1}{2}(2\pi r)s = \pi r s,$$

where s is the "slant height." Finally, consider the frustum of a cone with slant height s and radii r_1 and r_2 shown in Figure 9(a). Completing this to a cone, as in Figure 9(b), we have

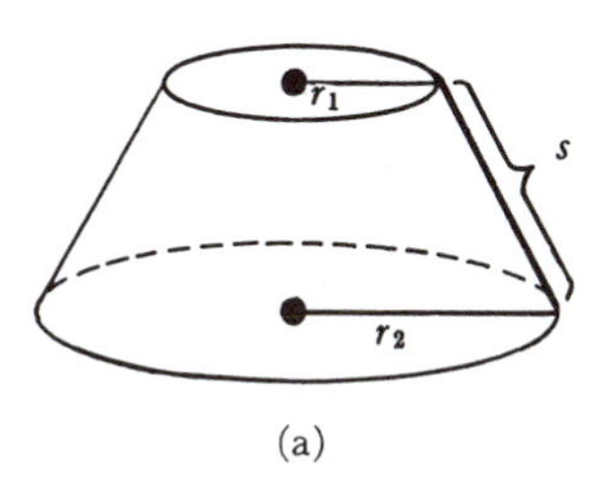

(a)

$$\frac{s_1}{r_1} = \frac{s_1 + s}{r_2},$$

so

$$s_1 = \frac{r_1 s}{r_2 - r_1}, \qquad s_1 + s = \frac{r_2 s}{r_2 - r_1}.$$

Consequently, the surface area is

$$\pi r_2(s_1 + s) - \pi r_1 s_1 = \pi s \frac{{r_2}^2 - {r_1}^2}{r_2 - r_1} = \pi s(r_1 + r_2).$$

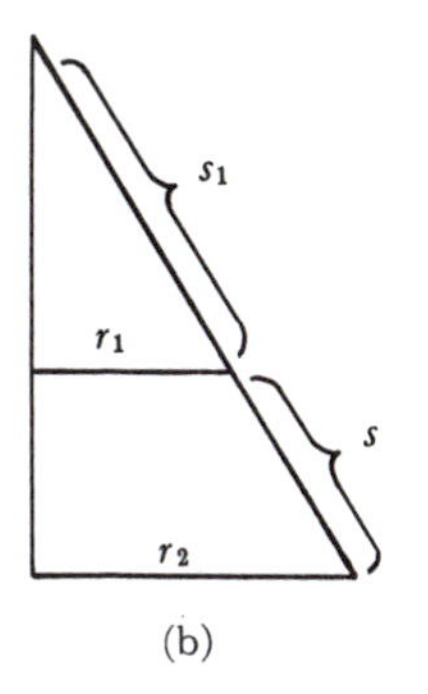

(b)

FIGURE 9

Now consider the surface formed by revolving the graph of f around the horizontal axis. For a partition $P = \{t_0, \dots, t_n\}$ we can inscribe a series of frusta of cones, as in Figure 10. The total surface area of these frusta is

$$\pi \sum_{i=1}^{n} [f(t_{i-1}) + f(t_i)] \sqrt{(t_i - t_{i-1})^2 + [f(t_i) - f(t_{i-1})]^2}$$

$$= \pi \sum_{i=1}^{n} [f(t_{i-1}) + f(t_i)] \sqrt{1 + \left(\frac{f(t_i) - f(t_{i-1})}{t_i - t_{i-1}}\right)^2} (t_i - t_{i-1}).$$

By the Mean Value Theorem, this is

$$\pi \sum_{i=1}^{n} [f(t_{i-1}) + f(t_i)] \sqrt{1 + f'(x_i)^2}\, (t_i - t_{i-1})$$

for some x_i in (t_{i-1}, t_i). Appealing to Problem 1 of the Appendix to Chapter 13, we conclude that the surface area is

$$2\pi \int_a^b f(x)\sqrt{1 + f'(x)^2}\, dx.$$

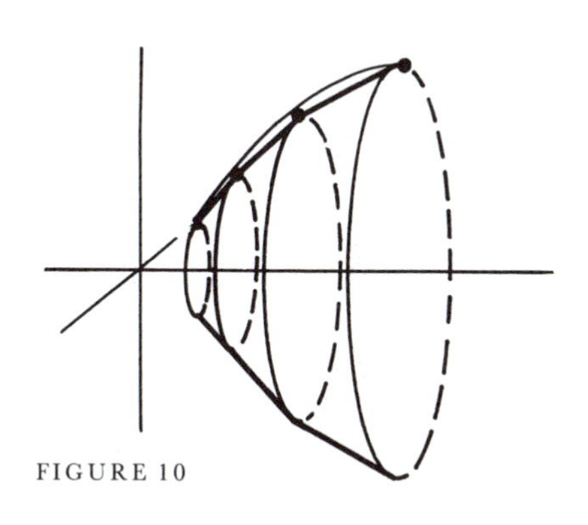

FIGURE 10

PROBLEMS

1. (a) Find the volume of the solid obtained by revolving the region bounded by the graphs of $f(x) = x$ and $f(x) = x^2$ around the horizontal axis.
(b) Find the volume of the solid obtained by revolving this same region around the vertical axis.

2. Find the volume of a sphere of radius r.

3. When the ellipse consisting of all points (x, y) with $x^2/a^2 + y^2/b^2 = 1$ is rotated around the horizontal axis we obtain an "ellipsoid of revolution" (Figure 11). Find the volume of the enclosed solid.

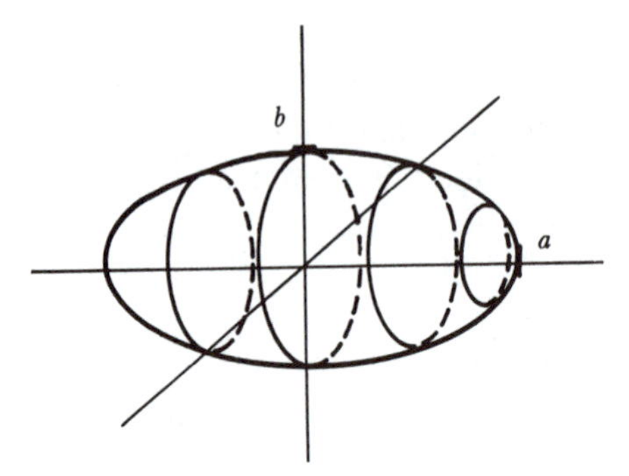

FIGURE 11

4. Find the volume of the "torus" (Figure 12), obtained by rotating the circle $(x - a)^2 + y^2 = b^2$ $(a > b)$ around the vertical axis.

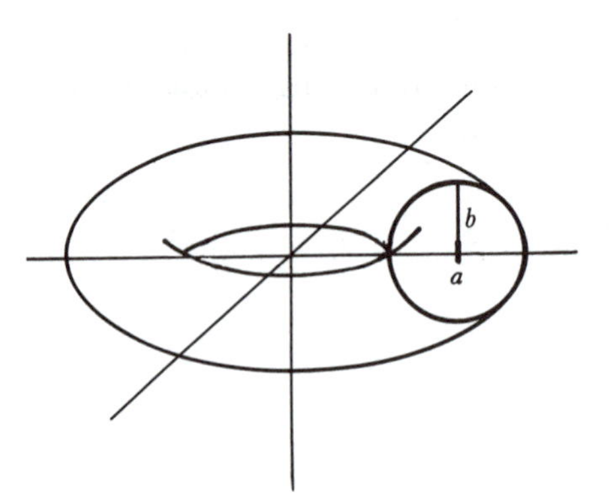

FIGURE 12

5. A cylindrical hole of radius a is bored through the center of a sphere of radius $2a$ (Figure 13). Find the volume of the remaining solid.

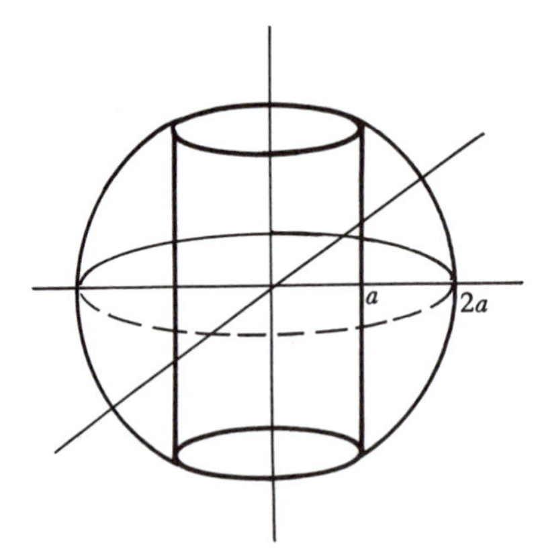

FIGURE 13

6. (a) For the solid shown in Figure 14, find the volume by the shell method.

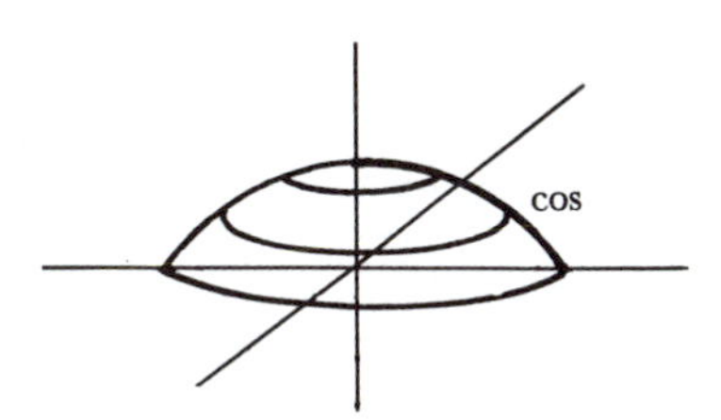

FIGURE 14

(b) This volume can also be evaluated by the disc method. Write down the integral which must be evaluated in this case; notice that it is more complicated. The next problem takes up a question which this might suggest.

7. Figure 15 shows a cylinder of height b and radius $f(b)$, divided into three solids, one of which, V_1, is a cylinder of height a and radius $f(a)$. If f is one-one, then a comparison of the disk method and the shell method of computing volumes leads us to believe that

$$\begin{aligned}\pi b f(b)^2 - \pi a f(a)^2 - \pi \int_a^b f(x)^2\,dx &= \text{volume } V_2 \\ &= 2\pi \int_{f(a)}^{f(b)} y f^{-1}(y)\,dy.\end{aligned}$$

Prove this analytically, using the formula for $\displaystyle\int f^{-1}$ from Problem 19-15.

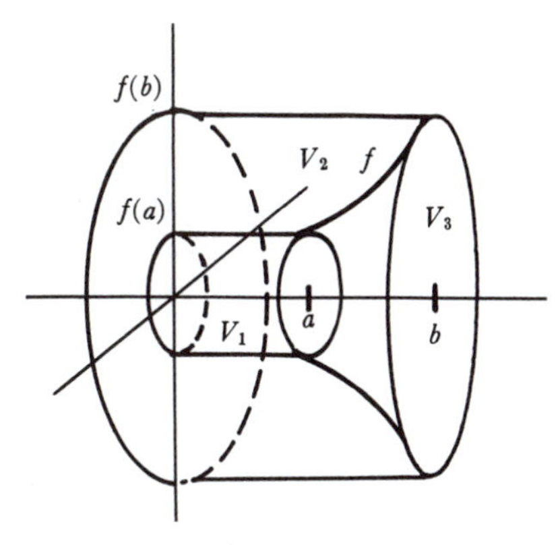

FIGURE 15

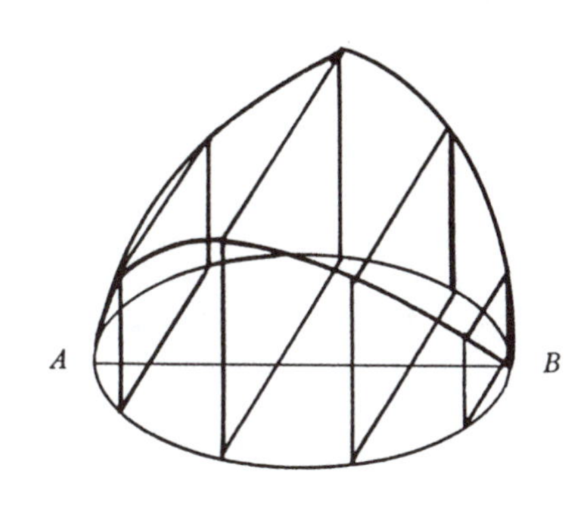

FIGURE 16

8. (a) Figure 16 shows a solid with a circular base of radius a. Each plane perpendicular to the diameter AB intersects the solid in a square. Using arguments similar to those already used in this Appendix, express the volume of the solid as an integral, and evaluate it.

(b) Same problem if each plane intersects the solid in an equilateral triangle.

9. Find the volume of a pyramid (Figure 17) in terms of its height h and the area A of its base.

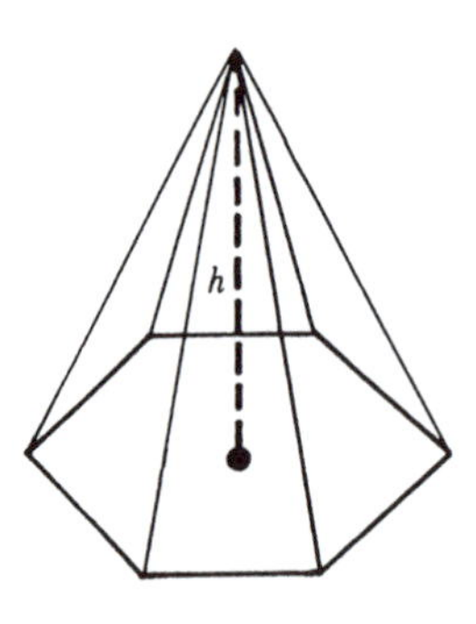

FIGURE 17

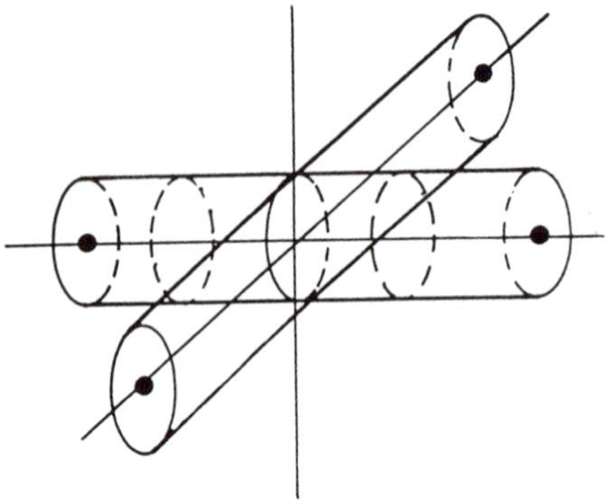

FIGURE 18

10. Find the volume of the solid which is the intersection of the two cylinders in Figure 18. Hint: Find the intersection of this solid with each horizontal plane.

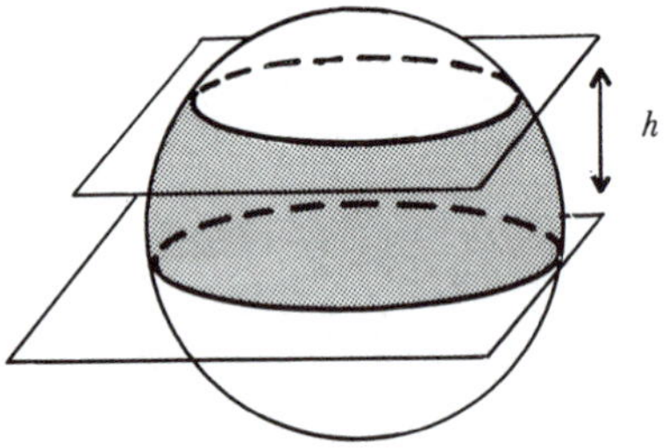

11. (a) Prove that the surface area of a sphere of radius r is $4\pi r^2$.

(b) Prove, more generally, that the area of the portion of the sphere shown in Figure 19 is $2\pi rh$. (Notice that this depends only on h, not on the position of the planes!)

FIGURE 19

12. (a) Find the surface area of the ellipsoid of revolution in Problem 19-3.

(b) Find the surface area of the torus in Problem 19-4.

13. The graph of $f(x) = 1/x$, $x \geq 1$ is revolved around the horizontal axis (Figure 20).

(a) Find the volume of the enclosed "infinite trumpet."

(b) Show that the surface area is infinite.

(c) Suppose that we fill up the trumpet with the finite amount of paint found in part (a). It would seem that we have thereby coated the infinite inside surface area with only a finite amount of paint! How is this possible?

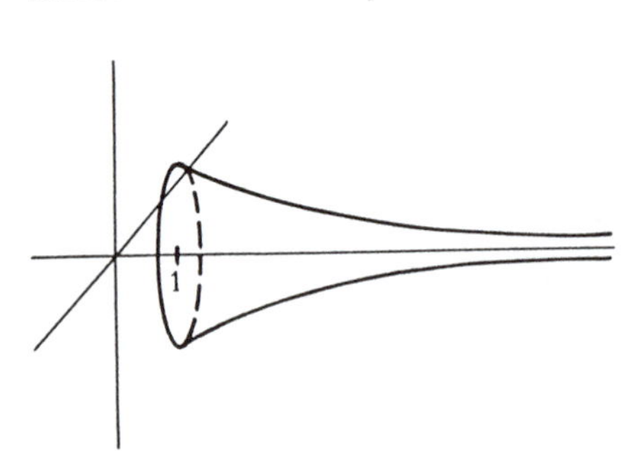

FIGURE 20

PART 4

INFINITE SEQUENCES AND INFINITE SERIES

One of the most remarkable series of algebraic analysis is the following:

$$1 + \frac{m}{1}x + \frac{m(m-1)}{1 \cdot 2}x^2 + \frac{m(m-1)(m-2)}{1 \cdot 2 \cdot 3}x^3 + \cdots + \frac{m(m-1)\cdots[m-(n-1)]}{1 \cdot 2 \cdots\cdots n}x^n + \cdots$$

When m is a positive whole number
the sum of the series,
which is then finite, can be expressed,
as is known, by $(1 + x)^m$*.*
When m is not an integer,
the series goes on to infinity, and it will
converge or diverge according
as the quantities
m and x have this or that value.
In this case, one writes the same equality

$$(1 + x)^m = 1 + \frac{m}{1}x + \frac{m(m-1)}{1 \cdot 2}x^2 + \cdots \textit{ etc.}$$

. . . It is assumed that
the numerical equality will always occur
whenever the series is convergent, but
this has never yet been proved.

NIELS HENRIK ABEL

CHAPTER 20 APPROXIMATION BY POLYNOMIAL FUNCTIONS

There is one sense in which the "elementary functions" are not elementary at all. If p is a polynomial function,

$$p(x) = a_0 + a_1x + \cdots + a_nx^n,$$

then $p(x)$ can be computed easily for any number x. This is not at all true for functions like sin, log, or exp. At present, to find $\log x = \int_1^x 1/t\,dt$ approximately, we must compute some upper or lower sums, and make certain that the error involved in accepting such a sum for $\log x$ is not too great. Computing $e^x = \log^{-1}(x)$ would be even more difficult: we would have to compute $\log a$ for many values of a until we found a number a such that $\log a$ is approximately x—then a would be approximately e^x.

In this chapter we will obtain important theoretical results which reduce the computation of $f(x)$, for many functions f, to the evaluation of polynomial functions. The method depends on finding polynomial functions which are close approximations to f. In order to guess a polynomial which is appropriate, it is useful to first examine polynomial functions themselves more thoroughly.

Suppose that

$$p(x) = a_0 + a_1x + \cdots + a_nx^n.$$

It is interesting, and for our purposes very important, to note that the coefficients a_i can be expressed in terms of the value of p and its various derivatives at 0. To begin with, note that

$$p(0) = a_0.$$

Differentiating the original expression for $p(x)$ yields

$$p'(x) = a_1 + 2a_2x + \cdots + na_nx^{n-1}.$$

Therefore,

$$p'(0) = p^{(1)}(0) = a_1.$$

Differentiating again we obtain

$$p''(x) = 2a_2 + 3 \cdot 2 \cdot a_3x + \cdots + n(n-1) \cdot a_nx^{n-2}.$$

Therefore,

$$p''(0) = p^{(2)}(0) = 2a_2.$$

In general, we will have

$$p^{(k)}(0) = k!\,a_k \quad \text{or} \quad a_k = \frac{p^{(k)}(0)}{k!}.$$

If we agree to define $0! = 1$, and recall the notation $p^{(0)} = p$, then this formula holds for $k = 0$ also.

If we had begun with a function p that was written as a "polynomial in $(x-a)$,"

$$p(x) = a_0 + a_1(x-a) + \cdots + a_n(x-a)^n,$$

then a similar argument would show that

$$a_k = \frac{p^{(k)}(a)}{k!}.$$

Suppose now that f is a function (not necessarily a polynomial) such that

$$f^{(1)}(a), \dots, f^{(n)}(a)$$

all exist. Let

$$a_k = \frac{f^{(k)}(a)}{k!}, \quad 0 \le k \le n,$$

and define

$$P_{n,a}(x) = a_0 + a_1(x-a) + \cdots + a_n(x-a)^n.$$

The polynomial $P_{n,a}$ is called the **Taylor polynomial of degree n for f at a.** (Strictly speaking, we should use an even more complicated expression, like $P_{n,a,f}$, to indicate the dependence on f; at times this more precise notation will be useful.) The Taylor polynomial has been defined so that

$$P_{n,a}{}^{(k)}(a) = f^{(k)}(a) \quad \text{for } 0 \le k \le n;$$

in fact, it is clearly the only polynomial *of degree* $\le n$ with this property.

Although the coefficients of $P_{n,a,f}$ seem to depend upon f in a fairly complicated way, the most important elementary functions have extremely simple Taylor polynomials. Consider first the function sin. We have

$$\begin{aligned} \sin(0) &= 0, \\ \sin'(0) &= \cos 0 = 1, \\ \sin''(0) &= -\sin 0 = 0, \\ \sin'''(0) &= -\cos 0 = -1, \\ \sin^{(4)}(0) &= \sin 0 = 0. \end{aligned}$$

From this point on, the derivatives repeat in a cycle of 4. The numbers

$$a_k = \frac{\sin^{(k)}(0)}{k!}$$

are

$$0,\ 1,\ 0,\ -\frac{1}{3!},\ 0,\ \frac{1}{5!},\ 0,\ -\frac{1}{7!},\ 0,\ \frac{1}{9!},\ \dots.$$

Therefore the Taylor polynomial $P_{2n+1,0}$ of degree $2n+1$ for sin at 0 is

$$P_{2n+1,0}(x) = x - \frac{x^3}{3!} + \frac{x^5}{5!} - \frac{x^7}{7!} + \cdots + (-1)^n \frac{x^{2n+1}}{(2n+1)!}.$$

(Of course, $P_{2n+1,0} = P_{2n+2,0}$).

The Taylor polynomial $P_{2n,0}$ of degree $2n$ for cos at 0 is (the computations are left to you)

$$P_{2n,0}(x) = 1 - \frac{x^2}{2!} + \frac{x^4}{4!} - \frac{x^6}{6!} + \cdots + (-1)^n \frac{x^{2n}}{(2n)!}.$$

The Taylor polynomial for exp is especially easy to compute. Since $\exp^{(k)}(0) = \exp(0) = 1$ for all k, the Taylor polynomial of degree n at 0 is

$$P_{n,0}(x) = 1 + \frac{x}{1!} + \frac{x^2}{2!} + \frac{x^3}{3!} + \frac{x^4}{4!} + \cdots + \frac{x^n}{n!}.$$

The Taylor polynomial for log must be computed at some point $a \neq 0$, since log is not even defined at 0. The standard choice is $a = 1$. Then

$$\begin{aligned} \log'(x) &= \frac{1}{x}, & \log'(1) &= 1; \\ \log''(x) &= -\frac{1}{x^2}, & \log''(1) &= -1; \\ \log'''(x) &= \frac{2}{x^3}, & \log'''(1) &= 2; \end{aligned}$$

in general

$$\log^{(k)}(x) = \frac{(-1)^{k-1}(k-1)!}{x^k}, \quad \log^{(k)}(1) = (-1)^{k-1}(k-1)!\,.$$

Therefore the Taylor polynomial of degree n for log at 1 is

$$P_{n,1}(x) = (x-1) - \frac{(x-1)^2}{2} + \frac{(x-1)^3}{3} + \cdots + \frac{(-1)^{n-1}(x-1)^n}{n}.$$

It is often more convenient to consider the function $f(x) = \log(1+x)$. In this case we can choose $a = 0$. We have

$$f^{(k)}(x) = \log^{(k)}(1+x),$$

so

$$f^{(k)}(0) = \log^{(k)}(1) = (-1)^{k-1}(k-1)!\,.$$

Therefore the Taylor polynomial of degree n for f at 0 is

$$P_{n,0}(x) = x - \frac{x^2}{2} + \frac{x^3}{3} - \frac{x^4}{4} + \cdots + \frac{(-1)^{n-1}x^n}{n}.$$

There is one other elementary function whose Taylor polynomial is important—arctan. The computations of the derivatives begin

$$\begin{aligned} \arctan'(x) &= \frac{1}{1+x^2} & \arctan'(0) &= 1; \\ \arctan''(x) &= \frac{-2x}{(1+x^2)^2}, & \arctan''(0) &= 0; \\ \arctan'''(x) &= \frac{(1+x^2)^2 \cdot (-2) + 2x \cdot 2(1+x^2) \cdot 2x}{(1+x^2)^4}, & \arctan'''(0) &= -2. \end{aligned}$$

It is clear that this brute force computation will never do. However, the Taylor polynomials of arctan will be easy to find after we have examined the properties of Taylor polynomials more closely—although the Taylor polynomial $P_{n,a,f}$ was simply defined so as to have the same first n derivatives at a as f, the connection between f and $P_{n,a,f}$ will actually turn out to be much deeper.

One line of evidence for a closer connection between f and the Taylor polynomials for f may be uncovered by examining the Taylor polynomial of degree 1, which is

$$P_{1,a}(x) = f(a) + f'(a)(x - a).$$

Notice that

$$\frac{f(x) - P_{1,a}(x)}{x - a} = \frac{f(x) - f(a)}{x - a} - f'(a).$$

Now, by the definition of $f'(a)$ we have

$$\lim_{x \to a} \frac{f(x) - P_{1,a}(x)}{x - a} = 0.$$

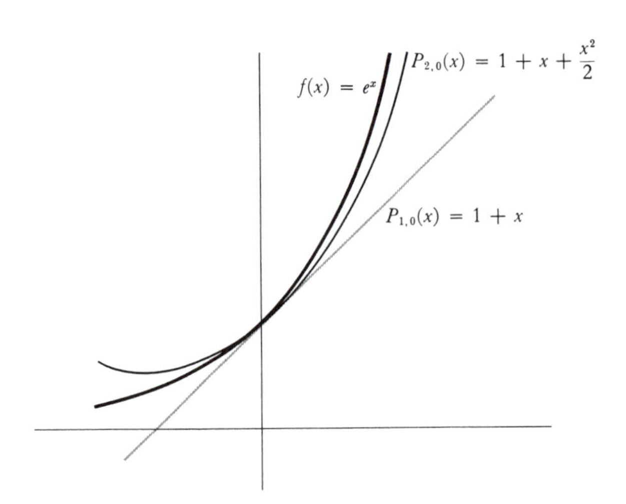

FIGURE 1

In other words, as x approaches a the difference $f(x) - P_{1,a}(x)$ not only becomes small, but actually becomes small even compared to $x - a$. Figure 1 illustrates the graph of $f(x) = e^x$ and of

$$P_{1,0}(x) = f(0) + f'(0)x = 1 + x,$$

which is the Taylor polynomial of degree 1 for f at 0. The diagram also shows the graph of

$$P_{2,0}(x) = f(0) + f'(0) + \frac{f''(0)}{2!}x^2 = 1 + x + \frac{x^2}{2},$$

which is the Taylor polynomial of degree 2 for f at 0. As x approaches 0, the difference $f(x) - P_{2,0}(x)$ seems to be getting small even faster than the difference

$f(x) - P_{1,0}(x)$. As it stands, this assertion is not very precise, but we are now prepared to give it a definite meaning. We have just noted that in general

$$\lim_{x \to a} \frac{f(x) - P_{1,a}(x)}{x - a} = 0.$$

For $f(x) = e^x$ and $a = 0$ this means that

$$\lim_{x \to 0} \frac{f(x) - P_{1,0}(x)}{x} = \lim_{x \to 0} \frac{e^x - 1 - x}{x} = 0.$$

On the other hand, an easy double application of l'Hôpital's Rule shows that

$$\lim_{x \to 0} \frac{e^x - 1 - x}{x^2} = \frac{1}{2} \neq 0.$$

Thus, although $f(x) - P_{1,0}(x)$ becomes small compared to x, as x approaches 0, it does *not* become small compared to x^2. For $P_{2,0}(x)$ the situation is quite different; the extra term $x^2/2$ provides just the right compensation:

$$\begin{aligned} \lim_{x \to 0} \frac{e^x - 1 - x - \dfrac{x^2}{2}}{x^2} &= \lim_{x \to 0} \frac{e^x - 1 - x}{2x} \\ &= \lim_{x \to 0} \frac{e^x - 1}{2} = 0. \end{aligned}$$

This result holds in general—if $f'(a)$ and $f''(a)$ exist, then

$$\lim_{x \to a} \frac{f(x) - P_{2,a}(x)}{(x - a)^2} = 0;$$

in fact, the analogous assertion for $P_{n,a}$ is also true.

THEOREM 1 Suppose that f is a function for which

$$f'(a), \ldots, f^{(n)}(a)$$

all exist. Let

$$a_k = \frac{f^{(k)}(a)}{k!}, \quad 0 \leq k \leq n,$$

and define

$$P_{n,a}(x) = a_0 + a_1(x - a) + \cdots + a_n(x - a)^n.$$

Then

$$\lim_{x \to a} \frac{f(x) - P_{n,a}(x)}{(x - a)^n} = 0.$$

PROOF Writing out $P_{n,a}(x)$ explicitly, we obtain

$$\frac{f(x) - P_{n,a}(x)}{(x-a)^n} = \frac{f(x) - \displaystyle\sum_{i=0}^{n-1} \frac{f^{(i)}(a)}{i!}(x-a)^i}{(x-a)^n} - \frac{f^{(n)}(a)}{n!}.$$

It will help to introduce the new functions

$$Q(x) = \sum_{i=0}^{n-1} \frac{f^{(i)}(a)}{i!}(x-a)^i \quad \text{and} \quad g(x) = (x-a)^n;$$

now we must prove that

$$\lim_{x \to a} \frac{f(x) - Q(x)}{g(x)} = \frac{f^{(n)}(a)}{n!}.$$

Notice that

$$\begin{aligned} Q^{(k)}(a) &= f^{(k)}(a), \quad k \le n-1, \\ g^{(k)}(x) &= n!(x-a)^{n-k}/(n-k)!. \end{aligned}$$

Thus

$$\begin{aligned} &\lim_{x \to a} [f(x) - Q(x)] = f(a) - Q(a) = 0, \\ &\lim_{x \to a} [f'(x) - Q'(x)] = f'(a) - Q'(a) = 0, \\ &\quad\vdots \\ &\lim_{x \to a} [f^{(n-2)}(x) - Q^{(n-2)}(x)] = f^{(n-2)}(a) - Q^{(n-2)}(a) = 0. \end{aligned}$$

and

$$\lim_{x \to a} g(x) = \lim_{x \to a} g'(x) = \cdots = \lim_{x \to a} g^{(n-2)}(x) = 0.$$

We may therefore apply l'Hôpital's Rule $n-1$ times to obtain

$$\lim_{x \to a} \frac{f(x) - Q(x)}{(x-a)^n} = \lim_{x \to a} \frac{f^{(n-1)}(x) - Q^{(n-1)}(x)}{n!\,(x-a)}.$$

Since Q is a polynomial of degree $n-1$, its $(n-1)$st derivative is a constant; in fact, $Q^{(n-1)}(x) = f^{(n-1)}(a)$. Thus

$$\lim_{x \to a} \frac{f(x) - Q(x)}{(x-a)^n} = \lim_{x \to a} \frac{f^{(n-1)}(x) - f^{(n-1)}(a)}{n!\,(x-a)}$$

and this last limit is $f^{(n)}(a)/n!$ by definition of $f^{(n)}(a)$. ▌

One simple consequence of Theorem 1 allows us to perfect the test for local maxima and minima which was developed in Chapter 11. If a is a critical point of f, then, according to Theorem 11-5, the function f has a local minimum at a if $f''(a) > 0$, and a local maximum at a if $f''(a) < 0$. If $f''(a) = 0$ no conclusion was possible, but it is conceivable that the sign of $f'''(a)$ might give further information; and if $f'''(a) = 0$, then the sign of $f^{(4)}(a) = 0$ might be significant. Even more generally, we can ask what happens when

$$(*) \qquad \begin{aligned} &f'(a) = f''(a) = \cdots = f^{(n-1)}(a) = 0, \\ &f^{(n)}(a) \neq 0. \end{aligned}$$

The situation in this case can be guessed by examining the functions

$$\begin{aligned} f(x) &= (x-a)^n, \\ g(x) &= -(x-a)^n, \end{aligned}$$

which satisfy $(*)$. Notice (Figure 2) that if n is odd, then a is neither a local maximum nor a local minimum point for f or g. On the other hand, if n is even, then f, with a positive nth derivative, has a local minimum at a, while g, with a negative nth derivative, has a local maximum at a. Of all functions satisfying $(*)$, these are about the simplest available; nevertheless they indicate the general situation exactly. In fact, the whole point of the next proof is that any function satisfying $(*)$ looks very much like one of these functions, in a sense that is made precise by Theorem 1.

THEOREM 2 Suppose that

$$\begin{aligned} &f'(a) = \cdots = f^{(n-1)}(a) = 0, \\ &f^{(n)}(a) \neq 0. \end{aligned}$$

(1) If n is even and $f^{(n)}(a) > 0$, then f has a local minimum at a.
(2) If n is even and $f^{(n)}(a) < 0$, then f has a local maximum at a.
(3) If n is odd, then f has neither a local maximum nor a local minimum at a.

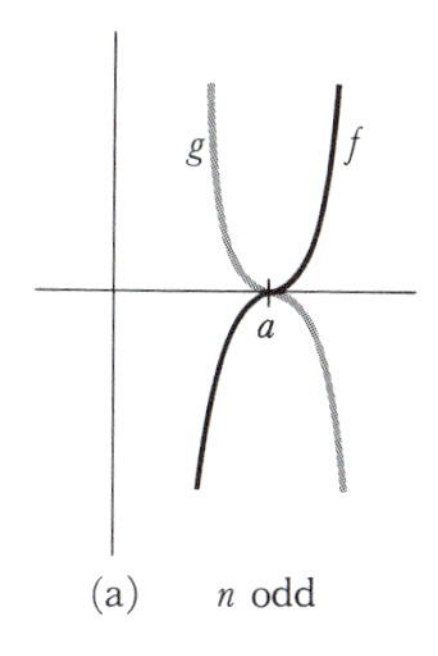

(a) n odd

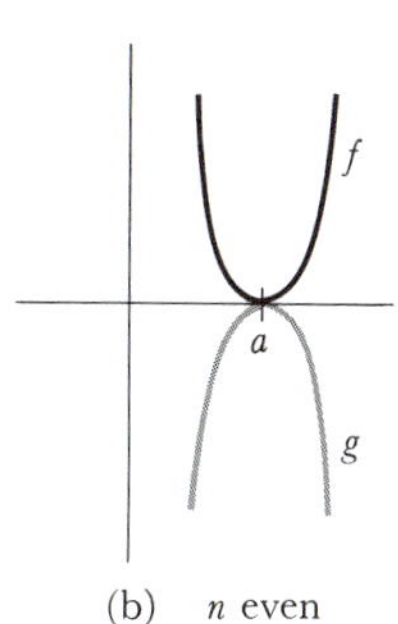

(b) n even

FIGURE 2

PROOF There is clearly no loss of generality in assuming that $f(a) = 0$, since neither the hypotheses nor the conclusion are affected if f is replaced by $f - f(a)$. Then, since the first $n-1$ derivatives of f at a are 0, the Taylor polynomial $P_{n,a}$ of f is

$$\begin{aligned} P_{n,a}(x) &= f(a) + \frac{f'(a)}{1!}(x-a) + \cdots + \frac{f^{(n)}(a)}{n!}(x-a)^n \\ &= \frac{f^{(n)}(a)}{n!}(x-a)^n. \end{aligned}$$

Thus, Theorem 1 states that

$$0 = \lim_{x \to a} \frac{f(x) - P_{n,a}(x)}{(x-a)^n} = \lim_{x \to a} \left[\frac{f(x)}{(x-a)^n} - \frac{f^{(n)}(a)}{n!} \right].$$

Consequently, if x is sufficiently close to a, then

$$\frac{f(x)}{(x-a)^n} \quad \text{has the same sign as} \quad \frac{f^{(n)}(a)}{n!}.$$

Suppose now that n is even. In this case $(x-a)^n > 0$ for all $x \neq a$. Since $f(x)/(x-a)^n$ has the same sign as $f^{(n)}(a)/n!$ for x sufficiently close to a, it follows

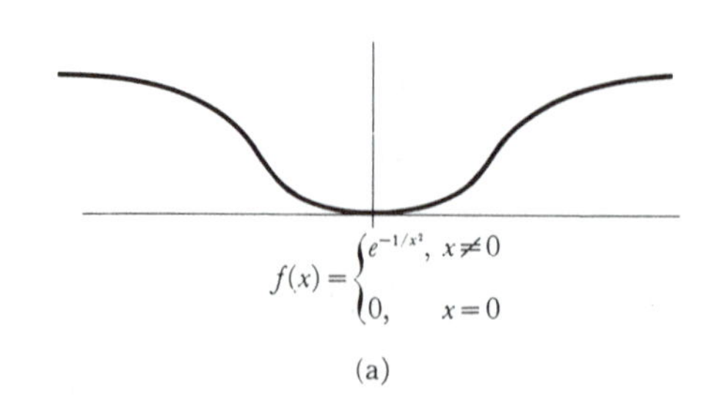

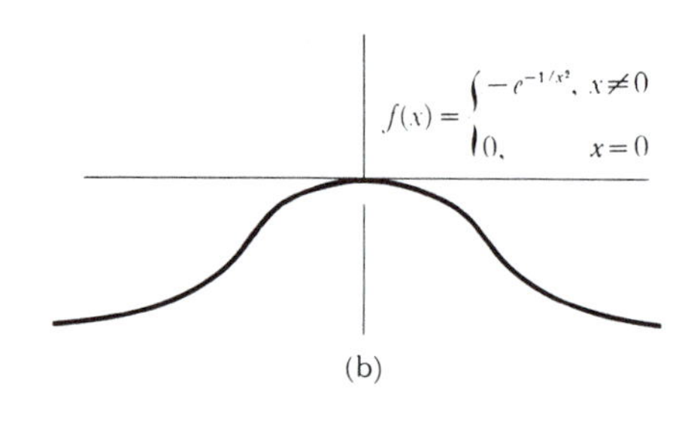

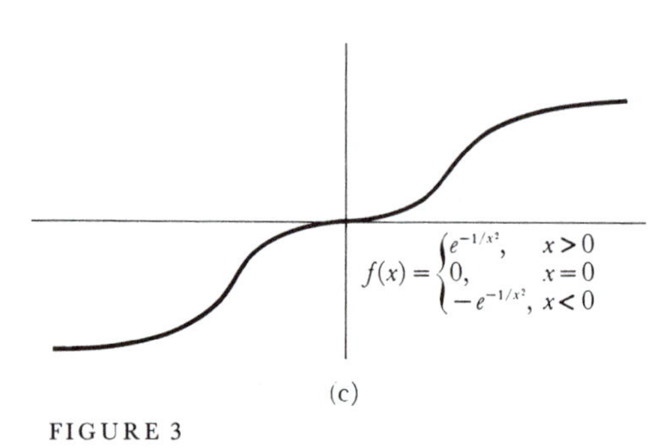

FIGURE 3

that $f(x)$ itself has the same sign as $f^n(a)/n!$ for x sufficiently close to a. If $f^{(n)}(a) > 0$, this means that

$$f(x) > 0 = f(a)$$

for x close to a. Consequently, f has a local minimum at a. A similar proof works for the case $f^{(n)}(a) < 0$.

Now suppose that n is odd. The same argument as before shows that if x is sufficiently close to a, then

$$\frac{f(x)}{(x-a)^n} \quad \text{always has the same sign.}$$

But $(x-a)^n > 0$ for $x > a$ and $(x-a)^n < 0$ for $x < a$. Therefore $f(x)$ has *different* signs for $x > a$ and $x < a$. This proves that f has neither a local maximum nor a local minimum at a. ▌

Although Theorem 2 will settle the question of local maxima and minima for just about any function which arises in practice, it does have some theoretical limitations, because $f^{(k)}(a)$ may be 0 for *all* k. This happens (Figure 3(a)) for the function

$$f(x) = \begin{cases} e^{-1/x^2}, & x \neq 0 \\ 0, & x = 0, \end{cases}$$

which has a minimum at 0, and also for the negative of this function (Figure 3(b)), which has a maximum at 0. Moreover (Figure 3(c)), if

$$f(x) = \begin{cases} e^{-1/x^2}, & x > 0 \\ 0, & x = 0 \\ -e^{-1/x^2}, & x < 0, \end{cases}$$

then $f^{(k)}(0) = 0$ for all k, but f has neither a local minimum nor a local maximum at 0.

The conclusion of Theorem 1 is often expressed in terms of an important concept of "order of equality." Two functions f and g are **equal up to order n at a** if

$$\lim_{x \to a} \frac{f(x) - g(x)}{(x-a)^n} = 0.$$

In the language of this definition, Theorem 1 says that the Taylor polynomial $P_{n,a,f}$ equals f up to order n at a. The Taylor polynomial might very well have been designed to make this fact true, because there is at most one polynomial of degree $\leq n$ with this property. This assertion is a consequence of the following elementary theorem.

THEOREM 3 Let P and Q be two polynomials in $(x-a)$, of degree $\leq n$, and suppose that P and Q are equal up to order n at a. Then $P = Q$.

PROOF Let $R = P - Q$. Since R is a polynomial of degree $\leq n$, it is only necessary to prove that if

$$R(x) = b_0 + \cdots + b_n(x-a)^n$$

satisfies

$$\lim_{x \to a} \frac{R(x)}{(x-a)^n} = 0,$$

then $R = 0$. Now the hypotheses on R surely imply that

$$\lim_{x \to a} \frac{R(x)}{(x-a)^i} = 0 \quad \text{for } 0 \le i \le n.$$

For $i = 0$ this condition reads simply $\lim_{x \to a} R(x) = 0$; on the other hand,

$$\begin{aligned} \lim_{x \to a} R(x) &= \lim_{x \to a} [b_0 + b_1(x-a) + \cdots + b_n(x-a)^n] \\ &= b_0. \end{aligned}$$

Thus $b_0 = 0$ and

$$R(x) = b_1(x-a) + \cdots + b_n(x-a)^n.$$

Therefore,

$$\frac{R(x)}{x-a} = b_1 + b_2(x-a) + \cdots + b_n(x-a)^{n-1}$$

and

$$\lim_{x \to a} \frac{R(x)}{x-a} = b_1.$$

Thus $b_1 = 0$ and

$$R(x) = b_2(x-a)^2 + \cdots + b_n(x-a)^n.$$

Continuing in this way, we find that

$$b_0 = \cdots = b_n = 0. \blacksquare$$

COROLLARY Let f be n-times differentiable at a, and suppose that P is a polynomial in $(x-a)$ of degree $\le n$, which equals f up to order n at a. Then $P = P_{n,a,f}$.

PROOF Since P and $P_{n,a,f}$ both equal f up to order n at a, it is easy to see that P equals $P_{n,a,f}$ up to order n at a. Consequently, $P = P_{n,a,f}$ by the Theorem. $\blacksquare$

At first sight this corollary appears to have unnecessarily complicated hypotheses; it might seem that the existence of the polynomial P would automatically imply that f is sufficiently differentiable for $P_{n,a,f}$ to exist. But in fact this is not so. For example (Figure 4), suppose that

$$f(x) = \begin{cases} x^{n+1}, & x \text{ irrational} \\ 0, & x \text{ rational.} \end{cases}$$

If $P(x) = 0$, then P is certainly a polynomial of degree $\le n$ which equals f up to order n at 0. On the other hand, $f'(a)$ does not exist for any $a \ne 0$, so $f''(0)$ is undefined.

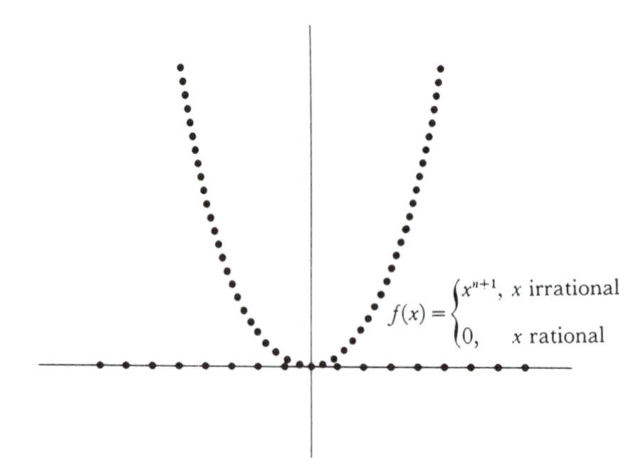

FIGURE 4

When f does have n derivatives at a, however, the corollary may provide a useful method for finding the Taylor polynomial of f. In particular, remember that our first attempt to find the Taylor polynomial for arctan ended in failure. The equation

$$\arctan x = \int_0^x \frac{1}{1+t^2}\,dt$$

suggests a promising method of finding a polynomial close to arctan—divide 1 by $1+t^2$, to obtain a polynomial plus a remainder:

$$\frac{1}{1+t^2} = 1 - t^2 + t^4 - t^6 + \cdots + (-1)^n t^{2n} + \frac{(-1)^{n+1}t^{2n+2}}{1+t^2}.$$

This formula, which can be checked easily by multiplying both sides by $1+t^2$, shows that

$$\begin{aligned}\arctan x &= \int_0^x 1 - t^2 + t^4 - \cdots + (-1)^n t^{2n}\,dt + (-1)^{n+1}\int_0^x \frac{t^{2n+2}}{1+t^2}\,dt \\ &= x - \frac{x^3}{3} + \frac{x^5}{5} - \cdots + (-1)^n \frac{x^{2n+1}}{2n+1} + (-1)^{n+1}\int_0^x \frac{t^{2n+2}}{1+t^2}\,dt.\end{aligned}$$

According to our corollary, the polynomial which appears here will be the Taylor polynomial of degree $2n+1$ for arctan at 0, provided that

$$\lim_{x\to 0} \frac{\displaystyle\int_0^x \frac{t^{2n+2}}{1+t^2}\,dt}{x^{2n+1}} = 0.$$

Since

$$\left|\int_0^x \frac{t^{2n+2}}{1+t^2}\,dt\right| \le \left|\int_0^x t^{2n+2}\,dt\right| = \frac{|x|^{2n+3}}{2n+3},$$

this is clearly true. Thus we have found that the Taylor polynomial of degree $2n+1$ for arctan at 0 is

$$P_{2n+1,0}(x) = x - \frac{x^3}{3} + \frac{x^5}{5} - \cdots + (-1)^n \frac{x^{2n+1}}{2n+1}.$$

By the way, now that we have discovered the Taylor polynomials of arctan, it is possible to work backwards and find $\arctan^{(k)}(0)$ for all k: Since

$$P_{2n+1,0}(x) = x - \frac{x^3}{3} + \frac{x^5}{5} - \cdots + (-1)^n \frac{x^{2n+1}}{2n+1},$$

and since this polynomial is, by definition,

$$\arctan^{(0)}(0) + \arctan^{(1)}(0) + \frac{\arctan^{(2)}(0)}{2!}x^2 + \cdots + \frac{\arctan^{(2n+1)}(0)}{(2n+1)!}x^{2n+1},$$

we can find $\arctan^{(k)}(0)$ by simply equating the coefficients of x^k in these two polynomials:

$$\frac{\arctan^{(k)}(0)}{k!} = 0 \quad \text{if } k \text{ is even},$$

$$\frac{\arctan^{(2l+1)}(0)}{(2l+1)!} = \frac{(-1)^l}{2l+1} \quad \text{or} \quad \arctan^{(2l+1)}(0) = (-1)^l \cdot (2l)!\,.$$

A much more interesting fact emerges if we go back to the original equation

$$\arctan x = x - \frac{x^3}{3} + \frac{x^5}{5} - \cdots + (-1)^n \frac{x^{2n+1}}{2n+1} + (-1)^{n+1} \int_0^x \frac{t^{2n+2}}{1+t^2}\,dt,$$

and remember the estimate

$$\left| \int_0^x \frac{t^{2n+2}}{1+t^2}\,dt \right| \le \frac{|x|^{2n+3}}{2n+3}.$$

When $|x| \le 1$, this expression is at most $1/(2n+3)$, and we can make this as small as we like simply by choosing n large enough. In other words, for $|x| \le 1$ *we can use the Taylor polynomials for* arctan *to compute* $\arctan x$ *as accurately as we like.* The most important theorems about Taylor polynomials extend this isolated result to other functions, and the Taylor polynomials will soon play quite a new role. The theorems proved so far have always examined the behavior of the Taylor polynomial $P_{n,a}$ for *fixed* n, as x approaches a. Henceforth we will compare Taylor polynomials $P_{n,a}$ for *fixed* x, and different n. In anticipation of the coming theorem we introduce some new notation.

If f is a function for which $P_{n,a}(x)$ exists, we define the **remainder term** $R_{n,a}(x)$ by

$$\begin{aligned} f(x) &= P_{n,a}(x) + R_{n,a}(x) \\ &= f(a) + f'(a)(x-a) + \cdots + \frac{f^{(n)}(a)}{n!}(x-a)^n + R_{n,a}(x). \end{aligned}$$

We would like to have an expression for $R_{n,a}(x)$ whose size is easy to estimate. There is such an expression, involving an integral, just as in the case for arctan. One way to guess this expression is to begin with the case $n = 0$:

$$f(x) = f(a) + R_{0,a}(x).$$

The Fundamental Theorem of Calculus enables us to write

$$f(x) = f(a) + \int_a^x f'(t)\,dt,$$

so that

$$R_{0,a}(x) = \int_a^x f'(t)\,dt.$$

A similar expression for $R_{1,a}(x)$ can be derived from this formula using integration by parts in a rather tricky way: Let

$$u(t) = f'(t) \quad \text{and} \quad v(t) = t - x$$

(notice that x represents some fixed number in the expression for $v(t)$, so $v'(t) = 1$); then

$$\begin{aligned} \int_a^x f'(t)\,dt &= \int_a^x \underset{\displaystyle u(t)}{\underset{\downarrow}{f'(t)}} \cdot \underset{\displaystyle v'(t)}{\underset{\downarrow}{1}}\,dt \\ &= u(t)v(t)\Big|_a^x - \int_a^x \underset{\displaystyle u'(t)}{\underset{\downarrow}{f''(t)}}\underset{\displaystyle v(t)}{\underset{\downarrow}{(t-x)}}\,dt. \end{aligned}$$

Since $v(x) = 0$, we obtain

$$\begin{aligned} f(x) &= f(a) + \int_a^x f'(t)\,dt \\ &= f(a) - u(a)v(a) + \int_a^x f''(t)(x-t)\,dt \\ &= f(a) + f'(a)(x-a) + \int_a^x f''(t)(x-t)\,dt. \end{aligned}$$

Thus

$$R_{1,a}(x) = \int_a^x f''(t)(x-t)\,dt.$$

It is hard to give any motivation for choosing $v(t) = t - x$, rather than $v(t) = t$. It just happens to be the choice which works out, the sort of thing one might discover after sufficiently many similar but futile manipulations. However, it is now easy to guess the formula for $R_{2,a}(x)$. If

$$u(t) = f''(t) \quad \text{and} \quad v(t) = \frac{-(x-t)^2}{2},$$

then $v'(t) = (x - t)$, so

$$\begin{aligned} \int_a^x f''(t)(x-t)\,dt &= u(t)v(t)\Big|_a^x - \int_a^x f'''(t) \cdot \frac{-(x-t)^2}{2}\,dt \\ &= \frac{f''(a)(x-a)^2}{2} + \int_a^x \frac{f'''(t)}{2}(x-t)^2\,dt. \end{aligned}$$

This shows that

$$R_{2,a}(x) = \int_a^x \frac{f^{(3)}(t)}{2}(x-t)^2\,dt.$$

You should now have little difficulty giving a rigorous proof, by induction, that if $f^{(n+1)}$ is continuous on $[a,x]$, then

$$R_{n,a}(x)=\int_a^x \frac{f^{(n+1)}(t)}{n!}(x-t)^n\,dt.$$

From this formula, which is called the integral form of the remainder, it is possible (Problem 15) to derive two other important expressions for $R_{n,a}(x)$: the Cauchy form of the remainder,

$$R_{n,a}(x)=\frac{f^{(n+1)}(t)}{n!}(x-t)^n(x-a) \quad \text{for some } t \text{ in } (a,x),$$

and the Lagrange form of the remainder,

$$R_{n,a}(x)=\frac{f^{(n+1)}(t)}{(n+1)!}(x-a)^{n+1} \quad \text{for some } t \text{ in } (a,x).$$

In the proof of the next theorem (Taylor's Theorem) we will derive all three forms of the remainder in an entirely different way. One virtue of this proof (aside from its cleverness) is the fact that the Cauchy and Lagrange forms of the remainder will be proved without assuming the extra hypothesis that $f^{(n+1)}$ is continuous. In this way Taylor's Theorem appears as a direct generalization of the Mean Value Theorem, to which it reduces for $n=0$, and which is the crucial tool used in the proof.

These remarks may suggest a strategy for proving Taylor's Theorem. Since $R_{n,a}(a)=0$, we might try to apply the Mean Value Theorem to the expression

$$\frac{R_{n,a}(x)}{x-a}=\frac{R_{n,a}(x)-R_{n,a}(a)}{x-a}.$$

On second thought, however, this idea does not look very promising, since it is not at all clear how $f^{(n+1)}(t)$ is ever going to be involved in the answer. Indeed, if we take the most straightforward route, and differentiate both sides of the equation which defines $R_{n,a}$, we obtain

$$f'(x)=f'(a)+f''(a)(x-a)+\cdots+\frac{f^{(n)}(a)}{(n-1)!}(x-a)^{n-1}+R_{n,a}{}'(x),$$

which is useless. The proper application of the Mean Value Theorem has a lot in common with the integration by parts proof outlined above. This proof involved the derivative of a function in which x *denoted a number which was fixed.* This is just how x will be treated in the following proof.

THEOREM 4 (TAYLOR'S THEOREM) Suppose that $f',\dots,f^{(n+1)}$ are defined on $[a,x]$, and that $R_{n,a}(x)$ is defined by

$$f(x)=f(a)+f'(a)(x-a)+\cdots+\frac{f^{(n)}(a)}{n!}(x-a)^n+R_{n,a}(x).$$

Then

$$(1)\quad R_{n,a}(x) = \frac{f^{(n+1)}(t)}{n!}(x-t)^n(x-a) \quad \text{for some } t \text{ in } (a,x).$$

$$(2)\quad R_{n,a}(x) = \frac{f^{(n+1)}(t)}{(n+1)!}(x-a)^{n+1} \quad \text{for some } t \text{ in } (a,x).$$

Moreover, if $f^{(n+1)}$ is integrable on $[a,x]$, then

$$(3)\quad R_{n,a}(x) = \int_a^x \frac{f^{(n+1)}(t)}{n!}(x-t)^n\,dt.$$

(If $x < a$, then the hypothesis should state that f is $(n+1)$-times differentiable on $[x,a]$; the number t in (1) and (2) will then be in (x,a), while (3) will remain true as stated, provided that $f^{(n+1)}$ is integrable on $[x,a]$.)

PROOF For each number t in $[a,x]$ we have

$$f(x) = f(t) + f'(t)(x-t) + \cdots + \frac{f^{(n)}(t)}{n!}(x-t)^n + R_{n,t}(x).$$

Let us denote the number $R_{n,t}(x)$ by $S(t)$; the function S is defined on $[a,x]$, and we have

$$(*)\quad f(x) = f(t) + f'(t)(x-t) + \cdots + \frac{f^{(n)}(t)}{n!}(x-t)^n + S(t) \qquad \text{for all } t \text{ in } [a,x].$$

We will now differentiate both sides of this equation, which asserts the equality of two functions: the one whose value at t is $f(x)$, and the one whose value at t is

$$f(t) + \cdots + \frac{f^{(n)}(t)}{n!}(x-t)^n + S(t).$$

(In common parlance we are considering both sides of $(*)$ "as a function of t".) Just to make sure that the letter x causes no confusion, notice that if

$$g(t) = f(x) \quad \text{for all } t,$$

then

$$g'(t) = 0 \quad \text{for all } t;$$

and if

$$g(t) = \frac{f^{(k)}(t)}{k!}(x-t)^k,$$

then

$$\begin{aligned} g'(t) &= \frac{f^{(k)}(t)}{k!}k(x-t)^{k-1}(-1) + \frac{f^{(k+1)}(t)}{k!}(x-t)^k \\ &= -\frac{f^{(k)}(t)}{(k-1)!}(x-t)^{k-1} + \frac{f^{(k+1)}(t)}{k!}(x-t)^k. \end{aligned}$$

Applying these formulas to each term of (*), we obtain

$$0 = f'(t) + \left[-f'(t) + \frac{f''(t)}{1!}(x-t)\right] + \left[\frac{-f''(t)}{1!}(x-t) + \frac{f^{(3)}(t)}{2!}(x-t)^2\right]$$
$$+ \cdots + \left[\frac{-f^{(n)}(t)}{(n-1)!}(x-t)^{n-1} + \frac{f^{(n+1)}(t)}{n!}(x-t)^n\right] + S'(t).$$

In this beautiful formula practically everything in sight cancels out, and we obtain

$$S'(t) = -\frac{f^{(n+1)}(t)}{n!}(x-t)^n.$$

Now we can apply the Mean Value Theorem to the function S on $[a, x]$: there is some t in (a, x) such that

$$\frac{S(x) - S(a)}{x - a} = S'(t) = -\frac{f^{(n+1)}(t)}{n!}(x-t)^n.$$

Remember that

$$S(t) = R_{n,t}(x);$$

this means in particular that

$$S(x) = R_{n,x}(x) = 0,$$
$$S(a) = R_{n,a}(x).$$

Thus

$$\frac{0 - R_{n,a}(x)}{x - a} = -\frac{f^{(n+1)}(t)}{n!}(x-t)^n$$

or

$$R_{n,a}(x) = \frac{f^{(n+1)}(t)}{n!}(x-t)^n(x-a);$$

this is the Cauchy form of the remainder.

To derive the Lagrange form we apply the Cauchy Mean Value Theorem to the functions S and $g(t) = (x-t)^{n+1}$: there is some t in (a, x) such that

$$\frac{S(x) - S(a)}{g(x) - g(a)} = \frac{S'(t)}{g'(t)} = \frac{-\dfrac{f^{(n+1)}(t)}{n!}(x-t)^n}{-(n+1)(x-t)^n}.$$

Thus

$$\frac{R_{n,a}(x)}{(x-a)^{n+1}} = \frac{f^{(n+1)}(t)}{(n+1)!}$$

or

$$R_{n,a}(x) = \frac{f^{(n+1)}(t)}{(n+1)!}(x-a)^{n+1},$$

which is the Lagrange form.

Finally, if $f^{(n+1)}$ is integrable on $[a, x]$, then

$$S(x) - S(a) = \int_a^x S'(t) = -\int_a^x \frac{f^{(n+1)}(t)}{n!}(x-t)^n\,dt$$

or

$$R_{n,a}(x) = \int_a^x \frac{f^{(n+1)}(t)}{n!}(x-t)^n\,dt. \blacksquare$$

Although the Lagrange and Cauchy forms of the remainder are more than theoretical curiosities (see, e.g., Problem 23-18), the integral form of the remainder will usually be quite adequate. If this form is applied to the functions sin, cos, and exp, with $a = 0$, Taylor's Theorem yields the following formulas:

$$\sin x = x - \frac{x^3}{3!} + \frac{x^5}{5!} - \cdots + (-1)^n \frac{x^{2n+1}}{(2n+1)!} + \int_0^x \frac{\sin^{(2n+2)}(t)}{(2n+1)!}(x-t)^{2n+1}\,dt,$$

$$\cos x = 1 - \frac{x^2}{2!} + \frac{x^4}{4!} - \cdots + (-1)^n \frac{x^{2n}}{(2n)!} + \int_0^x \frac{\cos^{(2n+1)}(t)}{(2n)!}(x-t)^{2n}\,dt,$$

$$e^x = 1 + x + \frac{x^2}{2!} + \cdots + \frac{x^n}{n!} + \int_0^x \frac{e^t}{n!}(x-t)^n\,dt.$$

To evaluate any of these integrals explicitly would be supreme foolishness—the answer of course will be exactly the difference of the left side and all the other terms on the right side! To *estimate* these integrals, however, is both easy and worthwhile.

The first two integrals are especially easy. Since

$$|\sin^{(2n+2)}(t)| \leq 1 \quad \text{for all } t,$$

we have

$$\left| \int_0^x \frac{\sin^{(2n+2)}(t)}{(2n+1)!}(x-t)^{2n+1}\,dt \right| \leq \frac{1}{(2n+1)!} \left| \int_0^x (x-t)^{2n+1}\,dt \right|.$$

Since

$$\int_0^x (x-t)^{2n+1}\,dt = \left. \frac{-(x-t)^{2n+2}}{2n+2} \right|_{t=0}^{t=x} = \frac{x^{2n+2}}{2n+2}$$

we conclude that

$$\left| \int_0^x \frac{\sin^{(2n+2)}(t)}{(2n+1)!}(x-t)^{2n+1}\,dt \right| \leq \frac{|x|^{2n+2}}{(2n+2)!}.$$

Similarly, we can show that

$$\left| \int_0^x \frac{\cos^{(2n+1)}(t)}{(2n)!}(x-t)^{2n}\,dt \right| \le \frac{|x|^{2n+1}}{(2n+1)!}.$$

These estimates are particularly interesting, because (as proved in Chapter 16) for any $\varepsilon > 0$ we can make

$$\frac{x^n}{n!} < \varepsilon$$

by choosing n large enough (how large n must be will depend on x). This enables us to compute $\sin x$ to any degree of accuracy desired simply by evaluating the proper Taylor polynomial $P_{n,0}(x)$. For example, suppose we wish to compute $\sin 2$ with an error of less than 10^{-4}. Since

$$\sin 2 = P_{2n+1,0}(2) + R, \quad \text{where } |R| \le \frac{2^{2n+2}}{(2n+2)!},$$

we can use $P_{2n+1,0}(2)$ as our answer, provided that

$$\frac{2^{2n+2}}{(2n+2)!} < 10^{-4}.$$

A number n with this property can be found by a straightforward search—it obviously helps to have a table of values for $n!$ and 2^n (see page 428). In this case it happens that $n = 5$ works, so that

$$\begin{aligned} \sin 2 &= P_{11,0}(2) + R \\ &= 2 - \frac{2^3}{3!} + \frac{2^5}{5!} - \frac{2^7}{7!} + \frac{2^9}{9!} - \frac{2^{11}}{11!} + R, \\ &\qquad \text{where } |R| < 10^{-4}. \end{aligned}$$

It is even easier to calculate $\sin 1$ approximately, since

$$\sin 1 = P_{2n+1,0}(1) + R, \quad \text{where } |R| < \frac{1}{(2n+2)!}.$$

To obtain an error less than ε we need only find an n such that

$$\frac{1}{(2n+2)!} < \varepsilon,$$

and this requires only a brief glance at a table of factorials. (Moreover, the individual terms of $P_{2n+1,0}(1)$ will be easier to handle.)

For very small x the estimates will be even easier. For example,

$$\sin\frac{1}{10} = P_{2n+1,0}\left(\frac{1}{10}\right) + R, \quad \text{where } |R| < \frac{1}{10^{2n+2}(2n+2)!}.$$

To obtain $|R| < 10^{-10}$ we can clearly take $n = 4$ (and we could even get away with $n = 3$). These methods are actually used to compute tables of sin and cos. A high-speed computer can compute $P_{2n+1,0}(x)$ for many different x in almost no time at all.

Estimating the remainder for e^x is only slightly harder. For simplicity assume that $x \geq 0$ (the estimates for $x \leq 0$ are obtained in Problem 10). On the interval $[0, x]$ the maximum value of e^t is e^x, since exp is increasing, so

$$\int_0^x \frac{e^t}{n!}(x-t)^n \, dt \leq \frac{e^x}{n!} \int_0^x (x-t)^n \, dt = \frac{e^x x^{n+1}}{(n+1)!}.$$

Since we already know that $e < 4$, we have

$$\frac{e^x x^{n+1}}{(n+1)!} < \frac{4^x x^{n+1}}{(n+1)!},$$

which can be made as small as desired by choosing n sufficiently large. How large n must be will depend on x (and the factor 4^x will make things more difficult). Once again, the estimates are easier for small x. If $0 \leq x \leq 1$, then

$$e^x = 1 + x + \frac{x^2}{2!} + \cdots + \frac{x^n}{n!} + R, \quad \text{where } 0 < R < \frac{4}{(n+1)!}.$$

(The inequality $0 < R$ follows immediately from the integral form for R.) In particular, if $n = 4$, then

$$0 < R < \frac{4}{5!} < \frac{1}{10},$$

so

$$\begin{aligned} e = e^1 &= 1 + 1 + \frac{1}{2!} + \frac{1}{3!} + \frac{1}{4!} + R, \quad \text{where } 0 < R < \frac{1}{10} \\ &= \frac{65}{24} + R \\ &= 2 + \frac{17}{24} + R, \end{aligned}$$

which shows that

$$2 < \varepsilon < 3.$$

(This then shows that

$$0 < R < \frac{3^x x^{n+1}}{(n+1)!},$$

allowing us to improve our estimate of R slightly.) By taking $n = 7$ you can compute that the first 3 decimals for e are

$$e = 2.718\ldots$$

(you should check that $n = 7$ does give this degree of accuracy, but it would be cruel to insist that you actually do the computations).

The function arctan is also important but, as you may recall, an expression for $\arctan^{(k)}(x)$ is hopelessly complicated, so that the integral form of the remainder is useless. On the other hand, our derivation of the Taylor polynomial for arctan automatically provided a formula for the remainder:

$$\arctan x = x - \frac{x^3}{3} + \cdots + \frac{(-1)^n x^{2n+1}}{2n+1} + \int_0^x \frac{(-1)^{n+1} t^{2n+2}}{1+t^2} \, dt.$$

As we have already estimated,

$$\left| \int_0^x \frac{(-1)^{n+1} t^{2n+2}}{1+t^2}\, dt \right| \le \left| \int_0^x t^{2n+2}\, dt \right| = \frac{|x|^{2n+3}}{2n+3}.$$

For the moment we will consider only numbers x with $|x| \le 1$. In this case, the remainder term can clearly be made as small as desired by choosing n sufficiently large. In particular,

$$\arctan 1 = 1 - \frac{1}{3} + \frac{1}{5} - \cdots + \frac{(-1)^n}{2n+1} + R, \quad \text{where } |R| < \frac{1}{2n+3}.$$

With this estimate it is easy to find an n which will make the remainder less than any preassigned number; on the other hand, n will usually have to be so large as to make computations hopelessly long. To obtain a remainder $< 10^{-4}$, for example, we must take $n > (10^4 - 3)/2$. This is really a shame, because $\arctan 1 = \pi/4$, so the Taylor polynomial for arctan should allow us to compute π. Fortunately, there are some clever tricks which enable us to surmount these difficulties. Since

$$|R_{2n+1,0}(x)| < \frac{|x|^{2n+3}}{2n+3},$$

much smaller n's will work for only somewhat smaller x's. The trick for computing π is to express $\arctan 1$ in terms of $\arctan x$ for smaller x; Problem 6 shows how this can be done in a convenient way.

The Taylor polynomial for the function $f(x) = \log(x+1)$ at $a = 1$ is best handled in the same manner as the Taylor polynomial for arctan. Although the integral form of the remainder for f is not hard to write down, it is difficult to estimate. On the other hand, we obtain a simple formula if we begin with the equation

$$\frac{1}{1+t} = 1 - t + t^2 - \cdots + (-1)^{n-1} t^{n-1} + \frac{(-1)^n t^n}{1+t};$$

this implies that

$$\log(1+x) = \int_0^x \frac{1}{1+t}\, dt = x - \frac{x^2}{2} + \frac{x^3}{3} - \cdots + (-1)^{n-1}\frac{x^n}{n} + (-1)^n \int_0^x \frac{t^n}{1+t}\, dt,$$

for all $x > -1$. If $x \ge 0$, then

$$\int_0^x \frac{t^n}{t+1}\, dt \le \int_0^x t^n\, dt = \frac{x^{n+1}}{n+1},$$

and there is a slightly more complicated estimate when $-1 < x < 0$ (Problem 11). For this function the remainder term can be made as small as desired by choosing n sufficiently large, provided that $-1 < x \le 1$.

The behavior of the remainder terms for arctan and $f(x) = \log(x+1)$ is quite another matter when $|x| > 1$. In this case, the estimates

$$|R_{2n+1,0}(x)| < \frac{|x|^{2n+3}}{2n+3} \quad \text{for arctan,}$$

$$|R_{n,0}(x)| < \frac{x^{n+1}}{n+1} \quad (x > 0) \text{ for } f,$$

are of no use, because when $|x| > 1$ the bounds x^m/m become large as m becomes large. This predicament is unavoidable, and is not just a deficiency of our estimates. It is easy to get estimates in the other direction which show that the remainders actually do remain large. To obtain such an estimate for arctan, note that if t is in $[0, x]$ (or in $[x, 0]$ if $x < 0$), then

$$1 + t^2 \le 1 + x^2 \le 2x^2, \quad \text{if } |x| \ge 1,$$

so

$$\left| \int_0^x \frac{t^{2n+2}}{1+t^2}\,dt \right| \ge \frac{1}{2x^2} \left| \int_0^x t^{2n+2}\,dt \right| = \frac{|x|^{2n+1}}{4n+6}.$$

Similarly, if $x > 0$, then for t in $[0, x]$ we have

$$1 + t \le 1 + x \le 2x, \quad \text{if } x \ge 1,$$

so

$$\int_0^x \frac{t^n}{t+1}\,dt \ge \frac{1}{2x} \int_0^x t^n\,dt = \frac{x^n}{2n+2}.$$

These estimates show that if $|x| > 1$, then the remainder terms become large as n becomes large. In other words, for $|x| > 1$, the Taylor polynomials for arctan and f *are of no use whatsoever in computing* $\arctan x$ *and* $\log(x+1)$. This is no tragedy, because the values of these functions can be found for any x once they are known for all x with $|x| < 1$.

This same situation occurs in a spectacular way for the function

$$f(x) = \begin{cases} e^{-1/x^2}, & x \ne 0 \\ 0, & x = 0. \end{cases}$$

We have already seen that $f^{(k)}(0) = 0$ for every natural number k. This means that the Taylor polynomial $P_{n,0}$ for f is

$$\begin{aligned} P_{n,0}(x) &= f(0) + f'(0)x + \frac{f''(0)}{2!}x^2 + \cdots + \frac{f^{(n)}(0)}{n!}x^n \\ &= 0. \end{aligned}$$

In other words, the remainder term $R_{n,0}(x)$ always equals $f(x)$, and the Taylor polynomial is useless for computing $f(x)$, except for $x = 0$. Eventually we will be able to offer some explanation for the behavior of this function, which is such a disconcerting illustration of the limitations of Taylor's Theorem.

The word "compute" has been used so often in connection with our estimates for the remainder term, that the significance of Taylor's Theorem might be misconstrued. It is true that Taylor's Theorem is an almost ideal computational aid (despite its ignominious failure in the previous example), but it has equally important theoretical consequences. Most of these will be developed in succeeding chapters, but two proofs will illustrate some ways in which Taylor's Theorem may be used. The first illustration will be particularly impressive to those who have waded through the proof, in Chapter 16, that π is irrational.

THEOREM 5 e is irrational.

PROOF We know that, for any n,

$$e = e^1 = 1 + \frac{1}{1!} + \frac{1}{2!} + \cdots + \frac{1}{n!} + R_n, \quad \text{where } 0 < R_n < \frac{3}{(n+1)!}.$$

Suppose that e were rational, say $e = a/b$, where a and b are positive integers. Choose $n > b$ and also $n > 3$. Then

$$\frac{a}{b} = 1 + 1 + \frac{1}{2!} + \cdots + \frac{1}{n!} + R_n,$$

so

$$\frac{n!\,a}{b} = n! + n! + \frac{n!}{2!} + \cdots + \frac{n!}{n!} + n!R_n.$$

Every term in this equation other than $n!R_n$ is an integer (the left side is an integer because $n > b$). Consequently, $n!R_n$ must be an integer also. But

$$0 < R_n < \frac{3}{(n+1)!},$$

so

$$0 < n!R_n < \frac{3}{n+1} < \frac{3}{4} < 1,$$

which is impossible for an integer. ▮

The second illustration is merely a straightforward demonstration of a fact proved in Chapter 15: If

$$\begin{aligned} f'' + f &= 0, \\ f(0) &= 0, \\ f'(0) &= 0, \end{aligned}$$

then $f = 0$. To prove this, observe first that $f^{(k)}$ exists for every k; in fact

$$\begin{aligned} f^{(3)} &= (f'')' = -f', \\ f^{(4)} &= (f^3)' = (-f')' = -f'' = f, \\ f^{(5)} &= (f^{(4)})' = f', \\ &\text{etc.} \end{aligned}$$

This shows, not only that all $f^{(k)}$ exist, but also that there are at most 4 different ones: $f, f', -f, -f'$. Since $f(0) = f'(0) = 0$, all $f^{(k)}(0)$ are 0. Now Taylor's Theorem states, for any n, that

$$f(x) = \int_0^x \frac{f^{(n+1)}(t)}{n!}(x-t)^n \, dt.$$

Each function $f^{(n+1)}$ is continuous (since $f^{(n+2)}$ exists), so for any particular x there is a number M such that

$$|f^{(n+1)}(t)| \le M \quad \text{for } 0 \le t \le x, \text{ and } all\ n$$

(we can add the phrase "and *all* n" because there are only four different $f^{(k)}$). Thus

$$|f(x)| \le M \left| \int_0^x \frac{(x-t)^n}{n!} \, dt \right| = \frac{M|x|^{n+1}}{(n+1)!}.$$

Since this is true for every n, and since $|x|^n/n!$ can be made as small as desired by choosing n sufficiently large, this shows that $|f(x)| \le \varepsilon$ for any $\varepsilon > 0$; consequently, $f(x) = 0$.

The other uses to which Taylor's Theorem will be put in succeeding chapters are closely related to the computational considerations which have concerned us for much of this chapter. If the remainder term $R_{n,a}(x)$ can be made as small as desired by choosing n sufficiently large, then $f(x)$ can be computed to any degree of accuracy desired by using the polynomials $P_{n,a}(x)$. As we require greater and greater accuracy we must add on more and more terms. If we are willing to add up infinitely many terms (in theory at least!), then we ought to be able to ignore the remainder completely. There should be "infinite sums" like

$$\begin{aligned}
\sin x &= x - \frac{x^3}{3!} + \frac{x^5}{5!} - \frac{x^7}{7!} + \cdots, \\
\cos x &= 1 - \frac{x^2}{2!} + \frac{x^4}{4!} - \frac{x^6}{6!} + \frac{x^8}{8!} - \cdots, \\
e^x &= 1 + x + \frac{x^2}{2!} + \frac{x^3}{3!} + \frac{x^4}{4!} + \cdots, \\
\arctan x &= x - \frac{x^3}{3} + \frac{x^5}{5} - \frac{x^7}{7} + \cdots \qquad \text{if } |x| \le 1, \\
\log(1+x) &= x - \frac{x^2}{2} + \frac{x^3}{3} - \frac{x^4}{4} + \cdots \qquad \text{if } -1 < x \le 1.
\end{aligned}$$

We are almost completely prepared for this step. Only one obstacle remains—we have never even defined an infinite sum. Chapters 22 and 23 contain the necessary definitions.

PROBLEMS

1. Find the Taylor polynomials (of the indicated degree, and at the indicated point) for the following functions.

(i) $f(x) = e^{e^x}$; degree 3, at 0.

(ii) $f(x) = e^{\sin x}$ degree 3, at 0.

(iii) sin; degree $2n$, at $\frac{\pi}{2}$.

(iv) cos; degree $2n$, at π.

(v) exp; degree n, at 1.

(vi) log; degree n, at 2.

(vii) $f(x) = x^5 + x^3 + x$; degree 4, at 0.

(viii) $f(x) = x^5 + x^3 + x$; degree 4, at 1.

(ix) $f(x) = \dfrac{1}{1+x^2}$; degree $2n+1$, at 0.

(x) $f(x) = \dfrac{1}{1+x}$; degree n, at 0.

2. Write each of the following polynomials in x as a polynomial in $(x-3)$. (It is only necessary to compute the Taylor polynomial at 3, of the same degree as the original polynomial. Why?)

(i) $x^2 - 4x - 9$.

(ii) $x^4 - 12x^3 + 44x^2 + 2x + 1$.

(iii) x^5.

(iv) $ax^2 + bx + c$.

3. Write down a sum (using $\sum$ notation) which equals each of the following numbers to within the specified accuracy. To minimize needless computation, consult the tables for 2^n and $n!$ on the next page.

(i) $\sin 1$; error $< 10^{-17}$.

(ii) $\sin 2$; error $< 10^{-12}$.

(iii) $\sin \frac{1}{2}$; error $< 10^{-20}$.

(iv) e; error $< 10^{-4}$.

(v) e^2; error $< 10^{-5}$.

n	2^n	$n!$
1	2	1
2	4	2
3	8	6
4	16	24
5	32	120
6	64	720
7	128	5,040
8	256	40,430
9	512	362,880
10	1,024	3,628,800
11	2,048	39,916,800
12	4,096	479,001,600
13	8,192	6,227,020,800
14	16,384	87,178,291,200
15	32,768	1,307,674,368,000
16	65,536	20,922,789,888,000
17	131,072	355,687,428,096,000
18	262,144	6,402,373,705,728,000
19	524,888	121,645,100,408,832,000
20	1,048,576	2,432,902,008,176,640,000

***4.** This problem is similar to the previous one, except that the errors demanded are so small that the tables cannot be used. You will have to do a little thinking, and in some cases it may be necessary to consult the proof, in Chapter 16, that $x^n/n!$ can be made small by choosing n large—the proof actually provides a method for finding the appropriate n. In the previous problem it was possible to find rather short sums; in fact, it was possible to find the smallest n which makes the estimate of the remainder given by Taylor's Theorem less than the desired error. But in this problem finding *any* specific sum is a moral victory (provided you can demonstrate that the sum works).

(i) $\sin 1$; error $< 10^{-(10^{10})}$.
(ii) e; error $< 10^{-1,000}$.
(iii) $\sin 10$; error $< 10^{-20}$.
(iv) e^{10}; error $< 10^{-30}$.
(v) $\arctan \frac{1}{10}$; error $< 10^{-(10^{10})}$.

5. (a) In Problem 11-38 you showed that the equation $x^2 = \cos x$ has precisely two solutions. Use the Taylor polynomial of cos to show that the solutions are approximately $\pm\sqrt{2/3}$, and find bounds on the error.
(b) Similarly, estimate the solutions of the equation $2x^2 = x \sin x + \cos^2 x$.

6. (a) Prove, using Problem 15-9, that

$$\frac{\pi}{4} = \arctan\frac{1}{2} + \arctan\frac{1}{3},$$
$$\frac{\pi}{4} = 4\arctan\frac{1}{5} - \arctan\frac{1}{239}.$$

(b) Show that $\pi = 3.14159\ldots$. (Every budding mathematician should verify a few decimals of π, but the purpose of this exercise is not to set you off on an immense calculation. If the second expression in part (a) is used, the first 5 decimals for π can be computed with remarkably little work.)

7. For every number α, and every natural number n, we define the "binomial coefficient"

$$\binom{\alpha}{n} = \frac{\alpha(\alpha-1)\cdot\ldots\cdot(\alpha-n+1)}{n!},$$

and we define $\binom{\alpha}{0} = 1$, as usual. If α is not an integer, then $\binom{\alpha}{n}$ is never 0, and alternates in sign for $n > \alpha$. Show that the Taylor polynomial of degree n for $f(x) = (1+x)^\alpha$ at 0 is $P_{n,0}(x) = \sum_{k=0}^{n}\binom{\alpha}{k}x^k$, and that the Cauchy and Lagrange forms of the remainder are the following:

Cauchy form:

$$\begin{aligned}R_{n,0}(x) &= \frac{\alpha(\alpha-1)\cdot\ldots\cdot(\alpha-n)}{n!}x(x-t)^n(1+t)^{\alpha-n-1}\\ &= \frac{\alpha(\alpha-1)\cdot\ldots\cdot(\alpha-n)}{n!}x(1+t)^{\alpha-1}\left(\frac{x-t}{1+t}\right)^n\\ &= (n+1)\binom{\alpha}{n+1}x(1+t)^{\alpha-1}\left(\frac{x-t}{1+t}\right)^n, \qquad t \text{ in } [0,x] \text{ or } [x,0].\end{aligned}$$

Lagrange form:

$$\begin{aligned}R_{n,0}(x) &= \frac{\alpha(\alpha-1)\cdot\ldots\cdot(\alpha-n)}{(n+1)!}x^{n+1}(1+t)^{\alpha-n-1}\\ &= \binom{\alpha}{n+1}x^{n+1}(1+t)^{\alpha-n-1}, \qquad t \text{ in } [0,x] \text{ or } [x,0].\end{aligned}$$

Estimates for these remainder terms are rather difficult to handle, and are postponed to Problem 23-18.

8. Suppose that a_i and b_i are the coefficients in the Taylor polynomials at a of f and g, respectively. In other words, $a_i = f^{(i)}(a)/i!$ and $b_i = g^{(i)}(a)/i!$. Find the coefficients c_i of the Taylor polynomials at a of the following functions, in terms of the a_i's and b_i's.

(i) $f+g$.
(ii) fg.

(iii) f'.

(iv) $h(x) = \int_a^x f(t)\,dt$.

(v) $k(x) = \int_0^x f(t)\,dt$.

9. (a) Prove that the Taylor polynomial of $f(x) = \sin(x^2)$ of degree $4n+2$ at 0 is

$$x^2 - \frac{x^6}{3!} + \frac{x^{10}}{5!} - \cdots + (-1)^n \frac{x^{4n+2}}{(2n+1)!}.$$

Hint: If P is the Taylor polynomial of degree $2n+1$ for sin at 0, then $\sin x = P(x) + R(x)$, where $\lim_{x\to 0} R(x)/x^{2n+1} = 0$. What does this imply about $\lim_{x\to 0} R(x^2)/x^{4n+2}$?

(b) Find $f^{(k)}(0)$ for all k.

(c) In general, if $f(x) = g(x^n)$, find $f^{(k)}(0)$ in terms of the derivatives of g at 0.

10. Prove that if $x \le 0$, then

$$\left| \int_0^x \frac{e^t}{n!}(x-t)^n\,dt \right| \le \frac{|x|^{n+1}}{(n+1)!}.$$

11. Prove that if $-1 < x \le 0$, then

$$\left| \int_0^x \frac{t^n}{1+t}\,dt \right| \le \frac{|x|^{n+1}}{(1+x)(n+1)}.$$

***12.** (a) Show that if $|g'(x)| \le M|x-a|^n$ for $|x-a| < \delta$, then $|g(x) - g(a)| \le M|x-a|^{n+1}/(n+1)$ for $|x-a| < \delta$.

(b) Use part (a) to show that if $\lim_{x\to a} g'(x)/(x-a)^n = 0$, then

$$\lim_{x\to a} \frac{g(x)}{(x-a)^{n+1}} = 0.$$

(c) Show that if $g(x) = f(x) - P_{n,a,f}(x)$, then $g'(x) = f'(x) - P_{n-1,a,f'}(x)$.

(d) Give an inductive proof of Theorem 1, without using l'Hôpital's Rule.

13. Deduce Theorem 1 as a corollary of Taylor's Theorem, with any form of the remainder. (The catch is that it will be necessary to assume one more derivative than in the hypotheses for Theorem 1.)

14. Deduce the Cauchy and Lagrange forms of the remainder from the integral form, using Problem 13-23. There will be the same catch as in Problem 13.

15. (a) Suppose that f is twice differentiable on $(0, \infty)$ and that $|f(x)| \le M_0$ for all $x > 0$, while $|f''(x)| \le M_2$ for all $x > 0$. Prove that for all $x > 0$ we have

$$|f'(x)| \le \frac{2}{h}M_0 + \frac{h}{2}M_2 \qquad \text{for all } h > 0.$$

(b) Show that for all $x > 0$ we have

$$|f'(x)| \le 2\sqrt{M_0 M_2}.$$

(c) If f is twice differentiable on $(0, \infty)$, f'' is bounded, and $f(x)$ approaches 0 as $x \to \infty$, then also $f'(x)$ approaches 0 as $x \to \infty$.

(d) If $\lim_{x\to\infty} f(x)$ exists and $\lim_{x\to\infty} f''(x)$ exists, then $\lim_{x\to\infty} f''(x) = \lim_{x\to\infty} f'(x) = 0$. (Compare Problem 11-31.)

16. (a) Prove that if $f''(a)$ exists, then

$$f''(a) = \lim_{h\to 0} \frac{f(a+h) + f(a-h) - 2f(a)}{h^2}.$$

The limit on the right is called the *Schwarz second derivative* of f at a. Hint: Use the Taylor polynomial $P_{2,a}(x)$ with $x = a + h$ and with $x = a - h$.

(b) Let $f(x) = x^2$ for $x \ge 0$, and $-x^2$ for $x \le 0$. Show that

$$\lim_{h\to 0} \frac{f(0+h) + f(0-h) - 2f(0)}{h^2}$$

exists, even though $f''(0)$ does not.

(c) Prove that if f has a local maximum at a, and the Schwarz second derivative of f at a exists, then it is ≤ 0.

(d) Prove that if $f'''(a)$ exists, then

$$\frac{f'''(a)}{3} = \lim_{h\to 0} \frac{f(a+h) - f(a-h) - 2hf'(x)}{h^3}.$$

17. Use the Taylor polynomial $P_{1,a,f}$, together with the remainder, to prove a weak form of Theorem 2 of the Appendix to Chapter 11: If $f'' > 0$, then the graph of f always lies above the tangent line of f, except at the point of contact.

***18.** Problem 18-43 presented a rather complicated proof that $f = 0$ if $f'' - f = 0$ and $f(0) = f'(0) = 0$. Give another proof, using Taylor's Theorem. (This problem is really a preliminary skirmish before doing battle with the general case in Problem 19, and is meant to convince you that Taylor's Theorem is a good tool for tackling such problems, even though tricks work out more neatly for special cases.)

****19.** Consider a function f which satisfies the differential equation

$$f^{(n)} = \sum_{j=0}^{n-1} a_j f^{(j)},$$

for certain numbers $a_0, \dots, a_{n-1}$. Several special cases have already received detailed treatment, either in the text or in other problems; in particular, we have found all functions satisfying $f' = f$, or $f''+f = 0$, or $f''-f = 0$. The trick in Problem 18-42 enables us to find many solutions for such equations, but doesn't say whether these are the only solutions. This requires a *uniqueness* result, which will be supplied by this problem. At the end you will find some (necessarily sketchy) remarks about the general solution.

(a) Derive the following formula for $f^{(n+1)}$ (let us agree that "a_{-1}" will be 0):

$$f^{(n+1)} = \sum_{j=0}^{n-1} (a_{j-1} + a_{n-1}a_j) f^{(j)}.$$

(b) Deduce a formula for $f^{(n+2)}$.

The formula in part (b) is not going to be used; it was inserted only to convince you that a general formula for $f^{(n+k)}$ is out of the question. On the other hand, as part (c) shows, it is not very hard to obtain estimates on the size of $f^{(n+k)}(x)$.

(c) Let $N = \max(1, |a_0|, \dots, |a_{n-1}|)$. Then $|a_{j-1} + a_{n-1}a_j| \le 2N^2$; this means that

$$f^{(n+1)} = \sum_{j=0}^{n-1} b_j{}^1 f^{(j)}, \quad \text{where } |b_j{}^1| \le 2N^2.$$

Show that

$$f^{(n+2)} = \sum_{j=0}^{n-1} b_j{}^2 f^{(j)}, \quad \text{where } |b_j{}^2| \le 4N^3,$$

and, more generally,

$$f^{(n+k)} = \sum_{j=0}^{n-1} b_j{}^k f^{(j)}, \quad \text{where } |b_j{}^k| \le 2N^{k+1}.$$

(d) Conclude from part (c) that, for any particular number x, there is a number M such that

$$|f^{(n+k}(x)| \le M \cdot 2^k N^{k+1} \quad \text{for all } k.$$

(e) Now suppose that $f(0) = f'(0) = \cdots = f^{(n-1)}(0) = 0$. Show that

$$|f(x)| \le \frac{M \cdot 2^{k+1} N^{k+2} |x|^{n+k+1}}{(n+k+1)!} \le \frac{M \cdot |2Nx|^{n+k+1}}{(n+k+1)!},$$

and conclude that $f = 0$.

(f) Show that if f_1 and f_2 are both solutions of the differential equation

$$f^{(n)} = \sum_{j=0}^{n-1} a_j f^{(j)},$$

and $f_1{}^{(j)}(0) = f_2{}^{(j)}(0)$ for $0 \le j \le n-1$, then $f_1 = f_2$.

In other words, the solutions of this differential equation are determined by the "initial conditions" (the values $f^{(j)}(0)$ for $0 \le j \le n-1$). This means that we can find *all* solutions once we can find enough solutions to obtain any given set of initial conditions. If the equation

$$x^n - a_{n-1}x^{n-1} - \cdots - a_0 = 0$$

has n distinct roots $\alpha_1, \dots, \alpha_n$, then any function of the form

$$f(x) = c_1 e^{\alpha_1 x} + \cdots + c_n e^{\alpha_n x}$$

is a solution, and

$$\begin{aligned} f(0) &= c_1 + \cdots + c_n, \\ f'(0) &= \alpha_1 c_1 + \cdots + \alpha_n c_n, \\ &\ \ \vdots \\ f^{(n-1)}(0) &= \alpha_1{}^{n-1} c_1 + \cdots + \alpha_n{}^{n-1} c_n. \end{aligned}$$

As a matter of fact, every solution is of this form, because we can obtain any set of numbers on the left side by choosing the c's properly, but we will not try to prove this last assertion. (It is a purely algebraic fact, which you can easily check for $n = 2$ or 3.) These remarks are also true if some of the roots are multiple roots, and even in the more general situation considered in Chapter 27.

****20.** (a) Suppose that f is a continuous function on $[a, b]$ with $f(a) = f(b)$ and that for all x in (a, b) the Schwarz second derivative of f at x is 0 (Problem 16). Show that f is constant on $[a, b]$. Hint: Suppose that $f(x) > f(a)$ for some x in (a, b). Consider the function

$$g(x) = f(x) - \varepsilon(x-a)(b-x)$$

with $g(a) = g(b) = f(a)$. For sufficiently small $\varepsilon > 0$ we will have $g(x) > g(a)$, so g will have a maximum point y in (a, b). Now use Problem 16(c) (the Schwarz second derivative of $(x-a)(b-x)$ is simply its ordinary second derivative).

(b) If f is a continuous function on $[a, b]$ whose Schwarz second derivative is 0 at all points of (a, b), then f is linear.

***21.** (a) Let $f(x) = x^4 \sin 1/x^2$ for $x \neq 0$, and $f(0) = 0$. Show that $f = 0$ up to order 2 at 0, even though $f''(0)$ does not exist.

This example is slightly more complex, but also slightly more impressive, than the example in the text, because both $f'(a)$ and $f''(a)$ exist for $a \neq 0$. Thus, for each number a there is another number $m(a)$ such that

$$(*) \quad f(x) = f(a) + f'(a)(x-a) + \frac{m(a)}{2}(x-a)^2 + R_a(x),$$

$$\text{where } \lim_{x \to a} \frac{R_a(x)}{(x-a)^2} = 0;$$

namely, $m(a) = f''(a)$ for $a \neq 0$, and $m(0) = 0$. Notice that the function m defined in this way is not continuous.

(b) Suppose that f is a differentiable function such that $(*)$ holds for all a, with $m(a) = 0$. Use Problem 20 to show that $f''(a) = m(a) = 0$ for all a.

(c) Now suppose that $(*)$ holds for all a, and that m is continuous. Prove that for all a the second derivative $f''(a)$ exists and equals $m(a)$.

*CHAPTER 21 e IS TRANSCENDENTAL

The irrationality of e was so easy to prove that in this optional chapter we will attempt a more difficult feat, and prove that the number e is not merely irrational, but actually much worse. Just how a number might be even worse than irrational is suggested by a slight rewording of definitions. A number x is irrational if it is not possible to write $x = a/b$ for any integers a and b, with $b \neq 0$. This is the same as saying that x does not satisfy any equation

$$bx - a = 0$$

for integers a and b, except for $a = 0$, $b = 0$. Viewed in this light, the irrationality of $\sqrt{2}$ does not seem to be such a terrible deficiency; rather, it appears that $\sqrt{2}$ just barely manages to be irrational—although $\sqrt{2}$ is not the solution of an equation

$$a_1 x + a_0 = 0,$$

it *is* the solution of the equation

$$x^2 - 2 = 0,$$

of one higher degree. Problem 2-18 shows how to produce many irrational numbers x which satisfy higher-degree equations

$$a_n x^n + a_{n-1} x^{n-1} + \cdots + a_0 = 0,$$

where the a_i are integers and $a_0 \neq 0$ (this condition rules out the possibility that all $a_i = 0$). A number which satisfies an "algebraic" equation of this sort is called an **algebraic number**, and practically every number we have ever encountered is defined in terms of solutions of algebraic equations (π and e are the great exceptions in our limited mathematical experience). All roots, such as

$$\sqrt{2}, \quad \sqrt[10]{3}, \quad \sqrt[4]{7},$$

are clearly algebraic numbers, and even complicated combinations, like

$$\sqrt[3]{3 + \sqrt{5} + \sqrt[4]{1 + \sqrt{2}} + \sqrt[5]{6}}$$

are algebraic (although we will not try to prove this). Numbers which cannot be obtained by the process of solving algebraic equations are called **transcendental**; the main result of this chapter states that e is a number of this anomalous sort.

The proof that e is transcendental is well within our grasp, and was theoretically possible even before Chapter 20. Nevertheless, with the inclusion of this proof, we can justifiably classify ourselves as something more than novices in the study of higher mathematics; while many irrationality proofs depend only on elementary properties of numbers, the proof that a number is transcendental usually involves some really high-powered mathematics. Even the dates connected with the transcendence of e are impressively recent—the first proof that e is transcendental, due to Hermite, dates from 1873. The proof that we will give is a simplification, due to Hilbert.

Before tackling the proof itself, it is a good idea to map out the strategy, which depends on an idea used even in the proof that e is irrational. Two features of the expression

$$e = 1 + \frac{1}{1!} + \frac{1}{2!} + \cdots + \frac{1}{n!} + R_n$$

were important for the proof that e is irrational: On the one hand, the number

$$1 + \frac{1}{1!} + \cdots + \frac{1}{n!}$$

can be written as a fraction p/q with $q \le n!$ (so that $n!\,(p/q)$ is an integer); on the other hand, $0 < R_n < 3/(n+1)!$ (so $n!R_n$ is not an integer). These two facts show that e can be approximated particularly well by rational numbers. Of course, every number x can be approximated arbitrarily closely by rational numbers—if $\varepsilon > 0$ there is a rational number r with $|x - r| < \varepsilon$; the catch, however, is that it may be necessary to allow a very large denominator for r, as large as $1/\varepsilon$ perhaps. For e we are assured that this is not the case: there is a fraction p/q within $3/(n+1)!$ of e, whose denominator q is at most $n!$. If you look carefully at the proof that e is irrational, you will see that only this fact about e is ever used. The number e is by no means unique in this respect: generally speaking, the *better* a number can be approximated by rational numbers, the *worse* it is (some evidence for this assertion is presented in Problem 3). The proof that e is transcendental depends on a natural extension of this idea: not only e, but any finite number of powers $e, e^2, \dots, e^n$, can be simultaneously approximated especially well by rational numbers. In our proof we will begin by assuming that e is algebraic, so that

$$(*) \quad a_n e^n + \cdots + a_1 e + a_0 = 0, \qquad a_0 \neq 0$$

for some integers $a_0, \dots, a_n$. In order to reach a contradiction we will than find certain integers $M, M_1, \dots, M_n$ and certain "small" numbers $\epsilon_1, \dots, \epsilon_n$ such that

$$e^1 = \frac{M_1 + \epsilon_1}{M},$$
$$e^2 = \frac{M_2 + \epsilon_2}{M},$$
$$\vdots$$
$$e^n = \frac{M_n + \epsilon_n}{M}.$$

Just how small the ϵ's must be will appear when these expressions are substituted into the assumed equation $(*)$. After multiplying through by M we obtain

$$[a_0 M + a_1 M_1 + \cdots + a_n M_n] + [\epsilon_1 a_1 + \cdots + \epsilon_n a_n] = 0.$$

The first term in brackets is an integer, and we will choose the M's so that it will necessarily be a *nonzero* integer. We will also manage to find ϵ's so small that

$$|\epsilon_1 a_1 + \cdots + \epsilon_n a_n| < \tfrac{1}{2};$$

this will lead to the desired contradiction—the sum of a nonzero integer and a number of absolute value less than $\frac{1}{2}$ cannot be zero!

As a basic strategy this is all very reasonable and quite straightforward. The remarkable part of the proof will be the way that the M's and ϵ's are defined. In order to read the proof you will need to know about the gamma function! (This function was introduced in Problem 19-39.)

THEOREM 1 e is transcendental.

PROOF Suppose there were integers $a_0, \dots, a_n$, with $a_0 \neq 0$, such that

$$(*) \quad a_n e^n + a_{n-1}e^{n-1} + \cdots + a_0 = 0.$$

Define numbers $M, M_1, \dots, M_n$ and $\epsilon_1, \dots, \epsilon_n$ as follows:

$$M = \int_0^\infty \frac{x^{p-1}[(x-1)\cdot \ldots \cdot (x-n)]^p e^{-x}}{(p-1)!}\,dx,$$

$$M_K = e^k \int_k^\infty \frac{x^{p-1}[(x-1)\cdot \ldots \cdot (x-n)]^p e^{-x}}{(p-1)!}\,dx,$$

$$\epsilon_k = e^k \int_0^k \frac{x^{p-1}[(x-1)\cdot \ldots \cdot (x-n)]^p e^{-x}}{(p-1)!}\,dx.$$

The unspecified number p represents a prime number* which we will choose later. Despite the forbidding aspect of these three expressions, with a little work they will appear much more reasonable. We concentrate on M first. If the expression in brackets,

$$[(x-1)\cdot \ldots \cdot (x-n)],$$

is actually multiplied out, we obtain a polynomial

$$x^n + \cdots \pm n!$$

*The term "prime number" was defined in Problem 2-17. An important fact about prime numbers will be used in the proof, although it is not proved in this book: If p is a prime number which does not divide the integer a, and which does not divide the integer b, then p also does not divide ab. The Suggested Reading mentions references for this theorem (which is crucial in proving that the factorization of an integer into primes is unique). We will also use the result of Problem 2-17(d), that there are infinitely many primes—the reader is asked to determine at precisely which points this information is required.

with integer coefficients. When raised to the pth power this becomes an even more complicated polynomial

$$x^{np} + \cdots \pm (n!)^p.$$

Thus M can be written in the form

$$M = \sum_{\alpha=0}^{np} \frac{1}{(p-1)!} C_\alpha \int_0^\infty x^{p-1+\alpha} e^{-x}\, dx,$$

where the C_α are certain integers, and $C_0 = \pm(n!)^p$. But

$$\int_0^\infty x^k e^{-x}\, dx = k!.$$

Thus

$$M = \sum_{\alpha=0}^{np} C_\alpha \frac{(p-1+\alpha)!}{(p-1)!}.$$

Now, for $\alpha = 0$ we obtain the term

$$\pm(n!)^p \frac{(p-1)!}{(p-1)!} = \pm(n!)^p.$$

We will now consider only primes $p > n$; then this term is an integer which is *not* divisible by p. On the other hand, if $\alpha > 0$, then

$$C_\alpha \frac{(p-1+\alpha)!}{(p-1)!} = C_\alpha (p+\alpha-1)(p+\alpha-2) \cdot \ldots \cdot p,$$

which *is* divisible by p. Therefore M itself is an integer which is *not* divisible by p.

Now consider M_k. We have

$$\begin{aligned} M_k &= e^k \int_k^\infty \frac{x^{p-1}[(x-1) \cdot \ldots \cdot (x-n)]^p e^{-x}}{(p-1)!}\, dx \\ &= \int_k^\infty \frac{x^{p-1}[(x-1) \cdot \ldots \cdot (x-n)]^p e^{-(x-k)}}{(p-1)!}\, dx. \end{aligned}$$

This can be transformed into an expression looking very much like M by the substitution

$$\begin{aligned} u &= x - k \\ du &= dx. \end{aligned}$$

The limits of integration are changed to 0 and ∞, and

$$M_k = \int_0^\infty \frac{(u+k)^{p-1}[(u+k-1) \cdot \ldots \cdot u \cdot \ldots \cdot (u+k-n)]^p e^{-u}}{(p-1)!}\, du.$$

There is one very significant difference between this expression and that for M. The term in brackets contains the factor u in the kth place. Thus the pth power contains the factor u^p. This means that the entire expression

$$(u+k)^{p-1}[(u+k-1) \cdot \ldots \cdot (u+k-n)]^p$$

is a polynomial with integer coefficients, *every term of which* has degree at least p. Thus

$$M_k = \sum_{\alpha=1}^{np} \frac{1}{(p-1)!} D_\alpha \int_0^\infty u^{p-1+\alpha} e^{-u}\, du = \sum_{\alpha=1}^{np} D_\alpha \frac{(p-1+\alpha)!}{(p-1)!},$$

where the D_α are certain integers. Notice that the summation begins with $\alpha = 1$; in this case *every* term in the sum is divisible by p. Thus each M_k is an integer which *is* divisible by p.

Now it is clear that

$$e^k = \frac{M_k + \epsilon_k}{M}, \quad k = 1, \ldots, n.$$

Substituting into (∗) and multiplying by M we obtain

$$[a_0 M + a_1 M + \cdots + a_n M_n] + [a_1\epsilon_1 + \cdots + a_n\epsilon_n] = 0.$$

In addition to requiring that $p > n$ let us also stipulate that $p > |a_0|$. This means that both M and a_0 are not divisible by p, so $a_0 M$ is also not divisible by p. Since each M_k is divisible by p, it follows that

$$a_0 M + a_1 M_1 + \cdots + a_n M_n$$

is *not* divisible by p. In particular it is a *nonzero* integer.

In order to obtain a contradiction to the assumed equation (∗), and thereby prove that e is transcendental, it is only necessary to show that

$$|a_1\epsilon_1 + \cdots + a_n\epsilon_n|$$

can be made as small as desired, by choosing p large enough; it is clearly sufficient to show that each $|\epsilon_k|$ can be made as small as desired. This requires nothing more than some simple estimates; for the remainder of the argument remember that n is a certain fixed number (the degree of the assumed polynomial equation (∗)). To begin with, if $1 \le k \le n$, then

$$\begin{aligned}
|\epsilon_k| &\le e^k \int_0^k \frac{|x^{p-1}[(x-1)\cdot\ldots\cdot(x-n)]^p|\, e^{-x}}{(p-1)!}\, dx \\
&\le e^n \int_0^n \frac{n^{p-1}|[(x-1)\cdot\ldots\cdot(x-n)]^p|\, e^{-x}}{(p-1)!}\, dx.
\end{aligned}$$

Now let A be the maximum of $|(x-1)\cdot\ldots\cdot(x-n)|$ for x in $[0, n]$. Then

$$\begin{aligned}
|\epsilon_k| &\le \frac{e^n n^{p-1} A^p}{(p-1)!} \int_0^n e^{-x}\, dx \\
&\le \frac{e^n n^{p-1} A^p}{(p-1)!} \int_0^\infty e^{-x}\, dx \\
&= \frac{e^n n^{p-1} A^p}{(p-1)!} \\
&\le \frac{e^n n^p A^p}{(p-1)!} = \frac{e^n (nA)^p}{(p-1)!}.
\end{aligned}$$

But n and A are fixed; thus $(nA)^p/(p-1)!$ can be made as small as desired by making p sufficiently large. ▮

This proof, like the proof that π is irrational, deserves some philosophic afterthoughts. At first sight, the argument seems quite "advanced"—after all, we use integrals, and integrals from 0 to ∞ at that. Actually, as many mathematicians have observed, integrals can be eliminated from the argument completely; the only integrals essential to the proof are of the form

$$\int_0^\infty x^k e^{-x}\,dx$$

for integral k, and these integrals can be replaced by $k!$ whenever they occur. Thus M, for example, could have been defined initially as

$$M = \sum_{\alpha=0}^{np} C_\alpha \frac{(p-1+\alpha)!}{(p-1)!},$$

where C_α are the coefficients of the polynomial

$$[(x-1)\cdot \ldots \cdot (x-n)]^p.$$

If this idea is developed consistently, one obtains a "completely elementary" proof that e is transcendental, depending only on the fact that

$$e = 1 + \frac{1}{1!} + \frac{1}{2!} + \frac{1}{3!} + \cdots.$$

Unfortunately, this "elementary" proof is harder to understand than the original one—the whole structure of the proof must be hidden just to eliminate a few integral signs! This situation is by no means peculiar to this specific theorem—"elementary" arguments are frequently more difficult than "advanced" ones. Our proof that π is irrational is a case in point. You probably remember nothing about this proof except that it involves quite a few complicated functions. There is actually a more advanced, but much more conceptual proof, which shows that π is *transcendental*, a fact which is of great historical, as well as intrinsic, interest. One of the classical problems of Greek mathematics was to construct, with compass and straightedge alone, a square whose area is that of a circle of radius 1. This requires the construction of a line segment whose length is $\sqrt{\pi}$, which can be accomplished if a line segment of length π is constructible. The Greeks were totally unable to decide whether such a line segment could be constructed, and even the full resources of modern mathematics were unable to settle this question until 1882. In that year Lindemann proved that π is transcendental; since the length of any segment that can be constructed with straightedge and compass can be written in terms of $+, \cdot, -, \div$, and $\sqrt{\ }$, and is therefore algebraic, this proves that a line segment of length π cannot be constructed.

The proof that π is transcendental requires a sizable amount of mathematics which is too advanced to be reached in this book. Nevertheless, the proof is not much more difficult than the proof that e is transcendental. In fact, the proof for π is practically the same as the proof for e. This last statement should certainly surprise you. The proof that e is transcendental seems to depend so thoroughly on particular properties of e that it is almost inconceivable how any modifications could ever be used for π; after all, what does e have to do with π? Just wait and see!

PROBLEMS

1. (a) Prove that if $\alpha > 0$ is algebraic, then $\sqrt{\alpha}$ is algebraic.
(b) Prove that if α is algebraic and r is rational, then $\alpha + r$ and αr are algebraic.

Part (b) can actually be strengthened considerably: the sum, product, and quotient of algebraic numbers is algebraic. This fact is too difficult for us to prove here, but some special cases can be examined:

2. Prove that $\sqrt{2} + \sqrt{3}$ and $\sqrt{2}(1 + \sqrt{3})$ are algebraic, by actually finding algebraic equations which they satisfy. (You will need equations of degree 4.)

***3.** (a) Let α be an algebraic number which is not rational. Suppose that α satisfies the polynomial equation

$$f(x) = a_n x^n + a_{n-1} x^{n-1} + \cdots + a_0 = 0,$$

and that no polynomial function of lower degree has this property. Show that $f(p/q) \neq 0$ for any rational number p/q. Hint: Use Problem 3-7(b).
(b) Now show that $|f(p/q)| \geq 1/q^n$ for all rational numbers p/q with $q > 0$. Hint: Write $f(p/q)$ as a fraction over the common denominator q^n.
(c) Let $M = \sup\{|f'(x)| : |x - \alpha| < 1\}$. Use the Mean Value Theorem to prove that if p/q is a rational number with $|\alpha - p/q| < 1$, then $|\alpha - p/q| > 1/Mq^n$. (It follows that for $c = \max(1, 1/M)$ we have $|\alpha - p/q| > c/q^n$ for all rational p/q.)

***4.** Let

$$\alpha = 0.110001000000000000000001000\ldots,$$

where the 1's occur in the $n!$ place, for each n. Use Problem 3 to prove that α is transcendental. (For each n, show that α is not the root of an equation of degree n.)

Although Problem 4 mentions only one specific transcendental number, it should be clear that one can easily construct infinitely many other numbers α which do not satisfy $|\alpha - p/q| > c/q^n$ for any c and n. Such numbers were first considered by Liouville (1809–1882), and the inequality in Problem 3 is often called Liouville's inequality. None of the transcendental numbers constructed in this way happens to be particularly interesting, but for a long time Liouville's transcendental numbers were the only ones known. This situation was changed quite radically by the work of Cantor (1845–1918), who showed, without exhibiting a single transcendental number, that *most* numbers are transcendental. The next two problems provide an introduction to the ideas that allow us to make sense of such statements. The basic definition with which we must work is the following: A set A is called **countable** if its elements can be arranged in a sequence

$$a_1, a_2, a_3, a_4, \ldots .$$

The obvious example (in fact, more or less the Platonic ideal of) a countable set is $\mathbf{N}$, the set of natural numbers; clearly the set of even natural numbers is also countable:

$$2, 4, 6, 8, \ldots .$$

It is a little more surprising to learn that $\mathbf{Z}$, the set of all integers (positive, negative and 0) is also countable, but seeing is believing:

$$0, 1, -1, 2, -2, 3, -3, \ldots .$$

The next two problems, which outline the basic features of countable sets, are really a series of examples to show that (1) a lot more sets are countable than one might think and (2) nevertheless, some sets are not countable.

***5.** (a) Show that if A and B are countable, then so is $A \cup B = \{x : x \text{ is in } A \text{ or } x \text{ is in } B\}$. Hint: Use the same trick that worked for $\mathbf{Z}$.

(b) Show that the set of positive rational numbers is countable. (This is really quite startling, but the figure below indicates the path to enlightenment.)

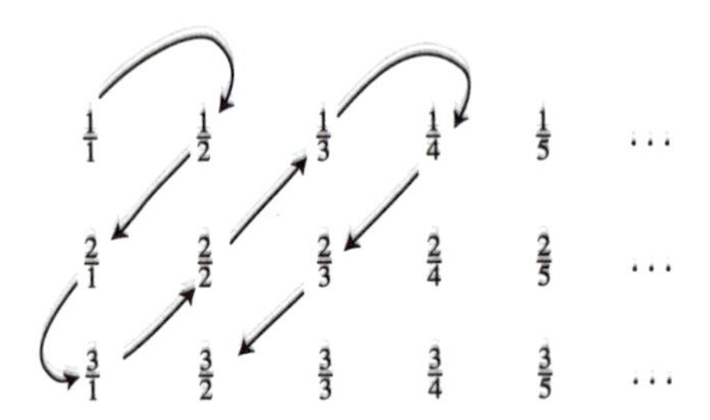

(c) Show that the set of all pairs (m, n) of integers is countable. (This is practically the same as part (b).)

(d) If $A_1, A_2, A_3, \ldots$ are each countable, prove that

$$A_1 \cup A_2 \cup A_3 \cup \ldots$$

is also countable. (Again use the same trick as in part (b).)

(e) Prove that the set of all triples (l, m, n) of integers is countable. (A triple (l, m, n) can be described by a pair (l, m) and a number n.)

(f) Prove that the set of all n-tuples $(a_1, a_2, \ldots, a_n)$ is countable. (If you have done part (e), you can do this, using induction.)

(g) Prove that the set of all roots of polynomial functions of degree n with integer coefficients is countable. (Part (f) shows that the set of all these polynomial functions can be arranged in a sequence, and each has at most n roots.)

(h) Now use parts (d) and (g) to prove that the set of all algebraic numbers is countable.

***6.** Since so many sets turn out to be countable, it is important to note that the set of all real numbers between 0 and 1 is *not* countable. In other words, there is no way of listing all these real numbers in a sequence

$$\begin{aligned} \alpha_1 &= 0.a_{11}a_{12}a_{13}a_{14}\dots \\ \alpha_2 &= 0.a_{21}a_{22}a_{23}a_{24}\dots \\ \alpha_3 &= 0.a_{31}a_{32}a_{33}a_{34}\dots \\ &\dots \end{aligned}$$

(decimal notation is being used on the right). To prove that this is so, suppose such a list were possible and consider the decimal

$$0.\bar{a}_{11}\bar{a}_{22}\bar{a}_{33}\bar{a}_{44}\dots,$$

where $\bar{a}_{nn} = 5$ if $a_{nn} \neq 5$ and $\bar{a}_{nn} = 6$ if $a_{nn} = 5$. Show that this number cannot possibly be in the list, thus obtaining a contradiction.

Problems 5 and 6 can be summed up as follows. The set of algebraic numbers is countable. If the set of transcendental numbers were also countable, then the set of all real numbers would be countable, by Problem 5(a), and consequently the set of real numbers between 0 and 1 would be countable. But this is false. Thus, the set of algebraic numbers is countable and the set of transcendental numbers is not ("there are more transcendental numbers than algebraic numbers"). The remaining two problems illustrate further how important it can be to distinguish between sets which are countable and sets which are not.

***7.** Let f be a nondecreasing function on $[0, 1]$. Recall (Problem 8-8) that $\lim_{x\to a^+} f(x)$ and $\lim_{x\to a^-} f(x)$ both exist.

(a) For any $\varepsilon > 0$ prove that there are only finitely many numbers a in $[0, 1]$ with $\lim_{x\to a^+} f(x) - \lim_{x\to a^-} f(x) > \varepsilon$. Hint: There are, in fact, at most $[f(1) - f(0)]/\varepsilon$ of them.

(b) Prove that the set of points at which f is discontinuous is countable. Hint: If $\lim_{x\to a^+} f(x) - \lim_{x\to a^-} f(x) > 0$, then it is $> 1/n$ for some natural number n.

This problem shows that a nondecreasing function is automatically continuous at most points. For differentiability the situation is more difficult to analyze and also more interesting. A nondecreasing function can fail to be differentiable at a set of points which is not countable, but it is still true that nondecreasing functions are differentiable at most points (in a different sense of the word "most"). Reference [32] of the Suggested Reading gives a beautiful proof, using the Rising Sun Lemma of

Problem 8-20. For those who have done Problem 10 of the Appendix to Chapter 11, it is possible to provide at least one application to differentiability of the ideas already developed in this problem set: If f is convex, then f is differentiable except at those points where its right-hand derivative ${f_+}'$ is discontinuous; but the function ${f_+}'$ is increasing, so a convex function is automatically differentiable except at a countable set of points.

***8.** (a) Problem 11-66 showed that if every point is a local maximum point for a *continuous* function f, then f is a constant function. Suppose now that the hypothesis of continuity is dropped. Prove that f takes on only a countable set of values. Hint: For each x choose *rational* numbers a_x and b_x such that $a_x < x < b_x$ and x is a maximum point for f on (a_x, b_x). Then every value $f(x)$ is the maximum value of f on some interval (a_x, b_x). How many such intervals are there?

(b) Deduce Problem 11-66(a) as a corollary.

(c) Prove the result of Problem 11-66(b) similarly.

CHAPTER 22 INFINITE SEQUENCES

The idea of an infinite sequence is so natural a concept that it is tempting to dispense with a definition altogether. One frequently writes simply "an infinite sequence

$$a_1, a_2, a_3, a_4, a_5, \ldots,"$$

the three dots indicating that the numbers a_i continue to the right "forever." A rigorous definition of an infinite sequence is not hard to formulate, however; the important point about an infinite sequence is that for each natural number, n, there is a real number a_n. This sort of correspondence is precisely what functions are meant to formalize.

DEFINITION An **infinite sequence** of real numbers is a function whose domain is **N**.

From the point of view of this definition, a sequence should be designated by a single letter like a, and particular values by

$$a(1), a(2), a(3), \ldots,$$

but the subscript notation

$$a_1, a_2, a_3, \ldots$$

is almost always used instead, and the sequence itself is usually denoted by a symbol like $\{a_n\}$. Thus $\{n\}$, $\{(-1)^n\}$, and $\{1/n\}$ denote the sequences α, β, and γ defined by

$$\begin{aligned} \alpha_n &= n, \\ \beta_n &= (-1)^n, \\ \gamma_n &= \frac{1}{n}. \end{aligned}$$

A sequence, like any function, can be graphed (Figure 1) but the graph is usually rather unrevealing, since most of the function cannot be fit on the page.

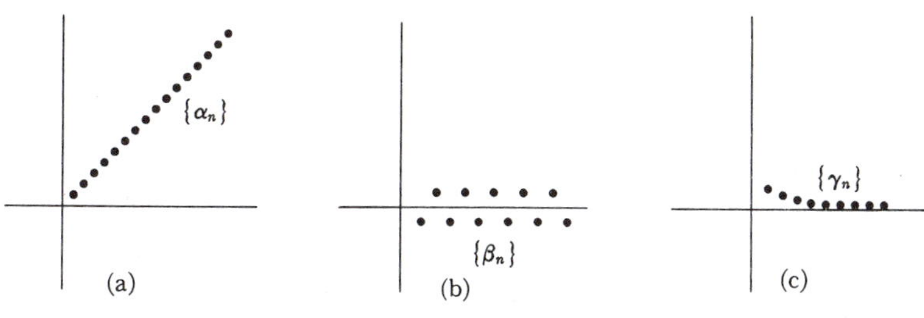

FIGURE 1

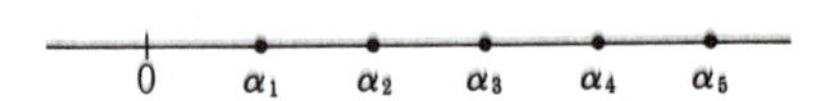

$\cdots = \beta_5 = \beta_3 = \beta_1$ $\quad 0 \quad$ $\beta_2 = \beta_4 = \beta_6 = \cdots$

$\gamma_4 \quad \gamma_2$

$0 \quad \gamma_5 \; \gamma_3 \qquad \gamma_1$

FIGURE 2

A more convenient representation of a sequence is obtained by simply labeling the points $a_1, a_2, a_3, \ldots$ on a line (Figure 2). This sort of picture shows where the sequence "is going." The sequence $\{\alpha_n\}$ "goes out to infinity," the sequence $\{\beta_n\}$ "jumps back and forth between -1 and 1," and the sequence $\{\gamma_n\}$ "converges to 0." Of the three phrases in quotation marks, the last is the crucial concept associated with sequences, and will be defined precisely (the definition is illustrated in Figure 3).

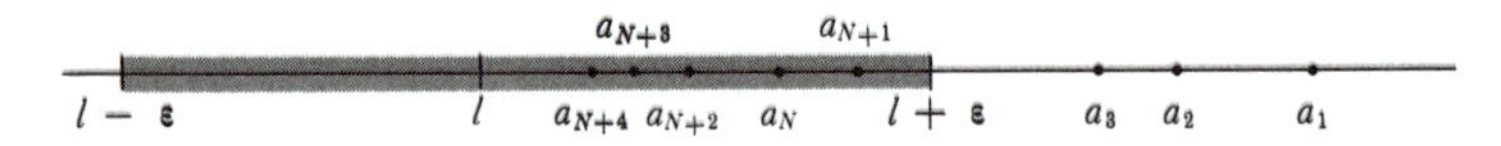

FIGURE 3

DEFINITION

A sequence $\{a_n\}$ **converges to l** (in symbols $\lim_{n\to\infty} a_n = l$) if for every $\varepsilon > 0$ there is a natural number N such that, for all natural numbers n,

$$\text{if } n > N, \text{ then } |a_n - l| < \varepsilon.$$

In addition to the terminology introduced in this definition, we sometimes say that the sequence $\{a_n\}$ **approaches l** or has the **limit l**. A sequence $\{a_n\}$ is said to **converge** if it converges to l for some l, and to **diverge** if it does not converge.

To show that the sequence $\{\gamma_n\}$ converges to 0, it suffices to observe the following. If $\varepsilon > 0$, there is a natural number N such that $1/N < \varepsilon$. Then, if $n > N$ we have

$$\gamma_n = \frac{1}{n} < \frac{1}{N} < \varepsilon, \qquad \text{so } |\gamma_n - 0| < \varepsilon.$$

The limit

$$\lim_{n\to\infty} \sqrt{n+1} - \sqrt{n} = 0$$

will probably seem reasonable after a little reflection (it just says that $\sqrt{n+1}$ is practically the same as $\sqrt{n}$ for large n), but a mathematical proof might not be so obvious. To estimate $\sqrt{n+1}-\sqrt{n}$ we can use an algebraic trick:

$$\begin{aligned}\sqrt{n+1}-\sqrt{n} &= \frac{(\sqrt{n+1}-\sqrt{n})(\sqrt{n+1}+\sqrt{n})}{\sqrt{n+1}+\sqrt{n}}\\ &= \frac{n+1-n}{\sqrt{n+1}+\sqrt{n}}\\ &= \frac{1}{\sqrt{n+1}+\sqrt{n}}.\end{aligned}$$

It is also possible to estimate $\sqrt{n+1}-\sqrt{n}$ by applying the Mean Value Theorem to the function $f(x)=\sqrt{x}$ on the interval $[n, n+1]$. We obtain

$$\begin{aligned}\frac{\sqrt{n+1}-\sqrt{n}}{1} &= f'(x)\\ &= \frac{1}{2\sqrt{x}}, \qquad \text{for some } x \text{ in } (n, n+1)\\ &< \frac{1}{2\sqrt{n}}.\end{aligned}$$

Either of these estimates may be used to prove the above limit; the detailed proof is left to you, as a simple but valuable exercise.

The limit

$$\lim_{n\to\infty}\frac{3n^3+7n^2+1}{4n^3-8n+63}=\frac{3}{4}$$

should also seem reasonable, because the terms involving n^3 are the most important when n is large. If you remember the proof of Theorem 7-9 you will be able to guess the trick that translates this idea into a proof—dividing top and bottom by n^3 yields

$$\frac{3n^3+7n^2+1}{4n^3-8n+63}=\frac{3+\dfrac{7}{n}+\dfrac{1}{n^3}}{4-\dfrac{8}{n^2}+\dfrac{63}{n^3}}.$$

Using this expression, the proof of the above limit is not difficult, especially if one uses the following facts:

If $\lim_{n\to\infty} a_n$ and $\lim_{n\to\infty} b_n$ both exist, then

$$\begin{aligned}\lim_{n\to\infty}(a_n+b_n) &= \lim_{n\to\infty} a_n + \lim_{n\to\infty} b_n,\\ \lim_{n\to\infty}(a_n\cdot b_n) &= \lim_{n\to\infty} a_n \cdot \lim_{n\to\infty} b_n;\end{aligned}$$

moreover, if $\lim_{n\to\infty} b_n \neq 0$, then $b_n \neq 0$ for all n greater than some N, and

$$\lim_{n\to\infty} a_n/b_n = \lim_{n\to\infty} a_n / \lim_{n\to\infty} b_n.$$

(If we wanted to be utterly precise, the third statement would have to be even more complicated. As it stands, we are considering the limit of the sequence $\{c_n\} = \{a_n/b_n\}$, where the numbers c_n might not even be defined for certain $n < N$. This doesn't really matter—we could define c_n any way we liked for such n—because the limit of a sequence is not changed if we change the sequence at a finite number of points.)

Although these facts are very useful, we will not bother stating them as a theorem—you should have no difficulty proving these results for yourself, because the definition of $\lim_{n\to\infty} a_n = l$ is so similar to previous definitions of limits, especially $\lim_{x\to\infty} f(x) = l$.

The similarity between the definitions of $\lim_{n\to\infty} a_n = l$ and $\lim_{x\to\infty} f(x) = l$ is actually closer than mere analogy; it is possible to define the first in terms of the second. If f is the function whose graph (Figure 4) consists of line segments joining

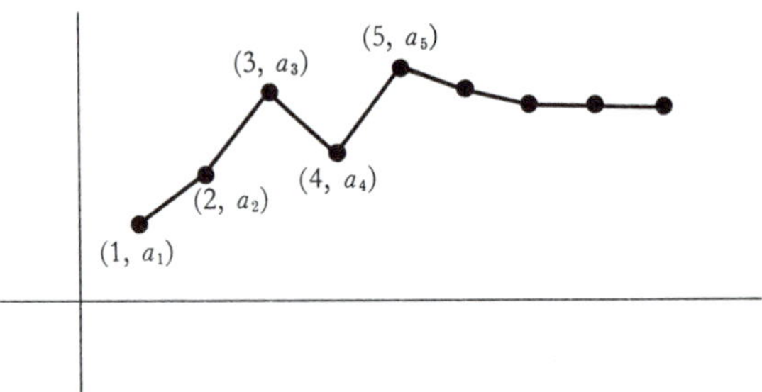

FIGURE 4

the points in the graph of the sequence $\{a_n\}$, so that

$$f(x) = (a_{n+1} - a_n)(x - n) + a_n \qquad n \le x \le n+1,$$

then

$$\lim_{n\to\infty} a_n = l \quad \text{if and only if} \quad \lim_{x\to\infty} f(x) = l.$$

Conversely, if f satisfies $\lim_{x\to\infty} = l$, and we set $a_n = f(n)$, then $\lim_{x\to\infty} a_n = l$.

This second observation is frequently very useful. For example, suppose that $0 < a < 1$. Then

$$\lim_{n\to\infty} a^n = 0.$$

To prove this we note that

$$\lim_{x\to\infty} a^x = \lim_{x\to\infty} e^{x \log a} = 0,$$

since $\log a < 0$, so that $x \log a$ is a negative and large in absolute value for large x. Notice that we actually have

$$\lim_{n\to\infty} a^n = 0 \quad \text{if } |a| < 1;$$

for if $a < 0$ we can write

$$\lim_{n\to\infty} a^n = \lim_{n\to\infty} (-1)^n |a|^n = 0.$$

The behavior of the logarithm function also shows that if $a > 1$, then a^n becomes arbitrarily large as n becomes large. This assertion is often written

$$\lim_{n\to\infty} a^n = \infty, \quad a > 1,$$

and it is sometimes even said that $\{a^n\}$ approaches ∞. We also write equations like

$$\lim_{n\to\infty} -a^n = -\infty,$$

and say that $\{-a^n\}$ approaches $-\infty$. Notice, however, that if $a < -1$, then $\lim_{n\to\infty} a^n$ does not exist, even in this extended sense.

Despite this connection with a familiar concept, it is more important to visualize convergence in terms of the picture of a sequence as points on a line (Figure 3). There is another connection between limits of functions and limits of sequences which is related to *this* picture. This connection is somewhat less obvious, but considerably more interesting, than the one previously mentioned—instead of defining limits of sequences in terms of limits of functions, it is possible to reverse the procedure.

THEOREM 1 Let f be a function defined in an open interval containing c, except perhaps at c itself, with

$$\lim_{x\to c} f(x) = l.$$

Suppose that $\{a_n\}$ is a sequence such that

$$\begin{aligned} &(1) \quad \text{each } a_n \text{ is in the domain of } f,\\ &(2) \quad \text{each } a_n \neq c,\\ &(3) \quad \lim_{n\to\infty} a_n = c. \end{aligned}$$

Then the sequence $\{f(a_n)\}$ satisfies

$$\lim_{n\to\infty} f(a_n) = l.$$

Conversely, if this is true for every sequence $\{a_n\}$ satisfying the above conditions, then $\lim_{x\to c} f(x) = l$.

PROOF Suppose first that $\lim_{x\to c} f(x) = l$. Then for every $\varepsilon > 0$ there is a $\delta > 0$ such that, for all x,

$$\text{if } 0 < |x - c| < \delta, \text{ then } |f(x) - l| < \varepsilon.$$

If the sequence $\{a_n\}$ satisfies $\lim_{n\to\infty} a_n = c$, then (Figure 3) there is a natural number N such that,

$$\text{if } n > N, \text{ then } |a_n - c| < \delta.$$

By our choice of δ, this means that

$$|f(a_n) - l| < \varepsilon,$$

showing that

$$\lim_{n\to\infty} f(a_n) = l.$$

Suppose, conversely, that $\lim_{n\to\infty} f(a_n) = l$ for every sequence $\{a_n\}$ with $\lim_{n\to\infty} a_n = c$. If $\lim_{x\to c} f(x) = l$ were *not* true, there would be some $\varepsilon > 0$ such that for *every* $\delta > 0$ there is an x with

$$0 < |x - c| < \delta \quad \text{but} \quad |f(x) - l| > \varepsilon.$$

In particular, for each n there would be a number x_n such that

$$0 < |x_n - c| < \frac{1}{n} \quad \text{but} \quad |f(x_n) - l| > \varepsilon.$$

Now the sequence $\{x_n\}$ clearly converges to c but, since $|f(x_n) - l| > \varepsilon$ for all n, the sequence $\{f(x_n)\}$ does not converge to l. This contradicts the hypothesis, so $\lim_{x\to c} f(x) = l$ must be true. ▌

Theorem 1 provides many examples of convergent sequences. For example, the sequences $\{a_n\}$ and $\{b_n\}$ defined by

$$a_n = \sin\left(13 + \frac{1}{n^2}\right)$$

$$b_n = \cos\left(\sin\left(1 + (-1)^n \cdot \frac{1}{n}\right)\right),$$

clearly converge to $\sin(13)$ and $\cos(\sin(1))$, respectively. It is important, however, to have some criteria guaranteeing convergence of sequences which are not obviously of this sort. There is one important criterion which is very easy to prove, but which is the basis for all other results. This criterion is stated in terms of concepts defined for functions, which therefore apply also to sequences: a sequence $\{a_n\}$ is **increasing** if $a_{n+1} > a_n$ for all n, **nondecreasing** if $a_{n+1} \geq a_n$ for all n, and **bounded above** if there is a number M such that $a_n \leq M$ for all n; there are similar definitions for sequences which are decreasing, nonincreasing, and bounded below.

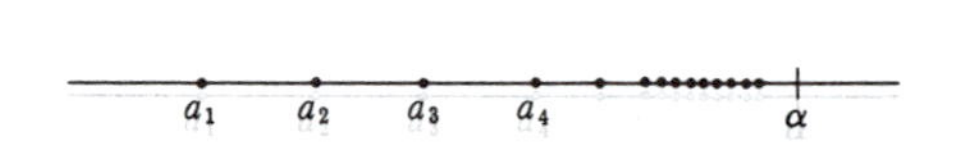

FIGURE 5

THEOREM 2 If $\{a_n\}$ is nondecreasing and bounded above, then $\{a_n\}$ converges (a similar statement is true if $\{a_n\}$ is nonincreasing and bounded below).

PROOF The set A consisting of all numbers a_n is, by assumption, bounded above, so A has a least upper bound α. We claim that $\lim_{n\to\infty} a_n = \alpha$ (Figure 5). In fact, if $\varepsilon > 0$, there is some a_N satisfying $\alpha - a_N < \varepsilon$, since α is the least upper bound of A. Then if $n > N$ we have

$$a_n \geq a_N, \quad \text{so} \quad \alpha - a_n \leq \alpha - a_N < \varepsilon.$$

This proves that $\lim_{n\to\infty} a_n = \alpha$. ▌

The hypothesis that $\{a_n\}$ is bounded above is clearly essential in Theorem 2: if $\{a_n\}$ is not bounded above, then (whether or not $\{a_n\}$ is nondecreasing) $\{a_n\}$ clearly diverges. Upon first consideration, it might appear that there should be little trouble deciding whether or not a given nondecreasing sequence $\{a_n\}$ is bounded above, and consequently whether or not $\{a_n\}$ converges. In the next chapter such sequences will arise very naturally and, as we shall see, deciding whether or not they converge is hardly a trivial matter. For the present, you might try to decide whether or not the following (obviously increasing) sequence is bounded above:

$$1,\ 1+\tfrac{1}{2},\ 1+\tfrac{1}{2}+\tfrac{1}{3},\ 1+\tfrac{1}{2}+\tfrac{1}{3}+\tfrac{1}{4},\ \dots.$$

Although Theorem 2 treats only a very special class of sequences, it is more useful than might appear at first, because it is always possible to extract from an arbitrary sequence $\{a_n\}$ another sequence which is either nonincreasing or else nondecreasing. To be precise, let us define a **subsequence** of the sequence $\{a_n\}$ to be a sequence of the form

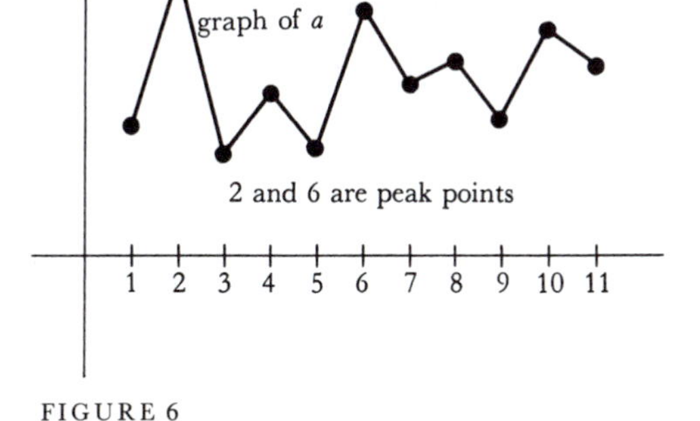

FIGURE 6

$$a_{n_1}, a_{n_2}, a_{n_3}, \dots,$$

where the n_j are natural numbers with

$$n_1 < n_2 < n_3 \cdots.$$

Then every sequence contains a subsequence which is either nondecreasing or nonincreasing. It is possible to become quite befuddled trying to prove this assertion, although the proof is very short if you think of the right idea; it is worth recording as a lemma.

LEMMA Any sequence $\{a_n\}$ contains a subsequence which is either nondecreasing or nonincreasing.

PROOF Call a natural number n a "peak point" of the sequence $\{a_n\}$ if $a_m < a_n$ for all $m > n$ (Figure 6).

Case 1. The sequence has infinitely many peak points. In this case, if $n_1 < n_2 < n_3 < \cdots$ are the peak points, then $a_{n_1} > a_{n_2} > a_{n_3} > \cdots$, so $\{a_{n_k}\}$ is the desired (nonincreasing) subsequence.

Case 2. The sequence has only finitely many peak points. In this case, let n_1 be greater than all peak points. Since n_1 is not a peak point, there is some $n_2 > n_1$ such that $a_{n_2} \geq a_{n_1}$. Since n_2 is not a peak point (it is greater than n_1, and hence greater than all peak points) there is some $n_3 > n_2$ such that $a_{n_3} \geq a_{n_2}$. Continuing in this way we obtain the desired (nondecreasing) subsequence. ▮

If we assume that our original sequence $\{a_n\}$ is bounded, we can pick up an extra corollary along the way.

COROLLARY (THE BOLZANO-WEIERSTRASS THEOREM) Every bounded sequence has a convergent subsequence.

Without some additional assumptions this is as far as we can go: it is easy to construct sequences having many, evenly infinitely many, subsequences converging to different numbers (see Problem 3). There is a reasonable assumption to add, which yields a necessary and sufficient condition for convergence of any sequence. Although this condition will not be crucial for our work, it does simplify many proofs. Moreover, this condition plays a fundamental role in more advanced investigations, and for this reason alone it is worth stating now.

If a sequence converges, so that the individual terms are eventually all close to the same number, then the difference of any two such individual terms should be very small. To be precise, if $\lim_{n\to\infty} a_n = l$ for some l, then for any $\varepsilon > 0$ there is an N such that $|a_n - l| < \varepsilon/2$ for $n > N$; now if both $n > N$ and $m > N$, then

$$|a_n - a_m| \le |a_n - l| + |l - a_m| < \frac{\varepsilon}{2} + \frac{\varepsilon}{2} = \varepsilon.$$

This final inequality, $|a_n - a_m| < \varepsilon$, which eliminates mention of the limit l, can be used to formulate a condition (the Cauchy condition) which is clearly necessary for convergence of a sequence.

DEFINITION

A sequence $\{a_n\}$ is a **Cauchy sequence** if for every $\varepsilon > 0$ there is a natural number N such that, for all m and n,

$$\text{if } m, n > N, \text{ then } |a_n - a_m| < \varepsilon.$$

(This condition is usually written $\lim_{m,n\to\infty} |a_m - a_n| = 0$.)

The beauty of the Cauchy condition is that it is also sufficient to ensure convergence of a sequence. After all our preliminary work, there is very little left to do in order to prove this.

THEOREM 3 A sequence $\{a_n\}$ converges if and only if it is a Cauchy sequence.

PROOF We have already shown that $\{a_n\}$ is a Cauchy sequence if it converges. The proof of the converse assertion contains only one tricky feature: showing that every Cauchy sequence $\{a_n\}$ is bounded. If we take $\varepsilon = 1$ in the definition of a Cauchy sequence we find that there is some N such that

$$|a_m - a_n| < 1 \qquad \text{for } m, n > N.$$

In particular, this means that

$$|a_m - a_{N+1}| < 1 \qquad \text{for all } m > N.$$

Thus $\{a_m : m > N\}$ is bounded; since there are only finitely many other a_i's the whole sequence is bounded.

The corollary to the Lemma thus implies that some subsequence of $\{a_n\}$ converges.

Only one point remains, whose proof will be left to you: if a subsequence of a Cauchy sequence converges, then the Cauchy sequence itself converges. ▮

PROBLEMS

1. Verify each of the following limits.

(i) $\displaystyle\lim_{n\to\infty} \frac{n}{n+1} = 1.$

(ii) $\displaystyle\lim_{n\to\infty} \frac{n+3}{n^3+4} = 0.$

(iii) $\displaystyle\lim_{n\to\infty} \sqrt[8]{n^2+1} - \sqrt[4]{n+1} = 0.$ Hint: You should at least be able to prove that $\displaystyle\lim_{n\to\infty} \sqrt[8]{n^2+1} - \sqrt[8]{n^2} = 0.$

(iv) $\displaystyle\lim_{n\to\infty} \frac{n!}{n^n} = 0.$ Hint: $n! = n(n-1)\cdot\ldots\cdot k!$ for $k < n$, in particular, for $k < n/2$.

(v) $\displaystyle\lim_{n\to\infty} \sqrt[n]{a} = 1, \qquad a > 0.$

(vi) $\displaystyle\lim_{n\to\infty} \sqrt[n]{n} = 1.$

(vii) $\displaystyle\lim_{n\to\infty} \sqrt[n]{n^2+n} = 1.$

(viii) $\displaystyle\lim_{n\to\infty} \sqrt[n]{a^n+b^n} = \max(a,b), \qquad a, b \geq 0.$

(ix) $\displaystyle\lim_{n\to\infty} \frac{\alpha(n)}{n} = 0$, where $\alpha(n)$ is the number of primes which divide n. Hint: The fact that each prime is ≥ 2 gives a very simple estimate of how small $\alpha(n)$ must be.

*(x) $$\lim_{n\to\infty} \frac{\displaystyle\sum_{k=1}^{n} k^p}{n^{p+1}} = \frac{1}{p+1}.$$

2. Find the following limits.

(i) $\displaystyle\lim_{n\to\infty} \frac{n}{n+1} - \frac{n+1}{n}.$

(ii) $\displaystyle\lim_{n\to\infty} n - \sqrt{n+a}\sqrt{n+b}.$

(iii) $\displaystyle\lim_{n\to\infty} \frac{2^n + (-1)^n}{2^{n+1} + (-1)^{n+1}}.$

(iv) $\displaystyle\lim_{n\to\infty} \frac{(-1)^n\sqrt{n}\sin(n^n)}{n+1}.$

(v) $\displaystyle\lim_{n\to\infty} \frac{a^n - b^n}{a^n + b^n}.$

(vi) $\displaystyle\lim_{n\to\infty} nc^n, \qquad |c| < 1.$

(vii) $\lim_{n\to\infty} \dfrac{2^{n^2}}{n!}$.

3. (a) What can be said about the sequence $\{a_n\}$ if it converges and each a_n is an integer?

(b) Find all convergent subsequences of the sequence $1, -1, 1, -1, 1, -1, \dots$. (There are infinitely many, although there are only two limits which such subsequences can have.)

(c) Find all convergent subsequences of the sequence $1, 2, 1, 2, 3, 1, 2, 3, 4, 1, 2, 3, 4, 5, \dots$. (There are infinitely many limits which such subsequences can have.)

(d) Consider the sequence

$$\tfrac{1}{2}, \tfrac{1}{3}, \tfrac{2}{3}, \tfrac{1}{4}, \tfrac{2}{4}, \tfrac{3}{4}, \tfrac{1}{5}, \tfrac{2}{5}, \tfrac{3}{5}, \tfrac{4}{5}, \tfrac{1}{6}, \dots.$$

For which numbers α is there a subsequence converging to α?

4. (a) Prove that if a subsequence of a Cauchy sequence converges, then so does the original Cauchy sequence.

(b) Prove that any subsequence of a convergent sequence converges.

5. (a) Prove that if $0 < a < 2$, then $a < \sqrt{2a} < 2$.

(b) Prove that the sequence

$$\sqrt{2},\ \sqrt{2\sqrt{2}},\ \sqrt{2\sqrt{2\sqrt{2}}},\ \dots$$

converges.

(c) Find the limit. Hint: Notice that if $\lim_{n\to\infty} a_n = l$, then $\lim_{n\to\infty} \sqrt{2a_n} = \sqrt{2l}$, by Theorem 1.

6. Let $0 < a_1 < b_1$ and define

$$a_{n+1} = \sqrt{a_n b_n}, \qquad b_{n+1} = \frac{a_n + b_n}{2}.$$

(a) Prove that the sequences $\{a_n\}$ and $\{b_n\}$ each converge.

(b) Prove that they have the same limit.

7. In Problem 2-16 we saw that any rational approximation m/n to $\sqrt{2}$ can be replaced by a better approximation $(m+2n)/(m+n)$. In particular, starting with $m = n = 1$, we obtain

$$1, \frac{3}{2}, \frac{7}{5}, \dots.$$

(a) Prove that this sequence is given recursively by

$$a_1 = 1, \qquad a_{n+1} = 1 + \frac{1}{1+a_n}.$$

(b) Prove that $\lim_{n\to\infty} a_n = \sqrt{2}$. This gives the so-called *continued fraction expansion*

$$\sqrt{2} = 1 + \cfrac{1}{2 + \cfrac{1}{2 + \cdots}}.$$

Hint: Consider separately the subsequences $\{a_{2n}\}$ and $\{a_{2n+1}\}$.

(c) Prove that for any natural numbers a and b,

$$\sqrt{a^2 + b} = a + \cfrac{b}{2a + \cfrac{b}{2a + \cdots}}.$$

8. Identify the function $f(x) = \lim_{n\to\infty} (\lim_{k\to\infty} (\cos n!\pi x)^{2k})$. (It has been mentioned many times in this book.)

9. Many impressive looking limits can be evaluated easily (especially by the person who makes them up), because they are really lower or upper sums in disguise. With this remark as hint, evaluate each of the following. (Warning: the list contains one red herring which can be evaluated by elementary considerations.)

(i) $\displaystyle \lim_{n\to\infty} \frac{\sqrt[n]{e} + \sqrt[n]{e^2} + \cdots + \sqrt[n]{e^n}}{n}.$

(ii) $\displaystyle \lim_{n\to\infty} \frac{\sqrt[n]{e} + \sqrt[n]{e^2} + \cdots + \sqrt[n]{e^{2n}}}{n}.$

(iii) $\displaystyle \lim_{n\to\infty} \left(\frac{1}{n+1} + \cdots + \frac{1}{2n}\right).$

(iv) $\displaystyle \lim_{n\to\infty} \left(\frac{1}{n^2} + \frac{1}{(n+1)^2} + \cdots + \frac{1}{(2n)^2}\right).$

(v) $\displaystyle \lim_{n\to\infty} \left(\frac{n}{(n+1)^2} + \frac{n}{(n+2)^2} + \cdots + \frac{n}{(n+n)^2}\right).$

(vi) $\displaystyle \lim_{n\to\infty} \left(\frac{n}{n^2+1} + \frac{n}{n^2+2^2} + \cdots + \frac{n}{n^2+n^2}\right).$

10. Although limits like $\lim_{n\to\infty} \sqrt[n]{n}$ and $\lim_{n\to\infty} a^n$ can be evaluated using facts about the behavior of the logarithm and exponential functions, this approach is vaguely dissatisfying, because integral roots and powers can be defined without using the exponential function. Some of the standard "elementary" arguments for such limits are outlined here; the basic tools are inequalities derived from the binomial theorem, notably

$$(1+h)^n \geq 1 + nh, \qquad \text{for } h > 0;$$

and, for part (e),

$$(1+h)^n \geq 1 + nh + \frac{n(n-1)}{2}h^2 \geq \frac{n(n-1)}{2}h^2, \qquad \text{for } h > 0.$$

(a) Prove that $\lim_{n\to\infty} a^n = \infty$ if $a > 1$, by setting $a = 1 + h$, where $h > 0$.
(b) Prove that $\lim_{n\to\infty} a^n = 0$ if $0 < a < 1$.
(c) Prove that $\lim_{n\to\infty} \sqrt[n]{a} = 1$ if $a > 1$, by setting $\sqrt[n]{a} = 1+h$ and estimating h.
(d) Prove that $\lim_{n\to\infty} \sqrt[n]{a} = 1$ if $0 < a < 1$.
(e) Prove that $\lim_{n\to\infty} \sqrt[n]{n} = 1$.

11. (a) Prove that a convergent sequence is always bounded.
(b) Suppose that $\lim_{n\to\infty} a_n = 0$, and that each $a_n > 0$. Prove that the set of all numbers a_n actually has a maximum member.

12. (a) Prove that

$$\frac{1}{n+1} < \log(n+1) - \log n < \frac{1}{n}.$$

(b) If

$$a_n = 1 + \frac{1}{2} + \frac{1}{3} + \cdots + \frac{1}{n} - \log n,$$

show that the sequence $\{a_n\}$ is decreasing, and that each $a_n \geq 0$. It follows that there is a number

$$\gamma = \lim_{n\to\infty} \left(1 + \cdots + \frac{1}{n} - \log n\right).$$

This number, known as Euler's number, has proved to be quite refractory; it is not even known whether γ is rational.

13. (a) Suppose that f is increasing on $[1, \infty)$. Show that

$$f(1) + \cdots + f(n-1) < \int_1^n f(x)\,dx < f(2) + \cdots + f(n).$$

(b) Now choose $f = \log$ and show that

$$\frac{n^n}{e^{n-1}} < n! < \frac{(n+1)^{n+1}}{e^n};$$

it follows that

$$\lim_{n\to\infty} \frac{\sqrt[n]{n!}}{n} = \frac{1}{e}.$$

This result shows that $\sqrt[n]{n!}$ is approximately n/e, in the sense that the ratio of these two quantities is close to 1 for large n. But we cannot conclude that $n!$ is close to $(n/e)^n$ in this sense; in fact, this is false. An estimate for $n!$ is very desirable, even for concrete computations, because $n!$ cannot be calculated easily even with logarithm tables. The standard (and difficult) theorem which provides the right information will be found in Problem 27-19.

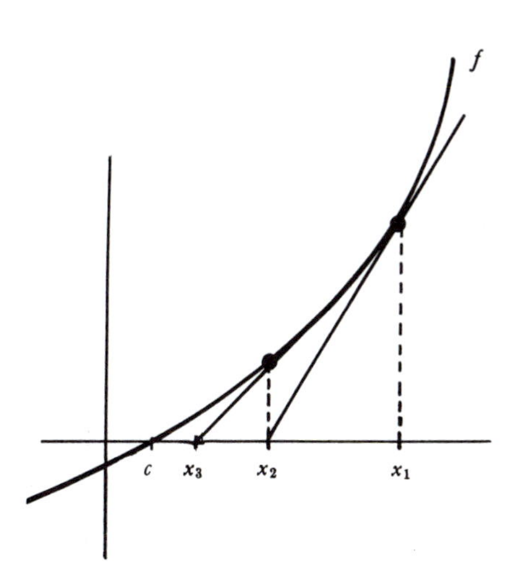

FIGURE 7

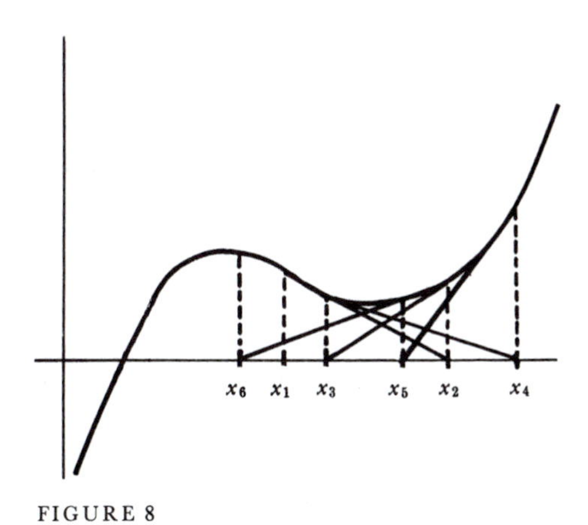

FIGURE 8

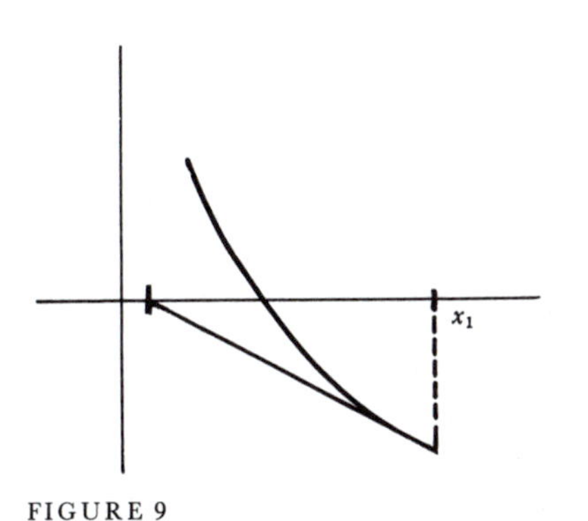

FIGURE 9

14. (a) Show that the tangent line to the graph of f at $(x_0, f(x_0))$ intersects the horizontal axis at $(x_1, 0)$, where

$$x_1 = x_0 - \frac{f(x_0)}{f'(x_0)}.$$

This intersection point may be regarded as a rough approximation to the point where the graph of f intersects the horizontal axis. If we now start at x_1 and repeat the process to get x_2, then use x_2 to get x_3, etc., we have a sequence $\{x_n\}$ defined inductively by

$$x_{n+1} = x_n - \frac{f(x_n)}{f'(x_n)}.$$

Figure 7 suggests that $\{x_n\}$ will converge to a number c with $f(c) = 0$; this is called *Newton's method* for finding a zero of f. In the remainder of this problem we will establish some conditions under which Newton's method works (Figures 8 and 9 show two cases where it doesn't). A few facts about convexity may be found useful; see Chapter 11, Appendix.

(b) Suppose that $f,' f'' > 0$, and that we choose x_0 with $f(x_0) > 0$. Show that $x_0 \geq x_1 \geq x_2 \geq \cdots \geq c$.

(c) Let $\delta_k = x_k - c$. Then

$$\delta_k = \frac{f(x_k)}{f'(\xi_k)}$$

for some ξ_k in (c, x_k). Show that

$$\delta_{k+1} = \frac{f(x_k)}{f'(\xi_k)} - \frac{f(x_k)}{f'(x_k)}.$$

Conclude that

$$\delta_{k+1} = \frac{f(x_k)}{f'(\xi_k)f'(x_k)} \cdot f''(\eta)(x_k - \xi_k)$$

for some η_k in (c, x_k), and then that

$$(*) \qquad \delta_{k+1} \leq \frac{f''(\eta_k)}{f'(x_k)}{\delta_k}^2.$$

(d) Let $m = \inf f'$ on $[c, x_1]$ and let $M = \sup |f''|$ on $[c, x_1]$. Show that Newton's method works if $x_0 - c < m/M$.

(e) What is the formula for x_{n+1} when $f(x) = x^2 - A$?

If we take $A = 2$ and $x_0 = 1.4$ we get

$$\begin{aligned} x_0 &= 1.4 \\ x_1 &= 1.4142857 \\ x_2 &= 1.4142136 \\ x_3 &= 1.4142136, \end{aligned}$$

which is already correct to 7 decimals! Notice that the number of correct decimals at least doubled each time. This is essentially guaranteed by the inequality $(*)$ when $M/m < 1$.

15. Use Newton's method to estimate the zeros of the following functions.

(i) $f(x) = \tan x - \cos^2 x$ near 0.
(ii) $f(x) = \cos x - x^2$ near 0.
(iii) $f(x) = x^3 + x - 1$ on $[0, 1]$.
(iv) $f(x) = x^3 - 3x^2 + 1$ on $[0, 1]$.

***16.** Prove that if $\lim_{n\to\infty} a_n = l$, then

$$\lim_{n\to\infty} \frac{(a_1 + \cdots + a_n)}{n} = l.$$

Hint: This problem is very similar to (in fact it is a special case of) Problem 13-40.

17. Suppose that f is continuous and $\lim_{x\to\infty} f(x+1) - f(x) = 0$. Prove that $\lim_{x\to\infty} f(x)/x = 0$. Hint: See the previous problem.

***18.** Suppose that $a_n > 0$ for each n and that $\lim_{n\to\infty} a_{n+1}/a_n = l$. Prove that $\lim_{n\to\infty} \sqrt[n]{a_n} = l$. Hint: This requires the same sort of argument that works in Problem 16, together with the fact that $\lim_{n\to\infty} \sqrt[n]{a} = 1$, for $a > 0$.

19. (a) Suppose that $\{a_n\}$ is a convergent sequence of points all in $[0, 1]$. Prove that $\lim_{n\to\infty} a_n$ is also in $[0, 1]$.
(b) Find a convergent sequence $\{a_n\}$ of points all in $(0, 1)$ such that $\lim_{n\to\infty} a_n$ is not in $(0, 1)$.

20. Suppose that f is continuous and that the sequence

$$x,\ f(x),\ f(f(x)),\ f(f(f(x))),\ \ldots$$

converges to l. Prove that l is a "fixed point" for f, i.e., $f(l) = l$. Hint: Two special cases have occurred already.

21. (a) Suppose that f is continuous on $[0, 1]$ and that $0 \le f(x) \le 1$ for all x in $[0, 1]$. Problem 7-11 shows that f has a fixed point (in the terminology of Problem 20). If f is *increasing*, a much stronger statement can be made: For any x in $[0, 1]$, the sequence

$$x,\ f(x),\ f(f(x)), \ldots$$

has a limit (which is necessarily a fixed point, by Problem 20). Prove this assertion, by examining the behavior of the sequence for $f(x) > x$ and $f(x) < x$, or by looking at Figure 10. A diagram of this sort is used in Littlewood's *Mathematician's Miscellany* to preach the value of drawing pictures: "For the professional the only proof needed is [this Figure]."

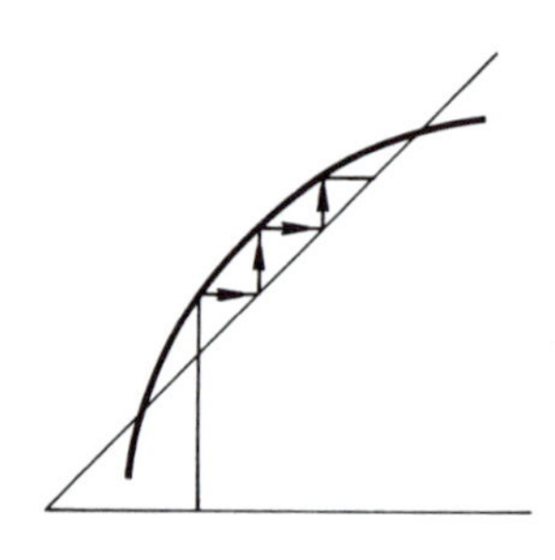

FIGURE 10

*(b) Suppose that f and g are two continuous functions on $[0, 1]$, with $0 \le f(x) \le 1$ and $0 \le g(x) \le 1$ for all x in $[0, 1]$, which satisfy $f \circ g = g \circ f$. Suppose, moreover, that f is increasing. Show that f and g have

a common fixed point; in other words, there is a number l such that $f(l) = l = g(l)$. Hint: Begin by choosing a fixed point for g.

For a long time mathematicians amused themselves by asking whether the conclusion of part (b) holds without the assumption that f is increasing, but two independent announcements in the *Notices* of the American Mathematical Society, Volume 14, Number 2 give counterexamples, so it was probably a pretty silly problem all along.

The trick in Problem 20 is really much more valuable than Problem 20 might suggest, and some of the most important "fixed point theorems" depend upon looking at sequences of the form $x, f(x), f(f(x)), \ldots$. A special, but representative, case of one such theorem is treated in Problem 23 (for which the next problem is preparation).

22. (a) Use Problem 2-5 to show that if $c \neq 1$, then

$$c^m + c^{m+1} + \cdots + c^n = \frac{c^m - c^{n+1}}{1 - c}.$$

(b) Suppose that $|c| < 1$. Prove that

$$\lim_{m,n\to\infty} c^m + \cdots + c^n = 0.$$

(c) Suppose that $\{x_n\}$ is a sequence with $|x_n - x_{n+1}| \leq c^n$, where $c < 1$. Prove that $\{x_n\}$ is a Cauchy sequence.

***23.** Suppose that f is a function on $\mathbf{R}$ such that

$$(*) \qquad |f(x) - f(y)| \leq c|x - y|, \qquad \text{for all } x \text{ and } y,$$

where $c < 1$. (Such a function is called a *contraction*.)

(a) Prove that f is continuous.
(b) Prove that f has at most one fixed point.
(c) By considering the sequence

$$x,\ f(x),\ f(f(x)),\ \ldots,$$

for any x, prove that f does have a fixed point. (This result, in a more general setting, is known as the "contraction lemma.")

24. (a) Prove that if f is differentiable and $|f'| < 1$, then f has at most one fixed point.
(b) Prove that if $|f'(x)| \leq c < 1$ for all x, then f has a fixed point.
(c) Give an example to show that the hypothesis $|f'(x)| \leq 1$ is not sufficient to insure that f has a fixed point.

25. This problem is a sort of converse to the previous problem. Let b_n be a sequence defined by $b_1 = a$, $b_{n+1} = f(b_n)$. Prove that if $b = \lim_{n\to\infty} b_n$ exists and f' is continuous at b, then $|f'(b)| \leq 1$. Hint: If $|f'(b)| > 1$, then

$|f'(x)| > 1$ for all x in an interval around b, and b_n will be in this interval for large enough n. Now consider f on the interval $[b, b_n]$.

26. This problem investigates for which $a > 0$ the symbol

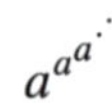

makes sense. In other words, if we define $b_1 = a$, $b_{n+1} = a^{b_n}$, when does $b = \lim_{n\to\infty} b_n$ exist?

(a) Prove that if b exists, then $a^b = b$. (The situation is similar to that in Problem 5.)

(b) According to part (a), if b exists, then a can be written in the form $y^{1/y}$ for some y. Describe the graph of $g(y) = y^{1/y}$ and conclude that $0 < a \le e^{1/e}$.

(c) Suppose that $1 \le a \le e^{1/e}$. Show that $\{b_n\}$ is increasing and also $b_n \le e$. This proves that b exists (and also that $b \le e$).

The analysis for $a < 1$ is more difficult.

(d) Using Problem 25, show that if b exists, then $e^{-1} \le b \le e$. Then show that $e^{-e} \le a \le e^{1/e}$.

From now on we will suppose that $e^{-e} \le a < 1$.

(e) Show that the function

$$f(x) = \frac{a^x}{\log x}$$

is decreasing on the interval $(0, 1)$.

(f) Let b be the unique number such that $a^b = b$. Show that $a < b < 1$. Using part (e), show that if $0 < x < b$, then $x < a^{a^x} < b$. Conclude that $l = \lim_{n\to\infty} a_{2n+1}$ exists and that $a^{a^l} = l$.

(g) Using part (e) again, show that $l = b$.

(h) Finally, show that $\lim_{n\to\infty} a_{2n+2} = b$, so that $\lim_{n\to\infty} b_n = b$.

27. Let $\{x_n\}$ be a sequence which is bounded, and let

$$y_n = \sup\{x_n, x_{n+1}, x_{n+2}, \dots\}.$$

(a) Prove that the sequence $\{y_n\}$ converges. The limit $\lim_{n\to\infty} y_n$ is denoted by $\overline{\lim_{n\to\infty}}\, x_n$ or $\limsup_{n\to\infty} x_n$, and called the **limit superior**, or **upper limit**, of the sequence $\{x_n\}$.

(b) Find $\overline{\lim_{n\to\infty}}\, x_n$ for each of the following:

(i) $x_n = \dfrac{1}{n}$.

(ii) $x_n = (-1)^n \dfrac{1}{n}$.

(iii) $x_n = (-1)^n \left[1 + \frac{1}{n}\right]$.

(iv) $x_n = \sqrt[n]{n}$.

(c) Define $\underline{\lim}_{n\to\infty} x_n$ (or $\liminf_{n\to\infty} x_n$) and prove that

$$\underline{\lim}_{n\to\infty} x_n \le \overline{\lim}_{n\to\infty} x_n.$$

(d) Prove that $\lim_{n\to\infty} x_n$ exists if and only if $\overline{\lim}_{n\to\infty} x_n = \underline{\lim}_{n\to\infty} x_n$ and that in this case $\lim_{n\to\infty} x_n = \overline{\lim}_{n\to\infty} x_n = \underline{\lim}_{n\to\infty} x_n$.

(e) Recall the definition, in Problem 8-18, of $\overline{\lim}\, A$ for a bounded set A. Prove that if the numbers x_n are distinct, then $\overline{\lim}_{n\to\infty} x_n = \overline{\lim}\, A$, where $A = \{x_n : n \text{ in } \mathbf{N}\}$.

28. In the Appendix to Chapter 8 we defined uniform continuity of a function on an interval. If $f(x)$ is defined only for rational x, this concept still makes sense: we say that f is uniformly continuous on an interval if for every $\varepsilon > 0$ there is some $\delta > 0$ such that, if x and y are rational numbers in the interval and $|x - y| < \delta$, then $|f(x) - f(y)| < \varepsilon$.

(a) Let x be any (rational or irrational) point in the interval, and let $\{x_n\}$ be a sequence of *rational* points in the interval such that $\lim_{n\to\infty} x_n = x$. Show that the sequence $\{f(x_n)\}$ converges.

(b) Prove that the limit of the sequence $\{f(x_n)\}$ doesn't depend on the choice of the sequence $\{x_n\}$.

We will denote this limit by $\bar{f}(x)$, so that $\bar{f}$ is an extension of f to the whole interval.

(c) Prove that the extended function $\bar{f}$ is uniformly continuous on the interval.

29. Let $a > 0$, and for rational x let $f(x) = a^x$, as defined in the usual elementary algebraic way. This problem shows directly that f can be extended to a continuous function $\bar{f}$ on the whole line. Problem 28 provides the necessary machinery.

(a) Show that $a^x < a^y$ for rational $x < y$.

(b) Using Problem 10, show that for any $\varepsilon > 0$ we have $|a^x - 1| < \varepsilon$ for rational numbers x close enough to 0.

(c) Using the equation $a^x - a^y = a^y(a^{x+y} - 1)$, prove that on any closed interval f is uniformly continuous, in the sense of Problem 28.

(d) Show that the extended function $\bar{f}$ of Problem 28 is increasing and satisfies $\bar{f}(x + y) = \bar{f}(x)\bar{f}(y)$.

***30.** The Bolzano-Weierstrass Theorem is usually stated, and also proved, quite differently than in the text—the classical statement uses the notion of limit points. A point x is a **limit point** of the set A if for every $\varepsilon > 0$ there is a point a in A with $|x - a| < \varepsilon$ but $x \neq a$.

(a) Find all limit points of the following sets.

(i) $\left\{\frac{1}{n} : n \text{ in } \mathbf{N}\right\}$.

(ii) $\left\{\frac{1}{n} + \frac{1}{m} : n \text{ and } m \text{ in } \mathbf{N}\right\}$.

(iii) $\left\{(-1)^n\left[1 + \frac{1}{n}\right] : n \text{ in } \mathbf{N}\right\}$.

(iv) $\mathbf{Z}$.

(v) $\mathbf{Q}$.

(b) Prove that x is a limit point of A if and only if for every $\varepsilon > 0$ there are infinitely many points a of A satisfying $|x - a| < \varepsilon$.

(c) Prove that $\overline{\lim}\, A$ is the largest limit point of A, and $\underline{\lim}\, A$ the smallest.

The usual form of the Bolzano-Weierstrass Theorem states that if A is an infinite set of numbers contained in a closed interval $[a, b]$, then some point of $[a, b]$ is a limit point of A. Prove this in two ways:

(d) Using the form already proved in the text. Hint: Since A is infinite, there are distinct numbers $x_1, x_2, x_3, \ldots$ in A.

(e) Using the Nested Intervals Theorem. Hint: If $[a, b]$ is divided into two intervals, at least one must contain infinitely many points of A.

31. (a) Use the Bolzano-Weierstrass Theorem to prove that if f is continuous on $[a, b]$, then f is bounded above on $[a, b]$. Hint: If f is not bounded above, then there are points x_n in $[a, b]$ with $f(x_n) > n$.

(b) Also use the Bolzano-Weierstrass Theorem to prove that if f is continuous on $[a, b]$, then f is uniformly continuous on $[a, b]$ (see Chapter 8, Appendix).

****32.** (a) Let $\{a_n\}$ be the sequence

$$\frac{1}{2}, \frac{1}{3}, \frac{2}{3}, \frac{1}{4}, \frac{2}{4}, \frac{3}{4}, \frac{1}{5}, \frac{2}{5}, \frac{3}{5}, \frac{4}{5}, \frac{1}{6}, \frac{2}{6}, \ldots.$$

Suppose that $0 \le a < b \le 1$. Let $N(n; a, b)$ be the number of integers $j \le n$ such that a_j is in $[a, b]$. (Thus $N(2; \frac{1}{3}, \frac{2}{3}) = 2$, and $N(4; \frac{1}{3}, \frac{2}{3}) = 3$.) Prove that

$$\lim_{n\to\infty} \frac{N(n; a, b)}{n} = b - a.$$

(b) A sequence $\{a_n\}$ of numbers in $[0, 1]$ is called **uniformly distributed** in $[0, 1]$ if

$$\lim_{n\to\infty} \frac{N(n; a, b)}{n} = b - a$$

for all a and b with $0 \le a < b \le 1$. Prove that if s is a step function defined on $[0, 1]$, and $\{a_n\}$ is uniformly distributed in $[0, 1]$, then

$$\int_0^1 s = \lim_{n\to\infty} \frac{s(a_1) + \cdots + s(a_n)}{n}.$$

(c) Prove that if $\{a_n\}$ is uniformly distributed in $[0, 1]$ and f is integrable on $[0, 1]$, then

$$\int_0^1 f = \lim_{n\to\infty} \frac{f(a_1) + \cdots + f(a_n)}{n}.$$

****33.** (a) Let f be a function defined on $[0, 1]$ such that $\lim_{y\to a} f(y)$ exists for all a in $[0, 1]$. For any $\varepsilon > 0$ prove that there are only finitely many points a in $[0, 1]$ with $|\lim_{y\to a} f(y) - f(a)| > \varepsilon$. Hint: Show that the set of such points cannot have a limit point x, by showing that $\lim_{y\to x} f(y)$ could not exist.

(b) Prove that, in the terminology of Problem 21-5, the set of points where f is discontinuous is countable. This finally answers the question of Problem 6-16: If f has only removable discontinuities, then f is continuous except at a countable set of points, and in particular, f cannot be discontinuous everywhere.

CHAPTER 23 INFINITE SERIES

Infinite sequences were introduced in the previous chapter with the specific intention of considering their "sums"

$$a_1 + a_2 + a_3 + \cdots$$

in this chapter. This is not an entirely straightforward matter, for the sum of infinitely many numbers is as yet completely undefined. What can be defined are the "partial sums"

$$s_n = a_1 + \cdots + a_n,$$

and the infinite sum must presumably be defined in terms of these partial sums. Fortunately, the mechanism for formulating this definition has already been developed in the previous chapter. If there is to be any hope of computing the infinite sum $a_1 + a_2 + a_3 + \cdots$, the partial sums s_n should represent closer and closer approximations as n is chosen larger and larger. This last assertion amounts to little more than a sloppy definition of limits: the "infinite sum" $a_1 + a_2 + a_3 + \cdots$ ought to be $\lim_{n\to\infty} s_n$. This approach will necessarily leave the "sum" of many sequences undefined, since the sequence $\{s_n\}$ may easily fail to have a limit. For example, the sequence

$$1,\ -1,\ 1,\ -1,\ \ldots$$

with $a_n = (-1)^{n+1}$ yields the new sequence

$$\begin{aligned} s_1 &= a_1 = 1, \\ s_2 &= a_1 + a_2 = 0, \\ s_3 &= a_1 + a_2 + a_3 = 1, \\ s_4 &= a_1 + a_2 + a_3 + a_4 = 0, \\ &\ldots, \end{aligned}$$

for which $\lim_{n\to\infty} s_n$ does not exist. Although there happen to be some clever extensions of the definition suggested here (see Problems 9 and 24-20) it seems unavoidable that some sequences will have no sum. For this reason, an acceptable definition of the sum of a sequence should contain, as an essential component, terminology which distinguishes sequences for which sums can be defined from less fortunate sequences.

DEFINITION

The sequence $\{a_n\}$ is **summable** if the sequence $\{s_n\}$ converges, where

$$s_n = a_1 + \cdots + a_n.$$

In this case, $\lim_{n \to \infty} s_n$ is denoted by

$$\sum_{n=1}^{\infty} a_n \qquad \text{(or, less formally, } a_1 + a_2 + a_3 + \cdots\text{)}$$

and is called the **sum** of the sequence $\{a_n\}$.

The terminology introduced in this definition is usually replaced by less precise expressions; indeed the title of this chapter is derived from such everyday language. An infinite sum $\sum_{n=1}^{\infty} a_n$ is usually called an *infinite series*, the word "series" emphasizing the connection with the infinite sequence $\{a_n\}$. The statement that $\{a_n\}$ is, or is not, summable is conventionally replaced by the statement that the series $\sum_{n=1}^{\infty} a_n$ does, or does not, converge. This terminology is somewhat peculiar, because at best the symbol $\sum_{n=1}^{\infty} a_n$ denotes a number (so it can't "converge"), and it doesn't denote anything at all unless $\{a_n\}$ is summable. Nevertheless, this informal language is convenient, standard, and unlikely to yield to attacks on logical grounds.

Certain elementary arithmetical operations on infinite series are direct consequences of the definition. It is a simple exercise to show that if $\{a_n\}$ and $\{b_n\}$ are summable, then

$$\sum_{n=1}^{\infty} (a_n + b_n) = \sum_{n=1}^{\infty} a_n + \sum_{n=1}^{\infty} b_n,$$

$$\sum_{n=1}^{\infty} c \cdot a_n = c \cdot \sum_{n=1}^{\infty} a_n.$$

As yet these equations are not very interesting, since we have no examples of summable sequences (except for the trivial examples in which the terms are eventually all 0). Before we actually exhibit a summable sequence, some general conditions for summability will be recorded.

There is one necessary and sufficient condition for summability which can be stated immediately. The sequence $\{a_n\}$ is summable if and only if the sequence $\{s_n\}$ converges, which happens, according to Theorem 22-3, if and only if $\lim_{m,n \to \infty} s_m - s_n = 0$; this condition can be rephrased in terms of the original sequence as follows.

THE CAUCHY CRITERION

The sequence $\{a_n\}$ is summable if and only if

$$\lim_{m,n\to\infty} a_{n+1} + \cdots + a_m = 0.$$

Although the Cauchy criterion is of theoretical importance, it is not very useful for deciding the summability of any particular sequence. However, one simple consequence of the Cauchy criterion provides a *necessary* condition for summability which is too important not to be mentioned explicitly.

THE VANISHING CONDITION

If $\{a_n\}$ is summable, then

$$\lim_{n\to\infty} a_n = 0.$$

This condition follows from the Cauchy criterion by taking $m = n + 1$; it can also be proved directly as follows. If $\lim_{n\to\infty} s_n = l$, then

$$\begin{aligned}\lim_{n\to\infty} a_n &= \lim_{n\to\infty}(s_n - s_{n-1}) = \lim_{n\to\infty} s_n - \lim_{n\to\infty} s_{n-1}\\ &= l - l = 0.\end{aligned}$$

Unfortunately, this condition is far from sufficient. For example, $\lim_{n\to\infty} 1/n = 0$, but the sequence $\{1/n\}$ is not summable; in fact, the following grouping of the numbers $1/n$ shows that the sequence $\{s_n\}$ is not bounded:

$$1 + \tfrac{1}{2} + \underbrace{\tfrac{1}{3} + \tfrac{1}{4}}_{\substack{\geq \frac{1}{2}\\ \text{(2 terms,}\\ \text{each } \geq \frac{1}{4})}} + \underbrace{\tfrac{1}{5} + \tfrac{1}{6} + \tfrac{1}{7} + \tfrac{1}{8}}_{\substack{\geq \frac{1}{2}\\ \text{(4 terms,}\\ \text{each } \geq \frac{1}{8})}} + \underbrace{\tfrac{1}{9} + \cdots + \tfrac{1}{16}}_{\substack{\geq \frac{1}{2}\\ \text{(8 terms,}\\ \text{each } \geq \frac{1}{16})}} + \cdots.$$

The method of proof used in this example, a clever trick which one might never see, reveals the need for some more standard methods for attacking these problems. These methods shall be developed soon (one of them will give an alternate proof that $\sum_{n=1}^{\infty} 1/n$ does not converge) but it will be necessary to first procure a few examples of convergent series.

The most important of all infinite series are the "geometric series"

$$\sum_{n=0}^{\infty} r^n = 1 + r + r^2 + r^3 + \cdots.$$

Only the cases $|r| < 1$ are interesting, since the individual terms do not approach 0 if $|r| \geq 1$. These series can be managed because the partial sums

$$s_n = 1 + r + \cdots + r^n$$

can be evaluated in simple terms. The two equations

$$\begin{aligned} s_n &= 1 + r + r^2 + \cdots + r^n\\ rs_n &= \phantom{1 +{}} r + r^2 + \cdots + r^n + r^{n+1}\end{aligned}$$

lead to

$$s_n(1-r) = 1 - r^{n+1}$$

or

$$s_n = \frac{1-r^{n+1}}{1-r}$$

(division by $1 - r$ is valid since we are not considering the case $r = 1$). Now $\lim_{n\to\infty} r^n = 0$, since $|r| < 1$. It follows that

$$\sum_{n=0}^{\infty} r^n = \lim_{n\to\infty} \frac{1-r^{n+1}}{1-r} = \frac{1}{1-r}, \qquad |r| < 1.$$

In particular,

$$\sum_{n=1}^{\infty} \left(\frac{1}{2}\right)^n = \sum_{n=0}^{\infty} \left(\frac{1}{2}\right)^n - 1 = \frac{1}{1-\frac{1}{2}} - 1 = 1,$$

that is,

$$\tfrac{1}{2} + \tfrac{1}{4} + \tfrac{1}{8} + \tfrac{1}{16} + \cdots = 1,$$

an infinite sum which can always be remembered from the picture in Figure 1.

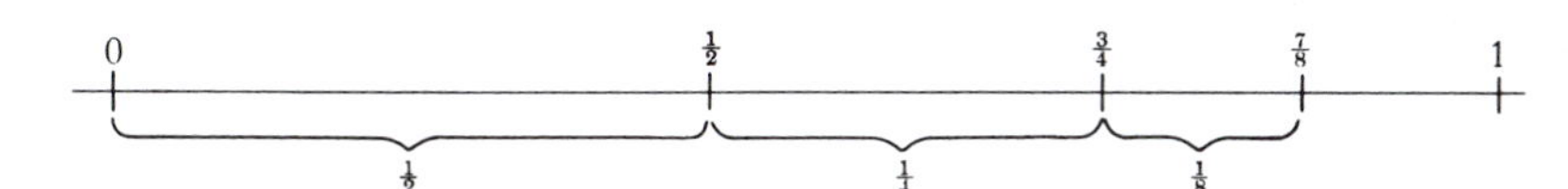

FIGURE 1

Special as they are, geometric series are standard examples from which important tests for summability will be derived.

For a while we shall consider only sequences $\{a_n\}$ with each $a_n \geq 0$; such sequences are called **nonnegative**. If $\{a_n\}$ is a nonnegative sequence, then the sequence $\{s_n\}$ is clearly nondecreasing. This remark, combined with Theorem 22-2, provides a simple-minded test for summability:

THE BOUNDEDNESS CRITERION

A nonnegative sequence $\{a_n\}$ is summable if and only if the set of partial sums s_n is bounded.

By itself, this criterion is not very helpful—deciding whether or not the set of all s_n is bounded is just what we are unable to do. On the other hand, if some convergent series are already available for comparison, this criterion can be used to obtain a result whose simplicity belies its importance (it is the basis for almost all other tests).

THEOREM 1 (THE COMPARISON TEST)

Suppose that

$$0 \leq a_n \leq b_n \qquad \text{for all } n.$$

Then if $\sum_{n=1}^{\infty} b_n$ converges, so does $\sum_{n=1}^{\infty} a_n$.

PROOF If

$$s_n = a_1 + \cdots + a_n,$$
$$t_n = b_1 + \cdots + b_n,$$

then

$$0 \leq s_n \leq t_n \qquad \text{for all } n.$$

Now $\{t_n\}$ is bounded, since $\sum_{n=1}^{\infty} b_n$ converges. Therefore $\{s_n\}$ is bounded; consequently, by the boundedness criterion $\sum_{n=1}^{\infty} a_n$ converges. ▮

Quite frequently the comparison test can be used to analyze very complicated looking series in which most of the complication is irrelevant. For example,

$$\sum_{n=1}^{\infty} \frac{2 + \sin^3(n+1)}{2^n + n^2}$$

converges because

$$0 \leq \frac{2 + \sin^3(n+1)}{2^n + n^2} < \frac{3}{2^n},$$

and

$$\sum_{n=1}^{\infty} \frac{3}{2^n} = 3 \sum_{n=1}^{\infty} \frac{1}{2^n}$$

is a convergent (geometric) series.

Similarly, we would expect the series

$$\sum_{n=1}^{\infty} \frac{1}{2^n - 1 + \sin^2 n^3}$$

to converge, since the nth term of the series is practically $1/2^n$ for large n, and we would expect the series

$$\sum_{n=1}^{\infty} \frac{n+1}{n^2+1}$$

to diverge, since $(n+1)/(n^2+1)$ is practically $1/n$ for large n. These facts can be derived immediately from the following theorem, another kind of "comparison test."

THEOREM 2 If $a_n, b_n > 0$ and $\lim_{n \to \infty} a_n/b_n = c \neq 0$, then $\sum_{n=1}^{\infty} a_n$ converges if and only if $\sum_{n=1}^{\infty} b_n$ converges.

PROOF Suppose $\sum_{n=1}^{\infty} b_n$ converges. Since $\lim_{n\to\infty} a_n/b_n = c$, there is some N such that

$$a_n \le 2cb_n \qquad \text{for } n \ge N.$$

But the sequence $2c\sum_{n=N}^{\infty} b_n$ certainly converges. Then Theorem 1 shows that $\sum_{n=N}^{\infty} a_n$ converges, and this implies convergence of the whole series $\sum_{n=1}^{\infty} a_n$, which has only finitely many additional terms.

The converse follows immediately, since we also have $\lim_{n\to\infty} b_n/a_n = 1/c \neq 0$. ▮

The comparison test yields other important tests when we use previously analyzed series as catalysts. Choosing the geometric series $\sum_{n=0}^{\infty} r^n$, the convergent series *par excellence*, we obtain the most important of all tests for summability.

THEOREM 3 (THE RATIO TEST) Let $a_n > 0$ for all n, and suppose that

$$\lim_{n\to\infty} \frac{a_{n+1}}{a_n} = r.$$

Then $\sum_{n=1}^{\infty} a_n$ converges if $r < 1$. On the other hand, if $r > 1$, then the terms a_n do not approach 0, so $\sum_{n=1}^{\infty} a_n$ diverges. (Notice that it is therefore essential to compute $\lim_{n\to\infty} a_{n+1}/a_n$ and not $\lim_{n\to\infty} a_n/a_{n+1}$!)

PROOF Suppose first that $r < 1$. Choose any number s with $r < s < 1$. The hypothesis

$$\lim_{n\to\infty} \frac{a_{n+1}}{a_n} = r < 1$$

implies that there is some N such that

$$\frac{a_{n+1}}{a_n} \le s \qquad \text{for } n \ge N.$$

This can be written

$$a_{n+1} \le sa_n \qquad \text{for } n \ge N.$$

Thus

$$\begin{aligned}
&a_{N+1} \le sa_N,\\
&a_{N+2} \le sa_{N+1} \le s^2 a_N,\\
&\quad\vdots\\
&a_{N+k} \le s^k a_N.
\end{aligned}$$

Since $\sum_{k=0}^{\infty} a_N s^k = a_N \sum_{k=0}^{\infty} s^k$ converges, the comparison test shows that

$$\sum_{n=N}^{\infty} a_n = \sum_{k=0}^{\infty} a_{N+k}$$

converges. This implies the convergence of the whole series $\sum_{n=1}^{\infty} a_n$.

The case $r > 1$ is even easier. If $1 < s < r$, then there is a number N such that

$$\frac{a_{n+1}}{a_n} \geq s \qquad \text{for } n \geq N,$$

which means that

$$a_{N+k} \geq a_N s^k \geq a_N \qquad k = 0, 1, \ldots .$$

This shows that the individual terms of $\{a_n\}$ do not approach 0, so $\{a_n\}$ is not summable. ▮

As a simple application of the ratio test, consider the series $\sum_{n=1}^{\infty} 1/n!$. Letting $a_n = 1/n!$ we obtain

$$\frac{a_{n+1}}{a_n} = \frac{\dfrac{1}{(n+1)!}}{\dfrac{1}{n!}} = \frac{n!}{(n+1)!} = \frac{1}{n+1}.$$

Thus

$$\lim_{n\to\infty} \frac{a_{n+1}}{a_n} = 0,$$

which shows that the series $\sum_{n=1}^{\infty} 1/n!$ converges. If we consider instead the series $\sum_{n=1}^{\infty} r^n/n!$, where r is some fixed positive number, then

$$\lim_{n\to\infty} \frac{\dfrac{r^{n+1}}{(n+1)!}}{\dfrac{r^n}{n!}} = \lim_{n\to\infty} \frac{r}{n+1} = 0,$$

so $\sum_{n=1}^{\infty} r^n/n!$ converges. It follows that

$$\lim_{n\to\infty} \frac{r^n}{n!} = 0,$$

a result already proved in Chapter 16 (the proof given there was based on the same ideas as those used in the ratio test). Finally, if we consider the series $\sum_{n=1}^{\infty} nr^n$ we have

$$\lim_{n\to\infty} \frac{(n+1)r^{n+1}}{nr^n} = \lim_{n\to\infty} r \cdot \frac{n+1}{n} = r,$$

since $\lim_{n\to\infty} (n+1)/n = 1$. This proves that if $0 \le r < 1$, then $\sum_{n=1}^{\infty} nr^n$ converges, and consequently

$$\lim_{n\to\infty} nr^n = 0.$$

(This result clearly holds for $-1 < r \le 0$, also.) It is a useful exercise to provide a direct proof of this limit, without using the ratio test as an intermediary.

Although the ratio test will be of the utmost theoretical importance, as a practical tool it will frequently be found disappointing. One drawback of the ratio test is the fact that $\lim_{n\to\infty} a_{n+1}/a_n$ may be quite difficult to determine, and may not even exist. A more serious deficiency, which appears with maddening regularity, is the fact that the limit might equal 1. The case $\lim_{n\to\infty} a_{n+1}/a_n = 1$ is precisely the one which is inconclusive: $\{a_n\}$ might not be summable (for example, if $a_n = 1/n$), but then again it might be. In fact, our very next test will show that $\sum_{n=1}^{\infty}(1/n)^2$ converges, even though

$$\lim_{n\to\infty} \frac{\left(\dfrac{1}{n+1}\right)^2}{\left(\dfrac{1}{n}\right)^2} = 1.$$

This test provides a quite different method for determining convergence or divergence of infinite series—like the ratio test, it is an immediate consequence of the comparison test, but the series chosen for comparison is quite novel.

THEOREM 4 (THE INTEGRAL TEST) Suppose that f is positive and decreasing on $[1, \infty)$, and that $f(n) = a_n$ for all n. Then $\sum_{n=1}^{\infty} a_n$ converges if and only if the limit

$$\int_1^{\infty} f = \lim_{A\to\infty} \int_1^A f$$

exists

PROOF The existence of $\lim_{A\to\infty} \int_1^A f$ is equivalent to convergence of the series

$$\int_1^2 f + \int_2^3 f + \int_3^4 f + \cdots.$$

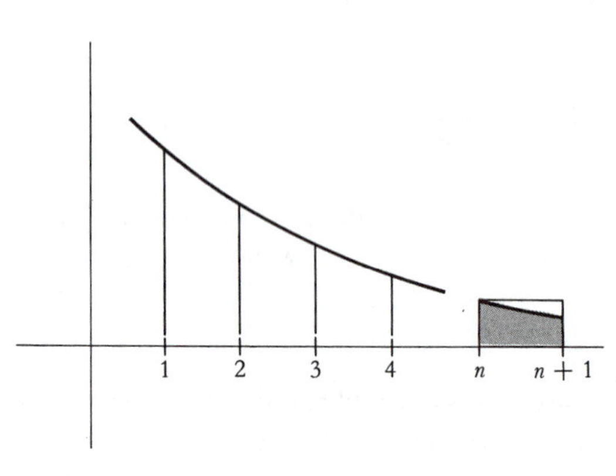

FIGURE 2

Now, since f is decreasing we have (Figure 2)

$$f(n+1) < \int_n^{n+1} f < f(n).$$

The first half of this double inequality shows that the series $\sum_{n=1}^{\infty} a_{n+1}$ may be compared to the series $\sum_{n=1}^{\infty} \int_n^{n+1} f$, proving that $\sum_{n=1}^{\infty} a_{n+1}$ (and hence $\sum_{n=1}^{\infty} a_n$) converges if $\lim_{A\to\infty} \int_1^A$ exists.

The second half of the inequality shows that the series $\sum_{n=1}^{\infty} \int_n^{n+1} f$ may be compared to the series $\sum_{n=1}^{\infty} a_n$, proving that $\lim_{A\to\infty} \int_1^A f$ must exist if $\sum_{n=1}^{\infty} a_n$ converges. ▌

Only one example using the integral test will be given here, but it settles the question of convergence for infinitely many series at once. If $p > 0$, the convergence of $\sum_{n=1}^{\infty} 1/n^p$ is equivalent, by the integral test, to the existence of

$$\int_1^{\infty} \frac{1}{x^p}\,dx.$$

Now

$$\int_1^A \frac{1}{x^p}\,dx = \begin{cases} -\dfrac{1}{(p-1)} \cdot \dfrac{1}{A^{p-1}} + \dfrac{1}{p-1}, & p \neq 1 \\ \log A, & p = 1. \end{cases}$$

This shows that $\lim_{A\to\infty} \int_1^A 1/x^p\,dx$ exists if $p > 1$, but not if $p \leq 1$. Thus $\sum_{n=1}^{\infty} 1/n^p$ converges precisely for $p > 1$. In particular, $\sum_{n=1}^{\infty} 1/n$ diverges.

The tests considered so far apply only to nonnegative sequences, but nonpositive sequences may be handled in precisely the same way. In fact, since

$$\sum_{n=1}^{\infty} a_n = -\left(\sum_{n=1}^{\infty} -a_n\right),$$

all considerations about nonpositive sequences can be reduced to questions involving nonnegative sequences. Sequences which contain both positive and negative terms are quite another story.

If $\sum_{n=1}^{\infty} a_n$ is a sequence with both positive and negative terms, one can consider instead the sequence $\sum_{n=1}^{\infty} |a_n|$, all of whose terms are nonnegative. Cheerfully ignoring the possibility that we may have thrown away all the interesting information about the original sequence, we proceed to eulogize those sequences which are converted by this procedure into convergent sequences.

DEFINITION

The series $\sum_{n=1}^{\infty} a_n$ is **absolutely convergent** if the series $\sum_{n=1}^{\infty} |a_n|$ is convergent. (In more formal language, the sequence $\{a_n\}$ is **absolutely summable** if the sequence $\{|a_n|\}$ is summable.)

Although we have no right to expect this definition to be of any interest, it turns out to be exceedingly important. The following theorem shows that the definition is at least not entirely useless.

THEOREM 5 Every absolutely convergent series is convergent. Moreover, a series is absolutely convergent if and only if the series formed from its positive terms and the series formed from its negative terms both converge.

PROOF If $\sum_{n=1}^{\infty} |a_n|$ converges, then, by the Cauchy criterion,

$$\lim_{m,n\to\infty} |a_{n+1}| + \cdots + |a_m| = 0.$$

Since

$$|a_{n+1} + \cdots + a_m| \le |a_{n+1}| + \cdots + |a_m|,$$

it follows that

$$\lim_{m,n\to\infty} a_{n+1} + \cdots + a_m = 0,$$

which shows that $\sum_{n=1}^{\infty} a_n$ converges.

To prove the second part of the theorem, let

$$a_n{}^+ = \begin{cases} a_n, & \text{if } a_n \ge 0 \\ 0, & \text{if } a_n \le 0, \end{cases}$$

$$a_n{}^- = \begin{cases} a_n, & \text{if } a_n \le 0 \\ 0, & \text{if } a_n \ge 0, \end{cases}$$

so that $\sum_{n=1}^{\infty} a_n{}^+$ is the series formed from the positive terms of $\sum_{n=1}^{\infty} a_n$, and $\sum_{n=1}^{\infty} a_n{}^-$ is the series formed from the negative terms.

If $\sum_{n=1}^{\infty} a_n{}^+$ and $\sum_{n=1}^{\infty} a_n{}^-$ both converge, then

$$\sum_{n=1}^{\infty} |a_n| = \sum_{n=1}^{\infty} [a_n{}^+ - (a_n{}^-)] = \sum_{n=1}^{\infty} a_n{}^+ - \sum_{n=1}^{\infty} a_n{}^-$$

also converges, so $\sum_{n=1}^{\infty} a_n$ converges absolutely.

On the other hand, if $\sum_{n=1}^{\infty} |a_n|$ converges, then, as we have just shown, $\sum_{n=1}^{\infty} a_n$ also converges. Therefore

$$\sum_{n=1}^{\infty} a_n{}^+ = \frac{1}{2}\left(\sum_{n=1}^{\infty} a_n + \sum_{n=1}^{\infty} |a_n|\right)$$

and

$$\sum_{n=1}^{\infty} a_n{}^- = \frac{1}{2}\left(\sum_{n=1}^{\infty} a_n - \sum_{n=1}^{\infty} |a_n|\right)$$

both converge. ▮

It follows from Theorem 5 that every convergent series with positive terms can be used to obtain infinitely many other convergent series, simply by putting in minus signs at random. Not every convergent series can be obtained in this way, however—there are series which are convergent but not absolutely convergent (such series are called **conditionally convergent**). In order to prove this statement we need a test for convergence which applies specifically to series with positive and negative terms.

THEOREM 6 (LEIBNIZ'S THEOREM) Suppose that

$$a_1 \geq a_2 \geq a_3 \geq \cdots \geq 0,$$

and that

$$\lim_{n\to\infty} a_n = 0.$$

Then the series

$$\sum_{n=1}^{\infty} (-1)^{n+1} a_n = a_1 - a_2 + a_3 - a_4 + a_5 - \cdots$$

converges.

PROOF Figure 3 illustrates relationships between the partial sums which we will establish:

(1) $s_2 \leq s_4 \leq s_6 \leq \cdots$,
(2) $s_1 \geq s_3 \geq s_5 \geq \cdots$,
(3) $s_k \leq s_l$ if k is even and l is odd.

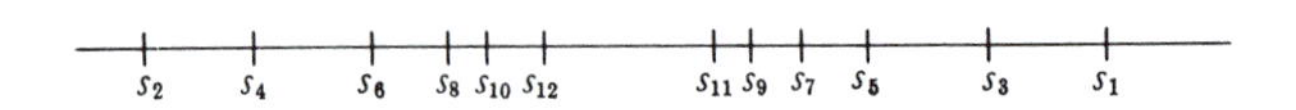

FIGURE 3

To prove the first two inequalities, observe that

$$\begin{aligned}
(1)\quad s_{2n+2} &= s_{2n} + a_{2n+1} - a_{2n+2} \\
&\geq s_{2n}, \qquad \text{since } a_{2n+1} \geq a_{2n+2} \\
(2)\quad s_{2n+3} &= s_{2n+1} - a_{2n+2} + a_{2n+3} \\
&\geq s_{2n+1}, \qquad \text{since } a_{2n+2} \geq a_{2n+3}.
\end{aligned}$$

To prove the third inequality, notice first that

$$\begin{aligned}
s_{2n} &= s_{2n_1} - a_{2n} \\
&\leq s_{2n-1} \qquad \text{since } a_{2n} \geq 0.
\end{aligned}$$

This proves only a special case of (3), but in conjunction with (1) and (2) the general case is easy: if k is even and l is odd, choose n such that

$$2n \geq k \quad \text{and} \quad 2n - 1 \geq l;$$

then

$$s_k \leq s_{2n} \leq s_{2n-1} \leq s_l,$$

which proves (3).

Now, the sequence $\{s_{2n}\}$ converges, because it is nondecreasing and is bounded above (by s_l for any odd l). Let

$$\alpha = \sup\{s_{2n}\} = \lim_{n\to\infty} s_{2n}.$$

Similarly, let

$$\beta = \inf\{s_{2n+1}\} = \lim_{n\to\infty} s_{2n+1}.$$

It follows from (3) that $\alpha \leq \beta$; since

$$s_{2n+1} - s_{2n} = a_{2n+1} \quad \text{and} \quad \lim_{n\to\infty} a_n = 0$$

it is actually the case that $\alpha = \beta$. This proves that $\alpha = \beta = \lim_{n\to\infty} s_n$. ▮

The standard example derived from Theorem 6 is the series

$$1 - \tfrac{1}{2} + \tfrac{1}{3} - \tfrac{1}{4} + \tfrac{1}{5} - \cdots,$$

which is convergent, but *not* absolutely convergent (since $\sum_{n=1}^{\infty} 1/n$ does not converge). If the sum of this series is denoted by x, the following manipulations lead to quite a paradoxical result:

$$\begin{aligned}
x &= 1 - \tfrac{1}{2} + \tfrac{1}{3} - \tfrac{1}{4} + \tfrac{1}{5} - \tfrac{1}{6} + \cdots \\
&= 1 - \tfrac{1}{2} - \tfrac{1}{4} + \tfrac{1}{3} - \tfrac{1}{6} - \tfrac{1}{8} + \tfrac{1}{5} - \tfrac{1}{10} - \tfrac{1}{12} + \tfrac{1}{7} - \tfrac{1}{14} - \tfrac{1}{16} + \cdots \\
&\qquad \text{(the pattern here is one positive term followed by two negative ones)} \\
&= (1 - \tfrac{1}{2}) - \tfrac{1}{4} + (\tfrac{1}{3} - \tfrac{1}{6}) - \tfrac{1}{8} + (\tfrac{1}{5} - \tfrac{1}{10}) - \tfrac{1}{12} + (\tfrac{1}{7} - \tfrac{1}{14}) - \tfrac{1}{16} + \cdots \\
&= \tfrac{1}{2} - \tfrac{1}{4} + \tfrac{1}{6} - \tfrac{1}{8} + \tfrac{1}{10} - \tfrac{1}{12} + \tfrac{1}{14} - \tfrac{1}{16} + \cdots \\
&= \tfrac{1}{2}(1 - \tfrac{1}{2} + \tfrac{1}{3} - \tfrac{1}{4} + \tfrac{1}{5} - \tfrac{1}{6} + \tfrac{1}{7} - \tfrac{1}{8} + \cdots) \\
&= \tfrac{1}{2}x,
\end{aligned}$$

so $x = x/2$, implying that $x = 0$. On the other hand, it is easy to see that $x \neq 0$: the partial sum s_2 equals $\frac{1}{2}$, and the proof of Leibniz's Theorem shows that $x \geq s_2$.

This contradiction depends on a step which takes for granted that operations valid for finite sums necessarily have analogues for infinite sums. It is true that the sequence

$$\{b_n\} = 1,\ -\tfrac{1}{2},\ -\tfrac{1}{4},\ \tfrac{1}{3},\ -\tfrac{1}{6},\ -\tfrac{1}{8},\ \tfrac{1}{5},\ -\tfrac{1}{10},\ -\tfrac{1}{12},\ \dots$$

contains all the numbers in the sequence

$$\{a_n\} = 1,\ -\tfrac{1}{2},\ \tfrac{1}{3},\ -\tfrac{1}{4},\ \tfrac{1}{5},\ -\tfrac{1}{6},\ \tfrac{1}{7},\ -\tfrac{1}{8},\ \tfrac{1}{9},\ -\tfrac{1}{10},\ \tfrac{1}{11},\ -\tfrac{1}{12},\ \dots.$$

In fact, $\{b_n\}$ is a **rearrangement** of $\{a_n\}$ in the following precise sense: each $b_n = a_{f(n)}$ where f is a certain function which "permutes" the natural numbers, that is, every natural number m is $f(n)$ for precisely one n. In our example

$f(2m+1) = 3m+1$ (the terms $1, \frac{1}{3}, \frac{1}{5}, \dots$ go into the 1st, 4th, 7th, ... places),

$f(4m) = 3m$ (the terms $-\frac{1}{4}, -\frac{1}{8}, -\frac{1}{12}, \dots$ go into the 3rd, 6th, 9th, ... places),

$f(4m+2) = 3m+2$ (the terms $-\frac{1}{2}, -\frac{1}{6}, -\frac{1}{10}, \dots$ go into the 2nd, 5th, 8th, ... places).

Nevertheless, there is no reason to assume that $\sum_{n=1}^{\infty} b_n$ should equal $\sum_{n=1}^{\infty} a_n$: these sums are, by definition, $\lim_{n\to\infty} b_1 + \dots + b_n$ and $\lim_{n\to\infty} a_1 + \dots + a_n$, so the particular order of the terms can quite conceivably matter. The series $\sum_{n=1}^{\infty} (-1)^{n+1}/n$ is not special in this regard; indeed, its behavior is typical of series which are not absolutely convergent—the following result (really more of a grand counterexample than a theorem) shows how bad conditionally convergent series are.

THEOREM 7 If $\sum_{n=1}^{\infty} a_n$ converges, but does not converge absolutely, then for any number α there is a rearrangement $\{b_n\}$ of $\{a_n\}$ such that $\sum_{n=1}^{\infty} b_n = \alpha$.

PROOF Let $\sum_{n=1}^{\infty} p_n$ denote the series formed from the positive terms of $\{a_n\}$ and let $\sum_{n=1}^{\infty} q_n$ denote the series of negative terms. It follows from Theorem 5 that at least one of these series does not converge. As a matter of fact, both must fail to converge, for if one had bounded partial sums, and the other had unbounded partial sums, then the original series $\sum_{n=1}^{\infty} a_n$ would also have unbounded partial sums, contradicting the assumption that it converges.

Now let α be any number. Assume, for simplicity, that $\alpha > 0$ (the proof for $\alpha < 0$ will be a simple modification). Since the series $\sum_{n=1}^{\infty} p_n$ is not convergent, there is a number N such that

$$\sum_{n=1}^{N} p_n > \alpha.$$

We will choose N_1 to be the *smallest* N with this property. This means that

$$(1) \quad \sum_{n=1}^{N_1-1} p_n \leq \alpha,$$

$$\text{but} \quad (2) \quad \sum_{n=1}^{N_1} p_n > \alpha.$$

Then if

$$S_1 = \sum_{n=1}^{N_1} p_n,$$

we have

$$S_1 - \alpha \leq p_{N_1}.$$

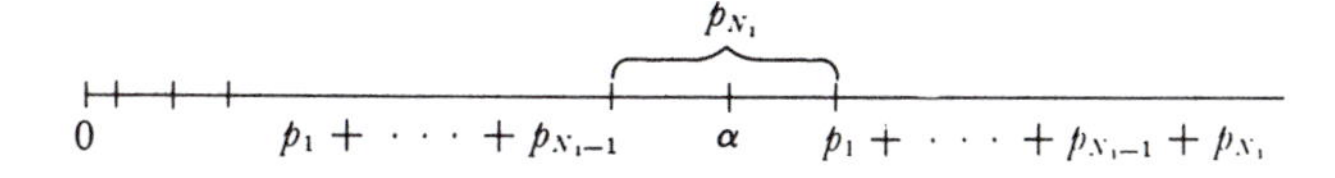

FIGURE 4

This relation, which is clear from Figure 4, follows immediately from equation (1):

$$S_1 - \alpha \leq S_1 - \sum_{n=1}^{N_1-1} p_n = p_{N_1}.$$

To the sum S_1 we now add on just enough negative terms to obtain a new sum T_1 which is less than α. In other words, we choose the smallest integer M_1 for which

$$T_1 = S_1 + \sum_{n=1}^{M_1} q_n < \alpha.$$

As before, we have

$$\alpha - T_1 \leq -q_{M_1}.$$

We now continue this procedure indefinitely, obtaining sums alternately larger and smaller than α, each time choosing the smallest N_k or M_k possible. The sequence

$$p_1, \ldots, p_{N_1}, q_1, \ldots, q_{M_1}, p_{N_1+1}, \ldots, p_{N_2}, \ldots$$

is a rearrangement of $\{a_n\}$. The partial sums of this rearrangement increase to S_1, then decrease to T_1, then increase to S_2, then decrease to T_2, etc. To complete the proof we simply note that $|S_k - \alpha|$ and $|T_k - \alpha|$ are less than or equal to p_{N_k} or $-q_{M_k}$, respectively, and that these terms, being members of the original sequence $\{a_n\}$, must decrease to 0, since $\sum_{n=1}^{\infty} a_n$ converges. ∎

Together with Theorem 7, the next theorem establishes conclusively the distinction between conditionally convergent and absolutely convergent series.

THEOREM 8 If $\sum_{n=1}^{\infty} a_n$ converges absolutely, and $\{b_n\}$ is any rearrangement of $\{a_n\}$, then $\sum_{n=1}^{\infty} b_n$ also converges (absolutely), and

$$\sum_{n=1}^{\infty} a_n = \sum_{n=1}^{\infty} b_n.$$

PROOF Let us denote the partial sums of $\{a_n\}$ by s_n, and the partial sums of $\{b_n\}$ by t_n.

Suppose that $\varepsilon > 0$. Since $\sum_{n=1}^{\infty} a_n$ converges, there is some N such that

$$\left| \sum_{n=1}^{\infty} a_n - s_N \right| < \varepsilon.$$

Moreover, since $\sum_{n=1}^{\infty} |a_n|$ converges, we can also choose N so that

$$\sum_{n=1}^{\infty} |a_n| - (|a_1| + \cdots + |a_N|) < \varepsilon,$$

i.e., so that

$$|a_{N+1}| + |a_{N+2}| + |a_{N+3}| + \cdots < \varepsilon.$$

Now choose M so large that each of $a_1, \ldots, a_N$ appear among $b_1, \ldots, b_M$. Then whenever $m > M$, the difference $t_m - s_N$ is the sum of certain a_i, *where* $a_1, \ldots, a_N$ *are definitely excluded.* Consequently,

$$|t_m - s_N| \le |a_{N+1}| + |a_{N+2}| + |a_{N+3}| + \cdots.$$

Thus, if $m > M$, then

$$\begin{aligned}\left|\sum_{n=1}^{\infty} a_n - t_m\right| &= \left|\sum_{n=1}^{\infty} a_n - s_N - (t_m - s_N)\right| \\ &\le \left|\sum_{n=1}^{\infty} a_n - s_N\right| + |t_m - s_N| \\ &< \varepsilon + \varepsilon.\end{aligned}$$

Since this is true for every $\varepsilon > 0$, the series $\sum_{n=1}^{\infty} b_n$ converges to $\sum_{n=1}^{\infty} a_n$.

To show that $\sum_{n=1}^{\infty} b_n$ converges absolutely, note that $\{\,|b_n|\,\}$ is a rearrangement of $\{\,|a_n|\,\}$; since $\sum_{n=1}^{\infty} |a_n|$ converges absolutely, $\sum_{n=1}^{\infty} |b_n|$ converges by the first part of the theorem. ▌

Absolute convergence is also important when we want to multiply two infinite series. Unlike the situation for addition, where we have the simple formula

$$\sum_{n=1}^{\infty} a_n + \sum_{n=1}^{\infty} b_n = \sum_{n=1}^{\infty} (a_n + b_n),$$

there isn't quite so obvious a candidate for the product

$$\left(\sum_{n=1}^{\infty} a_n\right) \cdot \left(\sum_{n=1}^{\infty} b_N\right) = (a_1 + a_2 + \cdots) \cdot (b_1 + b_2 + \cdots).$$

It would seem that we ought to sum all the products $a_i b_j$. The trouble is that these form a two-dimensional array, rather than a sequence:

$$\begin{array}{cccc} a_1b_1 & a_1b_2 & a_1b_3 & \dots \\ a_2b_1 & a_2b_2 & a_2b_3 & \dots \\ a_3b_1 & a_3b_2 & a_3b_3 & \dots \\ \vdots & \vdots & \vdots & \end{array}$$

Nevertheless, all the elements of this array can be arranged in a sequence. The picture below shows one way of doing this, and of course, there are (infinitely) many other ways.

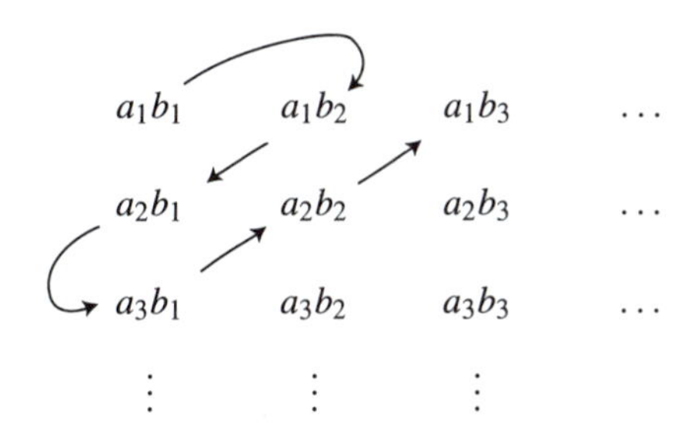

Suppose that $\{c_n\}$ is some sequence of this sort, containing each product $a_i b_j$ just once. Then we might naively expect to have

$$\sum_{n=1}^{\infty} c_n = \sum_{n=1}^{\infty} a_n \cdot \sum_{n=1}^{\infty} b_n.$$

But this *isn't* true (see Problem 8), nor is this really so surprising, since we've said nothing about the specific arrangement of the terms. The next theorem shows that the result does hold when the arrangement of terms is irrelevant.

THEOREM 9 If $\sum_{n=1}^{\infty} a_n$ and $\sum_{n=1}^{\infty} b_n$ converge absolutely, and $\{c_n\}$ is any sequence containing the products $a_i b_j$ for each pair (i, j), then

$$\sum_{n=1}^{\infty} c_n = \sum_{n=1}^{\infty} a_n \cdot \sum_{n=1}^{\infty} b_n.$$

PROOF Notice first that the sequence

$$p_L = \sum_{i=1}^{L} |a_i| \cdot \sum_{j=1}^{L} |b_j|$$

converges, since $\{a_n\}$ and $\{b_n\}$ are absolutely convergent, and since the limit of a product is the product of the limits. So $\{p_L\}$ is a Cauchy sequence, which means that for any $\varepsilon > 0$, if L and L' are large enough, then

$$\left| \sum_{i=1}^{L'} |a_i| \cdot \sum_{j=1}^{L'} |b_j| - \sum_{i=1}^{L} |a_i| \cdot \sum_{j=1}^{L} |b_j| \right| < \frac{\varepsilon}{2}.$$

It follows that

$$(1) \qquad \sum_{i \text{ or } j > L} |a_i| \cdot |b_j| \le \frac{\varepsilon}{2} < \varepsilon.$$

Now suppose that N is any number so large that the terms c_n for $n \le N$ include every term $a_i b_j$ for $i, j \le L$. Then the difference

$$\sum_{n=1}^{N} c_n - \sum_{i=1}^{L} a_i \cdot \sum_{j=1}^{L} b_j$$

consists of terms $a_i b_j$ with $i > L$ or $j > L$, so

$$(2) \qquad \left| \sum_{n=1}^{N} c_n - \sum_{i=1}^{L} a_i \cdot \sum_{j=1}^{L} b_j \right| \le \sum_{i \text{ or } j > L} |a_i| \cdot |b_j|$$

$$< \varepsilon \qquad \text{by (1).}$$

But since the limit of a product is the product of the limits, we also have

$$(3) \qquad \left| \sum_{i=1}^{\infty} a_i \cdot \sum_{j=1}^{\infty} b_j - \sum_{i=1}^{L} a_i \cdot \sum_{j=1}^{L} b_j \right| < \varepsilon$$

for large enough L. Consequently, if we choose L, and then N, large enough, we will have

$$\begin{aligned}\left|\sum_{i=1}^{\infty} a_i \cdot \sum_{j=1}^{\infty} b_j - \sum_{i=1}^{N} c_n\right| &\le \left|\sum_{i=1}^{\infty} a_i \cdot \sum_{j=1}^{\infty} b_j - \sum_{i=1}^{L} a_i \cdot \sum_{j=1}^{L} b_j\right| \\ &\quad + \left|\sum_{i=1}^{L} a_i \cdot \sum_{j=1}^{L} b_j - \sum_{n=1}^{N} c_n\right| \\ &< 2\varepsilon \qquad \text{by (2) and (3),}\end{aligned}$$

which proves the theorem. ∎

Unlike our previous theorems, which were merely concerned with summability, this result says something about the actual sums. Generally speaking, there is no reason to presume that a given infinite sum can be "evaluated" in any simpler terms. However, many simple expressions can be equated to infinite sums by using Taylor's Theorem. Chapter 20 provides many examples of functions for which

$$f(x) = \sum_{i=0}^{n} \frac{f^{(i)}(a)}{i!}(x-a)^i + R_{n,a}(x),$$

where $\lim_{n\to\infty} R_{n,a}(x) = 0$. This is precisely equivalent to

$$f(x) = \lim_{n\to\infty} \sum_{i=0}^{n} \frac{f^{(i)}(a)}{i!}(x-a)^i,$$

which means, in turn, that

$$f(x) = \sum_{i=0}^{\infty} \frac{f^{(i)}(a)}{i!}(x-a)^i.$$

As particular examples we have

$$\begin{aligned}\sin x &= x - \frac{x^3}{3!} + \frac{x^5}{5!} - \frac{x^7}{7!} + \cdots, \\ \cos x &= 1 - \frac{x^2}{2!} + \frac{x^4}{4!} - \frac{x^6}{6!} + \cdots, \\ e^x &= 1 + \frac{x}{1!} + \frac{x^2}{2!} + \frac{x^3}{3!} + \frac{x^4}{4!} + \cdots, \\ \arctan x &= x - \frac{x^3}{3} + \frac{x^5}{5} - \frac{x^7}{7} + \cdots, \qquad |x| \le 1, \\ \log(1+x) &= x - \frac{x^2}{2} + \frac{x^3}{3} - \frac{x^4}{4} + \frac{x^5}{5} + \cdots, \qquad -1 < x \le 1.\end{aligned}$$

(Notice that the series for $\arctan x$ and $\log(1+x)$ do not even converge for $|x| > 1$; in addition, when $x = -1$, the series for $\log(1+x)$ becomes

$$-1 - \tfrac{1}{2} - \tfrac{1}{3} - \tfrac{1}{4} - \cdots$$

which does not converge.)

Some pretty impressive results are obtained with particular values of x:

$$\begin{aligned} 0 &= \pi - \frac{\pi^3}{3!} + \frac{\pi^5}{5!} - \frac{\pi^7}{7!} + \cdots, \\ e &= 1 + \frac{1}{1!} + \frac{1}{2!} + \frac{1}{3!} + \cdots, \\ \frac{\pi}{4} &= 1 - \frac{1}{3} + \frac{1}{5} - \frac{1}{7} + \cdots, \\ \log 2 &= 1 - \frac{1}{2} + \frac{1}{3} - \frac{1}{4} + \cdots. \end{aligned}$$

More significant developments may be anticipated if we compare the series for $\sin x$ and $\cos x$ a little more carefully. The series for $\cos x$ is just the one we would have obtained if we had enthusiastically differentiated both sides of the equation

$$\sin x = x - \frac{x^3}{3!} + \frac{x^5}{5!} - \cdots$$

term-by-term, ignoring the fact that we have never proved anything about the derivatives of infinite sums. Likewise, if we differentiate both sides of the formula for $\cos x$ formally (i.e., without justification) we obtain the formula $\cos'(x) = -\sin x$, and if we differentiate the formula for e^x we obtain $\exp'(x) = \exp(x)$. In the next chapter we shall see that such term-by-term differentiation of infinite sums is indeed valid in certain important cases.

PROBLEMS

1. Decide whether each of the following infinite series is convergent or divergent. The tools which you will need are Leibniz's Theorem and the comparison, ratio, and integral tests. A few examples have been picked with malice aforethought; two series which look quite similar may require different tests (and then again, they may not). The hint below indicates which tests may be used.

(i) $\displaystyle\sum_{n=1}^{\infty} \frac{\sin n\theta}{n^2}$.

(ii) $1 - \tfrac{1}{3} + \tfrac{1}{5} - \tfrac{1}{7} + \cdots$.

(iii) $1 - \tfrac{1}{2} + \tfrac{2}{3} - \tfrac{1}{3} + \tfrac{2}{4} - \tfrac{1}{4} + \tfrac{2}{5} - \tfrac{1}{5} + \cdots$.

(iv) $\displaystyle\sum_{n=1}^{\infty} (-1)^n \frac{\log n}{n}$.

(v) $\displaystyle\sum_{n=2}^{\infty} \frac{1}{\sqrt[3]{n^2-1}}$. (The summation begins with $n = 2$ simply to avoid the meaningless term obtained for $n = 1$).

(vi) $\displaystyle\sum_{n=1}^{\infty} \frac{1}{\sqrt[3]{n^2+1}}$.

(vii) $\displaystyle\sum_{n=1}^{\infty} \frac{n^2}{n!}$.

(viii) $\displaystyle\sum_{n=1}^{\infty} \frac{\log n}{n}$.

(ix) $\displaystyle\sum_{n=2}^{\infty} \frac{1}{\log n}$.

(x) $\displaystyle\sum_{n=2}^{\infty} \frac{1}{(\log n)^k}$.

(xi) $\displaystyle\sum_{n=2}^{\infty} \frac{1}{(\log n)^n}$.

(xii) $\displaystyle\sum_{n=2}^{\infty} (-1)^n \frac{1}{(\log n)^n}$.

(xiii) $\displaystyle\sum_{n=1}^{\infty} \frac{n^2}{n^3+1}$.

(xiv) $\displaystyle\sum_{n=1}^{\infty} \sin \frac{1}{n}$.

(xv) $\displaystyle\sum_{n=2}^{\infty} \frac{1}{n \log n}$.

(xvi) $\displaystyle\sum_{n=2}^{\infty} \frac{1}{n(\log n)^2}$.

(xvii) $\displaystyle\sum_{n=2}^{\infty} \frac{1}{n^2(\log n)}$.

(xviii) $\displaystyle\sum_{n=1}^{\infty} \frac{n!}{n^n}$.

(xix) $\displaystyle\sum_{n=1}^{\infty} \frac{2^n n!}{n^n}$.

(xx) $\displaystyle\sum_{n=1}^{\infty} \frac{3^n n!}{n^n}$.

Hint: Use the comparison test for (i), (ii), (v), (vi), (ix), (x), (xi), (xiii), (xiv), (xvii); the ratio test for (vii), (xviii), (xix), (xx); the integral test for (viii), (xv), (xvi).

The next two problems examine, with hints, some infinite series that require more delicate analysis than those in Problem 1.

***2.** (a) If you have successfully solved examples (xix) and (xx) from Problem 1, it should be clear that $\displaystyle\sum_{n=1}^{\infty} a^n n!/n^n$ converges for $a < e$ and diverges for $a > e$. For $a = e$ the ratio test fails; show that $\displaystyle\sum_{n=1}^{\infty} e^n n!/n^n$ actually diverges, by using Problem 22-13.

(b) Decide when $\displaystyle\sum_{n=1}^{\infty} n^n/a^n n!$ converges, again resorting to Problem 22-13 when the ratio test fails.

***3.** Problem 1 presented the two series $\displaystyle\sum_{n=2}^{\infty} (\log n)^{-k}$ and $\displaystyle\sum_{n=2}^{\infty} (\log n)^{-n}$, of which the first diverges while the second converges. The series

$$\sum_{n=2}^{\infty} \frac{1}{(\log n)^{\log n}},$$

which lies between these two, is analyzed in parts (a) and (b).

(a) Show that $\int_0^\infty e^y/y^y\,dy$ exists, by considering the series $\displaystyle\sum_{n=1}^{\infty} (e/n)^n$.

(b) Show that

$$\sum_{n=2}^{\infty} \frac{1}{(\log n)^{\log n}}$$

converges, by using the integral test. Hint: Use an appropriate substitution and part (a).

(c) Show that

$$\sum_{n=2}^{\infty} \frac{1}{(\log n)^{\log(\log n)}}$$

diverges, by using the integral test. Hint: Use the same substitution as in part (b), and show directly that the resulting integral diverges.

4. Decide whether or not $\sum_{n=1}^{\infty} \frac{1}{n^{1+1/n}}$ converges.

5. (a) Let $\{a_n\}$ be a sequence of integers with $0 \le a_n \le 9$. Prove that $\sum_{n=1}^{\infty} a_n 10^{-n}$ exists (and lies between 0 and 1). (This, of course, is the number which we usually denote by $0.a_1a_2a_3a_4\dots$.)

(b) Suppose that $0 \le x \le 1$. Prove that there is a sequence of integers $\{a_n\}$ with $0 \le a_n \le 9$ and $\sum_{n=1}^{\infty} a_n 10^{-n} = x$. Hint: For example, $a_1 = [10x]$ (where $[y]$ denotes the greatest integer which is $\le y$).

(c) Show that if $\{a_n\}$ is repeating, i.e., is of the form $a_1, a_2, \dots, a_k, a_1, a_2, \dots, a_k, a_1, a_2, \dots$, then $\sum_{n=1}^{\infty} a_n 10^{-n}$ is a rational number (and find it). The same result naturally holds if $\{a_n\}$ is eventually repeating, i.e., if the sequence $\{a_{N+k}\}$ is repeating for some N.

(d) Prove that if $x = \sum_{n=1}^{\infty} a_n 10^{-n}$ is rational, then $\{a_n\}$ is eventually repeating. (Just look at the process of finding the decimal expansion of p/q—dividing q into p by long division.)

6. Suppose that $\{a_n\}$ satisfies the hypothesis of Leibniz's Theorem. Use the proof of Leibniz's Theorem to obtain the following estimate:

$$\left| \sum_{n=1}^{\infty} (-1)^{n+1} a_n - [a_1 - a_2 + \cdots \pm a_N] \right| < a_N.$$

7. Prove that if $a_n \ge 0$ and $\lim_{n\to\infty} \sqrt[n]{a_n} = r$, then $\sum_{n=1}^{\infty} a_n$ converges if $r < 1$, and diverges if $r > 1$. (The proof is very similar to that of the ratio test.) This result is known as the "root test." It is easy to construct series for which the ratio test fails, while the root test works. For example, the root test shows that the series

$$\tfrac{1}{2} + \tfrac{1}{3} + (\tfrac{1}{2})^2 + (\tfrac{1}{3})^2 + (\tfrac{1}{2})^3 + (\tfrac{1}{3})^3 + \cdots$$

converges, even though the ratios of successive terms do not approach a limit. Most examples are of this rather artificial nature, but the root test is nevertheless quite an important theoretical tool, and if the ratio test works the root test will also (by Problem 22-18). It is possible to eliminate limits from the root test; a simple modification of the proof shows that $\sum_{n=1}^{\infty} a_n$ converges if there is some $s < 1$ such that all but finitely many $\sqrt[n]{a_n}$ are $\le s$, and that $\sum_{n=1}^{\infty} a_n$ diverges if infinitely many $\sqrt[n]{a_n}$ are ≥ 1. This result is known as the "delicate root test" (there is a similar delicate ratio test). It follows, using the notation of Problem 22-27, that $\sum_{n=1}^{\infty} a_n$ converges if $\overline{\lim_{n\to\infty}} \sqrt[n]{a_n} < 1$ and diverges if $\overline{\lim_{n\to\infty}} \sqrt[n]{a_n} > 1$; no conclusion is possible if $\overline{\lim_{n\to\infty}} \sqrt[n]{a_n} = 1$.

8. For two sequences $\{a_n\}$ and $\{b_n\}$, let $c_n = \sum_{k=1}^{n} a_k b_{n+1-k}$. (Then c_n is the sum of the terms on the nth diagonal in the picture on page 479.) The series $\sum_{n=1}^{\infty} c_n$ is called the *Cauchy product* of $\sum_{n=1}^{\infty} a_n$ and $\sum_{n=1}^{\infty} b_n$. If $a_n = b_n = (-1)^n/\sqrt{n}$, show that $|c_n| \geq 1$, so that the Cauchy product does not converge.

9. A sequence $\{a_n\}$ is called **Cesaro summable**, with Cesaro sum l, if

$$\lim_{n\to\infty} \frac{s_1 + \cdots + s_n}{n} = l$$

(where $s_k = a_1 + \cdots + a_k$). Problem 22-16 shows that a summable sequence is automatically Cesaro summable, with sum equal to its Cesaro sum. Find a sequence which is *not* summable, but which *is* Cesaro summable.

10. Suppose that $a_n > 0$ and $\{a_n\}$ is Cesaro summable. Suppose also that the sequence $\{na_n\}$ is bounded. Prove that the series $\sum_{n=1}^{\infty} a_n$ converges. Hint: If $s_n = \sum_{i=1}^{n} a_i$ and $\sigma_n = \dfrac{1}{n}\sum_{i=1}^{n} s_i$, prove that $s_n - \dfrac{n}{n+1}\sigma_n$ is bounded.

11. This problem outlines an alternative proof of Theorem 8 which does not rely on the Cauchy criterion.

(a) Suppose that $a_n \geq 0$ for each n. Let $\{b_n\}$ be a rearrangement of $\{a_n\}$, and let $s_n = a_1 + \cdots + a_n$ and $t_n = b_1 + \cdots + b_n$. Show that for each n there is some m with $s_n \leq t_m$.

(b) Show that $\sum_{n=1}^{\infty} a_n \leq \sum_{n=1}^{\infty} b_n$ if $\sum_{n=1}^{\infty} b_n$ exists.

(c) Show that $\sum_{n=1}^{\infty} a_n = \sum_{n=1}^{\infty} b_n$.

(d) Now replace the condition $a_n \geq 0$ by the hypothesis that $\sum_{n=1}^{\infty} a_n$ converges absolutely, using the second part of Theorem 5.

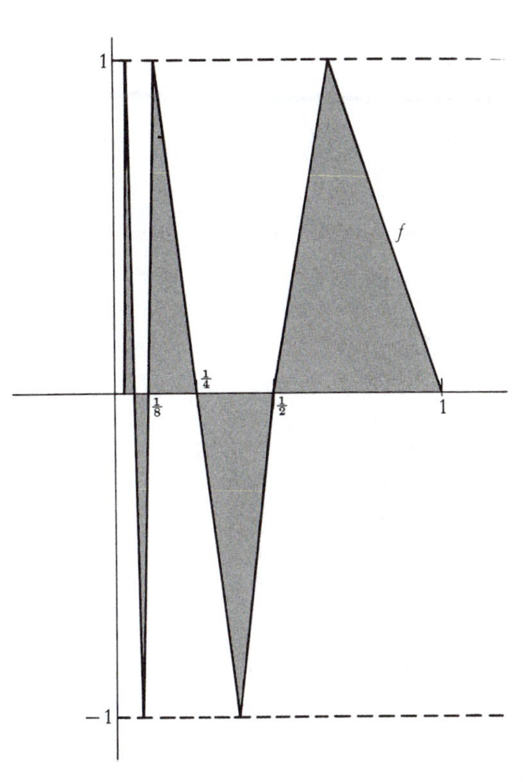

FIGURE 5

12. (a) Prove that if $\sum_{n=1}^{\infty} a_n$ converges absolutely, and $\{b_n\}$ is any subsequence of $\{a_n\}$, then $\sum_{n=1}^{\infty} b_n$ converges (absolutely).

(b) Show that this is false if $\sum_{n=1}^{\infty} a_n$ does not converge absolutely.

*(c) Prove that if $\sum_{n=1}^{\infty} a_n$ converges absolutely, then

$$\sum_{n=1}^{\infty} a_n = (a_1 + a_3 + a_5 + \cdots) + (a_2 + a_4 + a_6 + \cdots).$$

13. Prove that if $\sum_{n=1}^{\infty} a_n$ is absolutely convergent, then $\left|\sum_{n=1}^{\infty} a_n\right| \le \sum_{n=1}^{\infty} |a_n|$.

***14.** Problem 19-42 shows that $\int_0^\infty (\sin x)/x\,dx$ converges. Prove that $\int_0^\infty |(\sin x)/x|\,dx$ diverges.

***15.** Find a continuous function f with $f(x) \ge 0$ for all x such that $\int_0^\infty f(x)\,dx$ exists, but $\lim_{x\to\infty} f(x)$ does not exist.

***16.** Let $f(x) = x \sin 1/x$ for $0 < x \le 1$, and let $f(0) = 0$. Recall the definition of $\ell(f, P)$ from Problem 13-25. Show that the set of all $\ell(f, P)$ for P a partition of $[0, 1]$ is not bounded (thus f has "infinite length"). Hint: Try partitions of the form

$$P = \left\{0, \frac{2}{(2n+1)\pi}, \dots, \frac{2}{7\pi}, \frac{2}{5\pi}, \frac{2}{3\pi}, \frac{2}{\pi}, 1\right\}.$$

17. Let f be the function shown in Figure 5. Find $\int_0^1 f$, and also the area of the shaded region in Figure 5.

***18.** In this problem we will establish the "binomial series"

$$(1+x)^\alpha = \sum_{k=0}^{\infty} \binom{\alpha}{k} x^k, \qquad |x| < 1,$$

for any α, by showing that $\lim_{n\to\infty} R_{n,0}(x) = 0$. The proof is in several steps, and uses the Cauchy and Lagrange forms as found in Problem 20-7.

(a) Use the ratio test to show that the series $\sum_{k=0}^{\infty} \binom{\alpha}{k} r^k$ does indeed converge for $|r| < 1$ (this is not to say that it necessarily converges to $(1+r)^\alpha$). It follows in particular that $\lim_{n\to\infty} \binom{\alpha}{n} r^n = 0$ for $|r| < 1$.

(b) Suppose first that $0 \le x < 1$. Show that $\lim_{n\to\infty} R_{n,0}(x) = 0$, by using Lagrange's form of the remainder, noticing that $(1+t)^{\alpha-n-1} \le 1$ for $n+1 > \alpha$.

(c) Now suppose that $-1 < x < 0$; the number t in Cauchy's form of the remainder satisfies $-1 < x < t \le 0$. Show that

$$|x(1+t)^{\alpha-1}| \le |x|M, \qquad \text{where } M = \max(1, (1+x)^{\alpha-1}),$$

and

$$\left|\frac{x-t}{1+t}\right| = |x|\left(\frac{1-t/x}{1+t}\right) \le |x|.$$

Using Cauchy's form of the remainder, and the fact that

$$(n+1)\binom{\alpha}{n+1} = \alpha\binom{\alpha-1}{n},$$

show that $\lim_{n\to\infty} R_{n,0}(x) = 0$.

19. (a) Suppose that the partial sums of the sequence $\{a_n\}$ are bounded and that $\{b_n\}$ is a sequence with $b_n \ge b_{n+1}$ and $\lim_{n\to\infty} b_n = 0$. Prove that $\sum_{n=1}^{\infty} a_n b_n$ converges. This is known as *Dirichlet's test.* Hint: Use Abel's Lemma (Problem 19-35) to check the Cauchy criterion.

(b) Derive Leibniz's Theorem from this result.

(c) Prove, using Problem 15-33, that the series $\sum_{n=1}^{\infty} (\cos nx)/n$ converges if x is not of the form $2k\pi$ for any integer k (in which case it clearly diverges).

(d) Prove *Abel's test*: If $\sum_{n=1}^{\infty} a_n$ converges and $\{b_n\}$ is a sequence which is either nondecreasing or nonincreasing and which is bounded, then $\sum_{n=1}^{\infty} a_n b_n$ converges. Hint: Consider $b_n - b$, where $b = \lim_{n\to\infty} b_n$.

***20.** Suppose $\{a_n\}$ is decreasing and $\lim_{n\to\infty} a_n = 0$. Prove that if $\sum_{n=1}^{\infty} a_n$ converges, then $\sum_{n=1}^{\infty} 2^n a_{2^n}$ also converges (the "Cauchy Condensation Theorem"). Notice that the divergence of $\sum_{n=1}^{\infty} 1/n$ is a special case, for if $\sum_{n=1}^{\infty} 1/n$ converged, then $\sum_{n=1}^{\infty} 2^n(1/2^n)$ would also converge; this remark may serve as a hint.

***21.** (a) Prove that if $\sum_{n=1}^{\infty} a_n{}^2$ and $\sum_{n=1}^{\infty} b_n{}^2$ converge, then $\sum_{n=1}^{\infty} a_n b_n$ converges.

(b) Prove that if $\sum_{n=1}^{\infty} a_n{}^2$ converges, then $\sum_{n=1}^{\infty} a_n/n^{\alpha}$ converges for any $\alpha > \frac{1}{2}$.

***22.** Suppose $\{a_n\}$ is decreasing and each $a_n \geq 0$. Prove that if $\sum_{n=1}^{\infty} a_n$ converges, then $\lim_{n\to\infty} na_n = 0$. Hint: Write down the Cauchy criterion and be sure to use the fact that $\{a_n\}$ is decreasing.

***23.** If $\sum_{n=1}^{\infty} a_n$ converges, then the partial sums s_n are bounded, and $\lim_{n\to\infty} a_n = 0$. It is tempting to conjecture that boundedness of the partial sums, together with the condition $\lim_{n\to\infty} a_n = 0$, implies convergence of $\sum_{n=1}^{\infty} a_n$. This is *not* true, but finding a counterexample requires a little ingenuity. As a hint, notice that some *subsequence* of the partial sums will have to converge; you must somehow allow this to happen, without letting the sequence itself converge.

24. Prove that if $a_n \geq 0$ and $\sum_{n=1}^{\infty} a_n$ diverges, then $\sum_{n=1}^{\infty} \frac{a_n}{1+a_n}$ also diverges. Hint: Compare the partial sums. Does the converse hold?

25. Let $b_n \neq 0$. We say that the infinite product $\prod_{n=1}^{\infty} b_n$ *converges* if the sequence

$$p_n = \prod_{i=1}^{n} b_i \text{ converges, and also } \lim_{n\to\infty} p_n \neq 0.$$

(a) Prove that if $\prod_{n=1}^{\infty}(1+a_n)$ converges, then a_n approaches 0.

(b) Prove that $\prod_{n=1}^{\infty}(1+a_n)$ converges if and only if $\sum_{n=1}^{\infty}\log(1+a_n)$ converges.

(c) For $a_n \geq 0$, prove that $\prod_{n=1}^{\infty}(1+a_n)$ converges if and only if $\sum_{n=1}^{\infty} a_n$ converges. Hint: Use Problem 24 for one implication, and a simple estimate for $\log(1+a)$ for the reverse implication.

26. (a) Compute $\prod_{n=2}^{\infty}\left(1-\frac{1}{n^2}\right)$.

(b) Compute $\prod_{n=1}^{\infty}(1+x^{2^n})$ for $|x| < 1$.

27. The divergence of $\sum_{n=1}^{\infty} 1/n$ is related to the following remarkable fact: Any positive rational number x can be written as a *finite* sum of *distinct* numbers of the form $1/n$. The idea of the proof is shown by the following calculation for $\frac{27}{31}$: Since

$$\begin{aligned} \tfrac{27}{31} - \tfrac{1}{2} &= \tfrac{23}{62} \\ \tfrac{23}{62} - \tfrac{1}{3} &= \tfrac{7}{186} \\ \tfrac{7}{186} &< \tfrac{1}{4}, \dots, \tfrac{1}{26} \\ \tfrac{7}{186} - \tfrac{1}{27} &= \tfrac{1}{1674} \end{aligned}$$

we have

$$\frac{27}{31} = \frac{1}{2} + \frac{1}{3} + \frac{1}{27} + \frac{1}{1674}.$$

Notice that the numerators 23, 7, 1 of the differences are decreasing.

(a) Prove that if $1/(n+1) < x < 1/n$ for some n, then the numerator in this sort of calculation must always decrease; conclude that x can be written as a finite sum of distinct numbers $1/k$.

(b) Now prove the result for all x by using the divergence of $\sum_{n=1}^{\infty} 1/n$.

CHAPTER 24 UNIFORM CONVERGENCE AND POWER SERIES

The considerations at the end of the previous chapter suggest an entirely new way of looking at infinite series. Our attention will shift from particular infinite sums to equations like

$$e^x = 1 + \frac{x}{1!} + \frac{x^2}{2!} + \cdots$$

which concern sums of quantities that depend on x. In other words, we are interested in *functions* defined by equations of the form

$$f(x) = f_1(x) + f_2(x) + f_3(x) + \cdots$$

(in the previous example $f_n(x) = x^{n-1}/(n-1)!$). In such a situation $\{f_n\}$ will be some sequence of functions; for each x we obtain a sequence of numbers $\{f_n(x)\}$, and $f(x)$ is the sum of this sequence. In order to analyze such functions it will certainly be necessary to remember that each sum

$$f_1(x) + f_2(x) + f_3(x) + \cdots$$

is, by definition, the limit of the sequence

$$f_1(x),\ f_1(x) + f_2(x),\ f_1(x) + f_2(x) + f_3(x),\ \ldots.$$

If we define a new sequence of functions $\{s_n\}$ by

$$s_n = f_1 + \cdots + f_n,$$

then we can express this fact more succinctly by writing

$$f(x) = \lim_{n\to\infty} s_n(x).$$

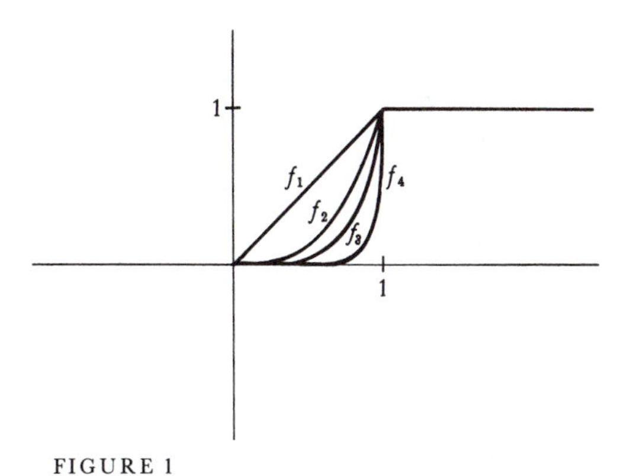

FIGURE 1

For some time we shall therefore concentrate on functions defined as limits,

$$f(x) = \lim_{n\to\infty} f_n(x),$$

rather than on functions defined as infinite sums. The total body of results about such functions can be summed up very easily: nothing one would hope to be true actually is—instead we have a splendid collection of counterexamples. The first of these shows that even if each f_n is continuous, the function f may not be! Contrary to what you may expect, the functions f_n will be very simple. Figure 1 shows the graphs of the functions

$$f_n(x) = \begin{cases} x^n, & 0 \le x \le 1 \\ 1, & x \ge 1. \end{cases}$$

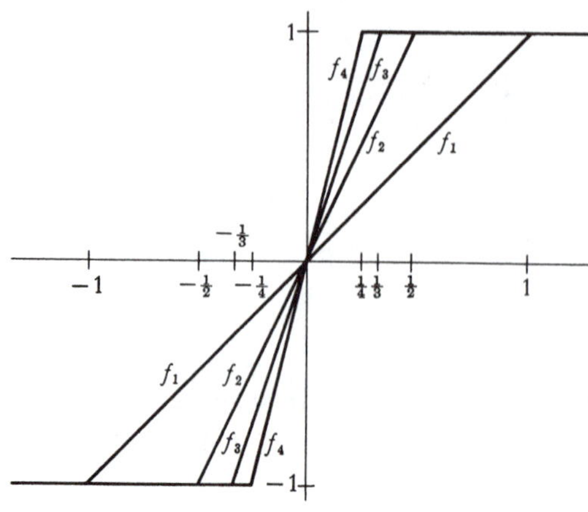

FIGURE 2

These functions are all continuous, but the function $f(x) = \lim_{n\to\infty} f_n(x)$ is not continuous; in fact,

$$\lim_{n\to\infty} f_n(x) = \begin{cases} 0, & 0 \le x < 1 \\ 1, & x \ge 1. \end{cases}$$

Another example of this same phenomenon is illustrated in Figure 2; the functions f_n are defined by

$$f_n(x) = \begin{cases} -1, & x \le -\dfrac{1}{n} \\ nx, & -\dfrac{1}{n} \le x \le \dfrac{1}{n} \\ 1, & \dfrac{1}{n} \le x. \end{cases}$$

In this case, if $x < 0$, then $f_n(x)$ is eventually (i.e., for large enough n) equal to -1, and if $x > 0$, then $f_n(x)$ is eventually 1, while $f_n(0) = 0$ for all n. Thus

$$\lim_{n\to\infty} f_n(x) = \begin{cases} -1, & x < 0 \\ 0, & x = 0 \\ 1, & x > 0; \end{cases}$$

so, once again, the function $f(x) = \lim_{n\to\infty} f_n(x)$ is not continuous.

By rounding off the corners in the previous examples it is even possible to produce a sequence of *differentiable* functions $\{f_n\}$ for which the function $f(x) = \lim_{n\to\infty} f_n(x)$ is not continuous. One such sequence is easy to define explicitly:

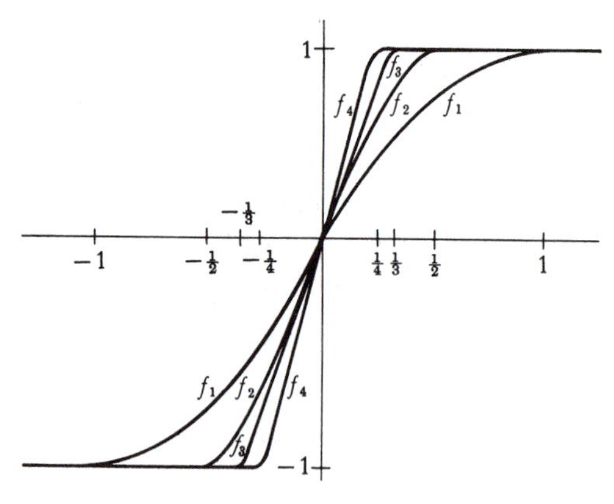

FIGURE 3

$$f_n(x) = \begin{cases} -1, & x \le -\dfrac{1}{n} \\ \sin\left(\dfrac{n\pi x}{2}\right), & -\dfrac{1}{n} \le x \le \dfrac{1}{n} \\ 1, & \dfrac{1}{n} \le x. \end{cases}$$

These functions are differentiable (Figure 3), but we still have

$$\lim_{n\to\infty} f_n(x) = \begin{cases} -1, & x < 0 \\ 0, & x = 0 \\ 1, & x > 0. \end{cases}$$

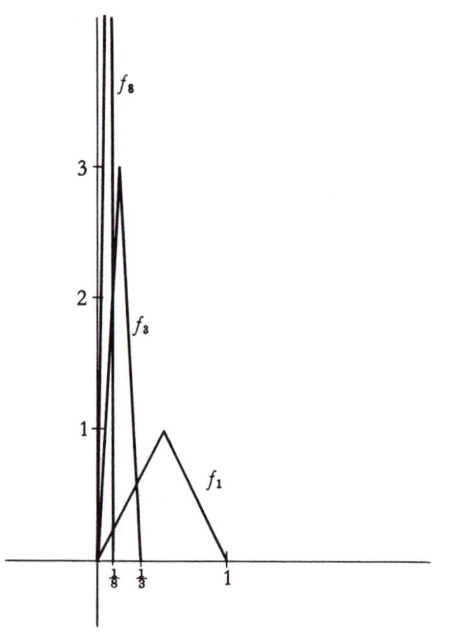

FIGURE 4

Continuity and differentiability are, moreover, not the only properties for which problems arise. Another difficulty is illustrated by the sequence $\{f_n\}$ shown in Figure 4; on the interval $[0, 1/n]$ the graph of f_n forms an isosceles triangle of altitude n, while $f_n(x) = 0$ for $x \ge 1/n$. These functions may be defined explicitly as follows:

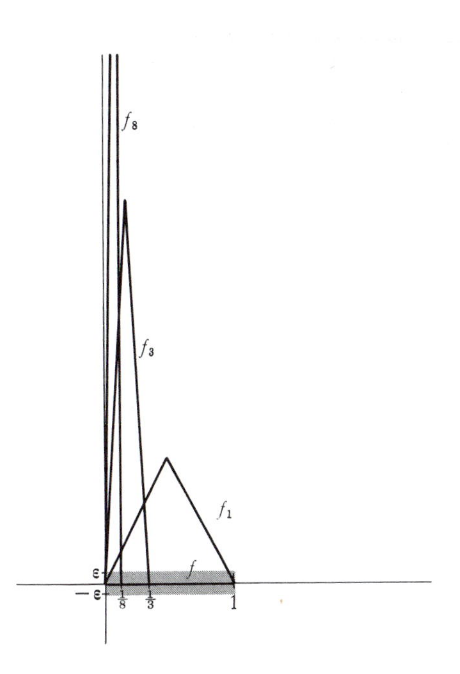

FIGURE 5

$$f_n(x) = \begin{cases} 2n^2x, & 0 \le x \le \dfrac{1}{2n} \\ 2n - 2n^2x, & \dfrac{1}{2n} \le x \le \dfrac{1}{n} \\ 0, & \dfrac{1}{n} \le x \le 1. \end{cases}$$

Because this sequence varies so erratically near 0, our primitive mathematical instincts might suggest that $\lim_{n\to\infty} f_n(x)$ does not always exist. Nevertheless, this limit does exist for all x, and the function $f(x) = \lim_{n\to\infty} f_n(x)$ is even continuous. In fact, if $x > 0$, then $f_n(x)$ is eventually 0, so $\lim_{n\to\infty} f_n(x) = 0$; moreover, $f_n(0) = 0$ for all n, so that we certainly have $\lim_{n\to\infty} f_n(0) = 0$. In other words, $f(x) = \lim_{n\to\infty} f_n(x) = 0$ for all x. On the other hand, the integral quickly reveals the strange behavior of this sequence; we have

$$\int_0^1 f_n(x)\,dx = \tfrac{1}{2},$$

but

$$\int_0^1 f(x)\,dx = 0.$$

Thus,

$$\lim_{n\to\infty} \int_0^1 f_n(x)\,dx \ne \int_0^1 \lim_{n\to\infty} f_n(x)\,dx.$$

This particular sequence of functions behaves in a way that we really never imagined when we first considered functions defined by limits. Although it is true that

$$f(x) = \lim_{n\to\infty} f_n(x) \qquad \text{for each } x \text{ in } [0, 1],$$

the graphs of the functions f_n do not "approach" the graph of f in the sense of lying close to it—if, as in Figure 5, we draw a strip around f of total width 2ε (allowing a width of ε above and below), then the graphs of f_n do not lie completely within this strip, no matter how large an n we choose. Of course, for each x there is some N such that the point $(x, f_n(x))$ lies in this strip for $n > N$; this assertion just amounts to the fact that $\lim_{n\to\infty} f_n(x) = f(x)$. But it is necessary to choose larger and larger N's as x is chosen closer and closer to 0, and no one N will work for all x at once.

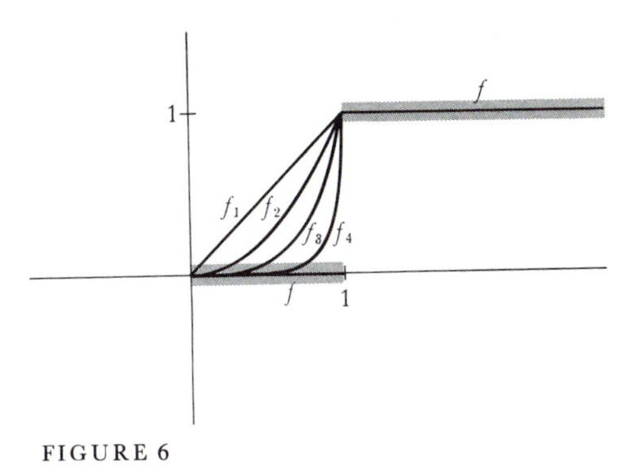

FIGURE 6

The same situation actually occurs, though less blatantly, for each of the other examples given previously. Figure 6 illustrates this point for the sequence

$$f_n(x) = \begin{cases} x^n, & 0 \le x \le 1 \\ 1, & x \ge 1. \end{cases}$$

A strip of total width 2ε has been drawn around the graph of $f(x) = \lim_{n\to\infty} f_n(x)$. If $\varepsilon < \frac{1}{2}$, this strip consists of two pieces, which contain no points with second

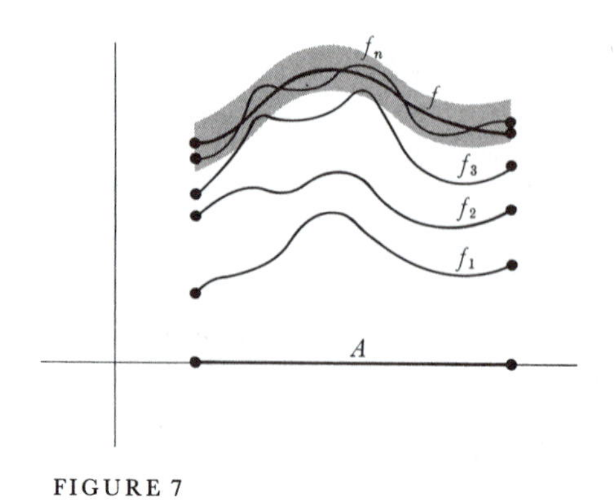

FIGURE 7

coordinate equal to $\frac{1}{2}$; since each function f_n takes on the value $\frac{1}{2}$, the graph of each f_n fails to lie within this strip. Once again, for each point x there is some N such that $(x, f_n(x))$ lies in the strip for $n > N$; but it is not possible to pick one N which works for all x at once.

It is easy to check that precisely the same situation occurs for each of the other examples. In each case we have a function f, and a sequence of functions $\{f_n\}$, all defined on some set A, such that

$$f(x) = \lim_{n\to\infty} f_n(x) \qquad \text{for all } x \text{ in } A.$$

This means that

> for all $\varepsilon > 0$, and for all x in A, there is some N such that if $n > N$, then $|f(x) - f_n(x)| < \varepsilon$.

But in each case different N's must be chosen for different x's, and in each case it is *not* true that

> for all $\varepsilon > 0$ there is some N such that for all x in A, if $n > N$, then $|f(x) - f_n(x)| < \varepsilon$.

Although this condition differs from the first only by a minor displacement of the phrase "for all x in A," it has a totally different significance. If a sequence $\{f_n\}$ satisfies this second condition, then the graphs of f_n eventually lie close to the graph of f, as illustrated in Figure 7. This condition turns out to be just the one which makes the study of limit functions feasible.

DEFINITION

Let $\{f_n\}$ be a sequence of functions defined on A, and let f be a function which is also defined on A. Then f is called the **uniform limit of** $\{f_n\}$ **on** A if for every $\varepsilon > 0$ there is some N such that for all x in A,

$$\text{if } n > N, \text{ then } |f(x) - f_n(x)| < \varepsilon.$$

We also say that $\{f_n\}$ **converges uniformly to** f **on** A, or that f_n **approaches** f **uniformly on** A.

As a contrast to this definition, if we know only that

$$f(x) = \lim_{n\to\infty} f_n(x) \qquad \text{for each } x \text{ in } A,$$

then we say that $\{f_n\}$ **converges pointwise to** f **on** A. Clearly, uniform convergence implies pointwise convergence (but not conversely!).

Evidence for the usefulness of uniform convergence is not at all difficult to amass. Integrals represent a particularly easy topic; Figure 7 makes it almost obvious that if $\{f_n\}$ converges uniformly to f, then the integral of f_n can be made as close to the integral of f as desired. Expressed more precisely, we have the following theorem.

THEOREM 1 Suppose that $\{f_n\}$ is a sequence of functions which are integrable on $[a,b]$, and that $\{f_n\}$ converges uniformly on $[a,b]$ to a function f which is integrable on $[a,b]$. Then

$$\int_a^b f = \lim_{n\to\infty} \int_a^b f_n.$$

PROOF Let $\varepsilon > 0$. There is some N such that for all $n > N$ we have

$$|f(x) - f_n(x)| < \varepsilon \qquad \text{for all } x \text{ in } [a,b].$$

Thus, if $n > N$ we have

$$\begin{aligned}\left|\int_a^b f(x)\,dx - \int_a^b f_n(x)\,dx\right| &= \left|\int_a^b [f(x) - f_n(x)]\,dx\right| \\ &\le \int_a^b |f(x) - f_n(x)|\,dx \\ &\le \int_a^b \varepsilon\,dx \\ &= \varepsilon(b-a).\end{aligned}$$

Since this is true for any $\varepsilon > 0$, it follows that

$$\int_a^b f = \lim_{n\to\infty} \int_a^b f_n. \blacksquare$$

The treatment of continuity is only a little more difficult, involving an "$\varepsilon/3$-argument," a three-step estimate of $|f(x) - f(x+h)|$. If $\{f_n\}$ is a sequence of continuous functions which converges uniformly to f, then there is some n such that

$$(1) \qquad |f(x) - f_n(x)| < \frac{\varepsilon}{3},$$

$$(2) \qquad |f(x+h) - f_n(x+h)| < \frac{\varepsilon}{3}.$$

Moreover, since f_n is continuous, for sufficiently small h we have

$$(3) \qquad |f_n(x) - f_n(x+h)| < \frac{\varepsilon}{3}.$$

It will follow from (1), (2), and (3) that $|f(x) - f(x+h)| < \varepsilon$. In order to obtain (3), however, we must restrict the size of $|h|$ in a way that cannot be predicted until n has already been chosen; it is therefore quite essential that there be some fixed n which makes (2) true, no matter how small $|h|$ may be—it is precisely at this point that uniform convergence enters the proof.

THEOREM 2 Suppose that $\{f_n\}$ is a sequence of functions which are continuous on $[a,b]$, and that $\{f_n\}$ converges uniformly on $[a,b]$ to f. Then f is also continuous on $[a,b]$.

PROOF For each x in $[a,b]$ we must prove that f is continuous at x. We will deal only with x in (a,b); the cases $x = a$ and $x = b$ require the usual simple modifications.

Let $\varepsilon > 0$. Since $\{f_n\}$ converges uniformly to f on $[a, b]$, there is some n such that

$$|f(y) - f_n(y)| < \frac{\varepsilon}{3} \qquad \text{for all } y \text{ in } [a, b].$$

In particular, for all h such that $x + h$ is in $[a, b]$, we have

$$(1) \qquad |f(x) - f_n(x)| < \frac{\varepsilon}{3},$$

$$(2) \qquad |f(x+h) - f_n(x+h)| < \frac{\varepsilon}{3}.$$

Now f_n is continuous, so there is some $\delta > 0$ such that for $|h| < \delta$ we have

$$(3) \qquad |f_n(x) - f_n(x+h)| < \frac{\varepsilon}{3}.$$

Thus, if $|h| < \delta$, then

$$\begin{aligned}
|f(x+h) - f(x)| & \\
&= |f(x+h) - f_n(x+h) + f_n(x+h) - f_n(x) + f_n(x) - f(x)| \\
&\leq |f(x+h) - f_n(x+h)| + |f_n(x+h) - f_n(x)| + |f_n(x) - f(x)| \\
&< \frac{\varepsilon}{3} + \frac{\varepsilon}{3} + \frac{\varepsilon}{3} \\
&= \varepsilon.
\end{aligned}$$

This proves that f is continuous at x. ▌

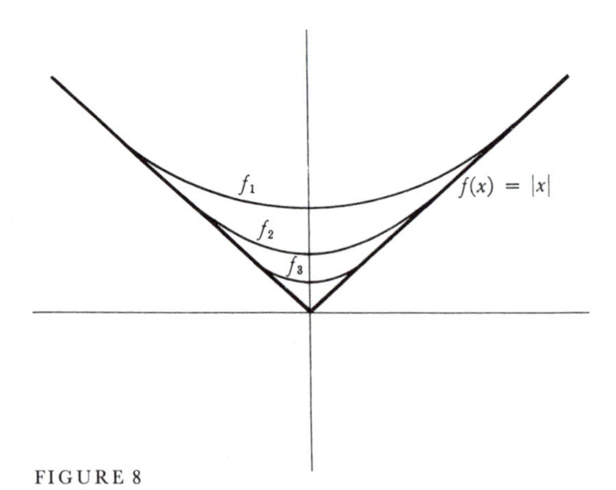

FIGURE 8

After the two noteworthy successes provided by Theorem 1 and Theorem 2, the situation for differentiability turns out to be very disappointing. If each f_n is differentiable, and if $\{f_n\}$ converges uniformly to f, it is still not necessarily true that f is differentiable. For example, Figure 8 shows that there is a sequence of differentiable functions $\{f_n\}$ which converges uniformly to the function $f(x) = |x|$.

Even if f *is* differentiable, it may not be true that

$$f'(x) = \lim_{n\to\infty} f_n{}'(x);$$

this is not at all surprising if we reflect that a smooth function can be approximated by very rapidly oscillating functions. For example (Figure 9), if

$$f_n(x) = \frac{1}{n}\sin(n^2x),$$

then $\{f_n\}$ converges uniformly to the function $f(x) = 0$, but

$$f_n{}'(x) = n\cos(n^2x),$$

and $\lim_{n\to\infty} n\cos(n^2x)$ does not always exist (for example, it does not exist if $x = 0$).

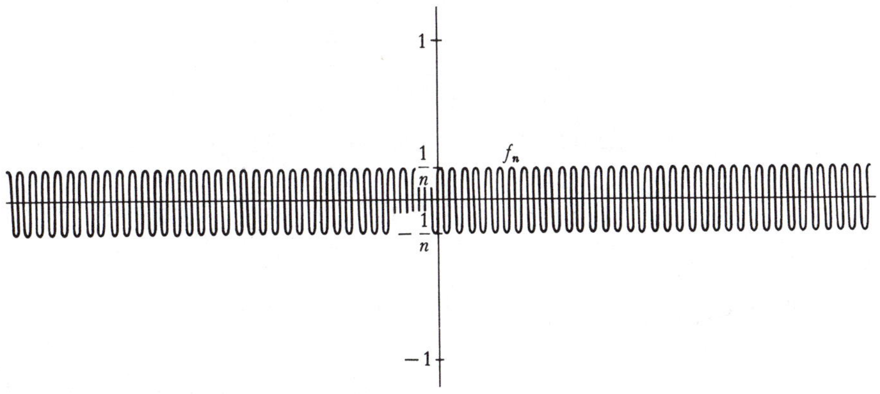

FIGURE 9

Despite such examples, the Fundamental Theorem of Calculus practically guarantees that some sort of theorem about derivatives will be a consequence of Theorem 1; the crucial hypothesis is that $\{f_n{}'\}$ converge uniformly (to *some* continuous function).

THEOREM 3 Suppose that $\{f_n\}$ is a sequence of functions which are differentiable on $[a, b]$, with integrable derivatives $f_n{}'$, and that $\{f_n\}$ converges (pointwise) to f. Suppose, moreover, that $\{f_n{}'\}$ converges uniformly on $[a, b]$ to some continuous function g. Then f is differentiable and

$$f'(x) = \lim_{n\to\infty} f_n{}'(x).$$

PROOF Applying Theorem 1 to the interval $[a, x]$, we see that for each x we have

$$\begin{aligned}\int_a^x g &= \lim_{n\to\infty}\int_a^x f_n{}' \\ &= \lim_{n\to\infty}[f_n(x) - f_n(a)] \\ &= f(x) - f(a).\end{aligned}$$

Since g is continuous, it follows that $f'(x) = g(x) = \lim_{n\to\infty} f_n{}'(x)$ for all x in the interval $[a, b]$. ▮

Now that the basic facts about uniform limits have been established, it is clear how to treat functions defined as infinite sums,

$$f(x) = f_1(x) + f_2(x) + f_3(x) + \cdots .$$

This equation means that

$$f(x) = \lim_{n\to\infty} f_1(x) + \cdots + f_n(x);$$

our previous theorems apply when the new sequence

$$f_1,\ f_1 + f_2,\ f_1 + f_2 + f_3,\ \ldots$$

converges uniformly to f. Since this is the only case we shall ever be interested in, we single it out with a definition.

DEFINITION

The series $\sum_{n=1}^{\infty} f_n$ **converges uniformly** (more formally: the sequence $\{f_n\}$ is **uniformly summable**) **to** $\boldsymbol{f}$ **on** $\boldsymbol{A}$, if the sequence

$$f_1,\ f_1 + f_2,\ f_1 + f_2 + f_3,\ \ldots$$

converges uniformly to f on A.

We can now apply each of Theorems 1, 2, and 3 to uniformly convergent series; the results may be stated in one common corollary.

COROLLARY

Let $\sum_{n=1}^{\infty} f_n$ converge uniformly to f on $[a, b]$.

(1) If each f_n is continuous on $[a, b]$, then f is continuous on $[a, b]$.
(2) If f and each f_n is integrable on $[a, b]$, then

$$\int_a^b f = \sum_{n=1}^{\infty} \int_a^b f_n.$$

Moreover, if $\sum_{n=1}^{\infty} f_n$ converges (pointwise) to f on $[a, b]$, each f_n has an integrable derivative $f_n{}'$ and $\sum_{n=1}^{\infty} f_n{}'$ converges uniformly on $[a, b]$ to some continuous function, then

$$(3)\quad f'(x) = \sum_{n=1}^{\infty} f_n{}'(x) \qquad \text{for all } x \text{ in } [a, b].$$

PROOF (1) If each f_n is continuous, then so is each $f_1+\cdots+f_n$, and f is the uniform limit of the sequence $f_1, f_1+f_2, f_1+f_2+f_3, \dots$, so f is continuous by Theorem 2.

(2) Since $f_1, f_1+f_2, f_1+f_2+f_3, \dots$ converges uniformly to f, it follows from Theorem 1 that

$$\begin{aligned}\int_a^b f &= \lim_{n\to\infty}\int_a^b (f_1+\cdots+f_n)\\ &= \lim_{n\to\infty}\left(\int_a^b f_1+\cdots+\int_a^b f_n\right)\\ &= \sum_{n=1}^{\infty}\int_a^b f_n.\end{aligned}$$

(3) Each function $f_1+\cdots+f_n$ is differentiable, with derivative $f_1{}'+\cdots+f_n{}'$, and $f_1{}', f_1{}'+f_2{}', f_1{}'+f_2{}'+f_3{}', \dots$ converges uniformly to a continuous function, by hypothesis. It follows from Theorem 3 that

$$\begin{aligned}f'(x) &= \lim_{n\to\infty}[f_1{}'(x)+\cdots+f_n{}'(x)]\\ &= \sum_{n=1}^{\infty} f_n{}'(x). \blacksquare\end{aligned}$$

At the moment this corollary is not very useful, since it seems quite difficult to predict when the sequence $f_1, f_1+f_2, f_1+f_2+f_3, \dots$ will converge uniformly. The most important condition which ensures such uniform convergence is provided by the following theorem; the proof is almost a triviality because of the cleverness with which the very simple hypotheses have been chosen.

THEOREM 4 (THE WEIERSTRASS M-TEST) Let $\{f_n\}$ be a sequence of functions defined on A, and suppose that $\{M_n\}$ is a sequence of numbers such that

$$|f_n(x)| \le M_n \qquad \text{for all } x \text{ in } A.$$

Suppose moreover that $\sum_{n=1}^{\infty} M_n$ converges. Then for each x in A the series $\sum_{n=1}^{\infty} f_n(x)$ converges (in fact, it converges absolutely), and $\sum_{n=1}^{\infty} f_n$ converges uniformly on A to the function

$$f(x) = \sum_{n=1}^{\infty} f_n(x).$$

PROOF For each x in A the series $\sum_{n=1}^{\infty} |f_n(x)|$ converges, by the comparison test; consequently $\sum_{n=1}^{\infty} f_n(x)$ converges (absolutely). Moreover, for all x in A we have

$$\begin{aligned} \left| f(x) - [f_1(x) + \cdots + f_n(x)] \right| &= \left| \sum_{n=N+1}^{\infty} f_n(x) \right| \\ &\leq \sum_{n=N+1}^{\infty} |f_n(x)| \\ &\leq \sum_{n=N+1}^{\infty} M_n. \end{aligned}$$

Since $\sum_{n=1}^{\infty} M_n$ converges, the number $\sum_{n=N+1}^{\infty} M_n$ can be made as small as desired, by choosing N sufficiently large. ▌

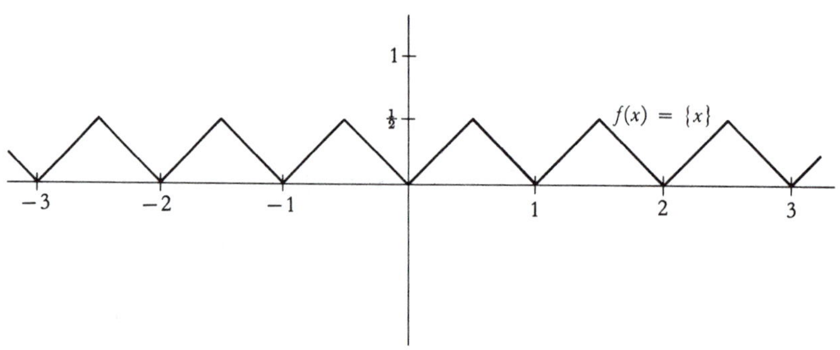

FIGURE 10

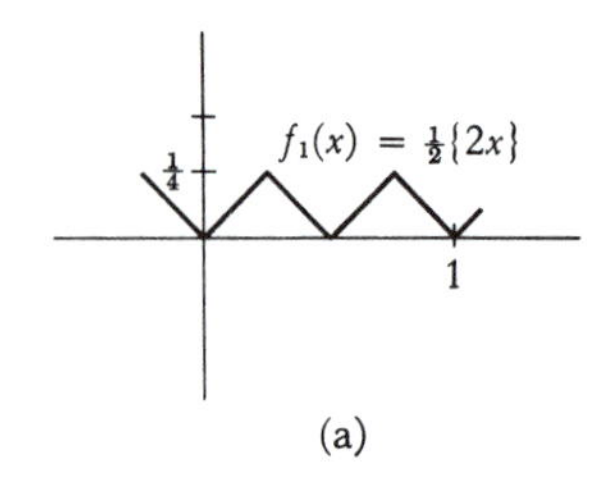

(a)

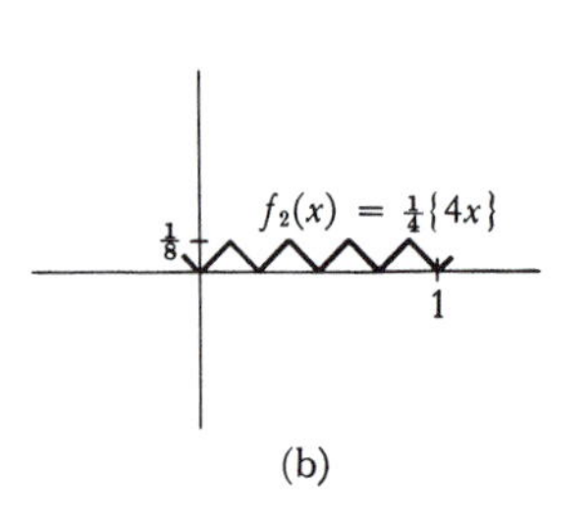

(b)

FIGURE 11

The following sequence $\{f_n\}$ illustrates a simple application of the Weierstrass M-test. Let $\{x\}$ denote the distance from x to the nearest integer (the graph of $f(x) = \{x\}$ is illustrated in Figure 10). Now define

$$f_n(x) = \frac{1}{10^n}\{10^n x\}.$$

The functions f_1 and f_2 are shown in Figure 11 (but to make the drawings simpler, 10^n has been replaced by 2^n). This sequence of functions has been defined so that the Weierstrass M-test automatically applies: clearly

$$|f_n(x)| \leq \frac{1}{10^n} \qquad \text{for all } x,$$

and $\sum_{n=1}^{\infty} 1/10^n$ converges. Thus $\sum_{n=1}^{\infty} f_n$ converges uniformly; since each f_n is continuous, the corollary implies that the function

$$f(x) = \sum_{n=1}^{\infty} f_n(x) = \sum_{n=1}^{\infty} \frac{1}{10^n}\{10^n x\}$$

is also continuous. Figure 12 shows the graph of the first few partial sums $f_1 + \cdots + f_n$. As n increases, the graphs become harder and harder to draw, and the infinite sum $\sum_{n=1}^{\infty} f_n$ is quite undrawable, as shown by the following theorem (included mainly as an interesting sidelight, to be skipped if you find the going too rough).

THEOREM 5 The function

$$f(x) = \sum_{n=1}^{\infty} \frac{1}{10^n}\{10^n x\}$$

is continuous everywhere and differentiable nowhere!

PROOF We have just shown that f is continuous; this is the only part of the proof which uses uniform convergence. We will prove that f is not differentiable at a, for any a, by the straightforward method of exhibiting a particular sequence $\{h_m\}$ approaching 0 for which

$$\lim_{m\to\infty} \frac{f(a+h_m) - f(a)}{h_m}$$

does not exist. It obviously suffices to consider only those numbers a satisfying $0 < a \le 1$.

Suppose that the decimal expansion of a is

$$a = 0.a_1a_2a_3a_4\ldots.$$

Let $h_m = 10^{-m}$ if $a_m \neq 4$ or 9, but let $h_m = -10^{-m}$ if $a_m = 4$ or 9 (the reason for these two exceptions will appear soon). Then

$$\begin{aligned}\frac{f(a+h_m) - f(a)}{h_m} &= \sum_{n=1}^{\infty} \frac{1}{10^n} \cdot \frac{\{10^n(a+h_m)\} - \{10^n a\}}{\pm 10^{-m}} \\ &= \sum_{n=1}^{\infty} \pm 10^{m-n}[\{10^n(a+h_m)\} - \{10^n a\}].\end{aligned}$$

This infinite series is really a finite sum, because if $n \ge m$, then $10^n h_m$ is an integer, so

$$\{10^n(a+h_m)\} - \{10^n a\} = 0.$$

On the other hand, for $n < m$ we can write

$$\begin{aligned}10^n a &= \text{integer} + 0.a_{n+1}a_{n+2}a_{n+3}\ldots a_m\ldots \\ 10^n(a+h_m) &= \text{integer} + 0.a_{n+1}a_{n+2}a_{n+3}\ldots(a_m \pm 1)\ldots\end{aligned}$$

(a)

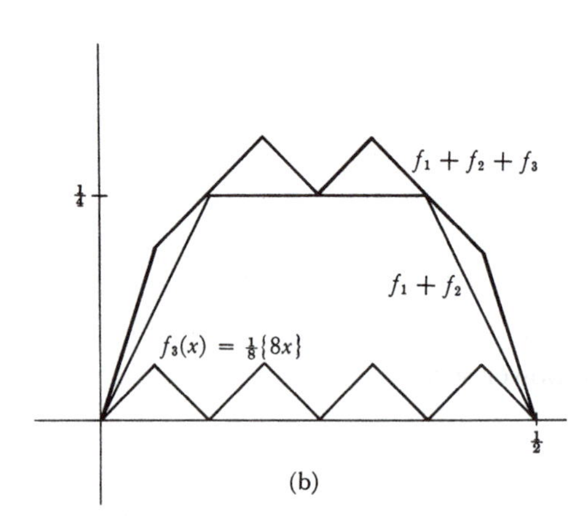

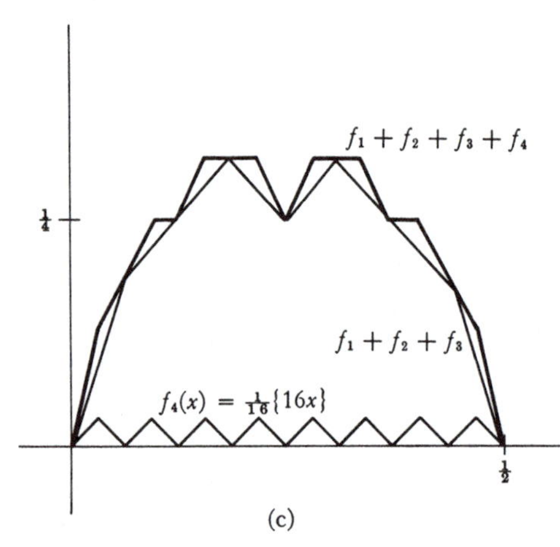

FIGURE 12

(in order for the second equation to be true it is essential that we choose $h_m = -10^{-m}$ when $a_m = 9$). Now suppose that

$$0.a_{n+1}a_{n+2}a_{n+3}\dots a_m \dots \le \tfrac{1}{2}.$$

Then we also have

$$0.a_{n+1}a_{n+2}a_{n+3}\dots(a_m \pm 1)\dots \le \tfrac{1}{2}$$

(in the special case $m = n + 1$ the second equation is true because we chose $h_m = -10^{-m}$ when $a_m = 4$). This means that

$$\{10^n(a + h_m)\} - \{10^n a\} = \pm 10^{n-m},$$

and exactly the same equation can be derived when $0.a_{n+1}a_{n+2}a_{n+3}\dots > \frac{1}{2}$. Thus, for $n < m$ we have

$$10^{m-n}[\{10^n(a + h_m)\} - \{10^n a\}] = \pm 1.$$

In other words,

$$\frac{f(a + h_m) - f(a)}{h_m}$$

is the sum of $m - 1$ numbers, each of which is ± 1. Now adding $+1$ or -1 to a number changes it from odd to even, and vice versa. The sum of $m - 1$ numbers each ± 1 is therefore an *even integer* if m is odd, and an *odd integer* if m is even. Consequently the sequence of ratios

$$\frac{f(a + h_m) - f(a)}{h_m}$$

cannot possibly converge, since it is a sequence of integers which are alternately odd and even. ▮

In addition to its role in the previous theorem, the Weierstrass M-test is an ideal tool for analyzing functions which are very well behaved. We will give special attention to functions of the form

$$f(x) = \sum_{n=0}^{\infty} a_n(x - a)^n,$$

which can also be described by the equation

$$f(x) = \sum_{n=0}^{\infty} f_n(x),$$

for $f_n(x) = a_n(x - a)^n$. Such an infinite sum, of functions which depend only on powers of $(x - a)$, is called a **power series centered at *a***. For the sake of simplicity, we will usually concentrate on power series centered at 0,

$$f(x) = \sum_{n=0}^{\infty} a_n x^n.$$

One especially important group of power series are those of the form

$$\sum_{n=0}^{\infty} \frac{f^{(n)}(a)}{n!}(x-a)^n,$$

where f is some function which has derivatives of all orders at a; this series is called the **Taylor series for f at a**. Of course, it is not necessarily true that

$$f(x) = \sum_{n=0}^{\infty} \frac{f^{(n)}(a)}{n!}(x-a)^n;$$

this equation holds only when the remainder terms satisfy $\lim_{n\to\infty} R_{n,a}(x) = 0$.

We already know that a power series $\sum_{n=0}^{\infty} a_n x^n$ does not necessarily converge for all x. For example, the power series

$$x - \frac{x^3}{3} + \frac{x^5}{5} - \frac{x^7}{7} + \cdots$$

converges only for $|x| \le 1$, while the power series

$$x - \frac{x^2}{2} + \frac{x^3}{3} - \frac{x^4}{4} + \frac{x^5}{5} + \cdots$$

converges only for $-1 < x \le 1$. It is even possible to produce a power series which converges only for $x = 0$. For example, the power series

$$\sum_{n=0}^{\infty} n!\, x^n$$

does not converge for $x \neq 0$; indeed, the ratios

$$\frac{(n+1)!\,(x^{n+1})}{n!\,x^n} = (n+1)x$$

are unbounded for any $x \neq 0$. If a power series $\sum_{n=0}^{\infty} a_n x^n$ does converge for some $x_0 \neq 0$ however, then a great deal can be said about the series $\sum_{n=0}^{\infty} a_n x^n$ for $|x| < |x_0|$.

THEOREM 6 Suppose that the series

$$f(x_0) = \sum_{n=0}^{\infty} a_n {x_0}^n$$

converges, and let a be any number with $0 < a < |x_0|$. Then on $[-a, a]$ the series

$$f(x) = \sum_{n=0}^{\infty} a_n x^n$$

converges uniformly (and absolutely). Moreover, the same is true for the series

$$g(x) = \sum_{n=1}^{\infty} na_n x^{n-1}.$$

Finally, f is differentiable and

$$f'(x) = \sum_{n=1}^{\infty} na_n x^{n-1}$$

for all x with $|x| < |x_0|$.

PROOF Since $\sum_{n=0}^{\infty} a_n {x_0}^n$ converges, the terms $a_n {x_0}^n$ approach 0. Hence they are surely bounded: there is some number M such that

$$|a_n {x_0}^n| = |a_n| \cdot |{x_0}^n| \le M \qquad \text{for all } n.$$

Now if x is in $[-a, a]$, then $|x| \le |a|$, so

$$\begin{aligned} |a_n x^n| &= |a_n| \cdot |x^n| \\ &\le |a_n| \cdot |a^n| \\ &= |a_n| \cdot |x_0|^n \cdot \left|\frac{a}{x_0}\right|^n \qquad \text{(this is the clever step)} \\ &\le M \left|\frac{a}{x_0}\right|^n . \end{aligned}$$

But $|a/x_0| < 1$, so the (geometric) series

$$\sum_{n=0}^{\infty} M \left|\frac{a}{x_0}\right|^n = M \sum_{n=0}^{\infty} \left|\frac{a}{x_0}\right|^n$$

converges. Choosing $M \cdot |a/x_0|^n$ as the number M_n in the Weierstrass M-test, it follows that $\sum_{n=0}^{\infty} a_n x^n$ converges uniformly on $[-a, a]$.

To prove the same assertion for $g(x) = \sum_{n=1}^{\infty} na_n x^{n-1}$ notice that

$$\begin{aligned} |na_n x^{n-1}| &= n|a_n| \cdot |x^{n-1}| \\ &\le n|a_n| \cdot |a^{n-1}| \\ &= \frac{|a_n|}{|a|} \cdot |x_0|^n \, n \left|\frac{a}{x_0}\right|^n \\ &\le \frac{M}{|a|} n \left|\frac{a}{x_0}\right|^n . \end{aligned}$$

Since $|a/x_0| < 1$, the series

$$\sum_{n=1}^{\infty} \frac{M}{|a|} n \left|\frac{a}{x_0}\right|^n = \frac{M}{|a|} \sum_{n=1}^{\infty} n \left|\frac{a}{x_0}\right|^n$$

converges (this fact was proved in Chapter 23 as an application of the ratio test).

Another appeal to the Weierstrass M-test proves that $\sum_{n=1}^{\infty} na_n x^{n-1}$ converges uniformly on $[-a, a]$.

Finally, our corollary proves, first that g is continuous, and then that

$$f'(x) = g(x) = \sum_{n=1}^{\infty} na_n x^{n-1} \qquad \text{for } x \text{ in } [-a, a].$$

Since we could have chosen any number a with $0 < a < |x_0|$, this result holds for all x with $|x| < |x_0|$. ▌

We are now in a position to manipulate power series with ease. Most algebraic manipulations are fairly straightforward consequences of general theorems about infinite series. For example, suppose that $f(x) = \sum_{n=0}^{\infty} a_n x^n$ and $g(x) = \sum_{n=0}^{\infty} b_n x^n$, where the two power series both converge for some x_0. Then for $|x| < |x_0|$ we have

$$\sum_{n=0}^{\infty} a_n x^n + \sum_{n=0}^{\infty} b_n x^n = \sum_{n=0}^{\infty} (a_n x^n + b_n x^n) = \sum_{n=0}^{\infty} (a_n + b_n) x^n.$$

So the series $h(x) = \sum_{n=0}^{\infty} (a_n + b_n) x^n$ also converges for $|x| < |x_0|$, and $h = f + g$ for these x.

The treatment of products is just a little more involved. If $|x| < |x_0|$, then we know that the series $\sum_{n=0}^{\infty} a_n x^n$ and $\sum_{n=0}^{\infty} b_n x^n$ converge *absolutely*. So it follows from Theorem 23-9 that the product $\sum_{n=0}^{\infty} a_n x^n \cdot \sum_{n=0}^{\infty} b_n x^n$ is given by

$$\sum_{i=0}^{\infty} \sum_{j=0}^{\infty} a_i x^i b_j x^j,$$

where the elements $a_i x^i b_j x^j$ are arranged in any order. In particular, we can choose the arrangement

$$a_0 b_0 + (a_0 b_1 + a_1 b_0)x + (a_0 b_2 + a_1 b_1 + a_2 b_0)x^2 + \cdots$$

which can be written as

$$\sum_{n=0}^{\infty} c_n x^n \qquad \text{for } c_n = \sum_{k=0}^{n} a_k b_{n-k}.$$

This is the "Cauchy product" that was introduced in Problem 23-8. Thus, the Cauchy product $h(x) = \sum_{n=0}^{\infty} c_n x^n$ also converges for $|x| < |x_0|$ and $h = fg$ for these x.

Finally, suppose that $f(x) = \sum_{n=0}^{\infty} a_n x^n$, where $a_0 \neq 0$, so that $f(0) = a_0 \neq 0$. Then we can try to find a power series $\sum_{n=0}^{\infty} b_n x^n$ which represents $1/f$. This means that we want to have

$$\sum_{n=0}^{\infty} a_n x^n \cdot \sum_{n=0}^{\infty} b_n x^n = 1 = 1 + 0 \cdot x + 0 \cdot x^2 + \cdots .$$

Since the left side of this equation will be given by the Cauchy product, we want to have

$$\begin{aligned} &a_0 b_0 = 1 \\ &a_0 b_1 + a_1 b_0 = 0 \\ &a_0 b_2 + a_1 b_1 + a_2 b_0 = 0 \\ &\cdots \end{aligned}$$

Since $a_0 \neq 0$, we can solve the first of these equations for b_0. Then we can solve the second for b_1, etc. Of course, we still have to prove that the new series $\sum_{n=0}^{\infty} b_n x^n$ does converge for some $x \neq 0$. This is left as an exercise (Problem 17).

For derivatives, Theorem 6 gives us all the information we need. In particular, when we apply Theorem 6 to the infinite series

$$\begin{aligned} \sin x &= x - \frac{x^3}{3!} + \frac{x^5}{5!} - \frac{x^7}{7!} + \frac{x^9}{9!} - \cdots, \\ \cos x &= 1 - \frac{x^2}{2!} + \frac{x^4}{4!} - \frac{x^6}{6!} + \frac{x^8}{8!} - \cdots, \\ e^x &= 1 + \frac{x}{1!} + \frac{x^2}{2!} + \frac{x^3}{3!} + \frac{x^4}{4!} + \cdots, \end{aligned}$$

we get precisely the results which are expected. Each of these converges for any x_0, hence the conclusions of Theorem 6 apply for any x:

$$\begin{aligned} \sin'(x) &= 1 - \frac{3x^2}{3!} + \frac{5x^4}{5!} - \cdots = \cos x, \\ \cos'(x) &= -\frac{2x}{2!} + \frac{4x^3}{4!} - \frac{6x^5}{6!} + \cdots = -\sin x, \\ \exp'(x) &= 1 + \frac{2x}{2!} + \frac{3x^2}{3!} + \cdots = \exp(x). \end{aligned}$$

For the functions arctan and $f(x) = \log(1+x)$ the situation is only slightly more complicated. Since the series

$$\arctan x = x - \frac{x^3}{3} + \frac{x^5}{5} - \frac{x^7}{7} + \cdots$$

converges for $x_0 = 1$, it also converges for $|x| < 1$, and

$$\arctan'(x) = 1 - x^2 + x^4 - x^6 + \cdots = \frac{1}{1+x^2} \qquad \text{for } |x| < 1.$$

In this case, the series happens to converge for $x = -1$ also. However, the formula for the derivative is not correct for $x = 1$ or $x = -1$; indeed the series

$$1 - x^2 + x^4 - x^6 + \cdots$$

diverges for $x = 1$ and $x = -1$. Notice that this does not contradict Theorem 6, which proves that the derivative is given by the expected formula only for $|x| < |x_0|$.

Since the series

$$\log(1+x) = x - \frac{x^2}{2} + \frac{x^3}{3} - \frac{x^4}{4} + \frac{x^5}{5} - \cdots$$

converges for $x_0 = 1$, it also converges for $|x| < 1$, and

$$\frac{1}{1+x} = \log'(1+x) = 1 - x + x^2 - x^3 + \cdots \qquad \text{for } |x| < 1.$$

In this case, the original series does not converge for $x = -1$; moreover, the differentiated series does not converge for $x = 1$.

All the considerations which apply to a power series will automatically apply to its derivative, at the points where the derivative is represented by a power series. If

$$f(x) = \sum_{n=0}^{\infty} a_n x^n$$

converges for all x in some interval $(-R, R)$, then Theorem 6 implies that

$$f'(x) = \sum_{n=1}^{\infty} n a_n x^{n-1}$$

for all x in $(-R, R)$. Applying Theorem 6 once again we find that

$$f''(x) = \sum_{n=2}^{\infty} n(n-1) a_n x^{n-2},$$

and proceeding by induction we find that

$$f^{(k)}(x) = \sum_{n=k}^{\infty} n(n-1) \cdot \ldots \cdot (n-k+1) a_n x^{n-k}.$$

Thus, a function defined by a power series which converges in some interval $(-R, R)$ is automatically infinitely differentiable in that interval. Moreover, the previous equation implies that

$$f^{(k)}(0) = k!\, a_k,$$

so that

$$a_k = \frac{f^{(k)}(0)}{k!}.$$

In other words, *a convergent power series centered at* 0 *is always the Taylor series at* 0 *of the function which it defines.*

On this happy note we could easily end our study of power series and Taylor series. A careful assessment of our situation will reveal some unexplained facts, however.

The Taylor series of sin, cos, and exp are as satisfactory as we could desire; they converge for all x, and can be differentiated term-by-term for all x. The Taylor series of the function $f(x) = \log(1 + x)$ is slightly less pleasing, because it converges only for $-1 < x \leq 1$, but this deficiency is a necessary consequence of the basic nature of power series. If the Taylor series for f converged for any x_0 with $|x_0| > 1$, then it would converge on the interval $(-|x_0|, |x_0|)$, and on this interval the function which it defines would be differentiable, and thus continuous. But this is impossible, since it is unbounded on the interval $(-1, 1)$, where it equals $\log(1 + x)$.

The Taylor series for arctan is more difficult to comprehend—there seems to be no possible excuse for the refusal of this series to converge when $|x| > 1$. This mysterious behavior is exemplified even more strikingly by the function $f(x) = 1/(1 + x^2)$, an infinitely differentiable function which is the next best thing to a polynomial function. The Taylor series of f is given by

$$f(x) = \frac{1}{1 + x^2} = 1 - x^2 + x^4 - x^6 + x^8 - \cdots .$$

If $|x| \geq 1$ the Taylor series does not converge at all. Why? What unseen obstacle prevents the Taylor series from extending past 1 and -1? Asking this sort of question is always dangerous, since we may have to settle for an unsympathetic answer: it happens because it happens—that's the way things are! In this case there does happen to be an explanation, but this explanation is impossible to give at the present time; although the question is about real numbers, it can be answered intelligently only when placed in a broader context. It will therefore be necessary to devote two chapters to quite new material before completing our discussion of Taylor series in Chapter 27.

PROBLEMS

1. For each of the following sequences $\{f_n\}$, determine the pointwise limit of $\{f_n\}$ (if it exists) on the indicated interval, and decide whether $\{f_n\}$ converges uniformly to this function.

(i) $f_n(x) = \sqrt[n]{x}$, on $[0, 1]$.

(ii) $f_n(x) = \begin{cases} 0, & x \leq n \\ x - n, & x \geq n, \end{cases}$ on $[a, b]$, and on $\mathbf{R}$.

(iii) $f_n(x) = \dfrac{e^x}{x^n}$, on $(1, \infty)$.

(iv) $f_n(x) = e^{-nx^2}$, on $[-1, 1]$.

(v) $f_n(x) = \frac{e^{-x^2}}{n}$, on $\mathbf{R}$.

2. This problem asks for the same information as in Problem 1, but the functions are not so easy to analyze. Some hints are given at the end.

(i) $f_n(x) = x^n - x^{2n}$ on $[0, 1]$.

(ii) $f_n(x) = \dfrac{nx}{1+n+x}$ on $[0, \infty)$.

(iii) $f_n(x) = \sqrt{x^2 + \dfrac{1}{n^2}}$ on $[a, \infty)$, $a > 0$.

(iv) $f_n(x) = \sqrt{x^2 + \dfrac{1}{n^2}}$ on $\mathbf{R}$.

(v) $f_n(x) = \sqrt{x + \dfrac{1}{n}} - \sqrt{x}$ on $[a, \infty)$, $a > 0$.

(vi) $f_n(x) = \sqrt{x + \dfrac{1}{n}} - \sqrt{x}$ on $\mathbf{R}$.

(vii) $f_n(x) = n\left(\sqrt{x + \dfrac{1}{n}} - \sqrt{x}\right)$ on $[a, \infty)$, $a > 0$.

(viii) $f_n(x) = n\left(\sqrt{x + \dfrac{1}{n}} - \sqrt{x}\right)$ on $[0, \infty)$.

Hints: (i) For each n, find the maximum of $|f - f_n|$ on $[0, 1]$. (ii) For each n, consider $|f(x) - f_n(x)|$ for x large. (iii) Express $f(x) - f_n(x)$ as a fraction and estimate $|f(x) - f_n(x)|$ for $x \geq a$. (iv) Give a separate estimate of $|f(x) - f_n(x)|$ for small $|x|$. (vii) Use (v).

3. Find the Taylor series at 0 for each of the following functions.

(i) $f(x) = \dfrac{1}{x-a}$, $\quad a \neq 0$.

(ii) $f(x) = \log(x - a)$, $\quad a \neq 0$.

(iii) $f(x) = \dfrac{1}{\sqrt{1-x}} = (1-x)^{-1/2}$. $\quad$ (Use Problem 20-7.)

(iv) $f(x) = \dfrac{1}{\sqrt{1-x^2}}$.

(v) $f(x) = \arcsin x$.

4. Find each of the following infinite sums.

(i) $1 - x + \dfrac{x^2}{2!} - \dfrac{x^3}{3!} + \dfrac{x^4}{4!} - \cdots$.

(ii) $1 - x^3 + x^6 - x^9 + \cdots$. Hint: What is $1 - x + x^2 - x^3 + \cdots$?

(iii) $\dfrac{x^2}{2} - \dfrac{x^3}{3 \cdot 2} + \dfrac{x^4}{4 \cdot 3} - \dfrac{x^5}{5 \cdot 4} + \cdots$ for $|x| < 1$. Hint: Differentiate.

5. Evaluate the following infinite sums. (In most cases they are $f(a)$ where a is some obvious number and $f(x)$ is given by some power series. To evaluate the various power series, manipulate them until some well-known power series emerge.)

(i) $\displaystyle\sum_{n=0}^{\infty} \frac{(-1)^n 2^{2n} \pi^{2n}}{(2n)!}$.

(ii) $\displaystyle\sum_{n=0}^{\infty} \frac{1}{(2n)!}$.

(iii) $\displaystyle\sum_{n=0}^{\infty} \frac{1}{2n+1}\left(\frac{1}{2}\right)^{2n+1}$

(iv) $\displaystyle\sum_{n=0}^{\infty} \frac{n}{2^n}$.

(v) $\displaystyle\sum_{n=0}^{\infty} \frac{1}{3^n(n+1)}$.

(vi) $\displaystyle\sum_{n=0}^{\infty} \frac{2n+1}{2^n n!}$.

6. If $f(x) = (\sin x)/x$ for $x \neq 0$ and $f(0) = 1$, find $f^{(k)}(0)$. Hint: Find the power series for f.

7. In this problem we deduce the binomial series $(1+x)^\alpha = \displaystyle\sum_{n=0}^{\infty} \binom{\alpha}{n} x^n$, $|x| < 1$ without all the work of Problem 23-18, although we will use a fact established in part (a) of that problem—the series $f(x) = \displaystyle\sum_{n=0}^{\infty} \binom{\alpha}{n} x^n$ does converge for $|x| < 1$.

(a) Prove that $(1+x)f'(x) = \alpha f(x)$ for $|x| < 1$.

(b) Now show that any function f satisfying part (a) is of the form $f(x) = c(1+x)^\alpha$ for some constant c, and use this fact to establish the binomial series. Hint: Consider $g(x) = f(x)/(1+x)^\alpha$.

8. Prove that the series

$$\sum_{n=1}^{\infty} \frac{x}{n(1+nx^2)}$$

converges uniformly on $\mathbf{R}$.

9. (a) Prove that the series

$$\sum_{n=0}^{\infty} 2^n \sin \frac{1}{3^n x}$$

converges uniformly on $[a, \infty)$ for $a > 0$. Hint: $\displaystyle\lim_{h\to 0} (\sin h)/h = 1$.

(b) By considering the sum from N to ∞ for $x = 2/(\pi 3^N)$, show that the series does not converge uniformly on $(0, \infty)$.

10. (a) Prove that the series

$$f(x) = \sum_{n=0}^{\infty} \frac{nx}{1+n^4x^2}$$

converges uniformly on $[a, \infty)$ for $a > 0$. Hint: First find the maximum of $nx/(1+n^4x^2)$ on $[0, \infty)$.

(b) Show that

$$f\left(\frac{1}{N}\right) \geq \frac{N}{2} \sum_{n \geq \sqrt{N}} \frac{1}{n^3},$$

and by using an integral to estimate the sum, show that $f\left(\frac{1}{N}\right) \geq 1/4$. Conclude that the series does not converge uniformly on $\mathbf{R}$.

(c) What about the series

$$\sum_{n=0}^{\infty} \frac{nx}{1+n^5x^2}?$$

11. (a) Use Problem 15-33 and the method of proof used for Dirichlet's test (Problem 23-19) to obtain a uniform Cauchy condition for the series

$$\sum_{n=1}^{\infty} \frac{\sin nx}{n}$$

uniformly on $[\varepsilon, 2\pi - \varepsilon]$, $\varepsilon > 0$, and conclude that the series converges uniformly there.

(b) For $x = \pi/N$, with N large, show that

$$\left| \sum_{k=N}^{2N} \sin kx \right| = \left| \sum_{k=0}^{N} \sin kx \right| \geq \frac{N}{\pi}.$$

Conclude that

$$\left| \sum_{k=N}^{2N} \frac{\sin kx}{k} \right| \geq \frac{1}{2\pi},$$

and that the series does not converge uniformly on $[0, 2\pi]$.

12. (a) Suppose that $f(x) = \sum_{n=0}^{\infty} a_n x^n$ converges for all x in some interval $(-R, R)$ and that $f(x) = 0$ for all x in $(-R, R)$. Prove that each $a_n = 0$. (If you remember the formula for a_n this is easy.)

(b) Suppose we know only that $f(x_n) = 0$ for some sequence $\{x_n\}$ with $\lim_{n \to \infty} x_n = 0$. Prove again that each $a_n = 0$. Hint: First show that $f(0) = a_0 = 0$; then that $f'(0) = a_1 = 0$, etc.

This result shows that if $f(x) = e^{-1/x^2} \sin 1/x$ for $x \neq 0$, then f cannot possibly be written as a power series. It also shows that a function defined by a power series cannot be 0 for $x \leq 0$ but nonzero for $x > 0$—thus a power series cannot describe the motion of a particle which has remained at rest until time 0, and then begins to move!

(c) Suppose that $f(x) = \sum_{n=0}^{\infty} a_n x^n$ and $g(x) = \sum_{n=0}^{\infty} b_n x^n$ converge for all x in some interval containing 0 and that $f(t_m) = g(t_m)$ for some sequence $\{t_m\}$ converging to 0. Show that $a_n = b_n$ for each n. In particular, *a function can have only one representation as a power series centered at* 0.

13. Prove that if $f(x) = \sum_{n=0}^{\infty} a_n x^n$ is an even function, then $a_n = 0$ for n odd, and if f is an odd function, then $a_n = 0$ for n even.

14. Show that the power series for $f(x) = \log(1 - x)$ converges only for $-1 \leq x < 1$, and that the power series for $g(x) = \log[(1 + x)/(1 - x)]$ converges only for x in $(-1, 1)$.

***15.** Recall that the Fibonacci sequence $\{a_n\}$ is defined by $a_1 = a_2 = 1$, $a_{n+1} = a_n + a_{n-1}$.

(a) Show that $a_{n+1}/a_n \leq 2$.
(b) Let

$$f(x) = \sum_{n=1}^{\infty} a_n x^{n-1} = 1 + x + 2x^2 + 3x^3 + \cdots.$$

Use the ratio test to prove that $f(x)$ converges if $|x| < 1/2$.
(c) Prove that if $|x| < 1/2$, then

$$f(x) = \frac{-1}{x^2 + x - 1}.$$

Hint: This equation can be written $f(x) - xf(x) - x^2 f(x) = 1$.
(d) Use the partial fraction decomposition for $1/(x^2 + x - 1)$, and the power series for $1/(x - a)$, to obtain another power series for f.
(e) It follows from Problem 12 that the two power series obtained for f must be the same. Use this fact to show that

$$a_n = \frac{\left(\dfrac{1+\sqrt{5}}{2}\right)^n - \left(\dfrac{1-\sqrt{5}}{2}\right)^n}{\sqrt{5}}.$$

16. Let $f(x) = \sum_{n=0}^{\infty} a_n x^n$ and $g(x) = \sum_{n=0}^{\infty} b_n x^n$. Suppose we merely knew that $f(x)g(x) = \sum_{n=0}^{\infty} c_n x^n$ for some c_n, but we didn't know how to multiply series in general. Use Leibniz's formula (Problem 10-18) to show directly that this series for fg must indeed be the Cauchy product of the series for f and g.

17. Suppose that $f(x) = \sum_{n=0}^{\infty} a_n x^n$ converges for some x_0, and that $a_0 \neq 0$; for simplicity, we'll assume that $a_0 = 1$. Let $\{b_n\}$ be the sequence defined recursively by

$$b_0 = 1$$
$$b_n = -\sum_{k=0}^{n-1} b_k a_{n-k}.$$

The aim of this problem is to show that $\sum_{n=0}^{\infty} b_n x^n$ also converges for some $x \neq 0$, so that it represents $1/f$ for small enough $|x|$.

(a) If all $|a_n x_0{}^n| \le M$, show that

$$|b_n x^n| \le M \sum_{k=0}^{n-1} |b_k x^k|.$$

(b) Choose $M \ge \sqrt{2}$ with all $|a_n x_0{}^n| \le M$. Show that

$$|b_n x_0{}^n| \le M^{2n}.$$

(c) Conclude that $\sum_{n=0}^{\infty} b_n x^n$ converges for $|x|$ sufficiently small.

18. Show that the series

$$\sum_{n=0}^{\infty} \frac{x^{2n+1}}{2n+1} - \frac{x^{n+1}}{2n+2}$$

converges uniformly to $\frac{1}{2}\log(x+1)$ on $[-a, a]$ for $0 < a < 1$, but that at 1 it converges to $\log 2$!

***19.** Suppose that $\sum_{n=0}^{\infty} a_n$ converges. We know that the series $f(x) = \sum_{n=0}^{\infty} a_n x^n$ must converge uniformly on $[-a, a]$ for $0 < a < 1$, but it may not converge uniformly on $[-1, 1]$; in fact, it may not even converge at the point -1 (for example, if $f(x) = \log(1+x)$). However, a beautiful theorem of Abel shows that the series *does* converge uniformly on $[0, 1]$. Consequently, f is continuous on $[0, 1]$ and, in particular, $\sum_{n=0}^{\infty} a_n = \lim_{x \to 1^-} \sum_{n=0}^{\infty} a_n x^n$. Prove Abel's Theorem by noticing that if $|a_m + \cdots + a_n| < \varepsilon$, then $|a_m x^m + \cdots + a_n x^n| < \varepsilon$, by Abel's Lemma (Problem 19-35).

20. A sequence $\{a_n\}$ is called **Abel summable** if $\lim_{x \to 1^-} \sum_{n=0}^{\infty} a_n x^n$ exists; Problem 19 shows that a summable sequence is necessarily Abel summable. Find a sequence which is Abel summable, but which is not summable. Hint: Look over the list of Taylor series until you find one which does not converge at 1, even though the function it represents is continuous at 1.

21. (a) Using Problem 19, find the following infinite sums.

(i) $\dfrac{1}{2 \cdot 1} - \dfrac{1}{3 \cdot 2} + \dfrac{1}{4 \cdot 3} - \dfrac{1}{5 \cdot 4} + \cdots.$

(ii) $1 - \frac{1}{4} + \frac{1}{7} - \frac{1}{10} + \cdots.$

(b) Let $\sum_{n=0}^{\infty} c_n$ be the Cauchy product of two convergent power series $\sum_{n=0}^{\infty} a_n$ and $\sum_{n=0}^{\infty} b_n$, and suppose merely that $\sum_{n=0}^{\infty} c_n$ converges. Prove that, in fact, it converges to the product $\sum_{n=0}^{\infty} a_n \cdot \sum_{n=0}^{\infty} b_n$.

22. (a) Suppose that $\{f_n\}$ is a sequence of bounded (not necessarily continuous) functions on $[a, b]$ which converge uniformly to f on $[a, b]$. Prove that f is bounded on $[a, b]$.

(b) Find a sequence of continuous functions on $[a, b]$ which converge pointwise to an unbounded function on $[a, b]$.

***23.** Suppose that f is differentiable. Prove that the function f' is the pointwise limit of a sequence of continuous functions. (Since we already know examples of discontinuous derivatives, this provides another example where the pointwise limit of continuous functions is not continuous.)

24. Find a sequence of integrable functions $\{f_n\}$ which converges to the (nonintegrable) function f that is 1 on the rationals and 0 on the irrationals. Hint: Each f_n will be 0 except at a few points.

25. (a) Prove that if f is the uniform limit of $\{f_n\}$ on $[a, b]$ and each f_n is integrable on $[a, b]$, then so is f. (So one of the hypotheses in Theorem 1 was unnecessary.)

(b) In Theorem 3 we assumed only that $\{f_n\}$ converges pointwise to f. Show that the remaining hypotheses ensure that $\{f_n\}$ actually converges uniformly to f.

(c) Suppose that in Theorem 3 we do not assume $\{f_n\}$ converges to a function f, but instead assume only that $f_n(x_0)$ converges for some x_0 in $[a, b]$. Show that f_n does converge (uniformly) to some f (with $f' = g$).

(d) Prove that the series

$$\sum_{n=1}^{\infty} \frac{(-1)^n}{x + n}$$

converges uniformly on $[0, \infty)$.

26. Suppose that f_n are continuous functions on $[0, 1]$ that converge uniformly to f. Prove that

$$\lim_{n\to\infty} \int_0^{1-1/n} f_n = \int_0^1 f.$$

Is this true if the convergence isn't uniform?

27. (a) Suppose that $\{f_n\}$ is a sequence of continuous functions on $[a, b]$ which approach 0 pointwise. Suppose moreover that we have $f_n(x) \ge f_{n+1}(x) \ge 0$ for all n and all x in $[a, b]$. Prove that $\{f_n\}$ actually approaches 0 uniformly on $[a, b]$. Hint: Suppose not, choose an appropriate sequence of points x_n in $[a, b]$, and apply the Bolzano-Weierstrass theorem.

(b) Prove Dini's Theorem: If $\{f_n\}$ is a nonincreasing sequence of continuous functions on $[a, b]$ which approach the continuous function f pointwise, then $\{f_n\}$ also approaches f uniformly on $[a, b]$. (The same result holds if $\{f_n\}$ is a nondecreasing sequence.)

(c) Does Dini's Theorem hold if f isn't continuous? How about if $[a, b]$ is replaced by the open interval (a, b)?

28. (a) Suppose that $\{f_n\}$ is a sequence of continuous functions on $[a, b]$ that converges uniformly to f. Prove that if x_n approaches x, then $f_n(x_n)$ approaches $f(x)$.

(b) Is this statement true without assuming that the f_n are continuous?

(c) Prove the converse of part (a): If f is continuous on $[a, b]$ and $\{f_n\}$ is a sequence with the property that $f_n(x_n)$ approaches $f(x)$ whenever x_n approaches x, then f_n converges uniformly to f on $[a, b]$. Hint: If not, there is an $\varepsilon > 0$ and a sequence x_n with $|f_n(x_n) - f(x_n)| > \varepsilon$. Then use the Bolzano-Weierstrass theorem.

29. This problem outlines a completely different approach to the integral; consequently, it is unfair to use any facts about integrals learned previously.

(a) Let s be a step function on $[a, b]$, so that s is constant on (t_{i-1}, t_i) for some partition $\{t_0, \dots, t_n\}$ of $[a, b]$. Define $\int_a^b s$ as $\sum_{i=1}^{n} s_i \cdot (t_i - t_{i-1})$ where s_i is the (constant) value of s on (t_{i-1}, t_i). Show that this definition does not depend on the partition $\{t_0, \dots, t_n\}$.

(b) A function f is called a **regulated** function on $[a, b]$ if it is the uniform limit of a sequence of step functions $\{s_n\}$ on $[a, b]$. Show that in this case there is, for every $\varepsilon > 0$, some N such that for $m, n > N$ we have $|s_n(x) - s_m(x)| < \varepsilon$ for all x in $[a, b]$.

(c) Show that the sequence of numbers $\left\{\int_a^b s_n\right\}$ will be a Cauchy sequence.

(d) Suppose that $\{t_n\}$ is another sequence of step functions on $[a, b]$ which converges uniformly to f. Show that for every $\varepsilon > 0$ there is an N such that for $n > N$ we have $|s_n(x) - t_n(x)| < \varepsilon$ for x in $[a, b]$.

(e) Conclude that $\lim_{n\to\infty} \int_a^b s_n = \lim_{n\to\infty} \int_a^b t_n$. This means that we can *define* $\int_a^b f$ to be $\lim_{n\to\infty} s_n$ for any sequence of step functions $\{s_n\}$ converging uniformly to f. The only remaining question is: Which functions are regulated? Here is a partial answer.

*(f) Prove that a continuous function is regulated. Hint: To find a step function s on $[a, b]$ with $|f(x) - s(x)| < \varepsilon$ for all x in $[a, b]$, consider all y for which there is such a step function on $[a, y]$.

***30.** Find a sequence $\{f_n\}$ approaching f uniformly on $[0, 1]$ for which $\lim_{n\to\infty}$ (length of f_n on $[0, 1]$) $\neq$ length of f on $[0, 1]$. (Length is defined in Problem 13-25, but the simplest example will involve functions the length of whose graphs will be obvious.)

CHAPTER 25 COMPLEX NUMBERS

With the exception of the last few paragraphs of the previous chapter, this book has presented unremitting propaganda for the real numbers. Nevertheless, the real numbers do have a great deficiency—not every polynomial function has a root. The simplest and most notable example is the fact that no number x can satisfy $x^2 + 1 = 0$. This deficiency is so severe that long ago mathematicians felt the need to "invent" a number i with the property that $i^2 + 1 = 0$. For a long time the status of the "number" i was quite mysterious: since there is no number x satisfying $x^2 + 1 = 0$, it is nonsensical to say "let i be the number satisfying $i^2 + 1 = 0$." Nevertheless, admission of the "imaginary" number i to the family of numbers seemed to simplify greatly many algebraic computations, especially when "complex numbers" $a + bi$ (for a and b in $\mathbf{R}$) were allowed, and all the laws of arithmetical computation enumerated in Chapter 1 were assumed to be valid. For example, every quadratic equation

$$ax^2 + bx + c = 0 \qquad (a \neq 0)$$

can be solved formally to give

$$x = \frac{-b + \sqrt{b^2 - 4ac}}{2a} \quad \text{or} \quad x = \frac{-b - \sqrt{b^2 - 4ac}}{2a}.$$

If $b^2 - 4ac \geq 0$, these formulas give correct solutions; when complex numbers are allowed the formulas seem to make sense in all cases. For example, the equation

$$x^2 + x + 1 = 0$$

has no real root, since

$$x^2 + x + 1 = (x + \tfrac{1}{2})^2 + \tfrac{3}{4} > 0, \quad \text{for all } x.$$

But the formula for the roots of a quadratic equation suggest the "solutions"

$$x = \frac{-1 + \sqrt{-3}}{2} \quad \text{and} \quad x = \frac{-1 - \sqrt{-3}}{2};$$

if we understand $\sqrt{-3}$ to mean $\sqrt{3 \cdot (-1)} = \sqrt{3} \cdot \sqrt{-1} = \sqrt{3}\, i$, then these numbers would be

$$-\frac{1}{2} + \frac{\sqrt{3}}{2} i \quad \text{and} \quad -\frac{1}{2} - \frac{\sqrt{3}}{2} i.$$

It is not hard to check that these, as yet purely formal, numbers do indeed satisfy the equation

$$x^2 + x + 1 = 0.$$

It is even possible to "solve" quadratic equations whose coefficients are themselves complex numbers. For example, the equation

$$x^2 + x + 1 + i = 0$$

ought to have the solutions

$$x = \frac{-1 \pm \sqrt{1 - 4(1 + i)}}{2} = \frac{-1 \pm \sqrt{-3 - 4i}}{2},$$

where the symbol $\sqrt{-3 - 4i}$ means a complex number $\alpha + \beta i$ whose square is $-3 - 4i$. In order to have

$$(\alpha + \beta i)^2 = \alpha^2 - \beta^2 + 2\alpha\beta i = -3 - 4i$$

we need

$$\begin{aligned} \alpha^2 - \beta^2 &= -3, \\ 2\alpha\beta &= -4. \end{aligned}$$

These two equations can easily be solved for real α and β; in fact, there are two possible solutions:

$$\begin{matrix} \alpha = 1 \\ \beta = -2 \end{matrix} \qquad \text{and} \qquad \begin{matrix} \alpha = -1 \\ \beta = 2. \end{matrix}$$

Thus the two "square roots" of $-3 - 4i$ are $1 - 2i$ and $-1 + 2i$. There is no reasonable way to decide which one of these should be called $\sqrt{-3 - 4i}$, and which $-\sqrt{-3 - 4i}$; the conventional usage of $\sqrt{x}$ makes sense only for real $x \geq 0$, in which case $\sqrt{x}$ denotes the (real) nonnegative root. For this reason, the solution

$$x = \frac{-1 \pm \sqrt{-3 - 4i}}{2}$$

must be understood as an abbreviation for:

$$x = \frac{-1 + r}{2}, \qquad \text{where } r \text{ is one of the square roots of } -3 - 4i.$$

With this understanding we arrive at the solutions

$$\begin{aligned} x &= \frac{-1 + 1 - 2i}{2} = -i, \\ x &= \frac{-1 - 1 + 2i}{2} = -1 + i; \end{aligned}$$

as you can easily check, these numbers do provide formal solutions for the equation

$$x^2 + x + 1 + i = 0.$$

For cubic equations complex numbers are equally useful. Every cubic equation

$$ax^3 + bx^2 + cx + d = 0 \qquad (a \neq 0)$$

with real coefficients a, b, c, and d, has, as we know, a real root α, and if we divide $ax^3 + bx^2 + cx + d$ by $x - \alpha$ we obtain a second-degree polynomial whose roots are the other roots of $ax^3 + bx^2 + cx + d = 0$; the roots of this second-degree polynomial may be complex numbers. Thus a cubic equation will have either three real roots or one real root and 2 complex roots. The existence of the real root is guaranteed by our theorem that every odd degree equation has a real root, but it is not really necessary to appeal to this theorem (which is of no use at all if the coefficients are complex); in the case of a cubic equation we can, with sufficient cleverness, actually find a formula for all the roots. The following derivation is presented not only as an interesting illustration of the ingenuity of early mathematicians, but as further evidence for the importance of complex numbers (whatever they may be).

To solve the most general cubic equation, it obviously suffices to consider only equations of the form

$$x^3 + bx^2 + cx + d = 0.$$

It is even possible to eliminate the term involving x^2, by a fairly straight-forward manipulation. If we let

$$x = y - \frac{b}{3},$$

then

$$\begin{aligned} x^3 &= y^3 - by^2 + \frac{b^2y}{3} - \frac{b^3}{27}, \\ x^2 &= y^2 - \frac{2by}{3} + \frac{b^2}{9}, \end{aligned}$$

so

$$\begin{aligned} 0 &= x^3 + bx^2 + cx + d \\ &= \left(y^3 - by^2 + \frac{b^2y}{3} - \frac{b^3}{27}\right) + \left(by^2 - \frac{2b^2y}{3} + \frac{b^3}{9}\right) + \left(cy - \frac{bc}{3}\right) + d \\ &= y^3 + \left(\frac{b^2}{3} - \frac{2b^2}{3} + c\right)y + \left(\frac{b^3}{9} - \frac{b^3}{27} - \frac{bc}{3} + d\right). \end{aligned}$$

The right-hand side now contains no term with y^2. If we can solve the equation for y we can find x; this shows that it suffices to consider in the first place only equations of the form

$$x^3 + px + q = 0.$$

In the special case $p = 0$ we obtain the equation $x^3 = -q$. We shall see later on that every complex number does have a cube root, in fact it has three, so that this equation has three solutions. The case $p \neq 0$, on the other hand, requires quite an ingenious step. Let

$$(*) \qquad x = w - \frac{p}{3w}$$

Then

$$\begin{aligned}0 = x^3 + px + q &= \left(w - \frac{p}{3w}\right)^3 + p\left(w - \frac{p}{3w}\right) + q \\ &= w^3 - \frac{3w^2p}{3w} + \frac{3wp^2}{9w^2} - \frac{p^3}{27w^3} + pw - \frac{p^2}{3w} + q \\ &= w^3 - \frac{p^3}{27w^3} + q.\end{aligned}$$

This equation can be written

$$27(w^3)^2 + 27q(w^3) - p^3 = 0,$$

which is a quadratic equation in w^3 (!!).

Thus

$$\begin{aligned}w^3 &= \frac{-27q \pm \sqrt{(27)^2q^2 + 4\cdot 27p^3}}{2\cdot 27} \\ &= -\frac{q}{2} \pm \sqrt{\frac{q^2}{4} + \frac{p^3}{27}}.\end{aligned}$$

Remember that this really means:

$$w^3 = -\frac{q}{2} + r, \quad \text{where } r \text{ is a square root of } \frac{q^2}{4} + \frac{p^3}{27}.$$

We can therefore write

$$w = \sqrt[3]{\frac{-q}{2} \pm \sqrt{\frac{q^2}{4} + \frac{p^3}{27}}};$$

this equation means that w is some cube root of $-q/2 + r$, where r is some square root of $q^2/4 + p^3/27$. This allows six possibilities for w, but when these are substituted into (*), yielding

$$x = \sqrt[3]{\frac{-q}{2} \pm \sqrt{\frac{q^2}{4} + \frac{p^3}{27}}} - \frac{p}{3\cdot\sqrt[3]{\frac{-q}{2} \pm \sqrt{\frac{q^2}{4} + \frac{p^3}{27}}}},$$

it turns out that only 3 different values for x will be obtained! An even more surprising feature of this solution arises when we consider a cubic equation all of whose roots are real; the formula derived above may still involve complex numbers in an essential way. For example, the roots of

$$x^3 - 15x - 4 = 0$$

are 4, $-2+\sqrt{3}$, and $-2-\sqrt{3}$. On the other hand, the formula derived above (with $p=-15$, $q=-4$) gives as one solution

$$\begin{aligned} x &= \sqrt[3]{2+\sqrt{4-125}} - \frac{-15}{3\cdot\sqrt[3]{2+\sqrt{4-125}}} \\ &= \sqrt[3]{2+11i} + \frac{15}{3\cdot\sqrt[3]{2+11i}}. \end{aligned}$$

Now,

$$\begin{aligned} (2+i)^3 &= 2^3 + 3\cdot 2^2 i + 3\cdot 2\cdot i^2 + i^3 \\ &= 8 + 12i - 6 - i \\ &= 2 + 11i, \end{aligned}$$

so one of the cube roots of $2+11i$ is $2+i$. Thus, for one solution of the equation we obtain

$$\begin{aligned} x &= 2+i+\frac{15}{6+3i} \\ &= 2+i+\frac{15}{6+3i}\cdot\frac{6-3i}{6-3i} \\ &= 2+i+\frac{90-45i}{36+9} \\ &= 4\ (!). \end{aligned}$$

The other roots can also be found if the other cube roots of $2+11i$ are known. The fact that even one of these real roots is obtained from an expression which depends on complex numbers is impressive enough to suggest that the use of complex numbers cannot be entirely nonsense. As a matter of fact, the formulas for the solutions of the quadratic and cubic equations can be interpreted entirely in terms of real numbers.

Suppose we agree, for the moment, to write all complex numbers as $a+bi$, writing the real number a as $a+0i$ and the number i as $0+1i$. The laws of ordinary arithmetic and the relation $i^2=-1$ show that

$$\begin{aligned} (a+bi)+(c+di) &= (a+c)+(b+d)i \\ (a+bi)\cdot(c+di) &= (ac-bd)+(ad+bc)i. \end{aligned}$$

Thus, an equation like

$$(1+2i)\cdot(3+1i) = 1+7i$$

may be regarded simply as an abbreviation for the *two* equations

$$\begin{aligned} 1\cdot 3 - 2\cdot 1 &= 1, \\ 1\cdot 1 + 2\cdot 3 &= 7. \end{aligned}$$

The solution of the quadratic equation $ax^2 + bx + c = 0$ with real coefficients could be paraphrased as follows:

$$\text{If } \begin{cases} u^2 - v^2 = b^2 - 4ac, \\ uv = 0, \end{cases}$$
$$\text{(i.e., if } (u + vi)^2 = b^2 - 4ac\text{)},$$

$$\text{then } \begin{cases} a\left[\left(\dfrac{-b+u}{2a}\right)^2 - \left(\dfrac{v}{2a}\right)^2\right] + b\left[\dfrac{-b+u}{2a}\right] + c = 0, \\ a\left[2\left(\dfrac{-b+u}{2a}\right)\left(\dfrac{v}{2a}\right)\right] + b\left[\dfrac{v}{a}\right] = 0, \end{cases}$$

$$\left(\text{i.e., then } a\left(\frac{-b+u+vi}{2a}\right)^2 + b\left(\frac{-b+u+vi}{2a}\right) + c = 0\right).$$

It is not very hard to check this assertion about real numbers without writing down a single "i," but the complications of the statement itself should convince you that equations about complex numbers are worthwhile as abbreviations for pairs of equations about real numbers. (If you are still not convinced, try paraphrasing the solution of the cubic equation.) If we really intend to use complex numbers consistently, however, it is going to be necessary to present some reasonable definition.

One possibility has been implicit in this whole discussion. All mathematical properties of a complex number $a + bi$ are determined completely by the real numbers a and b; any mathematical object with this same property may reasonably be used to define a complex number. The obvious candidate is the ordered pair (a, b) of real numbers; we shall accordingly *define* a complex number to be a pair of real numbers, and likewise *define* what addition and multiplication of complex numbers is to mean.

DEFINITION

A **complex number** is an ordered pair of real numbers; if $z = (a, b)$ is a complex number, then a is called the **real part** of z, and b is called the **imaginary part** of z. The set of all complex numbers is denoted by **C**. If (a, b) and (c, d) are two complex numbers we define

$$(a, b) + (c, d) = (a + c, b + d)$$
$$(a, b) \cdot (c, d) = (a \cdot c - b \cdot d, a \cdot d + b \cdot c).$$

(The $+$ and $\cdot$ appearing on the left side are new symbols being defined, while the $+$ and $\cdot$ appearing on the right side are the familiar addition and multiplication for real numbers.)

When complex numbers were first introduced, it was understood that real numbers were, in particular, complex numbers; if our definition is taken seriously this is not true—a real number is not a pair of real numbers, after all. This difficulty is only a minor annoyance, however. Notice that

$$(a,0)+(b,0)=(a+b,0+0)=(a+b,0),$$
$$(a,0)\cdot(b,0)=(a\cdot b-0\cdot 0, a\cdot 0+0\cdot b)=(a\cdot b,0);$$

this shows that the complex numbers of the form $(a,0)$ behave precisely the same with respect to addition and multiplication of complex numbers as real numbers do with their own addition and multiplication. For this reason we will adopt the convention that $(a,0)$ will be denoted simply by a. The familiar $a+bi$ notation for complex numbers can now be recovered if one more definition is made.

DEFINITION

$$\boldsymbol{i}=(0,1).$$

Notice that $i^2=(0,1)\cdot(0,1)=(-1,0)=-1$ (the last equality sign depends on our convention). Moreover

$$\begin{aligned}(a,b)&=(a,0)+(0,b)\\&=(a,0)+(b,0)\cdot(0,1)\\&=a+bi.\end{aligned}$$

You may feel that our definition was merely an elaborate device for defining complex numbers as "expressions of the form $a+bi$." That is essentially correct; it is a firmly established prejudice of modern mathematics that new objects must be defined as something specific, not as "expressions." Nevertheless, it is interesting to note that mathematicians were sincerely worried about using complex numbers until the modern definition was proposed. Moreover, the precise definition emphasizes one important point. Our aim in introducing complex numbers was to avoid the necessity of paraphrasing statements about complex numbers in terms of their real and imaginary parts. This means that we wish to work with complex numbers in the same way that we worked with rational or real numbers. For example, the solution of the cubic equation required writing $x=w-p/3w$, so we want to know that $1/w$ makes sense. Moreover, w^2 was found by solving a quadratic equation, which requires numerous other algebraic manipulations. In short, we are likely to use, at some time or other, any manipulations performed on real numbers. We certainly do not want to stop each time and justify every step. Fortunately this is not necessary. Since all algebraic manipulations performed on real numbers can be justified by the properties listed in Chapter 1, it is only necessary to check that these properties are also true for complex numbers. In most cases this is quite easy, and these facts will not be listed as formal theorems. For example, the proof of P1,

$$[(a,b)+(c,d)]+(e,f)=(a,b)+[(c,d)+(e,f)]$$

requires only the application of the definition of addition for complex numbers. The left side becomes

$$([a+c]+e,[b+d]+f),$$

and the right side becomes

$$(a + [c + e], b + [d + f]);$$

these two are equal because P1 is true for real numbers. It is a good idea to check P2–P6 and P8 and P9. Notice that the complex numbers playing the role of 0 and 1 in P2 and P6 are $(0, 0)$ and $(1, 0)$, respectively. It is not hard to figure out what $-(a, b)$ is, but the multiplicative inverse for (a, b) required in P7 is a little trickier: if $(a, b) \neq (0, 0)$, then $a^2 + b^2 \neq 0$ and

$$(a, b) \cdot \left(\frac{a}{a^2 + b^2}, \frac{-b}{a^2 + b^2} \right) = (1, 0).$$

This fact could have been guessed in two ways. To find (x, y) with

$$(a, b) \cdot (x, y) = (1, 0)$$

it is only necessary to solve the equations

$$\begin{aligned} ax - by &= 1, \\ bx + ay &= 0. \end{aligned}$$

The solutions are $x = a/(a^2 + b^2)$, $y = -b/(a^2 + b^2)$. It is also possible to reason that if $1/(a + bi)$ means anything, then it should be true that

$$\frac{1}{a + bi} = \frac{1}{a + bi} \cdot \frac{a - bi}{a - bi} = \frac{a - bi}{a^2 + b^2}.$$

Once the existence of inverses has actually been proved (after guessing the inverse by some method), it follows that this manipulation is really valid; it is the easiest one to remember when the inverse of a complex number is actually being sought—it was precisely this trick which we used to evaluate

$$\begin{aligned} \frac{15}{6 + 3i} &= \frac{15}{6 + 3i} \cdot \frac{6 - 3i}{6 - 3i} \\ &= \frac{90 - 45i}{36 + 9}. \end{aligned}$$

Unlike P1–P9, the rules P10–P12 do not have analogues: it is easy to prove that there is *no* set P of *complex* numbers such that P10–P12 are satisfied for all *complex* numbers. In fact, if there were, then P would have to contain 1 (since $1 = 1^2$) and also -1 (since $-1 = i^2$), and this would contradict P10. The absence of P10–P12 will not have disastrous consequences, but it does mean that we cannot define $z < w$ for complex z and w. Also, you may remember that for the real numbers, P10–P12 were used to prove that $1 + 1 \neq 0$. Fortunately, the corresponding fact for complex numbers can be reduced to this one: clearly $(1, 0) + (1, 0) \neq (0, 0)$.

Although we will usually write complex numbers in the form $a + bi$, it is worth remembering that the set of all complex numbers $\mathbf{C}$ is just the collection of all pairs of real numbers. Long ago this collection was identified with the plane, and for this reason the plane is often called the "complex plane." The horizontal axis, which consists of all points $(a, 0)$ for a in $\mathbf{R}$, is often called the *real axis*, and the vertical axis is called the *imaginary axis*. Two important definitions are also related to this geometric picture.

DEFINITION

If $z = x + iy$ is a complex number (with x and y real), then the **conjugate** $\bar{z}$ of z is defined as

$$\bar{z} = x - iy,$$

and the **absolute value** or **modulus** $|z|$ of z is defined as

$$|z| = \sqrt{x^2 + y^2}.$$

(Notice that $x^2 + y^2 \geq 0$, so that $\sqrt{x^2 + y^2}$ is defined unambiguously; it denotes the nonnegative real square root of $x^2 + y^2$.)

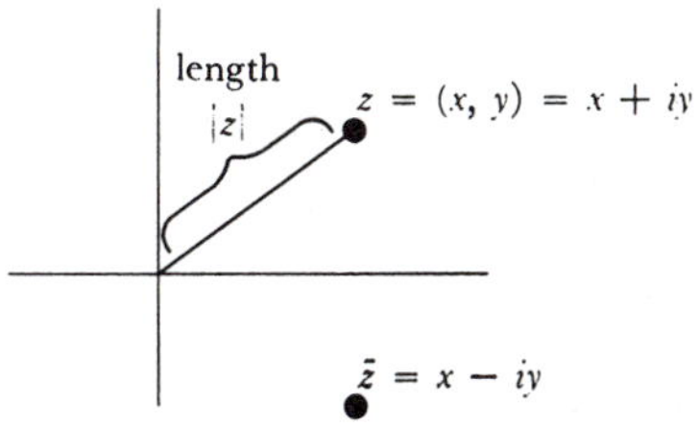

length $|z|$

$z = (x, y) = x + iy$

$\bar{z} = x - iy$

FIGURE 1

Geometrically, $\bar{z}$ is simply the reflection of z in the real axis, while $|z|$ is the distance from z to $(0, 0)$ (Figure 1). Notice that the absolute value notation for complex numbers is consistent with that for real numbers. The **distance** between two complex numbers z and w can be defined quite easily as $|z - w|$. The following theorem lists all the important properties of conjugates and absolute values.

THEOREM 1

Let z and w be complex numbers. Then

(1) $\bar{\bar{z}} = z$.
(2) $\bar{z} = z$ if and only if z is real (i.e., is of the form $a + 0i$, for some real number a).
(3) $\overline{z + w} = \bar{z} + \bar{w}$.
(4) $\overline{-z} = -\bar{z}$.
(5) $\overline{z \cdot w} = \bar{z} \cdot \bar{w}$.
(6) $\overline{z^{-1}} = (\bar{z})^{-1}$, if $z \neq 0$.
(7) $|z|^2 = z \cdot \bar{z}$.
(8) $|z \cdot w| = |z| \cdot |w|$.
(9) $|z + w| \leq |z| + |w|$.

PROOF

Assertions (1) and (2) are obvious. Equations (3) and (5) may be checked by straightforward calculations and (4) and (6) may then be proved by a trick:

$$0 = \bar{0} = \overline{z + (-z)} = \bar{z} + \overline{-z}, \qquad \text{so } \overline{-z} = -\bar{z},$$
$$1 = \bar{1} = \overline{z \cdot (z^{-1})} = \bar{z} \cdot \overline{z^{-1}}, \qquad \text{so } \overline{z^{-1}} = (\bar{z})^{-1}.$$

Equations (7) and (8) may also be proved by a straightforward calculation. The only difficult part of the theorem is (9). This inequality has, in fact, already occurred (Problem 4-9), but the proof will be repeated here, using slightly different terminology.

It is clear that equality holds in (9) if $z = 0$ or $w = 0$. It is also easy to see that (9) is true if $z = \lambda w$ for any real number λ (consider separately the cases $\lambda > 0$ and $\lambda < 0$). Suppose, on the other hand, that $z \neq \lambda w$ for any real number λ, and that

$w \neq 0$. Then, for all real numbers λ,

$$(*) \qquad \begin{aligned} 0 < |z - \lambda w|^2 &= (z - \lambda w) \cdot \overline{(z - \lambda w)} \\ &= (z - \lambda w) \cdot (\bar{z} - \lambda \bar{w}) \\ &= z\bar{z} + \lambda^2 w\bar{w} - \lambda(w\bar{z} + z\bar{w}) \\ &= \lambda^2 |w|^2 + |z|^2 - \lambda(w\bar{z} + z\bar{w}). \end{aligned}$$

Notice that $w\bar{z} + z\bar{w}$ is real, since

$$\overline{w\bar{z} + z\bar{w}} = \bar{w}\bar{\bar{z}} + \bar{z}\bar{\bar{w}} = \bar{w}z + \bar{z}w = w\bar{z} + z\bar{w}.$$

Thus the right side of $(*)$ is a quadratic equation in λ with real coefficients and no real solutions; its discriminant must therefore be negative. Thus

$$(w\bar{z} + z\bar{w})^2 - 4|w|^2 \cdot |z|^2 < 0;$$

it follows, since $w\bar{z} + \bar{z}w$ and $|w| \cdot |z|$ are real numbers, and $|w| \cdot |z| \geq 0$, that

$$(w\bar{z} + z\bar{w}) < 2|w| \cdot |z|.$$

From this inequality it follows that

$$\begin{aligned} |z + w|^2 &= (z + w) \cdot (\bar{z} + \bar{w}) \\ &= |z|^2 + |w|^2 + (w\bar{z} + z\bar{w}) \\ &< |z|^2 + |w|^2 + 2|w| \cdot |z| \\ &= (|z| + |w|)^2, \end{aligned}$$

which implies that

$$|z + w| < |z| + |w|. \blacksquare$$

The operations of addition and multiplication of complex numbers both have important geometric interpretations. The picture for addition is very simple (Figure 2). Two complex numbers $z = (a, b)$ and $w = (c, d)$ determine a parallelogram having for two of its sides the line segment from $(0, 0)$ to z, and the line segment from $(0, 0)$ to w; the vertex opposite $(0, 0)$ is $z + w$ (a proof of this geometric fact is left to you [compare Appendix 1 to Chapter 4]).

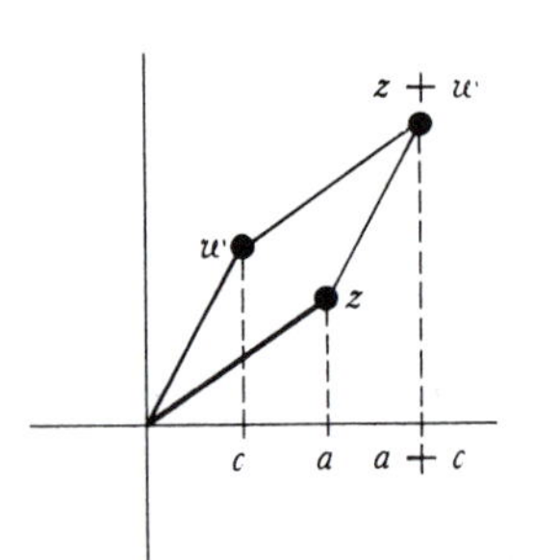

FIGURE 2

The interpretation of multiplication is more involved. If $z = 0$ or $w = 0$, then $z \cdot w = 0$ (a one-line computational proof can be given, but even this is unnecessary—the assertion has already been shown to follow from P1–P9), so we may restrict our attention to nonzero complex numbers. We begin by putting every nonzero complex number into a special form (compare Appendix 3 to Chapter 4).

For any complex number $z \neq 0$ we can write

$$z = |z| \frac{z}{|z|};$$

in this expression, $|z|$ is a positive real number, while

$$\left| \frac{z}{|z|} \right| = \frac{|z|}{|z|} = 1,$$

so that $z/|z|$ is a complex number of absolute value 1. Now any complex number $a = x + iy$ with $1 = |a| = x^2 + y^2$ can be written in the form

$$a = (\cos\theta, \sin\theta) = \cos\theta + i\sin\theta$$

for some number θ. Thus every nonzero complex number z can be written

$$z = r(\cos\theta + i\sin\theta)$$

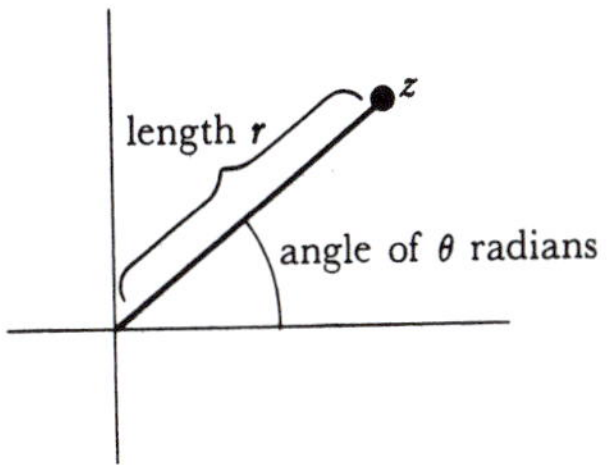

for some $r > 0$ and some number θ. The number r is unique (it equals $|z|$), but θ is not unique; if θ_0 is one possibility, then the others are $\theta_0 + 2k\pi$ for k in $\mathbf{Z}$—any one of these numbers is called an **argument** of z. Figure 3 shows z in terms of r and θ. (To find an argument θ for $z = x + iy$ we may note that the equation

$$x + iy = z = |z|(\cos\theta + i\sin\theta)$$

means that

$$x = |z|\cos\theta,$$
$$y = |z|\sin\theta.$$

So, for example, if $x > 0$ we can take $\theta = \arctan y/x$; if $x = 0$, we can take $\theta = \pi/2$ when $y > 0$ and $\theta = 3\pi/2$ when $y < 0$.)

FIGURE 3

Now the product of two nonzero complex numbers

$$z = r(\cos\theta + i\sin\theta),$$
$$w = s(\cos\phi + i\sin\phi),$$

is

$$\begin{aligned} z\cdot w &= rs(\cos\theta + i\sin\theta)(\cos\phi + i\sin\phi) \\ &= rs[(\cos\theta\cos\phi - \sin\theta\sin\phi) + i(\sin\theta\cos\phi + \cos\theta\sin\phi)] \\ &= rs[\cos(\theta+\phi) + i\sin(\theta+\phi)]. \end{aligned}$$

Thus, the absolute value of a product is the product of the absolute values of the factors, while the sum of any argument for each of the factors will be an argument for the product. For a nonzero complex number

$$z = r(\cos\theta + i\sin\theta)$$

it is now an easy matter to prove by induction the following very important formula (sometimes known as De Moivre's Theorem):

$$z^n = |z|^n(\cos n\theta + i\sin n\theta), \text{ for any argument } \theta \text{ of } z.$$

This formula describes z^n so explicitly that it is easy to decide just when $z^n = w$:

THEOREM 2 Every nonzero complex number has exactly n complex nth roots.

More precisely, for any complex number $w \neq 0$, and any natural number n, there are precisely n different complex numbers z satisfying $z^n = w$.

PROOF Let

$$w = s(\cos\phi + i\sin\phi)$$

for $s = |w|$ and some number ϕ. Then a complex number

$$z = r(\cos\theta + i\sin\theta)$$

satisfies $z^n = w$ if and only if

$$r^n(\cos n\theta + i\sin n\theta) = s(\cos\phi + i\sin\phi),$$

which happens if and only if

$$r^n = s,$$
$$\cos n\theta + i\sin n\theta = \cos\phi + i\sin\phi.$$

From the first equation it follows that

$$r = \sqrt[n]{s},$$

where $\sqrt[n]{s}$ denotes the positive real nth root of s. From the second equation it follows that for some integer k we have

$$\theta = \theta_k = \frac{\phi}{n} + \frac{2k\pi}{n}.$$

Conversely, if we choose $r = \sqrt[n]{s}$ and $\theta = \theta_k$ for some k, then the number $z = r(\cos\theta + i\sin\theta)$ will satisfy $z^n = w$. To determine the number of nth roots of w, it is therefore only necessary to determine which such z are distinct. Now any integer k can be written

$$k = nq + k'$$

for some integer q, and some integer k' between 0 and $n - 1$. Then

$$\cos\theta_k + i\sin\theta_k = \cos\theta_{k'} + i\sin\theta_{k'}.$$

This shows that every z satisfying $z^n = w$ can be written

$$z = \sqrt[n]{s}\,(\cos\theta_k + i\sin\theta_k) \quad k = 0, \dots, n-1.$$

Moreover, it is easy to see that these numbers are all different, since any two θ_k for $k = 0, \dots, n-1$ differ by less than 2π. ▮

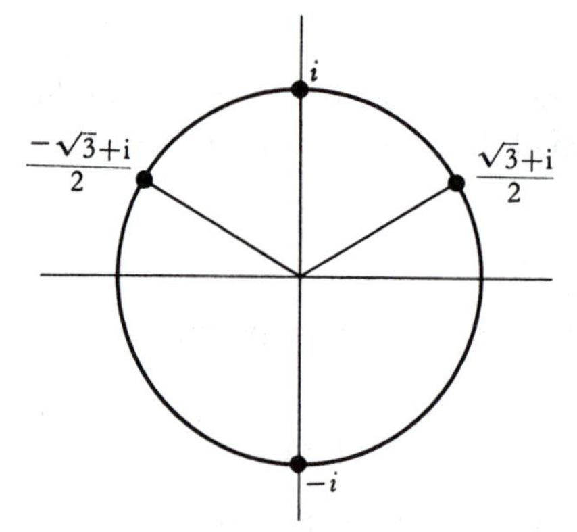

FIGURE 4

In the course of proving Theorem 2, we have actually developed a method for finding the nth roots of a complex number. For example, to find the cube roots of i (Figure 4) note that $|i| = 1$ and that $\pi/2$ is an argument for i. The cube roots of i are therefore

$$1\cdot\left[\cos\frac{\pi}{6} + i\sin\frac{\pi}{6}\right],$$
$$1\cdot\left[\cos\left(\frac{\pi}{6}+\frac{2\pi}{3}\right) + i\sin\left(\frac{\pi}{6}+\frac{2\pi}{3}\right)\right] = \cos\frac{5\pi}{6} + i\sin\frac{5\pi}{6},$$
$$1\cdot\left[\cos\left(\frac{\pi}{6}+\frac{4\pi}{3}\right) + i\sin\left(\frac{\pi}{6}+\frac{4\pi}{3}\right)\right] = \cos\frac{3\pi}{2} + i\sin\frac{3\pi}{2}.$$

Since

$$\begin{aligned} \cos\pi 6 &= \frac{\sqrt{3}}{2}, & \sin\frac{\pi}{6} &= \frac{1}{2}, \\ \cos\frac{5\pi}{6} &= -\frac{\sqrt{3}}{2}, & \sin\frac{5\pi}{6} &= \frac{1}{2}, \\ \cos\frac{3\pi}{2} &= 0, & \sin\frac{3\pi}{2} &= -1, \end{aligned}$$

the cube roots of i are

$$\frac{\sqrt{3}+i}{2}, \quad \frac{-\sqrt{3}+i}{2}, \quad -i.$$

In general, we cannot expect to obtain such simple results. For example, to find the cube roots of $2+11i$, note that $|2+11i| = \sqrt{2^2+11^2} = \sqrt{125}$ and that $\arctan\frac{11}{2}$ is an argument for $2+11i$. One of the cube roots of $2+11i$ is therefore

$$\begin{aligned} &\sqrt[6]{125}\left[\cos\left(\frac{\arctan\frac{11}{2}}{3}\right) + i\sin\left(\frac{\arctan\frac{11}{2}}{3}\right)\right] \\ &\qquad = \sqrt{5}\left[\cos\left(\frac{\arctan\frac{11}{2}}{3}\right) + i\sin\left(\frac{\arctan\frac{11}{2}}{3}\right)\right]. \end{aligned}$$

Previously we noted that $2+i$ is also a cube root of $2+11i$. Since $|2+i| = \sqrt{2^2+1^2} = \sqrt{5}$, and since $\arctan\frac{1}{2}$ is an argument of $2+i$, we can write this cube root as

$$2+i = \sqrt{5}(\cos\arctan\tfrac{1}{2} + i\sin\arctan\tfrac{1}{2}).$$

These two cube roots are actually the same number, because

$$\frac{\arctan\frac{11}{2}}{3} = \arctan\frac{1}{2}$$

(you can check this by using the formula in Problem 15-9), but this is hardly the sort of thing one might notice!

The fact that every complex number has an nth root for all n is just a special case of a very important theorem. The number i was originally introduced in order to provide a solution for the equation $x^2+1=0$. The *Fundamental Theorem of Algebra* states the remarkable fact that this one addition automatically provides solutions for all other polynomial equations: every equation

$$z^n + a_{n-1}z^{n-1} + \cdots + a_0 = 0 \qquad a_0, \ldots, a_{n-1} \text{ in } \mathbf{C}$$

has a complex root!

In the next chapter we shall give an almost complete proof of the Fundamental Theorem of Algebra; the slight gap left in the text can be filled in as an exercise (Problem 26-5). The proof of the theorem will rely on several new concepts which come up quite naturally in a more thorough investigation of complex numbers.

PROBLEMS

1. Find the absolute value and argument of each of the following.

(i) $3+4i$.
(ii) $(3+4i)^{-1}$.
(iii) $(1+i)^5$.
(iv) $\sqrt[7]{3+4i}$.
(v) $|3+4i|$.

2. Solve the following equations.

(i) $x^2+ix+1=0$.
(ii) $x^4+x^2+1=0$.
(iii) $x^2+2ix-1=0$.
(iv) $\begin{cases} ix-(1+i)y=3, \\ (2+i)x+iy=4 \end{cases}$.
(v) $x^3-x^2-x-2=0$.

3. Describe the set of all complex numbers z such that

(i) $\bar{z}=-z$.
(ii) $\bar{z}=z^{-1}$.
(iii) $|z-a|=|z-b|$.
(iv) $|z-a|+|z-b|=c$.
(v) $|z|<1-$ real part of z.

4. Prove that $|z|=|\bar{z}|$, and that the real part of z is $(z+\bar{z})/2$, while the imaginary part is $(z-\bar{z})/2i$.

5. Prove that $|z+w|^2+|z-w|^2=2(|z|^2+|w|^2)$, and interpret this statement geometrically.

6. What is the pictorial relation between z and $\sqrt{i}\cdot\overline{z\sqrt{-i}}$? Hint: Which line goes into the real axis under multiplication by $\sqrt{-i}$?

7. (a) Prove that if $a_0, \dots, a_{n-1}$ are *real* and $a+bi$ (for a and b real) satisfies the equation $z^n+a_{n-1}z^{n-1}+\dots+a_0=0$, then $a-bi$ also satisfies this equation. (Thus the nonreal roots of such an equation always occur in pairs, and the number of such roots is even.)
(b) Conclude that $z^n+a_{n-1}z^{n-1}+\dots+a_0$ is divisible by $z^2-2az+(a^2+b^2)$ (whose coefficients are real).

***8.** (a) Let c be an integer which is not the square of another integer. If a and b are integers we define the **conjugate** of $a+b\sqrt{c}$, denoted by $\overline{a+b\sqrt{c}}$, as $a-b\sqrt{c}$. Show that the conjugate is well defined by showing that a number can be written $a+b\sqrt{c}$, for integers a and b, in only one way.
(b) Show that for all α and β of the form $a+b\sqrt{c}$, we have $\bar{\bar{\alpha}}=\alpha$; $\bar{\alpha}=\alpha$ if

and only if α is an integer; $\overline{\alpha+\beta} = \bar{\alpha} + \bar{\beta}$; $\overline{-\alpha} = -\bar{\alpha}$; $\overline{\alpha \cdot \beta} = \bar{\alpha} \cdot \bar{\beta}$; and $\overline{\alpha^{-1}} = (\bar{\alpha})^{-1}$ if $\alpha \neq 0$.

(c) Prove that if $a_0, \ldots, a_{n-1}$ are *integers* and $z = a + b\sqrt{c}$ satisfies the equation $z^n + a_{n-1}z^{n-1} + \cdots + a_0 = 0$, then $\bar{z} = a - b\sqrt{c}$ also satisfies this equation.

9. Find all the 4th roots of i; express the one having smallest argument in a form that does not involve any trigonometric functions.

***10.** (a) Prove that if ω is an nth root of 1, then so is ω^k.

(b) A number ω is called a **primitive nth root** of 1 if $\{1, \omega, \omega^2, \ldots, \omega^{n-1}\}$ is the set of all nth roots of 1. How many primitive nth roots of 1 are there for $n = 3, 4, 5, 9$?

(c) Let ω be an nth root of 1, with $\omega \neq 1$. Prove that $\displaystyle\sum_{k=0}^{n-1} \omega^k = 0$.

***11.** (a) Prove that if $z_1, \ldots, z_k$ lie on one side of some straight line through 0, then $z_1 + \cdots + z_k \neq 0$. Hint: This is obvious from the geometric interpretation of addition, but an analytic proof is also easy: the assertion is clear if the line is the real axis, and a trick will reduce the general case to this one.

(b) Show further that $z_1{}^{-1}, \ldots, z_k{}^{-1}$ all lie on one side of a straight line through 0, so that $z_1{}^{-1} + \cdots + z_k{}^{-1} \neq 0$.

***12.** Prove that if $|z_1| = |z_2| = |z_3|$ and $z_1 + z_2 + z_3 = 0$, then z_1, z_2, and z_3 are the vertices of an equilateral triangle. Hint: It will help to assume that z_1 is real, and this can be done with no loss of generality. Why?

CHAPTER 26 COMPLEX FUNCTIONS

You will probably not be surprised to learn that a deeper investigation of complex numbers depends on the notion of functions. Until now a function was (intuitively) a rule which assigned real numbers to certain other real numbers. But there is no reason why this concept should not be extended; we might just as well consider a rule which assigns complex numbers to certain other complex numbers. A rigorous definition presents no problems (we will not even accord it the full honors of a formal definition): a function is a collection of pairs of complex numbers which does not contain two distinct pairs with the same first element. Since we consider real numbers to be certain complex numbers, the old definition is really a special case of the new one. Nevertheless, we will sometimes resort to special terminology in order to clarify the context in which a function is being considered. A function f is called **real-valued** if $f(z)$ is a real number for all z in the domain of f, and **complex-valued** to emphasize that it is not necessarily real-valued. Similarly, we will usually state explicitly that a function f is defined on [a subset of] **R** in those cases where the domain of f is [a subset of] **R**; in other cases we sometimes mention that f is defined on [a subset of] **C** to emphasize that $f(z)$ is defined for complex z as well as real z.

Among the multitude of functions defined on **C**, certain ones are particularly important. Foremost among these are the functions of the form

$$f(z) = a_n z^n + a_{n-1} z^{n-1} + \cdots + a_0,$$

where $a_0, \ldots, a_n$ are complex numbers. These functions are called, as in the real case, polynomial functions; they include the function $f(z) = z$ (the "identity function") and functions of the form $f(z) = a$ for some complex number a ("constant functions"). Another important generalization of a familiar function is the "absolute value function" $f(z) = |z|$ for all z in **C**.

Two functions of particular importance for complex numbers are Re (the "real part function") and Im (the "imaginary part function"), defined by

$$\begin{aligned} \operatorname{Re}(x+iy) &= x, \\ \operatorname{Im}(x+iy) &= y, \end{aligned} \qquad \text{for } x \text{ and } y \text{ real.}$$

The "conjugate function" is defined by

$$f(z) = \bar{z} = \operatorname{Re}(z) - i \operatorname{Im}(z).$$

Familiar real-valued functions defined on **R** may be combined in many ways to produce new complex-valued functions defined on **C**—an example is the function

$$f(x+iy) = e^y \sin(x-y) + ix^3 \cos y.$$

The formula for this particular function illustrates a decomposition which is always possible. Any complex-valued function f can be written in the form

$$f = u + iv$$

for some real-valued functions u and v—simply define $u(z)$ as the real part of $f(z)$, and $v(z)$ as the imaginary part. This decomposition is often very useful, but not always; for example, it would be inconvenient to describe a polynomial function in this way.

One other function will play an important role in this chapter. Recall that an *argument* of a nonzero complex number z is a (real) number θ such that

$$z = |z|(\cos\theta + i\sin\theta).$$

There are infinitely many arguments for z, but just one which satisfies $0 \le \theta < 2\pi$. If we call this unique argument $\theta(z)$, then θ is a (real-valued) function (the "argument function") on $\{z \text{ in } \mathbf{C} : z \neq 0\}$.

"Graphs" of complex-valued functions defined on $\mathbf{C}$, since they lie in 4-dimensional space, are presumably not very useful for visualization. The alternative picture of a function mentioned in Chapter 4 can be used instead: we draw two copies of $\mathbf{C}$, and arrows from z in one copy, to $f(z)$ in the other (Figure 1).

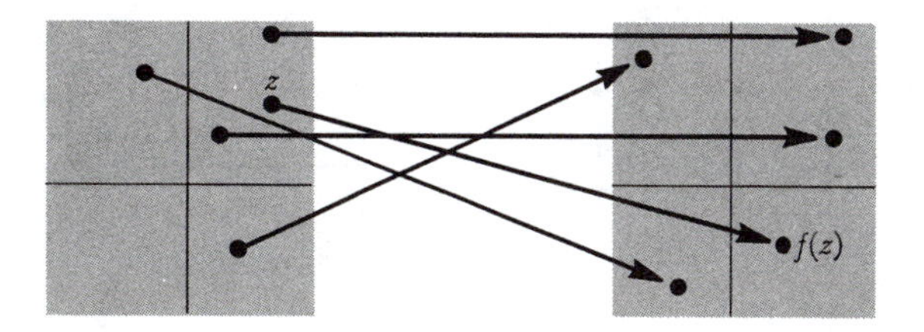

FIGURE 1

The most common pictorial representation of a complex-valued function is produced by labeling a point in the plane with the value $f(z)$, instead of with z (which can be estimated from the position of the point in the picture). Figure 2 shows this sort of picture for several different functions. Certain features of the function are illustrated very clearly by such a "graph." For example, the absolute value function is constant on concentric circles around 0, the functions Re and Im are constant on the vertical and horizontal lines, respectively, and the function $f(z) = z^2$ wraps the circle of radius r twice around the circle of radius r^2.

Despite the problems involved in visualizing complex-valued functions in general, it is still possible to define analogues of important properties previously defined for real-valued functions on $\mathbf{R}$, and in some cases these properties may be easier to visualize in the complex case. For example, the notion of limit can be defined as follows:

> $\lim\limits_{z \to a} f(z) = l$ means that for every (real) number $\varepsilon > 0$ there is a (real) number $\delta > 0$ such that, for all z, if $0 < |z - a| < \delta$, then $|f(z) - l| < \varepsilon$.

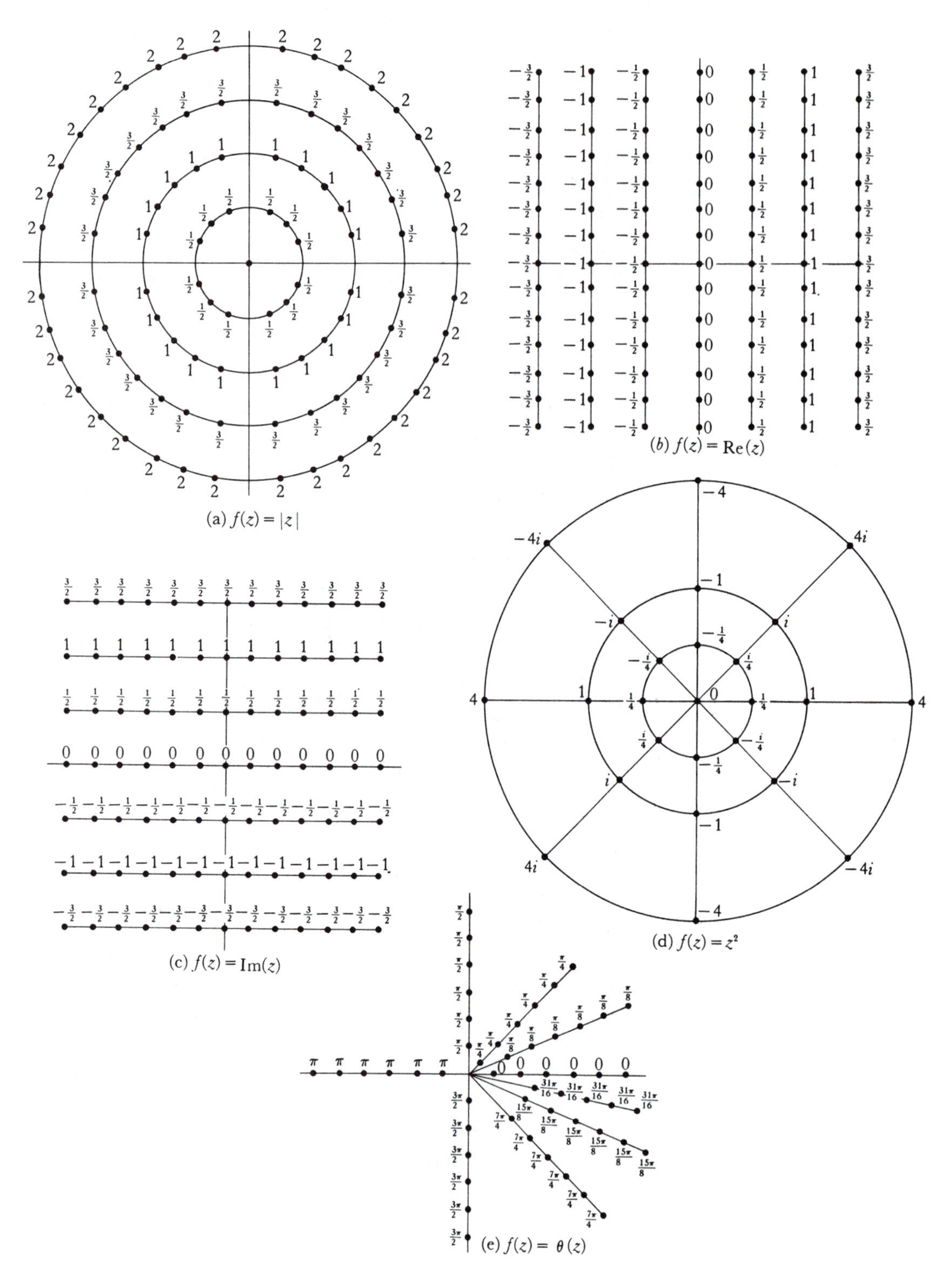

FIGURE 2

Although the definition reads precisely as before, the interpretation is slightly different. Since $|z-w|$ is the distance between the complex numbers z and w, the equation $\lim_{z\to a} f(z) = l$ means that the values of $f(z)$ can be made to lie inside any given circle around l, provided that z is restricted to lie inside a sufficiently small circle around a. This assertion is particularly easy to visualize using the "two copy" picture of a function (Figure 3).

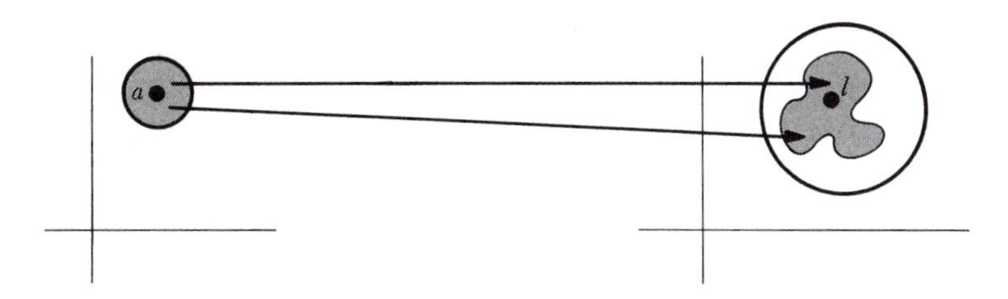

FIGURE 3

Certain facts about limits can be proved exactly as in the real case. In particular,

$$\lim_{z\to a} c = c,$$
$$\lim_{z\to a} z = a,$$
$$\lim_{z\to a}[f(z)+g(z)] = \lim_{z\to a} f(z) + \lim_{z\to a} g(z),$$
$$\lim_{z\to a} f(z)\cdot g(z) = \lim_{z\to a} f(z)\cdot \lim_{z\to a} g(z),$$
$$\lim_{z\to a}\frac{1}{g(z)} = \frac{1}{\lim_{z\to a} g(z)}, \qquad \text{if } \lim_{z\to a} g(z) \neq 0.$$

The essential property of absolute values upon which these results are based is the inequality $|z+w| \le |z| + |w|$, and this inequality holds for complex numbers as well as for real numbers. These facts already provide quite a few limits, but many more can be obtained from the following theorem.

THEOREM 1 Let $f(z) = u(z) + iv(z)$ for real-valued functions u and v, and let $l = \alpha + i\beta$ for real numbers α and β. Then $\lim_{z\to a} f(z) = l$ if and only if

$$\lim_{z\to a} u(z) = \alpha,$$
$$\lim_{z\to a} v(z) = \beta.$$

PROOF Suppose first that $\lim_{z\to a} f(z) = l$. If $\varepsilon > 0$, there is $\delta > 0$ such that, for all z,

$$\text{if } 0 < |z-a| < \delta, \text{ then } |f(z) - l| < \varepsilon.$$

The second inequality can be written

$$\big|[u(z)-\alpha] + i[v(z)-\beta]\big| < \varepsilon,$$

or

$$[u(z) - \alpha]^2 + [v(z) - \beta]^2 < \varepsilon^2.$$

Since $u(z) - \alpha$ and $v(z) - \beta$ are both real numbers, their squares are positive; this inequality therefore implies that

$$[u(z) - \alpha]^2 < \varepsilon^2 \quad \text{and} \quad [v(z) - \beta]^2 < \varepsilon^2,$$

which implies that

$$|u(z) - \alpha| < \varepsilon \quad \text{and} \quad |v(z) - \beta| < \varepsilon.$$

Since this is true for all $\varepsilon > 0$, it follows that

$$\lim_{z \to a} u(z) = \alpha \quad \text{and} \quad \lim_{z \to a} v(z) = \beta.$$

Now suppose that these two equations hold. If $\varepsilon > 0$, there is a $\delta > 0$ such that, for all z, if $0 < |z - a| < \delta$, then

$$|u(z) - \alpha| < \frac{\varepsilon}{2} \quad \text{and} \quad |v(z) - \alpha| < \frac{\varepsilon}{2},$$

which implies that

$$\begin{aligned} |f(z) - l| &= \big|[u(z) - \alpha] + i[v(z) - \beta]\big| \\ &\le |u(z) - \alpha| + |i| \cdot |v(z) - \beta| \\ &< \frac{\varepsilon}{2} + \frac{\varepsilon}{2} = \varepsilon. \end{aligned}$$

This proves that $\lim_{z \to a} f(z) = l$. ▮

In order to apply Theorem 1 fruitfully, notice that since we already know the limit $\lim_{z \to a} z = a$, we can conclude that

$$\begin{aligned} &\lim_{z \to a} \operatorname{Re}(z) = \operatorname{Re}(a), \\ &\lim_{z \to a} \operatorname{Im}(z) = \operatorname{Im}(a). \end{aligned}$$

A limit like

$$\lim_{z \to a} \sin(\operatorname{Re}(z)) = \sin(\operatorname{Re}(a))$$

follows easily, using continuity of sin. Many applications of these principles prove such limits as the following:

$$\begin{gathered} \lim_{z \to a} \bar{z} = \bar{a}, \\ \lim_{z \to a} |z| = |a|, \\ \lim_{(x+iy) \to a+bi} e^y \sin x + ix^3 \cos y = e^b \sin a + ia^3 \cos b. \end{gathered}$$

Now that the notion of limit has been extended to complex functions, the notion of continuity can also be extended: f is **continuous at *a*** if $\lim_{z \to a} f(z) = f(a)$, and

f is **continuous** if f is continuous at a for all a in the domain of f. The previous work on limits shows that all the following functions are continuous:

$$
\begin{aligned}
f(z) &= a_n z^n + a_{n-1} z^{n-1} + \cdots + a_0, \\
f(z) &= \bar{z}, \\
f(z) &= |z|, \\
f(x+iy) &= e^y \sin x + ix^3 \cos y.
\end{aligned}
$$

Examples of discontinuous functions are easy to produce, and certain ones come up very naturally. One particularly frustrating example is the "argument function" θ, which is discontinuous at all nonnegative real numbers (see the "graph" in Figure 2). By suitably redefining θ it is possible to change the discontinuities; for example (Figure 4), if $\theta'(z)$ denotes the unique argument of z with $\pi/2 \le \theta'(z) < 5\pi/2$, then θ' is discontinuous at ai for every nonnegative real number a. But, no matter how θ is redefined, some discontinuities will always occur.

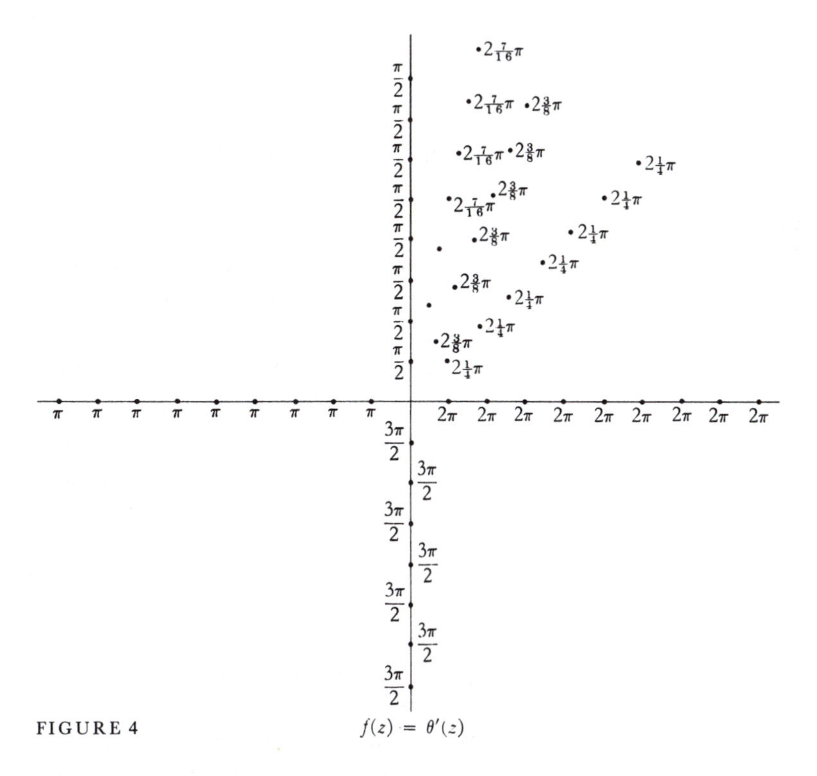

FIGURE 4

The discontinuity of θ has an important bearing on the problem of defining a "square-root function," that is, a function f such that $(f(z))^2 = z$ for all z. For real numbers the function $\sqrt{}$ had as domain only the nonnegative real numbers. If complex numbers are allowed, then every number has two square roots (except 0, which has only one). Although this situation may seem better, it is in some ways worse; since the square roots of z are complex numbers, there is no clear criterion for selecting one root to be $f(z)$, in preference to the other.

One way to define f is the following. We set $f(0) = 0$, and for $z \neq 0$ we set

$$f(z) = \sqrt{|z|}\left(\cos\frac{\theta(z)}{2} + i\sin\frac{\theta(z)}{2}\right).$$

Clearly $(f(z))^2 = z$, but the function f is discontinuous, since θ is discontinuous. As a matter of fact, it is impossible to find a continuous f such that $(f(z))^2 = z$ for all z. In fact, it is even impossible for $f(z)$ to be defined for all z with $|z| = 1$. To prove this by contradiction, we can assume that $f(1) = 1$ (since we could always replace f by $-f$). Then we claim that for all θ with $0 \leq \theta < 2\pi$ we have

$$(*) \qquad f(\cos\theta + i\sin\theta) = \cos\frac{\theta}{2} + i\sin\frac{\theta}{2}.$$

The argument for this is left to you (it is a standard type of least upper bound argument). But $(*)$ implies that

$$\begin{aligned}\lim_{\theta\to 2\pi} f(\cos\theta + i\sin\theta) &= \cos\pi + i\sin\pi \\ &= -1 \\ &\neq f(1),\end{aligned}$$

even though $\cos\theta + i\sin\theta \to 1$ as $\theta \to 2\pi$. Thus, we have our contradiction. A similar argument shows that it is impossible to define continuous "nth-root functions" for any $n \geq 2$.

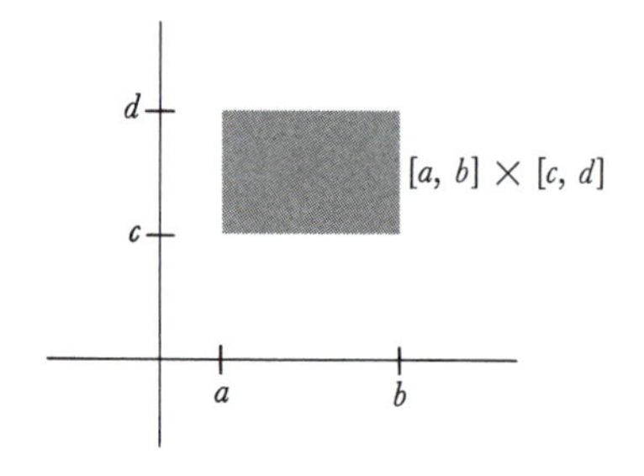

FIGURE 5

For continuous complex functions there are important analogues of certain theorems which describe the behavior of real-valued functions on closed intervals. A natural analogue of the interval $[a, b]$ is the set of all complex numbers $z = x + iy$ with $a \leq x \leq b$ and $c \leq y \leq d$ (Figure 5). This set is called a **closed rectangle**, and is denoted by $[a, b] \times [c, d]$.

If f is a continuous complex-valued function whose domain is $[a, b] \times [c, d]$, then it seems reasonable, and is indeed true, that f is bounded on $[a, b] \times [c, d]$. That is, there is some real number M such that

$$|f(z)| \leq M \quad \text{for all } z \text{ in } [a, b] \times [c, d].$$

It does not make sense to say that f has a maximum and a minimum value on $[a, b] \times [c, d]$, since there is no notion of order for complex numbers. If f is a real-valued function, however, then this assertion does make sense, and is true. In particular, if f is any complex-valued continuous function on $[a, b] \times [c, d]$, then $|f|$ is also continuous, so there is some z_0 in $[a, b] \times [c, d]$ such that

$$|f(z_0)| \leq |f(z)| \quad \text{for all } z \text{ in } [a, b] \times [c, d];$$

a similar statement is true with the inequality reversed. It is sometimes said that "f attains its maximum and minimum modulus on $[a, b] \times [c, d]$."

The various facts listed in the previous paragraph will not be proved here, although proofs are outlined in Problem 5. Assuming these facts, however, we can now give a proof of the Fundamental Theorem of Algebra, which is really quite surprising, since we have not yet said much to distinguish polynomial functions from other continuous functions.

THEOREM 2 (THE FUNDAMENTAL THEOREM OF ALGEBRA) Let $a_0, \dots, a_{n-1}$ be any complex numbers. Then there is a complex number z such that

$$z^n + a_{n-1}z^{n-1} + a_{n-2}z^{n-2} + \cdots + a_0 = 0.$$

PROOF Let

$$f(z) = z^n + a_{n-1}z^{n-1} + \cdots + a_0.$$

Then f is continuous, and so is the function $|f|$ defined by

$$|f|(z) = |f(z)| = |z^n + a_{n-1}z^{n-1} + \cdots + a_0|.$$

Our proof is based on the observation that a point z_0 with $f(z_0) = 0$ would clearly be a minimum point for $|f|$. To prove the theorem we will first show that $|f|$ does indeed have a smallest value on the *whole complex plane*. The proof will be almost identical to the proof, in Chapter 7, that a polynomial function of even degree (with real coefficients) has a smallest value on all of $\mathbf{R}$; both proofs depend on the fact that if $|z|$ is large, then $|f(z)|$ is large.

We begin by writing, for $z \neq 0$,

$$f(z) = z^n\left(1 + \frac{a_{n-1}}{z} + \cdots + \frac{a_0}{z^n}\right),$$

so that

$$|f(z)| = |z|^n \cdot \left|1 + \frac{a_{n-1}}{z} + \cdots + \frac{a_0}{z^n}\right|.$$

Let

$$M = \max(1, 2n|a_{n-1}|, \dots, 2n|a_0|).$$

Then for all z with $|z| \geq M$, we have $|z^k| \geq |z|$ and

$$\frac{|a_{n-k}|}{|z^k|} \leq \frac{|a_{n-k}|}{|z|} \leq \frac{|a_{n-k}|}{2n|a_{n-k}|} = \frac{1}{2n},$$

so

$$\left|\frac{a_{n-1}}{z} + \cdots + \frac{a_0}{z^n}\right| \leq \left|\frac{a_{n-1}}{z}\right| + \cdots + \left|\frac{a_0}{z^n}\right| \leq \frac{1}{2},$$

which implies that

$$\left|1 + \frac{a_{n-1}}{z} + \cdots + \frac{a_0}{z^n}\right| \geq 1 - \left|\frac{a_{n-1}}{z} + \cdots + \frac{a_0}{z^n}\right| \geq \frac{1}{2}.$$

This means that

$$|f(z)| \geq \frac{|z|^n}{2} \quad \text{for } |z| \geq M.$$

In particular, if $|z| \geq M$ and also $|z| \geq \sqrt[n]{2|f(0)|}$, then

$$|f(z)| \geq |f(0)|.$$

Now let $[a, b] \times [c, d]$ be a closed rectangle (Figure 6) which contains $\{z : |z| \le \max(M, \sqrt[n]{2|f(0)|})\}$, and suppose that the minimum of $|f|$ on $[a, b] \times [c, d]$ is attained at z_0, so that

$$(1) \quad |f(z_0)| \le |f(z)| \quad \text{for } z \text{ in } [a, b] \times [c, d].$$

It follows, in particular, that $|f(z_0)| \le |f(0)|$. Thus

$$(2) \quad \text{if } |z| \ge \max(M, \sqrt[n]{2|f(0)|}), \text{ then } |f(z)| \ge |f(0)| \ge |f(z_0)|.$$

Combining (1) and (2) we see that $|f(z_0)| \le |f(z)|$ for all z, so that $|f|$ attains its minimum value on the whole complex plane at z_0.

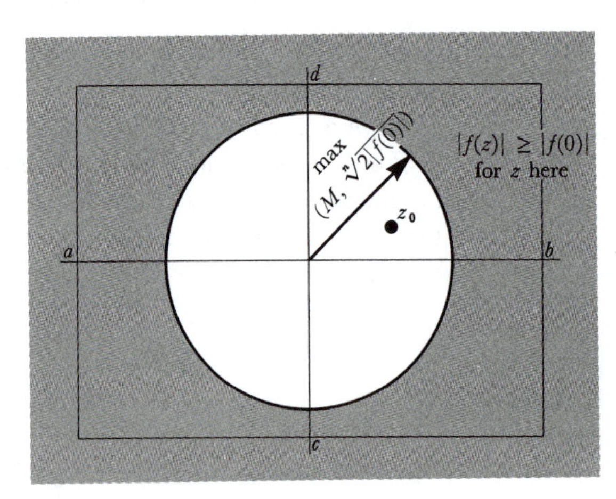

FIGURE 6

To complete the proof of the theorem we now show that $f(z_0) = 0$. It is convenient to introduce the function g defined by

$$g(z) = f(z + z_0).$$

Then g is a polynomial function of degree n, whose minimum absolute value occurs at 0. We want to show that $g(0) = 0$.

Suppose instead that $g(0) = \alpha \ne 0$. If m is the smallest positive power of z which occurs in the expression for g, we can write

$$g(z) = \alpha + \beta z^m + c_{m+1}z^{m+1} + \cdots + c_n z^n,$$

where $\beta \ne 0$. Now, according to Theorem 25-2 there is a complex number γ such that

$$\gamma^m = -\frac{\alpha}{\beta}.$$

Then, setting $d_k = c_k\gamma^k$, we have

$$\begin{aligned}
|g(\gamma z)| &= |\alpha + \beta\gamma^m z^m + d_{m+1}z^{m+1} + \cdots + d_n z^n| \\
&= |\alpha - \alpha z^m + d_{m+1}z^{m+1} + \cdots| \\
&= \left|\alpha\left(1 - z^m + \frac{d_{m+1}}{\alpha}z^{m+1} + \cdots\right)\right| \\
&= \left|\alpha\left(1 - z^m + z^m\left[\frac{d_{m+1}}{\alpha}z + \cdots\right]\right)\right| \\
&= |\alpha| \cdot \left|1 - z^m + z^m\left[\frac{d_{m+1}}{\alpha}z + \cdots\right]\right|.
\end{aligned}$$

This expression, so tortuously arrived at, will enable us to reach a quick contradiction. Notice first that if $|z|$ is chosen small enough, we will have

$$\left|\frac{d_{m+1}}{\alpha}z + \cdots\right| < 1.$$

If we choose, from among all z for which this inequality holds, some z which is *real and positive*, then

$$\left|z^m\left[\frac{d_{m+1}}{\alpha}z + \cdots\right]\right| < |z^m| = z^m.$$

Consequently, if $0 < z < 1$ we have

$$\begin{aligned}
\left|1 - z^m + z^m\left[\frac{d_{m+1}}{\alpha}z + \cdots\right]\right| &\le |1 - z^m| + \left|z^m\left[\frac{d_{m+1}}{\alpha}z + \cdots\right]\right| \\
&= 1 - z^m + \left|z^m\left[\frac{d_{m+1}}{\alpha}z + \cdots\right]\right| \\
&< 1 - z^m + z^m \\
&= 1.
\end{aligned}$$

This is the desired contradiction: for such a number z we have

$$|g(\gamma z)| < |\alpha|,$$

contradicting the fact that $|\alpha|$ is the minimum of $|g|$ on the whole plane. Hence, the original assumption must be incorrect, and $g(0) = 0$. This implies, finally, that $f(z_0) = 0$. ▮

Even taking into account our omission of the proofs for the basic facts about continuous complex functions, this proof verified a deep fact with surprisingly little work. It is only natural to hope that other interesting developments will arise if we pursue further the analogues of properties of real functions. The next obvious step is to define derivatives: a function f is **differentiable at *a*** if

$$\lim_{z\to 0}\frac{f(a+z) - f(a)}{z} \text{ exists,}$$

in which case the limit is denoted by $f'(a)$. It is easy to prove that

$$\begin{aligned}
f'(a) &= 0 \qquad \text{if } f(z) = c,\\
f'(a) &= 1 \qquad \text{if } f(z) = z,\\
(f+g)'(a) &= f'(a) + g'(a),\\
(f\cdot g)'(a) &= f'(a)g(a) + f(a)g'(a),\\
\left(\frac{1}{g}\right)'(a) &= \frac{-g'(a)}{[g(a)]^2} \qquad \text{if } g(a) \neq 0,\\
(f\circ g)'(a) &= f'(g(a))\cdot g'(a);
\end{aligned}$$

the proofs of all these formulas are exactly the same as before. It follows, in particular, that if $f(z) = z^n$, then $f'(z) = nz^{n-1}$. These formulas only prove the differentiability of rational functions however. Many other obvious candidates are *not* differentiable. Suppose, for example, that

$$f(x+iy) = x - iy \quad (\text{i.e., } f(z) = \bar{z}).$$

If f is to be differentiable at 0, then the limit

$$\lim_{(x+iy)\to 0} \frac{f(x+iy) - f(0)}{x+iy} = \lim_{(x+iy)\to 0} \frac{x-iy}{x+iy}$$

must exist. Notice however, that

$$\text{if } y = 0, \text{ then } \frac{x-iy}{x+iy} = 1,$$

and

$$\text{if } x = 0, \text{ then } \frac{x-iy}{x+iy} = -1;$$

therefore this limit cannot possibly exist, since the quotient has both the values 1 and -1 for $x+iy$ arbitrarily close to 0.

In view of this example, it is not at all clear where other differentiable functions are to come from. If you recall the definitions of sin and exp, you will see that there is no hope at all of generalizing these definitions to complex numbers. At the moment the outlook is bleak, but all our problems will soon be solved.

PROBLEMS

1. (a) For any real number y, define $\alpha(x) = x + iy$ (so that α is a complex-valued function defined on **R**). Show that α is continuous. (This follows immediately from a theorem in this chapter.) Show similarly that $\beta(y) = x + iy$ is continuous.

(b) Let f be a continuous function defined on **C**. For fixed y, let $g(x) = f(x+iy)$. Show that g is a continuous function (defined on **R**). Show similarly that $h(y) = f(x+iy)$ is continuous. Hint: Use part (a).

2. (a) Suppose that f is a continuous real-valued function defined on a closed rectangle $[a, b] \times [c, d]$. Prove that if f takes on the values $f(z)$ and $f(w)$

for z and w in $[a, b] \times [c, d]$, then f also takes all values between $f(z)$ and $f(w)$. Hint: Consider $g(t) = f(tz + (1 - t)w)$ for t in $[0, 1]$.

*(b) If f is a continuous complex-valued function defined on $[a, b] \times [c, d]$, the assertion in part (a) no longer makes any sense, since we cannot talk of complex numbers between $f(z)$ and $f(w)$. We might conjecture that f takes on all values on the line segment between $f(z)$ and $f(w)$, but even this is false. Find an example which shows this.

3. (a) Prove that if $a_0, \dots, a_{n-1}$ are any complex numbers, then there are complex numbers $z_1, \dots, z_n$ (not necessarily distinct) such that

$$z^n + a_{n-1}z^{n-1} + \cdots + a_0 = \prod_{i=1}^{n} (z - z_i).$$

(b) Prove that if $a_0, \dots, a_{n-1}$ are *real*, then $z^n + a_{n-1}z^{n-1} + \cdots + a_0$ can be written as a product of linear factors $z+a$ and quadratic factors z^2+az+b all of whose coefficients are real. (Use Problem 25-7.)

4. In this problem we will consider only polynomials with real coefficients. Such a polynomial is called a **sum of squares** if it can be written as $h_1{}^2 + \cdots + h_n{}^2$ for polynomials h_i with real coefficients.

(a) Prove that if f is a sum of squares, then $f(x) \ge 0$ for all x.
(b) Prove that if f and g are sums of squares, then so is $f \cdot g$.
(c) Suppose that $f(x) \ge 0$ for all x. Show that f is a sum of squares. Hint: First write $f(x) = x^k g(x)$, where $g(x) \ne 0$ for all x. Then k must be even (why?), and $g(x) > 0$ for all x. Now use Problem 3(b).

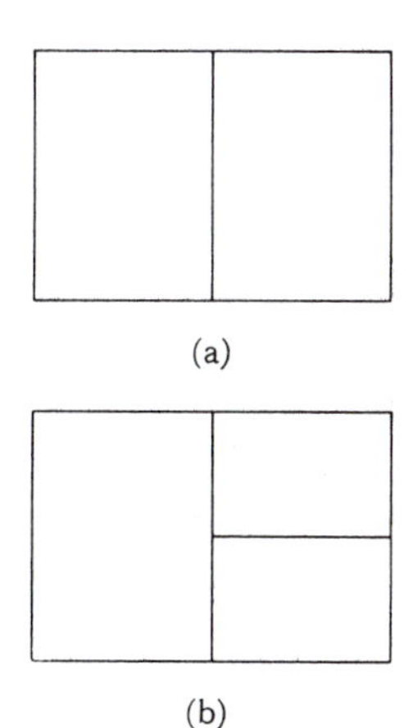

FIGURE 7

5. (a) Let A be a set of complex numbers. A number z is called, as in the real case, a **limit point** of the set A if for every (real) $\varepsilon > 0$, there is a point a in A with $|z - a| < \varepsilon$ but $z \ne a$. Prove the two-dimensional version of the Bolzano-Weierstrass Theorem: If A is an infinite subset of $[a, b] \times [c, d]$, then A has a limit point in $[a, b] \times [c, d]$. Hint: First divide $[a, b] \times [c, d]$ in half by a vertical line as in Figure 7(a). Since A is infinite, at least one half contains infinitely many points of A. Divide this in half by a horizontal line, as in Figure 7(b). Continue in this way, alternately dividing by vertical and horizontal lines.

(The two-dimensional bisection argument outlined in this hint is so standard that the title "Bolzano-Weierstrass" often serves to describe the method of proof, in addition to the theorem itself. See, for example, H. Petard, "A Contribution to the Mathematical Theory of Big Game Hunting," *Amer. Math. Monthly*, **45** (1938), 446–447.)

(b) Prove that a continuous (complex-valued) function on $[a, b] \times [c, d]$ is bounded on $[a, b] \times [c, d]$. (Imitate Problem 22-31.)
(c) Prove that if f is a real-valued continuous function on $[a, b] \times [c, d]$, then f takes on a maximum and minimum value on $[a, b] \times [c, d]$. (You can use the same trick that works for Theorem 7-3.)

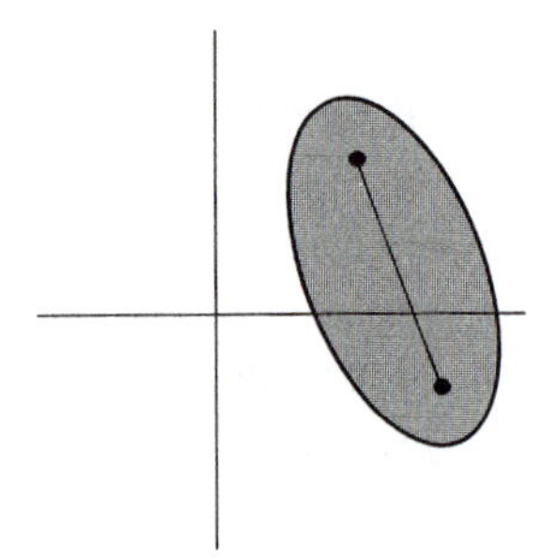

(a) a convex subset of the plane

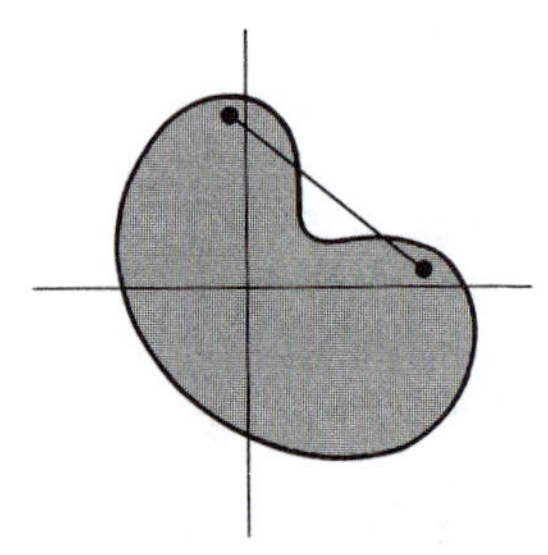

(b) a nonconvex subset of the plane

FIGURE 8

***6.** The proof of Theorem 2 cannot be considered to be completely elementary because the possibility of choosing γ with $\gamma^m = -\alpha/\beta$ depends on Theorem 25-2, and thus on the trigonometric functions. It is therefore of some interest to provide an elementary proof that there is a solution for the equation $z^n - c = 0$.

(a) Make an explicit computation to show that solutions of $z^2 - c = 0$ can be found for any complex number c.
(b) Explain why the solution of $z^n - c = 0$ can be reduced to the case where n is odd.
(c) Let z_0 be the point where the function $f(z) = z^n - c$ has its minimum absolute value. If $z_0 \neq 0$, show that the integer m in the proof of Theorem 2 is equal to 1; since we can certainly find γ with $\gamma^1 = -\alpha/\beta$, the remainder of the proof works for f. It therefore suffices to show that the minimum absolute value of f does not occur at 0.
(d) Suppose instead that f has its minimum absolute value at 0. Since n is odd, the points $\pm\delta$, $\pm\delta i$ go under f into $-c\pm\delta^n$, $-c\pm\delta^n i$. Show that for small δ at least one of these points has smaller absolute value than $-c$, thereby obtaining a contradiction.

7. Let $f(z) = (z - z_1)^{m_1} \cdot \ldots \cdot (z - z_k)^{m_k}$.

(a) Show that $f'(z) = (z - z_1)^{m_1} \cdot \ldots \cdot (z - z_k)^{m_k} \cdot \displaystyle\sum_{\alpha=1}^{k} m_\alpha (z - z_\alpha)^{-1}$.
(b) Let $g(z) = \displaystyle\sum_{\alpha=1}^{k} m_\alpha (z - z_\alpha)^{-1}$. Show that if $g(z) = 0$, then $z_1, \ldots, z_k$ cannot all lie on the same side of a straight line through z. Hint: Use Problem 25-11.
(c) A subset K of the plane is **convex** if K contains the line segment joining any two points in it (Figure 8). For any set A, there is a smallest convex set containing it, which is called the **convex hull** of A (Figure 9); if a

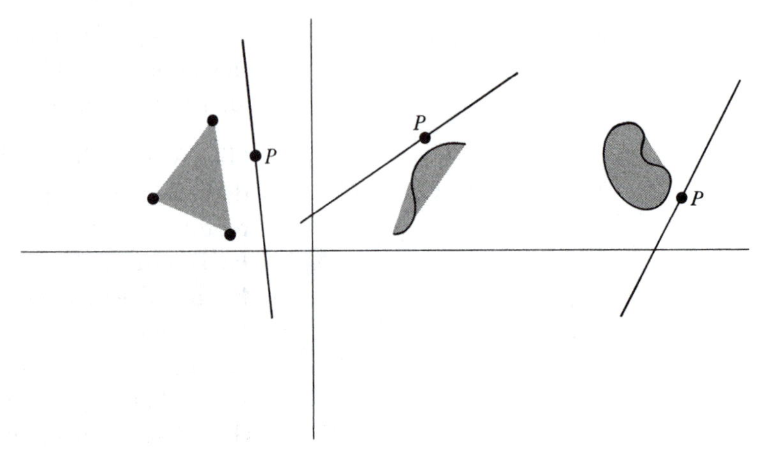

FIGURE 9

point P is not in the convex hull of A, then all of A is contained on one side of some straight line through P. Using this information, prove that the roots of $f'(z) = 0$ lie within the convex hull of the set $\{z_1, \dots, z_k\}$. Further information on convex sets will be found in reference [19] of the Suggested Reading.

8. Prove that if f is differentiable at z, then f is continuous at z.

***9.** Suppose that $f = u + iv$ where u and v are real-valued functions.

(a) For fixed y_0 let $g(x) = u(x + iy_0)$ and $h(x) = v(x + iy_0)$. Show that if $f'(x_0 + iy_0) = \alpha + i\beta$ for real α and β, then $g'(x_0) = \alpha$ and $h'(x_0) = \beta$.

(b) On the other hand, suppose that $k(y) = u(x_0 + iy)$ and $l(y) = v(x_0 + iy)$. Show that $l'(y_0) = \alpha$ and $k'(y_0) = -\beta$.

(c) Suppose that $f'(z) = 0$ for all z. Show that f is a constant function.

10. (a) Using the expression

$$f(x) = \frac{1}{1 + x^2} = \frac{1}{2i}\left(\frac{1}{x - i} - \frac{1}{x + i}\right),$$

find $f^{(k)}(x)$ for all k.

(b) Use this result to find $\arctan^{(k)}(0)$ for all k.

CHAPTER 27 COMPLEX POWER SERIES

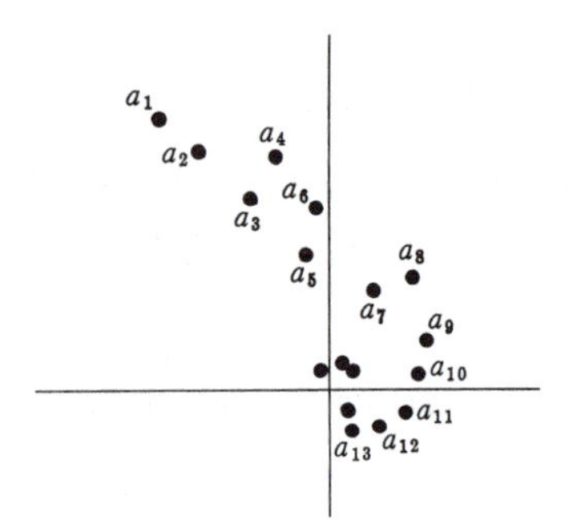

FIGURE 1

If you have not already guessed where differentiable complex functions are going to come from, the title of this chapter should give the secret away: we intend to define functions by means of infinite series. This will necessitate a discussion of infinite sequences of complex numbers, and sums of such sequences, but (as was the case with limits and continuity) the basic definitions are almost exactly the same as for real sequences and series.

An **infinite sequence** of complex numbers is, formally, a complex-valued function whose domain is $\mathbf{N}$; the convenient subscript notation for sequences of real numbers will also be used for sequences of complex numbers. A sequence $\{a_n\}$ of complex numbers is most conveniently pictured by labeling the points a_n in the plane (Figure 1).

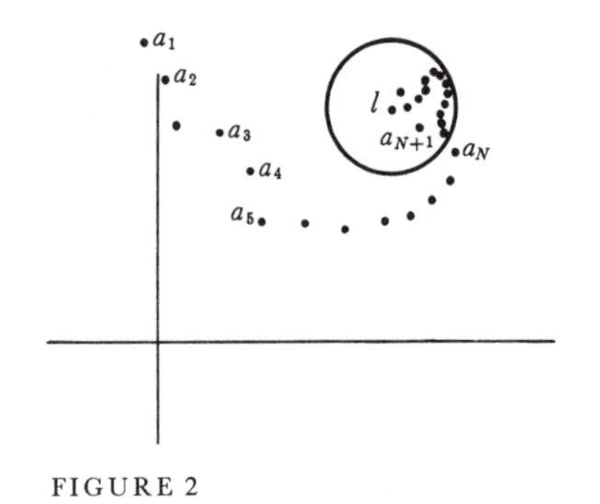

FIGURE 2

The sequence shown in Figure 1 converges to 0, "convergence" of complex sequences being defined precisely as for real sequences: the sequence $\{a_n\}$ **converges** to l, in symbols

$$\lim_{n\to\infty} a_n = l,$$

if for every $\varepsilon > 0$ there is a natural number N such that, for all n,

$$\text{if } n > N, \text{ then } |a_n - l| < \varepsilon.$$

This condition means that any circle drawn around l will contain a_n for all sufficiently large n (Figure 2); expressed more colloquially, the sequence is eventually inside any circle drawn around l.

Convergence of complex sequences is not only defined precisely as for real sequences, but can even be reduced to this familiar case.

THEOREM 1 Let

$$a_n = b_n + ic_n \qquad \text{for real } b_n \text{ and } c_n,$$

and let

$$l = \beta + i\gamma \qquad \text{for real } \beta \text{ and } \gamma.$$

Then $\lim_{n\to\infty} a_n = l$ if and only if

$$\lim_{n\to\infty} b_n = \beta \qquad \text{and} \qquad \lim_{n\to\infty} c_n = \gamma.$$

PROOF The proof is left as an easy exercise. If there is any doubt as to how to proceed, consult the similar Theorem 1 of Chapter 26. ▮

The **sum** of a sequence $\{a_n\}$ is defined, once again, as $\lim_{n\to\infty} s_n$, where

$$s_n = a_1 + \cdots + a_n.$$

Sequences for which this limit exists are **summable**; alternatively, we may say that the infinite series $\sum_{n=1}^{\infty} a_n$ **converges** if this limit exists, and **diverges** otherwise. It is unnecessary to develop any new tests for convergence of infinite series, because of the following theorem.

THEOREM 2 Let

$$a_n = b_n + ic_n \qquad \text{for real } b_n \text{ and } c_n.$$

Then $\sum_{n=1}^{\infty} a_n$ converges if and only if $\sum_{n=1}^{\infty} b_n$ and $\sum_{n=1}^{\infty} c_n$ both converge, and in this case

$$\sum_{n=1}^{\infty} a_n = \sum_{n=1}^{\infty} b_n + i\left(\sum_{n=1}^{\infty} c_n\right).$$

PROOF This is an immediate consequence of Theorem 1 applied to the sequence of partial sums of $\{a_n\}$. ▮

There is also a notion of absolute convergence for complex series: the series $\sum_{n=1}^{\infty} a_n$ **converges absolutely** if the series $\sum_{n=1}^{\infty} |a_n|$ converges (this is a series of real numbers, and consequently one to which our earlier tests may be applied). The following theorem is not quite so easy as the preceding two.

THEOREM 3 Let

$$a_n = b_n + ic_n \qquad \text{for real } b_n \text{ and } c_n.$$

Then $\sum_{n=1}^{\infty} a_n$ converges absolutely if and only if $\sum_{n=1}^{\infty} b_n$ and $\sum_{n=1}^{\infty} c_n$ both converge absolutely.

PROOF Suppose first that $\sum_{n=1}^{\infty} b_n$ and $\sum_{n=1}^{\infty} c_n$ both converge absolutely, i.e., that $\sum_{n=1}^{\infty} |b_n|$ and $\sum_{n=1}^{\infty} |c_n|$ both converge. It follows that $\sum_{n=1}^{\infty} |b_n| + |c_n|$ converges. Now,

$$|a_n| = |b_n + ic_n| \le |b_n| + |c_n|.$$

It follows from the comparison test that $\sum_{n=1}^{\infty} |a_n|$ converges (the numbers $|a_n|$ and $|b_n| + |c_n|$ are real and nonnegative). Thus $\sum_{n=1}^{\infty} a_n$ converges absolutely.

Now suppose that $\sum_{n=1}^{\infty} |a_n|$ converges. Since

$$|a_n| = \sqrt{b_n{}^2 + c_n{}^2},$$

it is clear that

$$|b_n| \le |a_n| \quad \text{and} \quad |c_n| \le |a_n|.$$

Once again, the comparison test shows that $\sum_{n=1}^{\infty} |b_n|$ and $\sum_{n=1}^{\infty} |c_n|$ converge. ∎

Two consequences of Theorem 3 are particularly noteworthy. If $\sum_{n=1}^{\infty} a_n$ converges absolutely, then $\sum_{n=1}^{\infty} b_n$ and $\sum_{n=1}^{\infty} c_n$ also converge absolutely; consequently $\sum_{n=1}^{\infty} b_n$ and $\sum_{n=1}^{\infty} c_n$ converge, by Theorem 23-5, so $\sum_{n=1}^{\infty} a_n$ converges by Theorem 2. In other words, absolute convergence implies convergence. Similar reasoning shows that any rearrangement of an absolutely convergent series has the same sum. These facts can also be proved directly, without using the corresponding theorems for real numbers, by first establishing an analogue of the Cauchy criterion (see Problem 13).

With these preliminaries safely disposed of, we can now consider **complex power series**, that is, functions of the form

$$f(z) = \sum_{n=0}^{\infty} a_n(z-a)^n = a_0 + a_1(z-a) + a_2(z-a)^2 + \cdots .$$

Here the numbers a and a_n are allowed to be complex, and we are naturally interested in the behavior of f for complex z. As in the real case, we shall usually consider power series centered at 0,

$$f(z) = \sum_{n=0}^{\infty} a_n z^n;$$

in this case, if $f(z_0)$ converges, then $f(z)$ will also converge for $|z| < |z_0|$. The proof of this fact will be similar to the proof of Theorem 24-6, but, for reasons that will soon become clear, we will not use all the paraphernalia of uniform convergence and the Weierstrass M-test, even though they have complex analogues. Our next theorem consequently generalizes only a small part of Theorem 24-6.

THEOREM 4 Suppose that

$$\sum_{n=0}^{\infty} a_n z_0{}^n = a_0 + a_1 z_0 + a_2 z_0{}^2 + \cdots$$

converges for some $z_0 \neq 0$. Then if $|z| < |z_0|$, the two series

$$\sum_{n=0}^{\infty} a_n z^n = a_0 + a_1 z + a_2 z^2 + \cdots$$

$$\sum_{n=1}^{\infty} n a_n z^{n-1} = a_1 + 2a_2 z + 3a_3 z^2 + \cdots$$

both converge absolutely.

PROOF As in the proof of Theorem 24-6, we will need only the fact that the set of numbers $a_n {z_0}^n$ is bounded: there is a number M such that

$$|a_n {z_0}^n| \leq M \qquad \text{for all } n.$$

We then have

$$\begin{aligned} |a_n z^n| &= |a_n {z_0}^n| \cdot \left|\frac{z}{z_0}\right|^n \\ &\leq M \left|\frac{z}{z_0}\right|^n, \end{aligned}$$

and, for $z \neq 0$,

$$\begin{aligned} |n a_n z^{n-1}| &= \frac{1}{|z|} n |a_n {z_0}^n| \cdot \left|\frac{z}{z_0}\right|^n \\ &\leq \frac{M}{|z|} n \left|\frac{z}{z_0}\right|^n. \end{aligned}$$

Since the series $\sum_{n=0}^{\infty} |z/z_0|^n$ and $\sum_{n=1}^{\infty} n\,|z/z_0|^n$ converge, this shows that both $\sum_{n=0}^{\infty} a_n z^n$ and $\sum_{n=1}^{\infty} n a_n z^{n-1}$ converge absolutely (the argument for $\sum_{n=1}^{\infty} n a_n z^{n-1}$ assumed that $z \neq 0$, but this series certainly converges for $z = 0$ also). ▮

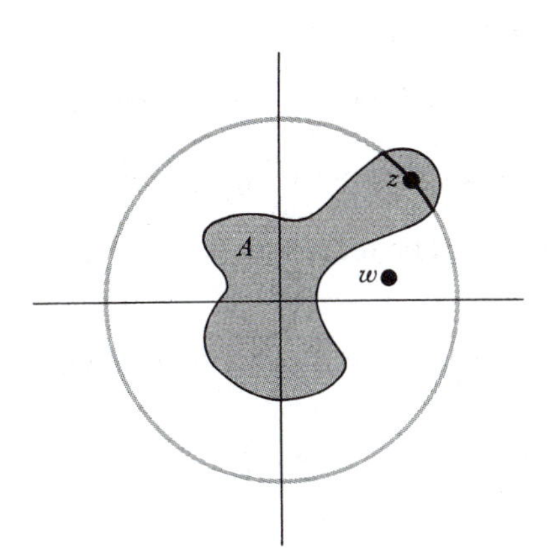

FIGURE 3

Theorem 4 evidently restricts greatly the possibilities for the set

$$\left\{z : \sum_{n=0}^{\infty} a_n z^n \text{ converges}\right\}.$$

For example, the shaded set A in Figure 3 cannot be the set of all z where $\sum_{n=0}^{\infty} a_n z^n$ converges, since it contains z, but not the number w satisfying $|w| < |z|$.

It seems quite unlikely that the set of points where a power series converges could be anything except the set of points inside a circle. If we allow "circles of radius 0" (when the power series converges only at 0) and "circles of radius ∞" (when the power series converges at all points), then this assertion is true (with one complication which we will soon mention); the proof requires only Theorem 4 and a knack for good organization.

THEOREM 5 For any power series

$$\sum_{n=0}^{\infty} a_n z^n = a_0 + a_1 z + a_2 z^2 + a_3 z^3 + \cdots$$

one of the following three possibilities must be true:

(1) $\sum_{n=0}^{\infty} a_n z^n$ converges only for $z = 0$.

(2) $\sum_{n=0}^{\infty} a_n z^n$ converges absolutely for all z in $\mathbf{C}$.

(3) There is a number $R > 0$ such that $\sum_{n=0}^{\infty} a_n z^n$ converges absolutely if $|z| < R$ and diverges if $|z| > R$. (Notice that we do not mention what happens when $|z| = R$.)

PROOF Let

$$S = \left\{ x \text{ in } \mathbf{R} : \sum_{n=0}^{\infty} a_n w^n \text{ converges for some } w \text{ with } |w| = x \right\}.$$

Suppose first that S is unbounded. Then for any complex number z, there is a number x in S such that $|z| < x$. By definition of S, this means that $\sum_{n=0}^{\infty} a_n w^n$ converges for some w with $|w| = x > |z|$. It follows from Theorem 4 that $\sum_{n=0}^{\infty} a_n z^n$ converges absolutely. Thus, in this case possibility (2) is true.

Now suppose that S is bounded, and let R be the least upper bound of S. If $R = 0$, then $\sum_{n=0}^{\infty} a_n z^n$ converges only for $z = 0$, so possibility (1) is true. Suppose, on the other hand, that $R > 0$. Then if z is a complex number with $|z| < R$, there is a number x in S with $|z| < x$. Once again, this means that $\sum_{n=0}^{\infty} a_n w^n$ converges for some w with $|z| < |w|$, so that $\sum_{n=0}^{\infty} a_n z^n$ converges absolutely. Moreover, if $|z| > R$, then $\sum_{n=0}^{\infty} a_n z^n$ does not converge, since $|z|$ is not in S. ▮

The number R which occurs in case (3) is called the **radius of convergence** of $\sum_{n=0}^{\infty} a_n z^n$. In cases (1) and (2) it is customary to say that the radius of convergence is 0 and ∞, respectively. When $0 < R < \infty$, the circle $\{z : |z| = R\}$ is called the **circle of convergence** of $\sum_{n=0}^{\infty} a_n z^n$. If z is outside the circle, then, of course,

$\sum_{n=0}^{\infty} a_n z^n$ does not converge, but actually a much stronger statement can be made: the terms $a_n z^n$ are not even bounded. To prove this, let w be any number with $|z| > |w| > R$; if the terms $a_n z^n$ were bounded, then the proof of Theorem 4 would show that $\sum_{n=0}^{\infty} a_n w^n$ converges, which is false. Thus (Figure 4), inside the circle of convergence the series $\sum_{n=0}^{\infty} a_n z^n$ converges in the best possible way (absolutely) and outside the circle the series diverges in the worst possible way (the terms $a_n z^n$ are not bounded).

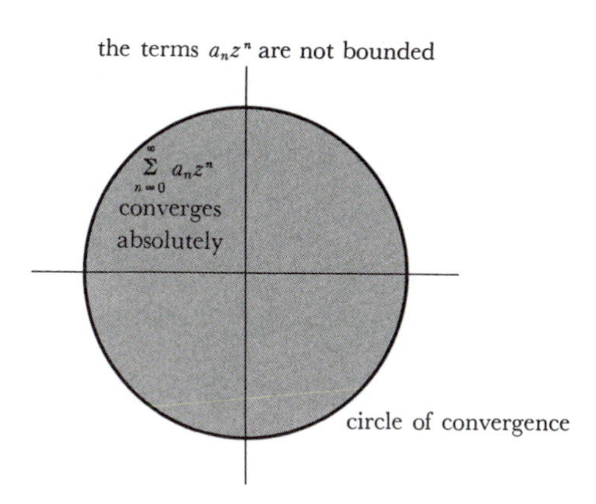

FIGURE 4

What happens *on* the circle of convergence is a much more difficult question. We will not consider that question at all, except to mention that there are power series which converge everywhere on the circle of convergence, power series which converge nowhere on the circle of convergence, and power series that do just about anything in between. (See Problem 5.)

Algebraic manipulations on complex power series can be justified just as in the real case. Thus, if $f(z) = \sum_{n=0}^{\infty} a_n z^n$ and $g(z) = \sum_{n=0}^{\infty} b_n z^n$ both have radius of convergence $\geq R$, then $h(z) = \sum_{n=0}^{\infty} (a_n + b_n) z^n$ also has radius of convergence $\geq R$ and $h = f + g$ inside the circle of radius R. Similarly, the Cauchy product $h(z) = \sum_{n=0}^{\infty} c_n z^n$, for $c_n = \sum_{k=0}^{n} a_k b_{n-k}$, has radius of convergence $\geq R$ and $h = fg$ inside the circle of radius R. And if $f(z) = \sum_{n=0}^{\infty} a_n z^n$ has radius of convergence > 0 and $a_0 \neq 0$, then we can find a power series $\sum_{n=0}^{\infty} b_n z^n$ with radius of convergence > 0 which represents $1/f$ inside its circle of convergence.

But our real goal in this chapter is to produce differentiable functions. We therefore want to generalize the result proved for real power series in Chapter 24, that a function defined by a power series can be differentiated term-by-term inside the circle of convergence. At this point we can no longer imitate the proof of Chapter 24, even if we were willing to introduce uniform convergence, because no analogue of Theorem 24-3 seems available. Instead we will use a direct argument (which could also have been used in Chapter 24). Before beginning the proof, we notice that at least there is no problem about the convergence of the series produced by term-by-term differentiation. If the series $\sum_{n=0}^{\infty} a_n z^n$ has radius of convergence R, then Theorem 4 immediately implies that the series $\sum_{n=1}^{\infty} n a_n z^{n-1}$ also converges for $|z| < R$. Moreover, if $|z| > R$, so that the terms $a_n z^n$ are unbounded, then the terms na_nz^{n-1} are surely unbounded, so $\sum_{n=1}^{\infty} na_nz^{n-1}$ does not converge. This shows that the radius of convergence of $\sum_{n=1}^{\infty} na_nz^{n-1}$ is also exactly R.

THEOREM 6 If the power series

$$f(z) = \sum_{n=0}^{\infty} a_nz^n$$

has radius of convergence $R > 0$, then f is differentiable at z for all z with $|z| < R$, and

$$f'(z) = \sum_{n=1}^{\infty} na_nz^{n-1}.$$

PROOF We will use another "$\varepsilon/3$ argument." The fact that the theorem is clearly true for polynomial functions suggests writing

$$(*) \quad \left|\frac{f(z+h)-f(z)}{h} - \sum_{n=1}^{\infty} na_nz^{n-1}\right| = \left|\sum_{n=0}^{\infty} a_n\frac{((z+h)^n - z^n)}{h} - \sum_{n=1}^{\infty} na_nz^{n-1}\right|$$
$$\leq \left|\sum_{n=0}^{\infty} a_n\frac{((z+h)^n - z^n)}{h} - \sum_{n=0}^{N} a_n\frac{((z+h)^n - z^n)}{h}\right|$$
$$+ \left|\sum_{n=0}^{N} a_n\frac{((z+h)^n - z^n)}{h} - \sum_{n=1}^{N} na_nz^{n-1}\right|$$
$$+ \left|\sum_{n=1}^{N} na_nz^{n-1} - \sum_{n=1}^{\infty} na_nz^{n-1}\right|.$$

We will show that for any $\varepsilon > 0$, each absolute value on the right side can be made $< \varepsilon/3$ by choosing N sufficiently large and h sufficiently small. This will clearly prove the theorem.

Only the first term in the right side of (*) will present any difficulties. To begin with, choose some z_0 with $|z| < |z_0| < R$; henceforth we will consider only h with $|z+h| \leq |z_0|$. The expression $((z+h)^n - z^n)/h$ can be written in a more convenient way if we remember that

$$\frac{x^n - y^n}{x - y} = x^{n-1} + x^{n-2}y + x^{n-3}y^2 + \cdots + y^{n-1}.$$

Applying this to

$$\frac{(z+h)^n - z^n}{h} = \frac{(z+h)^n - z^n}{(z+h) - z},$$

we obtain

$$\frac{(z+h)^n - z^n}{h} = (z+h)^{n-1} + z(z+h)^{n-2} + \cdots + z^{n-1}.$$

Since

$$|(z+h)^{n-1} + z(z+h)^{n-2} + \cdots + z^{n-1}| \leq n|z_0|^{n-1},$$

we have

$$\left| a_n \frac{((z+h)^n - z^n)}{h} \right| \leq n|a_n| \cdot |z_0|^{n-1}.$$

But the series $\sum_{n=1}^{\infty} n|a_n| \cdot |z_0|^{n-1}$ converges, so if N is sufficiently large, then

$$\sum_{n=N+1}^{\infty} n|a_n| \cdot |z_0|^{n-1} < \frac{\varepsilon}{3}.$$

This means that

$$\begin{aligned}
&\left| \sum_{n=0}^{\infty} a_n \frac{((z+h)^n - z^n)}{h} - \sum_{n=0}^{N} a_n \frac{((z+h)^n - z^n)}{h} \right| \\
&\qquad = \left| \sum_{n=N+1}^{\infty} a_n \frac{((z+h)^n - z^n)}{h} \right| \leq \sum_{n=N+1}^{\infty} \left| a_n \frac{((z+h)^n - z^n)}{h} \right| \\
&\qquad \leq \sum_{n=N+1}^{\infty} n|a_n| \cdot |z_0|^{n-1} < \frac{\varepsilon}{3}.
\end{aligned}$$

In short, if N is sufficiently large, then

$$(1) \qquad \left| \sum_{n=0}^{\infty} a_n \frac{((z+h)^n - z^n)}{h} - \sum_{n=0}^{N} a_n \frac{((z+h)^n - z^n)}{h} \right| < \frac{\varepsilon}{3},$$

for *all* h with $|z+h| \leq |z_0|$.

It is easy to deal with the third term on the right side of $(*)$: Since $\sum_{n=1}^{\infty} na_n z^{n-1}$ converges, it follows that if N is sufficiently large, then

$$(2) \qquad \left| \sum_{n=1}^{\infty} na_n z^{n-1} - \sum_{n=1}^{N} na_n z^{n-1} \right| < \frac{\varepsilon}{3}.$$

Finally, choosing an N such that (1) and (2) are true, we note that

$$\lim_{h \to 0} \sum_{n=0}^{N} a_n \frac{((z+h)^n - z^n)}{h} = \sum_{n=1}^{N} na_n z^{n-1},$$

since the polynomial function $g(z) = \sum_{n=0}^{N} a_n z^n$ is certainly differentiable. Therefore

$$(3) \qquad \left| \sum_{n=0}^{N} \frac{a_n((z+h)^n - z^n)}{h} - \sum_{n=1}^{N} na_n z^{n-1} \right| < \frac{\varepsilon}{3}.$$

for sufficiently small h.

As we have already indicated, (1), (2), and (3) prove the theorem. ▮

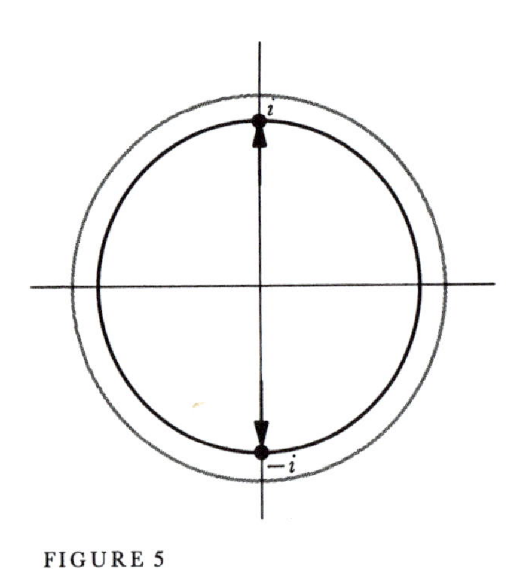

FIGURE 5

Theorem 6 has an obvious corollary: a function represented by a power series is infinitely differentiable inside the circle of convergence, and the power series is its Taylor series at 0. It follows, in particular, that f is continuous inside the circle of convergence, since a function differentiable at z is continuous at z (Problem 26-8).

The continuity of a power series inside its circle of convergence helps explain the behavior of certain Taylor series obtained for real functions, and gives the promised answers to the questions raised at the end of Chapter 24. We have already seen that the Taylor series for the function $f(z) = 1/(1+z^2)$, namely,

$$1 - z^2 + z^4 - z^6 + \cdots,$$

converges for real z only when $|z| < 1$, and consequently has radius of convergence 1. It is no accident that the circle of convergence contains the two points i and $-i$ at which f is undefined. If this power series converged in a circle of radius greater than 1, then (Figure 5) it would represent a function which was continuous in that circle, in particular at i and $-i$. But this is impossible, since it equals $1/(1+z^2)$ inside the unit circle, and $1/(1+z^2)$ does not approach a limit as z approaches i or $-i$ from inside the unit circle.

The use of complex numbers also sheds some light on the strange behavior of the Taylor series for the function

$$f(x) = \begin{cases} e^{-1/x^2}, & x \neq 0 \\ 0, & x = 0. \end{cases}$$

Although we have not yet defined e^z for complex z, it will presumably be true that if y is real and unequal to 0, then

$$f(iy) = e^{-1/(iy)^2} = e^{1/y^2}.$$

The interesting fact about this expression is that it becomes large as y becomes small. Thus f will not even be continuous at 0 when defined for complex numbers, so it is hardly surprising that it is equal to its Taylor series only for $z = 0$.

The method by which we will actually define e^z (as well as $\sin z$ and $\cos z$) for complex z should by now be clear. For real x we know that

$$\sin x = x - \frac{x^3}{3!} + \frac{x^5}{5!} - \cdots,$$
$$\cos x = 1 - \frac{x^2}{2!} + \frac{x^4}{4!} - \cdots,$$
$$e^x = 1 + \frac{x}{1!} + \frac{x^2}{2!} + \cdots$$

For complex z we therefore *define*

$$\sin z = z - \frac{z^3}{3!} + \frac{z^5}{5!} - \cdots,$$
$$\cos z = 1 - \frac{z^2}{2!} + \frac{z^4}{4!} + \cdots,$$
$$\exp(z) = e^z = 1 + \frac{z}{1!} + \frac{z^2}{2!} + \cdots$$

Then $\sin'(z) = \cos z$, $\cos'(z) = -\sin z$, and $\exp'(z) = \exp(z)$ by Theorem 6. Moreover, if we replace z by iz in the series for e^z, and make a rearrangement of the terms (justified by absolute convergence), something particularly interesting happens:

$$\begin{aligned} e^{iz} &= 1 + iz + \frac{(iz)^2}{2!} + \frac{(iz)^3}{3!} + \frac{(iz)^4}{4!} + \frac{(iz)^5}{5!} + \cdots \\ &= 1 + iz - \frac{z^2}{2!} - \frac{iz^3}{3!} + \frac{iz^4}{4!} + \frac{iz^5}{5!} + \cdots \\ &= \left(1 - \frac{z^2}{2!} + \frac{z^4}{4!} - \cdots\right) + i\left(z - \frac{z^3}{3!} + \frac{z^5}{5!} + \cdots\right), \end{aligned}$$

so

$$e^{iz} = \cos z + i \sin z.$$

It is clear from the definitions (i.e., the power series) that

$$\begin{aligned} \sin(-z) &= -\sin z, \\ \cos(-z) &= \cos z, \end{aligned}$$

so we also have

$$e^{-iz} = \cos z - i \sin z.$$

From the equations for e^{iz} and e^{-iz} we can derive the formulas

$$\begin{aligned} \sin z &= \frac{e^{iz} - e^{-iz}}{2i}, \\ \cos z &= \frac{e^{iz} + e^{-iz}}{2}. \end{aligned}$$

The development of complex power series thus places the exponential function at the very core of the development of the elementary functions—it reveals a connection between the trigonometric and exponential functions which was never imagined when these functions were first defined, and which could never have been discovered without the use of complex numbers. As a by-product of this relationship, we obtain a hitherto unsuspected connection between the numbers e and π: if in the formula

$$e^{iz} = \cos z + i \sin z$$

we take $z = \pi$, we obtain the remarkable result

$$e^{i\pi} = -1.$$

(More generally, $e^{2\pi i/n}$ is an nth root of 1.)

With these remarks we will bring to a close our investigation of complex functions. And yet there are still several basic facts about power series which have not been mentioned. Thus far, we have seldom considered power series centered at a,

$$f(z) = \sum_{n=0}^{\infty} a_n (z - a)^n,$$

except for $a = 0$. This omission was adopted partly to simplify the exposition. For power series centered at a there are obvious versions of all the theorems in this chapter (the proofs require only trivial modifications): there is a number R (possibly 0 or "∞") such that the series $\sum_{n=0}^{\infty} a_n(z-a)^n$ converges absolutely for z with $|z-a| < R$, and has unbounded terms for z with $|z-a| > R$; moreover, for all z with $|z-a| < R$ the function

$$f(z) = \sum_{n=0}^{\infty} a_n(z-a)^n$$

has derivative

$$f'(z) = \sum_{n=1}^{\infty} na_n(z-a)^{n-1}.$$

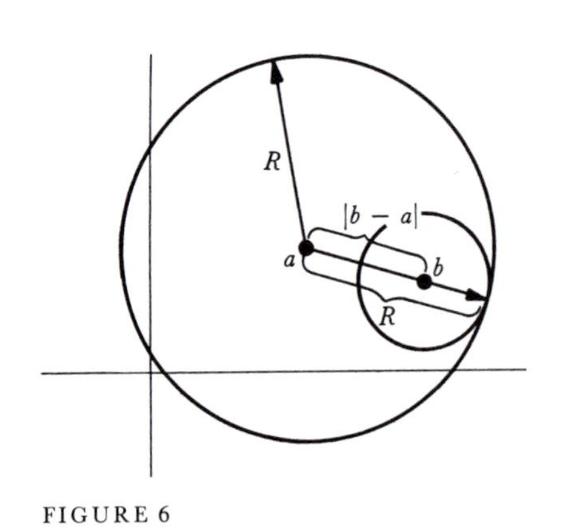

FIGURE 6

It is less straightforward to investigate the possibility of representing a function as a power series centered at b, if it is already written as a power series centered at a. If

$$f(z) = \sum_{n=0}^{\infty} a_n(z-a)^n$$

has radius of convergence R, and b is a point with $|b-a| < R$ (Figure 6), then it is true that $f(z)$ can also be written as a power series centered at b,

$$f(z) = \sum_{n=0}^{\infty} b_n(z-b)^n$$

(the numbers b_n are necessarily $f^{(n)}(b)/n!$); moreover, this series has radius of convergence at least $R - |b-a|$ (*it may be larger*).

We will *not* prove the facts mentioned in the previous paragraph, and there are several other important facts we shall not prove. For example, if

$$f(z) = \sum_{n=0}^{\infty} a_n(z-a)^n \qquad \text{and} \qquad g(z) = \sum_{n=0}^{\infty} b_n(z-b)^n,$$

and $g(b) = a$, then we would expect that $f \circ g$ can be written as a power series centered at b. All such facts could be proved now without introducing any basic new ideas, but the proofs would not be as easy as the proofs about sums, products and reciprocals of power series. The possibility of changing a power series centered at a into one centered at b is quite a bit more involved, and the treatment of $f \circ g$ requires still more skill. Rather than end this section with a *tour de force* of computations, we will instead give a preview of "complex analysis," one of the most beautiful branches of mathematics, where all these facts are derived as straightforward consequences of some fundamental results.

Power series were introduced in this chapter in order to provide complex functions which are differentiable. Since these functions are actually infinitely differentiable, it is natural to suppose that we have therefore selected only a very special collection of differentiable complex functions. The basic theorems of complex analysis show that this is not at all true:

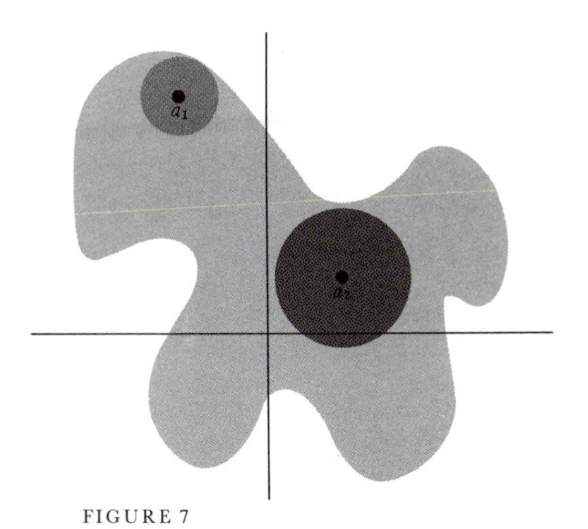

FIGURE 7

> *If a complex function is defined in some region A of the plane and is differentiable in A, then it is automatically infinitely differentiable in A. Moreover, for each point a in A the Taylor series for f at a will converge to f in any circle contained in A (Figure 7).*

These facts are among the first to be proved in complex analysis. It is impossible to give any idea of the proofs themselves—the methods used are quite different from anything in elementary calculus. If these facts are granted, however, then the facts mentioned before can be proved very easily.

Suppose, for example, that f and g are functions which can be written as power series. Then, as we have shown, f and g are differentiable—it then follows from easy general theorems that $f+g$, $f \cdot g$, $1/g$ and $f \circ g$ are also differentiable. Appealing to the results from complex analysis, it follows that they can be written as power series.

We already know how to compute the power series for $f+g$, $f \cdot g$ and $1/g$ from those for f and g. It is also easy to guess how one would compute an expression for $f \circ g$ as a power series in $(z-b)$ when we are given the power series expansions

$$f(z) = \sum_{n=0}^{\infty} a_n(z-a)^n$$

$$g(z) = \sum_{k=0}^{\infty} b_k(z-b)^k,$$

with $a = g(b) = b_0$, so that

$$g(z) - a = \sum_{k=1}^{\infty} b_k(z-b)^k.$$

First of all, we know how to compute the power series

$$(g(z)-a)^l = \left(\sum_{k=1}^{\infty} b_k(z-b)^k\right)^l,$$

and this power series will begin with $(z-b)^l$. Consequently, the coefficient of z^n in

$$f(g(z)) = \sum_{l=0}^{\infty} a_l(g(z)-a)^l$$

can be calculated as a finite sum, involving only coefficients arising from the first n powers of $g(z) - a$.

Similarly, if

$$f(z) = \sum_{n=0}^{\infty} a_n(z-a)^n$$

has radius of convergence R, then f is differentiable in the region $A = \{z : |z-a| < R\}$. Thus, if b is in A, it is possible to write f as a power series centered at b, which will converge in the circle of radius $R - |b - a|$. The coefficient of z^n will be $f^{(n)}(b)/n!$. This series may actually converge in a larger circle, because $\sum_{n=0}^{\infty} a_n(z-a)^n$ may be the series for a function differentiable in a larger region than A. For example, suppose that $f(z) = 1/(1+z^2)$. Then f is differentiable, except at i and $-i$, where it is not defined. Thus $f(z)$ can be written as a power series $\sum_{n=0}^{\infty} a_n z^n$ with radius of convergence 1 (as a matter of fact, we know that $a_{2n} = (-1)^n$ and $a_k = 0$ if k is odd). It is also possible to write

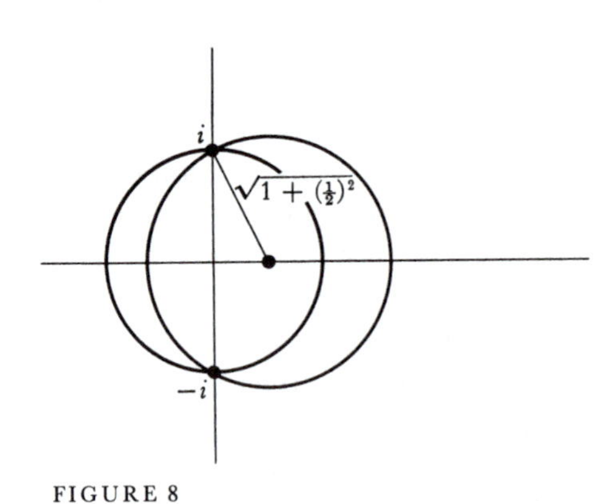

FIGURE 8

$$f(z) = \sum_{n=0}^{\infty} b_n (z - \tfrac{1}{2})^n,$$

where $b_n = f^{(n)}(\frac{1}{2})/n!$. We can easily predict the radius of convergence of this series: it is $\sqrt{1+(\frac{1}{2})^2}$, the distance from $\frac{1}{2}$ to i or $-i$ (Figure 8).

As an added incentive to investigate complex analysis further, one more result will be mentioned, which lies quite near the surface, and which will be found in any treatment of the subject.

For real z the values of $\sin z$ always lie between -1 and 1, but for complex z this is not at all true. In fact, if $z = iy$, for y real, then

$$\sin iy = \frac{e^{i(iy)} - e^{-i(iy)}}{2i} = \frac{e^{-y} - e^{y}}{2i}.$$

If y is large, then $\sin iy$ is also large in absolute value. This behavior of sin is typical of functions which are defined and differentiable on the whole complex plane (such functions are called *entire*). A result which comes quite early in complex analysis is the following:

Liouville's Theorem: The only bounded entire functions are the constant functions.

As a simple application of Liouville's Theorem, consider a polynomial function

$$f(z) = z^n + a_{n-1}z^{n-1} + \cdots + a_0,$$

where $n > 1$, so that f is not a constant. We already know that $f(z)$ is large for large z, so Liouville's Theorem tells us nothing interesting about f. But consider the function

$$g(z) = \frac{1}{f(z)}.$$

If $f(z)$ were never 0, then g would be entire; since $f(z)$ becomes large for large z, the function g would also be bounded, contradicting Liouville's Theorem. Thus $f(z) = 0$ for some z, and we have proved the Fundamental Theorem of Algebra.

PROBLEMS

1. Decide whether each of the following series converges, and whether it converges absolutely.

(i) $\displaystyle\sum_{n=1}^{\infty}\frac{(1+i)^n}{n!}$.

(ii) $\displaystyle\sum_{n=1}^{\infty}\frac{1+2i}{2^n}$.

(iii) $\displaystyle\sum_{n=1}^{\infty}\frac{i^n}{n}$.

(iv) $\displaystyle\sum_{n=1}^{\infty}(\tfrac{1}{2}+\tfrac{1}{2}i)^n$.

(v) $\displaystyle\sum_{n=2}^{\infty}\frac{\log n}{n}+i^n\frac{\log n}{n}$.

2. Use the ratio test to show that the radius of convergence of each of the following power series is 1. (In each case the ratios of successive terms will approach a limit < 1 if $|z| < 1$, but for $|z| > 1$ the ratios will tend to ∞ or to a limit > 1.)

(i) $\displaystyle\sum_{n=1}^{\infty}\frac{z^n}{n^2}$.

(ii) $\displaystyle\sum_{n=1}^{\infty}\frac{z^n}{n}$.

(iii) $\displaystyle\sum_{n=1}^{\infty}z^n$.

(iv) $\displaystyle\sum_{n=1}^{\infty}(n+2^{-n})z^n$.

(v) $\displaystyle\sum_{n=1}^{\infty}2^n z^{n!}$.

3. Use the root test (Problem 23-7) to find the radius of convergence of each of the following power series.

(i) $\displaystyle\frac{z}{2}+\frac{z^2}{3}+\frac{z^3}{2^2}+\frac{z^4}{3^2}+\frac{z^5}{2^3}+\frac{z^6}{3^3}+\cdots$.

(ii) $\displaystyle\sum_{n=1}^{\infty}\frac{n!\,z^n}{n^n}$.

(iii) $\displaystyle\sum_{n=1}^{\infty}\frac{n}{2^n}z^n$.

(iv) $\displaystyle\sum_{n=1}^{\infty}\frac{n^2}{2^n}z^n$.

(v) $\displaystyle\sum_{n=1}^{\infty}2^n z^{n!}$.

4. The root test can always be used, in theory at least, to find the radius of convergence of a power series; in fact, a close analysis of the situation leads to a formula for the radius of convergence, known as the "Cauchy-Hadamard formula." Suppose first that the set of numbers $\sqrt[n]{|a_n|}$ is bounded.

(a) Use Problem 23-7 to show that if $\overline{\lim_{n\to\infty}} \sqrt[n]{|a_n|}\,|z| < 1$, then $\sum_{n=0}^{\infty} a_n z^n$ converges.

(b) Also show that if $\overline{\lim_{n\to\infty}} \sqrt[n]{|a_n|}\,|z| > 1$, then $\sum_{n=0}^{\infty} a_n z^n$ has unbounded terms.

(c) Parts (a) and (b) show that the radius of convergence of $\sum_{n=0}^{\infty} a_n z^n$ is $1/\overline{\lim_{n\to\infty}} \sqrt[n]{|a_n|}$ (where "1/0" means "∞"). To complete the formula, define $\overline{\lim_{n\to\infty}} \sqrt[n]{|a_n|} = \infty$ if the set of all $\sqrt[n]{|a_n|}$ is unbounded. Prove that in this case, $\sum_{n=0}^{\infty} a_n z^n$ diverges for $z \neq 0$, so that the radius of convergence is 0 (which may be considered as "$1/\infty$").

5. Consider the following three series from Problem 2:

$$\sum_{n=1}^{\infty} \frac{z^n}{n^2}, \qquad \sum_{n=1}^{\infty} \frac{z^n}{n}, \qquad \sum_{n=1}^{\infty} z^n.$$

Prove that the first series converges everywhere on the unit circle; that the third series converges nowhere on the unit circle; and that the second series converges for at least one point on the unit circle and diverges for at least one point on the unit circle.

6. (a) Prove that $e^z \cdot e^w = e^{z+w}$ for all complex numbers z and w by showing that the infinite series for e^{z+w} is the Cauchy product of the series for e^z and e^w.

(b) Show that $\sin(z+w) = \sin z \cos w + \cos z \sin w$ and $\cos(z+w) = \cos z \cos w - \sin z \sin w$ for all complex z and w.

7. (a) Prove that every complex number of absolute value 1 can be written e^{iy} for some real number y.

(b) Prove that $|e^{x+iy}| = e^x$ for real x and y.

8. (a) Prove that exp takes on every complex value except 0.

(b) Prove that sin takes on every complex value.

9. For each of the following functions, compute the first three nonzero terms of the Taylor series centered at 0 by manipulating power series.

(i) $f(z) = \tan z$.

(ii) $f(z) = z(1-z)^{-1/2}$.

(iii) $f(z) = \dfrac{e^{\sin z} - 1}{z}$.

(iv) $f(z) = \log(1 - z^2)$.

(v) $f(z) = \dfrac{\sin^2 z}{z^2}$.

(vi) $f(z) = \dfrac{\sin(z^2)}{z \cos^2 z}$.

(vii) $f(z) = \dfrac{1}{z^4 - 2z^2 + 3}$.

(viii) $f(z) = \dfrac{1}{z}[e^{(\sqrt{1+z}-1)} - 1]$.

10. (a) Suppose that we write a differentiable complex function f as $f = u + iv$, where u and v are real-valued. Let $\bar{u}$ and $\bar{v}$ denote the restrictions of u and v to the real numbers. In other words, $\bar{u}(x) = u(x)$ for real numbers x (but $\bar{u}$ is not defined for other x). Using Problem 26-9, show that for real x we have

$$f'(x) = \bar{u}'(x) + i\bar{v}'(x),$$

where f' denotes the complex derivative, while $\bar{u}'$ and $\bar{v}'$ denote the ordinary derivatives of these real-valued functions on **R**.

(b) Show, more generally, that

$$f^{(k)}(x) = \bar{u}^{(k)}(x) + i\bar{v}^{(k)}(x).$$

(c) Suppose that f satisfies the equation

$$(*) \qquad f^{(n)} + a_{n-1}f^{(n-1)} + \cdots + a_0 f = 0,$$

where the a_i are real numbers, and where the $f^{(k)}$ denote higher-order complex derivatives. Show that $\bar{u}$ and $\bar{v}$ satisfy the same equation, where $\bar{u}^{(k)}$ and $\bar{v}^{(k)}$ now denote higher-order derivatives of real-valued functions on **R**.

(d) Show that if $a = b + ci$ is a complex root of the equation $z^n + a_{n-1}z^{n-1} + \cdots + a_0 = 0$, then $f(x) = e^{bx} \sin cx$ and $f(x) = e^{bx} \cos cx$ are both solutions of $(*)$.

11. (a) Show that exp is *not* one-one on **C**.

(b) Given $w \neq 0$, show that $e^z = w$ if and only if $z = x + iy$ with $x = \log|w|$ (here log denotes the real logarithm function), and y an argument of w.

*(c) Show that there does not exist a continuous function log defined for nonzero complex numbers, such that $\exp(\log(z)) = z$ for all $z \neq 0$. (Show that log cannot even be defined continuously for $|z| = 1$.)

Since there is no way to define a continuous logarithm function we cannot speak of *the* logarithm of a complex number, but only of "a logarithm for w," meaning one of the infinitely many numbers z with $e^z = w$. And for complex numbers a and b we define a^b to be a *set* of complex numbers, namely the set of all numbers $e^{b \log a}$ or, more precisely, the set of all numbers e^{bz} where z is a logarithm for a.

(d) If m is an integer, then a^m consists of only one number, the one given by the usual elementary definition of a^m.

(e) If m and n are integers, then the set $a^{m/n}$ coincides with the set of values given by the usual elementary definition, namely the set of all b^m where b is an nth root of a.

(f) If a and b are real and b is irrational, then a^b contains infinitely many members, even for $a > 0$.

(g) Find all logarithms of i, and find all values of i^i.

(h) By $(a^b)^c$ we mean the set of all numbers of the form z^c for some number z in the set a^b. Show that $(1^i)^i$ has infinitely many values, while $1^{i \cdot i}$ has only one.

(i) Show that all values of $a^{b \cdot c}$ are also values of $(a^b)^c$. Is $a^{b \cdot c} = (a^b)^c \cap (a^c)^b$?

12. (a) For real x show that we can choose $\log(x+i)$ and $\log(x-i)$ to be

$$\log(x+i) = \log(1+x^2) + i\left(\frac{\pi}{2} - \arctan x\right),$$

$$\log(x-i) = \log(1+x^2) - i\left(\frac{\pi}{2} - \arctan x\right).$$

(It will help to note that $\pi/2 - \arctan x = \arctan 1/x$ for $x \neq 0$.)

(b) The expression

$$\frac{1}{1+x^2} = \frac{1}{2i}\left(\frac{1}{x-i} - \frac{1}{x+i}\right)$$

yields, formally,

$$\int \frac{dx}{1+x^2} = \frac{1}{2i}[\log(x-i) - \log(x+i)].$$

Use part (a) to check that this answer agrees with the usual one.

13. (a) A sequence $\{a_n\}$ of complex numbers is called a **Cauchy sequence** if $\lim_{m,n \to \infty} |a_m - a_n| = 0$. Suppose that $a_n = b_n + ic_n$, where b_n and c_n are real. Prove that $\{a_n\}$ is a Cauchy sequence if and only if $\{b_n\}$ and $\{c_n\}$ are Cauchy sequences.

(b) Prove that every Cauchy sequence of complex numbers converges.

(c) Give direct proofs, without using theorems about real series, that an absolutely convergent series is convergent and that any rearrangement has the same sum. (It is permitted, and in fact advisable, to use the *proofs* of the corresponding theorems for real series.)

14. (a) Prove that

$$\sum_{k=1}^{n} e^{ikx} = e^{ix}\frac{1-e^{inx}}{1-e^{ix}} = \frac{\sin\left(\frac{n}{2}x\right)}{\sin\frac{x}{2}} e^{i(n+1)x/2}.$$

(b) Deduce the formulas for $\sum_{k=1}^{n} \cos kx$ and $\sum_{k=1}^{n} \sin kx$ that are given in Problem 15-33.

15. Let $\{a_n\}$ be the Fibonacci sequence, $a_1 = a_2 = 1, a_{n+2} = a_n + a_{n+1}$.

(a) If $r_n = a_{n+1}/a_n$, show that $r_{n+1} = 1 + 1/r_n$.
(b) Show that $r = \lim_{n\to\infty} r_n$ exists, and $r = 1 + 1/r$. Conclude that $r = (1+\sqrt{5})/2$.
(c) Show that $\sum_{n=1}^{\infty} a_n z^n$ has radius of convergence $2/(1+\sqrt{5})$. (Using the unproved theorems in this chapter and the fact that $\sum_{n=1}^{\infty} a_n z^n = -1/(z^2+z-1)$ from Problem 24-15 we could have predicted that the radius of convergence is the smallest absolute value of the roots of $z^2 + z - 1 = 0$; since the roots are $(-1 \pm \sqrt{5})/2$, the radius of convergence should be $(-1+\sqrt{5})/2$. Notice that this number is indeed equal to $2/(1+\sqrt{5})$.)

16. Since $(e^z - 1)/z$ can be written as the power series $1 + z/2! + z^2/3! + \cdots$ which is nonzero at 0, it follows that there is a power series

$$\frac{z}{e^z - 1} = \sum_{n=0}^{\infty} \frac{b_n}{n!} z^n$$

with nonzero radius of convergence. Using the unproved theorems in this chapter, we can even predict the radius of convergence; it is 2π, since this is the smallest absolute value of the numbers $z = 2k\pi i$ for which $e^z - 1 = 0$. The numbers b_n appearing here are called the **Bernoulli numbers.***

(a) Clearly $b_0 = 1$. Now show that

$$\frac{z}{e^z - 1} = -\frac{z}{2} + \frac{z}{2} \cdot \frac{e^z + 1}{e^z - 1},$$

$$\frac{e^{-z} + 1}{e^{-z} - 1} = -\frac{e^z + 1}{e^z - 1},$$

and deduce that

$$b_1 = -\tfrac{1}{2}, \quad b_n = 0 \quad \text{if } n \text{ is odd and } n > 1.$$

(b) By finding the coefficient of z^n in the right side of the equation

$$z = \left(\sum_{k=0}^{\infty} \frac{b_k}{k!}\right)\left(z + \frac{z^2}{2!} + \frac{z^3}{3!} + \cdots\right),$$

*Sometimes the numbers $B_n = (-1)^{n-1} b_{2n}$ are called the Bernoulli numbers, because $b_n = 0$ if n is odd and > 1 (see part (a)) and because the numbers b_{2n} alternate in sign, although we will not prove this. Other modifications of this nomenclature are also in use.

show that

$$\sum_{i=0}^{n-1} \binom{n}{i} b_i = 0 \quad \text{for } n > 1.$$

This formula allows us to compute any b_k in terms of previous ones, and shows that each is rational. Calculate two or three of the following:

$$b_2 = \tfrac{1}{6}, \quad b_4 = -\tfrac{1}{30}, \quad b_6 = \tfrac{1}{42}, \quad b_8 = -\tfrac{1}{30}.$$

*(c) Part (a) shows that

$$\sum_{n=0}^{\infty} \frac{b_{2n}}{(2n)!} z^{2n} = \frac{z}{2} \cdot \frac{e^z + 1}{e^z - 1} = \frac{z}{2} \cdot \frac{e^{z/2} + e^{-z/2}}{e^{z/2} - e^{-z/2}}.$$

Replace z by $2iz$ and show that

$$z \cot z = \sum_{n=0}^{\infty} \frac{b_{2n}}{(2n)!} (-1)^n 2^{2n} z^{2n}.$$

*(d) Show that

$$\tan z = \cot z - 2 \cot 2z.$$

*(e) Show that

$$\tan z = \sum_{n=1}^{\infty} \frac{b_{2n}}{(2n)!} (-1)^{n-1} 2^{2n} (2^{2n} - 1) z^{2n-1}.$$

(This series converges for $|z| < \pi/2$.)

17. The Bernoulli numbers play an important role in a theorem which is best introduced by some notational nonsense. Let us use D to denote the "differentiation operator," so that Df denotes f'. Then $D^k f$ will mean $f^{(k)}$ and $e^D f$ will mean $\sum_{n=0}^{\infty} f^{(n)}/n!$ (of course this series makes no sense in general, but it will make sense if f is a polynomial function, for example). Finally, let Δ denote the "difference operator" for which $\Delta f(x) = f(x+1) - f(x)$. Now Taylor's Theorem implies, disregarding questions of convergence, that

$$f(x+1) = \sum_{n=0}^{\infty} \frac{f^{(n)}(x)}{n!},$$

or

$$(*) \quad f(x+1) - f(x) = \sum_{n=1}^{\infty} \frac{f^{(n)}(x)}{n!};$$

we may write this symbolically as $\Delta f = (e^D - 1)f$, where 1 stands for the "identity operator." Even more symbolically this can be written $\Delta = e^D - 1$, which suggests that

$$D = \frac{D}{e^D - 1} \Delta.$$

Thus we obviously ought to have

$$D = \sum_{k=0}^{\infty} \frac{b_k}{k!} D^k \Delta,$$

i.e.,

$$(**) \quad f'(x) = \sum_{k=0}^{\infty} \frac{b_k}{k!} [f^{(k)}(x+1) - f^{(k)}(x)].$$

The beautiful thing about all this nonsense is that it works!

(a) Prove that $(**)$ is literally true if f is a polynomial function (in which case the infinite sum is really a finite sum). Hint: By applying $(*)$ to $f^{(k)}$, find a formula for $f^{(k)}(x+1) - f^{(k)}(x)$; then use the formula in Problem 16(b) to find the coefficient of $f^{(j)}(x)$ in the right side of $(**)$.

(b) Deduce from $(**)$ that

$$f'(0) + \cdots + f'(n) = \sum_{k=0}^{\infty} \frac{b_k}{k!} [f^{(k)}(n+1) - f^{(k)}(0)].$$

(c) Show that for any polynomial function g we have

$$g(0) + \cdots + g(n) = \int_0^{n+1} g(t)\,dt + \sum_{k=1}^{\infty} \frac{b_k}{k!} [g^{(k-1)}(n+1) - g^{(k-1)}(0)].$$

(d) Apply this to $g(x) = x^p$ to show that

$$\sum_{k=1}^{n} k^p = \frac{n^{p+1}}{p+1} + \sum_{k=1}^{p} \frac{b_k}{k} \binom{p}{k-1} n^{p-k+1}.$$

Using the fact that $b_1 = -\frac{1}{2}$, show that

$$\sum_{k=1}^{n} k^p = \frac{n^{p+1}}{p+1} + \frac{n^p}{2} + \sum_{k=2}^{p} \frac{b_k}{k} \binom{p}{k-1} n^{p-k+1}.$$

The first ten instances of this formula were written out in Problem 2-7, which offered as a challenge the discovery of the general pattern. This may now seem to be a preposterous suggestion, but the Bernoulli numbers were actually discovered in precisely this way! After writing out these 10 formulas, Bernoulli claims (in his posthumously printed work *Ars Conjectandi*, 1713): "Whoever will examine the series as to their regularity may be able to continue the table." He then writes down the above formula, offering no proof at all, merely noting that the coefficients b_k (which he denoted simply by $A, B, C, \ldots$) satisfy the equation in Problem 16(b). The relation between these numbers and the coefficients in the power series for $z/(e^z - 1)$ was discovered by Euler.

***18.** The formula in Problem 17(c) can be generalized to the case where g is not a polynomial function; the infinite sum must be replaced by a finite sum plus a remainder term. In order to find an expression for the remainder, it is useful to introduce some new functions.

(a) The *Bernoulli polynomials* φ_n are defined by

$$\varphi_n(x) = \sum_{k=0}^{n} \binom{n}{k} b_{n-k} x^k.$$

The first three are

$$\varphi_1(x) = x - \frac{1}{2},$$
$$\varphi_2(x) = x^2 - x + \frac{1}{6},$$
$$\varphi_3(x) = x^3 - \frac{3x^2}{2} + \frac{x}{2}.$$

Show that

$$\begin{aligned} \varphi_n(0) &= b_n, \\ \varphi_n(1) &= b_n \qquad \text{if } n > 1, \\ \varphi_n{}'(x) &= n\varphi_{n-1}(x), \\ \varphi_n(x) &= (-1)^n \varphi_n(1-x) \qquad \text{for } n > 1. \end{aligned}$$

Hint: Prove the last equation by induction on n, starting with $n = 2$.

(b) Let $R_N{}^k(x)$ be the remainder term in Taylor's Theorem for $f^{(k)}$, on the interval $[x, x+1]$, so that

$$(*) \qquad f^{(k)}(x+1) - f^{(k)}(x) = \sum_{n=1}^{N} \frac{f^{(k+n)}(x)}{n!} + R_N{}^k(x).$$

Prove that

$$f'(x) = \sum_{k=0}^{N} \frac{b_k}{k!}[f^{(k)}(x+1) - f^{(k)}(x)] - \sum_{k=0}^{N} \frac{b_k}{k!} R_{N-k}{}^k(x).$$

Hint: Imitate Problem 17(a). Notice the subscript $N - k$ on R.

(c) Use the integral form of the remainder to show that

$$\sum_{k=0}^{N} \frac{b_k}{k!} R_{N-k}{}^k(x) = \int_x^{x+1} \frac{\varphi_n(x+1-t)}{N!} f^{(N+1)}(t)\, dt.$$

(d) Deduce the "Euler-Maclaurin Summation Formula":

$$\begin{aligned} &g(x) + g(x+1) + \cdots + g(x+n) \\ &\quad = \int_x^{x+n+1} g(t)\, dt + \sum_{k=1}^{N} \frac{b_k}{k!}[g^{(k-1)}(x+n+1) - g^{(k-1)}(x)] + S_n(x, n), \end{aligned}$$

where

$$S_N(x,n) = -\sum_{j=0}^{n} \int_{x+j}^{x+j+1} \frac{\varphi_N(x+j+1-t)}{N!} g^{(N)}(t)\,dt.$$

(e) Let ψ_n be the periodic function, with period 1, which satisfies $\psi_n(t) = \varphi_n(t)$ for $0 \le t < 1$. (Part (a) implies that if $n > 1$, then ψ_n is continuous, since $\varphi_n(1) = \varphi_n(0)$, and also that ψ_n is even if n is even and odd if n is odd.) Show that

$$S_N(x,n) = -\int_x^{x+n+1} \frac{\psi_N(x-t)}{N!} g^{(N)}(t)\,dt$$

$$\left(= (-1)^{N+1} \int_x^{x+n+1} \frac{\psi_N(t)}{N!} g^{(N)}(t)\,dt \qquad \text{if } x \text{ is an integer}\right).$$

Unlike the remainder in Taylor's Theorem, the remainder $S_n(x,n)$ usually does not satisfy $\lim_{N\to\infty} S_N(x,n) = 0$, because the Bernoulli numbers and functions become large very rapidly (although the first few examples do not suggest this). Nevertheless, important information can often be obtained from the summation formula. The general situation is best discussed within the context of a specialized study ("asymptotic series"), but the next problem shows one particularly important example.

****19.** (a) Use the Euler-Maclaurin Formula, with $N = 2$, to show that

$$\log 1 + \cdots + \log(n-1)$$
$$= \int_1^n \log t\,dt - \frac{1}{2}\log n + \frac{1}{12}\left(\frac{1}{n} - 1\right) + \int_1^n \frac{\psi_2(t)}{2t^2}\,dt.$$

(b) Show that

$$\log\left(\frac{n!}{n^{n+1/2}e^{-n+1/12n}}\right) = \frac{11}{12} + \int_1^n \frac{\psi_2(t)}{2t^2}\,dt.$$

(c) Explain why the improper integral $\beta = \displaystyle\int_1^\infty \psi_2(t)/2t^2\,dt$ exists, and show that if $\alpha = \exp(\beta + 11/12)$, then

$$\log\left(\frac{n!}{\alpha n^{n+1/2}e^{-n+1/12n}}\right) = -\int_n^\infty \frac{\psi_2(t)}{2t^2}\,dt.$$

(d) Problem 19-40(d) shows that

$$\sqrt{\pi} = \lim_{n\to\infty} \frac{(n!)^2 2^{2n}}{(2n)!\sqrt{n}}.$$

Use part (c) to show that

$$\sqrt{\pi} = \lim_{n\to\infty} \frac{\alpha^2 n^{2n+1} e^{-2n} 2^{2n}}{\alpha(2n)^{2n+1/2} e^{-2n}\sqrt{n}},$$

and conclude that $\alpha = \sqrt{2\pi}$.

(e) Show that

$$\int_0^{1/2} \varphi_2(t)\,dt = \int_0^1 \varphi_2(t)\,dt = 0.$$

(You can do the computations explicitly, but the result also follows immediately from Problem 18(a).) Now what can be said about the graphs of $\bar{\psi}(x) = \int_0^x \psi_2(t)\,dt$ and $\bar{\bar{\psi}}(x) = \int_0^x \bar{\psi}(t)\,dt$? Use this information and integration by parts to show that

$$\int_n^\infty \frac{\psi_2(t)}{2t^2}\,dt > 0.$$

(f) Show that the maximum value of $|\varphi_2(x)|$ for x in $[0, 1]$ is $\frac{1}{6}$, and conclude that

$$\left|\int_n^\infty \frac{\psi_2(t)}{2t^2}\,dt\right| < \frac{1}{12n}.$$

(g) Finally, conclude that

$$\sqrt{2\pi}\,n^{n+1/2}e^{-n} < n! < \sqrt{2\pi}\,n^{n+1/2}e^{-n+1/12n}.$$

The final result of Problem 19, a strong form of Stirling's Formula, shows that $n!$ is approximately $\sqrt{2\pi}\,n^{n+1/2}e^{-n}$, in the sense that this expression differs from $n!$ by an amount which is small compared to n when n is large. For example, for $n = 10$ we obtain 3598696 instead of 3628800, with an error $< 1\%$.

A more general form of Stirling's Formula illustrates the "asymptotic" nature of the summation formula. The same argument which was used in Problem 19 can now be used to show that for $N \geq 2$ we have

$$\log\left(\frac{n!}{\sqrt{2\pi}\,n^{n+1/2}e^{-n}}\right) = \sum_{k=2}^{N} \frac{b_k}{k(k-1)n^{k-1}} \pm \int_n^\infty \frac{\psi_N(t)}{Nt^N}\,dt.$$

Since ψ_N is bounded, we can obtain estimates of the form

$$\left|\int_n^\infty \frac{\psi_N(t)}{Nt^N}\,dt\right| \leq \frac{M_N}{n^{N-1}}.$$

If N is large, the constant M_N will also be large; but for very large n the factor n^{1-N} will make the product very small. Thus, the expression

$$\sqrt{2\pi}\,n^{n+1/2}e^{-n} \cdot \exp\left(\sum_{k=2}^{N} \frac{b_k}{k(k-1)n^{k-1}}\right)$$

may be a very bad approximation for $n!$ when n is small, but for large n (*how* large depends on N) it will be an extremely good one (*how* good depends on N).

PART 5
EPILOGUE

There was a most ingenious Architect
who had contrived a new Method
for building Houses,
by beginning at the Roof, and working
downwards to the Foundation.

JONATHAN SWIFT

CHAPTER 28 FIELDS

Throughout this book a conscientious attempt has been made to define all important concepts, even terms like "function," for which an intuitive definition is often considered sufficient. But **Q** and **R**, the two main protagonists of this story, have only been named, never defined. What has never been defined can never be analyzed thoroughly, and "properties" P1–P13 must be considered assumptions, not theorems, about numbers. Nevertheless, the term "axiom" has been purposely avoided, and in this chapter the logical status of P1–P13 will be scrutinized more carefully.

Like **Q** and **R**, the sets **N** and **Z** have also remained undefined. True, some talk about all four was inserted in Chapter 2, but those rough descriptions are far from a definition. To say, for example, that **N** consists of 1, 2, 3, etc., merely names some elements of **N** without identifying them (and the "etc." is useless). The natural numbers *can* be defined, but the procedure is involved and not quite pertinent to the rest of the book. The Suggested Reading list contains references to this problem, as well as to the other steps that are required if one wishes to develop calculus from its basic logical starting point. The further development of this program would proceed with the definition of **Z**, in terms of **N**, and the definition of **Q** in terms of **Z**. This program results in a certain well-defined set **Q**, certain explicitly defined operations $+$ and $\cdot$, and properties P1–P12 as *theorems*. The final step in this program is the construction of **R**, in terms of **Q**. It is this last construction which concerns us. Assuming that **Q** has been defined, and that P1–P12 have been proved for **Q**, we shall ultimately *define* **R** and *prove* all of P1–P13 for **R**.

Our intention of proving P1–P13 means that we must define not only real numbers, but also addition and multiplication of real numbers. Indeed, the real numbers are of interest only as a set together with these operations: how the real numbers behave with respect to addition and multiplication is crucial; what the real numbers may actually be is quite irrelevant. This assertion can be expressed in a meaningful mathematical way, by using the concept of a "field," which includes as special cases the three important number systems of this book. This extraordinarily important abstraction of modern mathematics incorporates the properties P1–P9 common to **Q**, **R**, and **C**. A **field** is a set F (of objects of any sort whatsoever), together with two "binary operations" $\boldsymbol{+}$ and $\boldsymbol{\cdot}$ defined on F (that is, two rules which associate to elements a and b in F, other elements $a \boldsymbol{+} b$ and $a \boldsymbol{\cdot} b$ in F) for which the following conditions are satisfied:

(1) $(a \boldsymbol{+} b) \boldsymbol{+} c = a \boldsymbol{+} (b \boldsymbol{+} c)$ for all a, b, and c in F.

(2) There is some element $\mathbf{0}$ in F such that

  (i) $a \boldsymbol{+} \mathbf{0} = a$ for all a in F,

  (ii) for every a in F, there is some element b in F such that $a \boldsymbol{+} b = \mathbf{0}$.

(3) $a \boldsymbol{+} b = b \boldsymbol{+} a$ for all a and b in F.
(4) $(a \bullet b) \bullet c = a \bullet (b \bullet c)$ for all a, b, and c in F.
(5) There is some element $\mathbf{1}$ in F such that $\mathbf{1} \neq \mathbf{0}$ and

 (i) $a \bullet \mathbf{1} = a$ for all a in F,
 (ii) For every a in F with $a \neq \mathbf{0}$, there is some element b in F such that $a \bullet b = \mathbf{1}$.

(6) $a \bullet b = b \bullet a$ for all a and b in F.
(7) $a \bullet (b \boldsymbol{+} c) = a \bullet b \boldsymbol{+} a \bullet c$ for all a, b, and c in F.

The familiar examples of fields are, as already indicated, **Q**, **R**, and **C**, with $\boldsymbol{+}$ and $\bullet$ being the familiar operations of $+$ and $\cdot$. It is probably unnecessary to explain why these are fields, but the explanation is, at any rate, quite brief. When $\boldsymbol{+}$ and $\bullet$ are understood to mean the ordinary $+$ and $\cdot$, the rules (1), (3), (4), (6), (7) are simply restatements of P1, P4, P5, P8, P9; the elements which play the role of $\mathbf{0}$ and $\mathbf{1}$ are the numbers 0 and 1 (which accounts for the choice of the symbols $\mathbf{0}$, $\mathbf{1}$); and the number b in (2) or (5) is $-a$ or a^{-1}, respectively. (For this reason, in an arbitrary field F we denote by $\boldsymbol{-a}$ the element such that $a \boldsymbol{+} (\boldsymbol{-a}) = \mathbf{0}$, and by $\boldsymbol{a^{-1}}$ the element such that $a \bullet \boldsymbol{a^{-1}} = \mathbf{1}$, for $a \neq \mathbf{0}$.)

In addition to **Q**, **R**, and **C**, there are several other fields which can be described easily. One example is the collection F_1 of all numbers $a + b\sqrt{2}$ for a, b in **Q**. The operations $\boldsymbol{+}$ and $\bullet$ will, once again, be the usual $+$ and $\cdot$ for real numbers. It is necessary to point out that these operations really do produce new elements of F_1:

$$(a + b\sqrt{2}\,) + (c + d\sqrt{2}\,) = (a + c) + (b + d)\sqrt{2}, \quad \text{which is in } F_1;$$
$$(a + b\sqrt{2}\,) \cdot (c + d\sqrt{2}\,) = (ac + 2bd) + (bc + ad)\sqrt{2}, \quad \text{which is in } F_1.$$

Conditions (1), (3), (4), (6), (7) for a field are obvious for F_1: since these hold for all real numbers, they certainly hold for all real numbers of the form $a + b\sqrt{2}$. Condition (2) holds because the number $0 = 0 + 0\sqrt{2}$ is in F_1 and, for $\alpha = a + b\sqrt{2}$ in F_1 the number $\beta = (-a) + (-b)\sqrt{2}$ in F_1 satisfies $\alpha + \beta = 0$. Similarly, $1 = 1 + 0\sqrt{2}$ is in F_1, so (5i) is satisfied. The verification of (5ii) is the only slightly difficult point. If $a + b\sqrt{2} \neq 0$, then

$$a + b\sqrt{2} \cdot \frac{1}{a + b\sqrt{2}} = 1;$$

it is therefore necessary to show that $1/(a + b\sqrt{2}\,)$ is in F_1. This is true because

$$\frac{1}{a + b\sqrt{2}} = \frac{a - b\sqrt{2}}{(a - b\sqrt{2}\,)(a + b\sqrt{2}\,)} = \frac{a}{a^2 - 2b^2} + \frac{(-b)}{a^2 - 2b^2}\sqrt{2}.$$

(The division by $a - b\sqrt{2}$ is valid because the relation $a - b\sqrt{2} = 0$ could be true only if $a = b = 0$ (since $\sqrt{2}$ is irrational) which is ruled out by the hypothesis $a + b\sqrt{2} \neq 0$.)

The next example of a field, F_2, is considerably simpler in one respect: it contains only two elements, which we might as well denote by $\mathbf{0}$ and $\mathbf{1}$. The operations

$\boldsymbol{+}$ and $\boldsymbol{\cdot}$ are described by the following tables.

$\boldsymbol{+}$	0	1
0	0	1
1	1	0

$\boldsymbol{\cdot}$	0	1
0	0	0
1	0	1

The verification of conditions (1)–(7) are straightforward, case-by-case checks. For example, condition (1) may be proved by checking the 8 equations obtained by setting $a, b, c = \mathbf{0}$ or $\mathbf{1}$. Notice that in this field $\mathbf{1} \boldsymbol{+} \mathbf{1} = \mathbf{0}$; this equation may also be written $\mathbf{1} = \boldsymbol{-}\mathbf{1}$.

Our final example of a field is rather silly: F_3 consists of all pairs (a, a) for a in $\mathbf{R}$, and $\boldsymbol{+}$ and $\boldsymbol{\cdot}$ are defined by

$$(a, a) \boldsymbol{+} (b, b) = (a + b, a + b),$$
$$(a, a) \boldsymbol{\cdot} (b, b) = (a \cdot b, a \cdot b).$$

(The $+$ and $\cdot$ appearing on the right side are ordinary addition and multiplication for $\mathbf{R}$.) The verification that F_3 is a field is left to you as a simple exercise.

A detailed investigation of the properties of fields is a study in itself, but for our purposes, fields provide an ideal framework in which to discuss the properties of numbers in the most economical way. For example, the consequences of P1–P9 which were derived for "numbers" in Chapter 1 actually hold for any field; in particular, they are true for the fields $\mathbf{Q}$, $\mathbf{R}$, and $\mathbf{C}$.

Notice that certain common properties of $\mathbf{Q}$, $\mathbf{R}$, and $\mathbf{C}$ do not hold for all fields. For example, it is possible for the equation $\mathbf{1} \boldsymbol{+} \mathbf{1} = \mathbf{0}$ to hold in some fields, and consequently $a \boldsymbol{-} b = b \boldsymbol{-} a$ does not necessarily imply that $a = b$. For the field $\mathbf{C}$ the assertion $1+1 \neq 0$ was derived from the explicit description of $\mathbf{C}$; for the fields $\mathbf{Q}$ and $\mathbf{R}$, however, this assertion was derived from further properties which do not have analogues in the conditions for a field. There is a related concept which does use these properties. An **ordered field** is a field F (with operations $\boldsymbol{+}$ and $\boldsymbol{\cdot}$) together with a certain subset $\mathbf{P}$ of F (the "positive" elements) with the following properties:

(8) For all a in F, one and only one of the following is true:

- (i) $a = \mathbf{0}$,
- (ii) a is in $\mathbf{P}$,
- (iii) $\boldsymbol{-}a$ is in $\mathbf{P}$.

(9) If a and b are in $\mathbf{P}$, then $a \boldsymbol{+} b$ is in $\mathbf{P}$.

(10) If a and b are in $\mathbf{P}$, then $a \boldsymbol{\cdot} b$ is in $\mathbf{P}$.

We have already seen that the field $\mathbf{C}$ cannot be made into an ordered field. The field F_2, with only two elements, likewise cannot be made into an ordered field: in fact, condition (8), applied to $\mathbf{1} = \boldsymbol{-}\mathbf{1}$, shows that $\mathbf{1}$ must be in $\mathbf{P}$; then (9) implies that $\mathbf{1} \boldsymbol{+} \mathbf{1} = \mathbf{0}$ is in $\mathbf{P}$, contradicting (8). On the other hand, the field F_1, consisting of all numbers $a + b\sqrt{2}$ with a, b in $\mathbf{Q}$, certainly can be made into an ordered field: let $\mathbf{P}$ be the set of all $a + b\sqrt{2}$ which are positive real numbers (in the ordinary sense). The field F_3 can also be made into an ordered field; the description of $\mathbf{P}$ is left to you.

It is natural to introduce notation for an arbitrary ordered field which corresponds to that used for $\mathbf{Q}$ and $\mathbf{R}$: we define

$$\begin{aligned} a > b &\quad \text{if} \quad a - b \text{ is in } \mathbf{P}, \\ a < b &\quad \text{if} \quad b > a, \\ a \le b &\quad \text{if} \quad a < b \text{ or } a = b, \\ a \ge b &\quad \text{if} \quad a > b \text{ or } a = b. \end{aligned}$$

Using these definitions we can reproduce, for an arbitrary ordered field F, the definitions of Chapter 7:

> A set A of elements of F is **bounded above** if there is some x in F such that $x \ge a$ for all a in A. Any such x is called an **upper bound** for A. An element x of F is a **least upper bound** for A if x is an upper bound for A and $x \le y$ for every y in F which is an upper bound for A.

Finally, it is possible to state an analogue of property P13 for $\mathbf{R}$; this leads to the last abstraction of this chapter:

> A **complete ordered field** is an ordered field in which every nonempty set which is bounded above has a least upper bound.

The consideration of fields may seem to have taken us far from the goal of constructing the real numbers. However, we are now provided with an intelligible means of formulating this goal. There are two questions which will be answered in the remaining two chapters:

1. Is there a complete ordered field?
2. Is there only one complete ordered field?

Our starting point for these considerations will be $\mathbf{Q}$, assumed to be an ordered field, containing $\mathbf{N}$ and $\mathbf{Z}$ as certain subsets. At one crucial point it will be necessary to assume another fact about $\mathbf{Q}$:

> Let x be an element of $\mathbf{Q}$ with $x > 0$. Then for any y in $\mathbf{Q}$ there is some n in $\mathbf{N}$ such that $nx > y$.

This assumption, which asserts that the rational numbers have the Archimedian property of the real numbers, does not follow from the other properties of an ordered field (for the example that demonstrates this conclusively see reference [17] of the Suggested Reading). The important point for us is that when $\mathbf{Q}$ is explicitly constructed, properties P1–P12 appear as theorems, and so does this additional assumption; if we really began from the beginning, no assumptions about $\mathbf{Q}$ would be necessary.

PROBLEMS

1. Let F be the set $\{0, 1, 2\}$ and define operations $+$ and $\cdot$ on F by the following tables. (The rule for constructing these tables is as follows: add or multiply in the usual way, and then subtract the highest possible multiple of 3; thus $2 \cdot 2 = 4 = 3 + 1$, so $2 \cdot 2 = 1$.)

$+$	0	1	2
0	0	1	2
1	1	2	0
2	2	0	1

$\cdot$	0	1	2
0	0	0	0
1	0	1	2
2	0	2	1

Show that F is a field, and prove that it cannot be made into an ordered field.

2. Suppose now that we try to construct a field F having elements 0, 1, 2, 3 with operations $+$ and $\cdot$ defined as in the previous example, by adding or multiplying in the usual way, and then subtracting the highest possible multiple of 4. Show that F will *not* be a field.

3. Let $F = \{0, 1, \alpha, \beta\}$ and define operations $+$ and $\cdot$ on F by the following tables.

$+$	0	1	α	β
0	0	1	α	β
1	1	0	β	α
α	α	β	0	1
β	β	α	1	0

$\cdot$	0	1	α	β
0	0	0	0	0
1	0	1	α	β
α	0	α	β	1
β	0	β	1	α

Show that F is a field.

4. (a) Let F be a field in which $\mathbf{1} + \mathbf{1} = \mathbf{0}$. Show that $a + a = \mathbf{0}$ for all a (this can also be written $a = -a$).

(b) Suppose that $a + a = \mathbf{0}$ for some $a \neq \mathbf{0}$. Show that $\mathbf{1} + \mathbf{1} = \mathbf{0}$ (and consequently $b + b = \mathbf{0}$ for all b).

5. (a) Show that in any field we have

$$\underbrace{(\mathbf{1}+\cdots+\mathbf{1})}_{m \text{ times}}\cdot\underbrace{(\mathbf{1}+\cdots+\mathbf{1})}_{n \text{ times}} = \underbrace{\mathbf{1}+\cdots+\mathbf{1}}_{mn \text{ times}}$$

for all natural numbers m and n.

(b) Suppose that in the field F we have

$$\underbrace{\mathbf{1}+\cdots+\mathbf{1}}_{m \text{ times}} = 0$$

for some natural number m. Show that the smallest m with this property must be a prime number (this prime number is called the **characteristic** of F).

6. Let F be any field with only finitely many elements.

(a) Show that there must be distinct natural numbers m and n with

$$\underbrace{\mathbf{1}+\cdots+\mathbf{1}}_{m \text{ times}} = \underbrace{\mathbf{1}+\cdots+\mathbf{1}}_{n \text{ times}}.$$

(b) Conclude that there is some natural number k with

$$\underbrace{\mathbf{1}+\cdots+\mathbf{1}}_{k \text{ times}} = \mathbf{0}.$$

7. Let a, b, c, and d be elements of a field F with $a\cdot d - b\cdot c \neq \mathbf{0}$. Show that the equations

$$a\cdot x + b\cdot y = \alpha,$$
$$c\cdot x + d\cdot y = \beta,$$

can be solved for x and y.

8. Let a be an element of a field F. A "square root" of a is an element b of F with $b^2 = b\cdot b = a$.

(a) How many square roots does $\mathbf{0}$ have?

(b) Suppose $a \neq \mathbf{0}$. Show that if a has a square root, then it has two square roots, unless $\mathbf{1}+\mathbf{1} = \mathbf{0}$, in which case a has only one.

9. (a) Consider an equation $x^2 + b\cdot x + c = \mathbf{0}$, where b and c are elements of a field F. Suppose that $b^2 - 4\cdot c$ has a square root r in F. Show that $(-b+r)/2$ is a solution of this equation.

(b) In the field F_2 of the text, both elements clearly have a square root. On the other hand, it is easy to check that neither element satisfies the equation $x^2 + x + \mathbf{1} = \mathbf{0}$. Thus some detail in part (a) must be incorrect. What is it?

10. Let F be a field and a an element of F which does *not* have a square root. This problem shows how to construct a bigger field F', containing F, in which a does have a square root. (This construction has already been carried through in a special case, namely, $F = \mathbf{R}$ and $a = -1$; this special case should guide you through this example.)

Let F' consist of all pairs (x, y) with x and y in F. If the operations on F are $+$ and $\cdot$, define operations $\oplus$ and $\odot$ on F' as follows:

$$(x, y) \oplus (z, w) = (x + z, y + w),$$
$$(x, y) \odot (z, w) = (x \cdot z + a \cdot y \cdot w, y \cdot z + x \cdot w).$$

(a) Prove that F', with the operations $\oplus$ and $\odot$, is a field.

(b) Prove that

$$(x, \mathbf{0}) \oplus (y, \mathbf{0}) = (x + y, \mathbf{0}),$$
$$(x, \mathbf{0}) \odot (y, \mathbf{0}) = (x \cdot y, \mathbf{0}),$$

so that we may agree to abbreviate $(x, \mathbf{0})$ by x.

(c) Find a square root of $a = (a, \mathbf{0})$ in F'.

11. Let F be the set of all four-tuples (w, x, y, z) of real numbers. Define $+$ and $\cdot$ by

$$(s, t, u, v) + (w, x, y, z) = (s + w, t + x, u + y, v + z),$$
$$(s, t, u, v) \cdot (w, x, y, z) = (sw - tx - uy - vz, sx + tw + uz - vy,$$
$$sy + uw + vx - tz, sz + vw + ty - ux).$$

(a) Show that F satisfies all conditions for a field, except (6). At times the algebra will become quite ornate, but the existence of multiplicative inverses is the only point requiring any thought.

(b) It is customary to denote

$$(0, 1, 0, 0) \text{ by } i,$$
$$(0, 0, 1, 0) \text{ by } j,$$
$$(0, 0, 0, 1) \text{ by } k.$$

Find all 9 products of pairs i, j, and k. The results will show in particular that condition (6) is definitely false. This "skew field" F is known as the **quaternions**.

CHAPTER 29 CONSTRUCTION OF THE REAL NUMBERS

The mass of drudgery which this chapter necessarily contains is relieved by one truly first-rate idea. In order to prove that a complete ordered field exists we will have to explicitly describe one in detail; verifying conditions (1)–(10) for an ordered field will be a straightforward ordeal, but the description of the field itself, of the elements in it, is ingenious indeed.

At our disposal is the set of rational numbers, and from this raw material it is necessary to produce the field which will ultimately be called the real numbers. To the uninitiated this must seem utterly hopeless—if only the rational numbers are known, where are the others to come from? By now we have had enough experience to realize that the situation may not be quite so hopeless as that casual consideration suggests. The strategy to be adopted in our construction has already been used effectively for defining functions and complex numbers. Instead of trying to determine the "real nature" of these concepts, we settled for a definition that described enough about them to determine their mathematical properties completely.

A similar proposal for defining real numbers requires a description of real numbers in terms of rational numbers. The observation, that a real number ought to be determined completely by the set of rational numbers less than it, suggests a strikingly simple and quite attractive possibility: a real number might (and in fact eventually will) be described as a collection of rational numbers. In order to make this proposal effective, however, some means must be found for describing "the set of rational numbers less than a real number" without mentioning real numbers, which are still nothing more than heuristic figments of our mathematical imagination.

If A is to be regarded as the set of rational numbers which are less than the real number α, then A ought to have the following property: If x is in A and y is a rational number satisfying $y < x$, then y is in A. In addition to this property, the set A should have a few others. Since there should be some rational number $x < \alpha$, the set A should not be empty. Likewise, since there should be some rational number $x > \alpha$, the set A should not be all of $\mathbf{Q}$. Finally, if $x < \alpha$, then there should be another rational number y with $x < y < \alpha$, so A should not contain a greatest member.

If we temporarily regard the real numbers as known, then it is not hard to check (Problem 8-17) that a set A with these properties is indeed the set of rational numbers less than some real number α. Since the real numbers are presently in limbo, your proof, if you supply one, must be regarded only as an unofficial comment on these proceedings. It will serve to convince you, however, that we have not failed to notice any crucial property of the set A. There appears to be no reason for hesitating any longer.

DEFINITION

A **real number** is a set α, of rational numbers, with the following four properties:

(1) If x is in α and y is a rational number with $y < x$, then y is also in α.
(2) $\alpha \neq \emptyset$.
(3) $\alpha \neq \mathbf{Q}$.
(4) There is no greatest element in α; in other words, if x is in α, then there is some y in α with $y > x$.

The set of all real numbers is denoted by **R**.

Just to remind you of the philosophy behind our definition, here is an explicit example of a real number:

$$\alpha = \{x \text{ in } \mathbf{Q} : x < 0 \text{ or } x^2 < 2\}.$$

It should be clear that α is the real number which will eventually be known as $\sqrt{2}$, but it is not an entirely trivial exercise to show that α actually is a real number. The whole point of such an exercise is to prove this using only facts about **Q**; the hard part will be checking condition (4), but this has already appeared as a problem in a previous chapter (finding out which one is up to you). Notice that condition (4), although quite bothersome here, is really essential in order to avoid ambiguity; without it both

$$\{x \text{ in } \mathbf{Q} : x < 1\}$$

and

$$\{x \text{ in } \mathbf{Q} : x \leq 1\}$$

would be candidates for the "real number 1."

The shift from A to α in our definition indicates both a conceptual and a notational concern. Henceforth, a real number *is*, by definition, a set of rational numbers. This means, in particular, that a rational number (a member of **Q**) is *not* a real number; instead every rational number x has a natural counterpart which is a real number, namely, $\{y \text{ in } \mathbf{Q} : y < x\}$. After completing the construction of the real numbers, we can mentally throw away the elements of **Q** and agree that **Q** will henceforth denote these special sets. For the moment, however, it will be necessary to work at the same time with rational numbers, real numbers (sets of rational numbers) and even sets of real numbers (sets of sets of rational numbers). Some confusion is perhaps inevitable, but proper notation should keep this to a minimum. Rational numbers will be denoted by lower case Roman letters (x, y, z, a, b, c) and real numbers by lower case Greek letters (α, β, γ); capital Roman letters (A, B, C) will be used to denote sets of real numbers.

The remainder of this chapter is devoted to the definition of $\boldsymbol{+}$, $\boldsymbol{\cdot}$, and **P** for **R**, and a proof that with these structures **R** is indeed a complete ordered field.

We shall actually begin with the definition of **P**, and even here we shall work backwards. We first define $\alpha \boldsymbol{<} \beta$; later, when $\boldsymbol{+}$, $\boldsymbol{\cdot}$, and $\mathbf{0}$ are available, we shall define **P** as the set of all α with $\mathbf{0} \boldsymbol{<} \alpha$, and prove the necessary properties for **P**.

The reason for beginning with the definition of $<$ is the simplicity of this concept in our present setup:

Definition. If α and β are real numbers, then $\alpha < \beta$ means that α is contained in β (that is, every element of α is also an element of β), but $\alpha \neq \beta$.

A repetition of the definitions of $\leq, >, \geq$ would be stultifying, but it is interesting to note that $\leq$ can now be expressed more simply than $<$; if α and β are real numbers, then $\alpha \leq \beta$ if and only if α is contained in β.

If A is a bounded collection of real numbers, it is almost obvious that A should have a least upper bound. Each α in A is a collection of rational numbers; if these rational numbers are all put in one collection β, then β is presumably $\sup A$. In the proof of the following theorem we check all the little details which have not been mentioned, not least of which is the assertion that β *is* a real number. (We will not bother numbering theorems in this chapter, since they all add up to one big Theorem: There is a complete ordered field.)

THEOREM If A is a set of real numbers and $A \neq \emptyset$ and A is bounded above, then A has a least upper bound.

PROOF Let $\beta = \{x : x \text{ is in some } \alpha \text{ in } A\}$. Then β is certainly a collection of rational numbers; the proof that β is a real number requires checking four facts.

(1) Suppose that x is in β and $y < x$. The first condition means that x is in α for some α in A. Since α is a real number, the assumption $y < x$ implies that y is in α. Therefore it is certainly true that y is in β.
(2) Since $A \neq \emptyset$, there is some α in A. Since α is a real number, there is some x in α. This means that x is in β, so $\beta \neq \emptyset$.
(3) Since A is bounded above, there is some real number γ such that $\alpha < \gamma$ for every α in A. Since γ is a real number, there is some rational number x which is not in γ. Now $\alpha < \gamma$ means that α is contained in γ, so it is also true that x is not in α for any α in A. This means that x is not in β; so $\beta \neq \mathbf{Q}$.
(4) Suppose that x is in β. Then x is in α for some α in A. Since α does not have a greatest member, there is some rational number y with $x < y$ and y in α. But this means that y is in β; thus β does not have a greatest member.

These four observations prove that β is a real number. The proof that β is the least upper bound of A is easier. If α is in A, then clearly α is contained in β; this means that $\alpha \leq \beta$, so β is an upper bound for A. On the other hand, if γ is an upper bound for A, then $\alpha \leq \gamma$ for every α in A; this means that α is contained in γ, for every α in A, and this surely implies that β is contained in γ. *This*, in turn, means that $\beta \leq \gamma$; thus β is the least upper bound of A. ▌

The definition of $+$ is both obvious and easy, but is must be complemented with a proof that this "obvious" definition makes any sense at all.

Definition. If α and β are real numbers, then

$$\alpha \boldsymbol{+} \beta = \{x : x = y + z \text{ for some } y \text{ in } \alpha \text{ and some } z \text{ in } \beta\}.$$

THEOREM If α and β are real numbers, then $\alpha \boldsymbol{+} \beta$ is a real number.

PROOF Once again four facts must be verified.

(1) Suppose $w < x$ for some x in $\alpha \boldsymbol{+} \beta$. Then $x = y + z$ for some y in α and some z in β, which means that $w < y + z$, and consequently, $w - y < z$. This shows that $w - y$ is in β (since z is in β, and β is a real number). Since $w = y + (w - y)$, it follows that w is in $\alpha \boldsymbol{+} \beta$.
(2) It is clear that $\alpha \boldsymbol{+} \beta \neq \emptyset$, since $\alpha \neq \emptyset$ and $\beta \neq \emptyset$.
(3) Since $\alpha \neq \mathbf{Q}$ and $\beta \neq \mathbf{Q}$, there are rational numbers a and b with a not in α and b not in β. Any x in α satisfies $x < a$ (for if $a < x$, then condition (1) for a real number would imply that a is in α); similarly any y in β satisfies $y < b$. Thus $x + y < a + b$ for any x in α and y in β. This shows that $a + b$ is not in $\alpha \boldsymbol{+} \beta$, so $\alpha \boldsymbol{+} \beta \neq \mathbf{Q}$.
(4) If x is in $\alpha \boldsymbol{+} \beta$, then $x = y + z$ for y in α and z in β. There are y' in α and z' in β with $y < y'$ and $z < z'$; then $x < y' + z'$ and $y' + z'$ is in $\alpha \boldsymbol{+} \beta$. Thus $\alpha \boldsymbol{+} \beta$ has no greatest member. ▮

By now you can see how tiresome this whole procedure is going to be. Every time we mention a new real number, we must prove that it *is* a real number; this requires checking four conditions, and even when trivial they require concentration. There is really no help for this (except that it will be less boring if you check the four conditions for yourself). Fortunately, however, a few points of interest will arise now and then, and some of our theorems will be easy. In particular, two properties of $\boldsymbol{+}$ present no problems.

THEOREM If α, β, and γ are real numbers, then $(\alpha \boldsymbol{+} \beta) \boldsymbol{+} \gamma = \alpha \boldsymbol{+} (\beta \boldsymbol{+} \gamma)$.

PROOF Since $(x + y) + z = x + (y + z)$ for all rational numbers x, y, and z, every member of $(\alpha \boldsymbol{+} \beta) \boldsymbol{+} \gamma$ is also a member of $\alpha \boldsymbol{+} (\beta \boldsymbol{+} \gamma)$, and vice versa. ▮

THEOREM If α and β are real numbers, then $\alpha \boldsymbol{+} \beta = \beta \boldsymbol{+} \alpha$.

PROOF Left to you (even easier). ▮

To prove the other properties of $\boldsymbol{+}$ we first define $\mathbf{0}$.

Definition. $\mathbf{0} = \{x \text{ in } \mathbf{Q} : x < 0\}$.

It is, thank goodness, obvious that $\mathbf{0}$ is a real number, and the following theorem is also simple.

THEOREM If α is a real number, then $\alpha + \mathbf{0} = \alpha$.

PROOF If x is in α and y is in $\mathbf{0}$, then $y < 0$, so $x + y < x$. This implies that $x + y$ is in α. Thus every member of $\alpha + \mathbf{0}$ is also a member of α.

On the other hand, if x is in α, then there is a rational number y in α such that $y > x$. Since $x = y + (x - y)$, where y is in α, and $x - y < 0$ (so that $x - y$ is in $\mathbf{0}$), this shows that x is in $\alpha + \mathbf{0}$. Thus every member of α is also a member of $\alpha + \mathbf{0}$. ▌

The reasonable candidate for $-\alpha$ would seem to be the set

$$\{x \text{ in } \mathbf{Q} : -x \text{ is not in } \alpha\}$$

(since $-x$ not in α means, intuitively, that $-x > \alpha$, so that $x < -\alpha$). But in certain cases this set will not even be a real number. Although a real number α does not have a greatest member, the set

$$\mathbf{Q} - \alpha = \{x \text{ in } \mathbf{Q} : x \text{ is not in } \alpha\}$$

may have a *least* element x_0; when α is a real number of this kind, the set $\{x : -x \text{ is not in } \alpha\}$ will have a greatest element $-x_0$. It is therefore necessary to introduce a slight modification into the definition of $-\alpha$, which comes equipped with a theorem.

Definition. If α is a real number, then

$$-\alpha = \{x \text{ in } \mathbf{Q} : -x \text{ is not in } \alpha, \text{ but } -x \text{ is not the least element of } \mathbf{Q} - \alpha\}.$$

THEOREM If α is a real number, then $-\alpha$ is a real number.

PROOF

(1) Suppose that x is in $-\alpha$ and $y < x$. Then $-y > -x$. Since $-x$ is not in α, it is also true that $-y$ is not in α. Moreover, it is clear that $-y$ is not the smallest element of $\mathbf{Q} - \alpha$, since $-x$ is a smaller element. This shows that y is in $-\alpha$.

(2) Since $\alpha \neq \mathbf{Q}$, there is some rational number y which is not in α. We can assume that y is not the smallest rational number in $\mathbf{Q} - \alpha$ (since y can always be replaced by any $y' > y$). Then $-y$ is in $-\alpha$. Thus $-\alpha \neq \emptyset$.

(3) Since $\alpha \neq \emptyset$, there is some x in α. Then $-x$ cannot possibly be in $-\alpha$, so $-\alpha \neq \mathbf{Q}$.

(4) If x is in $-\alpha$, then $-x$ is not in α, and there is a rational number $y < -x$ which is also not in α. Let z be a rational number with $y < z < -x$. Then z is also not in α, and z is clearly not the smallest element of $\mathbf{Q} - \alpha$. So $-z$ is in $-\alpha$. Since $-z > x$, this shows that $-\alpha$ does not have a greatest element. ▌

The proof that $\alpha + (-\alpha) = \mathbf{0}$ is not entirely straightforward. The difficulties are not caused, as you might presume, by the finicky details in the definition of $-\alpha$. Rather, at this point we require the Archimedean property of $\mathbf{Q}$ stated on page 574, which does not follow from P1–P12. This property is needed to prove the following lemma, which plays a crucial role in the next theorem.

LEMMA Let α be a real number, and z a positive rational number. Then there are (Figure 1) rational numbers x in α, and y not in α, such that $y - x = z$. Moreover, we may assume that y is not the smallest element of $\mathbf{Q} - \alpha$.

PROOF Suppose first that z is in α. If the numbers

$$z, 2z, 3z, \ldots$$

were *all* in α, then *every* rational number would be in α, since every rational number w satisfies $w < nz$ for some n, by the additional assumption on page 574. This contradicts the fact that α is a real number, so there is some k such that $x = kz$ is in α and $y = (k+1)z$ is not in α. Clearly $y - x = z$.

Moreover, if y happens to be the smallest element of $\mathbf{Q} - \alpha$, let $x' > x$ be an element of α, and replace x by x', and y by $y + (x' - x)$.

If z is not in α, there is a similar proof, based on the fact that the numbers $(-n)z$ cannot all fail to be in α. ▮

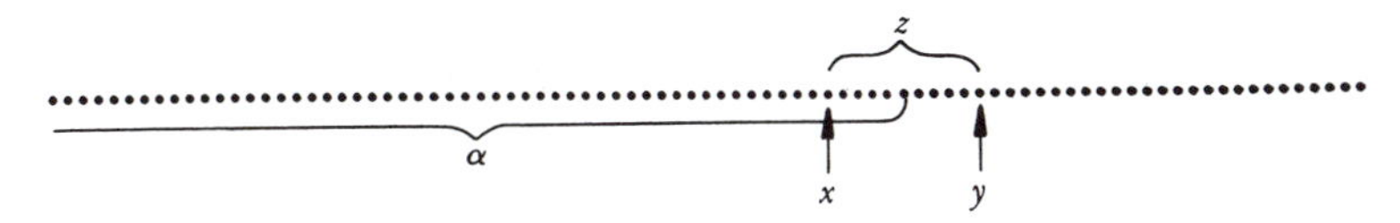

FIGURE 1

THEOREM If α is a real number, then

$$\alpha + (-\alpha) = \mathbf{0}.$$

PROOF Suppose x is in α and y is in $-\alpha$. Then $-y$ is not in α, so $-y > x$. Hence $x + y < 0$, so $x + y$ is in $\mathbf{0}$. Thus every member of $\alpha + (-\alpha)$ is in $\mathbf{0}$.

It is a little more difficult to go in the other direction. If z is in $\mathbf{0}$, then $-z > 0$. According to the lemma, there is some x in α, and some y not in α, with y not the smallest element of $\mathbf{Q} - \alpha$, such that $y - x = -z$. This equation can be written $x + (-y) = z$. Since x is in α, and $-y$ is in $-\alpha$, this proves that z is in $\alpha + (-\alpha)$. ▮

Before proceeding with multiplication, we define the "positive elements" and prove a basic property:

Definition. $\mathbf{P} = \{\alpha \text{ in } \mathbf{R} : \alpha > \mathbf{0}\}$.

Notice that $\alpha + \beta$ is clearly in $\mathbf{P}$ if α and β are.

THEOREM If α is a real number, then one and only one of the following conditions holds:

(i) $\alpha = \mathbf{0}$,
(ii) α is in $\mathbf{P}$,
(iii) $\mathbf{-}\alpha$ is in $\mathbf{P}$.

PROOF If α contains any positive rational number, then α certainly contains all negative rational numbers, so α contains $\mathbf{0}$ and $\alpha \neq \mathbf{0}$, i.e., α is in $\mathbf{P}$. If α contains no positive rational numbers, then one of two possibilities must hold:

(1) α contains all negative rational numbers; then $\alpha = \mathbf{0}$.
(2) there is some negative rational number x which is not in α; it can be assumed that x is not the least element of $\mathbf{Q} - \alpha$ (since x could be replaced by $x/2 > x$); then $\mathbf{-}\alpha$ contains the positive rational number $-x$, so, as we have just proved, $\mathbf{-}\alpha$ is in $\mathbf{P}$.

This shows that *at least one* of (i)–(iii) must hold. If $\alpha = \mathbf{0}$, it is clearly impossible for condition (ii) or (iii) to hold. Moreover, it is impossible that $\alpha \mathbf{>} \mathbf{0}$ and $\mathbf{-}\alpha \mathbf{>} \mathbf{0}$ both hold, since this would imply that $\mathbf{0} = \alpha \mathbf{+} (\mathbf{-}\alpha) \mathbf{>} \mathbf{0}$. ▌

Recall that $\alpha \mathbf{>} \beta$ was defined to mean that α contains β, but is unequal to β. This definition was fine for proving completeness, but now we have to show that it is equivalent to the definition which would be made in terms of $\mathbf{P}$. Thus, we must show that $\alpha \mathbf{-} \beta \mathbf{>} \mathbf{0}$ is equivalent to $\alpha \mathbf{>} \beta$. This is clearly a consequence of the next theorem.

THEOREM If α, β, and γ are real numbers and $\alpha \mathbf{>} \beta$, then $\alpha \mathbf{+} \gamma \mathbf{>} \beta \mathbf{+} \gamma$.

PROOF The hypothesis $\alpha \mathbf{>} \beta$ implies that β is contained in α; it follows immediately from the definition of $\mathbf{+}$ that $\beta \mathbf{+} \gamma$ is contained in $\alpha \mathbf{+} \gamma$. This shows that $\alpha \mathbf{+} \gamma \mathbf{\geq} \beta \mathbf{+} \gamma$. We can easily rule out the possibility of equality, for if

$$\alpha \mathbf{+} \gamma = \beta \mathbf{+} \gamma,$$

then

$$\alpha = (\alpha \mathbf{+} \gamma) \mathbf{+} (\mathbf{-}\gamma) = (\beta \mathbf{+} \gamma) \mathbf{+} (\mathbf{-}\gamma) = \beta,$$

which is false. Thus $\alpha \mathbf{+} \gamma \mathbf{>} \beta \mathbf{+} \gamma$. ▌

Multiplication presents difficulties of its own. If $\alpha, \beta \mathbf{>} \mathbf{0}$, then $\alpha \mathbf{\cdot} \beta$ can be defined as follows.

Definition. If α and β are real numbers and $\alpha, \beta \mathbf{>} \mathbf{0}$, then

$$\alpha \mathbf{\cdot} \beta = \{z : z \leq 0 \text{ or } z = x \cdot y \text{ for some } x \text{ in } \alpha \text{ and } y \text{ in } \beta \text{ with } x, y > 0\}.$$

THEOREM If α and β are real numbers with $\alpha, \beta \mathbf{>} \mathbf{0}$, then $\alpha \mathbf{\cdot} \beta$ is a real number.

PROOF As usual, we must check four conditions.

(1) Suppose $w < z$, where z is in $\alpha \cdot \beta$. If $w \le 0$, then w is automatically in $\alpha \cdot \beta$. Suppose that $w > 0$. Then $z > 0$, so $z = x \cdot y$ for some positive x in α and positive y in β. Now

$$w = \frac{wz}{z} = \frac{wxy}{z} = \left(\frac{w}{z} \cdot x\right) \cdot y.$$

Since $0 < w < z$, we have $w/z < 1$, so $(w/z) \cdot x$ is in α. Thus w is in $\alpha \cdot \beta$.

(2) Clearly $\alpha \cdot \beta \neq \emptyset$.

(3) If x is not in α, and y is not in β, then $x > x'$ for all x' in α, and $y > y'$ for all y' in β. Hence $xy > x'y'$ for all such positive x' and y'. So xy is not in $\alpha \cdot \beta$; thus $\alpha \cdot \beta \neq \mathbf{Q}$.

(4) Suppose w is in $\alpha \cdot \beta$, and $w \le 0$. There is some x in α with $x > 0$ and some y in β with $y > 0$. Then $z = xy$ is in $\alpha \cdot \beta$ and $z > w$. Now suppose $w > 0$. Then $w = xy$ for some positive x in α and some positive y in β. Moreover, α contains some $x' > x$; if $z = x'y$, then $z > xy = w$, and z is in $\alpha \cdot \beta$. Thus $\alpha \cdot \beta$ does not have a greatest element. ▮

Notice that $\alpha \cdot \beta$ is clearly in **P** if α and β are. This completes the verification of all properties of **P**. To complete the definition of $\cdot$ we first define $|\alpha|$.

Definition. If α is a real number, then

$$|\alpha| = \begin{cases} \alpha, & \text{if } \alpha \ge \mathbf{0} \\ -\alpha, & \text{if } \alpha \le \mathbf{0}. \end{cases}$$

Definition. If α and β are real numbers, then

$$\alpha \cdot \beta = \begin{cases} \mathbf{0}, & \text{if } \alpha = \mathbf{0} \text{ or } \beta = \mathbf{0} \\ |\alpha| \cdot |\beta|, & \text{if } \alpha > \mathbf{0}, \beta > \mathbf{0} \text{ or } \alpha < \mathbf{0}, \beta < \mathbf{0} \\ -(|\alpha| \cdot |\beta|), & \text{if } \alpha > \mathbf{0}, \beta < \mathbf{0} \text{ or } \alpha < \mathbf{0}, \beta > \mathbf{0}. \end{cases}$$

As one might suspect, the proofs of the properties of multiplication usually involve reduction to the case of positive numbers.

THEOREM If α, β, and γ are real numbers, then $\alpha \cdot (\beta \cdot \gamma) = (\alpha \cdot \beta) \cdot \gamma$.

PROOF This is clear if $\alpha, \beta, \gamma > \mathbf{0}$. The proof for the general case requires considering separate cases (and is simplified slightly if one uses the following theorem). ▮

THEOREM If α and β are real numbers, then $\alpha \cdot \beta = \beta \cdot \alpha$.

PROOF This is clear if $\alpha, \beta > \mathbf{0}$, and the other cases are easily checked. ∎

Definition. $\mathbf{1} = \{x \text{ in } \mathbf{Q} : x < 1\}$.
(It is clear that $\mathbf{1}$ is a real number.)

THEOREM If α is a real number, then $\alpha \cdot \mathbf{1} = \alpha$.

PROOF Let $\alpha > \mathbf{0}$. It is easy to see that every member of $\alpha \cdot \mathbf{1}$ is also a member of α. On the other hand, suppose x is in α. If $x \leq 0$, then x is automatically in $\alpha \cdot \mathbf{1}$. If $x > 0$, then there is some rational number y in α such that $x < y$. Then $x = y \cdot (x/y)$, and x/y is in $\mathbf{1}$, so x is in $\alpha \cdot \mathbf{1}$. This proves that $\alpha \cdot \mathbf{1} = \alpha$ if $\alpha > \mathbf{0}$.

If $\alpha < \mathbf{0}$, then, applying the result just proved, we have

$$\alpha \cdot \mathbf{1} = -(|\alpha| \cdot |\mathbf{1}|) = -(|\alpha|) = \alpha.$$

Finally, the theorem is obvious when $\alpha = \mathbf{0}$. ∎

Definition. If α is a real number and $\alpha > \mathbf{0}$, then
$\alpha^{-1} = \{x \text{ in } \mathbf{Q} : x \leq 0$, or $x > 0$ and $1/x$ is not in α, but $1/x$ is not the smallest member of $\mathbf{Q} - \alpha\}$;
if $\alpha < \mathbf{0}$, then $\alpha^{-1} = -(|\alpha|)^{-1}$.

THEOREM If α is a real number unequal to $\mathbf{0}$, then α^{-1} is a real number.

PROOF Clearly it suffices to consider only $\alpha > \mathbf{0}$. Four conditions must be checked.

(1) Suppose $y < x$, and x is in α^{-1}. If $y \leq 0$, then y is in α^{-1}. If $y > 0$, then $x > 0$, so $1/x$ is not in α. Since $1/y > 1/x$, it follows that $1/y$ is not in α, and $1/y$ is clearly not the smallest element of $\mathbf{Q} - \alpha$, so y is in α^{-1}.
(2) Clearly $\alpha^{-1} \neq \emptyset$.
(3) Since $\alpha > \mathbf{0}$, there is some positive rational number x in α. Then $1/x$ is not in α^{-1}, so $\alpha^{-1} \neq \mathbf{Q}$.
(4) Suppose x is in α^{-1}. If $x \leq 0$, there is clearly some y in α^{-1} with $y > x$ because α^{-1} contains some positive rationals. If $x > 0$, then $1/x$ is not in α. Since $1/x$ is not the smallest member of $\mathbf{Q} - \alpha$, there is a rational number y not in α, with $y < 1/x$. Choose a rational number z with $y < z < 1/x$. Then $1/z$ is in α, and $1/z > x$. Thus α^{-1} does not contain a largest member. ∎

In order to prove that α^{-1} is really the multiplicative inverse of α, it helps to have another lemma, which is the multiplicative analogue of our first lemma.

LEMMA Let α be a real number with $\alpha > \mathbf{0}$, and z a rational number with $z > 1$. Then there are rational numbers x in α, and y not in α, such that $y/x = z$. Moreover, we can assume that y is not the least element of $\mathbf{Q} - \alpha$.

PROOF Suppose first that z is in α. Since $z - 1 > 0$ and

$$z^n = (1 + (z-1))^n \geq 1 + n(z-1),$$

it follows that the numbers

$$z, z^2, z^3, \ldots$$

cannot all be in α. So there is some k such that $x = z^k$ is in α, and $y = z^{k+1}$ is not in α. Clearly $y/x = z$. Moreover, if y happens to be the least element of $\mathbf{Q} - \alpha$, let $x' > x$ be an element of α, and replace x by x' and y by yx'/x.

If z is not in α, there is a similar proof, based on the fact that the numbers $1/z^k$ cannot all fail to be in α. ▌

THEOREM If α is a real number and $\alpha \neq \mathbf{0}$, then $\alpha \bullet \alpha^{-1} = \mathbf{1}$.

PROOF It obviously suffices to consider only $\alpha > \mathbf{0}$, in which case $\alpha^{-1} > \mathbf{0}$. Suppose that x is a positive rational number in α, and y is a positive rational number in α^{-1}. Then $1/y$ is not in α, so $1/y > x$; consequently $xy < 1$, which means that xy is in $\mathbf{1}$. Since all rational numbers $x \leq 0$ are also in $\mathbf{1}$, this shows that every member of $\alpha \bullet \alpha^{-1}$ is in $\mathbf{1}$.

To prove the converse assertion, let z be in $\mathbf{1}$. If $z \leq 0$, then clearly z is in $\alpha \bullet \alpha^{-1}$. Suppose $0 < z < 1$. According to the lemma, there are positive rational numbers x in α, and y not in α, such that $y/x = 1/z$; and we can assume that y is not the smallest element of $\mathbf{Q} - \alpha$. But this means that $z = x \cdot (1/y)$, where x is in α, and $1/y$ is in α^{-1}. Consequently, z is in $\alpha \bullet \alpha^{-1}$. ▌

We are almost done! Only the proof of the distributive law remains. Once again we must consider many cases, but do not despair. The case when all numbers are positive contains an interesting point, and the other cases can all be taken care of very neatly.

THEOREM If α, β, and γ are real numbers, then $\alpha \bullet (\beta + \gamma) = \alpha \bullet \beta + \alpha \bullet \gamma$.

PROOF Assume first that $\alpha, \beta, \gamma > \mathbf{0}$. Then both numbers in the equation contain all rational numbers ≤ 0. A positive rational number in $\alpha \bullet (\beta + \gamma)$ is of the form $x \cdot (y + z)$ for positive x in α, y in β, and z in γ. Since $x \cdot (y + z) = x \cdot y + x \cdot z$, where $x \cdot y$ is a positive element of $\alpha \bullet \beta$, and $x \cdot z$ is a positive element of $\alpha \bullet \gamma$, this number is also in $\alpha \bullet \beta + \alpha \bullet \gamma$. Thus, every element of $\alpha \bullet (\beta + \gamma)$ is also in $\alpha \bullet \beta + \alpha \bullet \gamma$.

On the other hand, a positive rational number in $\alpha \bullet \beta + \alpha \bullet \gamma$ is of the form $x_1 \cdot y + x_2 \cdot z$ for positive x_1, x_2 in α, y in β, and z in γ. If $x_1 \leq x_2$, then $(x_1/x_2) \cdot y \leq y$, so $(x_1/x_2) \cdot y$ is in β. Thus

$$x_1 \cdot y + x_2 \cdot z = x_2[(x_1/x_2)y + z]$$

is in $\alpha \bullet (\beta + \gamma)$. Of course, the same trick works if $x_2 \leq x_1$.

To complete the proof it is necessary to consider the cases when α, β, and γ are not all $> \mathbf{0}$. If any one of the three equals $\mathbf{0}$, the proof is easy and the cases involving $\alpha < 0$ can be derived immediately once all the possibilities for β and γ have been accounted for. Thus we assume $\alpha > 0$ and consider three cases: $\beta, \gamma < 0$, and $\beta < 0$, $\gamma > 0$, and $\beta > 0$, $\gamma < 0$. The first follows immediately from the case already proved, and the third follows from the second by interchanging β and γ. Therefore we concentrate on the case $\beta < 0$, $\gamma > 0$. There are then two possibilities:

(1) $\beta + \gamma \geq 0$. Then

$$\alpha \cdot \gamma = \alpha \cdot ([\beta + \gamma] + |\beta|) = \alpha \cdot (\beta + \gamma) + \alpha \cdot |\beta|,$$

so

$$\begin{aligned} \alpha \cdot (\beta + \gamma) &= -(\alpha \cdot |\beta|) + \alpha \cdot \gamma \\ &= \alpha \cdot \beta + \alpha \cdot \gamma. \end{aligned}$$

(2) $\beta + \gamma \leq 0$. Then

$$\alpha \cdot |\beta| = \alpha \cdot (|\beta + \gamma| + \gamma) = \alpha \cdot |\beta + \gamma| + \alpha \cdot \gamma,$$

so

$$\alpha \cdot (\beta + \gamma) = -(\alpha \cdot |\beta + \gamma|) = -(\alpha \cdot |\beta|) + \alpha \cdot \gamma = \alpha \cdot \beta + \alpha \cdot \gamma. \blacksquare$$

This proof completes the work of the chapter. Although long and frequently tedious, this chapter contains results sufficiently important to be read in detail at least once (and preferably not more than once!). For the first time we know that we have not been operating in a vacuum—there is indeed a complete ordered field, the theorems of this book are not based on assumptions which can never be realized. One interesting and horrid possibility remains: there may be several complete ordered fields. If this is true, then the theorems of calculus are unexpectedly rich in content, but the properties P1–P13 are disappointingly incomplete. The last chapter disposes of this possibility; properties P1–P13 completely characterize the real numbers—anything that can be proved about real numbers can be proved on the basis of these properties alone.

PROBLEMS

There are only two problems in this set, but each asks for an entirely different construction of the real numbers! The detailed examination of another construction is recommended only for masochists, but the main idea behind these other constructions is worth knowing. The real numbers constructed in this chapter might be called "the algebraist's real numbers," since they were purposely defined so as to guarantee the least upper bound property, which involves the ordering $<$, an algebraic notion. The real number system constructed in the next problem might be called "the analyst's real numbers," since they are devised so that Cauchy sequences will always converge.

1. Since every real number ought to be the limit of some Cauchy sequence of rational numbers, we might try to *define* a real number to be a Cauchy sequence of rational numbers. Since two Cauchy sequences might converge to the same real number, however, this proposal requires some modifications.

(a) Define two Cauchy sequences of rational numbers $\{a_n\}$ and $\{b_n\}$ to be *equivalent* (denoted by $\{a_n\} \sim \{b_n\}$) if $\lim_{n\to\infty}(a_n - b_n) = 0$. Prove that $\{a_n\} \sim \{a_n\}$, that $\{b_n\} \sim \{a_n\}$ if $\{a_n\} \sim \{b_n\}$, and that $\{a_n\} \sim \{c_n\}$ if $\{a_n\} \sim \{b_n\}$ and $\{b_n\} \sim \{c_n\}$.

(b) Suppose that α is the set of all sequences equivalent to $\{a_n\}$, and β is the set of all sequences equivalent to $\{b_n\}$. Prove that either $\alpha \cap \beta = \emptyset$ or $\alpha = \beta$. (If $\alpha \cap \beta \neq \emptyset$, then there is some $\{c_n\}$ in both α and β. Show that in this case α and β both consist precisely of those sequences equivalent to $\{c_n\}$.)

Part (b) shows that the collection of all Cauchy sequences can be split up into disjoint sets, each set consisting of all sequences equivalent to some fixed sequence. We define a real number to be such a collection, and denote the set of all real numbers by **R**.

(c) If α and β are real numbers, let $\{a_n\}$ be a sequence in α, and $\{b_n\}$ a sequence in β. Define $\alpha + \beta$ to be the collection of all sequences equivalent to the sequence $\{a_n + b_n\}$. Show that $\{a_n + b_n\}$ is a Cauchy sequence and also show that this definition does not depend on the particular sequences $\{a_n\}$ and $\{b_n\}$ chosen for α and β. Check also that the analogous definition of multiplication is well defined.

(d) Show that **R** is a field with these operations; existence of a multiplicative inverse is the only interesting point to check.

(e) Define the positive real numbers P so that **R** will be an ordered field.

(f) Prove that every Cauchy sequence of real numbers converges. Remember that if $\{\alpha_n\}$ is a sequence of real numbers, then each α_n is itself a collection of Cauchy sequences of rational numbers.

2. This problem outlines a construction of "the high-school student's real numbers." We define a real number to be a pair $(a, \{b_n\})$, where a is an integer and $\{b_n\}$ is a sequence of natural numbers from 0 to 9, with the proviso that the sequence is not eventually 9; intuitively, this pair represents $a + \sum_{n=1}^{\infty} b_n 10^{-n}$. With this definition, a real number is a very concrete object, but the difficulties involved in defining addition and multiplication are formidable (how do you add infinite decimals without worrying about carrying digits infinitely far out?). A reasonable approach is outlined below; the trick is to use least upper bounds right from the start.

(a) Define $(a, \{b_n\}) \prec (c, \{d_n\})$ if $a < c$, or if $a = c$ and for some n we have $b_n < d_n$ but $b_j = d_j$ for $1 \le j < n$. Using this definition, prove the least upper bound property.

(b) Given $\alpha = (a, \{b_n\})$, define $\alpha_k = a + \sum_{n=1}^{k} b_n 10^{-n}$; intuitively, α_k is the rational number obtained by changing all decimal places after the kth to 0. Conversely, given a rational number r of the form $a + \sum_{n=1}^{k} b_n 10^{-n}$, let r' denote the real number $(a, \{b_n{}'\})$, where $b_n{}' = b_n$ for $1 \le n \le k$ and $b_n{}' = 0$ for $n > k$. Now for $\alpha = (a, \{b_n\})$ and $\beta = (c, \{d_n\})$ define

$$\alpha \mathbin{\boldsymbol{+}} \beta = \sup\{(\alpha_k + \beta_k)' : k \text{ a natural number}\}$$

(the least upper bound exists by part (a)). If multiplication is defined similarly, then the verification of all conditions for a field is a straightforward task, not highly recommended. Once more, however, existence of multiplicative inverses will be the hardest.

CHAPTER 30 UNIQUENESS OF THE REAL NUMBERS

We shall now revert to the usual notation for real numbers, reserving boldface symbols for other fields which may turn up. Moreover, we will regard integers and rational numbers as special kinds of real numbers, and forget about the specific way in which real numbers were defined. In this chapter we are interested in only one question: are there any complete ordered fields other than **R**? The answer to this question, if taken literally, is "yes." For example, the field F_3 introduced in Chapter 28 is a complete ordered field, and it is certainly not **R**. This field is a "silly" example because the pair (a, a) can be regarded as just another name for the real number a; the operations

$$(a, a) \mathbin{\boldsymbol{+}} (b, b) = (a + b, a + b),$$
$$(a, a) \mathbin{\boldsymbol{\cdot}} (b, b) = (a \cdot b, a \cdot b),$$

are consistent with this renaming. This sort of example shows that any intelligent consideration of the question requires some mathematical means of discussing such renaming procedures.

If the elements of a field F are going to be used to rename elements of **R**, then for each a in **R** there should correspond a "name" $f(a)$ in F. The notation $f(a)$ suggests that renaming can be formulated in terms of functions. In order to do this we will need a concept of function much more general than any which has occurred until now; in fact, we will require the most general notion of "function" used in mathematics. A function, in this general sense, is simply a rule which assigns to some things, other things. To be formal, a **function** is a collection of ordered pairs (of objects of any sort) which does not contain two distinct pairs with the same first element. The **domain** of a function f is the set A of all objects a such that (a, b) is in f for some b; this (unique) b is denoted by $\boldsymbol{f(a)}$. If $f(a)$ is in the set B for all a in A, then f is called a function **from** A **to** B. For example,

if $f(x) = \sin x$ for all x in **R** (and f is defined only for x in **R**), then f is a function from **R** to **R**; it is also a function from **R** to $[-1, 1]$;

if $f(z) = \sin z$ for all z in **C**, then f is a function from **C** to **C**;

if $f(z) = e^z$ for all z in **C**, then f is a function from **C** to **C**; it is also a function from **C** to $\{z \text{ in } \mathbf{C} : z \neq 0\}$;

θ is a function from $\{z \text{ in } \mathbf{C} : z \neq 0\}$ to $\{x \text{ in } \mathbf{R} : 0 \leq x < 2\pi\}$;

if f is the collection of all pairs $(a, (a, a))$ for a in **R**, then f is a function from **R** to F_3.

Suppose that F_1 and F_2 are two fields; we will denote the operations in F_1 by $\oplus$, $\odot$, etc., and the operations in F_2 by $+$, $\cdot$, etc. If F_2 is going to be considered as a collection of new names for elements of F_1, then there should be a function from F_1 to F_2 with the following properties:

(1) The function f should be one-one, that is, if $x \neq y$, then we should have $f(x) \neq f(y)$; this means that no two elements of F_1 have the same name.
(2) The function f should be "onto," that is, for every element z in F_2 there should be some x in F_1 such that $z = f(x)$; this means that every element of F_2 is used to name some element of F_1.
(3) For all x and y in F_1 we should have

$$f(x \oplus y) = f(x) + f(y),$$
$$f(x \odot y) = f(x) \cdot f(y);$$

this means that the renaming procedure is consistent with the operations of the field.

If we are also considering F_1 and F_2 as ordered fields, we add one more requirement:

(4) If $x \otimes y$, then $f(x) < f(y)$.

A function with these properties is called an *isomorphism* from F_1 to F_2. This definition is so important that we restate it formally.

DEFINITION

If F_1 and F_2 are two fields, an **isomorphism** from F_1 to F_2 is a function f from F_1 to F_2 with the following properties:

(1) If $x \neq y$, then $f(x) \neq f(y)$.
(2) If z is in F_2, then $z = f(x)$ for some x in F_1.
(3) If x and y are in F_1, then

$$f(x \oplus y) = f(x) + f(y),$$
$$f(x \odot y) = f(x) \cdot f(y).$$

If F_1 and F_2 are ordered fields we also require:

(4) If $x \otimes y$, then $f(x) < f(y)$.

The fields F_1 and F_2 are called **isomorphic** if there is an isomorphism between them. Isomorphic fields may be regarded as essentially the same—any important property of one will automatically hold for the other. Therefore, we can, and should, reformulate the question asked at the beginning of the chapter; if F is a complete ordered field it is silly to expect F to equal $\mathbf{R}$—rather, we would like to know if F is isomorphic to $\mathbf{R}$. In the following theorem, F will be a field, with operations $+$ and $\cdot$, and "positive elements" $\mathbf{P}$; we write $a < b$ to mean that $b - a$ is in $\mathbf{P}$, and so forth.

THEOREM If F is a complete ordered field, then F is isomorphic to $\mathbf{R}$.

PROOF Since two fields are defined to be isomorphic if there is an isomorphism between them, we must actually construct a function f from $\mathbf{R}$ to F which is an isomorphism. We begin by defining f on the integers as follows:

$$\begin{aligned} f(0) &= \mathbf{0}, \\ f(n) &= \underbrace{\mathbf{1} + \dots + \mathbf{1}}_{n \text{ times}} \qquad \text{for } n > 0, \\ f(n) &= -\underbrace{(\mathbf{1} + \dots + \mathbf{1})}_{|n| \text{ times}} \qquad \text{for } n < 0. \end{aligned}$$

It is easy to check that

$$\begin{aligned} f(m+n) &= f(m) + f(n), \\ f(m \cdot n) &= f(m) \cdot f(n), \end{aligned}$$

for all integers m and n, and it is convenient to denote $f(n)$ by $\boldsymbol{n}$. We then define f on the rational numbers by

$$f(m/n) = \boldsymbol{m}/\boldsymbol{n} = \boldsymbol{m} \cdot \boldsymbol{n}^{-1}$$

(notice that the n-fold sum $\mathbf{1} + \cdots + \mathbf{1} \neq \mathbf{0}$ if $n > 0$, since F is an ordered field). This definition makes sense because if $m/n = k/l$, then $ml = nk$, so $\boldsymbol{m} \cdot \boldsymbol{l} = \boldsymbol{k} \cdot \boldsymbol{n}$, so $\boldsymbol{m} \cdot \boldsymbol{n}^{-1} = \boldsymbol{k} \cdot \boldsymbol{l}^{-1}$. It is easy to check that

$$\begin{aligned} f(r_1 + r_2) &= f(r_1) + f(r_2), \\ f(r_1 \cdot r_2) &= f(r_1) \cdot f(r_2), \end{aligned}$$

for all rational numbers r_1 and r_2, and that $f(r_1) < f(r_2)$ if $r_1 < r_2$.

The definition of $f(x)$ for arbitrary x is based on the now familiar idea that any real number is determined by the rational numbers less than it. For any x in $\mathbf{R}$, let A_x be the subset of F consisting of all $f(r)$, for all rational numbers $r < x$. The set A_x is certainly not empty, and it is also bounded above, for if r_0 is a rational number with $r_0 > x$, then $f(r_0) > f(r)$ for all $f(r)$ in A_x. Since F is a complete ordered field, the set A_x has a least upper bound; we define $f(x)$ as $\sup A_x$.

We now have $f(x)$ defined in two different ways, first for rational x, and then for any x. Before proceeding further, it is necessary to show that these two definitions agree for rational x. In other words, if x is a rational number, we want to show that

$$\sup A_x = f(x),$$

where $f(x)$ here denotes $\boldsymbol{m}/\boldsymbol{n}$, for $x = m/n$. This is not automatic, but depends on the completeness of F; a slight digression is thus required.

Since F is complete, the elements

$$\underbrace{\mathbf{1} + \dots + \mathbf{1}}_{n \text{ times}} \qquad \text{for natural numbers } n$$

form a set which is not bounded above; the proof is exactly the same as the proof for **R** (Theorem 8-2). The consequences of this fact for **R** have exact analogues in F: in particular, if a and b are elements of F with $a \prec b$, then there is a rational number r such that

$$a \prec f(r) \prec b.$$

Having made this observation, we return to the proof that the two definitions of $f(x)$ agree for rational x. If y is a rational number with $y < x$, then we have already seen that $f(y) \prec f(x)$. Thus every element of A_x is $\prec f(x)$. Consequently,

$$\sup A_x \preceq f(x).$$

On the other hand, suppose that we had

$$\sup A_x \prec f(x).$$

Then there would be a rational number r such that

$$\sup A_x \prec f(r) \prec f(x).$$

But the condition $f(r) \prec f(x)$ means that $r < x$, which means that $f(r)$ is in the set A_x; this clearly contradicts the condition $\sup A_x \prec f(r)$. This shows that the original assumption is false, so

$$\sup A_x = f(x).$$

We thus have a certain well-defined function f from **R** to F. In order to show that f is an isomorphism we must verify conditions (1)–(4) of the definition. We will begin with (4).

If x and y are real numbers with $x < y$, then clearly A_x is contained in A_y. Thus

$$f(x) = \sup A_x \preceq \sup A_y = f(y).$$

To rule out the possibility of equality, notice that there are rational numbers r and s with

$$x < r < s < y.$$

We know that $f(r) \prec f(s)$. It follows that

$$f(x) \preceq f(r) \prec f(s) \preceq f(y).$$

This proves (4).

Condition (1) follows immediately from (4): If $x \neq y$, then either $x < y$ or $y < x$; in the first case $f(x) \prec f(y)$, and in the second case $f(y) \prec f(x)$; in either case $f(x) \neq f(y)$.

To prove (2), let a be an element of F, and let B be the set of all rational numbers r with $f(r) \prec a$. The set B is not empty, and it is also bounded above, because there is a rational number s with $f(s) \succ a$, so that $f(s) \succ f(r)$ for r in B, which implies that $s > r$. Let x be the least upper bound of B; we claim that $f(x) = a$. In order to prove this it suffices to eliminate the alternatives

$$f(x) \prec a,$$
$$a \prec f(x).$$

In the first case there would be a rational number r with

$$f(x) \prec f(r) \prec a.$$

But this means that $x < r$ and that r is in B, which contradicts the fact that $x = \sup B$. In the second case there would be a rational number r with

$$a \prec f(r) \prec f(x).$$

This implies that $r < x$. Since $x = \sup B$, this means that $r < s$ for some s in B. Hence

$$f(r) \prec f(s) \prec a,$$

again a contradiction. Thus $f(x) = a$, proving (2).

To check (3), let x and y be real numbers and suppose that $f(x + y) \neq f(x) \oplus f(y)$. Then either

$$f(x + y) \prec f(x) \oplus f(y) \qquad \text{or} \qquad f(x) \oplus f(y) \prec f(x + y).$$

In the first case there would be a rational number r such that

$$f(x + y) \prec f(r) \prec f(x) \oplus f(y).$$

But this would mean that

$$x + y \prec r.$$

Therefore r could be written as the sum of two rational numbers

$$r = r_1 + r_2, \qquad \text{where } x \prec r_1 \text{ and } y \prec r_2.$$

Then, using the facts checked about f for *rational* numbers, it would follow that

$$f(r) = f(r_1 + r_2) = f(r_1) \oplus f(r_2) \succ f(x) \oplus f(y),$$

a contradiction. The other case is handled similarly.

Finally, if x and y are positive real numbers, the same sort of reasoning shows that

$$f(x \cdot y) = f(x) \odot f(y);$$

the general case is then a simple consequence. ∎

This theorem brings to an end our investigation of the real numbers, and resolves any doubts about them: There *is* a complete ordered field and, up to isomorphism, only one complete ordered field. It is an important part of a mathematical education to follow a construction of the real numbers in detail, but it is not necessary to refer ever again to this particular construction. It is utterly irrelevant that a real number happens to be a collection of rational numbers, and such a fact should never enter the proof of any important theorem about the real numbers. Reasonable proofs should use only the fact that the real numbers are a complete ordered field, because this property of the real numbers characterizes them up to isomorphism, and any significant mathematical property of the real numbers will be true for all isomorphic fields. To be candid I should admit that this last assertion is just a prejudice of the author, but it is one shared by almost all other mathematicians.

PROBLEMS

1. Let f be an isomorphism from F_1 to F_2.

(a) Show that $f(\mathbf{0}) = \mathbf{0}$ and $f(\mathbf{1}) = \mathbf{1}$. (Here $\mathbf{0}$ and $\mathbf{1}$ on the left denote elements in F_1, while $\mathbf{0}$ and $\mathbf{1}$ on the right denote elements of F_2.)
(b) Show that $f(-a) = -f(a)$ and $f(a^{-1}) = f(a)^{-1}$, for $a \neq \mathbf{0}$.

2. Here is an opportunity to convince yourself that any significant property of a field is shared by any field isomorphic to it. The point of this problem is to write out very formal proofs until you are certain that all statements of this sort are obvious. F_1 and F_2 will be two fields which are isomorphic; for simplicity we will denote the operations in both by $+$ and $\cdot$. Show that:

(a) If the equation $x^2 + \mathbf{1} = \mathbf{0}$ has a solution in F_1, then it has a solution in F_2.
(b) If every polynomial equation $x^n + a_{n-1} \cdot x^{n-1} + \cdots + a_0 = \mathbf{0}$ with $a_0, \ldots, a_{n-1}$ in F_1, has a root in F_1, then every polynomial equation $x^n + b_{n-1} \cdot x^{n-1} + \cdots + b_0 = \mathbf{0}$ with $b_0, \ldots, b_{n-1}$ in F_2 has a root in F_2.
(c) If $1 + \cdots + 1$ (summed m times) $= \mathbf{0}$ in F_1, then the same is true in F_2.
(d) If F_1 and F_2 are ordered fields (and the isomorphism f satisfies $f(x) < f(y)$ for $x < y$) and F_1 is complete, then F_2 is complete.

3. Let f be an isomorphism from F_1 to F_2 and g an isomorphism from F_2 to F_3. Define the function $g \circ f$ from F_1 to F_3 by $(g \circ f)(x) = g(f(x))$. Show that $g \circ f$ is an isomorphism.

4. Suppose that F is a complete ordered field, so that there is an isomorphism f from $\mathbf{R}$ to F. Show that there is actually only *one* isomorphism from $\mathbf{R}$ to F. Hint: In case $F = \mathbf{R}$, this is Problem 3-17. Now if f and g are two isomorphisms from $\mathbf{R}$ to F consider $g^{-1} \circ f$.

5. Find an isomorphism from $\mathbf{C}$ to $\mathbf{C}$ other than the identity function.

SUGGESTED READING

A man ought to read
just as inclination leads him;
for what he reads as a task
will do him little good.

SAMUEL JOHNSON

One purpose of this bibliography is to guide the reader to other sources, but the most important function it can serve is to indicate the variety of mathematical reading available. Consequently, there is an attempt to achieve diversity, but no pretense of being complete. The present plethora of mathematics books would make such an undertaking almost hopeless in any case, and since I have tried to encourage independent reading, the more standard a text, the less likely it is to appear here. In some cases, this philosophy may seem to have been carried to extremes, as some entries in the list cannot be read by a student just finishing a first course of calculus until several years have elapsed. Nevertheless, there are many selections which can be read now, and I can't believe that it hurts to have some idea of what lies ahead.

Many of these books have gone through numerous editions and printings, which will be reflected in more recent publication dates. Many of the books with older publications dates are out of print, though that generally doesn't apply to books from the redoubtable Dover Publications, or from the Mathematical Association of America. Those that are no longer in print can still often be found in well-stocked academic libraries.

One of the most elementary unproved theorems mentioned in this book is the fact that every natural number can be written as a product of primes in only one way. A proof of this basic theorem will be found near the beginning of almost any book on elementary number theory. Few books have won so enthusiastic an audience as

[1] *An Introduction to the Theory of Numbers* (fifth edition), by G. H. Hardy and E. M. Wright; Oxford University Press, 1980.

The Pergamon Press published a series, Popular Lectures in Mathematics, with several titles worth investigating, among them

[2] *A Selection of Problems in the Theory of Numbers*, by W. Sierpinski; Macmillan (Pergamon), 1964.

Finally, I will mention an intriguing little book, now out of print I fear,

[3] *Three Pearls of Number Theory*, by A. Khinchin; Graylock Press, 1952.

The subject of irrational numbers straddles the fields of number theory and analysis. An excellent introduction will be found in

[4] *Irrational Numbers*, by I. M. Niven; Mathematical Association of America, 1956.

Together with many historical notes, there are references to some fairly elementary articles in journals. There is also a proof that π is transcendental (see also [51]) and, finally, a proof of the "Gelfond-Schneider theorem": If a and b are algebraic, with $a \neq 0$ or 1, and b is irrational, then a^b is transcendental.

All the books listed so far begin with natural numbers, but whenever necessary take for granted the irrational numbers, not to mention the integers and rational numbers. Several books present a construction of the rational numbers from the natural numbers, but one of the most lucid treatments is still to be found in

[5] *Foundations of Analysis* (second edition), by E. Landau; Chelsea, 1960.

Incidentally, the original German edition,

[6] *Grundlagen der Analysis* (fourth edition), by E. Landau; Chelsea, 1965.

has been printed in paper back, together with a complete German-English dictionary (of about 300 words) for the whole book—an excellent way to begin reading mathematical German. The basic idea for constructing the real numbers is derived from Dedekind, whose contributions can be found in

[7] *Essays on the Theory of Numbers*, by R. Dedekind; Dover, 1963.

While many mathematicians are content to accept the natural numbers as a natural starting point, numbers can be defined in terms of sets, the most basic starting point of all. A charming exposition of set theory can be found in a sophisticated little book called

[8] *Naive Set Theory*, by P. R. Halmos; Springer-Verlag, 1991.

Another very good introduction is

[9] *Theory of Sets*, by E. Kamke; Dover, 1950.

Perhaps it is necessary to assure some victims of the "new math" that set theory does have some mathematical content (in fact, some very deep theorems). Using these deep results, Kamke proves that there is a discontinuous function f such that $f(x+y) = f(x) + f(y)$ for all x and y. For those who enjoy reading the classics, the most important notions of set theory were first introduced by Cantor, whose work is reproduced in

[10] *Contributions to the Founding of the Theory of Transfinite Numbers*, by G. Cantor; Dover, 1952.

Inequalities, which were treated as an elementary topic in Chapters 1 and 2, actually form a specialized field. A good elementary introduction is provided by

[11] *Analytic Inequalities*, by N. Kazarinoff; Mathematical Association of America, 1961.

Twelve different proofs that the geometric mean is less than or equal to the arithmetic mean, each based on a different principle, can be found in the beginning of the more advanced book

[12] *An Introduction to Inequalities*, by E. Beckenbach and R. Bellman; Mathematical Association of America, 1961.

The classic work on inequalities is

[13] *Inequalities* (second edition), by G. H. Hardy, J. E. Littlewood, and G. Polya; Cambridge University Press, 1988.

Each of the authors of this triple collaboration has provided his own contribution to the sparse literature about the nature of mathematical thinking, written from a mathematician's point of view. My favorite is

[14] *A Mathematician's Apology*, by G. H. Hardy; Cambridge University Press, 1992.

Littlewood's anecdotal selections are entitled

[15] *A Mathematician's Miscellany*, by J. E. Littlewood; Methuen, 1953.

Polya's contribution is pedagogy at the highest level:

[16] *Mathematics and Plausible Reasoning* (Vol. I: *Induction and Analogy in Mathematics;* Vol. II: *Patterns of Plausible Inference*), by G. Polya; Princeton University Press, 1990.

Geometry is the other main field which can be considered as background for calculus. Euclid's *Elements* is still a masterful mathematical work, but should perhaps be postponed until some preparation has been made, with a modern work on "classical geometry," like

[17] *Elementary Geometry from an Advanced Standpoint* (second edition), by E. Moise; Addison-Wesley, 1974.

This beautiful book provides excellent historical perspectives and contains a thorough discussion of the role of the "Archimedean axiom" in geometry; in addition, Chapter 28 describes an ordered field in which the Archimedean axiom does not hold. Speaking of beautiful geometry books, all sorts of fascinating things can be found in

[18] *Introduction to Geometry* (second edition), by H. S. Coxeter; Wiley, 1989.

Almost all treatments of geometry at least mention convexity, which forms another specialized topic. I cannot imagine a better introduction to convexity, or a better mathematical experience in general, than reading and working through

[19] *Convex Figures*, by I. M. Yaglom and W. G. Boltyanskii; Holt, Rinehart and Winston, 1961.

This book contains a carefully arranged sequence of definitions and *statements* of theorems, whose proofs are to be supplied by the reader (worked-out proofs are supplied in the back of the book). Another geometry book has been modeled on the same principle:

[20] *Combinatorial Geometry in the Plane*, by H. Hadwiger and H. Debrunner; Holt, Rinehart and Winston, 1964.

Along with these two out-of-the-ordinary books, I might mention an extremely valuable little book, also of a specialized sort,

[21] *Counterexamples in Analysis*, by B. Gelbaum and J. Olmsted; Holden-Day, 1964.

Many of the example in this book come from more advanced topics in analysis, but quite a few can be appreciated by someone who knows calculus.

Of calculus books I will mention only two, each something of a classic:

[22] *A Course of Pure Mathematics* (tenth edition), by G. H. Hardy; Cambridge University Press, 1952.

[23] *Differential and Integral Calculus* (two volumes), by R. Courant; Wiley (Interscience), 1988.

Courant is especially strong on applications to physics. Speaking of such applications, an elegant exposition of the material in Chapter 17, together with much further discussion, can be found in the article

[24] *On the Geometry of the Kepler Problem*, by John Milnor; in *The American Mathematical Monthly*, Volume 90 (1983), pp. 353–365.

(In this paper the curve c' of Chapter 17 is denoted by $\mathbf{v}$, and the derivative of the important composition $\mathbf{v} \circ \theta^{-1}$ (page 331) is introduced quite off-handedly as $d\mathbf{v}/d\theta$.) A "straight-forward" derivation of Kepler's laws, together with numerous references, can be found in another article in this same journal,

[25] *The Mathematical Relationship Between Kepler's Laws and Newton's Laws*, by Andrew T. Hyman; in *The American Mathematical Monthly*, Volume 100 (1993), pp. 932–936.

The latter parts of Volume I of Courant contain material usually found in advanced calculus, including differential equations and Fourier series. An introduction to Fourier series (requiring a little advanced calculus) will also be found in

[26] *An Introduction to Fourier Series and Integrals*, by R. Seeley; W. A. Benjamin, 1966.

The second volume of Courant (advanced calculus in earnest) contains additional material on differential equations, as well as an introduction to the calculus of variations. A widely admired, though somewhat more advanced, book on differential equations is

[27] *Lectures on Ordinary Differential Equations*, by W. Hurewicz; Dover, 1990.

I will bypass the more or less standard advanced calculus books (which can easily be found by the reader) since nowadays there is a movement to revise the whole presentation of advanced calculus, basing it upon linear algebra. One of the first, and still one of the nicest, treatments of advanced calculus using linear algebra is

[28] *Calculus of Vector Functions*, by R. H. Crowell and R. E. Williamson; Prentice-Hall, 1962.

Several recent books on advanced calculus attempt to acquaint undergraduates with very large areas of modern mathematics. My favorite, of course, is

[29] *Calculus on Manifolds*, by M. Spivak; W. A. Benjamin, 1965.

There are three other topics which are somewhat out of place in this bibliography because they are rapidly becoming established as part of a standard undergraduate curriculum. The purposeful study of fields and related systems is the domain of "algebra." One of the favorite texts is

[30] *Topics in Algebra* (second edition), by I. N. Herstein; Wiley, 1975.

A more advanced book is the great classic:

[31] *Algebra*, by B. L. van der Waerden; Springer-Verlag, 1990.

By the way, this book contains a proof of the partial fraction decomposition of a rational function.

There are now several introductions to complex analysis, as well as many elementary books on topology. Although the latter subject has not been mentioned before, it has really been in the background of many discussions, since it is the natural generalization of the ideas about limits and continuity which play such a prominent role in Part II of this book.

The next few topics, ranging from elementary to very difficult, are included in this bibliography because they have been alluded to in the text. The proof that a nondecreasing function is differentiable at almost all points (and an explanation of just what this means) receives a beautiful exposition in

[32] *Functional Analysis*, by F. Riesz and B. Sz.-Nagy; Ungar, 1955.

(After this elementary beginning, the book moves on to quite advanced material.) The gamma function has an elegant little book devoted entirely to its properties, most of them proved by using the theorem of Bohr and Mollerup which was mentioned in Problem 19-39:

[33] *The Gamma Function*, by E. Artin; Holt, Rinehart and Winston, 1964.

The gamma function is only one of several important improper integrals in mathematics. In particular, the calculation of $\int_0^\infty e^{-x^2}\,dx$ (see Problem 19-41) is important in probability theory, where the "normal distribution function"

$$\Phi(x) = \frac{1}{\sqrt{2\pi}}\int_{-\infty}^{x} e^{-\frac{1}{2}y^2}\,dy$$

plays a fundamental role. A classic book on probability theory is

[34] *An Introduction to Probability Theory and Its Applications* (third edition), by W. Feller; Wiley, 1968.

The impossibility of integrating certain functions in elementary terms (among them $f(x) = e^{-x^2}$) is one of the most esoteric subjects in mathematics. An interesting discussion of the possibilities of integrating in elementary terms, with an outline of the impossibility proofs, and references to the original papers of Liouville, will be found in

[35] *The Integration of Functions of a Single Variable* (second edition), by G. H. Hardy; Cambridge University Press, 1958.

A complete presentation of the impossibility proofs will be found in

[36] *Integration in Finite Terms*, by J. Ritt; Columbia University Press, 1948.

Oddly enough, a related but seemingly more difficult problem has a much neater solution. There are simple differential equations ($y'' + xy = 0$ is a specific example) whose solutions cannot be expressed even in terms of indefinite integrals of elementary functions. This fact is proved on page 43 of the (60-page) book:

[37] *An Introduction to Differential Algebra*, by I. Kaplansky; Hermann, 1957.

To read this book you will need to know quite a bit of algebra, however.

A few words should also be said in defense of the process of integrating in elementary terms, which many mathematicians look upon as an art (unlike differentiation, which is merely a skill). You are probably already aware that the process of integration can be expedited by tables of indefinite integrals. For those who enjoy pursuing tables there is a really beautiful collection, that includes indefinite integrals, definite improper integrals, and a great deal more besides (if you should ever happen to need the value of the thirty-fourth Bernoulli number, this is the place to look):

[38] *Tables of Integrals, Series, and Products*, by I. S. Gradschteyn et al.; Academic Press, 1980.

For the thrifty, there is a paperback table of integrals:

[39] *Tables of Indefinite Integrals*, by G. Petit Bois; Dover, 1961.

The remaining references are of a somewhat different sort. They fall into three categories, of which the first is historical. The letter of H. A. Schwarz referred to in Problem 11-65 will be found in

[40] *Ways of Thought of Great Mathematicians*, by H. Meschkowski; Holden-Day, 1964.

Some historical remarks, and an attempt to incorporate them into the teaching of calculus, will be found in

[41] *The Calculus: A Genetic Approach*, by O. Toeplitz; University of Chicago Press, 1981.

An admirable textbook on the history of mathematics is

[42] *An Introduction to the History of Mathematics* (sixth edition), by H. Eves; Saunders College Publishing, 1990.

Three good scholarly works are

[43] *History of Analytic Geometry*, by C. Boyer; Scholar's Bookshelf, 1988.
[44] *A History of the Calculus, and Its Conceptual Development*, by C. Boyer; Dover, 1959.
[45] *The Mathematics of Great Amateurs*, by J. Coolidge; Oxford University Press, 1990

and extracts from original sources will be found in

[46] *A Source Book in Mathematics* (2 vols.), by D. Smith; Dover, 1959.

Despite the impression that might be given by the large number of books listed here, it is often hard to find specific concrete information about the origins of calculus. For example, it is almost impossible to find out who first proved the Mean Value Theorem (according to the *Encyklopädie der Mathematischen Wissenschaften*, Volume II, it was O. Bonnet, whose name is familiar to students of differential geometry from the "Gauss-Bonnet Theorem"). Similarly, though many history books tell us that Wallis proved Wallis' formula by a "complicated method of interpolation," most never bother to mention what it was, even though it inspired Euler's investigations of the gamma function (a description is given in the answer book, along with the solution to Problem 19-40).

The second category in this final group of books might be described as "popularizations." There are a surprisingly large number of first-rate ones by real mathematicians:

[47] *What is Mathematics?* (fourth edition), by R. Courant and H. Robbins; Oxford University Press, 1979.
[48] *Geometry and the Imagination*, by D. Hilbert and S. Cohn-Vossen; Chelsea, 1952.
[49] *The Enjoyment of Mathematics*, by H. Rademacher and O. Toeplitz; Dover, 1990.
[50] *Famous Problems of Mathematics* (second edition), by H. Tietze; Graylock Press, 1965.

One of the most renowned "popularizations" is especially concerned with the teaching of mathematics:

[51] *Elementary Mathematics from an Advanced Standpoint*, by F. Klein (vol. 1: *Arithmetic, Algebra, Analysis*; vol. 2: *Geometry*); Dover, 1948.

Volume 1 contains a proof of the transcendence of π which, although not so elementary as the one in [4], is a direct analogue of the proof that e is transcendental, replacing integrals with complex line integrals. It can be read as soon as the basic facts about complex analysis are known.

The third category is the very opposite extreme—original papers. The difficulties encountered here are formidable, and I have only had the courage to list one such paper, the source of the quotation for Part IV. It is not even in English, although you do have a choice of foreign languages. The article in the original French is in

[52] *Oeuvres Complètes d'Abel;* Christiania. Johnson Reprint Corporation, New York, 1965.

It first appeared in a German translation in the *Journal für die reine und angewandte Mathematik*, Volume 1, 1826. To compound the difficulties, these references will usually be available only in university libraries. Yet the study of this paper will probably be as valuable as any other reading mentioned here. The reason is suggested by a remark of Abel himself, who attributed his profound knowledge of mathematics to the fact that he read the masters, rather than the pupils.

ANSWERS TO SELECTED PROBLEMS

CHAPTER 1

1. (i) $1 = a^{-1}a = a^{-1}(ax) = (a^{-1}a)x = 1 \cdot x = x$.
   (iii) If $x^2 = y^2$, then $0 = x^2 - y^2 = (x - y)(x + y)$, so either $x - y = 0$ or $x + y = 0$, that is, either $x = -y$ or $x = y$.
   (vi) Replace y by $-y$ in (iv).
2. One step requires dividing by $x - y = 0$.
3. (i) $a/b = ab^{-1} = (ac)(b^{-1}c^{-1}) = (ac)(bc)^{-1}$ (by (iii)) $= ac/bc$.
   (ii) $(ad + bc)/(bd) = (ad + bc)(bd)^{-1} = (ad + bc)(b^{-1}d^{-1})$ (by (iii)) $= ab^{-1} + cd^{-1} = a/b + c/d$.
   (iii) $ab(a^{-1}b^{-1}) = (a \cdot a^{-1})(b \cdot b^{-1}) = 1$, so $a^{-1} \cdot b^{-1} = (ab)^{-1}$.
   (v) $(a/b)/(c/d) = (a/b)(c/d)^{-1} = (a \cdot b^{-1})(c \cdot d^{-1})^{-1} = (a \cdot b^{-1})(c^{-1} \cdot d) = ad(b^{-1} \cdot c^{-1}) = ad(bc)^{-1} = (ad)/(bc)$.
4. (i) $x < -1$.
   (iii) $x > \sqrt{7}$ or $x < -\sqrt{7}$.
   (v) All x, since $x^2 - 2x + 2 = (x - 1)^2 + 1$.
   (vii) $x > 3$ or $x < -2$, since 3 and -2 are the roots of $x^2 - x - 6 = 0$.
   (ix) $x > \pi$ or $-5 < x < 3$.
   (xi) $x < 3$.
   (xiii) $x > 1$ or $0 < x < 1$.
5. (i) $b - a$ and $d - c$ are in P, so $(b - a) + (d - c) = (b + d) - (a + c)$ is in P. Thus, $b + d > a + c$.
   (iii) Using (ii), $-c < -d$; then (i) implies that $a + (-c) < b + (-d)$.
   (v) $(b-a)$ and $-c$ are in P, so $-c(b-a) = ac - bc$ is in P, that is, $ac > bc$.
   (vii) Using (iv), $a > 0$ and $a < 1$, so $a^2 < a$.
   (ix) Substitute a for c and b for d in (viii).
9. (i) $\sqrt{2} + \sqrt{3} - \sqrt{5} + \sqrt{7}$.
   (iii) $|a + b| + |c| - |a + b + c|$.
   (v) $\sqrt{2} + \sqrt{3} + \sqrt{5} - \sqrt{7}$.
10. (i) a if $a \geq -b$ and $b \geq 0$;
    $-a$ if $a \leq -b$ and $b \leq 0$;
    $a + 2b$ if $a \geq -b$ and $b \leq 0$;
    $-a - 2b$ if $a \leq -b$ and $b \geq 0$.
    (iii) $x - x^2$ if $x \geq 0$;
    $-x - x^2$ if $x \leq 0$.
11. (i) $x = 11, -5$.
    (iii) $-6 < x < -2$.
    (v) No x (the distance from x to 1 plus the distance from x to -1 is at least 2).
    (vii) $x = 1, -1$.
12. (i) $(|xy|)^2 = (xy)^2 = x^2y^2 = |x|^2|y|^2 = (|x| \cdot |y|)^2$; since $|xy|$ and $|x| \cdot |y|$ are both ≥ 0, this proves that $|xy| = |x| \cdot |y|$.
    (iii) $|x|/|y| = |x| \cdot |y|^{-1} = |x| \cdot |y^{-1}|$ by (ii)) $= |xy^{-1}|$ (by (i)) $= |x/y|$.
    (v) It follows from (iv) that $|x| = |y - (y - x)| \leq |y| + |y - x|$, so $|x| - |y| \leq |x - y|$.
    (vii) $|x + y + z| \leq |x + y| + |z| \leq |x| + |y| + |z|$. If equality holds, then $|x + y| = |x| + |y|$, so x and y have the same sign. Moreover, z must

have the same sign as $x+y$, so x, y, and z must all have the same sign (unless one is 0).

CHAPTER 2

1. (i) Since $1^2 = 1\cdot(2)\cdot(2\cdot 1+1)/6$, the formula is true for $n=1$. Suppose that the formula is true for k. Then

$$\begin{aligned}1^2+\cdots+k^2+(k+1)^2 &= \frac{k(k+1)(2k+1)}{6}+(k+1)^2\\ &= \frac{(k+1)}{6}[k(2k+1)+6(k+1)]\\ &= \frac{(k+1)}{6}[(k+2)(2k+3)]\\ &= \frac{(k+1)(k+2)(2[k+1]+1)}{6},\end{aligned}$$

so the formula is true for $k+1$.

2. (i)

$$\begin{aligned}\sum_{i=1}^{n}(2i-1) &= 1+3+5+\cdots+(2n-1)\\ &= 1+2+3+\cdots+2n-2(1+\cdots+n)\\ &= \frac{(2n)(2n+1)}{2}-n(n+1)\\ &= n^2.\end{aligned}$$

5. (a) Since

$$1+r=\frac{1-r^2}{1-r},$$

the formula is true for $n=1$. Suppose that

$$1+r+\cdots+r^n=\frac{1-r^{n+1}}{1-r}.$$

Then

$$\begin{aligned}1+r+\cdots+r^n+r^{n+1} &= \frac{1-r^{n+1}}{1-r}+r^{n+1}\\ &= \frac{1-r^{n+1}+r^{n+1}(1-r)}{1-r}\\ &= \frac{1-r^{n+2}}{1-r}.\end{aligned}$$

(b)

$$\begin{aligned}S &= 1+r+\cdots+r^n\\ rS &= \phantom{1+{}} r+\cdots+r^n+r^{n+1}.\end{aligned}$$

Thus

$$S(1-r)=S-rS=1-r^{n+1},$$

so

$$S = \frac{1 - r^{n+1}}{1 - r}.$$

6\. (i) From

$$(k+1)^4 - k^4 = 4k^3 + 6k^2 + 4k + 1, \qquad k = 1, \dots, n$$

we obtain

$$(n+1)^4 - 1 = 4\sum_{k=1}^{n} k^3 + 6\sum_{k=1}^{n} k^2 + 4\sum_{k=1}^{n} k + n,$$

so

$$\sum_{k=1}^{n} k^3 = \frac{(n+1)^4 - 1 - 6\dfrac{n(n+1)(2n+1)}{6} - 4\dfrac{n(n+1)}{2} - n}{4}$$

$$= \frac{n^4}{4} + \frac{n^3}{2} + \frac{n^2}{4}.$$

(iii) From

$$\frac{1}{k} - \frac{1}{k+1} = \frac{1}{k(k+1)}, \qquad k = 1, \dots, n$$

we obtain

$$1 - \frac{1}{n+1} = \sum_{k=1}^{n} \frac{1}{k(k+1)}.$$

8\. 1 is either even or odd, in fact it is odd. Suppose n is either even or odd; then n can be written either as $2k$ or $2k+1$. In the first case $n+1 = 2k+1$ is odd; in the second case $n+1 = 2k+1+1 = 2(k+1)$ is even. In either case, $n+1$ is either even or odd. (Admittedly, this looks fishy, but it is really correct.)

9\. Let B be the set of all natural numbers l such that $n_0 - 1 + l$ is in A. Then 1 is in B, and $l+1$ is in B if l is in B, so B contains all natural numbers, which means that A contains all natural numbers $\geq n_0$.

12\. (a) Yes, for if $a+b$ were rational, then $b = (a+b) - a$ would be rational. If a and b are irrational, then $a+b$ could be rational, for b could be $r - a$ for some rational number a.

(b) If $a = 0$, then ab is rational. But if $a \neq 0$, then ab could not be rational, for then $b = (ab) \cdot a^{-1}$ would be rational.

(c) Yes; for example, $\sqrt[4]{2}$.

(d) Yes; for example, $\sqrt{2}$ and $-\sqrt{2}$.

13\. (a) Since

$$(3n+1)^2 = 9n^2 + 6n + 1 = 3(3n^2 + 2n) + 1,$$
$$(3n+2)^2 = 9n^2 + 12n + 4 = 3(3n^2 + 4n + 1) + 1,$$

it follows that if k^2 is divisible by 3, then k must also be divisible by 3. Now suppose that $\sqrt{3}$ were rational, and let $\sqrt{3} = p/q$ where p and

q have no common factor. Then $p^2 = 3q^2$, so p^2 is divisible by 3, so p must be. Thus, $p = 3p'$ for some natural number p', and consequently $(3p')^2 = 3q^2$, or $3(p')^2 = q^2$. Thus, q is also divisible by 3, a contradiction.

The same proofs work for $\sqrt{5}$ and $\sqrt{6}$, because the equations

$$\begin{aligned}(5n+1)^2 &= 25n^2 + 10n + 1 = 5(5n^2 + 2n) + 1,\\ (5n+2)^2 &= 25n^2 + 20n + 4 = 5(5n^2 + 4n) + 4,\\ (5n+3)^2 &= 25n^2 + 30n + 9 = 5(5n^2 + 6n + 1) + 4,\\ (5n+4)^2 &= 25n^2 + 40n + 16 = 5(5n^2 + 8n + 3) + 1,\end{aligned}$$

and the corresponding equations for numbers of the form $6n + m$, show that if k^2 is divisible by 5 or 6, then k must be. The proof fails for $\sqrt{4}$, because $(4n+2)^2$ is divisible by 4. (For precisely this reason this proof cannot be used to show that in general $\sqrt{a}$ is irrational if a is not a perfect square—we have no guarantee that $(an + m)^2$ might not be a multiple of a for some $m < a$. Actually, this assertion *is* true, but the proof requires the information in Problem 17.)

(b) Since

$$(2n+1)^3 = 8n^3 + 12n^2 + 6n + 1 = 2(4n^3 + 6n^2 + 3n) + 1,$$

it follows that if k^3 is even, then k is even. If $\sqrt[3]{2} = p/q$ where p and q have no common factors, then $p^3 = 2q^3$, so p^3 is divisible by 2, so p must be. Thus, $p = 2p'$ for some natural number p', and consequently $(2p')^3 = 2q^3$, or $4(p')^3 = q^3$. Thus, q is also even, a contradiction.

The proof for $\sqrt[3]{3}$ is similar, using the equations

$$\begin{aligned}(3n+1)^3 &= 27n^3 + 27n^2 + 9n + 1 = 3(9n^3 + 9n^2 + 3n) + 1,\\ (3n+2)^3 &= 27n^3 + 54n^2 + 36n + 8 = 3(9n^3 + 18n^2 + 12n + 2) + 2.\end{aligned}$$

19. If $n = 1$, then $(1+h)^n = 1 + nh$. Suppose that $(1+h)^n \geq 1 + nh$. Then

$$\begin{aligned}(1+h)^{n+1} &= (1+h)(1+h)^n \geq (1+h)(1+nh), \quad \text{since } 1+h > 0\\ &= 1 + (n+1)h + nh^2 \geq 1 + (n+1)h.\end{aligned}$$

For $h > 0$, the inequality follows directly from the binomial theorem, since all the other terms appearing in the expansion of $(1+h)^n$ are positive.

CHAPTER 3

1. (i) $(x+1)/(x+2)$; the expression $f(f(x))$ makes sense only when $x \neq -1$ and $x \neq -2$.
   (iii) $1/(1+cx)$ (for $x \neq -1/c$ if $c \neq 0$).
   (v) $(x+y+2)/(x+1)(y+1)$ (for $x, y \neq -1$).
   (vii) Only $c = 1$, since $f(x) = f(cx)$ implies that $x = cx$, and this must be true for at least one $x \neq 0$.
2. (i) $y \geq 0$ and rational, or $y \geq 1$.
   (iii) 0.
   (v) $-1, 0, 1$.

3. (i) $\{x : -1 \le x \le 1\}$.
   (iii) $\{x : x \ne 1 \text{ and } x \ne 2\}$.
   (v) $\emptyset$.
4. (i) 2^{2^y}.
   (iii) $2^{2\sin t} + \sin(2^t)$.
5. (i) $P \circ s$.
   (iii) $s \circ S$.
   (v) $P \circ P$.
   (vii) $s \circ s \circ s \circ P \circ P \circ P \circ s$.
11. (a) y.
   (b) $H(y)$.
   (c) $H(y)$.
12. (a)

	even	odd
even	even	neither
odd	neither	odd

(b)

	even	odd
even	even	odd
odd	odd	even

(c)

	f even	f odd
g even	even	even
g odd	even	odd

(d) Let $g(x) = f(x)$ for $x \ge 0$ and define g arbitrarily for $x < 0$.

21. (i) Let $g(x) = h(x) = 1$ and let f be a function for which $f(2) \ne f(1) + f(1)$. Then $f \circ (g + h) \ne f \circ g + f \circ h$.
   (ii) $[(g + h) \circ f](x) = (g + h)(f(x)) = g(f(x)) + h(f(x)) = (g \circ f)(x) + (h \circ f)(x) = [(g \circ f) + (h \circ f)](x)$.
   (iii) $\dfrac{1}{f \circ g}(x) = \dfrac{1}{f(g(x))} = \dfrac{1}{f}(g(x)) = \left(\dfrac{1}{f} \circ g\right)(x)$.
   (iv) Let $g(x) = 2$ and let f be a function for which $f(\frac{1}{2}) \ne 1/f(2)$. Then $1/(f \circ g) \ne f \circ (1/g)$.

CHAPTER 4

1. (i) $(2, 4)$.
   (iii) $[2, 4]$.
   (v) $(-2, 2)$.
   (vii) $(-\infty, 1] \cup [1, \infty)$.
3. (i) All points below the graph of $f(x) = x$.
   (iii) All points below the graph of $f(x) = x^2$.
   (v) All points between the graphs of $f(x) = x + 1$ and $f(x) = x - 1$.
   (vii) A collection of straight lines parallel to the graph of $f(x) = -x$, intersecting the horizontal axis at the points $(n, 0)$ for integers n.
   (ix) All points inside the circle of radius 1 and around $(1, 2)$.
4. (i) A square with vertices $(1, 0)$, $(0, 1)$, $(-1, 0)$, and $(0, -1)$.
   (iii) The union of the graph of $f(x) = x$ and of $f(x) = 2 - x$.
   (v) The point $(0, 0)$.
   (vii) The circle of radius $\sqrt{5}$ around $(1, 0)$, since $x^2 - 2x + y^2 = (x - 1)^2 + y^2 - 1$.
6. (a) Simply observe that the graph of $f(x) = m(x - a) + b = mx + (b - ma)$ *is* a straight line with slope m, which goes through the point (a, b). (The important point about this exercise is simply to remember the point slope form.)
   (b) The straight line through (a, b) and (c, d) has slope $(d - b)/(c - a)$, so the equation follows from part (a).
   (c) When $m = m'$ and $b \neq b'$. In that case, there is clearly no number x with $f(x) = g(x)$, while such a number x always exists if $m \neq m'$, namely, $x = (b' - b)/(m - m')$.
7. (a) If $B = 0$ and $A \neq 0$, then the set is the vertical straight line formed by all points (x, y) with $x = -C/A$. If $B \neq 0$, the set is the graph of $f(x) = (-A/B)x + (-C/A)$.
   (b) The points (x, y) on the vertical line with $x = a$ are precisely the ones which satisfy $1 \cdot x + 0 \cdot y + (-a) = 0$. The points (x, y) on the graph of $f(x) = mx + b$ are precisely the ones which satisfy $(-m)x + 1 \cdot y + (-b) = 0$.
11. (i) The graph of f is symmetric with respect to the vertical axis.
    (ii) The graph of f is symmetric with respect to the origin. Equivalently, the part of the graph to the left of the vertical axis is obtained by reflecting first through the vertical axis, and then through the horizontal axis.
    (iii) The graph of f lies above or on the horizontal axis.
    (iv) The graph of f repeats the part between 0 and a over and over.
21. (a) The square of the distance from (x, x^2) to $(0, \frac{1}{4})$ is

$$x^2 + \left(x^2 - \frac{1}{4}\right)^2 = x^2 + x^4 - \frac{x^2}{2} + \frac{1}{16}$$
$$= x^4 + \frac{x^2}{2} + \frac{1}{16}$$
$$= (x^2 + \tfrac{1}{4})^2,$$

which is the square of the distance from (x, x^2) to the graph of g.

(b) The point (x, y) satisfies this condition if and only if

$$(x-\alpha)^2 + (y-\beta)^2 = (y-\gamma)^2,$$

or

$$x^2 - 2\alpha x + \alpha^2 + y^2 - 2\beta y + \beta^2 = y^2 - 2\gamma y + \gamma^2,$$

or

$$y = \left(\frac{1}{2\beta - 2\gamma}\right)x^2 + \left(\frac{\alpha}{\gamma - \beta}\right)x + \left(\frac{\alpha^2 + \beta^2 - \gamma^2}{2\beta - 2\gamma}\right).$$

(This solution works only for $\beta \neq \gamma$, which is just the condition that P is not on L. If P is on L, then the solution is the vertical line through P.)

CHAPTER 5

1. (ii)

$$\lim_{x\to 2} \frac{x^3 - 8}{x - 2} = \lim_{x\to 2}(x^2 + 2x + 4) = 12.$$

(iv)

$$\begin{aligned}\lim_{x\to y} \frac{x^n - y^n}{x - y} &= \lim_{x\to y} x^{n-1} + x^{n-2}y + \cdots + xy^{n-2} + y^{n-1} \\ &= y^{n-1} + y^{n-1} + \cdots + y^{n-1} = ny^{n-1}.\end{aligned}$$

(vi)

$$\begin{aligned}\lim_{h\to 0} \frac{\sqrt{a+h} - \sqrt{a}}{h} &= \lim_{h\to 0} \frac{(\sqrt{a+h} - \sqrt{a})(\sqrt{a+h} + \sqrt{a})}{h(\sqrt{a+h} + \sqrt{a})} \\ &= \lim_{h\to 0} \frac{1}{\sqrt{a+h} + \sqrt{a}} \\ &= \frac{1}{2\sqrt{a}}.\end{aligned}$$

3. (i) It is possible to find δ by beginning with the equation

$$x^4 - a^4 = (x-a)(x^3 + ax^2 + a^2x + a^3).$$

If $|x - a| < 1$, then $|x| < 1 + |a|$, so

$$\begin{aligned}|x^3 + ax^2 + a^2x + a^3| &\le |x|^3 + |a| \cdot |x|^2 + |a|^2 \cdot |x| + |a|^3 \\ &< (1+|a|)^3 + |a|(1+|a|)^2 + |a|^2(1+|a|) + |a|^3;\end{aligned}$$

therefore we can choose

$$\delta = \min\left(1, \frac{\varepsilon}{(1+|a|)^3 + |a|(1+|a|)^2 + |a|^2(1+|a|) + |a|^3}\right).$$

It is instructive, and probably easier, to use part (2) of the lemma. This shows that $|x^4 - a^4| < \varepsilon$ when

$$|x^2 - a^2| < \min\left(1, \frac{\varepsilon}{2(|a|^2 + 1)}\right),$$

which is true when

$$|x-a| < \min\left(1, \frac{\min\left(1, \dfrac{\varepsilon}{2(|a|^2+1)}\right)}{2(|a|+1)}\right)$$
$$= \min\left(1, \frac{\varepsilon}{4(|a|^2+1)(|a|+1)}\right) = \delta.$$

(ii) By part (3) of the lemma, $|1/x - 1| < \varepsilon$ when

$$|x-1| < \min\left(\frac{1}{2}, \frac{\varepsilon}{2}\right) = \delta.$$

(iii) By part (1) of the lemma, $|(x^4 + 1/x) - 2| < \varepsilon$ when $|1/x - 1| < \varepsilon/2$ and $|x^4 - 1| < \varepsilon/2$. According to parts (i) and (ii) of this problem, this happens when

$$|x-1| < \min\left(\frac{1}{2}, \frac{\varepsilon}{4}, 1, \frac{\varepsilon}{8\cdot 2\cdot 2}\right) = \min\left(\frac{1}{2}, \frac{\varepsilon}{32}\right) = \delta.$$

(v) Let $\delta = \varepsilon^2$, since $0 < |x| < \varepsilon^2$ implies that $\sqrt{|x|} < \varepsilon$.

6. (i) We need $|f(x) - 2| < \varepsilon/2$ and $|g(x) - 4| < \varepsilon/2$, so we need

$$0 < |x-2| < \min\left(\sin^2\left(\frac{\varepsilon^2}{36}\right) + \frac{\varepsilon}{2}, \frac{\varepsilon^2}{4}\right) = \delta.$$

(iii) We need

$$|g(x) - 4| < \min\left(\frac{|4|}{2}, \frac{\varepsilon|4|^2}{2}\right),$$

so we need

$$0 < |x-2| < [\min(2, 8\varepsilon)]^2 = \delta.$$

9. Let $l = \lim_{x\to a} f(x)$ and define $g(h) = f(a+h)$. Then for every $\varepsilon > 0$ there is a $\delta > 0$ such that, for all x, if $0 < |x-a| < \delta$, then $|f(x) - l < \varepsilon|$. Now, if $0 < |h| < \delta$, then $0 < |(h+a)-a| < \delta$, so $|f(a+h) - l| < \varepsilon$. This inequality can be written $|g(h) - l| < \varepsilon$. Thus, $\lim_{h\to 0} g(h) = l$, which can also be written $\lim_{h\to 0} f(a+h) = l$. The same sort of argument shows that if $\lim_{h\to 0} f(a+h) = m$, then $\lim_{x\to a} f(x) = m$. So either limit exists if the other does, and in this case they are equal.

10. (a) Intuitively, we can get $f(x)$ as close to l as we like if and only if we can get $f(x) - l$ as close to 0 as we like. The formal proof is so trivial that it takes a bit of work to make it look like a proof at all. To be very precise, suppose $\lim_{x\to a} f(x) = l$ and let $g(x) = f(x) - l$. Then for all $\varepsilon > 0$ there is a $\delta > 0$ such that, for all x, if $0 < |x-a| < \delta$, then $|f(x) - l| < \varepsilon$. This last inequality can be written $|g(x) - 0| < \varepsilon$, so $\lim_{x\to a} g(x) = 0$. The argument in the other direction is similarly uninteresting.

(b) Intuitively, making x close to a is the same as making $x - a$ close to 0. Formally: Suppose that $\lim_{x \to a} f(x) = l$, and let $g(x) = f(x - a)$. Then for all $\varepsilon > 0$ there is a $\delta > 0$ such that, for all x, if $0 < |x - a| < \delta$, then $|f(x) - l| < \varepsilon$. Now, if $0 < |y| < \delta$, then $0 < |(y + a) - a| < \delta$, so $|f(y + a) - l| < \varepsilon$. But this last inequality can be written $|g(y) - l| < \varepsilon$. So $\lim_{y \to 0} g(y) = l$. The argument in the reverse direction is similar.

(c) Intuitively, x is close to 0 if and only if x^3 is. Formally: Let $\lim_{x \to 0} f(x) = l$. For every $\varepsilon > 0$ there is a $\delta > 0$ such that if $0 < |x| < \delta$, then $|f(x) - l| < \varepsilon$. Then if $0 < |x| < \min(1, \delta)$, we have $0 < |x^3| < \delta$, so $|f(x^3) - l| < \varepsilon$. Thus, $\lim_{x \to 0} f(x) = l$. On the other hand, if we assume that $\lim_{x \to 0} f(x^3)$ exists, say $\lim_{x \to 0} f(x^3) = m$, then for all $\varepsilon > 0$ there is a δ such that if $0 < |x| < \delta$, then $|f(x^3) - m| < \varepsilon$. Then if $0 < |x| < \delta^3$, we have $0 < |\sqrt[3]{x}| < \delta$, so $|f([\sqrt[3]{x}]^3) - m| < \varepsilon$, or $|f(x) - m| < \varepsilon$. Thus $\lim_{x \to 0} f(x) = m$.

(d) Let $f(x) = 1$ for $x \geq 0$, and $f(x) = -1$ for $x < 0$. Then $\lim_{x \to 0} f(x^2) = 1$, but $\lim_{x \to 0} f(x)$ does not exist.

17. (a) The function $f(x) = 1/x$ cannot approach a limit at 0, since it becomes arbitrarily large near 0. In fact, no matter what $\delta > 0$ may be, there is some x satisfying $0 < |x| < \delta$, but $1/x > |l| + \varepsilon$, namely, $x = \min(\delta, 1/(|l| + \varepsilon))$. This x does not satisfy $|1/(x - l)| < \varepsilon$.

(b) No matter what $\delta > 0$ may be, there is some x satisfying $0 < |x - 1| < \delta$, but $1/(x-1) > |l|+\varepsilon$, namely, $x = \min(1+\delta, 1+1/(|l|+\varepsilon))$. This x does not satisfy $|1/(x - 1) - l| < \varepsilon$. (It is also possible to apply Problem 10(b): $\lim_{x \to 0} 1/x = \lim_{x \to 1} 1/(x - 1)$ if the latter exists, so this limit does not exist, because of part (a).)

25. (i) This is the usual definition, simply calling the numbers δ and ε, instead of ε and δ.

(ii) This is a minor modification of (i): if the condition is true for *all* $\delta > 0$, then it applies to $\delta/2$, so there is an $\varepsilon > 0$ such that if $0 < |x - a| < \varepsilon$, then $|f(x) - l| \leq \delta/2 < \delta$.

(iii) This is a similar modification: apply it to $\delta/5$ to obtain (i).

(iv) This is also a modification: it says the same thing as (i), since $\varepsilon/10 > 0$, and it is only the existence of *some* $\varepsilon > 0$ that is in question.

29. If $\lim_{x \to a^+} f(x) = \lim_{x \to a^-} f(x) = l$, then for every $\varepsilon > 0$ there are $\delta_1, \delta_2 > 0$ such that, for all x,

$$\text{if } a < x < a + \delta_1, \text{ then } |f(x) - l| < \varepsilon,$$
$$\text{if } a - \delta_2 < x < a, \text{ then } |f(x) - l| < \varepsilon.$$

Let $\delta = \min(\delta_1, \delta_2)$. If $0 < |x - a| < \delta$, then either $a - \delta_2 < a - \delta < x < a$ or else $a < x < a + \delta < a + \delta_1$, so $|f(x) - l| < \varepsilon$.

30. (i) If $l = \lim_{x\to 0^+} f(x)$, then for all $\varepsilon > 0$ there is a $\delta > 0$ such that $|f(x) - l| < \varepsilon$ for $0 < x < \delta$. If $-\delta < x < 0$, then $0 < -x < \delta$, so $|f(-x) - l| < \varepsilon$. Thus $\lim_{x\to 0^-} f(-x) = l$. Similarly, if $\lim_{x\to 0^-} f(x)$ exists, then $\lim_{x\to 0^+} f(x)$ exists and has the same value. (Intuitively, x is close to 0 and positive if and only if $-x$ is close to 0 and negative.)
(ii) If $l = \lim_{x\to 0^+} f(x)$, then for all $\varepsilon > 0$ there is a $\delta > 0$ such that $|f(x) - l| < \varepsilon$ for $0 < x < \delta$. So if $0 < |x| < \delta$, then $|f(|x|) - l| < \varepsilon$. Thus $\lim_{x\to 0} f(|x|) = l$. The reverse direction is similar. (Intuitively, if x is close to 0, then $|x|$ is close to 0 and positive.)
(iii) If $l = \lim_{x\to 0^+} f(x)$, then for all $\varepsilon > 0$ there is a $\delta > 0$ such that $|f(x) - l| < \varepsilon$ for $0 < x < \delta$. If $0 < |x| < \sqrt{\delta}$, then $0 < x^2 < \delta$, so $|f(x^2) - l| < \varepsilon$. Thus $\lim_{x\to 0} f(x^2) = l$. The reverse direction is similar. (Intuitively, if x is close to 0, then x^2 is close to 0 and positive.)

34. If $l = \lim_{x\to\infty} f(x)$, then for every $\varepsilon > 0$ there is some N such that $|f(x) - l| < \varepsilon$ for $x > N$, and we can clearly assume that $N > 0$. Now, if $0 < x < 1/N$, then $1/x > N$, so $|f(1/x) - l| < \varepsilon$. Thus $\lim_{x\to 0^+} f(1/x) = l$. The reverse direction is similar.

CHAPTER 6

1. (i) $F(x) = x + 2$ for all x.
(iii) $F(x) = 0$ for all x.

CHAPTER 7

1. (i) Bounded above and below; minimum value 0; no maximum value.
(iii) Bounded below but not above; minimum value 0.
(v) Bounded above and below. It is understood that $a > -1$ (so that $-a - 1 < a + 1$). If $-1 < a \le -\frac{1}{2}$, then $a \le -a - 1$, so $f(x) = a + 2$ for all x in $(-a - 1, a + 1)$, so $a + 2$ is the maximum and minimum value. If $-\frac{1}{2} < a \le 0$, then f has the minimum value a^2, and if $a \ge 0$, then f has the minimum value 0. Since $a + 2 > (a + 1)^2$ only for $[-1 - \sqrt{5}]/2 < a < [1 + \sqrt{5}]/2$, when $a \ge -\frac{1}{2}$ the function f has a maximum value only for $a \le [1 + \sqrt{5}]/2$ (the maximum value being $a + 2$).
(vii) Bounded above and below; maximum value 1; minimum value 0.
(ix) Bounded above and below; maximum value 1; minimum value -1.
(xi) f has a maximum and minimum value, since f is continuous.

2. (i) $n = -2$, since $f(-2) < 0 < f(-1)$.
(iii) $n = -1$, since $f(-1) = -1 < 0 < f(0)$.

3. (i) If $f(x) = x^{179} + 163/(1 + x^2 + \sin^2 x)$, then f is continuous on $\mathbf{R}$ and $f(1) > 0$, while $f(-2) < 0$, so $f(x) = 0$ for some x in $(-2, 1)$.

5. f is constant, for if f took on two different values, then f would take on all values in between, which would include irrational values.

7. (1) $f(x) = x$;
(2) $f(x) = -x$;

(3) $f(x) = |x|$;
(4) $f(x) = -|x|$.

10. Apply Theorem 1 to $f - g$.
11. If $f(0) = 0$ or $f(1) = 1$, choose $x = 0$ or 1. If $f(0) > 0 = I(0)$ and $f(1) < 1 = I(1)$, then Problem 10 applied to f and I implies that $f(x) = x$ for some x.

CHAPTER 8

1. (i) 1 is the greatest element, and the greatest lower bound is 0, which is not in the set.
   (iii) 1 is the greatest element, and 0 is the least element.
   (v) Since $\{x : x^2+x+1 \geq 0\} = \mathbf{R}$, there is no least upper bound or greatest lower bound.
   (vii) Since $\{x : x < 0 \text{ and } x^2+x-1 < 0\} = ([-1-\sqrt{5}]/2, 0)$, the greatest lower bound is $[-1-\sqrt{5}]/2$, and the least upper bound is 0; neither belongs to the set.
2. (a) Since $A \neq \emptyset$, there is some x in A. Then $-x$ is in $-A$, so $-A \neq \emptyset$. Since A is bounded above, there is some y such that $y \geq x$ for all x in A. Then $-y \leq -x$ for all x in A, so $-y \leq z$ for all z in $-A$, so $-A$ is bounded below. Let $\alpha = \sup(-A)$. Then α is an upper bound for $-A$, so, reversing the argument just given, $-\alpha$ is a lower bound for A. Moreover, if β is any lower bound for A, then $-\beta$ is an upper bound for $-A$, so $-\beta \geq \alpha$, so $\beta \leq -\alpha$. Thus $-\alpha$ is the greatest lower bound for A.
5. (a) If l is the largest integer with $l \leq x$, then $l+1 > x$, but $l+1 \leq x+1 < y$. So we can let $k = l+1$. (Proof that a largest such integer l exists: Since $\mathbf{N}$ is not bounded above, there is some natural number n with $-n < x < n$. There are consequently only a finite number of integers l with $-n \leq l \leq x$. Pick the largest.)
   (b) Since $y - x > 0$, there is some natural number n with $1/n < y - x$. Since $ny - nx > 1$, there is, by part (a), an integer k with $nx < k < ny$, which means that $x < k/n < y$.
   (c) Choose $r + \sqrt{2}(s-r)/2$.
   (d) By part (b), there is a rational number r with $x < r < y$, and therefore a rational number s with $x < r < s < y$. Apply part (c) to $r < s$.
10. Let k be the largest integer $\leq x/\alpha$ (the solution to Problem 5 shows that such a k exists), and let $x' = x - k\alpha \geq 0$. If $x - k\alpha = x' \geq \alpha$, then $x \geq (k+1)\alpha$, so $k+1 \leq x/\alpha$, contradicting the choice of k. So $0 \leq x' < \alpha$.
12. (a) Since any y in B satisfies $y \geq x$ for all x in A, any y in B is an upper bound for A, so $y \geq \sup A$.
    (b) Part (a) shows that $\sup A$ is a lower bound for B, so $\sup A \leq \inf B$.
13. Since $x \leq \sup A$ and $y \leq \sup B$ for every x in A, and y in B, it follows that $x+y \leq \sup A + \sup B$. Thus, $\sup A + \sup B$ is an upper bound for $A+B$, so $\sup(A+B) \leq \sup A + \sup B$. If x and y are chosen in A and B, respectively, so that $\sup A - x < \varepsilon/2$ and $\sup B - y < \varepsilon/2$, then $\sup A + \sup B - (x+y) < \varepsilon$. Hence,

$$\sup(A+B) \geq x + y > \sup A + \sup B - \varepsilon.$$

CHAPTER 9

1. (a)

$$f'(a) = \lim_{h\to 0} \frac{f(a+h)-f(a)}{h} = \lim_{h\to 0} \frac{\dfrac{1}{a+h} - \dfrac{1}{a}}{h}$$
$$= \lim_{h\to 0} \frac{-1}{a(a+h)} = -\frac{1}{a^2}.$$

(b) The tangent line through $(a, 1/a)$ is the graph of

$$g(x) = \frac{-1}{a^2}(x-a) + \frac{1}{a}$$
$$= \frac{-x}{a^2} + \frac{2}{a}.$$

If $f(x) = g(x)$, then

$$\frac{1}{x} = -\frac{x}{a^2} + \frac{2}{a}$$

or

$$x^2 - 2ax + a^2 = 0,$$

so $x = a$.

2. (a)

$$f'(a) = \lim_{h\to 0} \frac{f(a+h)-f(a)}{h} = \lim_{h\to 0} \frac{\dfrac{1}{(a+h)^2} - \dfrac{1}{a^2}}{h}$$
$$= \lim_{h\to 0} \frac{(-2ah - h^2)}{ha^2(a+h)^2} = -\frac{2}{a^3}.$$

(b) The tangent line through $(a, 1/a^2)$ is the graph of

$$g(x) = -\frac{2}{a^3}(x-a) + \frac{1}{a^2}$$
$$= -\frac{2x}{a^3} + \frac{3}{a^2}.$$

If $f(x) = g(x)$, then

$$\frac{1}{x^2} = \frac{-2x}{a^3} + \frac{3}{a^2},$$

or

$$2x^3 - 3ax^2 + a^3 = 0,$$

or

$$0 = (x-a)(2x^2 - ax - a^2) = (x-a)(2x+a)(x-a).$$

So $x = a$ or $x = -a/2$; the point $(-a/2, 4/a^2)$ lies on the opposite side of the vertical axis from $(a, 1/a^2)$.

3.

$$\begin{aligned}
f'(a) &= \lim_{h\to 0} \frac{f(a+h) - f(a)}{h} = \lim_{h\to 0} \frac{\sqrt{a+h} - \sqrt{a}}{h} \\
&= \lim_{h\to 0} \frac{(\sqrt{a+h} - \sqrt{a})(\sqrt{a+h} + \sqrt{a})}{h(\sqrt{a+h} + \sqrt{a})} = \lim_{h\to 0} \frac{h}{h(\sqrt{a+h} + \sqrt{a})} \\
&= \frac{1}{2\sqrt{a}}.
\end{aligned}$$

4. Conjecture: ${S_n}'(x) = nx^{n-1}$. Proof:

$$\begin{aligned}
{S_n}'(x) &= \lim_{h\to 0} \frac{S_n(x+h) - S_n(x)}{h} = \lim_{h\to 0} \frac{(x+h)^n - x^n}{h} \\
&= \lim_{h\to 0} \frac{\displaystyle\sum_{j=0}^{n} \binom{n}{j} x^{n-j} h^j - x^n}{h} \\
&= \lim_{h\to 0} \sum_{j=1}^{n} \binom{n}{j} x^{n-j} h^{j-1} \\
&= \binom{n}{1} x^{n-1} = nx^{n-1}, \qquad \text{since } \lim_{h\to 0} h^{j-1} = 0 \text{ for } j > 1.
\end{aligned}$$

5. $f'(x) = 0$ for x not an integer, and $f'(x)$ is not defined if x is an integer.

6. (a)

$$\begin{aligned}
g'(x) &= \lim_{h\to 0} \frac{g(x+h) - g(x)}{h} = \lim_{h\to 0} \frac{[f(x+h) + c] - [f(x) + c]}{h} \\
&= \lim_{h\to 0} \frac{f(x+h) - f(x)}{h} = f'(x).
\end{aligned}$$

(b)

$$\begin{aligned}
g'(x) &= \lim_{h\to 0} \frac{g(x+h) - g(x)}{h} = \lim_{h\to 0} \frac{cf(x+h) - cf(x)}{h} \\
&= c \cdot \lim_{h\to 0} \frac{f(x+h) - f(x)}{h} = cf'(x).
\end{aligned}$$

7. (a) $f'(9) = 3 \cdot 9^2$; $f'(25) = 3 \cdot (25)^2$; $f'(36) = 3 \cdot (36)^2$.
   (b) $f'(3^2) = f'(9) = 3 \cdot 9^2$; $f'(5^2) = f'(25) = 3 \cdot (25)^2$; $f'(6^2) = f'(36) = 3 \cdot (36)^2$.
   (c) $f'(a^2) = 3(a^2)^2 = 3a^4$; $f'(x^2) = 3(x^2)^2 = 3x^4$.
   (d) $f'(x^2) = 3x^4$; but $g(x) = x^6$, so $g'(x) = 6x^5$.

8. (a)

$$\begin{aligned}
g'(x) &= \lim_{h\to 0} \frac{g(x+h) - g(x)}{h} = \lim_{h\to 0} \frac{f(x+h+c) - f(x+c)}{h} \\
&= \lim_{h\to 0} \frac{f([x+c]+h) - f(x+c)}{h} = f'(x+c).
\end{aligned}$$

(b)

$$\begin{aligned} g'(x) &= \lim_{h\to 0}\frac{g(x+h)-g(x)}{h} = \lim_{h\to 0}\frac{f(cx+ch)-f(cx)}{h} \\ &= \lim_{h\to 0}\frac{c[f(cx+ch)-f(cx)]}{ch} = \lim_{k\to 0}\frac{c[f(cx+k)-f(cx)]}{k} \\ &= c\cdot\lim_{k\to 0}\frac{f(cx+k)-f(cx)}{k} = c\cdot f'(cx). \end{aligned}$$

(Compare the manipulations in this calculation with Problem 5-14.)

(c) If $g(x) = f(x+a)$, then $g'(x) = f'(x+a)$, by part (a). But $g = f$, so $f'(x) = g'(x) = f'(x+a)$ for all x, which means that f' is periodic, with period a.

9. (i) If $g(x) = x^5$, then $g'(x) = 5x^4$. Now $f(x) = g(x+3)$, so by Problem 8(a), $f'(x) = g'(x+3) = 5(x+3)^4$. And $f'(x+3) = 5(x+6)^4$.
   (ii) $f(x) = (x-3)^5$, so $f'(x) = 5(x-3)^4$, as in part (i). And $f'(x+3) = 5\cdot 0^4 = 0$.
   (iii) $f(x) = (x+2)^7$, so $f'(x) = 7(x+2)^6$, as in part (i). And $f'(x+3) = 7(x+5)^6$.

10. If $f(x) = g(t+x)$, then $f'(x) = g'(t+x)$, by Problem 8(a). If $f(t) = g(t+x)$, then $f'(t) = g'(t+x)$, by Problem 8(a), so $f'(x) = g'(2x)$.

11. (a) If $s(t) = ct^2$, then $s'(t) = 2ct$, and there is no number k such that $s'(t) = ks(t)$ [that is, $2ct = kct^2$] for all t.
    (By the way, at this point we do not know any nonzero function f for which f' is proportional to f. After Chapter 18 it might be amusing to determine what the world would be like if Galileo were correct.)
    (b) (i) If $s(t) = (a/2)t^2$, then $s'(t) = at$, so $s''(t) = a$.
        (ii) $[s'(t)]^2 = (at)^2 = 2a\cdot(a/2)t^2 = 2as(t)$.
    (c) The chandelier falls $s(t) = 16t^2$ feet in t seconds, so it falls 400 feet in t seconds, if $400 = 16t^2$, or $t = 5$. After 5 seconds the velocity will be $s'(5) = 5a = 5\cdot 32 = 160$ feet per second. The speed was half this amount when $80 = s'(t) = 32t$, or $t = \frac{5}{2}$.

21. (a) This is another way of writing the definition (see Problem 5-9).
    (b) This follows from Problem 5-11, applied to the functions $\alpha(h) = [f(a+h)-f(a)]/h$ and $\beta(h) = [g(a+h)-g(a)]/h$.

26. (i) $f''(x) = 6x$.
    (iii) $f''(x) = 4x^3$.

30. (i) means that $f'(a) = na^{n-1}$ if $f(x) = x^n$.
    (iii) means that $g'(a) = f'(a)$ if $g(x) = f(x)+c$.
    (v) means the same as (iii).
    (vii) means that $g'(b) = f'(b+a)$ if $g(x) = f(x+a)$.
    (ix) means that $g'(b) = cf'(cb)$ if $g(x) = f(cx)$.

CHAPTER 10

1. (i) $(1+2x)\cdot\cos(x+x^2)$.
   (iii) $(-\sin x)\cdot\cos(\cos x)$.

(v) $\cos\left(\dfrac{\cos x}{x}\right) \cdot \dfrac{-x\sin x - \cos x}{x^2}$.

(vii) $(\cos(x + \sin x)) \cdot (1 + \cos x)$.

2. (i) $(\cos((x+1)^2(x+2))) \cdot [2(x+1)(x+2) + (x+1)^2]$.

(iii) $[2\sin((x+\sin x)^2)\cos((x+\sin x)^2)] \cdot 2(x+\sin x)(1+\cos x)$.

(v) $(\cos(x \sin x)) \cdot (\sin x + x\cos x) + (\cos(\sin x^2)(\cos x^2)) \cdot 2x$.

(vii) $(2\sin x\cos x\sin x^2\sin^2 x^2) + (2x\cos x^2\sin^2 x\sin^2 x^2)$
$+ (4x\sin x^2\cos x^2\sin^2 x\sin x^2)$.

(ix) $6(x+\sin^5 x)^5(1+5\sin^4 x\cos x)$.

(xi) $\cos(\sin^7 x^7+1)^7 \cdot 7(\sin^7 x^7+1)^6 \cdot (7\sin^6 x^7 \cdot \cos x^7 \cdot 7x^6)$.

(xiii) $\cos(x^2+\sin(x^2+\sin x^2)) \cdot [(2x+\cos(x^2+\sin x^2)\cdot(2x+2x\cos x^2))]$.

(xv) $\dfrac{(1+\sin x)(2x\cos x^2\cdot\sin^2 x+\sin x^2\cdot 2\sin x\cos x) - \cos x\sin x^2\sin^2 x}{(1+\sin x)^2}$.

(xvii) $\cos\left(\dfrac{x^3}{\sin\left(\dfrac{x^3}{\sin x}\right)}\right) \cdot$

$$\frac{3x^2\sin\left(\dfrac{x^3}{\sin x}\right) - x^3\cos\left(\dfrac{x^3}{\sin x}\right)\cdot\left(\dfrac{3x^2\sin x - x^3\cos x}{\sin^2 x}\right)}{\sin^2\left(\dfrac{x^3}{\sin x}\right)}.$$

4. (i) $-\dfrac{(x+1)^2}{(x+2)^2}$.

(iii) $2x^2$.

5. (i) $-x^2$.

(iii) 17.

6. (i) $f'(x) = g'(x+g(a))$.

(iii) $f'(x) = g'(x+g(x))\cdot(1+g'(x))$.

(v) $f'(x) = g(a)$.

7. (a) $A'(t) = 2\pi r(t)r'(t)$. Since $r'(t) = 4$ for that t with $r(t) = 6$, it follows that $A'(t) = 2\pi\cdot 6\cdot 4 = 48\pi$ when $r(t) = 6$.

(b) If $V(t)$ is the volume at time t, then $V(t) = 4\pi r(t)^3/3$, so $V'(t) = 4\pi r(t)^2 r'(t) = 4\pi\cdot 6^2\cdot 4 = 576\pi$ when $r(t) = 6$.

(c) First method: Since $A'(t) = 2\pi r(t)r'(t)$, and $A'(t) = 5$ for $r(t) = 3$, it follows that

$$r'(t) = \frac{A'(t)}{2\pi r(t)} = \frac{5}{6\pi} \quad \text{when } r(t) = 3.$$

Thus

$$\begin{aligned} V'(t) &= 4\pi r(t)^2 r'(t) \\ &= 4\pi \cdot 9 \cdot \frac{5}{6\pi} \\ &= 30 \quad \text{when } r(t) = 3. \end{aligned}$$

To apply the second method, we first note that if

$$f(t) = A(t)^{3/2} = \sqrt{A(t)^3},$$

then, using Problem 9-3 and the Chain Rule,

$$\begin{aligned} f'(t) &= \frac{1}{2\sqrt{A(t)^3}} \cdot 3A(t)^2 A'(t) \\ &= \frac{1}{2A(t)^{3/2}} \cdot 3A(t)^2 A'(t) \\ &= \frac{3}{2} A(t)^{1/2} A'(t) \quad \text{(just as we might have guessed).} \end{aligned}$$

Now

$$\begin{aligned} V(t) &= \frac{4\pi r(t)^3}{3} = \frac{4\pi [r(t)^2]^{3/2}}{3} \\ &= \frac{4[\pi r(t)^2]^{3/2}}{3\pi^{1/2}} \\ &= \frac{4A(t)^{3/2}}{3\pi^{1/2}}. \end{aligned}$$

So

$$\begin{aligned} V'(t) &= \frac{4}{3\pi^{1/2}} \cdot \frac{3}{2}\sqrt{A(t)}A'(t) \\ &= \frac{2}{\pi^{1/2}} \cdot \pi^{1/2} r(t) A'(t) \\ &= 2 \cdot 3 \cdot 5 = 30. \end{aligned}$$

10. (i) $(f \circ h)'(0) = f'(h(0)) \cdot h'(0) = f'(3) \cdot \sin^2(\sin 1) = [6 \sin \frac{1}{3} - \cos \frac{1}{3}] \sin^2(\sin 1)$.

    (iii) $\alpha'(x^2) = h'(x^4) \cdot 2x^2 = \sin^2(\sin(x^4 + 1)) \cdot 2x^2$.

12. The Chain Rule implies that

$$\begin{aligned} \left(\frac{1}{g}\right)'(x) &= (f \circ g)'(x) = f'(g(x)) \cdot g'(x) \\ &= -\frac{1}{g(x)^2} \cdot g'(x). \end{aligned}$$

33. (i) $\dfrac{dz}{dx} = \dfrac{dz}{dy} \cdot \dfrac{dy}{dx} = (\cos y) \cdot (1 + 2x) = (\cos(x + x^2)) \cdot (1 + 2x)$.

(iii) $\dfrac{dz}{dx} = \dfrac{dz}{du} \cdot \dfrac{du}{dx} = (-\sin u) \cdot (\cos x) = (-\cos(\sin x)) \cdot (\cos x).$

CHAPTER 11

1. (i) $0 = f'(x) = 3x^2 - 2x - 8$ for $x = 2$ and $x = -\frac{4}{3}$, both of which are in $[-2, 2]$;
$f(-2) = 5$, $f(2) = -11$, $f(-\frac{4}{3}) = \frac{203}{27}$;
maximum $= \frac{203}{27}$, minimum $= -11$.
(iii) $0 = f'(x) = 12x^3 - 24x^2 + 12x = 12x(x^2 - 2x + 1)$ for $x = 0$ and $x = 1$, of which only 0 is in $[-\frac{1}{2}, \frac{1}{2}]$;
$f(-\frac{1}{2}) = \frac{43}{16}$, $f(\frac{1}{2}) = \frac{11}{16}$, $f(0) = 0$;
maximum $= \frac{43}{16}$, minimum $= 0$.
(v) $0 = f'(x) =$

$$\frac{x^2 + 1 - (x+1)2x}{(x^2+1)^2} = \frac{1 - 2x - x^2}{(x^2+1)^2}$$

for $x = -1 + \sqrt{2}$ and $x = -1 - \sqrt{2}$, of which only $-1 + \sqrt{2}$ is in $[-1, \frac{1}{2}]$;
$f(-1) = 0$, $f(\frac{1}{2}) = \frac{6}{5}$, $f(-1 + \sqrt{2}) = (1 + \sqrt{2})/2$;
maximum $= (1 + \sqrt{2})/2$, minimum $= 0$.

2. (i) $-\frac{4}{3}$ is a local maximum point, and 2 is a local minimum point.
(iii) 0 is a local minimum point, and there are no local maximum points.
(v) $-1 + \sqrt{2}$ is a local maximum point, and $-1 - \sqrt{2}$ is a local minimum point.

4. (a) Notice that f actually has a minimum value, since f is a polynomial function of even degree. The minimum occurs at a point x with

$$0 = f'(x) = 2\sum_{i=1}^{n}(x - a_i),$$

so $x = (a_1 + \cdots + a_n)/n$.

5. (i) 3 and 7 are local maximum points, and 5 and 9 are local minimum points.
(iii) All irrational $x > 0$ are local minimum points, and all irrational $x < 0$ are local maximum points.
(v) x is a local maximum (minimum) point if the decimal expansion contains (does not contain) a 5.

8. If $f(x)$ is the total length of the path, then

$$f(x) = \sqrt{x^2 + a^2} + \sqrt{(1-x)^2 + b^2}.$$

The positive function f clearly has a minimum, since $\lim_{x\to\infty} f(x) = \lim_{x\to-\infty} f(x) = \infty$, and f is differentiable everywhere, so the minimum occurs at a point x with $f'(x) = 0$. Now, $f'(x) = 0$ when

$$\frac{x}{\sqrt{x^2 + a^2}} - \frac{(1-x)}{\sqrt{(1-x)^2 + b^2}} = 0.$$

This equation says that $\cos\alpha = \cos\beta$.

It is also possible to notice that $f(x)$ is equal to the sum of the lengths of the dashed line segment and the line segment from $(x, 0)$ to $(1, b)$. This is shortest when the two line segments lie along a line (because of Problem 4-9(b), if a rigorous reason is required); a little plane geometry shows that this happens when $\alpha = \beta$.

9. If x is the length of one side of a rectangle of perimeter P, then the length of the other side is $(P - 2x)/2$, so the area is

$$A(x) = \frac{x(P - 2x)}{2}.$$

So the rectangle with greatest area occurs when x is the maximum point for f on $(0, P/2)$. Since A is continuous on $[0, P/2]$, and $A(0) = A(P/2) = 0$, and $A(x) > 0$ for x in $(0, P/2)$, the maximum exists. Since A is differentiable on $(0, P/2)$, the minimum point x satisfies

$$\begin{aligned} 0 = A'(x) &= \frac{P - 2x}{2} - x \\ &= \frac{P - 4x}{2}, \end{aligned}$$

so $x = P/4$.

10. Let $S(r)$ be the surface area of the right circular cylinder of volume V with radius r. Since

$$V = \pi r^2 h \quad \text{where } h \text{ is the height,}$$

we have $h = V/\pi r^2$, so

$$\begin{aligned} S(r) &= 2\pi r^2 + 2\pi r h \\ &= 2\pi r^2 + \frac{2V}{r}. \end{aligned}$$

We want the minimum point of S on $(0, \infty)$; this exists, since $\lim\limits_{r \to 0} S(r) = \lim\limits_{r \to \infty} S(r) = \infty$. Since S is differentiable on $(0, \infty)$, the minimum point r satisfies

$$\begin{aligned} 0 = S'(r) &= 4\pi r - \frac{2V}{r^2} \\ &= \frac{4\pi r^3 - 2V}{r^2}, \end{aligned}$$

or

$$r = \sqrt[3]{\frac{V}{2\pi}}.$$

19. 1 is a local maximum point, and 3 is a local minimum point.

25. (a) We have

$$\begin{aligned} \frac{f(b) - f(a)}{b - a} &= f'(x) \quad \text{for some } x \text{ in } (a, b) \\ &\geq M, \end{aligned}$$

so $f(b) - f(a) \geq M(b-a)$.

(b) We have

$$\begin{aligned}\frac{f(b)-f(a)}{b-a} &= f'(x) \quad \text{for some } x \text{ in } (a,b)\\ &\leq m,\end{aligned}$$

so $f(b) - f(a) \leq m(b-a)$.

(c) If $|f'(x)| \leq M$ for all x in $[a,b]$, then $-M \leq f(x) \leq M$, so

$$f(a) - M(b-a) \leq f(b) \leq f(a) + M(b-a),$$

or

$$|f(b) - f(a)| \leq M(b-a).$$

28. (a) $f(x) = -\cos x + a$ for some number a (because $f(x) = -\cos x$ is one such function, and any two such functions differ by a constant function).

(b) $f'(x) = x^4/4 + a$ for some number a, so $f(x) = x^5/20 + ax + b$ for some numbers a and b.

(c) $f''(x) = x^2 + x^3/3 + a$ for some a, so $f'(x) = x^3/6 + x^4/12 + ax + b$ for some a and b, so $f(x) = x^4/24 + x^5/60 + ax^2/2 + bx + c$ for some numbers a, b, and c. Equivalently, and more simply, $f(x) = x^4/24 + x^5/60 + ax^2 + bx + c$ for some numbers a, b, and c.

29. (a) Since $s''(t) = -32$, we have $s'(t) = -32t + \alpha$ for some α, so $s(t) = -16t^2 + \alpha t + \beta$ for some α and β.

(b) Clearly, $s(0) = 0 + 0 + \beta$ and $s'(0) = 0 + \alpha$. Thus, $\alpha = v_0$ and $\beta = s_0$.

(c) In this case, $s_0 = 0$ and $v_0 = v$, so $s(t) = -16t^2 + vt$. The maximum value of s occurs when $0 = s'(t) = -32t + v$, or $t = v/32$, so the maximum value is

$$\begin{aligned} s\left(\frac{v}{32}\right) &= -16\left(\frac{v}{32}\right)^2 + v \cdot \left(\frac{v}{32}\right)\\ &= \frac{-v^2}{64} + \frac{v^2}{32}\\ &= \frac{v^2}{64}.\end{aligned}$$

At that moment the velocity is clearly 0, but the acceleration is -32 (as at any time). The weight hits the ground at time $t > 0$ when

$$0 = s(t) = -16t^2 + vt,$$

or $t = v/16$ (it takes as long to fall back down as it took to reach the top). The velocity is then

$$\begin{aligned} s'(v/16) &= -32\left(\frac{v}{16}\right) + v\\ &= -v\end{aligned}$$

(the same velocity with which it was initially moving upward).

44. Apply the Mean Value Theorem to $f(x) = \sqrt{x}$ on $[64, 66]$:

$$\frac{\sqrt{66} - \sqrt{64}}{66 - 64} = f'(x) = \frac{1}{2\sqrt{x}} \quad \text{for some } x \text{ in } [64, 66].$$

Since $64 < x < 81$, we have $8 < \sqrt{x} < 9$, so

$$\frac{1}{2 \cdot 9} < \frac{\sqrt{66} - 8}{2} < \frac{1}{2 \cdot 8}.$$

48. l'Hôpital's Rule does not lead to the equation

$$\lim_{x \to 1} \frac{3x^2 + 1}{2x - 3} = \lim_{x \to 1} \frac{6x}{2}$$

because $\lim_{x \to 1} 3x^2 + 1 \neq 0$.

49. (i)

$$\lim_{x \to 0} \frac{x}{\tan x} = \lim_{x \to 0} \frac{1}{\sec^2 x} = \lim_{x \to 0} \cos^2 x = 1.$$

(ii)

$$\lim_{x \to 0} \frac{\cos^2 x - 1}{x^2} = \lim_{x \to 0} \frac{-2 \sin x \cos x}{2x} = -1.$$

CHAPTER 12

1. (i) $f^{-1}(x) = (x - 1)^{1/3}$. (If $y = f^{-1}(x)$, then $x = f(y) = y^3 + 1$, so $y = (x - 1)^{1/3}$.)

   (iii) $f^{-1} = f$. (If $y = f^{-1}(x)$, then

$$x = f(y) = \begin{cases} y, & y \text{ rational} \\ -y, & y \text{ irrational}; \end{cases}$$

since $\pm y$ is rational or irrational if and only if y is, we have $y = x$ if x is rational and $y = -x$ if x is irrational, so $y = f(x)$.)

   (v)

$$f^{-1}(x) = \begin{cases} x, & x \neq a_1, \dots, a_n \\ a_{i-1}, & x = a_i, \quad i = 2, \dots, n \\ a_n, & x = a_1. \end{cases}$$

   (vii) $f^{-1} = f$.

2. (i) f^{-1} is increasing and $f^{-1}(x)$ is not defined for $x \leq 0$.

   (iii) f^{-1} is decreasing and $f^{-1}(x)$ is not defined for $x \leq 0$.

3. Suppose f is increasing. Let $a < b$. Then $f^{-1}(a) \neq f^{-1}(b)$, since f^{-1} is one-one. So either $f^{-1}(a) < f^{-1}(b)$ or $f^{-1}(a) > f^{-1}(b)$. But if $f^{-1}(a) > f^{-1}(b)$, then

$$b = f(f^{-1}(b)) < f(f^{-1}(a)) = a,$$

a contradiction. The proof is similar for decreasing f, or one can consider $-f$ instead.

4. Clearly, $f + g$ is increasing, for if $f(a) < f(b)$ and $g(a) < g(b)$, then $(f + g)(a) = f(a) + g(a) < f(b) + g(b) = (f + g)(b)$.
   $f \cdot g$ is not necessarily increasing; for example, if $f(x) = g(x) = x$. (But $f \cdot g$

is increasing if $f(x) \geq 0$ for all x.)
$f \circ g$ is increasing, for if $a < b$, then $g(a) < g(b)$, so $f(g(a)) < f(g(b))$.

5. (a) If $(f \circ g)(x) = (f \circ g)(y)$, so that $f(g(x)) = f(g(y))$, then $g(x) = g(y)$, since f is one-one, so $x = y$, since g is one-one.
$(f \circ g)^{-1} = g^{-1} \circ f^{-1}$: for if $y = (f \circ g)^{-1}(x)$, then $x = (f \circ g)(y) = f(g(y))$, so $g(y) = f^{-1}(x)$, so $y = g^{-1}(f^{-1}(x))$.

6. If $f(x) = f(y)$, then

$$\frac{ax+b}{cx+d} = \frac{ay+b}{cy+d},$$

so

$$acxy + bcy + adx + bd = acxy + ady + bcx + bd,$$

or

$$ad(x-y) = bc(x-y).$$

If $ad \neq bc$, this implies that $x - y = 0$. (But if $ad = bc$, then $f(x) = f(y)$ for all x and y in the domain of f.)
If $y = f^{-1}(x)$, then $x = f(y)$, so

$$x = \frac{ay+b}{cy+d}$$

so

$$f^{-1}(x) = y = \frac{-dx+b}{cx-a} \qquad \text{for } x \neq a/c.$$

7. (i) Those intervals $[a, b]$ which are contained in $(-\infty, 0]$ or $[0, 2]$ or $[2, \infty)$, since f is increasing on $(-\infty, 0]$ and $[2, \infty)$, and decreasing on $[0, 2]$.
(iii) Those intervals $[a, b]$ which are contained in $(-\infty, 0]$ or $[0, \infty)$, since f is increasing on $(-\infty, 0]$ and decreasing on $[0, \infty)$.

17. The formula for the derivative reads:

$$\frac{dx}{dy} = \frac{1}{\dfrac{dy}{dx}}.$$

(In this formula, it is understood that dx/dy means $(f^{-1})'(y)$, while dy/dx is an "expression involving x," and in the final answer x must be replaced by y, by means of the equation $y = f(x)$.)
The computation in Problem 17, when completed, shows that

$$\frac{dx^{1/n}}{dx} = \frac{1}{n(x^{1/n})^{n-1}} = \frac{1}{nx^{1-(1/n)}}$$
$$= \frac{1}{n}x^{(1/n)-1}.$$

18.

$$\begin{aligned} G'(x) &= x(f^{-1})'(x) + f^{-1}(x) - F'(f^{-1}(x)) \cdot (f^{-1})'(x) \\ &= x(f^{-1})'(x) + f^{-1}(x) - f(f^{-1}(x)) \cdot (f^{-1})'(x) \\ &= x(f^{-1})'(x) + f^{-1}(x) - x(f^{-1})'(x) \\ &= f^{-1}(x). \end{aligned}$$

19. (i)

$$(h^{-1})'(3) = \frac{1}{h'(h^{-1}(3))} = \frac{1}{h'(0)} = \frac{1}{\sin^2(\sin 1)}.$$

20. Since

$$(f^{-1})'(x) = \frac{1}{f'(f^{-1}(x))},$$

we have

$$\begin{aligned} (f^{-1})''(x) &= \frac{-f''(f^{-1}(x)) \cdot (f^{-1})'(x)}{[f'(f^{-1}(x))]^2} \\ &= \frac{-f''(f^{-1}(x))}{[f'(f^{-1}(x))]^3}. \end{aligned}$$

CHAPTER 13

1. If $P_n = \{t_0, \dots, t_n\}$ is the partition with $t_i = ib/n$, then

$$\begin{aligned} L(f, P_n) &= \sum_{i=1}^{n} (t_{i-1})^3 \cdot (t_i - t_{i-1}) \\ &= \sum_{i=1}^{n} (i-1)^3 \cdot \frac{b_3}{n^3} \cdot \frac{b}{n} \\ &= \frac{b^4}{n^4} \sum_{j=0}^{n-1} j^3 \\ &= \frac{b^4}{n^4} \left[\frac{(n-1)^4}{4} + \frac{(n-1)^3}{2} + \frac{(n-1)^2}{4} \right], \end{aligned}$$

and similarly

$$\begin{aligned} U(f, P_n) &= \frac{b^4}{n^4} \sum_{j=1}^{n} j^3 \\ &= \frac{b^4}{n^4} \left[\frac{n^4}{4} + \frac{n^3}{2} + \frac{n^2}{4} \right]. \end{aligned}$$

Clearly $L(f, P_n)$ and $U(f, P_n)$ can be made as close to $b^4/4$ as desired by choosing n sufficiently large, so $U(f, P_n) - L(f, P_n)$ can be made as small as desired, by choosing n large enough. This shows that f is integrable. Moreover, there is only one number a with $L(f, P_n) \le a \le U(f, P_n)$ for all n; since $\int_0^b x^3\,dx$ has this property, the proof that $\int_0^b x^3\,dx = b^4/4$ will be complete once we show that $L(f, P_n) \le b^4/4 \le U(f, P_n)$ for all n. This can be done by a straightforward computation, but it actually follows from the fact that $L(f, P_n)$ and $U(f, P_n)$ can be made as close to $b^4/4$ as desired by choosing n sufficiently large. In fact, if it were true that $b^4/4 < \int_0^b x^3\,dx$, then it would not be possible to make $U(f, P_n)$ as close as desired to $b^4/4$ by choosing n large enough, since each $U(f, P_n) \geq \int_0^b x^3\,dx$, and similarly we cannot have $b^4/4 > \int_0^b x^3\,dx$.

2. We have

$$L(f, P_n) = \frac{b^5}{n^5}\left[\frac{(n-1)^5}{5} + \frac{(n-1)^4}{2} + \frac{(n-1)^3}{3} - \frac{(n-1)}{30}\right],$$

$$U(f, P_n) = \frac{b^5}{n^5}\left[\frac{n^5}{5} + \frac{n^4}{2} + \frac{n^3}{3} - \frac{n}{30}\right].$$

Clearly $L(f, P_n)$ and $U(f, P_n)$ can be made as close to $b^5/5$ as desired by choosing n large enough. As in Problem 1, this implies that $\int_0^b x^4\,dx = b^5/5$.

7. (i) $\int_0^2 f = 0$.
(iii) $\int_0^2 f = 3$.
(v) f is not integrable.
(vii) $\int_0^2 f = 1$.
(For a rigorous proof that the functions in (i), (iii), and (vii) are integrable, see Problem 20. The values of the integrals, which are clear from the geometric picture, can also be deduced rigorously by using the ideas in the proof of Problem 20, together with known integrals.)

8. (i)

$$\int_{-2}^{2}\left[\left(\frac{x^2}{2}+2\right) - x^2\right] dx = \frac{16}{3}.$$

(iii)

$$\int_{-\sqrt{2}/2}^{\sqrt{2}/2} [(1-x^2) - x^2]\,dx = \frac{2\sqrt{2}}{3}.$$

(v)

$$\int_0^2 [(x^2 - 2x + 4) - x^2]\,dx = 4.$$

9.

$$\int_a^b \left(\int_c^d f(x)g(y)\,dy\right) dx = \int_a^b \left(f(x)\int_c^d g(y)\,dy\right) dx \quad \text{(here } f(x) \text{ is the constant)}$$

$$= \int_c^d g(y)\,dy \cdot \int_a^b f(x)\,dx$$

$$\text{(here } \int_c^d g(y)\,dy \text{ is the constant).}$$

13. (a) Clearly $L(f, P) \geq 0$ for every partition P.

(b) Apply part (a) to $f - g$, and use the fact that

$$\int_a^b (f-g) = \int_a^b f - \int_a^b g.$$

23. (a) Clearly

$$m(b-a) \le L(f,P) \le U(f,P) \le M(b-a)$$

for all partitions P of $[a,b]$. Consequently,

$$m(b-a) \le \int_a^b f(x)\,dx \le M(b-a).$$

Thus

$$\mu = \frac{\displaystyle\int_a^b f(x)\,dx}{b-a}$$

satisfies $m \le \mu \le M$.

(b) Let m and M be the minimum and maximum values of f on $[a,b]$. Since f is continuous, it takes on the values m and M, and consequently the number μ of part (a).

33. (a) 0.
(b) $\frac{1}{2}$.

37. Since

$$-|f| \le f \le |f|,$$

we have

$$-\int_a^b |f| \le \int_a^b f \le \int_a^b |f|,$$

so

$$\left|\int_a^b f\right| \le \int_a^b |f|.$$

(Problem 36 implies that $\displaystyle\int_a^b |f|$ makes sense.)

CHAPTER 14

1. (i) $(\sin^3 x^3)\cdot 3x^2$.

(iii) $\displaystyle\int_8^x \frac{1}{1+t^2+\sin^2 t}\,dt$.

(v) $\displaystyle\int_a^b \frac{1}{1+t^2+\sin^2 t}\,dt$.

(vii) $\displaystyle (F^{-1})'(x) = \frac{1}{F'(F^{-1}(x))} = F^{-1}(x)$.

2. (i) All $x \ne 1$.
(iii) All $x \ne 1$.
(v) All x.
(vii) All $x \ne 0$. (F is not differentiable at 0 because $F(x) = 0$ for $x \le 0$, but there are $x > 0$ arbitrarily close to 0 with $\frac{F(x)}{x} = \frac{1}{2}$.)

5. (i)

$$(f^{-1})'(0) = \frac{1}{f'(f^{-1}(0))} = \frac{1}{1+\sin(\sin(f^{-1}(0)))}$$
$$= \frac{1}{1+\sin(\sin 0)} = 1.$$

11. $F(x) = x\int_0^x f(t)\,dt$, so

$$F'(x) = xf(x) + \int_0^x f(t)\,dt.$$

14.

$$f(x) = \int_0^x \left(\int_0^y \left(\int_0^x \frac{1}{\sqrt{1+\sin^2 t}}\,dt\right) dz\right) dy.$$

16. We can choose

$$f(x) = \frac{x^{(1/n)+1}}{\dfrac{1}{n}+1}.$$

Then

$$\int_0^b \sqrt[n]{x}\,dx = f(b) - f(0) = \frac{b^{(1/n)+1}}{\dfrac{1}{n}+1}.$$

CHAPTER 15

1. (i)

$$\frac{1}{1+\arctan^2(\arctan x)} \cdot \frac{1}{1+\arctan^2 x} \cdot \frac{1}{1+x^2}.$$

(iii)

$$\frac{1}{1+(\tan x \arctan x)^2} \cdot \left(\sec^2 x \arctan x + \frac{\tan x}{1+x^2}\right).$$

2. (i) 0.
   (iii) 0.
   (v) 0.
7. (a)

$$\sin 2x = \sin(x+x) = \sin x \cos x + \cos x \sin x = 2\sin x \cos x.$$
$$\cos 2x = \cos^2 x - \sin^2 x = 2\cos^2 x - 1 = 1 - 2\sin^2 x.$$
$$\begin{aligned}\sin 3x &= \sin(2x+x) = \sin 2x \cos x + \cos 2x \sin x\\ &= 2\sin x \cos^2 x + (\cos^2 x - \sin^2 x)\sin x\\ &= 3\sin x \cos^2 x - \sin^3 x.\end{aligned}$$
$$\begin{aligned}\cos 3x = \cos(2x+x) &= \cos 2x \cos x - \sin 2x \sin x\\ &= (\cos^2 x - \sin^2 x)\cos x = 2\sin^2 x \cos x\\ &= \cos^3 x - \sin^2 x \cos x - 2\sin^2 x \cos x\\ &= \cos^3 x - 3\sin^2 x \cos x\\ &= 4\cos^3 x - 3\cos x.\end{aligned}$$

(b) Since $\cos \pi/4 > 0$ and

$$0 = \cos\frac{\pi}{2} = \cos 2\cdot\frac{\pi}{4} = 2\cos^2\frac{\pi}{4} - 1,$$

we have $\cos \pi/4 = \sqrt{2}/2$. It follows, since $\sin \pi/4 > 0$ and $\sin^2 + \cos^2 = 1$, that $\sin \pi/4 = \sqrt{2}/2$, and consequently $\tan \pi/4 = 1$. Similarly, since $\cos \pi/6 > 0$ and

$$0 = \cos\frac{\pi}{2} = \cos 3\cdot\frac{\pi}{6} = 4\cos^3\frac{\pi}{6} - 3\cos\frac{\pi}{6},$$

we have $\cos \pi/6 = \sqrt{3}/2$. It follows, since $\sin \pi/6 > 0$, that $\sin \pi/6 = \sqrt{1-(\sqrt{3}/2)^2} = \frac{1}{2}$.

9. (a)

$$\begin{aligned}\tan(x+y) &= \frac{\sin(x+y)}{\cos(x+y)}\\ &= \frac{\sin x\cos y + \cos x \sin y}{\cos x \cos y - \sin x \sin y}\\ &= \frac{\dfrac{\sin x\cos y}{\cos x\cos y} + \dfrac{\cos x\sin y}{\cos x \cos y}}{\dfrac{\cos x\cos y}{\cos x\cos y} - \dfrac{\sin x \sin y}{\cos x\cos y}}\\ &= \frac{\tan x + \tan y}{1 - \tan x \tan y}.\end{aligned}$$

(b) From part (a) we have

$$\begin{aligned}\tan(\arctan x + \arctan y) &= \frac{\tan(\arctan x) + \tan(\arctan y)}{1 - \tan(\arctan x)\tan(\arctan y)}\\ &= \frac{x+y}{1-xy},\end{aligned}$$

provided that $\arctan x$, $\arctan y$, and $\arctan x + \arctan y \neq k\pi + \pi/2$. Since $-\pi/2 < \arctan x, \arctan y < \pi/2$, this is always the case except when $\arctan x + \arctan y = \pm\pi/2$, which is equivalent to $xy = 1$. From this equation we can conclude that

$$\arctan x + \arctan y = \arctan\left(\frac{x+y}{1-xy}\right)$$

provided that $\arctan x + \arctan y$ lies in $(-\pi/2, \pi/2)$, which is true whenever $xy < 1$. (If $x, y > 0$ and $xy > 1$, so that $\arctan x + \arctan y > \pi/2$, then we must add π to the right side, and if $x, y < 0$ and $xy > 1$, so that $\arctan x + \arctan y < -\pi/2$, then we must subtract π.)

11. The first formula is derived by subtracting the second of the following two equations from the first:

$$\begin{aligned}\cos(m-n)x &= \cos(mx-nx) = \cos mx \cos(-nx) - \sin mx \sin(-nx)\\ &= \cos mx \cos nx + \sin mx \sin nx,\\ \cos(m+n)x &= \cos mx \cos nx - \sin mx \sin nx.\end{aligned}$$

The other formulas are derived similarly.

12. It follows from Problem 11 that if $m \neq n$, then

$$\begin{aligned}\int_{-\pi}^{\pi} \sin mx \sin nx\, dx &= \frac{1}{2}\int_{-\pi}^{\pi} [\cos(m-n)x - \cos(m+n)x]\, dx\\ &= \frac{1}{2}\left\{\left[\frac{\sin(m-n)\pi}{m-n} - \frac{\sin(m+n)\pi}{m+n}\right]\right.\\ &\qquad \left. - \left[\frac{\sin(m-n)\pi}{m-n} - \frac{\sin(m+n)\pi}{m+n}\right]\right\}\\ &= 0.\end{aligned}$$

But if $m = n$, then

$$\begin{aligned}\int_{-\pi}^{\pi} \sin mx \sin nx\, dx &= \frac{1}{2}\int_{-\pi}^{\pi} 1 - \cos(m+n)x\, dx\\ &= \frac{1}{2}\{[\pi - \cos(m+n)\pi] - [-\pi - \cos(m+n)\pi]\}\\ &= \pi.\end{aligned}$$

The other formulas are proved similarly.

15. (a) We have

$$\begin{aligned}\cos 2x &= \cos^2 x - \sin^2 x\\ &= 1 - 2\sin^2 x\\ &= 2\cos^2 x - 1.\end{aligned}$$

So

$$\begin{aligned}\sin^2 x &= \frac{1-\cos 2x}{2},\\ \cos^2 x &= \frac{1+\cos 2x}{2}.\end{aligned}$$

(b) These formulas follow from part (a), because $\cos x/2 \geq 0$ and $\sin x/2 \geq 0$ (since $0 \leq x \leq \pi/2$).

(c)

$$\int_a^b \sin^2 x\, dx = \int_a^b \frac{1-\cos 2x}{2}\, dx = \frac{1}{2}(b-a) - \frac{1}{4}(\sin 2b - \sin 2a).$$

$$\int_a^b \cos^2 x\, dx = \int_a^b \frac{1+\cos 2x}{2}\, dx = \frac{1}{2}(b-a) + \frac{1}{4}(\sin 2b - \sin 2a).$$

19. (a) $\arctan 1 - \arctan 0 = \pi/4$.

(b) $\lim_{x\to\infty} \arctan x - \arctan 0 = \pi/2$.

20. $\lim_{x\to\infty} x \sin\frac{1}{x} = \lim_{x\to 0^+} \frac{1}{x}\sin x = 1$.

21. (a)
$$(\sin^\circ)'(x) = \frac{\pi}{180}\cos\left(\frac{\pi x}{180}\right) = \frac{\pi}{180}\cos^\circ(x).$$
$$(\cos^\circ)'(x) = \frac{\pi}{180}\cdot -\sin\left(\frac{\pi x}{180}\right) = \frac{-\pi}{180}\sin^\circ(x).$$

(b) $\lim_{x\to 0}\frac{\sin^\circ x}{x} = \lim_{x\to 0}\frac{\sin(\pi x/180)}{x} = \lim_{x\to 0}\frac{\pi}{180}\cdot\frac{\sin(\pi x/180)}{\pi x/180} = \frac{\pi}{180}$.

$\lim_{x\to\infty} x\sin^\circ\frac{1}{x} = \lim_{x\to 0^+}\frac{\sin^\circ x}{x} = \frac{\pi}{180}$.

CHAPTER 18

1. (i) $e^{e^{e^{e^x}}}\cdot e^{e^{e^x}}\cdot e^{e^x}\cdot e^x$.

(iii) $(\sin x)^{\sin(\sin x)}[(\log(\sin x))\cdot\cos(\sin x)\cdot\cos x + (\cos x/\sin x)\cdot\sin(\sin x)]$

(v) $\sin x^{\sin x^{\sin x}}[(\log(\sin x))\cdot\sin x^{\sin x} \cdot\{(\log(\sin x))\cdot\cos x + (\cos x/\sin x)\cdot\sin x\} + (\cos x/\sin x)\cdot\sin x^{\sin x}]$.

(vii)
$$\left[\arcsin\left(\frac{x}{\sin x}\right)\right]^{\log(\sin e^x)}\left[\left(\log\left(\arcsin\left(\frac{x}{\sin x}\right)\right)\right)\cdot\frac{(\cos e^x)e^x}{\sin e^x} + \log(\sin e^x)\cdot\frac{\sin x - x\cos x}{\arcsin\left(\frac{x}{\sin x}\right)\sqrt{1-\left(\frac{x}{\sin x}\right)^2}\cdot\sin^2 x}\right].$$

(ix) $(\log x)^{\log x}\cdot\left[\log(\log x)\cdot\frac{1}{x} + \log x\cdot\frac{1}{\log x}\cdot\frac{1}{x}\right]$.

5. (i) 0.

(iii) $\frac{1}{6}$.

(v) $\frac{1}{3}$.

7. (a)
$$\begin{aligned}\cosh^2 x - \sinh^2 x &= \left(\frac{e^x+e^{-x}}{2}\right)^2 - \left(\frac{e^x-e^{-x}}{2}\right)^2\\ &= \left[\frac{e^{2x}}{4}+\frac{1}{2}+\frac{e^{-2x}}{4}\right] - \left[\frac{e^{2x}}{4}-\frac{1}{2}+\frac{e^{-2x}}{4}\right]\\ &= 1.\end{aligned}$$

(c)

$$\sinh x \cosh y + \cosh x \sinh y = \left(\frac{e^x - e^{-x}}{2}\right)\left(\frac{e^y + e^{-y}}{2}\right) + \left(\frac{e^x + e^{-x}}{2}\right)\left(\frac{e^y - e^{-y}}{2}\right)$$

$$= \left[\frac{e^{x+y}}{4} + \frac{e^{-x-y}}{4} - \frac{e^{-x+y}}{4} + \frac{e^{x-y}}{4}\right] + \left[\frac{e^{x+y}}{4} + \frac{e^{-x-y}}{4} + \frac{e^{-x+y}}{4} - \frac{e^{-x-y}}{4}\right]$$

$$= \frac{e^{x+y} + e^{-(x+y)}}{2} = \sinh(x+y).$$

(e) Since

$$\sinh x = \frac{e^x + e^{-x}}{2},$$

we have

$$\sinh'(x) = \frac{e^x - e^{-x}}{2} = \cosh x.$$

(g) Since

$$\tanh x = \frac{\sinh x}{\cosh x},$$

we have

$$\tanh'(x) = \frac{(\cosh x)^2 - (\sinh x)^2}{\cosh^2 x}$$

$$= \frac{1}{\cosh^2 x} \quad \text{by part (a).}$$

8. (a) If $y = \arg\cosh x$, then $x \geq 0$ and

$$x = \cosh y = \sqrt{1 + \sinh^2 y} \quad \text{by part (a).}$$

So

$$\sinh(\arg\cosh x) = \sinh y = \sqrt{x^2 - 1} \quad \text{since } \sinh y \geq 0 \text{ for } y \geq 0.$$

(c)

$$(\arg\sinh)'(x) = \frac{1}{\sinh'(\arg\sinh(x))}$$

$$= \frac{1}{\cosh(\arg\sinh(x))}$$

$$= \frac{1}{\sqrt{1+x^2}} \quad \text{by part (b).}$$

(e)

$$(\arg\tanh)'(x) = \frac{1}{\tanh'(\arg\tanh(x))},$$

$$= \cosh^2(\arg\tanh(x)).$$

Now,

$$\tanh^2 y + \frac{1}{\cosh^2 y} = 1 \quad \text{by Problem 7(b),}$$

so

$$\tanh^2(\text{arg tanh}(x)) + \frac{1}{\cosh^2(\text{arg tanh}(x))} = 1,$$

or

$$\cosh^2(\text{arg tanh}(x)) = \frac{1}{1-x^2}.$$

9. (a) If $y = \text{arg sinh}\, x$, then

$$x = \sinh y = \frac{e^y - e^{-y}}{2}$$

so

$$\begin{aligned} e^y - e^{-y} &= 2x, \\ e^{2y} - 2xe^y - 1 &= 0, \\ e^y &= \frac{2x \pm \sqrt{4x^2+4}}{2} \end{aligned}$$

so

$$e^y = x + \sqrt{1+x^2} \quad \text{since } e^y > 0$$

or

$$y = \text{arg sinh}\, x = \log(x + \sqrt{1+x^2}).$$

Similarly,

$$\begin{aligned} \text{arg cosh}\, x &= \log(x + \sqrt{x^2-1}), \\ \text{arg tanh}\, x &= \tfrac{1}{2}\log(1+x) - \tfrac{1}{2}\log(1-x). \end{aligned}$$

(b)

$$\begin{aligned} \int_a^b \frac{1}{\sqrt{1+x^2}}\,dx &= \text{arg sinh}\, b - \text{arg sinh}\, a \quad \text{by Problem 8(c)} \\ &= \log(b + \sqrt{1+b^2}) - \log(a + \sqrt{1+a^2}). \\ \int_a^b \frac{1}{\sqrt{x^2-1}}\,dx &= \log(b + \sqrt{b^2-1}) - \log(a + \sqrt{a^2-1}). \end{aligned}$$

$$\int_a^b \frac{1}{1-x^2}\,dx = \frac{1}{2}[\log(1+b) - \log(1-b) - \log(1+a) + \log(1-a)].$$

12. (a) $\lim_{x\to\infty} a^x = \lim_{x\to\infty} e^{x\log a}$. Since $\log a < 0$, we have $\lim_{x\to\infty} x\log a = -\infty$, so $\lim_{x\to\infty} e^{x\log a} = 0$.

(c) $\displaystyle\lim_{x\to\infty} \frac{(\log x)^n}{x} = \lim_{y\to\infty} \frac{y^n}{e^y} = 0.$

(e) $\lim_{x\to 0^+} x^x = \lim_{x\to 0^+} e^{x\log x}$. Now, $\lim_{x\to 0^+} x\log x = 0$ by part (d), so $\lim_{x\to 0^+} x^x = 1$.

16. (a) $\lim_{y\to 0} \log(1+y)/y = \log'(1) = 1.$

(b) $\lim_{x\to\infty} x\log(1+1/x) = \lim_{y\to 0^+} \log(1+y)/y = 1.$

(c)

$$\begin{aligned} e = \exp(1) &= \exp(\lim_{x\to\infty} x\log(1+1/x)) \\ (*) &= \lim_{x\to\infty} \exp(x\log(1+1/x)) \\ &= \lim_{x\to\infty} (1+1/x)^x. \end{aligned}$$

(The starred equality depends on the continuity of exp at 1, and can be justified as follows. For every $\varepsilon > 0$ there is some $\delta > 0$ such that $|e - \exp y| < \varepsilon$ for $|y - 1| < \delta$. Moreover, there is some N such that $|x\log(1+1/x) - 1| < \delta$ for $x > N$. So $|e - \exp(x\log(1+1/x))| < \varepsilon$ for $x > N$.

(d)

$$\begin{aligned} e^a &= [\lim_{x\to\infty} (1+1/x)^x]^a = \lim_{x\to\infty} (1+1/x)^{ax} \\ &= \lim_{ax\to\infty} (1+1/x)^{ax} \\ &= \lim_{y\to\infty} (1+a/y)^y. \end{aligned}$$

18. After one year the number of dollars yielded by an initial investment of one dollar will be

$$\lim_{x\to\infty} (1+a/100x)^x = e^{a/100}.$$

19. (a) Clearly $f'(x) = 1/x$ for $x > 0$. If $x < 0$, then $f(x) = \log(-x)$, so $f'(x) = (-1)\cdot 1/(-x) = 1/x$.

(b) We can write $\log|f|$ as $g \circ f$ where $g(x) = \log|x|$ is the function of part (a). So $(\log|f|)' = (g' \circ f)\cdot f' = 1/f \cdot f'$.

20. (c) Let $g(x) = f(x)/e^{cx}$. Then

$$g'(x) = \frac{e^{cx}f'(x) - f(x)ce^{cx}}{e^{2cx}} = 0,$$

so there is some number k such that $g(x) = k$ for all x.

21. (a) According to Problem 20, there is some k such that $A(t) = ke^{ct}$. Then $k = ke^{0\cdot t} = A_0$. So $A(t) = A_0e^{ct}$.

(b) If $A(t+\tau) = A(t)/2$, then

$$A_0e^{ct+c\tau} = \frac{A_0e^{ct}}{2},$$

so $e^{ct}e^{c\tau} = e^{ct}/2$ or $e^{c\tau} = \frac{1}{2}$, so $\tau = -(\log 2)/c$. It is easy to check that this τ does work.

22. Newton's law states that, for a certain (positive) number c,

$$T'(t) = c(T - M),$$

which can be written

$$(T - M)' = c(T - M).$$

So by Problem 20 there is some number k such that

$$T(t) - M = ke^{ct},$$

and $k = ke^{0 \cdot t} = T(0) = T_0$. So $T(t) = M + T_0 e^{ct}$.

CHAPTER 19

1. (i) $\left(\sqrt[5]{x^3} + \sqrt[6]{x}\right)/\sqrt{x} = x^{1/10} + x^{-1/3}$.

(ii) $\dfrac{1}{\sqrt{x-1}+\sqrt{x+1}} = \dfrac{\sqrt{x-1}-\sqrt{x+1}}{-2}$.

(iii) $(e^x + e^{2x} + e^{3x})/e^{4x} = e^{-3x} + e^{-2x} + e^{-x}$.

(iv) $a^x/b^x = (a/b)^x = e^{x\log(a/b)}$.

(v) $\tan^2 x = \sec^2 x - 1$.

(vi) $\dfrac{1}{a^2+x^2} = \dfrac{1/a^2}{1+\left(\dfrac{x}{a}\right)^2}$.

(vii) $\dfrac{1}{\sqrt{a^2-x^2}} = \dfrac{1/a}{\sqrt{1-(x/a)^2}}$.

(viii) $\dfrac{1}{1+\sin x} = \dfrac{1-\sin x}{1-\sin^2 x} = \dfrac{1-\sin x}{\cos^2 x} = \sec^2 x - \sec x \tan x$.

(ix) $\dfrac{8x^2+6x+4}{x+1} = 8x - 2 + \dfrac{6}{x+1}$.

(x) $\dfrac{1}{\sqrt{2x-x^2}} = \dfrac{1}{\sqrt{1-(x-1)^2}}$.

2. (i) $-\cos e^x$. (Let $u = e^x$.)

(iii) $(\log x)^2/2$. (Let $u = \log x$.)

(v) e^{e^x}. (Let $u = e^{e^x}$.)

(vii) $2e^{\sqrt{x}}$. (Let $u = \sqrt{x}$.)

(ix) $-(\log(\cos x))^2/2$. (Let $u = \log(\cos x)$.)

3. (i)
$$\begin{aligned}\int x^2 e^x\,dx &= x^2e^x - \int 2xe^x\,dx = x^2e^x - \left[2xe^x - \int e^x\,dx\right]\\ &= x^2e^x - 2xe^x + 2e^x.\end{aligned}$$

(iii) We have

$$\begin{aligned}\int e^{ax}\sin bx\,dx &= \frac{e^{ax}\sin bx}{a} - \frac{b}{a}\int e^{ax}\cos bx\,dx\\ &= \frac{e^{ax}\sin bx}{a} - \frac{b}{a}\left[\frac{e^{ax}\cos bx}{a} - \frac{b}{a}\int e^{ax}(-\sin bx)\,dx\right],\end{aligned}$$

so

$$\int e^{ax}\sin bx\,dx = \frac{a}{a^2+b^2}e^{ax}\sin bx - \frac{b}{a^2+b^2}e^{ax}\cos bx.$$

(v) Using the result $\int (\log x)^2\,dx = x(\log x)^2 - 2x(\log x) + 2x$ from the text, we have

$$\begin{aligned}\int (\log x)^3\,dx &= [x(\log x)^2 - 2x(\log x) + 2x]\log x \\ &\qquad - \int \frac{1}{x}[x(\log x)^2 - 2x(\log x) + 2x]\,dx \\ &= x(\log x)^3 - 2x(\log x)^2 + 2x\log x \\ &\qquad - \int (\log x)^2\,dx + 2[x\log x - x] - 2x \\ &= x(\log x)^3 - 2x(\log x)^2 + 2x\log x \\ &\qquad - [x(\log x)^2 - 2x(\log x) + 2x] + 2[x\log x - x] - 2x \\ &= x(\log x)^3 - 3x(\log x)^2 + 6x\log x - 6x.\end{aligned}$$

(vii)

$$\begin{aligned}\int \sec^3 x\,dx &= \int (\sec^2 x)(\sec x)\,dx = \tan x\sec x - \int (\tan x)(\sec x\tan x)\,dx \\ &= \tan x\sec x - \int \sec x(\sec^2 x - 1)\,dx \\ &= \tan x\sec x - \int \sec^3 x\,dx + \int \sec x\,dx,\end{aligned}$$

so

$$\int \sec^3 x\,dx = \tfrac{1}{2}[\tan x\sec x + \log(\sec x + \tan x)].$$

(ix)

$$\begin{aligned}\int \sqrt{x}\log x\,dx &= \frac{2x^{3/2}}{3}\log x - \frac{2}{3}\int x^{3/2}\cdot\frac{1}{x}\,dx \\ &= \frac{2x^{3/2}}{3}\log x - \frac{2}{3}\int x^{1/2}\,dx \\ &= \frac{2x^{3/2}}{3}\log x - \frac{4}{9}x^{3/2}.\end{aligned}$$

4. (i) Let $x = \sin u$, $dx = \cos u\,du$. The integral becomes

$$\int \frac{\cos u\,du}{\sqrt{1 - \sin^2 u}} = \int 1\,du = u = \arcsin x.$$

(iii) Let $x = \sec u$, $dx = \sec u\tan u\,du$. The integral becomes

$$\begin{aligned}\int \frac{\sec u\tan u\,du}{\sqrt{\sec^2 u - 1}} &= \int \sec u\,du = \log(\sec u + \tan u) \\ &= \log(x + \sqrt{x^2 - 1}).\end{aligned}$$

(v) Let $x = \sin u$, $dx = \cos u\, du$. The integral becomes

$$\int \frac{\cos u\, du}{\sin u\sqrt{1-\sin^2 u}} = \int \csc u\, du = -\log(\csc u + \cot u)$$
$$= -\log\left(\frac{1}{x} + \frac{\sqrt{1-x^2}}{x}\right).$$

(vii) Let $x = \sin u$, $dx = \cos u\, du$. The integral becomes

$$\int (\sin^3 u \cos u)\cos u\, du = \int \sin^3 u \cos^2 u\, du = \int (\sin u)(1-\cos^2 u)\cos^2 u\, du$$
$$= \int (\sin u)(\cos^2 u - \cos^4 u)\, du = -\frac{\cos^3 u}{3} + \frac{\cos^5 u}{5}$$
$$= -\frac{(1-x^2)^{3/2}}{3} + \frac{(1-x^2)^{5/2}}{5}.$$

(ix) Let $x = \tan u$, $dx = \sec^2 u\, du$. The integral becomes

$$\int \sec u \sec^2 u\, du = \int \sec^3 u\, du$$
$$= \tfrac{1}{2}[\tan u \sec u + \log(\sec u + \tan u)] \quad \text{by Problem 3(vii)}$$
$$= \tfrac{1}{2}[x\sqrt{1+x^2} + \log(x + \sqrt{1+x^2})].$$

5\. (i) Let $u = \sqrt{x+1}$, $x = u^2 - 1$, $dx = 2u\, du$. The integral becomes

$$\int \frac{2u\, du}{1+u} = \int \left(2 + \frac{-2}{1+u}\right) du$$
$$= 2u - 2\log(1+u) = 2\sqrt{x+1} - 2\log(1+\sqrt{x+1}).$$

(iii) Let $u = x^{1/6}$, $x = u^6$, $dx = 6u^5\, du$. The integral becomes

$$\int \frac{6u^5\, du}{u^3+u^2} = 6\int \left(u^2 - u + 1 - \frac{1}{u+1}\right) du = 2u^3 - 3u^2 + 6u - 6\log(u+1)$$
$$= 2\sqrt{x} - 3\sqrt[3]{x} + 6\sqrt[6]{x} - 6\log(\sqrt[6]{x} + 1).$$

(v) Let $u = \tan x$, $x = \arctan u$, $dx = du/(1+u^2)$. The integral becomes

$$\int \frac{du}{(1+u^2)(2+u)} = \frac{1}{5}\int \left(\frac{1}{2+u} - \frac{u-2}{1+u^2}\right) du$$
$$= \frac{1}{5}\int \frac{du}{2+u} - \frac{1}{10}\int \frac{2u}{1+u^2}\, du + \frac{2}{5}\int \frac{du}{1+u^2}$$
$$= \frac{1}{5}\log(2+u) - \frac{1}{10}\log(1+u^2) + \frac{2}{5}\arctan u$$
$$= \frac{1}{5}\log(2+\tan x) - \frac{1}{10}\log(1+\tan^2 x) + \frac{2}{5}x.$$

(vii) Let $u = 2^x$, $x = (\log u)/(\log 2)$, $dx = du/(u\log 2)$. The integral becomes

$$\begin{aligned}\frac{1}{\log 2}\int \frac{u^2+1}{(u+1)u}\,du &= \frac{1}{\log 2}\int\left(1+\frac{1-u}{u(u+1)}\right)du\\ &= \frac{1}{\log 2}\int\left(1+\frac{1}{u}-\frac{2}{u+1}\right)du\\ &= \frac{1}{\log 2}[u+\log u - 2\log(u+1)]\\ &= \frac{1}{\log 2}[2^x + x\log 2 - 2\log(2^x+1)].\end{aligned}$$

(ix) Let $u = \sqrt{x}$, $x = u^2$, $dx = 2u$. The integral becomes

$$\int \frac{\sqrt{1-u^2}\,2u\,du}{1-u}.$$

Now let $u = \sin y$, $du = \cos y\,dy$. The integral becomes

$$\begin{aligned}\int \frac{2\cos y\sin y\cos y}{1-\sin y}\,dy &= 2\int\frac{(1-\sin^2 y)\sin y}{1-\sin y}\,dy\\ &= 2\int(1+\sin y)\sin y\,dy\\ &= 2\int \sin y\,dy + \int 1-\cos 2y\,dy\\ &= -2\cos y + y - \frac{\sin 2y}{2} = -2\cos y + y - \sin y\cos y\\ &= -2\sqrt{1-u^2} + \arcsin u - u\sqrt{1-u^2}\\ &= -2\sqrt{1-x} + \arcsin\sqrt{x} - \sqrt{x}\sqrt{1-x}.\end{aligned}$$

The substitution $u = \sqrt{1-x}$, $x = 1-u^2$, $dx = -2u\,du$ leads to

$$\int \frac{-2u^2\,du}{1-\sqrt{1-u^2}}$$

and the substitution $u = \sin y$ then leads to

$$\begin{aligned}\int \frac{-2\sin^2 y\cos y\,dy}{1-\cos y} &= -2\sin y - y - \sin y\cos y\\ &= -2u - \arcsin u - u\sqrt{1-u^2}\\ &= -2\sqrt{1-x} - \arcsin\sqrt{1-x} - \sqrt{1-x}\sqrt{x}.\end{aligned}$$

These answers agree, since

$$\arcsin\sqrt{x} = \frac{\pi}{2} - \arcsin\sqrt{1-x}$$

(check this by comparing their derivatives and their values for $x = 0$).

6. In these problems I will denote the original integral.

(i)

$$I = \int \frac{2}{x-1}\,dx + \int \frac{3}{(x+1)^2}\,dx$$
$$= 2\log(x-1) - \frac{3}{x+1}.$$

(iii)

$$I = \int \frac{1}{(x-1)^2}\,dx + \int \frac{4}{(x+1)^3}\,dx$$
$$= -\frac{1}{(x-1)} - \frac{2}{(x+1)^2}.$$

(v)

$$I = \frac{1}{2}\int \frac{2x}{x^2+1}\,dx + \int \frac{4}{x^2+1}\,dx$$
$$= \tfrac{1}{2}\log(x^2+1) + 4\arctan x.$$

(vii)

$$I = \int \frac{1}{(x+1)}\,dx + \int \frac{2x}{(x^2+x+1)}\,dx$$
$$= \int \frac{1}{x+1}\,dx + \int \frac{2x+1}{x^2+x+1}\,dx - \int \frac{1}{x^2+x+1}\,dx.$$

Now

$$\int \frac{1}{x^2+x+1}\,dx = \int \frac{1}{(x+\frac{1}{2})^2+\frac{3}{4}}\,dx$$
$$= \frac{4}{3}\int \frac{1}{\left[\frac{2}{\sqrt{3}}\left(x+\frac{1}{2}\right)\right]^2+1}\,dx$$
$$= \frac{4}{3}\cdot\frac{\sqrt{3}}{2}\arctan\left(\tfrac{2}{\sqrt{3}}\left(x+\tfrac{1}{2}\right)\right)$$
$$= \frac{2\sqrt{3}}{3}\arctan\left(\tfrac{2}{\sqrt{3}}\left(x+\tfrac{1}{2}\right)\right),$$

so

$$I = \log(x+1) + \log(x^2+x+1) - \frac{2\sqrt{3}}{3}\arctan\left(\tfrac{2}{\sqrt{3}}\left(x+\tfrac{1}{2}\right)\right).$$

(ix)

$$I = \int \frac{2x+1}{(x^2+x+1)^2}\,dx - \int \frac{1}{(x^2+x+1)^2}\,dx$$
$$= \int \frac{2x+1}{(x^2+x+1)^2}\,dx - \frac{16}{9}\int \frac{1}{\left(\left[\frac{2}{\sqrt{3}}\left(x+\frac{1}{2}\right)\right]^2+1\right)^2}\,dx.$$

Now the substitution

$$u = \tfrac{2}{\sqrt{3}}\left(x+\tfrac{1}{2}\right), \quad dx = \tfrac{\sqrt{3}}{2}\,du$$

changes the second integral to

$$-\frac{16}{9}\cdot\frac{\sqrt{3}}{2}\int\frac{du}{(u^2+1)^2}.$$

Using the reduction formula, this can be written

$$-\frac{8\sqrt{3}}{9}\left[\frac{u}{2(u^2+1)}+\frac{1}{2}\int\frac{du}{u^2+1}\right]=-\frac{8\sqrt{3}}{9}\left[\frac{\log(u^2+1)}{4}+\frac{1}{2}\arctan u\right],$$

so

$$I=-\frac{1}{x^2+x+1}-\frac{2\sqrt{3}}{9}\log\left(\frac{4}{3}(x^2+x+1)\right)$$
$$-\frac{4\sqrt{3}}{9}\arctan\left(\tfrac{2}{\sqrt{3}}\left(x+\tfrac{1}{2}\right)\right).$$

13. The equation $\int e^x\sin x\,dx=e^x\sin x-e^x\cos x-\int e^x\sin x\,dx$ means that any function F with $F'(x)=e^x\sin x$ can be written $F(x)=e^x\sin x-e^x\cos x-G(x)$ where G is another function with $G'(x)=e^x\sin x$. Of course, $G=F+c$ for some number c, but it is not necessarily true that $F=G$.

15. (a)

$$\int\arcsin x\,dx=\int 1\cdot\arcsin x\,dx=x\arcsin x-\int\frac{x}{\sqrt{1-x^2}}\,dx$$
$$=x\arcsin x+\sqrt{1-x^2}.$$

16. (a)

$$\int\sin^4 x\,dx=-\frac{\sin^3 x\cos x}{4}+\frac{3}{4}\int\sin^2 x\,dx$$
$$=-\frac{\sin^3 x\cos x}{4}+\frac{3}{4}\left[-\frac{\sin x\cos x}{2}+\frac{1}{2}\int 1\,dx\right]$$
$$=-\frac{\sin^3 x\cos x}{4}-\frac{3\sin x\cos x}{8}+\frac{3}{8}x.$$

$$\int\sin^4 x\,dx=\int\left(\frac{1-\cos 2x}{2}\right)^2dx=\int\left(\frac{1}{4}-\frac{\cos 2x}{2}+\frac{\cos^2 2x}{4}\right)dx$$
$$=\frac{x}{4}-\frac{\sin 2x}{4}+\frac{1}{4}\int\frac{1+\cos 4x}{2}\,dx$$
$$=\frac{x}{4}-\frac{\sin 2x}{4}+\frac{1}{4}\left[\frac{x}{2}+\frac{\sin 4x}{8}\right]$$
$$=\frac{3x}{8}-\frac{\sin 2x}{4}+\frac{\sin 4x}{32}.$$

(b) It follows that these two answers are the same, since they have the same value for $x=0$.

20. (a)

$$\begin{aligned}\sin^n x\,dx &= \int (\sin x)(\sin^{n-1} x)\,dx \\ &= -\cos x \sin^{n-1} x + (n-1)\int \cos x(\sin^{n-2} x)\cos x\,dx \\ &= -\cos x \sin^{n-1} x + (n-1)\int (\sin^{n-2} x - \sin^n x)\,dx,\end{aligned}$$

so

$$\int \sin^n x\,dx = -\frac{1}{n}\cos x \sin^{n-1} x + \frac{n-1}{n}\int \sin^{n-2} x\,dx.$$

(b)

$$\begin{aligned}\int \cos^n x\,dx &= \int (\cos x)(\cos^{n-1} x)\,dx \\ &= \sin x \cos^{n-1} x + (n-1)\int \sin x(\cos^{n-2} x)\sin x\,dx \\ &= \sin x \cos^{n-1} x + (n-1)\int (\cos^{n-2} x - \cos^n x)\,dx,\end{aligned}$$

so

$$\int \cos^n x\,dx = \frac{1}{n}\sin x \cos^{n-1} x + \frac{n-1}{n}\int \cos^{n-2} x\,dx.$$

(c)

$$\begin{aligned}\int \frac{dx}{(1+x^2)^n} &= \int \frac{dx}{(1+x^2)^{n-1}} - \int \frac{x^2\,dx}{(1+x^2)^n} \\ &= \int \frac{dx}{(1+x^2)^{n-1}} - \int x \cdot \frac{x}{(1+x^2)^n}\,dx \\ &= \int \frac{dx}{(1+x^2)^{n-1}} - \left[\frac{x}{2(1-n)(1+x^2)^{n-1}} \right. \\ &\qquad \left. - \int \frac{dx}{2(1-n)(1+x^2)^{n-1}}\right]\end{aligned}$$

so

$$\int \frac{dx}{(1+x^2)^n} = \frac{1}{2(n-1)}\,\frac{x}{(x^2+1)^{n-1}} - \frac{(2n-3)}{2(n-1)}\int \frac{1}{(x^2+1)^{n-1}}\,dx.$$

We can also use the substitution $x = \tan u$, $dx = \sec^2 u\,du$, which changes the integral to

$$\begin{aligned}\int \frac{\sec^2 u\,du}{\sec^{2n} u} &= \int \cos^{2n-2} u\,du \\ &= \frac{1}{2n-2}\cos^{2n-3} u \sin u + \frac{2n-3}{2n-2}\int \cos^{2n-4} u\,du \\ &= \frac{1}{2n-2}\cdot\frac{1}{(\sqrt{1+x^2})^{2n-3}}\cdot\frac{x}{\sqrt{1+x^2}} + \frac{2n-3}{2n-2}\int\frac{dx}{(1+x^2)^{n-1}} \\ &= \frac{1}{2(n-1)}\frac{x}{(1+x^2)^{n-1}} + \frac{2n-3}{2n-2}\int\frac{dx}{(1+x^2)^{n-1}}.\end{aligned}$$

CHAPTER 20

1. (i) $P_{3,0}(x) = e + ex + ex^2 + (5e/3!)x^3$.

   (iii) $P_{2n,\pi/2}(x) = 1 - \dfrac{(x-\pi/2)^2}{2!} + \dfrac{(x-\pi/2)^4}{4!} - \cdots + \dfrac{(-1)^n(x-\pi/2)^{2n}}{(2n)!}$.

   (v) $P_{n,1}(x) = e + e(x-1) + \dfrac{e(x-1)^2}{2!} + \cdots + \dfrac{e(x-1)^n}{n!}$.

   (vii) $P_{4,0}(x) = x + x^3$.

   (ix) $P_{2n+1,0}(x) = 1 - x^2 + x^4 - \cdots + (-1)^n x^{2n}$.

2. If f is a polynomial function of degree n, then $f^{(n+1)} = 0$. It follows from Taylor's Theorem that $R_{n,a}(x) = 0$, so $f(x) = P_{n,a}(x)$.

   (i) $-12 + 2(x-3) + (x-3)^2$.

   (iii) $243 + 405(x-3) + 270(x-3)^2 + 90(x-3)^3 + 15(x-3)^4 + (x-3)^5$.

3. (i) $\displaystyle\sum_{i=0}^{9}\frac{(-1)^i}{(2i+1)!}\left(\text{since } \frac{1}{(2n+2)!} < 10^{-17} \text{ for } 2n+2 \geq 19, \text{ or } n \geq 9\right)$.

   (iii) $\displaystyle\sum_{i=0}^{8}\frac{(-1)^i}{2^i(2i+1)!}\left(\text{since } \frac{1}{2^{2n+2}(2n+2)!} < 10^{-20} \text{ for } 2n+2 \geq 18, \text{ or } n \geq 8\right)$.

   (v) $\displaystyle\sum_{i=0}^{11}\frac{2^i}{i!}\left(\text{since } \frac{3^2\cdot 2^{n+1}}{(n+1)!} < 10^{-5} \text{ for } n+1 \geq 12, \text{ or } n \geq 11\right)$.

8. (i) $c_i = a_i + b_i$.

   (iii) $c_i = (i+1)a_i$.

   (v) $c_0 = \displaystyle\int_0^a f(t)\,dt$; $c_i = a_{i-1}/i$ for $i > 0$.

CHAPTER 22

1. (i) $1 - n/(n+1) = 1/(n+1) < \varepsilon$ for $n+1 > 1/\varepsilon$.

   (iii) $\displaystyle\lim_{n\to\infty}\sqrt[8]{n^2+1} - \sqrt[4]{n+1} = \lim_{n\to\infty}\left(\sqrt[8]{n^2+1} - \sqrt[8]{n^2}\right) + \lim_{n\to\infty}\left(\sqrt[4]{n} - \sqrt[4]{n+1}\right)$ $= 0 + 0 = 0$. (Each of these two limits can be proved in the same way that $\displaystyle\lim_{n\to\infty}\left(\sqrt{n+1} - \sqrt{n}\right) = 0$ was proved in the text.)

(v) Clearly $\lim_{n\to\infty}(\log a)/n = 0$. So $\lim_{n\to\infty}\sqrt[n]{a} = \lim_{n\to\infty} e^{(\log a)/n} = e^0$ (by Theorem 1) $= 1$.

(vii) $\sqrt[n]{n^2} \le \sqrt[n]{n^2+n} \le \sqrt[n]{2n^2}$, so $(\sqrt[n]{n})^2 \le \sqrt[n]{n^2+n} \le \sqrt[n]{2}(\sqrt[n]{n})^2$, and $\lim_{n\to\infty}(\sqrt[n]{n})^2 = \lim_{n\to\infty}\sqrt[n]{2}(\sqrt[n]{n})^2 = 1$ by parts (v) and (vi).

(ix) Clearly $\alpha(n) \le \log_2 n$, and $\lim_{n\to\infty}(\log_2 n)/n = 0$.

5. (a) If $0 < a < 2$, then $a^2 < 2a < 4$, so $a < \sqrt{2a} < 2$.

(b) Part (a) shows that

$$\sqrt{2} < \sqrt{2\sqrt{2}} < \sqrt{2\sqrt{2\sqrt{2}}} < \cdots < 2,$$

so the sequence converges by Theorem 2.

(c) If this sequence is denoted by $\{a_n\}$, then the sequence $\{\sqrt{2a_n}\}$ is the same as $\{a_{n+1}\}$. So the hint shows that $l = \sqrt{2l}$, or $l = 2$.

8. If x is rational, then $n!\pi x$ is a multiple of π for sufficiently large n, so $(\cos n!\pi x)^{2k} = 1$ for all such n, so $\lim_{n\to\infty}(\lim_{k\to\infty}(\cos n!\pi x)^{2k}) = 1$. If x is irrational, then $n!\pi x$ is not a multiple of π for any n, so $|\cos n!\pi x| < 1$, so $\lim_{k\to\infty}(\cos n!\pi x)^{2k} = 0$, so $f(x) = 0$.

9. (i) $\displaystyle\int_0^1 e^x\,dx = e - 1$. (Use partitions of $[0, 1]$ into n equal parts.)

(iii) $\displaystyle\int_0^1 \frac{1}{1+x}\,dx = \log 2.$

(v) $\displaystyle\int_0^1 \frac{1}{(1+x)^2}\,dx = \frac{1}{2}.$

CHAPTER 23

1. (i) (Absolutely) convergent, since $|(\sin n\theta)/n^2| \le 1/n^2$.

(iii) Divergent, since the first $2n$ terms have sum $\frac{1}{2} + \cdots + 1/n$. (Leibniz's Theorem does not apply since the terms are not decreasing in absolute value.)

(v) Divergent, since

$$\frac{1}{\sqrt[3]{n^2-1}} \ge \frac{1}{2n^{2/3}}$$

for sufficiently large n.

(vii) Convergent, since

$$\lim_{n\to\infty}\frac{(n+1)^2/(n+1)!}{n^2/n!} = \lim_{n\to\infty}\left(\frac{n+1}{n}\right)^2 \cdot \frac{1}{n+1} = 0.$$

(ix) Divergent, since $1/(\log n) > 1/n$.

(xi) Convergent, since $1/(\log n)^n < \dfrac{1}{2^n}$ for $n > 9$.

(xiii) Divergent, since

$$\frac{n^2}{n^3+1} > \frac{1}{2n}$$

for large enough n.

(xv) Divergent, since

$$\int_2^N \frac{1}{x \log x}\,dx = \log(\log N) - \log(\log 2) \to \infty \text{ as } N \to \infty.$$

(Notice that $f(x) = 1/(x \log x)$ is decreasing on $[2, \infty)$, since

$$f'(x) = \frac{-[1 + \log x]}{(x \log x)^2} < 0 \qquad \text{for } x > 1.$$

(xvii) Convergent, since $1/n^2(\log n) < 1/n^2$ for $n > 2$.

(xix) Convergent, since

$$\begin{aligned}\lim_{n\to\infty} \frac{2^{n+1}(n+1)!/(n+1)^{n+1}}{2^n n!/n^n} &= \lim_{n\to\infty} \frac{2(n+1)n^n}{(n+1)^{n+1}} \\ &= \lim_{n\to\infty} \frac{2}{\left(1+\dfrac{1}{n}\right)^n} = \frac{2}{e},\end{aligned}$$

by Problem 18-16.

5. (a) For each N we clearly have

$$0 \le \sum_{n=1}^{N} a_n 10^{-n} < 9 \sum_{n=1}^{\infty} 10^{-n} = 1,$$

so $\displaystyle\sum_{n=1}^{\infty} a_n 10^{-n}$ converges by the boundedness criterion, and lies between 0 and 1. (Actually, this number is denoted by $0.a_1a_2a_3a_4\dots$ only when the sequence $\{a_n\}$ is not eventually 0.)

17. The area of the shaded region is $\frac{1}{2}$. The integral is

$$\begin{aligned}&\tfrac{1}{2}([1 - \tfrac{1}{2}] + [\tfrac{1}{4} - \tfrac{1}{8}] + [\tfrac{1}{16} - \tfrac{1}{32}] + \cdots) - \tfrac{1}{2}([\tfrac{1}{2} - \tfrac{1}{4}] + [\tfrac{1}{8} - \tfrac{1}{16}] + \cdots) \\ &\quad = \tfrac{1}{2}(\tfrac{1}{2} + \tfrac{1}{8} + \tfrac{1}{32} + \tfrac{1}{128} + \cdots) - \tfrac{1}{2}(\tfrac{1}{4} + \tfrac{1}{16} + \tfrac{1}{64} + \tfrac{1}{256} + \cdots) \\ &\quad = \tfrac{1}{4}(1 + \tfrac{1}{4} + \tfrac{1}{16} + \tfrac{1}{64} + \cdots) - \tfrac{1}{8}(1 + \tfrac{1}{4} + \tfrac{1}{16} + \tfrac{1}{64} + \cdots) \\ &\quad = \frac{1}{8}\left(1 + \frac{1}{4} + \frac{1}{4^2} + \frac{1}{4^3} + \cdots\right) \\ &\quad = \frac{1}{8} \cdot \frac{1}{1 - \frac{1}{4}} \\ &\quad = \frac{1}{6}.\end{aligned}$$

CHAPTER 24

1. (i)

$$f(x) = \lim_{n\to\infty} f_n(x) = \begin{cases} 0, & x = 0 \\ 1, & 0 < x \le 1. \end{cases}$$

$\{f_n\}$ does not converge uniformly to f.

(iii) $f(x) = \lim_{n\to\infty} f_n(x) = 0$ (since $\lim_{n\to\infty} x^n = \infty$ for $x > 1$). The sequence $\{f_n\}$ does not converge uniformly to f; in fact, for any n we have $f_n(x)$ large for sufficiently large x.

(v) $f(x) = \lim_{n\to\infty} f_n(x) = 0$, and $\{f_n\}$ converges uniformly to f, since $|f_n(x)| \le 1/n$ for all x.

3. (i) $-\dfrac{1}{a} - \dfrac{x}{a^2} - \dfrac{x^2}{a^3} - \cdots.$

(iii) $\displaystyle\sum_{k=0}^{\infty}(-1)^k\binom{-\frac{1}{2}}{k}x^k.$

(v) $\displaystyle\sum_{k=0}^{\infty}\frac{(-1)^k\binom{-\frac{1}{2}}{k}}{2k+1}x^{2k+1}.$

4. (i) e^{-x}.

(iii) If

$$f(x) = \frac{x^2}{2} - \frac{x^3}{3\cdot 2} + \frac{x^4}{4\cdot 3} - \cdots, \qquad |x| \le 1$$

then

$$\begin{aligned} f'(x) &= x - \frac{x^2}{2} + \frac{x^3}{3} - \cdots \\ &= \log(1+x) \qquad |x| < 1, \end{aligned}$$

so for $|x| < 1$ we have $f(x) = (1+x)\log(1+x) - (1+x) + c$ for some number c. Since $f(0) = 0$, we have $c = 1$, so $f(x) = (1+x)\cdot\log(1+x) - x$ for $|x| < 1$.

6. Since

$$\sin x = \sum_{n=0}^{\infty}\frac{(-1)^n x^{2n+1}}{(2n+1)!}$$

we have

$$f(x) = \sum_{n=0}^{\infty}\frac{(-1)^n x^{2n}}{(2n+1)!}$$

(notice that the right side is 1 for $x = 0$). So

$$f^{(k)}(0) = \begin{cases} \dfrac{(-1)^n}{(2n+1)!}, & k = 2n \\ 0, & k \text{ odd}. \end{cases}$$

CHAPTER 25

1. (i) $|3+4i| = 5$; $\theta = \arctan\frac{4}{3}$.

(iii) $|(1+i)^5| = (|1+i|)^5 = (\sqrt{2})^5$; since $\pi/4 = \arctan 1/1$ is an argument for $1+i$, an argument for $(1+i)^5$ is $5\pi/4$.

(v) $|(|3+4i|)| = |5| = 5$; $\theta = 0$.

2. (i)

$$\begin{aligned} x &= \frac{-i \pm \sqrt{-1-4}}{2} \\ &= \frac{-i \pm \sqrt{5}\,i}{2} \\ &= \frac{(-1+\sqrt{5})i}{2} \quad \text{or} \quad \frac{(-1-\sqrt{5})i}{2}. \end{aligned}$$

(iii) $x^2 + 2ix - 1 = (x+i)^2$, so the only solution is $x = -i$.

(v) $x^3 - x^2 - x - 2 = (x-2)(x^2+x+1)$. The solutions are

$$2, \quad -\frac{1}{2} + \frac{\sqrt{3}}{2}i, \quad -\frac{1}{2} - \frac{\sqrt{3}}{2}i.$$

3. (i) All $z = iy$ with y real.

(iii) All z on the perpendicular bisector of the line segment between a and b.

(v) For $z = x + iy$, we clearly need $1 - x = 1 -$ real part of $z \geq 0$, or $x \leq 1$. For such x, our condition $\sqrt{x^2+y^2} < 1 - x$ is equivalent to $x^2 + y^2 < (1-x)^2$, or $x < (1-y^2)/2$. The set of points with $x = (1-y^2)/2$ is the parabola pointing along the second axis, with the point $(0, 1/2)$ closest to the origin, and intersecting the line $x = 1$ at $(1,1)$ and $(1,-1)$. The area bounded by this parabola and the line $x = 1$ is the desired set of points.

4. $|x+iy|^2 = x^2 + y^2 = x^2 + (-y)^2 = |x - iy|^2$.
$(z + \bar{z})/2 = [(x+iy) + (x-iy)]/2 = x$.
$(z - \bar{z})/2 = [(x+iy) - (x-iy)]/2i = y$.

5. $|z+w|^2 + |z-w|^2 = (z+w)(\bar{z}+\bar{w}) + (z-w)(\bar{z}-\bar{w}) = 2z\bar{z} + 2w\bar{w} = 2(|z|^2 + |w|^2)$. Geometrically, this says that the sum of the squares of the diagonals of a parallelogram equal the sum of the squares of the sides.

CHAPTER 27

1. (i) Converges absolutely, since $|(1+i)^n/n!| \leq (\sqrt{2})^n/n!$, and $\displaystyle\sum_{n=1}^{\infty} (\sqrt{2})^n/n!$ converges.

(iii) Converges, but not absolutely, since the real terms form the series

$$-\tfrac{1}{2} + \tfrac{1}{4} - \tfrac{1}{6} + \tfrac{1}{8} - \cdots$$

and the imaginary terms form the series

$$i(\tfrac{1}{1} - \tfrac{1}{3} + \tfrac{1}{5} - \tfrac{1}{7} + \cdots).$$

(v) Diverges, since the real terms form the series

$$\frac{\log 3}{3} + 2\frac{\log 4}{4} + \frac{\log 5}{5} + \frac{\log 7}{7} + 2\frac{\log 8}{8} + \frac{\log 9}{9} + \cdots.$$

2. (i) The limit

$$\lim_{n\to\infty} \frac{|z|^{n+1}/(n+1)^2}{|z|^n/n^2} = \lim_{n\to\infty} \left(\frac{n+1}{n}\right)^2 |z| = |z|$$

is < 1 for $|z| < 1$, but > 1 for $|z| > 1$.

(iii) The limit

$$\lim_{n\to\infty} \frac{|z|^{n+1}}{|z|^n} = |z|$$

is < 1 for $|z| < 1$ but > 1 for $|z| > 1$.

(v) The limit

$$\lim_{n\to\infty} \frac{2^{n+1}|z|^{(n+1)!}}{2^n|z|^{n!}} = \lim_{n\to\infty} 2|z|^{(n+1)!-n!}$$

is 0 for $|z| < 1$, but ∞ for $|z| > 1$.

3. (i) The limits

$$\lim_{n\to\infty} \sqrt[2n]{\frac{|z|^{2n}}{3^n}} = \frac{|z|}{\sqrt{3}} \quad \text{and} \quad \lim_{n\to\infty} \sqrt[2n+1]{\frac{|z|^{2n+1}}{2^{n+1}}} = \frac{|z|}{\sqrt{2}}$$

are < 1 for $|z| < \sqrt{2}$, so the series converges absolutely for $|z| < \sqrt{2}$. But the series does not converge absolutely for $|z| > \sqrt{2}$, so the radius of convergence is $\sqrt{2}$.

(iii) Since

$$\lim_{n\to\infty} \sqrt[n]{\frac{n!\,|z|^n}{n^n}} = \lim_{n\to\infty} \frac{|z|\sqrt[n]{n!}}{n} \le \lim_{n\to\infty} \frac{|z|}{n} = 0,$$

the series converges absolutely for all z, so the radius of convergence is ∞.

(v) The limit

$$\lim_{n\to\infty} \sqrt[n]{2^n z^{n!}} = 2 \lim_{n\to\infty} z^{(n-1)!}$$

is 0 for $|z| < 1$, but ∞ for $|z| > 1$, so the radius of convergence is 1.

GLOSSARY OF SYMBOLS